The Molecular Basis of Viral Replication

NATO ASI Series

Advanced Science Institutes Series

A series presenting the results of activities sponsored by the NATO Science Committee, which aims at the dissemination of advanced scientific and technological knowledge, with a view to strengthening links between scientific communities.

The series is published by an international board of publishers in conjunction with the NATO Scientific Affairs Division

A	**Life Sciences**	Plenum Publishing Corporation
B	**Physics**	New York and London
C	**Mathematical and Physical Sciences**	D. Reidel Publishing Company Dordrecht, Boston, and Lancaster
D	**Behavioral and Social Sciences**	Martinus Nijhoff Publishers
E	**Engineering and Materials Sciences**	The Hague, Boston, Dordrecht, and Lancaster
F	**Computer and Systems Sciences**	Springer-Verlag
G	**Ecological Sciences**	Berlin, Heidelberg, New York, London,
H	**Cell Biology**	Paris, and Tokyo

Recent Volumes in this Series

Volume 129—Cellular and Humoral Components of Cerebrospinal Fluid in Multiple Sclerosis
edited by A. Lowenthal and J. Raus

Volume 130—Individual Differences in Hemispheric Specialization
edited by A. Glass

Volume 131—Fat Production and Consumption: Technologies and Nutritional Implications
edited by C. Galli and E. Fedeli

Volume 132—Biomechanics of Cell Division
edited by Nuri Akkas

Volume 133—Membrane Receptors, Dynamics, and Energetics
edited by K. W. A. Wirtz

Volume 134—Plant Vacuoles: Their Importance in Solute Compartmentation in Cells and Their Applications in Plant Biotechnology
edited by B. Marin

Volume 135—Signal Transduction and Protein Phosphorylation
edited by L. M. G. Heilmayer

Volume 136—The Molecular Basis of Viral Replication
edited by R. Perez Bercoff

Series A: Life Sciences

The Molecular Basis of Viral Replication

Edited by

R. Perez Bercoff

The Institute for Virology
University of Rome
Rome, Italy

Springer Science+Business Media, LLC

Proceedings of a NATO Advanced Study Institute Summer School on
The Molecular Basis of Viral Replication,
held August 26–September 6, 1986,
in Maratea, Italy

Library of Congress Cataloging in Publication Data

NATO Advanced Study Institute Summer School on the Molecular Basis of Viral Replication (1986: Maratea, Italy)
The molecular basis of viral replication.

(NATO ASI series. Series A, life sciences; v. 136)
"Proceedings of a NATO Advanced Study Institute Summer School on the Molecular Basis of Viral Replication, held August 26–September 6, 1986, in Maratea, Italy"—T.p. verso.
"Published in cooperation with NATO Scientific Affairs Division."
Includes bibliographies and index.
1. Viruses—Reproduction—Congresses. 2. Molecular biology—Congresses. I. Pérez-Bercoff, R. II. North Atlantic Treaty Organization. Scientific Affairs Division. III. Title. IV. Series. [DNLM: 1. DNA Replication—congresses. 2. DNA, Viral—congresses. 3. Genes, Viral—congresses. 4. Molecular Biology—congresses. 5. RNA, Viral—congresses. 6. Virus Replication—congresses. QW 160 N279m 1986]
QR470.N38 1986 576′.64 87-15226

DOI 10.1007/978-1-4684-5350-8

Originally published by Plenum Press, New York in 1987
MyCopy version of the original edition 1987

To Minou and Kiki

R. P-B.

"From what I hear from our graduate students, anything before recombinant DNA is ancient virology and anything before polio vaccination is ancient-as-hell virology. Be that as it may, (...) it is well to recall that, before there were Picornaviridae, *there was poliovirus."*

Joseph L. Melnick
Intervirology (1983), **20**, *61-100*

THE MOLECULAR BASIS OF VIRAL REPLICATION

"In the beginning, it was poliovirus..."

And despite systematic vaccination campaigns poliovirus is still alive together with hepatitis (A and B), rotavirus, influenza, herpes, papilloma, yellow fever, and human immunodeficiency virus (HIV), to mention just a few.

The youngest offspring of microbiology (its 100th anniversary is still five years ahead), virology's contribution to build up our present understanding of the basic biochemical processes that characterize the living organisms count among the most incisive ones: From the milestone experiment by Chase and Hershey to the discovery of the "splicing" of the gene transcripts, the list would include the whole molecular genetics of the lambda bacteriophage, the notions of "promotor," "repressor," and "integration," the discovery of the reverse flow of genetic information, the very existence of oncogenes, the 5'-terminal "cap" structure of eukaryotic mRNAs,...

Electronmicroscopy, ultracentrifugation and tissue culture were the landmarks on the way of the young science. During the past few years, however, a major (and not so silent) revolution took place: recombinant DNA technology with all its might entered in our laboratories, and restriction mapping of cloned genomes and sequencing gels have replaced plaque counting and sucrose gradients.

The new techniques have made it possible to "dissect" the entire genome of a virus at the molecular level, and studies that would have been dreamt of just in the mid-seventies became the everyday experiments of our days. With new insight into the structure of viral genomes, and a deeper understanding of the mechanisms that regulate their expression, our view of viruses was bound to change: this volume bears witness to this impressive advance.

In organizing this NATO Advanced Study Institute we constantly kept in mind that the final product of this course (for a course it was, and held at the highest possible level) should be a reference book able to help outsiders willing to enter the field. Accordingly, we requested that our faculty contribute "extensive, interpretative reviews of the state of the art in the areas of their expertise." A most challenging task indeed, as we were actually demanding nothing less than to write at the same time for the (over) critical eyes of the expert colleague in need of an updated review, and to a broader (and probably more benevolent...) audience of virologists and molecular biologists at large.

Nobody would expect this to be easy, but it happened to be so thanks to the unlimited collaboration of the faculty members of this NATO Summer School: the reader will soon realize that they spared no efforts to secure that their lectures at this conference and their chapters for this book provided a comprehensive view of virology as we can see it today. Consequently, all the credit for this volume should go entirely to them. The editor, for his part, is totally accountable for any imperfections that may be found, and takes the blame for all of them.

The reader will also realize that although this book is mainly concerned with the molecular biology of *animal* viruses, the substantial contributions made by phagists and plant virologists have found their way into the proper chapters of this volume, and constant reference is made to the most recent developments in those areas.

The editor is especially indebted to the secretaries and typists from so many laboratories who went to great lengths to comply with his not-so-usual request of sending the manuscripts as hard copies and *as floppy disks of acceptable formats*: without such a help the fast editing, correcting, and printing of this book would have have been impossible. Unfortunately, most of them politely declined his invitation, and preferred to remain anonymous. He is left with no option but to thank them collectively. In the same vein, thanks are due to our computer expert, Mr. M.Toscano, whose skills and ability in "debugging" ruined programs or making compatible systems that originally were not so, saved the day time and again.

We gratefully acknowledge the financial support from the Italian National Research Council (C.N.R.), partly to defray the costs of preparing the edited manuscript for publication. And on behalf of all participants, it is my privilege to thank the local authorities of Maratea and the Chairman of its Tourist Board, Mr. B. Vitolo, who transformed our time in their wonderful seaside resort into an unforgettable experience.

R. Perez Bercoff
Rome, Fall 86

CONTENTS

SECTION I: VIRUSES AT THE MOLECULAR LEVEL

SECTION II: THE BASIC PROCESSES INVOLVED IN VIRAL REPLICATION

SECTION I

Viruses at the molecular level

CHAPTER 1

VIRUSES: AN OVERVIEW

MILTON W. TAYLOR and HOWARD V. HERSHEY

Program in Microbiology and Genetics, Department of Biology
Indiana University, Bloomington, IN 47405, U.S.A

INTRODUCTION

It is very difficult to be the opening author of a treatise of this type. It is even more difficult to write a chapter entitled "Viruses — An Overview." It is like a lecture on *"Animals — A Brief Summary"*.

The diversity in the animal kingdom — from sponges to insects to the primates, we, perhaps erroneously, call "sapiens" — make a brief summary impossible. The degree of diversity and variability within the viral "kingdom" which co-evolved with and infects all the other kingdoms and which is itself highly variable in morphology, basic chemistry, and molecular biology make it an extremely difficult topic to discuss in a short summary paper. So we will look for some commonalities among viruses by asking a few simple questions about virus structure, function, and evolution.

Perhaps the best way to start is to ask "what is a virus?" Early attempts to define a virus have been fraught with contradictions and confusion. This, of course, reflects the history of infectious diseases and their relationship to viruses. As "new" diseases were diagnosed, clinicians attempted to relate them to infectious agents, the majority of which were thought to be of bacterial origin. Microbiologists in the 1940-50's tried to find a common evolutionary pathway from the then recognized viruses to bacteria, hoping that viruses would prove to be the missing link in evolution at the procaryotic end of the scale. There have also been historical controversies as to whether viruses are "inert" molecules, living organisms, or autocatalytic proteins. None of these controversies seem relevant today. To repeat Lwoff's famous saying (1) "Viruses should be considered as viruses because viruses are viruses".

A more functional definition is the one presented by Luria and Darnell in their textbook, General Virology, in 1967 (2). They define viruses as "entities whose genome is an element of nucleic acid either DNA or RNA, which reproduce inside living cells and use their synthetic machinery to direct the synthesis of specialized particles, the virion, which contain the viral genome and transfer it to other cells". Thus, viruses are here defined by virtue of their obligate intracellular parasitism at the genetic level. Although this definition is now 20 years old, it is still generally acceptable. More recently, S. Harrison described a virus particle as "a structure for transferring nucleic acid from one cell to another", adding that "the nucleic acid may be either RNA or DNA and, in both cases particles of varying complexity are found. Observed structures reflect requirements for efficient and accurate assembly, for exit and re-entry, and for correctly localized disassembly" (55).

The concept of the virus as discussed by Lwoff in his 1957 paper (1), and the definitions above emphasize three characteristics of the virus particle: i) its infectivity, i.e., the ability to be transferred from cell to cell, ii) ability to exist in a non-cellular state, and iii) the obligate parasitism at the genetic level.

1. HISTORICAL BACKGROUND

Although the published history of virology begins with Jenner's experimental reports on vaccinia virus, (3) we know that the observation that transfer of pus from a lesion of an smallpox infected individual to a non-infected individual could result in immunity (variolation) was recognized among people in the Far and Middle East who suffered from periodic outbreaks of the disease centuries before Jenner's time. Of course neither they nor Jenner knew the nature of the causative agent of smallpox although Jenner does refer to it as a virus. In fact, the transfer of infectious smallpox material from one individual to another was introduced into England from Turkey a long time before Jenner's experiments. However, Jenner noted an inverse correlation between the severity of smallpox and exposure to cowpox (3). Jenner's paper makes interesting reading and is well documented. It is also noteworthy that Jenner's first attempts at publication were rejected ! The first vaccine used on a worldwide scale to eradicate a human viral diease — smallpox — contained live vaccinia virus developed by Jenner almost 200 years ago.

Continued research on immunization against disease has resulted in vaccines against a large number of other human viral pathogens, including such major diseases as polio, yellow fever, measles, mumps, and rubella, as well as many economically important viral pathogens which infect domestic animals. This list will surely become larger as more viral antigens are isolated by recombinant DNA technology, e.g., rabies and hepatitis B vaccines.

By the late 1800's, Koch and Pasteur had established the germ theory, which attributed disease to bacterial-like organisms. Such organisms were retained by the porcelain filters used at that time. However in 1892, Ivanovski comunicated to the Imperial Academy of Sciences of St. Petersbourg that the causative agent of the tobacco mosaic disease was filterable (Fig. 1). He proposed that the agent of the disease was some type of filterable toxin or a small microbe, and it was not until a few years later that Beijerinck proposed a living (reproducing) organism as the causative agent of tobacco mosaic disease, an organism smaller than all known bacteria. Beijerinck (4), who was unaware of Ivanovski's work, proposed that the

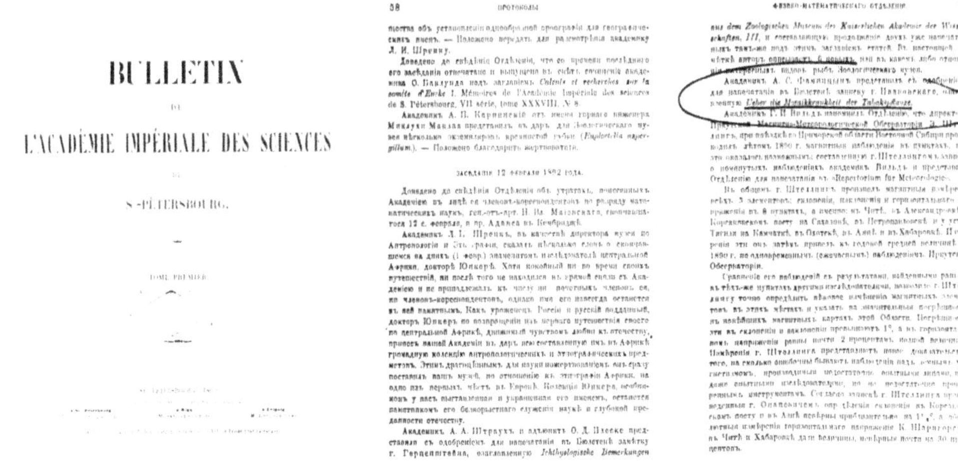
BULLETIN DE L'ACADÉMIE IMPÉRIALE DES SCIENCES DE ST.-PÉTERSBOURG

Figure 1. Ivanowki's communication to the Imperial Academy of Sciences of St. Petersburg.

tobacco mosaic disease was caused by a novel type of organism that existed in fluid or soluble form, and he called this *contagium vivum fluidum*. In essence, he rejected the idea that all infectious agents must be cellular in origin and proposed a non-cellular form (liquid!). This was a speculative jump that many feel opened up the field of virology. Beijerinck also recognized the obligatory parasitic nature of viruses by showing that there was no independent reproduction in the test tube. In order to reproduce "they must become incorporated into the living protoplasm of the cell", he writes (4). He also quantitated the amount of material necessary for infection, and showed a relationship between dilution and severity of the disease. Perhaps, Beijerinck should be regarded as the real founder of modern virology.

In 1898, Freidrich Loeffler and Paul Frosch investigated the outbreaks of Foot-and-Mouth (FAM) disease in German cattle. Like Ivanovski they discovered an infectious, filterable agent, and like Ivanovski they discussed the possibility that this substance might be a soluble toxin. However, they later rejected this conclusion and suggested that this was "an agent capable of reproducing... so small that the pores of a filter which will hold back the smallest bacterium will still allow it to pass" (5).

These pioneer virologists established working criteria, all negative, for identifying what we now know as viruses. Viruses i) unlike bacteria could not be seen through a light microscope, ii) could not be cultivated in cell-free medium, and iii) are not retained by filters known to prevent passage of bacteria. However, the concept that these were an entirely new class of biological entities was not yet considered. They were assumed to be "small" microbes, although called viruses by all the scientists at this time.

Perhaps one of the most important discoveries of modern virology was made at the turn of the century, and long ignored for 40 or so years. This was the discovery of the transmissability of avian leukemia by Ellerman and Bang (6) in

Denmark in 1908 and of a sarcoma of chickens by Peyton Rous (7) in the U.S. in 1911. Unfortunately these discoveries were relegated to the rank of avian curiosities, and their importance to virology and medicine was not recognized for many decades.

In 1915 and independently in 1917, the host range of viruses was expanded by the discovery of d'Herelle and Twort (8, 9) of bacterial viruses. The bacteriophage has since become one of the best studied organisms on earth. Modern molecular biology would not have developed without the work of the Cold Spring Harbor group of Hershey, Luria, and Delbruck in the 1940's who laid the groundwork for the quantitative aspects of virology (10). Much of this work was stimulated by the speculations of the physicist Schrodinger in his book "What Is Life?" which directed many people trained in the physical sciences to explore these small replicative "minimal" organisms (10).

Around the 1930's, two major discoveries were made that helped characterize the virus further. William Elford, of the National Institute for Medical Research, London, used a material called "collodion" to construct a range of membrane filters with different pore sizes (11). Using these filters, Elford estimated the size of several viruses. Two important results derive from his experiments, i) viruses were shown to be particulate entities with a definite size, and ii) viruses causing different diseases had different sizes, although viruses causing any specific diseases were identical in size. He estimated, for example, that the size of the Foot-and-Mouth disease virus was 10 nm. Thus, Elford's work gave some indication of how small viruses really were (12).

In 1935, Wendell Stanley, an organic chemist, reported the crystallization of tobacco mosaic virus (13). Although this led to controversy as to whether viruses were living organisms or auto-replicating proteins, it demonstrated the proteinous nature of viruses. However because of lack of knowledge of the nucleic acid component, it was difficult to explain the mechanism of viral replication. Stanley proposed that TMV was an "autocatalytic" protein which required the living cell for multiplication (13). More important — although not clearly understood at the time — was the demonstration a few years later that bacteriophage contain a nucleic acid (14). The concept that viruses were quite different from bacteria was beginning to be understood.

The importance of the bacteriophage research of the 1950's and 1960's by Luria, Hershey, Lwoff, and many others will never be too much stressed (10): Their research made virology into a quantitative science, gave birth to modern molecular biology, and led to the basic discoveries that opened up nucleic acid research and genetic engineering.

In parallel with the advances in virology, major advances were being made in the field of cell-culture. The art (for at first it was more art than science) of cell culture, began with the work of Alexis Carrel, who, in 1910, showed that it was possible to maintain chick tissues in culture by growing them in plasma clots supplemented with extracts from living chick embryos (15).

William Earle (a former student of Carrel's) established the first truly immortal cell line in the early 1940's (16). These cell lines were established by treating primary mouse fibroblasts with the chemical carcinogen methylcholanthrene. One of these cell lines, the L-cell, was established from mouse embryo fibroblasts in 1943. This cell line is widely used today and has proved invaluable in virology. A mutant derivative of this cell line has become a major tool in gene isolation experiments. In addition to the work of Earle, one must mention the work of George Gey, who established many human and rodent cell lines at about the same time (17).

Enders, in the late 1940's (18) showed that it was possible to culture poliomyelitis virus in various human embryonic tissues of non-neural origin. This led to the era of well-funded polio research and the development of methods for quantitating animal viruses, different tissue-culture media, and animal virus plaque assays (19). Dulbecco in 1963, demonstrated that viral transformation could be quantitated in a similar manner (20).

Closer to our own time, the work of Saul Spiegelman should be mentioned and in particular the *in vitro* replication of bacteriophage RNA (21). The discovery of the reverse transcriptase by Baltimore and Temin (22, 23) was a landmark in tumor virology and has profoundly altered all thinking in the area of cell biology and eukaryotic development. More recently the characterization of oncogenes by Bishop and Weinberg (23, 24), and the isolation of viruses of the HTLV/LAV series (Human Immune-deficiency Virus, HIV) by the groups of Gallo and Montagnier (26,27) have resulted in major insights into virus organization and replication.

2. VIRAL STRUCTURE

Since other chapters of this volume will describe in detail the molecular biology of individual viral species and virus-host interaction, we shall give here a simplified overview of viral structure and viral classification. Basically, virus are placed into three structural groupings based on electron microscopy. They are either

(a) *spherical* (''isometric''),

(b) *rod* shaped or filamentous (rigid or flexible) (Fig. 2) or

(c) *complex* (implying either a combination or neither of the above) (Fig. 3).

Many viruses posses lipid bilayer membranes, in part derived from the host cell, but usually with viral proteins inserted into the host lipid bilayer. As originally hypothesized by Crick and Watson (28), based on the limited coding potential of viral nucleic acids, the viral capsids are in most cases made up of repeating subunits. These capsid proteins protect the internalized nucleic acid from degradation, and may also act as means of cell attachment.

Viruses as we see them are symmetrical objects. It is important to remember that proteins themselves are *not* symmetrical and are irregular in shape. If a symmetrical arrangement did not occur, the same set of amino acids would have different patterns of noncovalent bonding in different places. Thus, because of the physical constraints of forming a symmetrical structure from asymmetrical proteins, spherical and rod-shaped structures fit the optimum energy requirements (Fig. 4a,

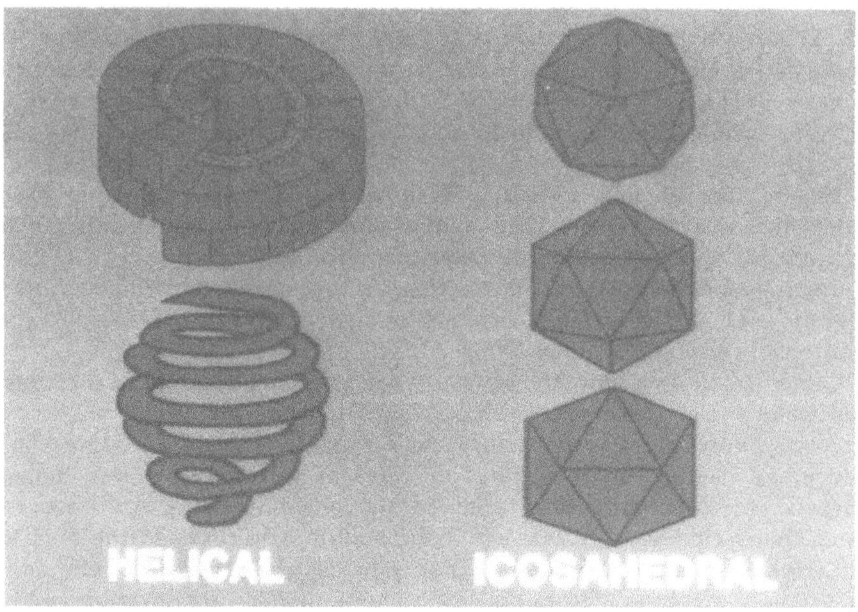

Figure 2. Hypothetical structure of a virus particle

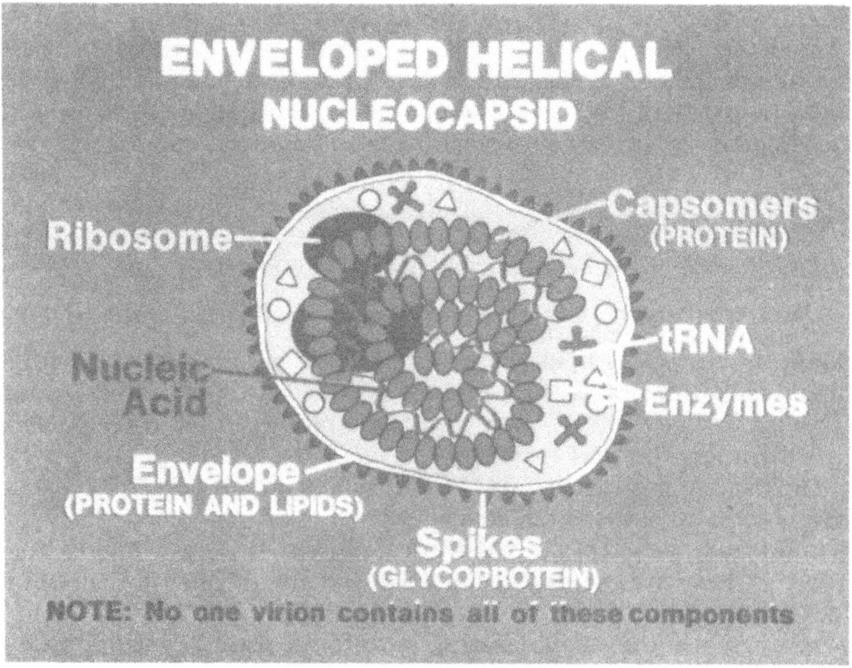

Figure 3. Hypothetical structure of a complex virus

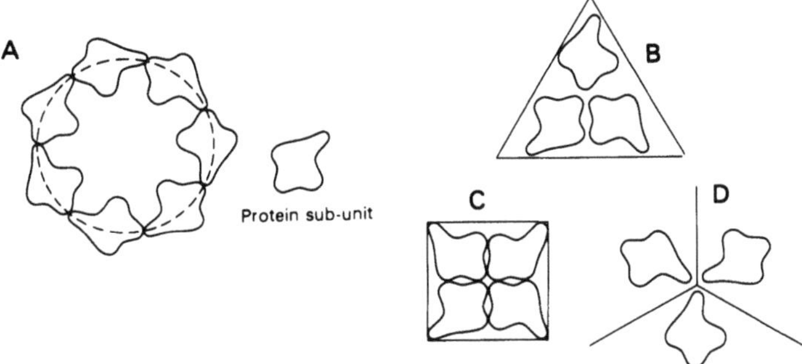

Figure 4. (A): Arrangement of identical asymmetrical components around the circumference of a circle to yield an asymmetrical structure. (B): Asymmetrical subunits located at the vertices of each triangular facet. (C): Asymmetrical subunits at each corner of a square with face represented in (D).
(Adapted from *Introduction to Modern Virology* (1974) S. B. Primrose, Halsted Press, with permission)

b, c, d). Although there may be slight deviations from symmetry this bonding of identical proteins is essential to self-assembly. Since viruses are fairly stable structures, the maximum number of bonds must be formed between the subunits and there are only limited ways in which this can be done.

Helical Structure

It turns out that one of the simplest ways of arranging non-symmetrical protein units is to place them around the circumference of a circle (Fig. 4a) to obtain a disc-like structure. If we examine the assembly process of TMV, we find disk-like structures as intermediates during self-assembly depending on the pH of the incubation buffer (Fig. 5) (29). However, because of the interaction between these discs and viral RNA the disc-like structures form a helical structure (29, 30). All filamentous viruses have helical protein structures, which probably reflects the constraints of wrapping disk structures around a long nucleic acid.

Icosahedral Viruses

Most spherical animal viruses have icosahedral symmetry. Multiplying the number of subunits per face by the number of faces gives the number of subunits that can be arranged around such a closed shell. For the icosahedron it turns out that 60 subunits or multiples of 60 are the number required (Fig. 6a, b). Since all spherical viruses are icosahedral, there must be some constraints on building other structures. Many viruses have more than 60 subunits (60N) but are still icosahedrons. The number of subunits does not have to equal the number of structural proteins (30).

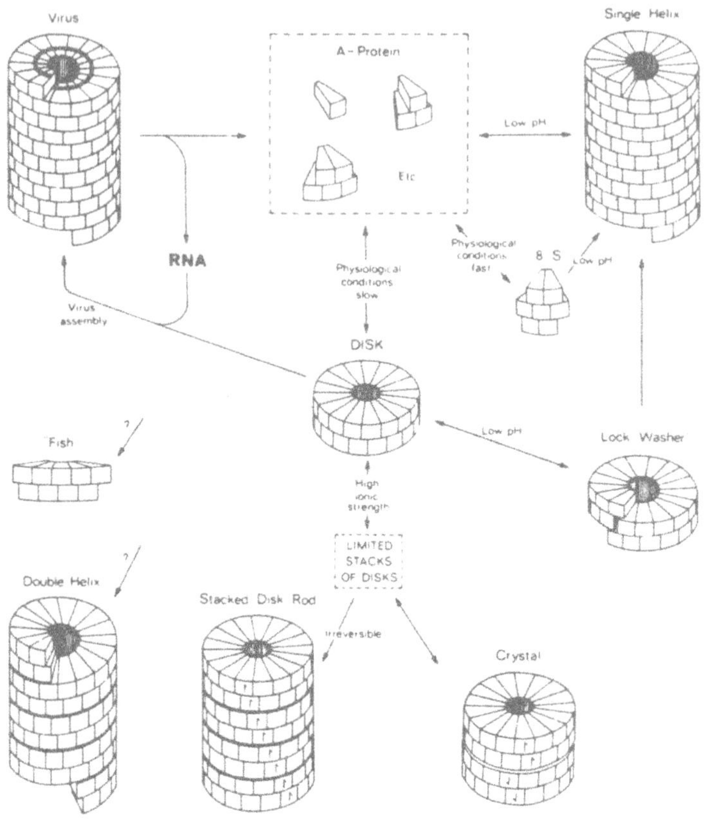

Figure 5. Diagram of interconversions that have been observed between some of the better aggregates of TMV proteins. (Reproduced from ref. 29, with permission).

Complex Viruses

Not all viruses are obviously helical or spherical. Viruses such as pox, herpes, rhabdo, T-phage, and λ-phage have complex morphology. In some cases this is due to the presence of a lipid membrane, and a helical/spherical basic structure (nucleocapsid) is found within the lipid membrane. Some of the plant viruses are flexible rods. These are basically helical but they have no straight axis of symmetry and so the subunits are quasi-equivalently related. Complex viruses such as the bacteriophage, are assembled independently from distinct sub-assemblies of icosahedral heads, rod-shaped tails, and tail fiber assemblies, and are then put together in the presence of a scaffolding protein.

Another important aspect of structure is the relationship between the viral nucleic acid and the capsid protein. In the case of helical viruses, such as TMV, there is a specific interaction between the viral nucleic acid and the protein subunits. In the case of isometric viruses, however, the condensation of nucleic acid is often independent of the protein structure and other viral and non-viral nucleic acids can be packaged into the capsid protein. In these cases it appears that the only restriction is that the viral RNA fit into the shell structure.

3. CLASSIFICATION OF VIRUSES

Because viruses contain either RNA or DNA, double-stranded or single-stranded, circular or linear, and these features can change quickly upon entry into the host, different viruses often have little in common with each other than their parasitic nature. The taxonomic scheme proposed by the ICTV (International Committee on Taxonomy of Viruses) (31) uses these structure and biochemical differences as the basis of its classification scheme. The hierarchy of this scheme subdivides viruses on the basis of their nucleic acid (RNA or DNA), viral structure (e.g., helical, isocohedral), whether they are enveloped, and genome structure (e.g., linear, d-s) (Fig. 7a,b).

Family names end in *viridae*, subfamily names in *virinae*, and genera, like species, in... *virus*. This taxonomic scheme is a mixture of the old and the new, since the names of some groups — such as adenoviridae and herpesviridae — refer to the original source of isolation or pathology of the virus, whereas the actual classification scheme is based on structure, type and character of nucleic acid, and in the case of retroviridae, on the presence of an enzyme, the reverse transcriptase.

Baltimore (32) has modified this scheme to use the mode of gene replication and expression to classify viruses (Fig. 8). In his classification scheme, mRNA (or + strand RNA) plays a pivotal role since protein synthesis occurs by the same mechanism for all viruses. All viruses are assigned to a numbered class based on the mode of synthesis of mRNA. All mRNA is designated (+) RNA. RNA which is complementary to the mRNA is designated as (−), and those which are non-

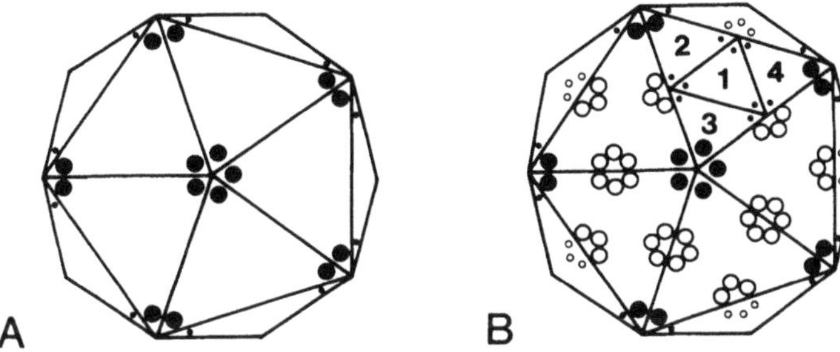

Figure 6. Arrangement of 60n identical subunits on the surface of an icosahedron. (A): n = 1, and the 60 subunits are distributed such that there is one subunit at the vertices of each triangular face. (B): n = 4, each triangular facet is divided into smaller (but identical) equilateral triangles.
(Reproduced from *Introduction to Modern Virology*, S. B. Primrose (1974), Halsted Press, with permission)

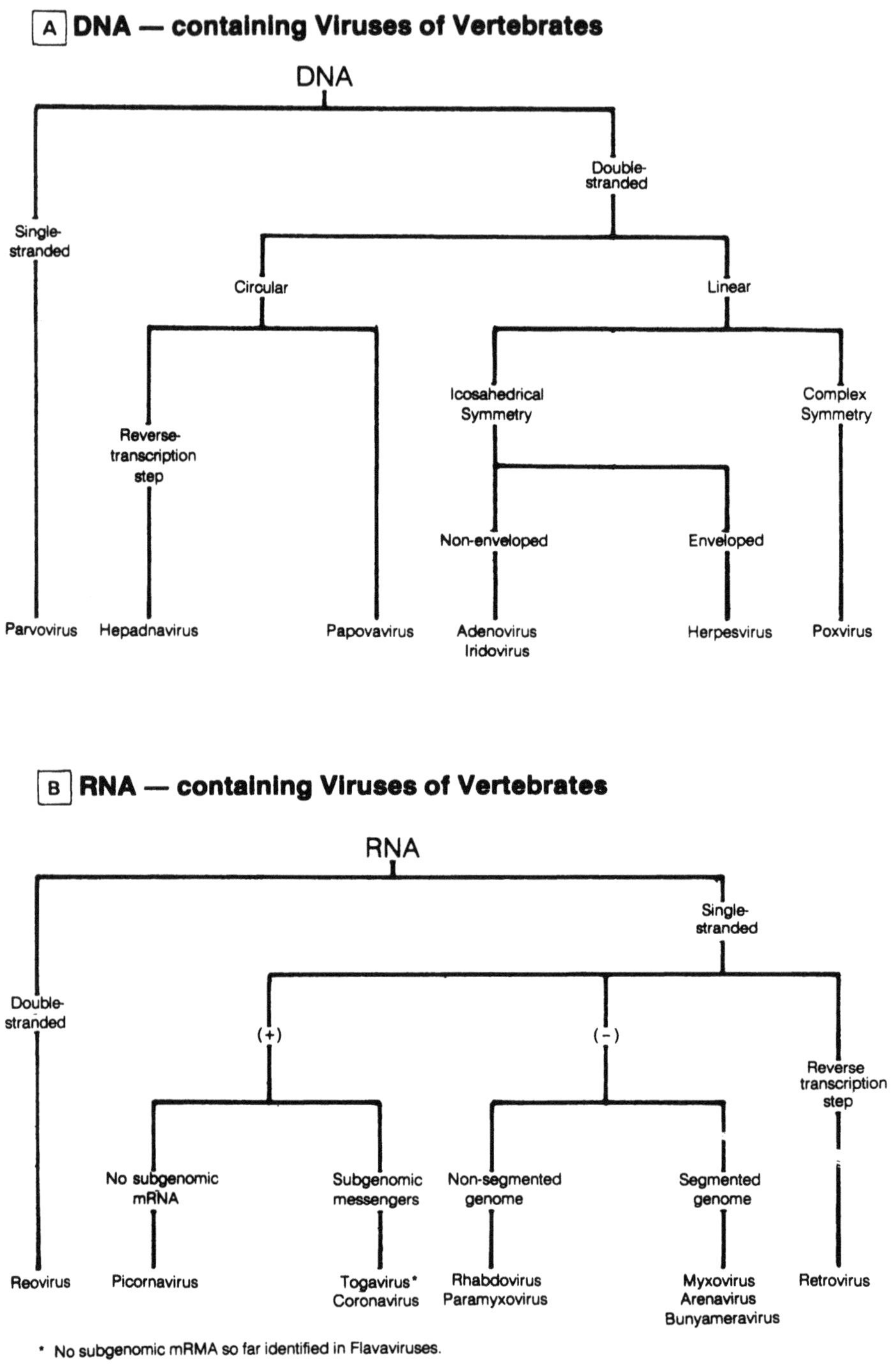

A DNA — containing Viruses of Vertebrates

DNA

Single-stranded

Double-stranded

Circular

Linear

Reverse-transcription step

Icosahedrical Symmetry

Complex Symmetry

Non-enveloped

Enveloped

Parvovirus

Hepadnavirus

Papovavirus

Adenovirus
Iridovirus

Herpesvirus

Poxvirus

B RNA — containing Viruses of Vertebrates

RNA

Double-stranded

Single-stranded

(+)

(−)

Reverse transcription step

No subgenomic mRNA

Subgenomic messengers

Non-segmented genome

Segmented genome

Reovirus

Picornavirus

Togavirus*
Coronavirus

Rhabdovirus
Paramyxovirus

Myxovirus
Arenavirus
Bunyameravirus

Retrovirus

* No subgenomic mRMA so far identified in Flavaviruses.

Figure 7. Classification of Animal Viruses

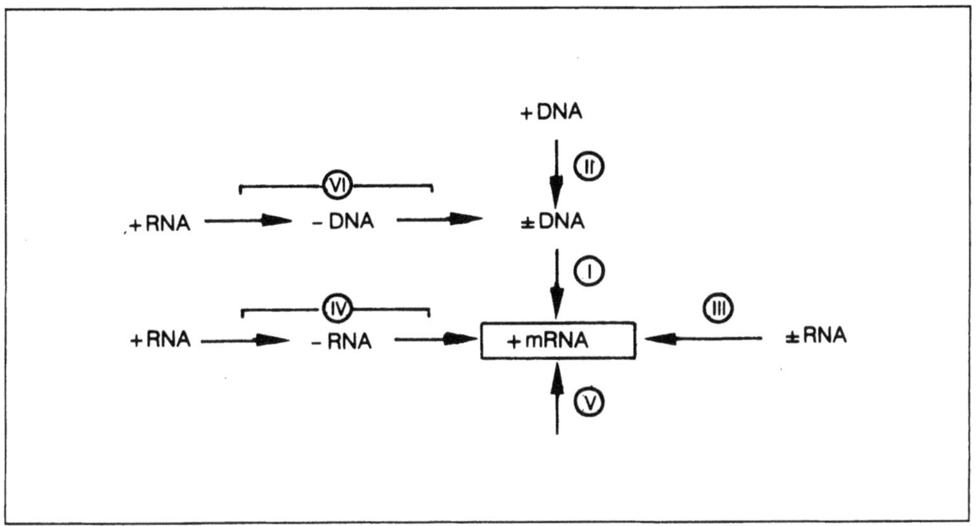

Figure 8. Functional Classification of Viruses according to D. Baltimore. Representative examples: I = T4 phage, vaccinia virus; II = ΦX174; III = reovirus; IV = RNA phages, poliovirus; V = vesicular stomatitis virus, Newcastle disease virus; VI = RNA tumor viruses.
(Adapted from ref. 32)

complementary as (+). Using this terminology, six classes of virus can be distinguished. We can subdivide the classes using other characteristics, such as enveloped, non-enveloped, segmented genome, etc.

Neither of these classification schemes do imply any phylogenetic relationship beyond those which can be imputed through nucleic acid hybridization, genetic recombination, and nucleic acid sequences of common regions. The presumption is that viruses which differ at few nucleotides have a more recent common ancestor than those that differ at a hundred. These techniques, however, are mute on the phylogenetic relationship between viral groups too disparate to have common features.

Since there is no fossil record of viruses, we cannot even consider viruses as a monophyletic group, since it is entirely possible (in fact, probable) that viruses do not always share a common ancestry, but arose independently more than once from different sources and by different mechanisms. We will discuss possible mechanisms later. Moreover, because viruses are parasitic, their further evolution is closely co-ordinated to the evolution of their hosts. Thus, the differences between for example, single-stranded circular DNA animal and plant viruses may reflect *either* common ancestry with subsequent divergence due to separate evolution of the host or may reflect two separate evolutionary events (by similar or different mechanisms) resulting in two independently-evolved viruses, one specific to plants and the other to animals. One must be cautious, moreover, in ascribing total linkage of viral evolution to a single host species since many plant and animal viruses are spread through the intervening vehicle of fungus or insect vectors, and these viruses must be coadapted for existence in both hosts.

4. VIRAL ONCOGENESIS

Probably one of the most exciting areas of virology is the role of oncogenic viruses. Again, as in so many other areas of virology, our model system is really derived from the bacteriophage work and the concept of provirus. The existence of bacteria that had the ability to generate phage *de novo* — i.e., were lysogenic — goes back to the 1920's with the work of Bordet (33) and Bail (34). However, the theory of lysogeny was really crystallized by the research of Lwoff in the 1950's (1). In temperate phage we have a system in which the phage DNA can be incorporated into the host genome, express unique repressor functions, prevent super-infection by the same phage, can be excised out, and on occasion carry (transduce) nearby genes.

The analogy of the prophage concept to RNA tumor viruses has proven very fruitful. The oncogene theory of Todaro and Huebner (35) is basically a restatement of the prophage theory with some modifications for animal viruses. The reader is referred to the proper section of chapter 16 of this volume for a detailed discussion of the oncogene theory. For the purpose of this introductory review enough to remind that the theory stated that many forms of neoplasia arise by the action of carcinogens on the expression of retrovirus oncogenes resident in the cell. Although this model is incorrect, the analogy between lysogeny and retrovirus integration proved to be a good one. The RNA tumor viruses integrate their DNA into the host chromosome, place nearby genes under the control of their Long Terminal Repeats (LTR's), and "transduce" oncogenes (Fig. 9). We can detect these rare, and often defective, transducing viruses by the characteristic of the oncogenes. Not only has the study of retroviruses given insight into the process of oncogenesis, but it has also contributed greatly to our understanding of cellular biology and cell differentiation (see chapter 16 of this volume).

5. ORIGINS OF VIRUSES

It is obvious that many of the early theories proposing viruses as very primitive organisms, as precursors of bacteria, or as precursors of lower eucaryotes, are difficult to accept in light of modern molecular biology. Viral evolution is ongoing, and the discovery of new viral strains affecting man or other animals, probably reflects ongoing recombination and evolution. Of course, one can not give a definitive answer to the question of where viruses came from; rather, one can only present a number of different hypotheses.

Three main theories have been advanced to explain the origin of viruses: i) viruses originated very early in evolution before the development of cellular life, that is, viral nucleic acid were among the first molecules replicating in the "primordial soup"; ii) viruses are the result of the degeneration of more complex parasitic organisms that have lost many of their key components, thus utilizing the protein synthetic and genetic apparatus of the cell; iii) viruses are derived from genetic

(a) TRANSFORMING VIRUS

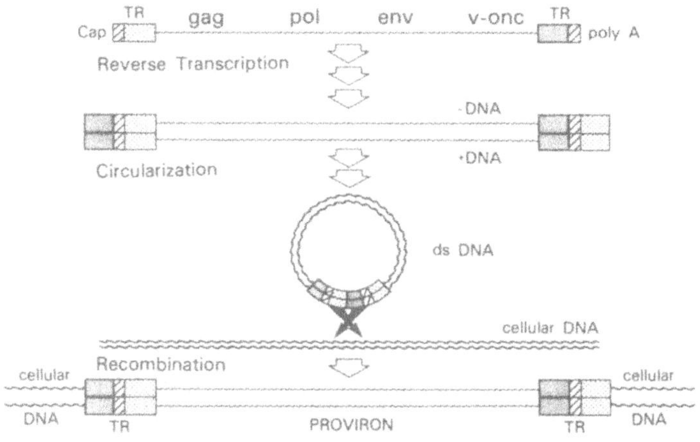

(b) LEUKEMIC VIRUS

Figure 9. Model to explain the formation and differences between (A) transforming viruses, (B) leukemic viruses, and expression of oncogenes.

elements of the cell which have evolved a semi-independent existence, presumably by "modular recombination" (36).

The more classical theory of parasitism (38, 39) is that viruses have originated by loss of essential functions from some type of free-living procaryotic organism. If we accept this basic premise, our candidate organisms for the original parasite would be eubacteria, archaeobacteria, rickettsia, mycoplasma, chlamydia or possibly a fungal eucaryote. Matthews (39) has argued that it is chlamydia-like organisms that have given rise to viruses such as pox virus. The chlamydia are obligate parasites, and lack an energy-generating system. They have two phases in their life cycle, outside the cells they exist as infectious *elementary* bodies about 300 nm in diameter. They have an outer cell wall, inner wall layer, and a plasma membrane. The genome DNA has a molecular weight of 4×10^8, and the chlamydia also contain RNA.

On entering the cell the elementary body is converted into the noninfectious *reticulate body*. This body is surrounded by a bilayer membrane derived from the host and divides by binary fission, giving rise to thousands of progeny within a few hours. Sometimes infection can lead to an immune state, and prevent superinfection.

What do chlamydia and poxviruses share in common, and what do they not?

a) Size of cell/viral particle — Approximately the same.

b) Genomic size — Approximately the same.

c) Cell Wall: Present in chlamydia, absent in pox virus.

d) Growth: Neither chlamydia or pox virus grow outside of cell. Both are obligate *intracellular* parasites.

e) Energy-yielding system: Absent in both.

f) Inability to synthesize amino-acids. Dependent on host cell.

g) Presence of two nucleic acids: Pox viruses do not contain RNA.

This theory is quite tenable for larger viruses such as T2, T4, herpes, pox viruses, etc., although it does not explain the formation of specialized organelles such as tail fibers, etc. Smaller viruses could be generated as defective interfering derivatives of a larger virus.

An alternate possibility, is that complex viruses may have resulted not by subtractive processes but from recombination type events, or physical joining/reordering of DNA segments of "primitive" viruses and host genes, an additive process. That herpes, T2 and T4, have genes for functions normally found in the host cell, such as thymidine kinase, tRNAs, and other enzymes of nucleotide biosynthesis may be explained as resulting from recombination, or transduction type events between virus and host cells or even by recombination between different viruses.

At the present time it is often speculated (and accepted) that certain groups of viruses arose from genetic elements of the cell or from chromosomal DNA (RNA?) which have somehow evolved an independent existence.

The late Herman Muller, the Nobel Prize winner who worked in our own department, suggested in 1922 (40) that bacteriophages may be derived from genes, "if these d'Herelle bodies were really genes, fundamentally like our chromosome genes, they would give us an utterly new angle from which to attack the gene problem. They are filterable, to some extent isolable, can be handled in test-tubes, and their properties, as shown by their effects on the bacteria, can then be studied after treatment. It would be very rash to call these bodies genes, and yet at present we must confess that there is no distinction known between the genes and them. Hence we cannot categorically deny that perhaps we may be able to grind genes in a mortar and cook them in a beaker after all." This statement was obviously farsighted and very modern.

By genetic elements we imply either segments of chromosomal DNA, transposons, insertion elements, or plasmid-like elements. As will be discussed below, we should also consider the reverse possibility: that such elements may be degenerate viruses. Let us now consider some possible examples.

During RNA processing, introns of varying length are spliced out of the heterogeneous nuclear RNA (HnRNA). It has been speculated that these RNAs might occasionally circularize and replicate autonomously. This has been suggested as a model for the generation of autonomously replicating RNAs. Viroids are small (250-400 base) closed circular single stranded RNAs, that are infectious to many

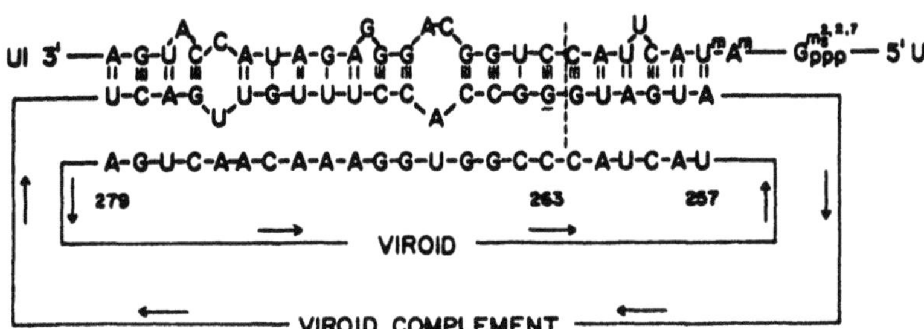

Figure 10. Possible relationship between viroid RNA and UIRNA (Reproduced from ref. 41, with permission).

plant species. Viroids can replicate in isolated nuclei and seem to have concatemeric precursors.

Small nuclear RNAs (snRNAs) associated with RNP particles are believed to be involved in the processing of primary transcription products (see chapter 4 of this volume). The 5' end of one such RNA, U1, has been shown to exhibit complementarity with the ends of many eucaryotic introns. Although no homologous sequences have been found between viroid and U1 sequences, some homologous sequences have been found between viroid RNA complements and U1 RNA. Figure 10 illustrates the possible base pairing interaction between the Potato Spindle Tuber viroid (PSTV) RNA complement and the 5' end of U1 RNA (41), (42). It should be pointed out that this is the only case in which such homologous regions have been detected.

Zimmern (43) has proposed a model in which intron-like RNA's (termed signal RNAs in the model), interact with "antenna RNAs", derived from structural or other genes. The product of signal RNA integrated into an antenna RNA is a fuson. Such fusons could code for replicases involved in subsequent independent replication of the fuson. Antenna RNA could be activated in different ways by signal RNA. Integration of the fuson or signal sequence into the genome might lead to amplification. Zimmern (43) argues that RNA viruses might have evolved from the attachment of a signal RNA carrying an origin of replication, to the antenna mRNA for a polymerase recognizing that origin, thus perpetuating a self-replicating RNA molecule. Recombination between this molecule and other mRNAs might lead to the formation of a complete RNA virus.

Another candidate for the origin of viruses is the transposon, or transposable element. Transposable elements are generally integrated into the host DNA with short reiterations of cellular DNA at either ends, have inverted terminal repeats at either end which can be similar to retrovirus LTR's in the case of complex transposons, have a long open reading frame, and many other characteristics similar to a retrovirus.

Although we can argue that viruses evolved from transposons the reverse may also be true, i.e., transposons and Ty-elements may be derived from retroviruses, or all may have a common ancestor. The recent finding that yeast carrying Ty elements have reverse transcriptase and produce virus-like particles similar to the intracisternal A-like particles of the mouse suggests that this transposable element may be a defective retrovirus (44). Likewise the copia element of *Drosophila* produces a virus-like particle resembling a retrovirus (45). Thus, before concluding that transposons are precursors to viruses, one should consider the opposite direction of evolution, i.e., transposons may have evolved from defective retroviruses.

The best argument against the "transposon" origin, but in favor of transposons being relics of retroviruses, is the recent finding (46) that Hepatitis B virus, a DNA virus, is possibly derived from an "ancient" retrovirus. It is now a well established fact that Hepatitis B virus (Hepadnaviruses) replicates through an RNA intermediate by reverse transcription (see chapter 17 and ref.47), a process similar to retrovirus replication. Miller and Robinson (46) have compared the nucleotide sequences of thirteen viral genes by means of a computer search program, and have recently reported that extensive homology exists over a 100 nucleotide segment of the conserved region of hepatitis DNA and retroviral U5 sequences. An examination of retrovirus-like sequences in human and simian chromosomes shows a similar pattern of conserved sequences.

Although hepadnaviruses do not contain an integrase, they contain two 11 or 12 base-pair direct repeats in the region homologous to the retrovirus U5 sequence. Thus, hepadnaviruses may be capable of transposition. When the nucleocapsids of Hepatitis B viruses and retroviruses are compared, similar regions of homology are found.

The copia and 17.6 transposable element of *Drosophila*, cauliflower mosaic viruses of plants, the Ty element of yeast, endogenous retrovirus-like elements of mammals, retroviruses, and hepatitis B virus share homology over several regions of their genome. The data indicate that all may have evolved from a common ancestor. Was this ancestor a large RNA virus, or a small transposon-like element that later incorporated host genes and become autonomous and infectious?

This last possibility can be explored by comparing the codon usage of viral genes and cellular genes. If one examines the codon usage of oncogenes, we find that there is a strong bias at the third position for cytidine rather than uridine, and guanine rather than adenine as in genes of eukaryotic cells. This bias is reversed in eukaryotic viruses. UUC is favored 3 to 1 over UUU for phenylalanine in the *src* gene, but UUU is favored for phenylalanine in the virus proteins. By examining the sequence in hepatitis viruses, it can be seen that in the X gene the codon usage preference is similar to that of eukaryotic genes, whereas the other genes definitely show a different pattern. It would thus seem that all of these viruses were derived from an ancestral retrovirus capable of undergoing recombination with its host cell.

It would be useful to do a similar computer-assisted analysis for herpes genes such as thymidine kinase.

6. VIRAL EVOLUTION

It is widely believed (48, 49) that RNA preceded DNA as the genetic material. The argument for this is that the 2' hydroxyl group in RNA tends to make the RNA more labile than DNA and thus selection would tend to favor DNA. RNA viruses are unique, in that they are the only self-reproducing organism in the biosphere that utilize an RNA genome.

DNA replication has associated proofreading mechanisms and repair systems. DNA replication in *E. coli* has an error rate of one false nucleotide incorporated for every 10^6-10^7 bases polymerized. Post-replicative repair systems may decrease this another thousand-fold.

The estimated error rate for RNA replicase, however, lies between 10^{-3}-10^{-6}. There are no known RNA repair enzymes. Thus RNA is intrinsically prone to error (referred to as noisiness by Reanney (38)). That this can lead to changes in the viral population emerged from experiments that used the small RNA-containing phage Qβ (50); 15% of the clones arising from a multiply-passaged population of Qβ had fingerprint patterns that deviated from those of the RNA of the population as a whole. Almost all progeny virus had at least one base sequence differing from wild type. This phenomenon is obviously the same as that referred to below as antigenic drift in foot-and-mouth-disease virus and influenza virus.

The noisiness of RNA replication has probably had an effect on the evolution of *RNA genomic size*. The largest known RNA genome is about 8×10^6 daltons. This may reflect the error rate, as the larger the genome the greater the possible error. Some RNA viruses may have better their performance by segmentation. Segmentation may allow for an escape from some of the deleterious effects of this high mutation rate by the ability to shuffle viral segments, and the greater likelihood of selection of the "correct" combination. Also segmentation substitutes for the type of recombination that might occur in DNA viruses. This is best illustrated by recombination (or antigenic shift) in influenza viruses (see chapter 13 of this volume).

There seems to be pressure toward smaller genomes in RNA viruses. *In vitro* experiments of Spiegelman and colleagues using Qβ replicase and a Qβ template (21) appeared to confirm that small RNA templates reproduce more progeny than large RNA templates. This also appears to be true, in general, of DI particles, where the small DI RNA out-replicates the intact viral RNA. Thus the evolutionary pressure would be for smaller RNA genomes.

Another aspect of evolution which one must consider is speciation; how do new species of virus arise, and how (or why?) do viruses change their host range? Many viruses can multiply in both insects and vertebrates, or insects and plants (but not in all three). Obviously this reflects ecological opportunity, the ability of insects to feed on vertebrates, or plants. Thus the spread of certain viruses will follow insect feeding mechanisms, and may have little to do with evolutionary pressures. This may explain why a particular rhabdoviruses can multiply in vertebrates or plants but not in both.

Studies with wound tumor viruses (51, 52) showed that when WTV was maintained in sweet clover plants for up to two years without passage through an insect vector, mutants arose that could no longer colonize the insect. Some of these mutants lacked segments of the genome, but retained their ability to replicate effectively in plants. Thus changing the ecology of the virus resulted in changing its host range. We know from studies of influenza viruses that antigenic shift results in large changes in the character of the virus (chapter 13).

Botstein (36) has proposed a generalized modular theory of viral evolution: He suggests that viral evolution is not so much the result of mutational events, but rather the result of recombination between individual blocks of genes, (functional blocks), here termed modules. Each module would specify a specific function. According to this theory, each viral type is the result of a combination of genes modules, which successfully occupy a niche in the environment. It is obvious that such a system does work in the generation of new lambdoid phage, where we can have the immunity of one phage, and host range of another (e.g., recombination between λ and 434, λ and Φ80).

Similar mechanisms may be at work in animal virus systems. Segmented genomes do undergo reassortment. Recombinants can be found between apparently non-related viruses, such as SV-40 and adenoviruses and, at the nucleic acid level, although possibly not at the functional level, between SV-40 and ΦX174 (53).

Similar mechanisms are also at work between defective viruses and infectious virus, endogenous retroviruses and exogenous retroviruses, and similar pathways

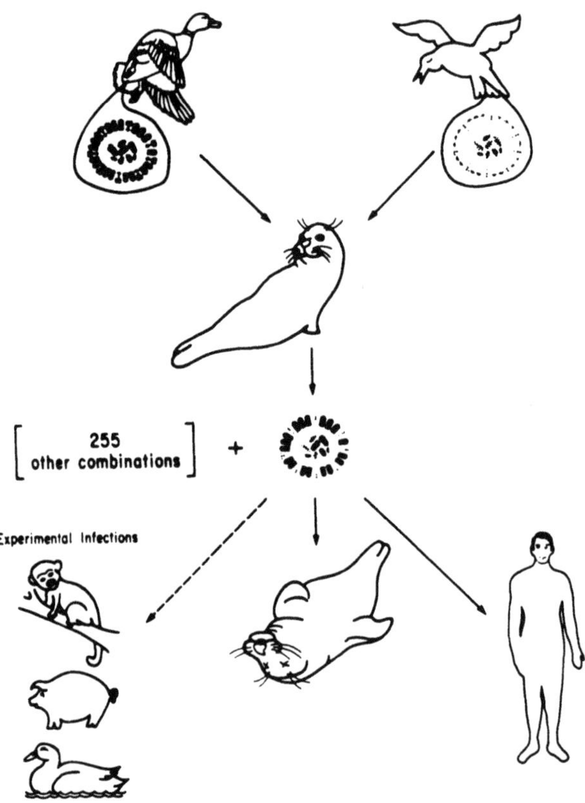

Figure 11. Diagramatic representation of the origin of seal influenza virus A/Seal/Mass/1/80 (H7N7). (Reproduced from ref. 54, with permission).

can be expanded to include the incorporation of host genetic material, such as oncogenes, into retroviruses.

At the microevolutionary scale, two other mechanisms leading to viral variation are antigenic drift, and antigenic shift. The best example of these microevolutionary mechanisms are presented by influenza viruses, although other examples could be mentioned — such as HTLV-3 and rhinoviruses. Antigenic drift involves minor changes in coat protein, or, in the case of influenza, in hemagglutinin and neuraminidase (54). Antigenic shift involves major antigenic changes.

Antigenic drift can be mimicked in the laboratory by growing influenza virus in the presence of monoclonal antibodies to the specific hemagglutinin (HA). Antigenic variants occur at a frequency of 1×10^{-5}/ml (54). Such variants have single amino acid changes in the HA polypeptide chain. In nature antigenic drift occurs by the accumulation of point mutations. Sequence analysis has shown that two mutations or more are necessary for a new strain to occur that escapes neutralization by the antibody to the parental virus. Thus, the accumulation of point mutation results in the evolution of new substrains of virus.

In *antigenic shift* large changes occur in the character of influenza virus. Sequence data indicates that the new subtypes have occurred by genetic reassortment. Genetic reassortment has been shown to occur between influenza A viruses of humans and lower animals *in vivo* (54). This type of evidence substantiates in part the Botstein (36) theory of viral evolution. This reassortment is diagrammed in Fig. 11 (55).

From the discussion above it is obvious the sources of viral evolution could be recombination between the viral genomes, recombination between viruses and the host genome; or segment reshuffling in segmented viruses.

7. REFERENCES

1) Lwoff, A. (1957) J. Gen. Microbiol. **17**,239-253.

2) Luria, S.E. and Darnell, J.E., Jr. (1967). In: *Virology*, John Wiley and Sons, Inc., New York.

3) Jenner, E. (1961) In: *Milestones in Microbiology*. T.D.Brock,ed., Prentice-Hall Inc., New Jersey.

4) Beijerinck, M.W. (1961) In: *Milestones in Microbiology*, T.D. Brock, ed., Prentice-Hall Inc., New Jersey.

5) Loeffler, F. and Frosch, P. (1961) In: *Milestones in Microbiology*. T.D. Brock, ed., Prentice-Hall Inc., New Jersey.

6) Ellerman, V. and O. Bang. (1908) Zentrablatt fur Bakteriologie, Parasitenkunde, Infektionskrank-heiten und Hygienes **46**, 595-609.

7) Rous, P. (1911) J. Exper. Med. **13**,397-411.

8) d'Herelle, F. (1961) In: *Milestones in Microbiology*, T.D. Brock, ed., Prentice-Hall Inc., New Jersey.

9) Twort, F. (1915) Lancet *11*,1241-1243.

10) Cairns, J., Stent, G.S., and Watson, J.D. (1966) In: *Phage and the Origins of Molecular Biology*, Cold Spring Harbor Laboratory, Cold Spring Harbor.

11) Elford, W.J. (1931) J. Pathol. Bacteriol. **34**, 505-521.

12) Elford, W.J. (1938) In: *Handbuch der Virusforschung*, R.Doerr and C. Hallaue eds., Springer-Verlag, Vienna.

13) Stanley, W.M. (1935) Science **81**, 644-645.

14) Schlessinger, M. (1935) Nature (London) **138**, 508-509.

15) Carrel, A. (1912) J. Exp. Med. **15**, 516-528.

16) Earle, W.R, Schilling, E.L., Staele, T.H., Straus, N.P., Brown, M.F., and Shelton, E. (1943). J. Nat. Cancer Inst. **4**, 165-212.

17) Gey, G., Coffman, W., and Kubiceck, M. (1952) Cancer Research **12**, 364-365.

18) Enders, J.F., Weller, T.H., and Robbins, F.C. (1949) Science **109**, 85-87.

19) Dulbecco, R. (1952) Proc. Natl. Acad. Sci. USA. **38**, 747-752.

20) Dulbecco, R. (1963) Science **142**, 932-36.

21) Spiegelman, S., Haruna, I., Holland, I.B., Beaudreau, G. and Mills, D. (1965) Proc. Natl. Acad. Sci. U.S.A. **54**, 919-927.

22) Baltimore, D. (1970) Nature (London) 13226, 1209-1211.

23) Temin, H.M. and Mizutani, S. (1970) Nature (London) **226**, 1211-1213.

24) Bishop, J.M. (1984). In: *The Microbe 1984. Part I Viruses*, Mahy, B.W.J. and J.R. Pattison, eds. Cambridge University Press, Cambridge, pp. 121-147.

25) Weinberg, R.A. (1982) Adv. in Cancer Res. **36**,149-63.

26) Gallo, R.C., Sarin, P.S., Gelmann, E.P., Robert-Guroff, M., Richardson, E., Kalyanaraman, V.S., Mann, D., Sidhu, G.D., Stahl, R.E., Zolla-Pazner, S., Leibowitch, J., and Popovic, M. (1983) Science **220**, 865-867.

27) Barre-Sinoussi, F., Chermann, J.C., Rey, F., Nugeyre, M.T., Chamaret, S., Gruest, J., Daguet, C., Axler-Blin, E., Vizenet-Grum, F., Rouzioux, C., Rozenbaum, W. and Montagnier, L. (1983) Science **220**, 868-871.

28) Crick, F.H.C. and Watson, J.D. (1956). Nature (London) **177**, 473-475.

29) Butler, P.J.G. and Durham, A.C.H. (1977) Adv. Protein Chem. **31**,187-251.

30) Klug, A. (1979) Harvey Lecture **74**, 141-172.

31) Wildy, P. (1971) In: *Monographs of Virology* , J.L.Melnick, ed., **5** Karger, Basel.

32) Baltimore, D. (1971) Bacteriol. Rev. **35**, 234-241.

33) Bordet, J. (1925) Annales de l'Institut Pasteur, **39**, 711-763.

34) Bail, O. (1925) Medizinische Klink (Munchen) **21**, 1271-1273.

35) Todaro, G.J. and Huebner, R.H. (1972) Proc. Natl. Acad. Sci. USA.**69**, 1009-1015.

36) Botstein, D. (1980) In: *Animal Virus Genetics*, Fields,B.W., Jaenisch, R., and Fox, C.C. , eds, Academic Press. New York pp. 363-384.

[20]37) Hoyle, F. and Wickramasinghe, C. (1978) New Scientist, pp. 946-948.

38) Reanny, D.C. (1982) Ann. Rev. Microbiol. **36**, 47-73.

39) Matthews, R.E.F. (1983) Int. Review of Cytology, Suppl. 15. J.F. Danielli, ed. Academic Press, New York, pp. 245-280.

40) Muller, H.J. (1922) Am. Naturalist **56**, 32-50.

41) Diener, T.O. (1981) Proc. Natl. Acad. Sci. U.S.A. **78**, 5014-5015.

42) Kiefer, M.C., Owens, B.A., and Diener, T.O. (1983) Proc. Natl. Acad. Sci. USA. **80**, 6234-6238.

43) Zimmern, D. (1982) Trends in Biochemical Sciences. **7**, 205-207.

44) Garfinkel, D.J., Boebe, J.D., and Fink, G.R. (1985) Cell **42**,507-517.

45) Shiba, T. and Saigo, R. (1983) Nature (London) **302**, 119-24.

46) Miller, R.H. and Robinson, W.S. (1986) Proc. Natl. Acad. Sci. USA. **83**, 2531-2535.

47) Summers, J. and Mason, W.S. (1982) Cell **29**, 403-415.

48) Eigen, M., Gardiner, W., Schuster, P. Winkler-Oswatisch, R. (1981) Sci. Am. **244**, 88-118.

49) Reanney, D.C. (1979) Nature (London) **288**, 598-600.

50) Domingo, E., Sabo, D., Taniguchi, T., & Weissman, C. (1978) Cell **13**, 735-744.

51) Reddy, D.V.R., Black, L.M. (1974) Virology **61**, 458-73.

52) Reddy, D.V.R., Black, L.M. (1977) Virology **80**, 336-46.

53) Dorsett, D.L., Keshet, I., and Winocour, E. (1983) J. Virol. **45**, 218-228.

54) Murphy, B.R. and Webster, R.G. (1985). In: *Virology*, B.N.Fields, ed. Raven Press, New York. pp. 1179-1240.

55) Harrison, S. C. (1984) In: *The Microbe 1984*, B.W.J. Mahy & J.R. Pattison, eds., Cambridge University Press, Cambridge, **1**, pp.29-73

CHAPTER 2

VIRAL PARTICLES AT ATOMIC RESOLUTION

MICHAEL G. ROSSMANN, EDWARD ARNOLD, GREG KAMER, MARCIA J. KREMER, MING LUO, THOMAS J. SMITH, & GERRIT VRIEND

Department of Biological Sciences, Purdue University, West Lafayette, Indiana 47907, USA

ROLAND R. RUECKERT, ANNE G. MOSSER, & BARBARA SHERRY

Biophysics Laboratory, University of Wisconsin, Madison, Wisconsin 53706, USA

ULRIKE BOEGE & DOUGLAS G. SCRABA

Department of Biochemistry, University of Alberta,Edmonton, Alberta T6 2H7, Canada

MARK A. McKINLAY & GUY D. DIANA

Microbiology and Medicinal Chemistry,Sterling-Winthrop Research Institute, Columbia Turnpike,Rensselaer NY 12144, USA

INTRODUCTION

Crick and Watson (ref. 1) first recognized that spherical viruses had to be regular polyhedra. Of these, the icosahedron has the largest number (60) of asymmetric units and was subsequently found to be the preferred envelope. The coding capacity of the enclosed genetic material could therefore be limited to coding only a relatively limited structural protein(s) for one-sixth of the virion shell. The assembly of viral particles from smaller, repeated subunits presents several defined advantages: such strategy of replication reduces considerably the amount of genetic information needed to code for the structural protein(s), and minimizes the risks of incurring in fatal errors, as faulty subunits inaccurately synthesized can be discarded at assembly time. The entire replication cycle of a virus, therefore, can be visualized as a two-step process:

a) the synthesis of viral components (nucleic acid and proteins), i.e: a template dependent, energy consuming process of polymerization of preformed blocks (nucleotides or amino acids), and

b) the self-assembly of the subunits into more complex structures, a process that does not involve the formation of stable chemical bonds but brings the the subunits (or the intermediate structures) to a thermodynamically stable configuration.

Caspar and Klug (2) extended these concepts by introducing quasi-equivalent triangulation number (**T**) and thus showed how larger viruses could be constituted of more than 60 subunits. The triangulation number T of viruses relates to their theory (2) in which a series of surfaces lattices were proposed for icosahedrical particles. These lattices would imply a quasi-equivalent environment for each identical subunit in the viral coat. A virus would have 60 T copies of each different type of subunit. However, the possible values of T are thought to be limited by the formula:

$$T = (h^2 + hk + k^2)\, f^2$$

where h, k, and f are integers. In practice, the proposed lattices have been found more or less as predicted but the number of subunits is not necessarily 60 T. For instance, in the case of polyoma virus $T = 7$, but the number of subunits is only 60×6 (3).

The first viruses to be studied to atomic detail were plant viruses as it is far easier, in general, to grow these in quantity and with relative ease. The first studies of spherical plant viruses were of two $T = 3$ particles, namely tomato bushy stunt virus (TBSV, ref. 4) and southern bean mosaic virus (SBMV, ref. 5), as well as one **T** = 1 structure, namely satellite tobacco necrosis virus (STNV, ref. 6). Surprisingly, TBSV and SBMV had closely related quaternary and tertiary protein shell structures and the STNV protein subunit was also similar to that of SBMV and the shell domain of TBSV. These results suggested a common evolutionary ancestor to many spherical RNA plant viruses. The concept of quasi-symmetry, however, needed a little modification (7), a situation which was especially obvious in the low resolution results for polyoma virus (3) where the anticipated hexamers were, in fact, pentamers.

With the development of many of the crystallographic techniques required in solving structures as complex as a virus, it became feasible to consider the even more challenging problem of animal viruses. The first three such structures to be solved were picornaviruses, namely a human rhinovirus (8), poliovirus (9) and Mengo virus (10). Not only were these three viruses strikingly similar, but the tertiary structure of each of the three larger viral proteins, VP1, VP2 and VP3 (see below), were similar to each other and similar to those found for the plant viruses. Furthermore, their quaternary arrangement in the coat was almost identical to that found in the **T** = 3 plant viruses. Thus, whereas the plant viruses had three covalent-

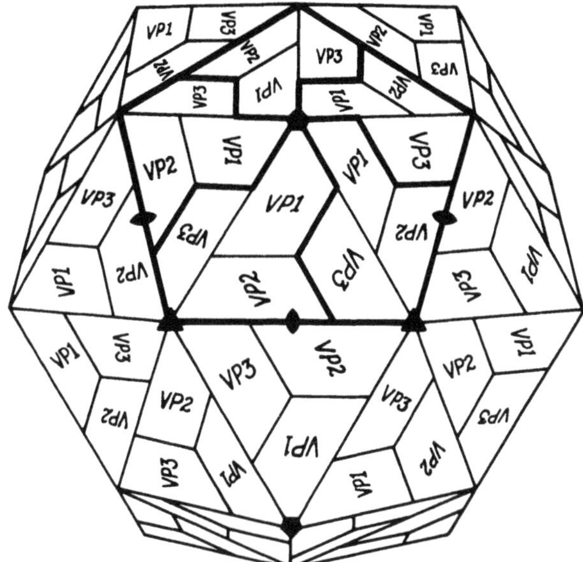

Figure 1. Relationship of the pseudo equivalent VP1, VP2 and VP3 subunits in the icosahedral capsid. The thick outline corresponds to the 6S (VP1, VP0, VP3) protomer and the 14S pentamer observed in assembly experiments.

ly identical protein subunits in quasi-equivalent evironments, picornaviruses have three subunits of similar structure but almost completely different amino acid sequences in pseudo-equivalent arrangements. The similarity of spherical RNA plant and picornaviruses makes it probable that these all have had a common evolutionary ancestor, possibly as a result of horizontal transfer of genetic material. Even more fascinating were the results on the DNA containing adenovirus hexon structure which was shown to have two successive β-barrels within a single polypeptide chain (11). The recent resolution of the structure of human common cold virus 14 at the atomic level has shed light into a number of important questions such as evolution, assembly of picornaviruses, sites for attachment of neutralizing monoclonal antibodies or the device used by the virus to maintain a constant receptor binding site. Accordingly, in this review we shall discuss primarily human rhinovirus 14 and refer to other structural results for comparisons.

1. VIRION STRUCTURE

HRV14 consists of an icosahedral protein shell (Fig. 1) surrounding an RNA core. 60 identical protomers arranged in groups of 5 form the protein capsid. Each protomer is composed of 4 structural peptides called VP1, VP2, VP3, and VP4 in decreasing order of size. The lack of visible structure in the central cavity results from the random orientation of the asymmetrical RNA molecule. VP1, VP2 and VP3 are each eight-stranded anti-parallel β-barrels closely similar to that observed

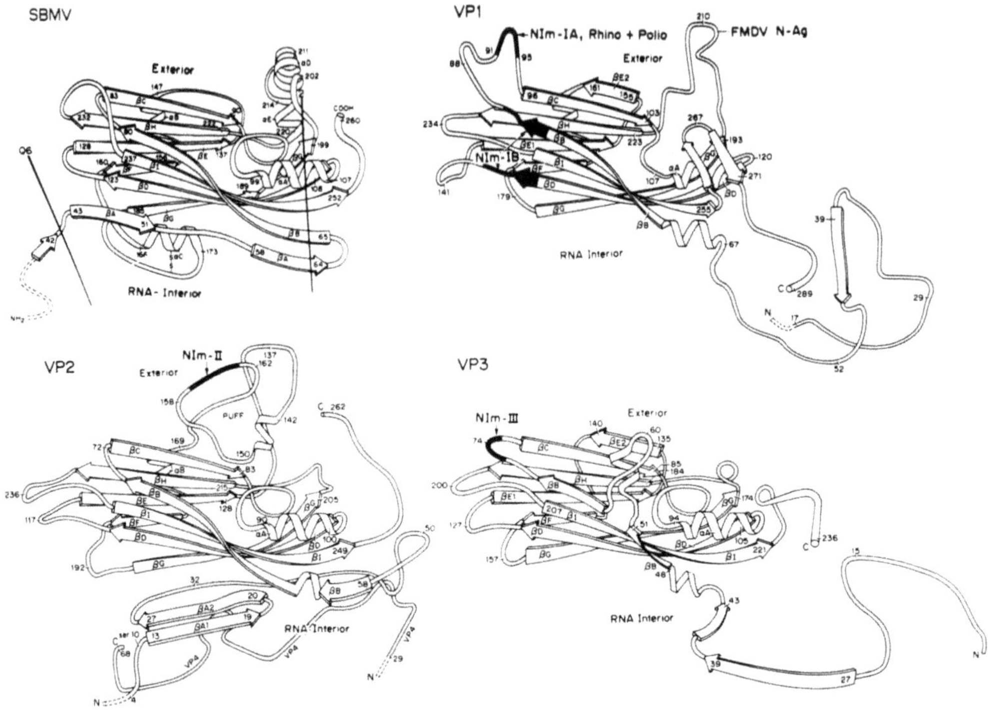

Figure 2. Comparison of the polypeptide folding of the structural proteins of HRV14 and SBMV coat protein.

in southern bean mosaic virus (5) and other plant viruses (Fig. 2): they have a pseudo threefold relation to each other as if the chains were identical and obeyed the rules of a T = 3 virus (2). For a more extensive discussion of the structure of the picornavirion, the reader is referred to the proper section of chapter 8 of this volume.

The Protein Component

The first 64 of the 73 amino-terminal residues of VP1 reside under VP3 (Fig. 3), whereas the first 42 of the 71 amino-terminal residues of VP3 are under VP1. Thus, the predominant positions of VP1 and VP3 at the RNA-protein interface are exchanged relative to their positions at the exterior surface. The first 25 of the 69 residues of the internal structural protein VP4 are not seen in the electron density map, implying that they lack icosahedral symmetry. VP4 is an extended polypeptide chain, positioned internally below, but in contact with, VP1 and VP2 (Fig. 2). Its visible amino end surrounds the fivefold axis. The carboxy ends of VP1 and VP3 are external and function in part to associate proteins within a protomer.

The "Canyon"

Large sequence insertions relative to the typical shell domain form protrusions on VP1, VP2 and VP3 and create a deep cleft or "canyon" on the HRV14 viral surface. The canyon separates the major part of five VP1 subunits (in the "north") clustered about a pentamer axis from the surrounding VP2 and VP3 subunits (in the "south"), thus forming a moat around the VP1 protrusions on the fivefold axis. The south canyon walls are lined with the carboxy-terminal ends of VP1 and a large sequence insertion in VP1 corresponding to helices αD and αE in the equivalent SBMV capsid protein. The north canyon wall is partially lined with the carboxy terminus of VP3. VP2 is hardly associated at all with the canyon, whereas VP1 is the major contributor to the residues lining the canyon. The canyon is 25 Å deep and 12-30 Å wide.

The Immunogenic Sites

Because of the additional elaborations which VP1 has on the surface relative to VP2 and VP3, its overall shape is that of a kidney (Fig. 2), with the depression forming a large part of the canyon (Fig.6). The first 16 residues of VP1 are not seen in the electron density map. The shell domain of VP1 in HRV14 starts at residue 74. The small sequence insertion between βB and βC in rhinovirus and poliovirus is not found in foot-and-mouth disease virus (FMDV). This loop forms a major immunogen in HRV14, NIm-IA, and poliovirus. The residues in VP1 of HRV14 that are analogous to αD and αE helices in SBMV protrude to the surface and form part of the south rim of the canyon, but do not form helices in HRV14. The most external portion of this segment contains the major antigenic site of FMDV and has been predicted to form an α-helix (12).

There is a large 43-residue insertion in the VP2 position, corresponding to βE2 of VP1 and VP3 (Fig. 2), forming an external mushroom-shaped "puff". This is positioned next to the VP1 elaborations, associated with the major antigenic site in FMDV, which line the south canyon wall. The most external residues of this puff correspond to NIm-II of HRV14. In contrast to VP1 and VP3, the carboxy-terminus of VP2 has no extensions beyond the shell domain.

All residues of VP3 can be seen in the electron density map. The 26 amino-terminal residues form a fivefold β-cylinder about the pentamer axis (Fig. 3). This fivefold cylinder extends down into the RNA to a radius of 111 Å. The polypeptide emerges from the β-cylinder and circles around the base of the VP1 shell domain, probably making extensive contact with the RNA in the central cavity. It then emerges on the viral surface near residue 61 and enters the shell domain at residue 72. The top corner of the VP3 shell domain, between βB and βC, is the NIm-III site of HRV14, structurally equivalent to NIm-IA in VP1.

RNA-triggered Cleavage of VP2/VP4

The capsid protein VP0 (the precursor to VP2 and VP4) is cleaved into its components only in the final stages of assembly. Since in HRV-14 the cleavage occurs at an Asn-Ser site it cannot be performed by the viral protease, specific for

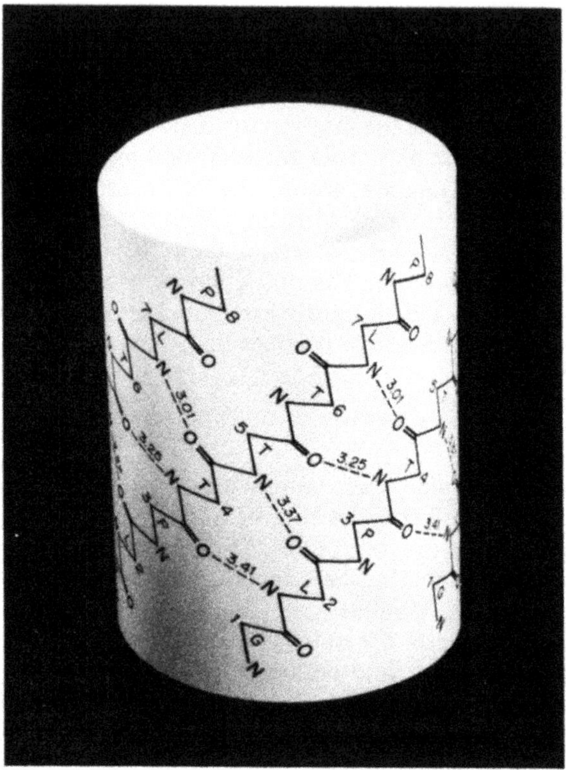

Figure 3. The fivefold β-cylinder formed by the amino-terminal ends of VP3 at the protein/RNA interface.

Glu-Gly peptides. The proximity of Ser 10 in VP2 to the carboxy-terminus of VP4 suggests that the cleavage of VP0 is autoproteolytic. There is no histidine next to Ser 10 and the carboxy end of VP4. However, nucleotide bases of the RNA could act as proton acceptors in the autocatalysis. Thus, the insertion of RNA into the growing capsids could trigger the cleavage of VP0 into VP2 and VP4. *This "enzyme" is, therefore, composed both of protein and RNA components.*

The Structure of Mengovirus

The Mengo virus structure shows greatest similarity to HRV14 in VP3. In VP2 the puff is deleted as was correctly predicted by sequence alignments (13). The major differences occur, however, in VP1. Here residues at the top corner corresponding to NIm-IA are deleted. The whole of the top βB-βC corner is folded over into the canyon between VP2 and VP1. The "FMDV-loop" is deleted as is also the carboxy-terminal end lining the "south canyon wall". The result is a totally changed virion surface (no puff, no FMDV loop, no NIm-IA protrusion), and a filling in of part of the canyon leaving only a deep "pit" into the deepest part by the highly conserved Pro-Pro-Gly-Ala-Tyr-Pro sequence of VP1.

Assembly

Assembly of picornaviruses (14) proceeds from 6S protomers of VP1, VP3 and VP0, via 14S pentamers of five 6S protomers, to mature virions. The final step involves inclusion of the RNA into empty capsids or partially assembled shells with simultaneous cleavage of VP0 into VP2 and VP4. Conversely, *in vitro* disassembly, produced by mild denaturation, proceeds via the expulsion of VP4 followed by the RNA (14).

Both the amino and carboxy ends of VP1 and $VP3_5$ are intertwined with each other. Furthermore, if VP4 and VP2 are considered as VP0, then VP0 is also intertwined with VP1 and $VP3_5$, which strongly suggests that the 6S protomer is as shown in Figure 1. These protomers are themselves intertwined because of the fivefold β-cylinder formed by the amino ends of the VP3's and the proximity of the observed amino ends of VP4's to the fivefold axis. Thus, the 14S pentamers closely correlate with the observed structure, shown diagrammatically in Figure 1.

The protein VP2, once cleaved from VP4, is globular and does not contact the other proteins extensively, although there are extensive solvent accessible regions around VP2 continuing into VP1. This, as well as the extraordinarily internal heavy atom sites on VP2, is consistent with loose binding of VP2 to the capsid and the channels that lead to the interior through a pore in the canyon floor. Disruption of pentamer-pentamer contacts, mediated by a slight reorientation of VP2 or its complete removal, could provide a port by which the VP4 and RNA can exit. Binding of a cell receptor in the canyon next to VP3 could facilitate this process, possibly accompanied by an isoelectric change.

Sequence Alignements

The three-dimensional coordinates of the three larger viral proteins have been superimposed on each other and on the structures of SBMV and tomato bushy stunt virus (TBSV) using techniques developed by Rossmann (15, 16). The resultant *structural* alignments are shown in Table 1 at the end of this chapter. The aligned residues have been systematically examined for common physical properties and some of these are also shown in Table 1. These characteristics may be essential for providing the common virus folding pattern in the absence of any obvious sequence homology. This "fingerprint" of the virus β-barrel fold can be used (17) to detect similar folds in other viral sequences.

Neutralizing Immunogenic Sites

The immunological response to a virus is a major defense against disease in animals. Antibodies can bind to viruses but they do not necessarily neutralize infectivity. Despite the sixtyfold equivalence of each potential binding site on the virus, as few as four neutralizing antibodies per virion can be sufficient to inhibit infectivity of poliovirus, (18). Neutralizing antibodies usually change the isoelectric point of the picornavirions (19, 20), indicating that a conformational change frequently accompanies neutralization. Antibodies may neutralize by interfering with cell attachment (antibodies to NIm-II and IA), membrane penetration or virus uncoating (antibodies to NIm-III) (21). Antibodies that bind to poliovirus may require bivalent attachment for neutralization of the virus (18, 22).

Amino acid residues within the major neutralization immunogens of HRV14 have been identified by Sherry and Rueckert (23) and by Sherry *et al.* (24). They found four major immunogenic neutralization sites NIm-IA, NIm-IB, NIm-II and NIm-III, each one composed of overlapping epitopes where a given mutant was resistant to many or all of the antibodies directed against that site. The residues, corresponding to escape mutants, invariably faced outward towards the viral exterior.

Many of the methods have been used to determine antibody binding sites depend on the use of synthetic peptides as antigens. Peptides associated with neutralizing antigenic regions do, in general, elicit antibodies that can neutralize the intact virus. In a significant number of cases, however, the sequence in question lies far below the viral surface or is even buried in the RNA. This suggests that some peptides can elicit antibodies which subsequently bind to totally unrelated portions of the native virus.

The Canyon as Receptor Binding Site

Despite the sequence and surface similarities of picornaviruses, they have different host and tissue specificity. This is particularly obvious in a comparison of HRV14 and Mengo virus. The presence of the 25 Å deep canyon, circulating around each of the 12 pentamer vertices, suggests that this is the site for cell receptor binding. An antibody molecule, whose Fab fragment would have a diameter of the order of 35 Å, would have difficulty in reaching the canyon floor, its entrance being blocked by the canyon rim. Thus, the residues in the deeper recesses of the canyon would not be under immune selection and could remain constant, permitting the virus to retain its ability to seek out the same cell receptor (Fig. 2).

Although retention of the canyon structure for all picornaviruses is to be expected, variation in the residues lining the canyon should be anticipated between viruses that attach themselves to different host cell receptors. Those parts of the carboxy-terminal ends of VP1 and VP3 which line the canyon walls are some of the least conserved amino acids among picornaviruses. As the topology of the canyon should be retained, the highly conserved, structurally equivalent sequences (Met-Tyr-Val-Pro-Pro starting at 151 of VP1 and Gly-Ala-Pro-Asn-Pro starting at 130 of VP3 for HRV14) in rhino, polio and FMD viruses situated near the floor of the canyon may be significant. The conserved sequence in VP1 in part provides the drug binding site mentioned below.

2. ANTIVIRAL DRUG BINDING

Arildone (25-28) and WIN 51711 (29-31) are examples of compounds (32, 32) which inhibit picornavirus replication by inhibiting viral uncoating without affecting cellular attachment. These compounds stabilize the virion against alkaline and

oxazoline phenyl aliphatic chain isoxazole

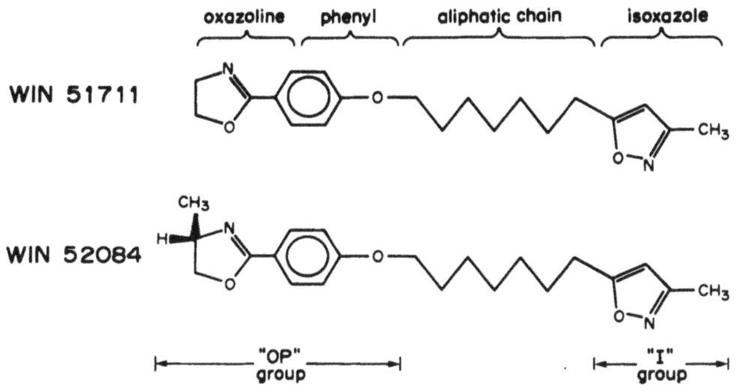

WIN 51711

WIN 52084

"OP" group "I" group

Figure 4. Structural formula of the antiviral compounds WIN 51711 and WIN 52084.

heat denaturation, loss of VP4 and induce a significant pI change. The effect of these substances on the virion properties in some ways resembles that of neutralizing antibodies or of sulfhydryl reagents (34, 35). By virtue of their conformation, these compounds can bind specifically to defined sites on the virion, and have proven a most useful tool to study the structure of the viral particle at the subatomic level.

WIN 51711 represents a class of compounds which inhibits picornavirus replication in tissue culture and in animal infected models. A number of compounds related to WIN 51711 have been synthesized (Sterling-Winthrop Research Institute) and tested against several different picornaviruses. The end point used in these studies was the minimal inhibition concentration (MIC), defined as the concentration which produces a 50% reduction of plaque count in cell culture. Although many of these compounds have a fairly wide spectrum of anti-picornavirus activity, the MIC values varied considerably between picornavirus pathogens, depending upon a number of factors such as the binding affinity of the compound to the virus, the ability of the bound compound to interfere with the uncoating process, and the incorporation of the compound into the host cell. The compounds that have been selected on the basis of their efficacy against human rhinoviruses, polioviruses and coxsackie A and B viruses were inactive against encephalomyocarditis and hepatitis A virus.

The site of binding on HRV14 was determined for two such compounds, namely WIN 51711 and WIN 52084 (Fig. 4). These compounds consist of a hydrophobic phenoxazole head, a seven membered aliphatic chain and an isoxazole functional tail. Two out of close to a thousand such compounds have been tested against a variety of picornaviruses. There were close to 60 drug molecules per virion as estimated both by the height of the electron density and the amount of radioactively labelled WIN 51711 bound. (36). The electron density for both WIN 51711 and WIN 52084 when bound to HRV14 was unequivocal (Fig. 5). The isoxazole (I) end binds into a hydrophobic pocket formed by the interior of the VP1 β-barrel (Fig. 2), in part created by the displacements of Phe 152 and Met 221. The oxazoline-phenolic end of the compounds binds into a hydrophilic pore on the base

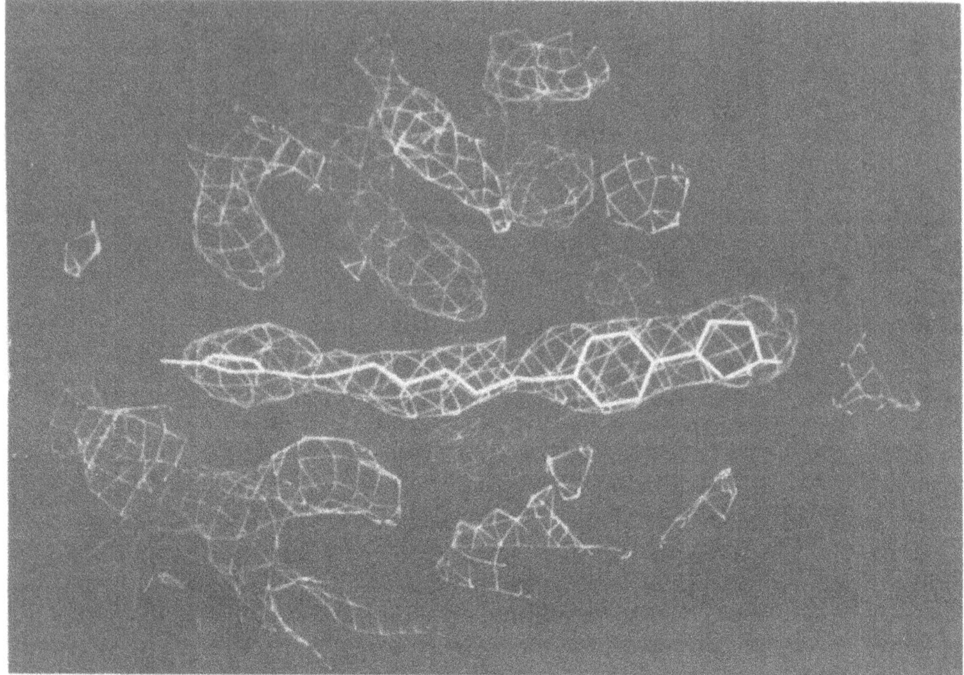

Figure 5. Model of WIN 52084 in its electron density.

of the canyon. The pore opens into large channels leading to the RNA interior (Fig. 6). The residues involved in drug binding are given in Table 2. A hydrogen bond between the nitrogen atom of the oxazoline ring and Asn 219 orients this end of the compound. Steric displacement of some of the residues by the drug affect considerable portions of βH leading from the "FMDV loop" across the canyon floor in VP1. The main chain is displaced as much as 3 Å in places and the side chains by even larger distances.

Table 2 lists not only the residues at the drug binding site, but also minimum inhibition concentrations required to reduce the plaque counts by one-half. The apparent insensitivity of HRV2 to WIN 51711 is explained in terms of the steric hindrance caused by the leucine, corresponding to valine 188 in HRV14.

Numerous substitutions of these compounds have been tested and the resultant MIC values can be explained on a qualitative basis in terms of the potential binding ability of the drug. The most important observation is that the drug becomes inactive on replacing the nitrogen and oxygen atoms in the oxazoline ring by carbon. Similarly, the compounds are inactive in those viruses (namely encephalomyocarditis, Mengo, hepatitis A and FMDV) which do not conserve Asn 219 of HRV14 as Asn or Asp (although the entrance to the pocket is far more constricted in Mengo virus). The hydrogen bond between the drug N or O atoms and Asn 219 is, therefore, probably essential to the drug function.

Table 2. – Residues within 3.6 Å of the bound WIN compounds

VP	Residue Number	HRV14 Amino Acid Type	Interaction	HRV2, HRV39, & HRV49	PV1 Mahoney	PV1 Sabin	PV2 Lansing	PV3 Leon	Mengo	FMDV (A10)
1	104	I	phenolic oxygen	I	I	I	I	I	L	L
1	106	L	phenyl	L	Y	Y	Y	Y	-	-
1	107	S	methyl on oxazoline	Q	K	K	K	K	-	-
1	116	L	oxazoline	F	L	L	L	L	S	A
1	128	Y	phenolic oxygen, C_3 in aliphatic chain	I	L	F	F	F	V	I
1	152	Y	C_4, C_2 & C_1 of aliphatic chain	Y	Y	Y	Y	Y	C	V
1	174	P	isoxazole	A	P	P	P	P	P	P
1	176	V	methyl of isoxazole	V	I	I	V	I	V	A
1	186	F	isoxazole	F	I	I	I	I	F	L
1	188	V	C_2 and C_3 of aliphatic chain	L	V	V	V	V	V	L
1	191	V	C_6 of aliphatic chain	L	V	V	V	V	N	T
1	197	Y	phenol C_7 & C_5 of aliphatic chain	Y	Y	Y	Y	Y	L	L
1	199	C	oxazoline	M	H	H	H	H	A	T
1	219	N	phenyl; N of oxazoline	N	N	N	N	D	S	G
1	221	M	phenyl	M	F	F	F	F	F	M
1	224	M	C_1 of aliphatic chain	L	L	L	L	L	L	A
3	24	A	isoxazole	A	A	A	A	A	I	V

Compound	MIC (μM)	HRV2	HRV39	HRV49	PV1 Mahoney	PV1 Sabin	PV2 Lansing	PV3 Leon	Mengo	FMDV (A10)
WIN 51711	0.4	3.8	9.3	4.7	8.7	1.1	0.8	1.1	inactive	inactive
WIN 52084	0.06	0.1	2.2	0.8	>9.3	6.9	5.7	1.2	no data	no data

Notes:
(i) Abbreviations for viruses: HRV, human rhinovirus; PV, poliovirus; FMDV (A10), foot-and-mouth disease virus strain A10.
(ii) Sequence alignments are based on work by A. Palmenberg (unpublished).

The effect of the compounds on the virion can be compared to that of the binding of an NAD co-factor to a dehydrogenase (37, 38): the isoxazole group (compared to the adenine end of NAD) binds into a hydrophobic pocket and stabilizes the virions against denaturation by heat or alkaline treatment. The binding of the aliphatic chain, corresponding to the pyrophosphate in NAD, causes essential conformational changes to permit the binding of the functional oxazoline-phenyl group (or nicotinamide group of NAD). The oxazoline-phenolic group covers a pore, thereby changing the pI, in the canyon floor which admits anions (e.g. $Au(CN)_2$ as used in the structure determination of HRV14) to the RNA, causing swelling and disassembly.

The structure of poliovirus Mahoney type I (9) shows electron density at the precise binding site of the WIN compounds. Thus a "co-factor" appears to bind to the same site as is occupied by the WIN compounds. This co-factor (perhaps a peptide or lipid component) would permit the virion to penetrate the membrane as a complete virion. The WIN compound might compete with the co-factor but binds more tenaciously, thus inhibiting disassembly.

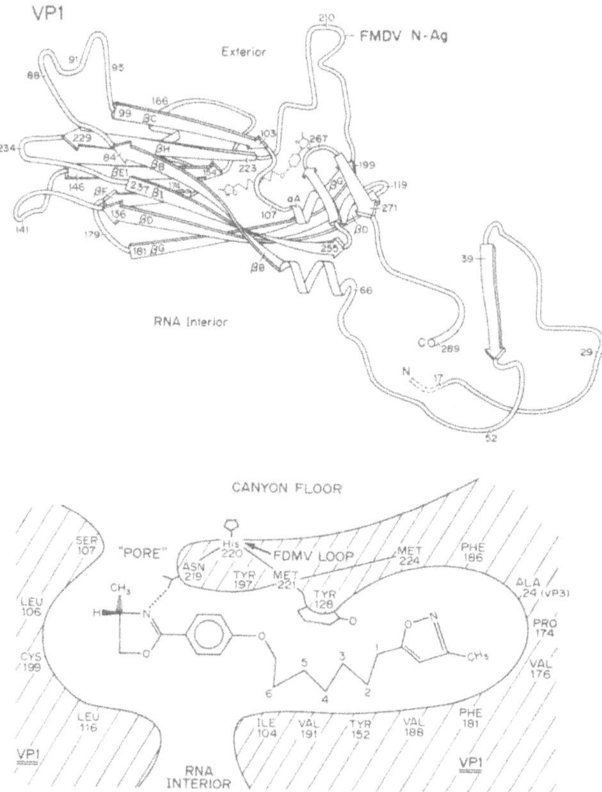

Figure 6. *Upper Panel*: Diagrammatic view of the polypeptide folding of VP1 of HRV14 showing the binding site of WIN 52084. *Lower panel*: WIN 52084.

3. CONCLUDING REMARKS

The determination of the structure of human common cold virus 14 has been rich in providing understanding of many phenomena. These studies have provided information on:

a) the evolution of picornaviruses as described above;

b) the assembly of picornaviruses: the 6S protomer formed by VP1, VP3 and VP0 was clearly recognized, and the larger 12S assembly unit (the pentamer), consisting of five protomers, was also an obvious intermediate deduced from the structure. The close association of VP1 and VP3 is consistent with the assembly processes of the RNA spherical plant viruses (7, 39, 40);

c) the autocatalytic process that mediates the cleavage of VP0 into VP2 and VP4 *involving both protein and RNA components*;

d) the sites for attachment of neutralizing monoclonal antibodies and how these sites may be interrelated for bivalent antibodies;

e) the significance of using synthetic peptides as antigens for mapping the antigenic surface of viruses or proteins;

f) the device used by the virus for maintaining a constant receptor binding site in a deep canyon on the virus surface, protected from antibody attachment: selective pressure would favor at the same time the hypervariability of the outside surface (exposed to neutralizing antibodies elicited by previous exposure to similar virions), and the conservation of the receptor binding site which, hidden in a deep canyon, can easily escape the immunological surveillance (the "canyon hypothesis");

g) the major differences and similarities of other picornaviruses such as foot-and-mouth disease virus or hepatitus A virus;

h) the site of binding and mode of action of a series of specific antiviral drugs (36).

Acknowledgements

We are grateful to Richard Colonno (Merck, Sharp and Dohme) for the initial analysis of $Au(CN)_2$ treated HRV14 and for stimulating discussions, Stan Lemon (UNC) for MIC measurements with respect to hepatitis A virus and R. K. Kulnig (Sterling-Winthrop Research Institute) for providing the crystal structure of WIN 51711. The authors' work described in this review was supported by a grant from the Sterling-Winthrop Research Institute to M.G.R. and by grants from the National Institutes of Health and the National Science Foundation to M.G.R. E.A. was supported by a National Institutes of Health Postdoctoral Fellowship during part of this work.

Table 1. – Alignment of HRV14, VP1, VP2, VP3, SBMV and TBSV based on structural superpositions

	HRV14								Exceptional
	VP1	VP2	VP3	SBMV	TBSV	STNV	ConA	Flu	Physical Properties
	66 D	55 K	42 N						13+
	67 V	56 P	43 L						
	68 E	57 D	44 L						12+,22-
α_Z	69 C	58 T	45 E						
	70 F	59 S	46 I						
	71 L	60 V	47 I						3+,19+
	72 G	61 C	48 Q						
	73 R		49 V	71 H	108 H	29 F			6+,8-
	74 A		50 D	72 C	109 R	30 A	2 D		6+
	75 A	63 F	51 T	73 E	110 E	31 L	3 T		22-
	76 C	64 Y	52 L	74 L	111 Y	32 I	4 I	164 L	3+,4+,13-,17-,18-,20+
	77 V	65 T	53 L	75 S	112 L	33 N	5 V	165 N	
βB	78 H	66 L		76 T	113 T	34 S	6 A	166 V	
	79 V						7 V	167 T	0,19+
	80 T	69 K	69 I	79 L	116 N	37 T	40 W	168 M	
	81 E	70 T	70 P	80 A	117 N	38 N	41 N	169 P	
	82 I	71 W	71 L	81 V	118 S		42 M	170 N	
	83 Q	72 T	72 N	82 T	119 S		43 Q	171 N	
	84 N	73 T	73 A		120 G		44 D	172 D	
	85 K						45 G		
	86 D						46 K		
	87 A						47 V		
	88 T						48 G		
	89 G								
	90 I								
Top	91 D								
corner	92 N								
	93 H								
	94 R								
	95 E								
	96 A								
	97 K								
	98 L								
	99 F	77 G	80 V				49 T	177 L	
βC	100 N	78 W	81 F	87 T		43 T	50 A	178 Y	23+
	101 D	79 C	82 G	88 S		44 V	51 H	179 I	
	102 W	80 W	83 T	89 E		45 Q	52 I	180 W	14+,23+
	103 K						53 I	181 G	
	104 I			92 M		48 S	54 L	182 I	
	105 N			93 P		49 N	55 F		
αA_0	106 L			94 F		50 G			
	107 S			95 T		51 I			
	108 S			96 V	141 L				
	109 L			97 G	142 F				
	110 V	90 G	95 T	98 T	143 S				
	111 Q	91 V	96 L	99 W	144 W				14+,23+
αA	112 L	92 F	97 L	100 L	145 L				0,3+,7+,9-,17-,21+
	113 R	93 G	98 G	101 R	146 P	54 G			15+

Table 1. – (continued)

	HRV14								
	VP1	VP2	VP3	SBMV	TBSV	STNV	ConA	Flu	Exceptional Physical Properties
	114 K	94 Q	99 E	102 G	147 A	55 D			13+
	115 K	95 N	100 I	103 V	148 L				
	116 L	96 M	101 V	104 A	149 A	58 N			
αA	117 E	97 F	102 Q	105 Q	150 S	59 Q			
	118 L	98 F	103 Y	106 N	151 N	60 R			
	119 F	99 H	104 Y	107 W	152 F	61 S			14+,23+
	120 T	100 S	105 T	108 S	153 D	62 G			4-,7-
	121 Y	101 L	106 H	109 K	154 Q	63 D			6+
	122 V	102 G	107 W	110 Y	155 Y				18-,20+,23+
	123 R	103 R	108 S	111 A	156 S				0,6+,10-,15+,16+,18+,22+
	124 F	104 S	109 G						
	125 D	105 G	110 S	114 A	159 S	69 S			7-,14-
	126 S	106 Y	111 L	115 I	160 V	70 H	89 V	198 A	
	127 E	107 T	112 R	116 R	161 V	71 K	90 R	199 S	6+,10-,15+,16+,18+,22+
βD	128 Y	108 V	113 F	117 Y	162 L	72 L	91 V	200 G	2+,3+,4+,6-,7+,14+,17+,18-,20+,23+
	129 T	109 H	114 S	118 T	163 D	73 H	92 G	201 R	
	130 I	110 V	115 L	119 Y	164 Y	74 V	93 L	202 V	2+,3+,4+,7+,13-,17-,18-,19+,20+
	131 L	111 Q	116 M	120 L	165 V	75 R	94 S	203 T	
	132 A	112 C	117 Y	121 P	166 P	76 G	95 A		9+
	133 T	113 N	118 T	122 S	167 L	77 T	96 S		
	134 A	114 A	119 G	123 C	168 C	78 A	97 T		9+,14-
	135 S	115 T	120 P	124 P	169 G		98 G		
	136 Q	116 K	121 A	125 T	170 T				
	143 S	117 F	122 L	126 T	171 T	80 T			
	144 S	118 H	123 S	127 T	172 E	81 V		206 T	
	145 N	119 S	124 S	128 S	173 V	82 S	104 N	207 R	4-
	146 L	120 G	125 A	129 G	174 G	83 Q	105 T	208 R	7-
	147 V	121 C	126 K	130 A	175 R	84 T	106 I	209 S	
	148 V	122 L	127 L	131 I	176 V	85 F	107 L	210 Q	3+,13-,17-,19+,21+
	149 Q	123 L	128 I	132 H	177 A	86 R	108 S	211 Q	
βE	150 A	124 V	129 L	133 M	178 L	87 F	109 W		
	151 M	125 V	130 A	134 G	179 Y	88 I	110 T		
	152 Y	126 V	131 Y	135 F	180 F	89 W	111 F		2+,4+,13-,14+,17-,18-,20+,23+
	153 V	127 I	132 T	136 Q	181 D	90 F	112 T		
	154 P	128 P	133 P	137 Y	182 K	91 R	113 S		10-,11+,15+,16+
	155 P	129 E	134 P	138 D	183 D				9+,11+,12+
	156 G	130 H	135 G						7-
	157 A	131 Q	136 A						
	158 P	132 L	137 R	142 T	187 D				
	159 N		138 G	143 I	188 E				
	160 P	PUFF	139 P	144 P	189 P				0,1-,2-,4+,9+,11+,20-
	161 K		140 Q	145 V	190 A	109 A			
	162 E			146 S	191 D	110 N			
	163 W			147 V	192 R	111 F			
αB	164 D			148 N	193 V	112 M			
	165 D			149 Q	194 E	113 S			

Table 1. – (continued)

	HRV14								
	VP1	VP2	VP3	SBMV	TBSV	STNV	ConA	Flu	Exceptional Physical Properties
↑	166 Y								
	167 T								
	168 W								
αB	169 Q			150 L	195 L	115 Y			
	170 S			151 S	196 F	116 N			
	171 A			152 N	197 N	117 P			
	172 S			153 L	198 F	118 I	127 H		17-
↓	173 N			154 K	199 G		128 F		
↑	174 P	185 P	149 T	155 G	200 V	129 K	129 M		
	175 S	186 H	150 H	156 Y	201 L	130 D	130 F		23+
	176 V	187 Q	151 V	157 V	202 K	131 V	131 N		19+
βF	177 F	188 F	152 V	158 T	203 E	132 T	154 L		
	178 F	189 I	153 W	159 G	204 T	133 L	155 E		23+
	179 K	190 N	154 D	160 P	205 A	134 N	156 L		2-,20-
↓	180 V	191 L	155 I	161 L	206 P	135 C	157 T		9+
Fourth corner down	181 G								
↑	182 D		159 S			143 K	171 G		7-
	183 T	196 T	160 T			144 D	172 R	212 T	6+,12+,16+,18+
	184 S	197 A	161 I	183 I	210 A	145 R	173 A	213 I	
	185 R	198 T	162 V	184 T	211 M	146 I	174 L	214 I	
βG1	186 F	199 I	163 M	185 I	212 L	147 I	175 F	215 P	0,2+,3+,4+,7+,10+, 11-,12-,13-,14+,16+, 17-,18-,19+,20+,21+
	187 S	200 V	164 T	186 A	213 R	148 N	176 Y	216 N	
↓	188 V	201 I	165 I	187 L	214 I	149 L	177 A	217 I	1+,3+,4+,5+,7+,8+,10+, 11-,13-,16-,17-,19+, 20+,21+
	189 P	202 P	166 P	188 D	215 P	150 P	178 P	218 G	0,1-,2-,3-,4+,8-,9+, 11+,14-,19-,20-
	190 Y	203 Y	167 W	189 T	216 T	151 G	179 V	219 S	23+
	191 V	204 I	168 T			152 Q	180 V	220 R	
	192 G	205 N	169 S	192 V			181 I	221 P	
	193 L	206 S	170 G	193 S				222 W	
	194 A	207 V	171 V	194 E	218 K			223 V	
	195 S	208 P	172 Q	195 K	219 V				
	196 A	209 I	173 F	196 R	220 K				15+,22+
↑	197 Y	210 D	174 R	197 Y	221 R	154 V			6+,10+,23+
βG2	198 N	211 S	175 Y	198 P	222 Y	155 N			23+
↓	199 C	212 M	176 T	199 F	223 C	156 Y			8-
	200 F			200 K	224 N	157 N			
	210 Y			201 T	225 D				
	FMDV loop								
	217 V		180 T	217 N	232 K			224 T	
	218 L		181 Y	218 I	233 L			225 G	17-
	219 N		182 T	219 L	234 I			226 L	
	220 H		183 S	220 V	235 D		189 S	227 S	
	221 M		184 A	221 P	236 L		190 A	228 S	

Table 1. – (continued)

	HRV14								Exceptional
	VP1	VP2	VP3	SBMV	TBSV	STNV	ConA	Flu	Physical Properties
	222 G		185 G	222 A	237 G	171 I	191 F	229 R	5+
	223 S	219 S	186 F	223 R	238 Q	172 F	192 E	230 I	
	224 M	220 L	187 L	224 L	239 L	173 M	193 A	231 I	0,1+,3+,5+,11-,13-, 16-,17-,21+
βH	225 A	221 M	188 S	225 V	240 G	174 L	194 T	232 I	
	226 F	222 V	189 C	226 T	241 I	175 Q	195 F	233 Y	
	227 R	223 I	190 W	227 A	242 A	176 I	196 A	234 W	
	228 I			228 M	243 T	177 G		235 T	
	229 V			229 E	244 Y	178 D			12+,15-,22-
	230 N			230 G	245 G	179 S			7-,11+,14-
	231 E			231 G	246 G				
Second	232 H			232 S	247 A		199 I		
corner	233 D						200 K		
down	234 E						201 S		
	235 H						202 P		
	236 K						204 S	240 G	
	237 T	236 P	203 G	236 A	249 R		205 H	241 D	
	238 L	237 S	204 Q	237 V	250 L		206 P	242 V	
	239 V	238 L	205 V	238 N	251 A	182 G	207 A	243 L	
	240 K	239 P	206 Y	239 T	252 V	183 L	208 D	244 V	
	241 I	240 I	207 L	240 G	253 G	184 W	210 I	245 I	3+,4+,8+,17-,19+
	242 R	241 T	208 L	241 R	254 E	185 D	211 A	246 N	6+,16+
	243 V	242 V	209 S	242 L	255 L	186 S	212 F	247 S	
	244 Y	243 T	210 F	243 Y	256 F	187 S	213 F	248 N	23+
	245 H	244 I	211 I	244 A	257 L	188 Y	214 I	249 G	19+
βI	246 R	245 A	212 S	245 S	258 A	189 E	215 S		
	247 A	246 P	213 C	246 Y	259 R	190 A	216 N		
	248 K	247 M		247 T	260 S	191 V			
	249 H	248 C				192 Y			23+
	250 V	249 T				193 T			
	251 E	250 E	218 K	249 R	262 T	194 D			5-,6+,10-,12+,13+, 16+,18+,21-
	252 A	251 F	219 L	250 L	263 L				17-
	253 W	252 S	220 R	251 I	264 Y				23+
	254 I	253 G	221 L	252 E	265 F				
	255 P	254 I	222 M	253 P	266 P				4+,9+
	256 R	255 R	223 K	254 I	267 Q				6+,10-,15+,16+,18+,22+
	257 A	256 S	224 D	255 A	268 P				
	258 P	257 K	225 T	256 A	269 T				
	259 R	258 S	226 Q	257 A	270 N				
	260 A	259 I	227 T	258 L	271 T				
	261 L								
	262 P								
	263 Y								

Notes:
(1) Comments indicate physical properties dominant at aligned positions which may be the determining factors that are required for producing the virus β-barrel fold.

Table 1. – (continued)

(2) The column headed "Exceptional physical properties" gives the particular property which is unusually large (+) or small (-) for the aligned set of amino acids.

0	Small minimum base change per codon
1	Helix forming
2	Sheet forming
3	Hydrophobicity (1) (Ponmanvolan)
4	Hydrophobicity (2) (Tanford)
5	Hydration potential
6	Polarity (1)
7	Bulkiness
8	pK (1)
9	pK (3)
10	Transfer of free energy
11	Turn forming
12	Hydrophilicity (Kuntz)
13	Polarity (2)
14	Volume
15	pI
16	Transfer free buried energy
17	Surface tension transfer energy
18	Hydrophilicity
19	Parallel sheet forming
20	Anti-parallel sheet forming
21	Hydropathy (Kyte & Doolittle)
22	Charge
23	Aromaticity

(3) All alignments were performed by superposition of the three-dimensional structures.

(4) Coordinates for the hemagglutinin spike of influenza virus ("Flu") were kindly supplied by Don Wiley.

4. REFERENCES

1) Crick, F.H.C. and Watson, J.D., (1956), Nature (London), **177**, 473-475.

2) Caspar, D.L.D. & Klug, A. (1962), Cold Spring Harbor Symp. Quant. Biol. **27**, 1-24.

3) Rayment, I., Baker, T.S., Caspar, D.L.D. & Murakami, W.T. (1982), Nature (London), **295**, 110-115.

Harrison, S.C., Olson, A.J., Schutt, C.E., Winkler, F.K. & Bricogne, G. (1978), Nature (London), **276**, 368-373.

5) Abad-Zapatero, C., Abdel-Meguid, S.S., Johnson, J.E., Leslie, A.G.W., Rayment, I., Rossmann, M.G., Suck, D. & Tsukihara, T. (1980), Nature (London), **286**, 33-39.

6) Liljas, L., Unge, T., Jones, T.A., Fridborg, K., Lovgren, S., Skoglund, U. & Strandberg, B. (1982), J. Mol. Biol. **159**, 93-108.

7) Rossmann, M.G. (1984), Virology, **134**, 1-11.

8) Rossmann, M.G., Arnold, E., Erickson, J.W., Frankenberger, E.A., Griffith, J.P., Hecht, H.J., Johnson, J.E., Kamer, G., Luo, M., Mosser, A.G., Rueckert, R.R., Sherry, B. & Vriend, G. (1985), Nature (London), **317**, 145-153.

9) Hogle, J.M., Chow, M. & Filman, D.J. (1985), Science, **229**, 1358-1365.

10) Luo, M., Boege, U., Vriend, G., Kamer, G., Minor, I., Arnold, E., Scraba, D.G. & Rossmann, M.G. (1986), manuscript in preparation.

11) Roberts, M.M., White, J.L., Gruetter, M.G. & Burnett, R.M. (1986, Science, **232**, 1148-1151.

12) Pfaff, E., Mussgay, M., Boehm, H.O., Schulz, G.E. & Schaller, H. (1982), EMBO J. **1**, 869-874.

13) Palmenberg, A.C. (1986), unpublished results.

14) Rueckert, R.R. (1976). In: *Comprehensive Virology*, H. Fraenkel-Conrat & R.R. Wagner, eds., Vol. 6, pp. 131-213, Plenum Press, New York

15) Rao, S.T. & Rossmann, M.G. (1973), J. Mol. Biol. **76**, 241-256.

16) Rossmann, M.G. & Argos, P. (1976), J. Mol. Biol. **105**, 75-96.

17) Rossmann, M.G. & Palmenberg, A.C. (1986), unpublished results.

18) Icenogle, J., Shiwen, H., Duke, G., Gilbert, S., Rueckert, R. & Anderegg, J. (1983), Virology, **127**, 412-425.

19) Mandel, B. (1976), Virology, **69**, 500-510.

20) Emini, E.A., Jameson, B.A. & Wimmer, E. (1983a), Nature (London), **304**, 699-703.

21) Diamond, D.C., Jameson, B.A., Bonin, J., Kohara, M., Abe, S., Itoh, H., Komatsu, T., Arita, M., Kuge, S., Nomoto, A., Osterhaus, A.D. M.E., Crainic, R. & Wimmer, E. (1985), Science, **229**, 1090-1093.

22) Emini, E.A., Ostapchuk, P. & Wimmer, E. (1983b), J. Virol. **48**, 547-550.

23) Sherry, B. & Rueckert, R.R. (1985), J. Virol. **53**, 137-143.

24) Sherry, B., Mosser, A.G., Colonno, R.J., & Rueckert, R.R. (1986), J. Virol. **57**, 246-257.

25) Caliguiri, L.A., McSharry, J.J. & Lawrence, G.W. (1980), Virology, **105**, 86-93.

26) McSharry, J.J., Caliguiri, L.A. & Effers, H.J. (1979), Virology, **97**, 307-315.

27) Diana, G.D., Salvador, U.H., Jonson, D., Henshaw, W.B., Lorenz, R.R., Thielking, W.H. & Pancic, F. (1977a), J. Med. Chem. **20**, 750-756.

28) Diana, G.D., Salvador, U.H., Zolay, E.S., Carabateas, P.M., Williams, G.L., Collins, J.C. & Pancic, F. (1977b), J. Med. Chem. **20**, 757-761.

29) McKinlay, M.A. & Steinberg, B.A. (1986), Antimicrobiol. Agents Chemother. **29**, 30-32.

30) Otto, M.J., Fox, M.P., Fancher, M.J., Kuhrt, M.F., Diana, G. & McKinlay, M.A. (1985), Antimicrobiol. Agents Chemother. **27**, 883-886.

31) Fox, M.P., Otto, M.J. & McKinlay, M.A. (1986), Antimicrobiol. Agents Chemother., manuscript submitted for publication.

32) Ninomiya, Y., Osborne, C., Aoyama, M., Umeda, I., Suhara, Y. & Ishitsuka, H. (1984), Virology, **134**, 269-276.

33) Lonberg-Holm, K., Gosser, L.B. & Kauer, J.C. (1975), J. Gen. Virol. **27**, 329-342.

34) Koch, F. & Koch, G. (1985). *The Molecular Biology of Poliovirus*, Springer-Verlag, New York.

35) Fenwick, M.L. & Cooper, P.D. (1962), Virology, **18**, 212-223.

36) Smith, T.J., Kremer, M.J., Luo, M., Vriend, G., Arnold, E., Kamer, G. & Rossmann, M.G. (1986), Science, manuscript submitted for publication.

37) Adams, M.J., Buehner, M., Chandrasekhar, K., Ford, G.C., Hackert, M.L., Liljas, A., Rossmann, M.G., Smiley, I.E., Allison, W.S., Everse, J., Kaplan, N.O. & Taylor, S.S. (1973), Proc. Natl. Acad. Sci. U.S. **70**, 1968-1972.

38) Holbrook, J.J., Liljas, A., Steindel, S.J. & Rossmann, M.G. (1975). In: *The Enzymes* Boyer, P.D., ed., 3rd edn., Vol. XI, pp. 191-292. Academic Press, New York.

39) Rossmann, M.G., Abad-Zapatero, C., Hermodson, M.A. & Erickson, J.W. (1983), J. Mol. Biol. **166**, 37-83.

40) Rossmann, M.G. & Erickson, J.W. (1985). In: *Virus Structure and Assembly*, S. Casjens, ed., pp. 29-73, Jones & Bartlett, Boston.

SECTION II

The basic processes involved in viral replication

CHAPTER 3

STRATEGY OF REPLICATION OF THE VIRAL GENOME

HOWARD V. HERSHEY AND MILTON W. TAYLOR
Department of Biology, Indiana University, Bloomington IN 47405, USA

Abbreviations:

ss = single-strand(ed); ds = double-strand(ed); (+) = nucleic acid strand with the same sequence as mRNA or anti-sense DNA; (−) = nucleic acid strand with the same sequence as sense DNA (or RNA equivalent); bp = base pair(s).

INTRODUCTION

Unlike higher organisms, viruses utilize a wide range of strategies in the replication of their genome. This is clearly a consequence of their need to deal with genomes that may be single-stranded (ss), double-stranded (ds), or partially ds; with genomes that may be composed of DNA, RNA, or both (at different stages in their life cycle); and with genomic structures which may be simple circles, simple linear molecules, single-stranded linear structures with terminal hairpin folds, ds linear structures in which the complementary strands are covalently linked at the ends, multipartite structures, and complex linear structures. In some cases the genome may also include covalently-linked proteins.

The variability of viral genomes (particularly the presence of RNA and single-strandedness) is probably an example of the selective influence of size and complexity in evolution. As a genome becomes larger and more complex, there is a greater need to minimize strand scission and to prevent copying errors in order to minimize the frequency of lethal defects per genome. Just as one cannot obtain man-sized insects because of inherent limitations in insect structure at large sizes, one probably cannot have a genome as complex as *E. coli* (or even vaccinia) composed or RNA or which is single-stranded. Evolution favors the more chemically stable DNA and the more structurally stable ds forms of genomes as genome size increases. Indeed, the larger single-stranded RNA viruses resort to segmented genomes (1, 2) to circumvent some of these difficulties.

Given the extreme variability of virus structures, it would be unrealistic to expect a unified mechanism of replication. However, there are common problems which all viruses must solve during nucleic acid replication; problems which are a consequence of polymerase enzymes which require a template, initiating factors (or primer in the case of DNA viruses), and the fact that all known polymerases can only add 5'nucleotide triphosphate to the 3' end of a growing chain.

After a general analysis of how viral replication is studied, we will examine the general problems and properties of replication of RNA viruses (Section 1), DNA viruses (Section 2), and viruses with both RNA and DNA forms (Section 3). In each section, we will discuss replication in specific selected viruses and bacteriophage. We will make particular note of the production of virus-specified replicative enzymes and the formation of viral, host, or mixed enzyme complexes, the mechanism of initiation (or priming) of replication, the formation of replicative structures and, finally, how the virus completes replication so as to end up with a full-length viral genome.

Viral One-Step Synchronized Growth Cycle

Before examining the molecular mechanism of viral replication one needs to standardize certain fundamental parameters of infection such as the type and physiological state of the host cells and the most suitable media and serum concentration for optimal viral production. One of the most important parameters affecting studies of viral replication is the multiplicity of infection (m.o.i.). Too high a m.o.i. can cause cell lysis from without, and too low a m.o.i. may result in undetectable levels of viral RNA or DNA at crucial early stages.

Before starting any studies on viral nucleic acid synthesis, it is important to carry out a single-step growth experiment in order to have a time-frame for further experiment. Typically, one wants to synchronize viral infection for all cells of the population to simplify analysis of viral products at various times during the infectious process.

An idealized one-step growth cycle for synchronized infection demonstrates several distinct stages in the infectious process. The first stage is attachment of the virus to a receptor site and penetration of the virus into the cell, or in the case of phage, the introduction of the nucleic acid into the bacterial host. Once this has been accomplished, and unattached virus removed, infectious particles are not usually detectable, even in the cell. This second stage is termed the *eclipse period* and ends with the earliest detection of infectious *intracellular* virus. This period should be distinguished from the *latent* period, the time at which extracellular virus is first released from the cell. From the end of the latent period, we see a large increase in the number of virus particles formed. This period is termed the growth period and is the time of maximum viral assembly, maturation, and release.

Starting the virus-cell interaction with a large m.o.i. (10-100) ensures a synchronous infection. After washing off unadsorbed virus, and if necessary adding anti-viral antibody to prevent a further round of adsorbtion and replication (which should not be a problem with cytolytic viruses introduced at a high m.o.i.), samples are assayed at various times (usually by plaque assay) for infectious material. Any free virus present early in the infectious process is typically due to unadsorbed virus.

Viral RNA or DNA synthesis is normally measured by labelling the cells with radioactive nucleic acid precursors, either continuously throughout the cycle, or in short pulses. In the case of many RNA viruses, one can also use an inhibitor of host-dependent RNA synthesis, such as actinomycin D, to analyze viral RNA synthesis independent of host RNA synthesis. For DNA viruses, the size of the viral DNA or specific hybridization probes can be used to distinguish viral DNA from host DNA. In some cases (e.g., T-even phages) one can use the presence of unusual nucleotides to identify viral DNA.

1. RNA VIRUSES

The most salient features of the replication of RNA viruses is that RNA viruses lack unwinding and ligating enzymes and must replicate in an assynchronous fashion from the viral ends. They cannot replicate *both* strands from an internal initiation site, nor replicate in a discontinuous fashion. Two general types of RNA replicative structures can be found:

i) the replicative intermediate (RI), which includes a single-stranded template with several newly-made transcripts peeling off (Fig. 1A), and
ii) the replicative form (RF), which involves the formation or utilization of a double-stranded structure (Fig. 1B).

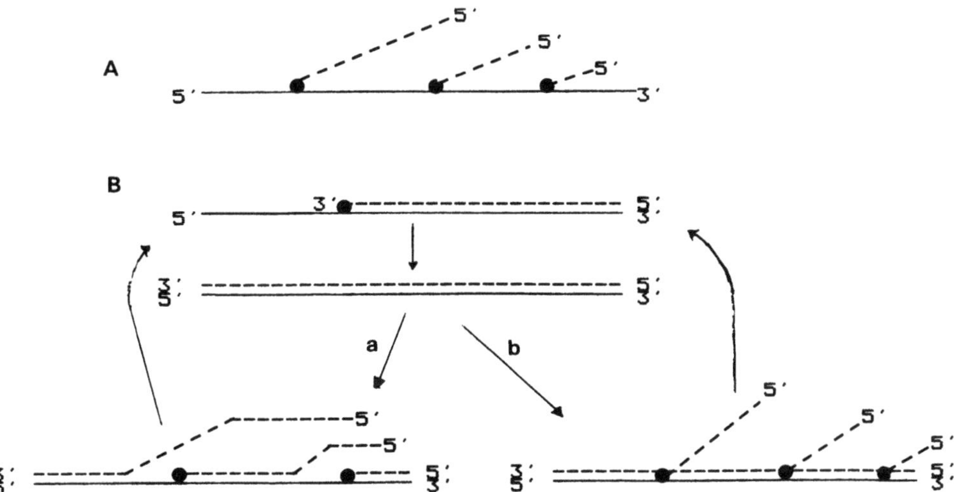

Figure 1. Mechanisms of RNA Replication. A. Replication via a replicative intermediate (RI). B. Replication via a double-stranded replicative form (RF). a. Semiconserative replication of RF. b. Conserative replication of RF.

The replication of RF can proceed either by semi-conservative displacement or in a conservative manner to produce ssRNA (+) strand, (−) strand, or mRNA. Note that the semi-conservative displacement replication of RF differs from RI only in the initial presence of ds RNA and in the extent of hybridization between newly-synthesized and template strands. Thus ds RNA viruses or RF structures produce displaced single strands which subsequently serve as template for formation of the complementary strand (3).

In general viral RNA replication exploits features of host transcription/translation systems rather than host replication systems. Specifically, some RNA viruses utilize tRNA-like structures (4, 5, 6), host translation factors (7, 8), host 3' polyadenylation enzymes, 5' capping enzymes, and even 5' fragments of host mRNA (9-14) in the replicative or transcription process. Viroids, the only RNA virus-like replicative entity which do not rely on viral replicase (plant virus satellites (15) do), rely on host DNA-dependent RNA polymerase (16, 17).

With the exception of a very few covalent circular single-stranded RNA species — not viruses but viral satellites — replication starts at the viral ends. However, many RNA viruses have strong intra-strand hydrogen-bonding giving a complex secondary structure typically having short hairpin or stem-loop structures. Some viruses have complementary ends or include proteins which interact to create a non-covalently closed circle. In many cases, the full-length virion and/or antivirion RNAs are always encapsidated.

The RNA replication/transcription mechanism must, at minimum, accomplish two discrete tasks. It must:

i) generate full-length genomic RNA apt to be encapsidated, and
ii) secure the synthesis of trnaslatable mRNAs.

Except for picornaviruses and the small RNA-containing phages (in which genomic RNA is identical with mRNA) *replication* (i.e: the synthesis of full-length genome-like RNA), and *transcription* (i.e: synthesis of viral messenger RNAs) are two totally distinct processes that take place at different times and under different conditions. The mechanism(s) that trigger the shift from *transcription* to *replication* will be discussed in detail in the following chapters. For the purpose of this review it is important to stress that in order to produce full-length genome-like RNA the replication machinery must generate a full-length anti-genomic strand and transcribe the template so generated into RNA copies of genomic polarity.

The requirement for complete viral information in both strands can be circumvented if there are non-template-directed mechanisms for regenerating complete viral information on the strand(s) which is/are to be packaged as virion (7, 21). In the case of multipartite viruses, the individual RNA segments typically use the same basic replication mechanism (not necessarily the case with contaminating satellite species) and can be treated as minor variations of unipartite genomes for purposes of examining the replication process. Multipartite genomes have more importance as a *transcriptional* mechanism than as a distinctive *replicative* one. Because of the highly assynchronous nature of RNA replication, one must examine the synthesis of (+) and (−) full-length transcripts separately, since the 3' ends of the (+) and (−) strands can (and do) differ in some viruses, thus requiring (in some cases) different initiation mechanisms.

In addition, because viral (−) strands are also used as the template for mRNA synthesis, a third task is often imposed on the replicative system, the production of mRNA.

Usually, the mRNA is distinctive from the full-length (+) transcript. Often the mRNA is subgenomic (initiated internally, terminated early, or both) or the mRNA may utilize distinctive 5' ends (e.g., by loss of VPg in picornaviruses, or the presence of host-derived 5'-capped sequences as in orthomyxoviruses or bunyaviruses). Often there is a sequential nature to mRNA, (−) strand, and (+) strand transcription, with a switch from greater mRNA production to greater virion transcription as infection proceeds. The distinctive nature of mRNA may play a role in preventing its encapsidation and directing it into polysomes.

In principle, RNA replication of both strands can proceed from the precise 3' end of the template strand. Since only a single virus-encoded RNA-directed RNA polymerase is present in RNA viruses, other factors (host- or virus-derived) are usually involved in directing the polymerase to initiate at the 3' end of (+) or (−) template, to produce mRNA, and, in some cases, not to terminate early during full-length (+) strand synthesis.

A. Positive-strand RNA Viruses

This category includes the bacteriophage group Leviviridae (e.g., Qβ, MS2, f2, R17). Animal (+) single-stranded RNA virus groups include the Caliciviridae, Picornaviridae, Alphaviridae, Flaviviridae, Coronaviridae, and Nodaviride. And most plant viruses are (+) single-stranded RNA viruses. These viruses exhibit considerable variation in the initiation of synthesis of (+) and (−) strands, often, but not always, using different mechanisms for each strand.

Qβ phage replication and tRNA-like ends

In addition to phage-specified core replicase, the complete replication complex of Qβ phage requires:

i) the 30S ribosomal S1 protein (22) complexed to replicase,
ii) the protein elongation factors EF-Tu and EF-Ts (23) acting as a EF-Tu.Ts complex, and
iii) another host factor protein (24) needed for initiation. This factor has single-stranded RNA-binding activity, and the level of factor determines the ratio of (+) to (−) strand synthesis.

The complete replicase complex recognizes several regions of the Qβ RNA sequence or structure (7) and binds, in the presence of GTP and Mn^+, to the 3' end of phage RNA, a 3' CCCA-OH sequence. Transcription of the (−) strand starts with a GTP opposite the penultimate C and ends with a C opposite the 5' terminal pppGGG of the (+) strand. At this point a termination event adds a 3'-terminal A (25, 26) in a non-template-directed fashion. A similar process occurs in the replication of (+) strand RNA, thus regenerating the terminal A lost during (-) strand replication. The non-templated terminal A may be added by the replicase since *in vitro* replication systems produce full-length virus.

Despite the nominal correspondence of the viral end to the 3' termini of tRNAs, which can have their 3' ends completed post-transcriptionally by tRNA nucleotidyltransferase, neither the replicase (7) nor tRNA nucleotidyltransferase (27) can add a terminal A to *free, complete* (+) or (-) strands which have had their terminal A removed. This may reflect an altered secondary structure of isolated complete strands (compared to newly-synthesized strands associated with replicase). Endonuclease-cleaved phage (as opposed to intact phage) treated with snake venom phosphodiesterase can have terminal CCA added to two specific fragments by tRNA nucleotidyltransferase-including what appears to be 3' terminus (27). Regardless of the possible enzymatic roles of the host factors in replication, it is evident that recognition of tRNA-like structures (presumably specific stem-loop structures) plays a role in assembly of replicase complex at the appropriate site.

The 3' ends of a number of plant (+) strand RNA viruses have a tRNA-like structure (21). These structures interact with host tRNA-related enzymes. Indeed, in some cases, the packaged virion, in at least part of the population, lacks the terminal A, which can be added post-infection by tRNA nucleotidyltransferase (21). The plant virus tRNA-like 3' structures, unlike the RNA phage 3' end, can also be aminoacylated by the appropriate tRNA aminoacylsynthetase (21). In analogy with the binding of EF-Tu and EF-Ts by RNA phage, the aminoacylated (but not the unacylated) tRNA-like structure binds EF-1 (4, 6, 21) in the presence of GTP. Recent analysis of the ability of the tRNA-like end of brome mosaic virus to recognize viral replicase or tRNA aminoacylsynthetase indicates that the recognition of viral end by replicase is affected by deletions in arm B and in the anticodon of arm C (Fig. 2). Base substitutions in the anticodon loop also causes loss of replicase-binding activity without affecting aminoacylation (29). In contrast, aminoacylation is affected by deletions of arm B in a manner generally proportional to its effect on replication (with some exceptions which affect aminoacylation without greatly affecting replication) but is not greatly affected by deletions in arm C. These plant viral terminal structures can form alternate secondary structures (4, 6, 21), which may also play a role in replication.

The ability to produce specific mutations, produce full-length viral RNA sequences (30), and to create *in vitro* replication systems (31) should help to resolve some of the details of the replicative role of tRNA-like struc in the near future. Of particular interest will be the differences in the replication complexes used in producing (-) antivirion and (+) virion.

Picornavirus replication and the role of VPg

Picornaviruses, using poliovirus (see chapter 8 and ref. 32, 33, for reviews) as an example, are single-stranded (+) RNA viruses which are polyadenylated at the 3' end. The 5' end is linked to a short peptide (VPg), which, however, is not essential for infection (34, 35). Other VPg-containing (+) RNA viruses, however, do re-

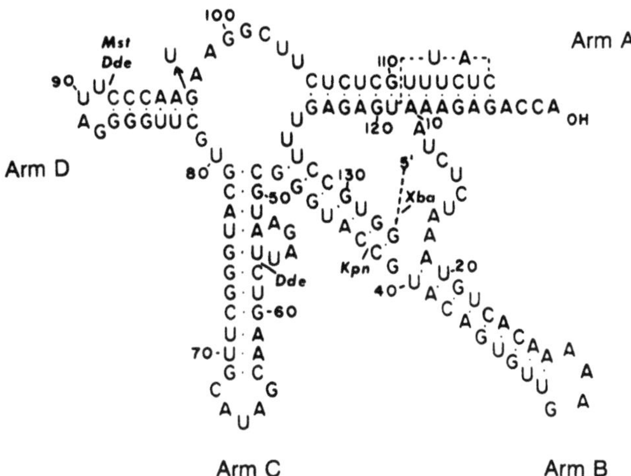

Figure 2. The 3' tRNA-like structure of brome mosaic virus. The role of arms B and C are described in the text. Reprinted with permission (reference 28).

quire VPg for infectivity. Although VPg-containing poliovirus RNA can act as mRNA *in vitro* (36) and are sometimes found in polysomes late in replication (37), most (+) strands found in polysomes end in 5'-pUU-3' and have had the VPg removed (34), presumably by a host enzyme found in uninfected cells (38). This enzyme has an unknown normal cellular role. The level of this enzyme and coat protein during the course of infection may play a role in regulating the fate of the (+) transcript. In contrast to the non-essential nature of VPg for initial infectivity, removal of the poly(A) tail greatly reduces infectivity (39).

The virion (+) RNA is mainly translated from an internal AUG as a polyprotein which is proteolytically cleaved to generate non-structural and structural proteins (chapters 5 and 8 of this volume, and ref. 32, 33). Among these peptides are the virally-encoded RNA-dependent RNA polymerase (40) and VPg.

The first step in viral replication, the synthesis of (-) antivirion from (+) virion, can be performed *in vitro* and requires 2 or 3 proteins: the viral polymerase, a host factor recently identified as a terminal uridylyl transferase (41, 42), and possibly, a VPg donor. Recent evidence (43-45) suggests that VPg is *not* required *in vitro* for initiation of (-) strand of the virion (+) strand.

Two models of replication have been proposed: according to the first one, the host factor would add an oligo-U sequence to the poly-A tail, leading to hairpin formation. The hairpin would then be extended by viral replicase. *In vitro*, this results in a 2-unit length fold-back genome with VPg attached only to the 5' end of the original (+) virion template. Since, *in vivo*, the 5' end of the (-), antivirion strand has VPg (46, 47), the hairpin, if formed *in vivo*, must be quickly cleaved, presumably with addition of VPg (see Fig. 3A). VPg, in such a model, might play a role in preventing the formation of ds RNA — the RF (replicative form) which has been considered a dead-end product.

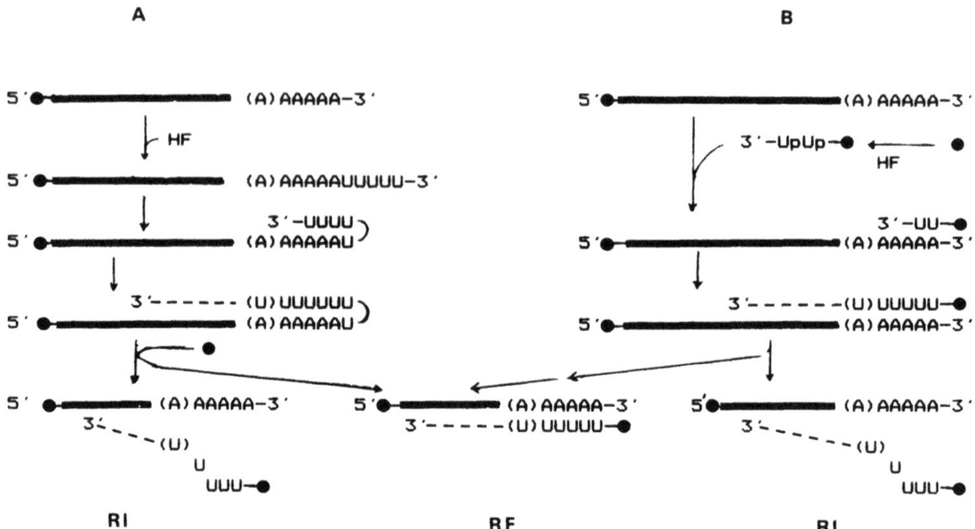

Figure 3. Two models of picornavirus replication. A. Replication via hairpin formation. B. Replication via use of VPg primer.

Most nascent virus RNA is in the form of replicative intermediates (RI) with the template being transcribed by multiple replicases (32, 33).

In a contrasting model of replication, VPg (or precursor) can itself be uridylated to VPg-pUpU (47, 48). Although this has not been chased into longer RNAs (47), it could act as a primer for either (–) or (+) synthesis (50), providing an alternative model for (–) strand replication (see Fig. 3B). At the present time, it is not possible to distinguish between these models.

Replication of (+) strand starts with a 5' UU sequence, which is highly conserved in all picornaviruses (Fig. 3). The 5' sequence is different from the complement of the sequence at the 3' viral (non-poly(A)) end (51). This, of course, may explain why poly(A) is necessary for replication. The 3' viral (non-poly(A) ends are conserved only within genera (51). However, using cloned cDNA, a small insertion in this 3' non-coding region produces a *ts* mutant for productive infection (52) indicating that these sequences are important. The importance of the natural 5'-UU sequence to virion production can be seen from *in vitro* synthesized rhinovirus RNA. The RNA produced *in vitro* has an additional 21 nucleotide 5' of the natural 5' end. This RNA is infectious, but when the progeny are recovered, the natural end has been regenerated, despite the presence of a second UU in the 5' added sequence (53). A similar process may occur in virus produced by transfection of cloned DNA copies of poliovirus (54). The regeneration of natural viral 5' ends is compatible with either model of picornavirus replication.

The role of VPg proteins in the replication of plant and other animal (+) single-stranded RNA viruses is even less clear, although in some cases the VPg is necessary for infection (51). Similarities to picornaviruses do exist in plant viruses (55).

Clearly, one must be cautious about ascribing a priming role to VPg proteins. RNA polymerases, unlike DNA polymerases, do *not* require a stabilized 3'0H primer and *can* start synthesis directly at the 3' end of a template. However, doing so usually involves more than the core polymerase itself, and one mechanism of initiation could involve a priming protein. In such a model, the difference between those viruses in which VPg is necessary for infection and those in which it is not may be similar to the distinction between adenovirus and Φ29 bacteriophage, the linear DNA viruses that utilize priming proteins (see Section 2). The recent development of cloned poliovirus, which can be used to produce poliovirus RNA (56), will allow the production of specific mutants which may resolve some of these questions.

Alphaviruses, Flaviviruses, Coronaviruses, and timing of (-) and (+) strand synthesis

Alphaviruses, Flaviviruses, and Coronaviruses have radically different mechanisms for transcription of mRNAs off of the anti-genomic, (-) strand.

Alphaviruses produce a single subgenomic mRNA (in addition to a genome-length (+) strand which acts as virion RNA *and* mRNA). The genomic RNA is directs the synthesis of a polyprotein encoding 4 nonstructural proteins, the last of which requires read-through of a stop codon (51) for translation. Both (+) strand RNAs are capped and have poly(A) (chapter 9, and ref.51).

Flavivirus (+) RNA, in contrast, is capped but not polyadenylated (51). ***No subgenomic mRNAs are produced.*** Rather, like picornavirus, a polycistronic protein is produced which is cleaved during translation in such a way that the polyprotein never actually has an independent existence (57). Little is known about flavivirus replication.

The genomic RNA of Coronaviruses is capped and polyadenylated and a nested series of subgenomic mRNAs has been identified in coronavirus-infected cells.

All of the mRNAs have the same 5' leader sequence (58-60), which apparently primes the nested subgenomic mRNA species. All the genomic and subgenomic (+) mRNAs terminate at the 3' terminus of the virion RNA (51) and are polyadenylated. Only the 5'-most open reading frame of each mRNA is translated (51).

However, these disparate viruses may replicate by similar mechanisms. In each case, the 3' end of the (-) and (+) strands, upon which the replicase complexes must be assembled, are quite different (51, 61, 62), both in sequence and in secondary structure (62), allowing temporal regulation of (+) strand, (-) strand, and mRNA synthesis.

We will examine the replication of alphaviruses, especially Sindbis virus, in detail. A model has been proposed (63) whereby the preferred template for replication is the (-) strand, with preferential initiation of (+) strand occuring at the internal conserved subgenomic mRNA site (51). Typically, (-) strand production occurs early in infection in both alpha (64-66) and coronaviruses (67), and is later shut off or greatly reduced. Late in infection, the rapid inclusion of full-length (+) strands into capsids may deplete available template for (−) strand synthesis. Recent work (68) with a cloned defective interfering (DI) particle of Sindbis virus shows that only sequences in the 5'-most 162 nucleotides and 3'-most 19 nucleotides of the DI are needed in *cis* for its replication and packaging.

The 3' end of the DI is the same as that of the native virus, and is strongly conserved in all alpha viruses (69) and natural DI particles.

The 5' end of this particular cloned DI, like many Sindbis DI particles generated in chicken embryo fibroblasts (but not baby hamster kidney cells) (70,71) have nucleotides 10-75 of cellular $tRNA^{asp}$, in this case replacing the 5'-most 30 nucleotides of native viral sequences. The common occurrence of $tRNA^{asp}$ in DIs from these cells may imply a role of tRNA aminoactylsynthetase in replicase assembly, perhaps by recognition of a viral template structure. The recognition need not be specific for $tRNA^{asp}$, if this tRNA is the only one whose complement permits replicase assembly.

In this respect, it is of interest that viral 5' *sequences* are not strongly conserved among alphaviruses (62), but a *structure* (which would also be present on the 3' end of the template (−) strand) appears to be conserved. Such a complex betweeen 3' template structure and host aminoacylsynthetase may, occasionally, put an uncharged tRNA in a position to be (aberrantly ?) extended by replicase.

Other DI particles have either native 5' ends, 5' ends which have undergone sequence rearrangement, and even one which has 100 nucleotides of the 5' end of the subgenomic mRNA added to the native 5' viral end (70). Clearly the ability to assemble replicase at the 3' end of antivirion RNA does not require a specific sequence, nor (given the ability of rearranged 5' ends to function) does replication require, necessarily, that recognition structures be positioned precisely at the 3' end of the template. This does not mean, however, that there is much flexibility at the 5' end: removal of even 7 nucleotides from the 5' end of the cloned $tRNA^{asp}$-containing DI eliminates its ability to replicate.

Another strongly conserved region in alphaviruses and their naturally occurring DI particles is a 51-nucleotide sequence (62) seen at 155-205 bases from the natural viral 5' end. This sequence can be removed in *in vitro* constructed DI particles without eliminating replication and packaging, implying that the sequence is *not* required in *cis* in avian cells. The sequence may, however, have a quantitative effect or play a role in the arthropod host. Little is known about replication of these viruses in their arthropod hosts.

Sindbis virus has two complementation groups in which temperature-sensitive mutations affect production of (−) strands:

i) complementation group B contains a temperature-sensitive mutant, *ts*-11 (65), whose normal product is required for production of (−) strands; and
ii) complementation group A including a second group of temperature-sensitive mutants (e.g: *ts*-24), whose active product is necessary for shutting off (−) strand replication. The assignment of complementation groups to specific polypeptide products is incomplete.

The group A product is probably not acting as a protease which inactivates the B product, since shifting to the non-permissive temperature late in replication in the absence of new protein synthesis results in (-) strand synthesis, indicating that functional B product still is present.

Since both the (+) genome-length RNA and the subgenomic mRNA include the same 3' sequences, and no truncated (-) strands are seen, the (-) strand replicase complex must recognize both the 3' and 5' end of the (+) strand simultaneously in order to assemble the replicase enzymes only on full-length (+) RNAs (51), presumably aided by the possibility that the virion, *in vivo*, may be present as a non-covalently linked circle (72). Perhaps B product is involved in directing the elongation factor — the presumed core polymerase (complementation group F) — to the 3' end of (+) strand full-length virion. Complementation group A product, then, may subsequently block binding of polymerase to the B product. Synthesis of the (-) strand starts within the poly(A) tail and produces a poly(U) 5' sequence on the (-) strand (73), which, like all (-) RNA strands is uncapped. Extensive production of (-) strand requires continued protein synthesis, indicating that one of the replication complex factors is unstable (64). Production of (+) strands (both virion and subgenomic mRNA) does not require continued protein synthesis (74).

Coronavirus replication also exhibits a time-dependent reduction of (-) strand synthesis, although the reduction is not as sharp as in alpha viruses. Here, also, synthesis of (-) strand is more sensitive to inhibition of protein synthesis than is (+) strand synthesis or mRNA synthesis (67), and the polymerase complexes involved in replication of (-) and (+) strands may differ from each other (75, 76).

In these viruses, the role of host factors is unclear but may be of major importance in assembling the replicase complex on either virion or antivirion RNA. Any use of host factors by alphaviruses, however, must be compatible with the divergent eucaryotic organisms (both arthropod and vertebrate) within which these viruses replicate.

The absence of similar 5' and 3' structures or sequences at the virion ends points out that replication recognition signals need not be the same at the 3' end of (-) and (+) templates, and that these differences can be exploited to time the production of virion RNA, mRNA, and antivirion RNA.

B. Negative-strand RNA Viruses

The replication of negative-strand, single-stranded RNA viruses can be considered in two distinctive groups:

i) the unipartite viruses, rhabdoviruses and paramyxoviruses, in which the process of transcription and replication are clearly distinct: in addition to producing a full-length anti-genomic (+) transcript which is uncapped and not polyadenylated (*replication*), the virus transcribes a number (5-7) of monocistronic mRNAs which either initiate internally or are processed to an internal start site and which typically terminate prior to the 5' end of the virion RNA (51, 77);

ii) the multipartite negative-strand single-stranded RNA virus groups orthomyxoviruses, bunyaviruses, and arenaviruses in which, in general, each RNA segment produces a single monocistronic mRNA (although a few RNA segments, via mRNA splicing, proteolytic processing, or via an ambisense RNA segment which encodes sense information for one gene at the 3' end of the (+) strand and sense information for a second gene at the 3' end of the (−) strand, do produce more than one protein).

The latter group of viruses — with the possible exception of arenaviruses, about which little is known (78) — utilize capped host mRNA 5' fragments as primers for transcription of their mRNA. The mRNA terminates at a polyadenylation site near, but not at, the 5' end of the template (−) strand. In contrast, the (+) strand full-length template is not capped, starts without host primer, is encapsidated like the (−) virion, and does not terminate at the polyadenylation site, but proceeds through to the end of the (−) template (77, 79).

In both types of virus, then, the (−) template is used to produce two distinctive products, capped and polyadenylated subgenomic mRNAs and uncapped, unpolyadenylated full-length (+) transcripts.

All of the negative-strand RNA viruses, unlike many positive-strand RNA viruses, contain complementary or nearly complementary sequences of between 10-20 nt at the extreme 3' ends of (+) and (−) templates indicating similar recognition signals for replicase assembly on both strands. In the case of the multipartite viruses, the 3' terminal sequences are very similar on the different segments (51, 77, 79).

However, other sequences at the 3' end of the templates probably play a role in both the switch between mRNA production and (+) strand synthesis and in the switch from producing more or less equal amounts of (+) and (−) strands to producing primarily (−) strands (virions) late in infection.

Rhabdoviruses and Paramyxovirus

The rhabdovirus Vesicular Stomatitis Virus (VSV) and the paramyxoviruses Sendai and New Castle disease virus (NDV) are cytoplasmic and do not require continued cellular mRNA synthesis in order to transcribe viral mRNA (80, 81). VSV is quite capable of replication in an enucleated cell, whereas the paramyxovirus require a nucleus for virion production, although the nucleus need not function biochemically (81). Transcription and replication of VSV (see Fig. 4) occurs

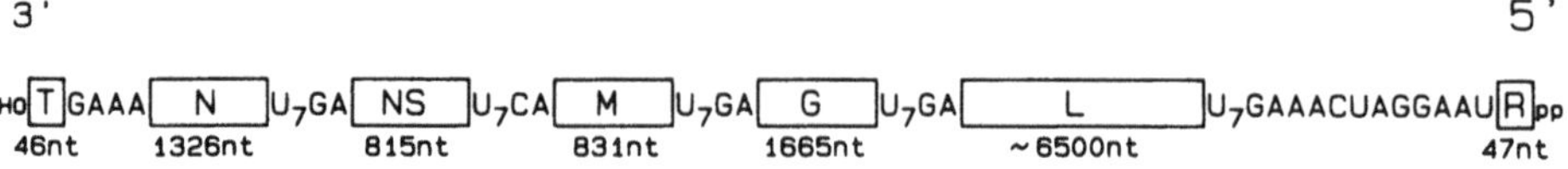

Figure 4. Structure of VSV. T and R represent the terminal leader sequences of (−) and (+) virion, respectively. N, NS, M, G, and L represent mRNA sequences of viral genes. Each mRNA is capped, starts at the 5' end of each box and ends within the oligo (U) sequence at the 3' end of each box. Reprinted with permission (reference 51).

on an RNA-N protein complex. Transcription requires NS and L proteins, both of which are required for polymerase activity. NS protein, but not L protein, can bind to the RNA-N protein complex (82). NS protein can be phosphorylated and the degree of phosphorylation — high (NS2) or low (NS1) — may influence whether replication or transcription occurs (51, 83, 85), perhaps related to the influence of pH (86, 87). Phosphorylation and dephosphorylation appears to be accomplished by host kinases and phosphatases.

Two other proteins have been implicated in viral replication:

i) the M structural protein, which may, by interacting with nucleocapsid at the 3' region of the (-) strand, prevent substantial transcription off of the (-) strand. *In vitro*, its presence permits synthesis of only the leader and 14 nt of the N gene. It may also, by helping to package the (-) strand, prevent continued synthesis of (+) template (88); and
ii) the N protein is also involved in replication since neither full-length strand is present as unencapsidated RNA.

It is clear that continued protein synthesis is necessary for VSV *replication* but not for *primary transcription* (89). This may reflect a requirement for N protein to bind the (+) strand leader (89), or it may reflect a need for newly-synthesized (altered ?) NS protein.

A model for N protein-mediated regulation has been proposed (91) which accounts for the role of the leader sequences in transcription and replication. Basically, it is proposed that both replication and transcription initiate at the extreme 3' end of (-) or (+) strands. This transcript is not capped, and transcription continues through to a decision point (45-47 nucleotides in the case of (+) strand leader synthesis (92, 93), 46 or 50 nt in the case of (-) strand leader synthesis (94).

At this point, some signal — either the level of N protein (91), the degree of phosphorylation of the NS protein of the replicase (51, 83, 87), or perhaps interaction with host protein (93, 93, 95) — will result in one of two choices:

i) there will be either termination (or possibly, processing) of leader transcript and re-initiation (or processing) of what is now mRNA transcription 4 nucleotides downstream from the termination site (96). This new mRNA transcript is capped (presumably, by viral enzymes); or
ii) transcription will continue through this decision point resulting, eventually, in full-length transcript.

To produce full-length (+) antivirion transcript, however, replicative transcription must continue through several internal transcription termination sites. In contrast, mRNA transcripts terminate at these signals (or are processed at them) and transcription usually, but not always, restarts 2 nucleotides downstream of each termination site. Each new mRNA transcript is capped and a polar effect is observed, with 5'-most transcripts present at higher levels than 3'-most transcripts. The mechanism which prevents termination in full-length transcripts is not known. Perhaps the presence of N protein on (+) replicative RNA or the presence of cap on mRNAs signal whether the replicase will proceed through these termination signals. Anti-termination, of course, could also represent a property of differentially-modified NS protein or even an indirect effect, such as changed cell pH. The (+) replicative RNA proceeds through the final mRNA transcription termination/polyadenylation site to the end of the (-) template and is not polyadenylated. It is also uncapped—although the 5'-terminal γ phosphate is usually removed to leave a ppA 5'-terminus.

In producing (-) virion RNA from full-length (+) antivirion RNA, it is proposed that a similar process occurs, since the inverted terminal repeats at the ends of the viral strands present the replicative machinery with similar end structures. Here also, the replicase appears to pause after producing a short leader sequence. However, signals indicating termination of nearly synthesized (-) strand are rare, although termination does occur. Typically, then, full-length (-) strand transcription continues.

This model would be strengthened if N protein were found to bind processively and co-operatively to RNA from a signal sequence within the 5' leader. Replication of the paramyxoviruses is probably quite similar (51, 77, 97) given the similarities in genome sequence, gene order and transcriptional processes.

Orthomyxoviruses and Bunyaviruses

The replication of segmented orthomyxoviruses (chapter 13) has been recently reviewed (14, 77) and, like the other negative-strand RNA viruses, the (+) antivirion RNA and (-) virion RNA are always present as nucleoprotein, in contrast to the mRNAs.

The mRNAs also differ in sequence from the (+) antivirion RNA. In the case of influenza (orthomyxoviruses), the virion contains has 8-9 RNA segments. The 3' ends of each of the segments are largely identical to each other within the first 13-22 nt, and very similar to the 3' ends of the (+) antivirion strands (Fig. 5), differing primarily in the presence of different nucleotides at position 3, 5, and 8 (98). These conserved sequences are probably involved in the recognition of the 3'-ends by the transcriptional enzyme complex. There are at least 3-4 proteins involved in the transcription/replication complex. Two of these, PB1 (P1) and PA, probably represent the core polymerase functions, required for both transcription and replication (19). A third protein, PB2 (P3), forms a complex with PB1 and PA during transcription (99). It is unclear whether PB2 plays a role in replication (100). The fourth protein, NP, is the nucleocapsid and complexes both (+) antigenome and (-) virion, with the mRNAs remaining unencapsidated.

Influenza virus transcription, and probably replication, occurs in the nucleus (101). Transcription/replication occurs early during infection (102) and cannot occur in the absence of continued host mRNA synthesis (81, 103). The PB2 protein acts as a cap-binding protein which binds new host mRNA and cleaves it 10-14 nucleotides downstream of the cap, preferentially after a purine (12, 104). This host mRNA stretch bound to PB2 then becomes complexed to the PB1 and PA proteins. The host mRNA oligonucleotide is positioned such that its 3' purine (typically an A) is opposite the 3' terminal U on the (-) template. The PB1 protein then binds GTP, complementary to the penultimate C on the (-) template (104) and polymerization proceeds. After initiation, the complex of all three proteins,

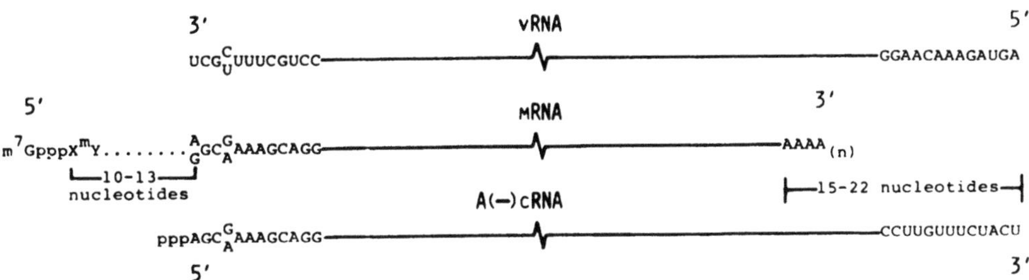

Figure 5. 5' and 3' sequences of influenza virus virion RNA, antivirion RNA, and mRNA. Reprinted with permission (reference 14).

with the cap still associated with PB2, continues transcription for at least 10 nucleotides before the 5' end dissociates from PB2. The replication complex continues until it reaches a tract of 5-7 U residues located about 20 nucleotides from the 5' end of the template. At this position, the capped influenza mRNA terminates (105) adding variable amounts of poly(A). The result is an mRNA transcript which includes non-template 5' sequences and lacks the final 15-22 nucleotides of (-) template (see Fig. 5).

In contrast, *replication* of both (+) and (-) full-length strands initiates directly at the 3' end of the template (HO-UpCp) leaving a triphosphate at the 5' end (pppApG). *Replication* does not terminate at the poly(U) stretch, but proceeds through to the end of the template 15-22 nucleotides downstream.

It is possible that nucleocapsid binding may play a role both in antitermination and, perhaps, in determining whether transcription or replication is initiated. There is essentially no knowledge of the mechanism by which (-) strands are preferentially exported or by which the different segments are appropriately packaged together.

The replication of bunyaviruses is probably, to a large extent, similar to that of the orthomyxoviruses (79). These viruses have a tripartite genome, the segments of which are rarely packaged in an equimolar fashion. The 3'- ends of (+) and (-) full-length transcripts are quite similar for both the S and M subunits of Snowshoe Hare bunyavirus (107) and the two segments also have considerable sequence homology at their ends.

The mRNAs of both the S and M fragment utilize host-derived 5'-capped oligonucleotides of 12-17 nucleotides to prime mRNA synthesis. Both terminate mRNA transcription before the 3' end of template — about 100 nucleotides before in the case of S segment, about 60 nt in the case of the M segment (107). Lacrosse virus has been shown to have a primer-stimulated RNA polymerase and a methylated cap-dependent nuclease (101). Full-length transcripts start precisely at the 3' end of template, are present as nucleoprotein and proceed through the mRNA termination sites. Unlike orthomyxoviruses, however, bunyavirus replication/transcription is cytoplasmic and does not require continued host mRNA synthesis. Presumably the source of host 5'-capped oligonucleotides is preformed host mRNAs (79).

In contrast to the many positive-strand RNA plant viruses, there are only two groups of negative-strand RNA plant viruses: plant rhabdoviruses and tomato spot wilt virus, which has been suggested to be a bunyavirus (109). There are no known negative-strand RNA bacteriophage.

C. Reovirus and Double-stranded RNA Replication

When reovirus enter lysosomes, the virus particle becomes uncoated with consequent activation of the double-stranded RNA-dependent RNA polymerase (110). The intermediate subviral particles probably then enter the cytoplasm where transcription occurs.

The virus contains 10 double-stranded RNA segments (111) which appear to be perfectly base-paired with no single-stranded RNA tails (112), and which are capped at the (+) strand 5' termini (113, 114).

Transcription, like replication, starts at the precise 3' end of a (-) strand and proceeds to the 5' end of template (115). No poly(A) tail is present. Only the (+) strand is transcribed off of ds RNA, and transcription is fully conservative (116) with neither parental strand appearing among the transcription products. Elongation, rather than initiation, appears to be rate limiting. Transcription from different RNA segments, each of which encodes a single message, does not all occur simultaneously, some segments being transcribed early and others later (117).

Replication proceeds initially like transcription, by conservative transcription of a capped (+) strand off the parental ds RNA template which remain in a subviral particle with active viral capping enzymes. Eventually these capped transcripts become associated with a new subviral particle where (-) strand synthesis occurs. The initiation, but not elongation, of (-) strand synthesis off of the (+) template in the new subviral particle requires protein synthesis (115). The newly-synthesized (-) strand remains associated with the (+) strand to form a new ds RNA molecule (118). Only a single round of (-) synthesis occurs on each (+) template (119). Like transcription, it appears that those (+) strands destined for the replication of certain of the ds segments appear earlier than those of other segments (120).

In contrast to the conservative replication seen in reoviruses and in the noninfectious ds RNA mycovirus of *Saccharomyces cerevisiae* (121), replication of the ds RNA bacteriophage Φ6 (122) and other mycoviruses (123, 124) appears to be by a semiconservative displacement mechanism (see Fig. 1B).

D. RNA Circles: Viroids, Virusoids and Plant Satellite Viruses

Although covalently closed circular genomes and multimer production are common features of many DNA viruses (see Section 2 of this review), they are only rarely seen in RNA. Indeed, they are mostly seen in plants — the delta "virus" associated with hepatitis B infection is an exception — and are associated with small satellite viruses, virusoids, and viroids (15, 16, 19, 125). These RNA elements typically encode few, if any, obvious proteins, but utilize either viral RNA-directed RNA polymerase (required for virusoid and satellite virus replication) or host DNA-directed RNA polymerase (capable of being utilized by viroids).

Multimeric RNAs of Satellite Tobacco Ring Spot Virus (STobRV) (126, 127), potatoe spindle tuber viroid (PSTV) (128), and possibly some virusoids (129) have the apparent ability to generate (in some cases spontaneously) monomeric subunits which are 5' HO-Ap and 3' C-2',3' cyclic phosphodiester (126-128). The monomer can, in some cases, also spontaneously dimerize (or circularize) due to the nature of the cyclic phosphodiester — the number of phosphoester bonds is unchanged in linear monomer, circular monomer, and dimer forms (126, 127). In other cases, RNA ligase may be involved in circularization (128, 129). That the most abundant form of STobRV are linear monomers whereas that of viroids is circular monomers may be irrelevent to their replicative scheme, since host cells may contain activities (such as RNA ligase) which permit, assist, or catalyze circularization (128-130) or oligomerization (131). One can envision production of multimeric or circular (-) strands off a circular (+) template by a rolling circle model, which, in turn, can act as a template for production of multimeric (+) strands. The multimeric (+) strands can then (possibly autocatalytically) produce circular or monomeric products. Alternatively, oligomerization might, in some cases, proceed in the absence of circularization by processes similar to splicing (131), producing 5' phosphates and 3'-OH linear monomers.

These subviral RNAs are invariably small and exhibit very high intrastrand complementarity. These features may be requirements for replication, cleavage, and ligation of these RNA molecules and may limit the ability of such RNAs to code for proteins. This may limit the use of a rolling circle or multimeric replication strategy in RNA to non-informational parasitic molecules, despite the obvious enzymatic capabilities of RNA molecules (132, 133).

One should note the similarity of some of these processes to those which are involved in processing tRNA precursors. The autocatalytic (132) or RNase F-assisted (134) cleavage of a tRNA-like T4 RNA (or of tRNA precursor) depends upon a stem-loop structure and produces a 5'-OH adenine and 2',3' cyclic phosphodiester C, as is the case in PSTV (128) and (+) strand of STobRV (126, 127). These ends can be annealed by RNA ligase, which normally acts on tRNA precursors, to initially produce a CpA dinucleotide in which the C has a 2' phosphate in addition to the 3',5' phosphodiester linkage to A. This unusual nucleotide is seen in some virusoids (129).

The complementarity of the (-) strand of PSTV to U1 RNA, discussed in the preceding chapter, (16) and the intron-like features of peanut stunt virus associated RNA 5 (131) may mean that other (splicing-derived?) features may be involved in the production of cyclic, monomeric, or oligomeric forms of these replicative RNA molecules.

E. RNA Replication: A Summary

RNA virus replication studies have (and will) be greatly aided by the recently acquired ability to clone DNA copies of the viral (or subviral) genome and produce RNA transcripts in vitro. This allows one to explore the structural and sequence requirements for replication in a precisely defined way. One is also struck by the extensive utilization of host factors and processes which normally utilize tRNA molecules and by the extent to which "ends" play a role in RNA replication. With the exception of circular RNAs, replication (as opposed to transcription) starts at one end of a virus and proceeds to the other. In many cases, the virus or its mRNA is capped, and this is true of those viruses which are cytoplasmic as well as those which are nuclear, indicating that the capping enzyme activity may be viral in origin in some cases. Only (+) strands are capped, and presumably 5' sequences (structure ?) of full-length and subgenomic (+) RNAs are important in determining whether the sequence is capped.

2. DNA VIRUSES

DNA viruses differ from RNA viruses in that they need not encode their own polymerase (although many of the larger ds DNA viruses do), in that many can integrate into the host genome (retroviruses (see Section 3), of course, do so as a double-stranded DNA), and in that recombination can play a role in replication. In general, whereas RNA viruses exploit host translation/transcription proteins as associated factors in their replication, DNA viruses exploit host DNA replication enzymes in their replication.

No known DNA polymerase, unlike RNA polymerases, initiates synthesis on a template in the absence of a stabilized primer providing a nucleotide 3'-OH end.

Typically, this primer is a short ribonucleotide synthesized by an RNA polymerase. Replication of ds DNA is always semi-conservative, like the replication of host DNA. Moreover, the presence of host and viral DNA ligases, topoisomerases, and recombinational enzymes — including transposases — provide a repertoire of alternative structures not seen in RNA replication, perhaps because such enzymes are not seen in transcriptional/translational processes.

The fact that DNA polymerases cannot initiate at the extreme 3' end of a linear template poses problems for the replication of *linear* DNA viruses: how to replace the 5'-most oligoribonucleotide primer with DNA, particularly in the absence of any enzyme (host or viral) capable of adding template-directed nucleotides to the 5' end of an oligonucleotide ? The structural mechanisms by which DNA viruses solve this problem will provide the basis for organizing the following sections.

A. Simple Circles:Papova Viruses

The simplest way to avoid problems of replicating the ends of a DNA molecule is not to have ends, but to replicate via a circular molecule. The simplest model is that of the papova viruses. These viruses are small ds DNA circles which in the nucleus are typically coated with about 24 host-derived nucleosomes containing histones H2a, H2b, H3, and H4 but not H1 (chapters 18 and 19, ref. 135).

The replication complex is largely host-derived; the only virally-encoded protein(s) required for replication are the multifunctional T-antigen (T-ag) proteins. The T-ag proteins are expressed early in viral replication. The large T-ag is found primarily in the nucleus (136) and is the only protein absolutely required for the lytic cycle, although polyoma middle and small T-ag may enhance replication in some, but not all, cell lines (137). T-ag can: 1) Stimulate G0/G1 arrested cells to enter S phase (138-140), thereby inducing the production of the host DNA-polymerizing system needed for viral replication. Induction of S-phase may be due to T-ag activity as a DNA-independent ATPase which is associated with a protein kinase activity (141, 142) or to its ability to bind to host p53 protein (143, 144). 2) Act as the primary regulatory molecule switching the ds DNA virus between early strand mRNA transcription, late-strand mRNA transcription, and initiation of replication, as a consequence of T-ag concentration and specific binding to several sites near the origin of replication (145-148).

Virtually all DNA replication in SV40 initiates within a 60-bp sequence almost entirely within T-ag binding site 2, and initiation requires the binding of T-ag to this site (149). Recent work indicates that the DNA between the pentanucleotide T-ag recognition sequences, to which the dimeric T-ag binds, is "bent" (150). A similar "bent" structure is seen at the λ origin of replication (151). *In vitro* studies (152) indicate that the interaction between bound T-ag and host replication complex is species specific and probably contributes to host-range limitations of the virus. The replicative role of T-ag other than initiation is not clear (153) although there is some evidence that it may remain associated with the replicase complex (154).

The leading (5' to 3') strand initiates with an oligoribonucleotide of 5-8 nucleotides starting with a ribo-A (155). Usually replication proceeds bi-directionally from the origin, with the 3' to 5' template strand being filled in by discontinuous synthesis in the retrograde (relative to fork movement) direction, in a manner similar to synthesis of other ds DNA molecules initiated internally.

Replication appears to pause after about 90% of the template duplication has occurred (156). This is an indication of a rate-limiting step at the final resolution of the two circles, perhaps a requirement for resolution of concatenated dimers (157). Duplex SV40 circles can be produced in in vitro replicative systems (158).

Many bacteriophage (both single-stranded phage in their double-stranded replicative forms, and double-stranded phage) also utilize this simple circular replication mechanism early in infection. Some later switch to a rolling circle mechanism. λ is, of course, a prime example (159).

B. Modified Circular Replication: Rolling Circles and Concatamer Formation

The bacterophage ΦX174, perhaps the most thoroughly analyzed virus in terms of its replicative scheme, is the model for rolling circle models of replication (see Fig. 6). Indeed, because of the extensive use of host enzymes, this phage has been used as a window to explore host replication (160-162).

The first step in replication of ΦX174 is the formation of a preprimosome complex, starting with host n, n', and n'' proteins.

n' binds to a specific 55 nucleotide fragment of phage single-stranded DNA which can form a 44-nucleotide hairpin structure. In the process of binding, ATP is hydrolyzed. The n' protein is capable of using the energy of ATP to propel the final primosome along DNA.

A

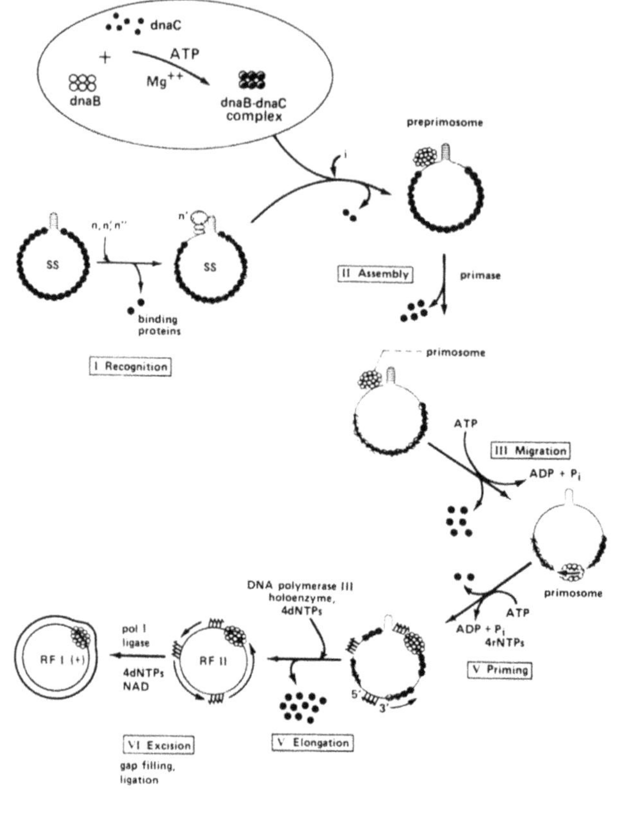

B

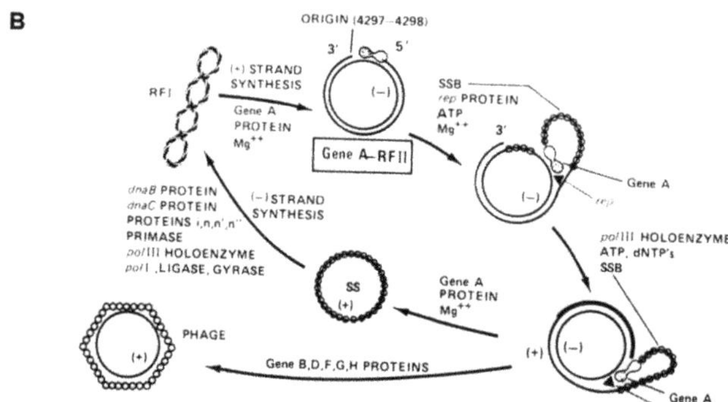

Figure 6. Replication of ΦX174. A. Formation of RF. B. Replication of RF to produce single-stranded virion. See text for discussion. Reprinted with permission (reference 160, 161).

Protein *n* serves to aid binding of dnaB protein, helping to form an initial complex containing *dnaB/dnaC* complex and *i* proteins to form the *preprimosome.*

Primase, the specialized RNA polymerase used for DNA initiation, is then added, forming the primosome, which can migrate to random sites on the ΦX174 genome to start DNA replication. Once a short oligoribonucleotide is synthesized, DNA polymerase activity proceeds, and, in conjunction with RNase H activity and ligase, produces a ds DNA circle — the replicative form (RF). At this point, viral transcription off the (−) strand occurs, producing the important multifunctional viral A protein (162).

The A protein can cleave the RF (+) strand at a unique site in the hairpin structure. The nick provides a 3'-OH end upon which DNA polymerase can proceed, displacing the old strand. The similarity to a replication fork is obvious (see Fig. 6) and, indeed, only synthesis in the retrograde direction is needed to produce a ds linear concatameric structure like phage λ. The viral concatamers, however, may never actually form. After one round of replication protein A typically cleaves unit length (+) ΦX174. Because the cleavage site is within the hairpin structure, a ds region with a nick can form. The ligase activity of protein A can subsequently seal the nick to produce circular (+) virion. This scheme, differing only in detail, accounts for the replication of filiamentous single-stranded DNA circular phages, many ds DNA phages and, perhaps, for porcine circovirus, a circular single-stranded DNA animal virus.

A similar mechanism, circularization and concatamer formation by rolling circle replication has been suggested for herpesvirus replication (163, 164). However, the herpes system is more complex because these viruses undergo extensive recombination, particularly in repeat regions. When the repeats are in an inverted orientation, recombination results in packaged viruses which are isomers of the original virus with different orientations of the unique sequences between the inverted repeats (165). When the repeated sequences are direct repeats (especially terminal direct repeats) packaged virions often differ in the iteration number of the repeats (165). Since recombinational mechanisms can also produce concatamers (see Section 2 C.), recombination may also be involved in herpesvirus replication.

The *geminiviruses*, one of the few DNA plant viruses, contain two single-stranded DNA circles per genome. The two single-stranded circles contain a 200-nucleotide common region (the rest of the sequence of each circle is unique). The common region can form a hairpin with a GC rich stem (166, 167). Moreover, the packaged virion of one strain of the virus, Maize Streak virus (MSV), contains a small segment of complementary strand which has a few ribonucleotides at its 5' end and which terminates after 80 or so nucleotides. Other geminiviruses (e.g: Cassava Latent Virus) do not appear to have this short complementary strand. *In vivo*, after infection, one finds primarily ds circular molecules, dimeric concatamers, and occasional concatameric single-stranded DNA. Although not proven, the most likely mechanism of replication is via a rolling circle. The short second strand in MSV may be an "Okazaki" fragment which can act as a primer for RF ds circle formation or as a signal for processing the concatamers or a signal for packaging the virion DNA (these are not, of course, mutually exclusive possibilities).

C. Concatamerization by Recombination: T-Phage and Iridoviruses

T4 phage is probably the best studied virus which emphasizes a recombinational strategy for concatamer formation (168). T4 only rarely forms circles, but is packaged as a terminally-redundant permuted circle. Fig. 7 shows how T4 produces long branched concatamers. After initiation at one of several possible origins, replication proceeds bidirectionally until the end of the linear molecule is reached. Because of the inability of DNA polymerases to initiate at the exact 3' end of the template DNA, this 3' end remains single-stranded . Such a single-stranded end is highly recombinogenic and can invade and displace the complementary region in another co-infecting phage, at the other end of a progeny molecule, or (rarely) at the other end of the same molecule producing a circular molecule. In any case, the invading strand provides a 3' end on a template molecule and is, in essence, a new replication fork (see boxed area in Fig. 7). As replication and/or branch migration continues, resolution of the branched structure, perhaps during the process of packaging, can occur by standard recombinational processes. The generation of replication forks by recombination is well-suited to rapidly accelerate overall DNA synthesis.

Recombinational processes may also play a role in the replication of herpes viruses, and may explain the finding that newly-synthesized and packaged virus contains a high proportion of nicks and gaps and explain the appearance of viral DNA in "tangles" (which have some similarity to the dense branching structure of T4) late in infection (163).

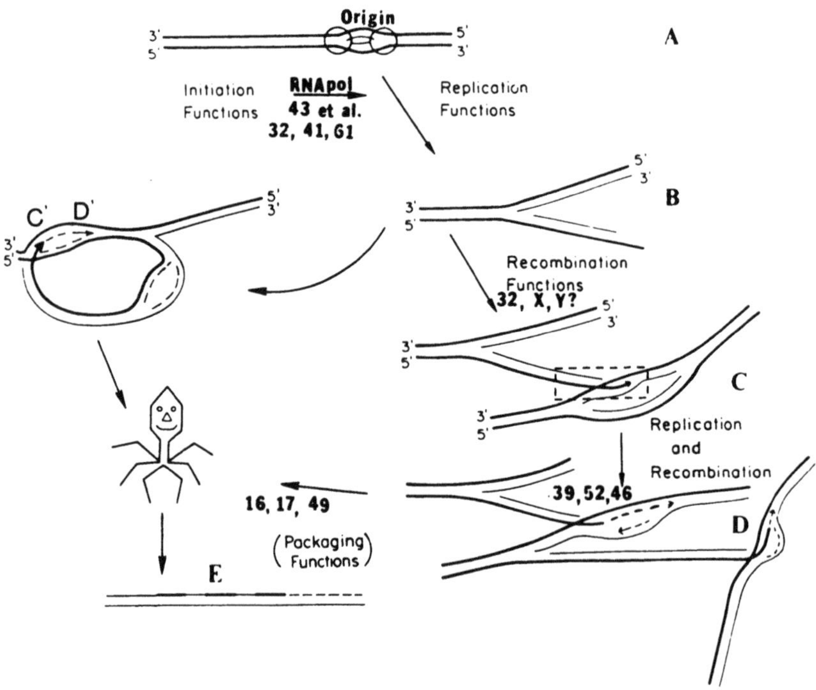

Figure 7. Replication of phage T4. A. Bidirectional replication from origin. B. Fork with single-stranded 3' end. C. Invasiveness of single-stranded end to form a new replicative fork (see boxed area). D. Further replication E. Resolution and packaging. Numbers and symbols in bold face indicate viral genes involved in that function. Reprinted with permission (reference 168).

The replication of the iridoviruses (169) is even more similar to T4. Iridoviruses are linear, terminally redundant, and circularly permuted (although, like phage P22 (170) the circular permutation is limited to certain regions — accounting for 20-28% — of the genome). Like many phage (e.g. ΦX174, λ) which replicate in two stages, an early stage producing monomeric RF and a later stage producing concatameric DNA, iridoviruses exhibit a staged replicative system. Early replication is nuclear and the size of replicating DNA is up to twice genome size. Late replication (>3hrs.) is cytoplasmic and produces DNA greater than 10X genome length (171). In addition to the displacement in time and space, the late-replicating DNA (but not the nuclear DNA) is heavily methylated (172) at the C position: apparently every CpG is methylated.

Another class of viruses in which recombination may play an important role are the poxviruses (discussed in detail chapter 21 and in Section 2 D. of this review). These viruses replicate entirely within the cytoplasm of the host in foci called *factory areas* (173), even in the absence of host nucleus (174, 175). The ability to produce recombinant viruses by site-specific recombination (176, 178) shows that recombination can, and does, occur in the cytoplasmic factory areas (presumably using viral enzymes). The extent to which recombination plays a role in poxvirus replication is unknown, although one model of viral replication involves recombination (179) as a means of resolving the termini. Concatameric DNA is observed (180, 181), but is consistent with other models of replication (180) as well.

D. Hairpin Structures and Viral Replication

The hairpin mechanism of dealing with the ends of linear DNA molecules was first proposed by Cavalier-Smith (182). Basically, for single-stranded linear DNA molecules, the end must be capable of folding-back to generate a self-priming hairpin structure with a 3'-OH end.

For linear ds molecules it is proposed that the end structure has complementary strands which are covalently linked at their ends upon which an endonuclease can generate a 3'OH end and a 5' hairpin. After polymerase activity, the terminal palindromes generated can reform into two terminal hairpins. This appears to be the mechanism which is used not only by the single-stranded DNA parvoviruses, and, possibly, the ds DNA poxviruses, but also by eucaryotic telomeres, e.g., in yeast (183).

Parvovirus replication

Two classes of the small (less than 6000 nucleotides) single-stranded linear DNA parvoviruses exist: the adeno-associated viruses (AAV) and the autonomously-replicating parvoviruses (ARPV). Since the AAV (184) and ARPV (185) exhibit significant differences in their replication process, they will be considered separately.

AAV requires co-infection with adenovirus or herpesvirus for productive infection. In the absence of the requisite helper functions from adenovirus or herpesvirus, AAV can integrate into host chromosomal DNA as a full-length or nearly full-length provirus (see Section IV. F). Rescue of the latent AAV requires adenovirus, and, specifically (in addition to other regions required for replication) requires a functional product located within the EIB region of adenovirus (186, 187). In addition, AAV replication appears to require:

i) the EIA region of adenovirus, which encodes early functions (188, 189) that have a cascade effect on other early adenovirus functions. Whether EIA acts directly (188) or indirectly via other adenovirus functions is unclear; and
ii) the E4 region (190, 191) appears to be required for full replicative activity. (3) The E2A and VA1 regions appear to be related to AAV capsid mRNA production (192) and have an effect on the level of single-stranded AAV virion produced but not on the level of ds RF (193). This implies a role of capsid protein in the regulation of the switch from RF to virion production.

The E2B region, which encodes adenovirus terminal protein and polymerase, is not vital for AAV replication. This does not mean that these proteins play no role in AAV replication — AAV replication complexes isolated from herpesvirus co-infected cells contain herpes polymerase (194) — only that host proteins can be utilized as well.

The sequence of AAV (195) shows a 5' and 3' terminal sequence which contain two inverted repeats imbedded in a longer inverted repeat (Fig. 8). This sequence can fold into a "T"-shaped structure and can provide, probably through an intermediate circular structure (184, 196), a 3' end which can act as a primer for synthesis of a ds hairpin RF. The existence of these structures (and others) *in vivo* has led to the model of AAV replication seen in Fig. 8 (197).

Briefly, after the ds hairpin RF is formed, it is nicked opposite the site of initiation. This nick provides a second 3'-OH end which can act as a primer for polymerase to act on, producing a non-hairpin, full-length ds-RF. The site of cleavage preferentially gives a 3'-GTT-5' sequence at the 5' end of the ds-RF. However, a substantial percentage of virions end in 3'-GT-5' or 3'-G-5', with these microdeletions (and even larger ones) being regenerated upon RF formation (196-198). The ds-RF intermediate can then, presumably under the influence of DNA binding proteins, restructure its terminal segments to generate hairpin structures on both strands. This rearranged ds-RF provides a 3'-OH end which can act as a primer for replication. Replication of ds-RF results in displacement of a single-stranded virion DNA, which is presumably — at least late in infection — packaged by capsid proteins. A ds hairpin RF is also generated and it can be cleaved again and extended to regenerate full-length ds-RF.

One consequence of such a model is that there is a "flip-flop" of terminal sequences, and virion DNA — both (+) and (−) strands are packaged in AAV — will have 4 different isomeric orientations of its ends. This prediction is, in fact, confirmed in viral preparations (199).

The ARPV (Fig. 9) differ from AAV in structure (200). The 3' end is similar in structure (but not sequence) to that of AAV, but does not undergo "flip-flop"

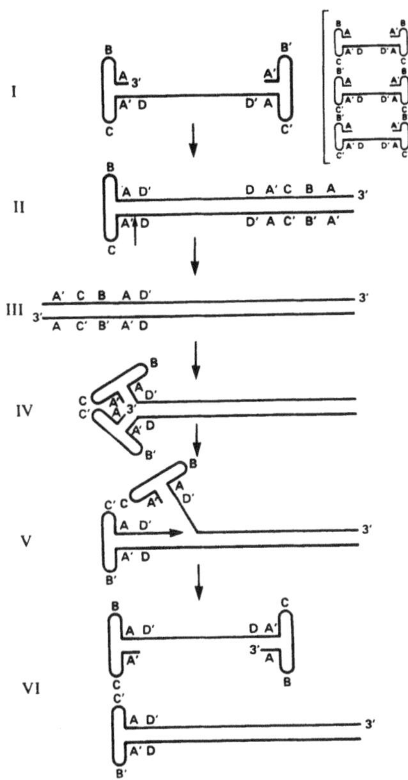

Figure 8. Replication of adeno-associated virus. See text for discussion. Reprinted with permission (reference 161).

isomerization, either in the packaged virion DNA or the several RF structures one can isolate (201). The 5' end of ARPV forms an imperfect simple hairpin which is different in structure and sequence from the 3' end. The 5' end is present as isomeric "flip-flops" in virion DNA and RF structures. Unlike AAV, typically only the (-) strand is packaged in virions (185). In addition, at least some of the RF 5' ends contain a covalently attached protein (of unknown, possibly host (202), origin) linked to an ~18-nucleotide extension beyond the viral 5' end. Again, the viral 5' end is not at a specific nucleotide but can be 1 or 2 nt shorter, indicating that the nuclease activity producing this end is probably cutting at a specific distance from a recognition structure rather than cleaving at a specific sequence. The extra 18 nt seen in some RF structures are viral in origin (201) and are removed before viral DNA is packaged. A model of ARPV replication, based on these constraints and the existence of appropriate intermediates has been proposed (185, 201) and is described in Fig. 9.

Replication of the ARPV requires concomitant host DNA synthesis (203) and appears to involve host DNA polymerases α and γ (204). It has been suggested (205) that production of ds RF involves polymeraseα, with polymerase γ — which is used in displacement replication of mitochondrial DNA in the cell (206) — being responsible for the displacement synthesis of virion DNA.

The source(s) of the three endonuclease activities required for the proposed replication model (one which cleaves the 5' hairpin 18 nt past the original viral se-

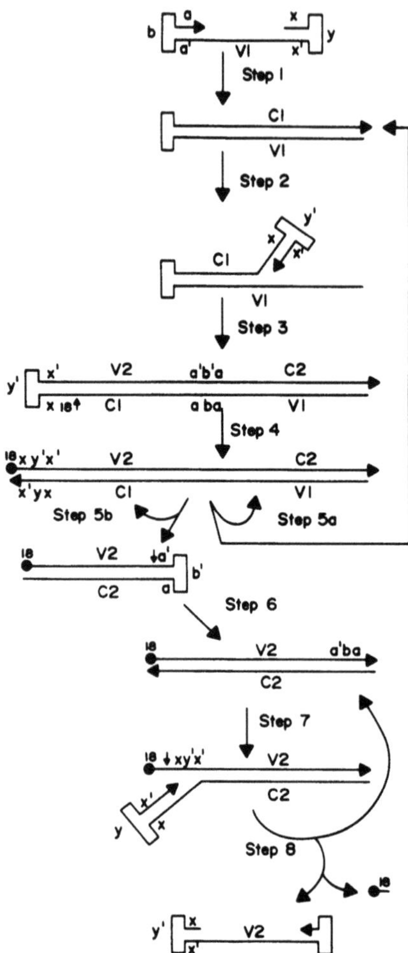

Figure 9. Replication of autonomous parvovirus. xyx' represents viral 5' and aba' represents viral 3' sequences. Solid circle represents 5' terminal protein. Reprinted with permission from *Virology* B.N. Fields, ed. Raven Press.

quence when present as a ds hairpin RF; one which cleaves the 3' hairpin only when it is present in the proper orientation relative to virion sequence, thereby avoiding isomerization; and one which removes the extra 18 nucleotides and terminal protein from the virion) are not known. Because both AAV and ARPV require structural (conformational) changes in the terminal regions involving melting duplex DNA to permit hairpin formation, DNA binding proteins are required and several strongly- (but not covalently) binding proteins of unknown, but probably host, origin (202) are found associated with RF structures.

Poxvirus replication

Poxviruses, unlike parvoviruses, replicate entirely in the cytoplasm utilizing viral enzymes for their replication. The viral DNA is nominally double-stranded, but both termini form covalently-linked imperfect hairpin structures which are present as isomeric inverted complements in progeny viruses (206, 207) — i.e: they flip-flop like the terminal structures of parvoviruses. In addition, the similar inverted terminal regions at either end of vaccinia include two sets of a 70-bp tandem repeat, the first set of 13 repeats starts 87 base pairs from the proximal end and is separated by 325 base pairs from a second set of 19 tandem repeats. This is followed by a somewhat overlapping pair of 125-bp repeats followed by 8 tandem 54-base pair repeats.

All the repeats have regions of homology (208-209). There is variation of the number of repeats both within isolates of the same strain and between vaccinia and rabbitpox (205) and some sequence divergence between cowpox and vaccinia (210).

Replication appears to initiate near the termini of the virus and involves discontinuous synthesis with RNA primers (211-213). Whether a specific viral origin of replication exists in poxviruses is not known, since, unlike other viruses (214-216), replication of plasmids introduced into cells co-infected with vaccinia does not require the presence on the plasmid of a specific viral origin of replication or, indeed, any viral sequence at all (217).

Nicks are introduced near the end of parental viral genomes soon after infection (218, 219) and are only sealed late in infection (220). This has led to the proposal that these nicks provide a 3'-OH end, which, when extended, provides a terminal hairpin palindromic sequence (see Fig. 10). When these ends are restructured

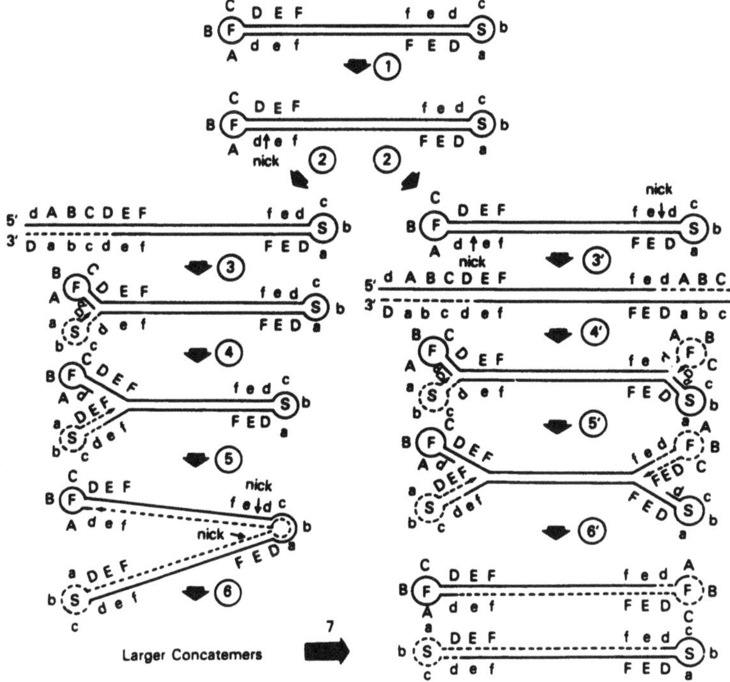

Figure 10. Poxvirus replication. S and F represent isomeric forms of termini. See text for discussion. Reprinted with permission (reference 180).

into two terminal hairpins, one of the hairpins provides a 3'-OH end which can act as a primer permitting synthesis from the resolved ends (205, 207, 221-223). Note that retrograde synthesis on the lagging strand of resolved vaccinia termini produces a ds replication fork and fills in single-stranded gaps.

Although vaccinia termini can be attached in vitro to a linear yeast plasmid, these ends apparently do not act as telomeres in yeast. DNA synthesis probably produces a dimeric circle which is resolved through recombination to produce a plasmid containing an inverted repeat structure of the termini (224). When circular plasmids containing such a head-to-head structure (isolated from concatenated replicative structures) are placed in virus-infected cells, the ends are replicatively resolved, converting the plasmid progeny to a linear structure with vaccinia termini and plasmid sequence in between (225). The ability of the concatenated end joint to resolve a circular plasmid to a linear structure is dependent upon the presence of at least 252 base pairs of concatamer joint (about 125 nucleotides from each terminus). Plasmids with 132 base pairs of concatamer joint are unable to resolve into linear structures. The role, then, of the terminal 70-bp repeat structures (which start at nucleotide 87 from the terminus) in replication is unclear (225).

E. The Direct Approach: Protein Primers

Perhaps the most direct method for supplying the requirement for a 3'OH primer is to utilize a nucleotide linked to a protein which recognizes the 3'-OH terminus of the template strand. Adenovirus uses such a system (reviewed in chapter 20 of this volume and ref. 226-229), as does the bacteriophage Φ29 (230).

Adenovirus

Adenovirus is packaged as a linear, double-stranded DNA molecule with a terminal protein covalently linked to the 5' deoxycytidine of the terminal inverted repeat ends through a phosphodiester bond to serine (231). The terminal protein (TP) is a proteolytic product of the precursor terminal protein (pTP) seen at the 5' end of newly synthesized strands. Proteolytic processing occurs late in infection (232) and uses a viral-encoded protease. The ends of adenovirus appear to be circular, but the linkage is usually not covalent. Rather, the TP at the 5' ends interact to circularize adenovirus. However, true covalently-closed circles do occur after infection (233, 234) and these head-to-tail circles are infectious (234). The role of these circles in replication is unclear. Their possible role in integration will be discussed in Section 2 F. of this review.

The initiation reaction adds dCTP to pTP in the presence of either end of the ds viral DNA and viral polymerase. The viral template DNA can either have TP (or pTP) on the to-be-displaced parental strand or can have the TP completely removed. Incomplete removal, however, leaving peptides on the parental strand inhibits initiation (235). The reaction also requires Mg^{++} and ATP (GTP or dATP can substitute) (236). The ability of deproteinized, and, hence, linear, double-stranded adenovirus DNA to support initiation, albeit poorly, indicates that circularization

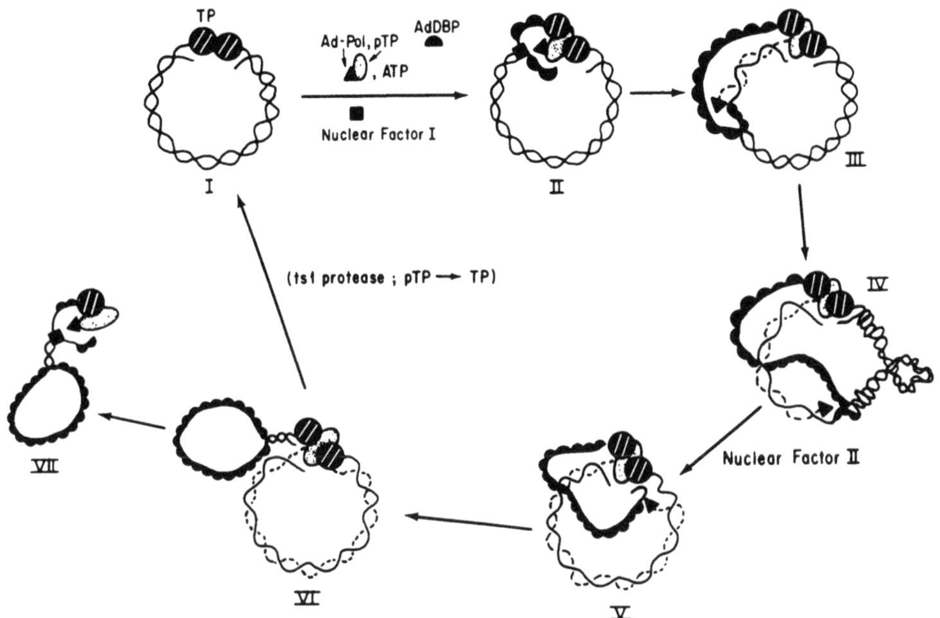

Figure 11. Adenovirus replication. See text for discussion. Reprinted with permission (reference 228).

is not crucial for replication initiation. However, the normal replicative form of the virus *in vivo* is probably circular (see Fig. 11).

Utilizing linearized plasmid constructs with variable amounts and organization of viral terminal sequences, it has been seen that the initiation event can tolerate two (but not three) additional G/C pairs (235, 237), can tolerate removal of up to eight terminal base pairs (238) but cannot tolerate termini which have less than 19 base pairs of viral sequence (235). Site-directed mutagenesis to the conserved region between positions 9-17 (239) destroyed all template activity. Although this region is an important recognition sequence for initiation and replication, it is probably not sufficient, since adenovirus AdF1 DNA/TP is inactive as a template in an *in vitro* system with proteins from Ad2 infected cells, even though AdF1 and Ad2 share identical terminal sequences through base 17 (237). The converse is also true, Ad2 DNA/TP is inactive in an in vitro system from AdF1 infected cells. This may reflect a strain-specific interaction between the TP on parental strands and new pTP/polymerase.

In addition to the pTP, viral replication utilizes a viral-encoded polymerase which can be found as the pTP-polymerase complex which binds to nucleotides 9-22 at the viral termini (240). Both pTP and polymerase are necessary to synthesize pTP-dCMP, and the viral polymerase cannot be replaced by other polymerases (241). This region slightly overlaps a region (nucleotides 17-48) protected from nuclease digestion by host-derived nuclear factor I (242), which, in the presence of ATP, stimulates initiation by the pTP-polymerase complex. A viral-encoded nonspecific single-stranded DNA-binding protein (DBP), although not necessary for the initiation phase, does greatly stimulate initiation when host nuclear factor I is present (243). In the absence of host nuclear factor I, DBP inhibits initiation (243). The host nuclear factor I does not affect the binding of pTP-polymerase complex to single-stranded viral DNA, such as a displaced parental strand, although this reaction is inhibited by viral DBP (244). These data suggest that nuclear factor I and DBP may interact in opening up a region of viral termini which can be recognized by pTP-polymerase (228). Significant elongation of the newly-initiated strand does, however, require viral DBP, and this function cannot be replaced by DNA binding protein from *E. coli* (228).

In addition to DBP, full-length elongation of viral DNA, which proceeds only in the direction of fork movement with displacement of parental strand, requires a second host factor (nuclear factor II) which has topoisomerase activity (228, 242). In the absence of nuclear factor II, replication stops at around 10,000 bp. This factor can be replaced by eucaryotic topoisomerase I. The activity of factor II requires the presence of factor I, as does eucaryotic topoisomerase I (245). *E. coli* topoisomerase will not replace either eucaryotic topoisomerase (245) in permiting full-length adenovirus replication.

Once initiation begins, apparently randomly, at one of the ends of the virus, it appears that this end is preferentially re-initiated, and molecules which have initiated at both ends are only rarely seen. This process, then, leads to displacement of a single-stranded parental DNA. Synthesis of the complementary strand is subsequently initiated on the displaced strand in a manner reminiscent of reovirus replication (see section 1 C.). Whether the normal structure of the single-stranded parental strand, upon separation from the replication complex, involves hybridization of the terminal inverted repeats and, hence, formation of a duplex terminus strictly analogous to the termini of duplex virus, is not known.

A similar, though slightly different, process is involved in Φ29 synthesis (230, 246, 247). The main difference is that the terminal protein on the parental strand is absolutely required for initiation (247) of DNA synthesis by a new terminal protein to which dAMP will be added — i.e: protein-protein interaction of the two terminal proteins is required for initiation. Full-length virus can be made *in vitro* with only the terminal protein and viral polymerase, although host factors and other viral factors may stimulate activity (246).

F. Viral Integration and Transposition

A number of DNA viruses in bacteria (e.g., mu, lambda) and in animals (including some papovaviruses, adeno-associated virus, adenovirus, the DNA forms of hepatitis B and retroviruses, and possibly herpesvirus) are capable of integrating into host DNA. In some cases, most notably in bacteriophage, retroviruses, and adeno-associated viruses (AAV), integration into host DNA can play a role in the viral life cycle. For some of these viruses — e.g., retroviruses and bacteriophage mu — integration is crucial to their productive infection.

In other bacteriophage and in AAV, integration represents an alternate life style to productive infection. Typically, once integrated, the virus is quiescent, expressing only a few, if any, viral functions, and replicating as part of the host chromosome. In this integrated structure, the virus can often escape host mechanisms which are designed to protect the host from viral infections. The virus remains in a cryptic state until an environmental signal rescues it and productive infection begins again. In some of the bacteriophage, the shut-down of most viral functions by integrated virus also protects the host from lytic infection by the same or related viruses. In the case of herpesviruses and hepatitis B viruses, viral latency occurs, as does at least partial integration. However, it is unclear whether there is any relationship between latency and integration.

In many other animal DNA viruses which integrate into host DNA (adenoviruses, polyoma, SV40, and, possibly, hepatitis B and herpesviruses), integration is probably a replicative dead-end, although integration is often (but not always — e.g., in papilloma viruses and, possibly, herpesviruses) required for oncogenicity. Typically integration of these viruses is a relatively rare event, occurs in hosts which are semipermissive or nonpermissive for productive infection, and often results (either initially or over time) in rearrangement or deletion of viral information.

In this section we will discuss viruses in terms of the specificity of the process of integration and excision. However, one should remember that integration is often a rare event, and little is known about it except from those cell lines transformed by the process, often long after secondary rearrangements may have occurred.

Site-specific integration

In the first class of integration events we will examine, integration occurs at specific homologous sequences in both the virus and host. The best characterized system of this type, of course, is bacteriophage lambda (see ref. 159, 160, 161, 248). Integration and excison are *conservative* events; that is, replication is *not* involved. The integration/excision events require viral and host functions.

In order to integrate, λ DNA must be in a supercoiled circular form. Integration requires the product of the *int* viral gene, which exhibits type I topoisomerase activity, and two host-encoded proteins (Him A and Him B).

The *int* product binds to several sites near and in the phage attachment site, and produces a staggered cleavage 7 base pairs apart, in the 15-bp core homologous region in both virus and host, with consequent sealing leading to physical integration.

Excision also requires the *int* topoisomerase, but, in addition, requires the product of the viral *xis* gene, probably to provide a directionality (excision *vs.* integration) to the *int* reaction. Although such a process, with specificity in both host and viral sequences, is common in bacteriophage, it is unknown in animal viruses.

Integration at non-specific sites

In the second type of integration event, integration does not involve specific sequences on either host or virus. The best examples of this type of integration are polyoma and SV40.

Integration sometimes results in (or from) extensive deletion of chromosomal DNA (249). The sites of integration show no extensive homology to viral sequences and there are no repeated flanking host sequences. There is no particular integration site on the virus. The integration process appears to be entirely host-directed and can probably utilize any input DNA. Occurring in nonpermissive hosts, this non-specific integration is a rare event (250) which is probably not normally a part of viral function. Typically, in such illegitimate recombination, some viral functions continue. In the case of polyoma and SV40, the continued synthesis of T-ag is a necessary (but not sufficient condition) for their ability to transform cells. Since many of the integrated viral sequences are present as multimers or partial multimers in a head-to-tail arrangement, a model of integration involving linear concatamers generated by a (probably aberrant — since it occurs in nonpermissive cells) rolling circle mechanism has been suggested (251). The presence of multimers probably is necessary for the regeneration of complete viral circles via homologous recombination.

Excision of viral DNA and production of circular virus appears to involve the replication of integrated viral sequences independently of host DNA (requiring a functional viral origin and T-ag sequence) in combination with homologous recombination (252,253). This process not only produces viral circles, but can eventually lead to the spontaneous elimination of integrated virus from cell lines.

Integration at preferred sites

The third type of viral integration in animal viruses involves no specific sites in host DNA, but there are preferred sites by which the virus integrates — namely sequences at or near one or both of the ends of a linear viral form. Examples are adenovirus, AAV, and hepadna viruses. Of these, the most interesting, in terms of replication, is AAV, which is the only one which can normally be rescued.

AAV provides an interesting model which may be suitable for understanding the integration of adenovirus and hepadna virus. Interestingly, the autonomously replicating parvoviruses (ARPV) have not been found to be integrated perhaps because of their limited host ranges. However, in the case of a plasmid construct containing an ARPV, the virus can be rescued if the 5' hairpin region is present (254), indicating that, in this case, rescue may be a distinct process from integration. Integration of AAV occurs naturally, although rarely, in the absence of helper virus (i.e., under nonpermissive conditions). Integration appears to preferentially involve the terminal inverted repeats (255). Often, a head-to-tail multimer (within the limit of restriction enzyme analysis) is seen to be integrated, although multimers are not necessary for rescue (186, 197). Rescue of integrated AAV can be initiated by infection with adenovirus which, of course, is necessary for productive infection. Rescue of virus from plasmid constructs with specific terminal deletions indicate that rescue involves a specialized form of replication and requires that at least one terminal palindrome remain either intact or be repairable by replication (see section 2 D.) (197, 256). It probably also requires sufficient sequence of the other palindrome so that repair synthesis of a panhandle structure would regenerate a complete end.

Adenovirus integration

Adenovirus integration into nonpermissive host cells has been recently reviewed (257). There is no evidence that adenovirus integration occurs at a specific site in host DNA. Although there are some short homologies between viral and cellular sequences which may play a role in the integration process, these do not exceed the statistically expected values. Like SV40 and polyoma, there is no duplication of cellular nucleotides at the junction between viral and host sequence (thus ruling out a transposition-like event), and, in some cases, no deletions of cellular nucleotides. Integration typically (but not always) occurs near, but not at, the termini of viral DNA. Deletions of 2, 8, 10, 45, 64, and 174 nucleotides of viral terminal sequences have been observed.

As we saw in Section 2 E. small deletions (less than 13 nucleotides) at the head-tail junction of the circular form of adenovirus do not prevent infectivity (234), and small deletions (less than 8 nucleotides) in the terminus of adenovirus do not prevent initiation of replication (238).

However, in contrast to the circular form of adenovirus in which the *viral* ends are joined (234), plasmids in which the viral ends are linked to plasmid sequences are not infectious unless the ends are cleaved (258) and free. This implies that integrated adenovirus would not normally be rescueable. The presence of nearly complete adenovirus integrates which end near the termini (257) and the occasional appearance of integrates with linked termini (259) indicate that circular forms may sometimes play a role in integration. Unlike SV40 and polyoma, multiple insertions are usually not true tandem repeats, but are separated by other (host) sequences (257). Secondary re-arrangements of adenovirus integrates are common, as is eventual loss of viral sequences.

Integration of Hepatitis B virus DNA

When integrated hepatitis B virus in hepatocellular carcinomas are examined, one often sees extensive rearrangement of both viral (260, 261) and host sequences (262). But in several cell lines without extensive rearrangement, one viral/cell junction is near the site of initiation of the reverse transcriptase while the other junction is found at various sites in the viral genome (263). In some cases, flanking host sequences are duplicated as direct repeats of various lengths; in other cases they are not. The mechanism of integration may be similar to that suggested as the way small nuclear RNA pseudogenes can be generated from cDNA copies via integration at a staggered cut in host DNA (264).

Hepatitis B at the stage of reverse transcription is a single-stranded DNA molecule, as is AAV, and as is, at least in some stages of the replicative cycle, adenovirus. It may be that this is a common feature of the integration of these viruses into host chromosomes, perhaps by a process of patch homology recognition to sites in host DNA which are temporarily single-stranded themselves.

In the case of adenovirus, it may be that single-stranded ends of a double-stranded virus rather than a true single-stranded replicative intermediate induces integration. This may account, because replication would be unnecessary, for the failure to find terminal direct repeats of host sequences at the junctions of adenovirus integrates (257) as is also the case with the double-stranded polyoma or SV40 integrates. Normally, only in the case of AAV does the integrated form play a role in the replication of virus. This may be because minor deletions at the viral termini in AAV can be regenerated during its replication and mechanisms for specific DNA cleavage are part of its replication mechanism.

Transposition and integration

The last type of viral integration involves transposition, whereby a specific viral transposase makes a staggered cut in host DNA, ligates the precise ends of the virus to the single-stranded extensions, and introduces viral DNA, either conservatively or by replication into the site which now has flanking host direct repeats. Retroviruses (see Section 3 of this review and chapter 16) are the animal viruses which transpose. In this section, however, we will examine bacteriophage mu.

Bacteriophage Mu

Bacteriophage mu replicates exclusively by a transposition-directed process, making this virus unique and also allowing it to be studied as a model transposable element (265). The ability to analyze mu-dependent replication in defined *in vitro* systems (265, 266) has permitted the trapping and analysis of the transposition intermediate (Fig. 12) (267). Transposition to the intermediate stage requires that both mu ends be present and in the proper orientation in a supercoiled structure

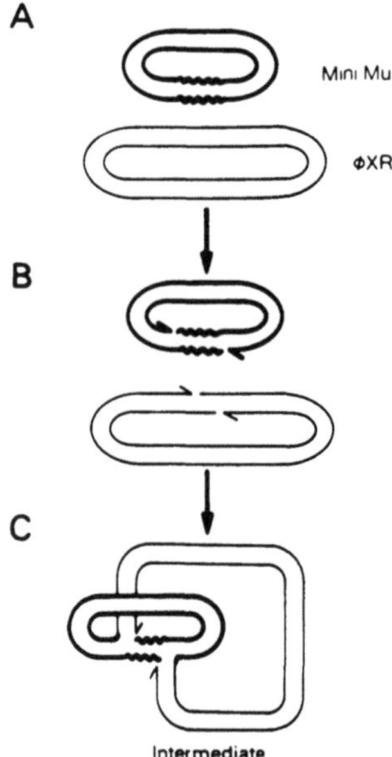

Figure 12. Formation of phage mu replicative intermediate. A. Plasmid with mini Mu transposon and ΦX174 RF as recipient. B. Introduction of staggered (5bp) cut in recipient DNA and cuts at termini of mu transposon. C. Formation of integration intermediate. Completion of transposition can either be replicative to yield a cointegrate or involve cleavage and loss of the non-mu donor DNA. Reprinted with permission (reference 268).

(268). Unlike most transposons, the ends of mu are not identical inverted repeats, but do contain sequences recognized by the viral transposase (269), the gene A product. The transpose specifically makes a staggered cut in the acceptor DNA, nicks the precise ends of the phage, and ligates the 3' ends of the phage to the 5' extensions of the acceptor DNA. This forms the *transposition intermediate.*

Once the transposition intermediate is formed, either co-integrate formation or simple insertion can proceed using only host enzymes (268). Co-integrates form by the initiation of DNA replication starting at the 3' end of the host DNA which is at the viral left-most end of the transposition intermediate; simple inserts require degradation of the DNA which originally flanked the transposon, followed by gap repair (270). In either case a 5-bp direct flanking repeat of host DNA is produced indicating a 5-nucleotide staggered cut in the case of the mu phage. To produce the transposition intermediate, one needs, for high efficiency, ATP, Mg^{++}, Mu A (the transposase) protein, Mu B protein (Mu B protein binds to DNA non-specifically (271) and is not absolutely required for mu transposition (272) but acts as a strong enhancer of transposition), and a host protein recently identified as HU protein (a nonspecific DNA-binding protein) (273). In addition, the mu phage must be supercoiled although the acceptor DNA can be supercoiled, relaxed, or linear (273). Since mu phage is packaged as a linear molecule with the viral information imbedded in random flanking host DNA, the phage should not be in the form of covalently closed circles. However, a circular, and thus potentially supercoiled form of mu in which the ends are held together by a protein bridge has been observed after infection (274).

3. VIRUSES THAT USE REVERSE TRANSCRIPTASE

The last group of viruses we will examine are unique in that genomic information switches between RNA and DNA, aided by a unique polymerase able to make a DNA copy of a pregenomic mRNA. Included are the retroviruses, the hepadna viruses, and cauliflower mosaic virus.

In the case of retroviruses, it is the pregenomic mRNA which is packaged in the virion. When this RNA virion enters the cell, it is converted to a double-stranded DNA copy with duplicated information at the termini (the long terminal direct repeats — the LTRs). These direct LTR repeats also include a small inverted repeat at their ends. Fig. 13 shows the proposed mechanism of the conversion of virion RNA of retroviruses to a double-stranded DNA, a process which is extensively discussed in chapter 16 of this volume. The reader is also referred to recent reviews (275, 276).

Virion reverse transcriptase activity requires a primer for initiation. In the case of retroviruses, the primer for (–) strand DNA synthesis is provided by the 3' end of a host tRNA which is complementary to a region of viral RNA. Synthesis proceeds to the 5' end of the virion RNA, which is terminally repeated at the 3' end (Fig. 13).

RNase H activity (associated with reverse transcriptase) removes 5' RNA sequences hybridized to newly-synthesized DNA (Fig. 13. 3), leaving a single-stranded DNA which is complementary to the direct repeat R region RNA at the 3'

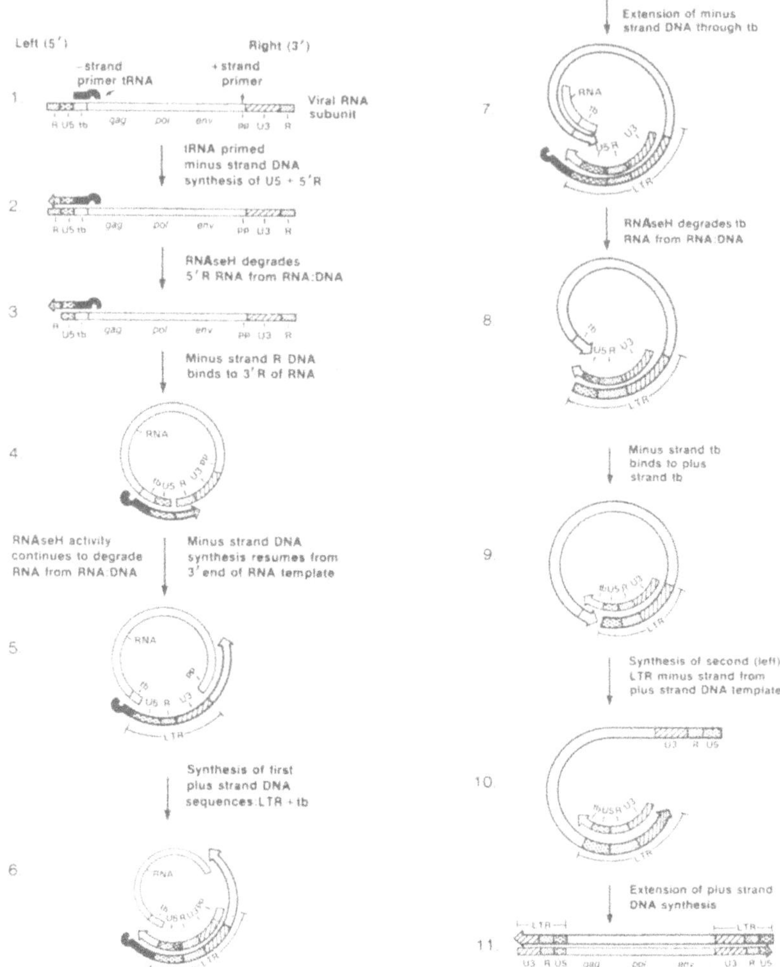

Figure 13. Replication of retrovirus. See text for description. Reprinted with permission (reference 276).

end of the same molecule (or of the other virion RNA packaged with it, since the genomic RNA of retroviruses is packaged as *dimers*). Hybridization of the free single-stranded DNA to the direct repeat produces a circular (or concatameric if to the other virion RNA) molecule (Fig. 13. 4), and transcription of (-) strand and RNase H activity continue (Fig. 13. 5). A short sequence of template RNA just past the LTR sequence (Fig. 13. 6), generated by its resistance to RNase H activity, is the primer for (+) DNA strand synthesis (277-280). Plus strand synthesis (using the primer) continues through at least 18 nucleotides of the tRNA primer of the (-) strand before stopping (281).

Minus strand synthesis, meanwhile, proceeds through to the 5' end of the RNA template (Fig 13.7), which, because of RNase H activity, has been trimmed back to the sequence complementary to tRNA (Fig. 13.8). This remaining viral RNA sequence was not removed by RNase H because it involved RNA:RNA pairing. Rather, the tRNA was displaced by (+) strand synthesis. This tRNA-binding region, where (−) DNA strand synthesis stops, is, after removal of RNA template, complementary to the 18 nucleotides or so extension of the (+) DNA strand, which is also single-stranded DNA because of RNase H removal of the displaced tRNA primer (Fig. 13. 9).

The (−) strand 3' end, then, will hybridize to the complementary (+) strand 3' end. The molecule is now a mostly single-stranded (−) DNA circle with a nick whose ends are held together by an overlapping (+) complementary DNA strand initiated at the LTR junction (Fig. 13. 9). Displacement synthesis from the (−) 3' end and continued synthesis at the (+) 3' end (Fig. 13. 10) will now produce a linear DNA with a direct repeat LTR at each end (Fig. 13. 11). This structure acts like a transposable element and will integrate into host DNA.

For retroviruses, integration is an important element in the replicative cycle. Integration results in the deletion of the 2 terminal nucleotides from the linear double-stranded DNA structure. Integration does not appear to depend upon host sequence (282), requires the terminal inverted repeats (283), and results in a 4-6 nucleotides (depending upon the virus) flanking direct repeat of host DNA. In addition to the endonuclease function of the *pol* viral gene, which recognizes specific sequences in the viral terminal inverted repeats (284), cellular functions are also required (285).

The integrated virus then initiates transcription at the left end of the R sequences in the left-hand LTR, utilizing host enzymes. There is a poly(A) termination signal about 20 nucleotides upstream from the right-hand end of the R sequences, which, however, is not utilized when first encountered after initiation of the transcript. It is utilized, however, when encountered at the right-hand end of the virus (that is, when full-length viral transcript is made) and poly(A) is added to sequences downstream of the R sequences producing a terminal redundancy of R sequences. This transcript is both the viral primary mRNA, which can be processed by splicing, and the virion RNA genome, which is encapsidated.

Hepadnaviruses

Hepadnaviruses (Hepatitis B) and cauliflower mosaic viruses, although packaged as DNA virions, share features which make them appear to be more closely related to the retroviruses than to the other DNA viruses (286, 287). In particular, all three groups utilize a reverse-transcription step in the production of DNA virus (288, 289). In turn, the DNA virus is the source of the RNA transcript used as template during the reverse transcription.

The analysis of hepatitis B viral replication has been hampered by the extremely hepatotropic nature of the virus. Infection of hepatocytes requires a cellular function which is quickly lost (within 7 days) from hepatocytes in culture (290), so most analysis must be done with *in vivo* infections rather than the more manipulable tissue culture infections. Although hepatocytes are the primary focus of infection, hepadnaviruses can infect other tissues *in vivo* including a small fraction of peripheral blood lymphocytes where they reside primarily in episomal (latent ?) form (291). In hepatocytes integration of virus is relatively common — although monomeric circles are much more common. The differential importance of integrated and non-integrated forms of hepadnaviruses is not clear.

Hepatitis B virus is packaged as a partially double-stranded DNA circle (for a detailed review, see chapted 17). The (-) strand is full-length and contains a 5'-terminal covalently-linked protein at a specific nick site. The (+) strand overlaps this nick and has heterogenous 3' ends. When the partially double-stranded DNA virion enters an hepatocyte, it is converted to a complete duplex supercoiled circular minichromosome. Multimeric circles and integrated virus are also produced.

The covalently-linked protein is assumed to act as a primer for reverse transcription of the pregenomic RNA (see Fig. 14). The pregenomic RNA contains a terminal redundancy of 270 nucleotides produced by synthesis of a more than genome length RNA, a process which requires that a polyadenylation signal in the double-stranded DNA circle be ignored (295) when first encountered. The initiation site for pregenomic RNA synthesis is just upstream of one of two direct repeat sequences (292) found near or at the ends of the cohesive overlap region — the region between the 5' end of the (-) DNA strand and the 5' end of the (+) DNA strand of the packaged virion.

Elongation of the (-) DNA strand off pregenomic RNA appears to occur concommitantly with degredation of the RNA template by RNase H activity, so that (-) strand DNA appears to be single-stranded rather than templated to RNA (288, 293). The complementary (+) strand DNA is initiated on single-stranded (-) DNA with a capped oligoribonucleotide of 18-19 bases which differs at its 5'-most 6 nucleotides from the sequence to which it is bound, one of the 12-13 nucleotide direct repeat sequences. This primer, however, is identical to the sequence at the 5' end of the (+) pregenomic RNA used as a template for reverse transcriptase (see Fig. 14). The suggestion is that, like retroviruses, this oligoribonucleotide primer is "borrowed" from pregenomic RNA (or another RNA which starts at this site) (294). The newly synthesized (+) DNA, which, at it's 3' end, is complementary to the sequence at the 5' end of the (-) DNA, then "jumps" to hybridize to that end. In this process, the terminal protein and some 7-8 nt are displaced. Thus the (-) DNA "circle" has a small 7-8 nucleotide single-stranded 5 "wisker" with terminal protein attached. At this point, replication of (+) strand DNA continues on the now "circularized" (-) DNA strand. Typically, (+) DNA replication terminates before the strand is completed, leaving a partial (+) DNA strand with heterogeneous 3'-termini (296).

Cauliflower Mosaic Virus

Cauliflower mosaic virus replication appears to be similar (289, see Fig. 15). The encapsidated virion is a circular double-stranded molecule with three interruptions and all have small single-stranded 5' duplications which, upon entry into host, are trimmed back, allowing the virus to become a covalently closed double-

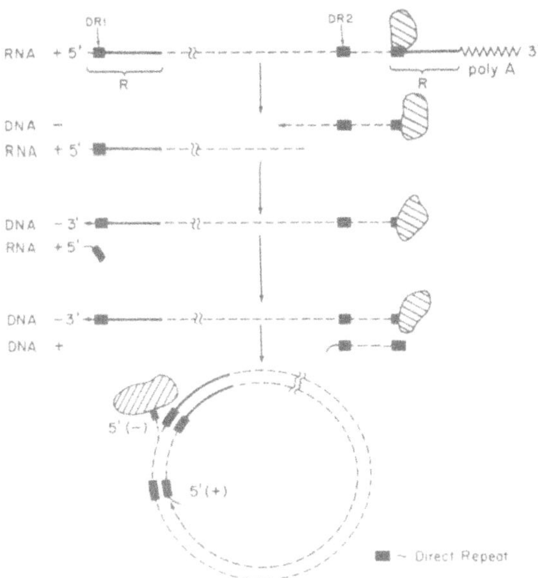

Figure 14. Replication of hepadna virus. See text for description. Reprinted with permission (reference 294).

stranded DNA circle. The virion is then found in the nucleus as a supercoiled minichromosome (297) where, along with subgenomic mRNAs, it produces a pregenomic RNA with a terminal redundancy of 180 nucleotides (298). Again a transcriptional termination signal must be ignored upon first being encountered. At this point, it is proposed that reverse transcription (using a viral polymerase (299) which has homology to retroviral reverse transcriptase (287) and exhibits reverse-transcriptase activity *in vitro*) produces (−) strand DNA. There is some evidence that (−) strand production occurs in virioplasms in an asymmetric fashion (300) and it has been suggested that, like retroviruses, the primer is a host tRNA (289). After synthesis of (−) DNA to the 5' end of the pregenome RNA, a switch to the terminally redundant 3' end of pregenome RNA is suggested with continued replication to produce a full-length (−) strand. The (−) DNA remains circular due, possibly, to hybridization to part of the RNA template not removed by RNase H activity. It is slightly redundant because a small amount of displacement synthesis occurs when the polymerase encounters the double-stranded DNA/RNA hybrid near the site of initiation on the circular RNA. Synthesis of (+) strand initiates at two sites, in regions which have short stretches of very high G content RNA, which may be RNase H resistant. It is suggested that these RNAs act as primers for (+) DNA strand synthesis. One of these (+) strands must "jump" across the slightly redundant gap in the (−) strand and continue synthesis until the 5'-end of another (+) strand is met. In all cases, the polymerase appears to continue, via displacement of the 5' end, through any remaining 5' RNA primer and 5,

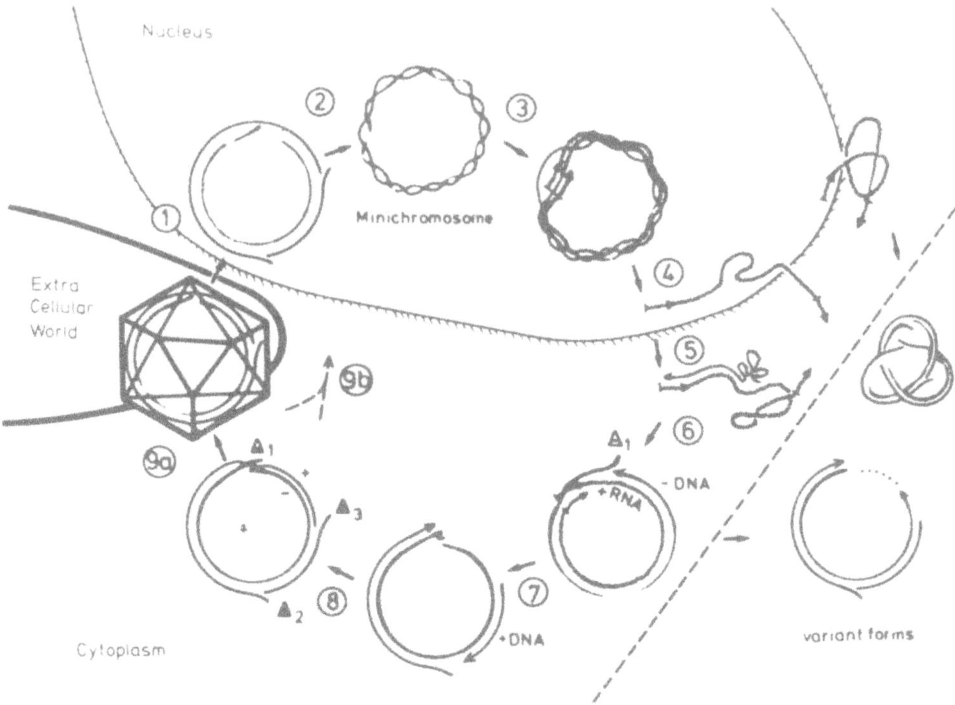

Figure 15. Replication of cauliflower mosaic virus. See text for description. Reprinted with permission (reference 289).

15, and 18 deoxynucleotides for nicks 1, 2 and 3 respectively, thereby producing a small single-stranded terminal 5' redundant region at each nick. The length of these single-stranded "wiskers" varies slightly.

Two remarkable features emerge from these data:

i) replication of all three viruses requires redundant regions in a pregenomic RNA which is produced by ignoring a transcriptional termination signal upon first encounter. This allows switching of DNA/RNA hybrids from one site on the pregenomic RNA to another when the RNA template is removed by RNase H activity; and

ii) displacement synthesis can occur when polymerase encounters a duplex region, again producing small redundancies useful for circularization or generating terminal duplications, as in retrovirus LTRs.

4. CONCLUDING REMARKS

Although there is no one mechanism of viral replication, one is struck by the extent to which RNA viruses utilize host transcriptional and translational proteins to perform the task of assisting the viral replicase recognize the 3' ends of the template. The presence of structural features which are tRNA-like at the ends of a number of RNA viruses are a further manifestation of the use of host translation mechanisms. The use of these proteins probably reflects the requirement, shared by RNA viruses (but not some viral satellites and viroids), that replication start at the 3'-OH end of a linear template molecule. Although RNA viruses can use various methods for initiation of replication, including primers such as protein primers and host mRNA sequences (although these latter are used only in transcriptional products), primers are not a necessity for all RNA viruses. Moreover, some RNA viruses take advantage of the ability to add ribonucleotides in a non-template-directed manner (adding terminal CCA-3' via tRNA-associated enzymes; 5-methylguanosine cap; poly (A) 3'-tail) to either complete a full-length transcript, to help in positioning transcription/replication proteins, or to distinguish between transcriptive or replicative molecules.

RNA viral replication always involves displacement synthesis (sometimes only after RF formation), a feature probably required by the lack of RNA unwinding enzymes. The ability to recognize and to initiate RNA synthesis directly at an end is probably the reason why no RNA virus uses a scheme similar to the parvoviruses, a scheme which otherwise is compatible with RNA replication.

DNA viruses, in contrast, require a 3'-OH primer for DNA polymerase activity. For most DNA viruses, the primer is supplied by an RNA oligonucleotide (which is subsequently replaced by DNA) in a manner similar to that observed in host DNA replication. The requirement for a primer can also be satisfied by 3'-OH sites in double-stranded DNA nicked at specific loci. Such 3'-OH DNA primers can either be generated in viral hairpins (e.g. poxvirus, parvovirus), at a junction with flanking host DNA (e.g. phage mu), in double-stranded circles, permitting concatamer formation (e.g: ΦX174), or in invasive recombination-like processes (e.g. T4). Finally, the primer requirement can be satisfied by a specific priming protein (e.g. adenovirus).

Because DNA unwinding enzymes are available, DNA viral replication can (but need not) initiate internally and proceed in a semi-conservative fashion on *both* strands (i.e. the displaced or lagging strand is also replicated) and in *both* directions from the initiation site. Internal initiation and the need for a 3'-OH primer requires either that there are no ends in the virus (i.e. the molecule is circular) or that other mechanisms are in place for resolving ends. Retrograde DNA synthesis on the lagging strand requires multiple initiations, primer removal, and ligation of DNA. This is an enzymatically more complex process than simple displacement synthesis and, often, host enzymes are utilized. The absence of a double-stranded RNA nicking/sealing activity similar to DNA ligase activity is probably one reason why RNA viruses only initiate at ends and by displacement of strands.

The viruses that utilize both RNA and DNA genomes, using the unique reverse-transcriptase activity to switch between them, all require production of RNA genomes (pregenomes) which have terminal redundancies. Although the RNA form is not circular, the DNA form becomes circular by hybridization of single-stranded DNA ends which have had their RNA templates removed by RNase H activity. The reverse transcriptase, like all DNA polymerases, requires a primer which can be supplied by either a host tRNA (retroviruses and cauliflower mosaic virus) or by a protein (hepadna virus). It can also produce strand displacement synthesis, a feature which can generate ssDNA redundancies necessary for hybridization or to generate a linear DNA with terminal redundancies.

The enzymology of viral replication appears to require enzymes which are either host enzymes or are modifications of host-like enzymatic activity. There are no truly unique enzymatic activities (e.g. a 3' to 5' polymerase) encoded by viruses. Although priming proteins do not appear, as such, in host cells, proteins with covalently-linked nucleotides are not unusual and some are also oligonucleotide-binding proteins. In many cases, the covalent-linkage is only transient and involved in energy transfer. However, no great leap is required to transform such a protein to a priming protein.

Thus, despite the wide variability in mechanisms of viral replication, one observes limitations in the range of possible mechanisms which are related to the nature of the genome (RNA or DNA) and the ways that cells deal with these molecules in their normal functioning. This, of course, is what one would expect in a parasitic entity which evolved from a free-living progenitor (although not necessarily evolved by loss of function — the acquisition of mobile, infectious status by a cell component is also possible), and which has co-evolved with its host by ex-ploiting its normal functions.

5. REFERENCES

1) D.C. Reanny (1982) Ann. Rev. Microbiol. **36**, 47-73.
2) Matthews, R.E.F. (1983) In *A Critical Appraisal of Viral Taxonomy* Matthews, R.E.F., ed. CRC Press, Boca Raton, 220-245.
3) Silverstein, S.C., Christmas, J.K., & Acs, G. (1976) Ann. Rev. Biochem) 45, 375-408.
4) Ahlquist, P., Dasgupta, R., & Kaesberg, P. (1981) Cell **23**, 183-189.
5) Goelet, P., Lomonossoff, G.P., Butler, P.J.G., Akam, M.E., Gait, M.J., & Karn, J. (1982) Proc. Natl. Acad. Sci. USA **79**, 5818-5822.
6) Ahlquist, P., Dasgupta, R., & Kaesberg, P. (1984) J. Mol. Biol. 172, 369-383.
7) Blumenthal, T. (1979) Ann. Rev. Biochem. **48**, 525-548.
8) Bastin, M. & Hall, T.C. (1976) J. Virol. **20**, 117-122.
9) Bishop, D.H.L., Gay, M.E., & Matsuoko, Y. (1983) Nucleic Acids Res) **11**, 6409-6418.
10) Patterson, J.L. & Kolakofsky, D. (1984) J. Virol. **49**, 680-685.
11) Krug, R.M., Broni, B.B., & Bouloy, M. (1979) Cell **18**, 329-334.
12) Plotch, S.J., Bouloy, M., Ulmanen, I., & Krug, R.M. (1981) Cell 23, 847-858.
13) Winter, G., Fields, S., Gait, M.J., & Brownlee, G.G. (1981) Nucleic Acids Res. **9**, 237-245.
14) Lamb, R.A. & Choppin, P.W. (1983) Ann. Rev. Biochem. **52**, 467-506.
15) Francki, R.I.B. (1985) Ann. Rev. Microbiol. **39**, 151-174.
16) Diener, T.O. (1979) Science **205**, 859-866.
17) Rackwitz, J.W., Davies, C., Hatta, T., Gould, A.R., & Francki, R.I.B) (1981) Virology **108**, 111-122.

18) Gerlach, W.L., Buzayan, J.M., Schneider, I.R., & Bruening, G. (1986) Virology **151**, 172-185.

19) Symons, R.J., Haseloff, J., Visvader, J.E., Keese, P., Murphy, P.J., Gordon, K.H.J., & Bruening, G. (1985) In *Subviral Pathogens of Plants and Animals: Viroids and Prions* Maramorosch, K. & McKelvey, J.J., eds) Academic Press, New York, 235-263.

20) Branch, A.D., Willis, K.K., Davatelis, G., & Robertson, H. (1985) In: *Subviral Pathogens of Plants and Animals: Viroids and Prions*, Maramorosch, K. & McKelvey, J.J., eds. Academic Press, New York, 201-234.

21) Haenni, A.-L., Joshi, S., & Chapeville, F. (1982) Prog. Nucl. Acid Res) Mol. Biol. **27**, 85-104.

22) Inouye, H., Pollack, Y., & Petre, J. (1974) Eur. J. Biochem. **45**, 109-117.

23) Blumenthal, T., Landers, T.A., & Weber, K. (1972) Proc. Natl. Acad. Sci) USA **69**, 1313-1317.

24) Franze de Fernandez, M.T., Eoyang, L., & August, J.T. (1968) Nature (London) **219**, 588-590.

25) Rensing, U. & August, J.T. (1969) Nature (London) **224**, 853-856.

26) Weber, H. & Weissman, C. (1970) J. Mol. Biol. **51**, 215-224.

27) Prochiantz, A., Benicourt, C., Carre, D., & Haenni, A.L. (1975) Eur. J) Biochem. **52**, 17-23.

28) Bujarski, J.J., Dreher, T.W., & Hall, T.C. (1985) Proc. Natl. Acad. Sci) USA **82**, 5636-5640.

29) Dreher, T.W., Bujarski, J.J., & Hall, T.C. (1984) Nature (London) **311**, 171-175.

30) Ahlquist, P., French, R., Janda, M., & Loesch-Fries, L.S. (1984) Proc) Natl. Acad. Sci. USA **81**, 7066-7070.

31) Miller, W.A., Dreher, T.W., & Hall, T.C. (1985) Nature (London) **313**, 68-70.

32) Rueckert, R.R. (1985) In *Virology* Fields, B.N., ed. Raven Press, New York, 705-738.

33) Perez-Bercoff, R. (1978) In *Molecular Biology of Picornaviruses* Perez- Bercoff, R., ed. Plenum Press, New York, 319-330.

34) Nomoto, A., Lee, Y.J., & Wimmer, E. (1976) Proc. Natl. Acad. Sci. USA **73**, 375-380.

35) Flanegan, J.B., Petterson, R.F., Ambros, V., Hewlett, M.J., & Baltimore, D. (1977) Proc. Natl. Acad. Sci. USA **74**, 961-965.

36) Golini, F., Semeler, B.L., Dorner, A.J., & Wimmer, E. (1980) Nature (London) **287**, 600-603.

37) Sanger, D.V. (1979) J. Gen. Virol. **45**, 1-11.

38) Ambrose, V. & Baltimore, D. (1980) J. Biol. Chem. **255**, 6739-6744.

39) Spector, D.H. & Baltimore, D. (1974) Proc. Natl. Acad. Sci. USA **71**, 2983-2987.

40) Van Dyke, T.A. & Flanegan, J.B. (1980) J. Virol. **35**, 732-740.

41) Andrews, N.C., Levin, D.H., & Baltimore, D. (1985) J. Biol. Chem) **260**, 7628-7635.

42) Andrews, N.C. & Baltimore, D. (1986) Proc. Natl. Acad. Sci. USA **83**, 221-225.

43) Andrews, N.C. & Baltimore, D. (1986) J. Virol. **58**, 212-215.

44) Young, D.C., Dunn, B.N., Tobin, G.J., & Flanegan, J.B. (1986) J. Virol) **58**, 715-723.

45) Hey, T.D., Richards, O.C., & Ehrenfeld, E. (1986) J. Virol. **58**, 790-796.

46) Nomoto, A., Detjer, B., Pozzati, R., & Wimmer, E. (1977) Nature (London) **268**, 208-213.

47) Pettersson, R.F., Ambros, V., & Baltimore, D. (1978) J. Virol) **27**, 357-365.

48) Takegani, T., Kuhn, R.J., Anderson, C.W., & Wimmer, E. (1983) Proc) Natl. Acad. Sci. USA **80**, 7447-7451.

49) Crawford, N.M. & Baltimore, D. (1983) Proc. Natl. Acad. Sci. USA **80**, 7452-7455.

50) Onata, T., Kohara, M., Sakai, Y., Kaneda, A., Imura, N., & Nomoto, A) (1984) Gene **32**, 1-10.

51) Strauss, E.G. & Strauss, J.H. (1983) Curr. Top. Micro. Immun. **105**, 1-98.

52) Sarnow, P., Bernstein, H.D., & Baltimore, D. (1986) Proc. Natl. Acad) Sci. USA **83**, 571-575.

53) Mizutani, S. & Colonno, R.J. (1985) J. Virol **56**, 628-632.

54) Semler, B.L., Dorner, A.J., & Wimmer, E. (1984) Nucleic Acids Res) **12**, 5123-5141.

55) Meyer, M., Hemmer, O., & Fritsch, C. (1984) J. Gen. Virol. **65**, 1575-1583.

56) van der Werf, S., Bradley, J., Wimmer, E., Studier, F.W., & Dunn, J.J) (1986) Proc. Natl. Acad. Sci. USA **83**, 2330-2334.

57) Rice, E.M., Lenches, E.M., Eddy, S.R., Shin, S.J., Sheets, R.L., & Strauss, J.H. (1985) Science **229**, 726-733.

58) Brown, T.D.K., Boursnell, M.E.G., Binns, M.M., & Tonley, F.M. (1986) J) Gen. Virol. **67**, 221-228.

59) Lai, M.M.C., Baric, R.S., Brayton, P.R., & Stohlman, S.A. (1984) Proc) Natl. Acad. Sci. USA **81**, 3626-3630.

60) Spaan, W., Delius, H., Skinner, M.A., Armstrong, J., Rottier, P., Smeekens, S., Siddell, S.G., & van der Zeijst, B. (1983) Adv. Exp. Biol) **173**, 173-186.

61) Ou, J.H., Trent, D.W., & Strauss, J.H. (1982) J. Mol. Biol. **156**, 719-730.

62) Ou, J.H., Strauss, E.G., & Strauss, J.H. (1983) J. Mol. Biol. **168**, 1-15.

63) Lewis, R., Weiss, B.G., Tsiang, M., Huang, A., & Schlesinger, S. (1986) Cell **44**, 137-145.

64) Sawicki, D.L. & Sawicki, S.G. (1980) J. Virol. **34**, 108-118.

65) Sawicki, S.G., Sawicki, D.L., Keranen, S., & Kaariainen, L. (1981) J) Virol. **39**, 348-358.

66) Sawicki, S.G., Sawicki, D.L., Kaariainen, L., & Keranen, S. (1981) Virology **115**, 161-172.

67) Sawicki, S.G. & Sawicki, D.L. (1986) Virology **151**, 339-349.

68) Sawicki, S.G. & Sawicki, D.L. (1986) J. Virol. **57**, 328-334.

69) Frey, T.K., Gard, D.L., & Strauss, J.H. (1979) J. Mol. Biol. **132**, 1-18.

70) Sawicki, D.L. & Gomatos, P.J. (1976) J. Virol. **20**, 446-464.

71) Wengler, G. & Wengler, G. (1975) Virology **66**, 322-326.

72) Monroe, S.S. & Schlesinger, S. (1983) Proc. Natl. Acad. Sci. USA **80**, 3279-3283.

73) Banerjee, A.K. (1980) Microbiol. Rev. **44**, 175-205.

74) Ishida, I., Simizu, B., Koizuni, S., Oya, A., & Yamada, M. (1981) Virology **108**, 13-20.

75) Brayton, P.R., Lai, M.M.C., Patton, C.D., & Stohlman, S.A. (1982) J) Virol. **42**, 847-853.

76) Lai, M.M.C., Baric, R.S., Brayton, P.R., & Stohlman, S.A. (1984) Adv) Exp. Med. Biol. **173**, 197-200.

77) Kingsbury, D.W. (1985) In *Virology* Fields, B.N., ed. Raven Press, New York, 1157-1178.

78) Compans, R.W. & Bishop, D.H.L. (1985) Curr. Topics Microbiol. Immun) **114**, 153-175.

79) Bishop, D.H.L. (1985) In *Virology* Fields, B.N., ed. Raven Press, New York, 1083-1110.

80) Bishop, D.H.L. & Smith, M.S. (1977) In *The Molecular Biology of Animal Viruses* Nyak, D.P., ed. Dekker, New York, 167-315.

81) Pennington, T.H. & Pringle, C.R. (1978) In *Negative Strand Viruses and the Host Cell* Mahy, B.W.J. & Berry, R.D., eds. Academic Press, London, 457-464.

82) Mellon, M.G. & Emerson, S.U. (1978) J. Virol. **27**, 560-567.

83) Emerson, S.U. (1985) In *Virology* Fields, B.N., ed. Raven Press, New York, 1119-1132.

84) Hsu, C.-H., Morgan, E.M., & Kingsburg, D.W. (1982) J. Virol. **43**, 104-112.

85) Kingsford, L. & Emerson S.U. (1980) J. Virol. **33**, 1097-1105.

86) Clinton, G.M., Burge, B.W., & Huang, A.S. (1978) J. Virol. **27**, 340-346.

87) Fiszman, M., Leaute, J.-B., Chany, C., & Girard, M. (1974) J. Virol) **13**, 801-808.

88) Wilson, T. & Lenard, J. (1981) Biochemistry **20**, 1349-1354.

89) Hill, V.M., Marnell, L., & Summers, D.F. (1981) Virology **113**, 109-118.

90) Blumberg, B.M. & Kolakofsky, D. (1981) J. Virol. **40**, 568-576.

91) Blumberg, B.M., Leppert, M., & Kolakofsky, D. (1981) Cell **23**, 837-845.

92) Kurilla, M.G., Cabradilla, C.D., Holloway, B.P., & Keene, J.D. (1984) J) Virol. **50**, 773-778.

93) Kurilla, M.G. & Keene, J.D. (1983) Cell **34**, 837-845.

94) Leppert, M., Rittenhouse, L., Perrault, J., Summers, D.F., & Kolakofsky, D.F. (1979) Cell **18**, 735-747.

95) Wilusz, J. & Keene, J.D. (1984) Virology **135**, 65-73.

96) Keene, J.D., Schubert, M., & Lazzarini, R.A. (1980) J. Virol. **33**, 789-794.

97) Kurilla, M.G., Stone, H.O., & Keene, J.D. (1985) Virology **145**, 203-212.

98) Robertson, J.S. (1979) Nucleic Acids Res. **6**, 3745-3757.

99) Braam, J., Ulmanen, I., & Krug, R.M. (1983) Cell **34**, 609-618.

100) Mahy, B.W.J., Barrett, T., Nichol, S.T., Penn, C.R., & Wolstenholme, A.J) (1981) In *The Replication of Negative Strand Viruses* Bishop, D.H.L. & Compans, R.W., eds. Elsevier/North-Holland, New York, 379-394.

101) Herz, C., Stavnezer, E., Krug, R.M., & Gurney, T. (1981) Cell **26**, 391-400.

102) Smith, G.L. & Hay, A.J. (1982) Virology **118**, 96-108.

103) Petri, T., Meier-Ewert, H., & Compans, R.W. (1979) J. Virol) **32**, 1037-1040.

104) Ulmanen, I., Broni, B.A., & Krug, R.M. (1981) Proc. Natl. Acad. Sci. USA **78**, 7355-7359.

105) Robertson, J.S., Caton, A.J., Schubert, M., & Lazzarini, R.A. (1981) In *The Replication of Negative Strand Viruses* Bishop, D.H.L. & Compans, R.W., eds. Elsevier/North Holland, New York, 303-308.

106) Hay, A.J., Skehel, J.J., & McCauley, J. (1982) Virology **116**, 517-522.

107) Eshita, Y., Ericson, B., Romanowski, V., & Bishop, D.H.L. (1985) J. Virol) **55**, 681-689.

108) Patterson, J.L., Holloway, B., & Kolakofsky, D. (1984) J. Virol) **52**, 215-222.

109) Milne, R.G. & Francki, R.I.B. (1984) Intervirology **22**, 72-76.

110) Silverstein, S.C., Astell, C., Levin, D.H., Schonberg, M., & Acs, G) (1972) Virology **47**, 797-806.

111) Shatkin, A.J., Sipe, J.D., & Loh, P.C. (1968) J. Virol. **2**, 986-991.

112) Muthukrishnan, S. & Shatkin, A.J. (1975) Virology **64**, 96-105.

113) Furuichi, Y., Muthukrishnan, S., & Shatkin, A.J. (1975) Proc. Natl. Acad) Sci. USA **72**, 742-745.

114) Chow, N.-L. & Shatkin, A.J. (1975) J. Virol. **15**, 1057-1064.

115) Shatkin, A.J. & Kozak, M. (1983) In *The Reoviridae* Joklik, W.K., ed) Plenum Press, New York, 79-106.

116) Silverstein, S.C., Christman, J.K., & Acs, G. (1976) Ann. Rev. Biochem) **45**, 375-408.

117) Tyler, K.L. & Fields, B.N. (1985) In *Virology* Fields, B.N., ed); Raven Press, New York, 823-862.

118) Sakuma, S. & Watanabe, Y. (1972) J. Virol. **10**, 628-638.

119) Zweerink, H.J. (1974) Nature (London) **247**, 313-315.

120) Zarble, H. & Millward, S. (1983) In *The Reoviridae* Joklik, W.K., ed) Plenum Press, New York, 108-196.

121) Nemeroff, M.E. & Bruenn, J.A. (1986) J. Virol. **57**, 754-758.

122) Van Etten, J.L., Burbank, D.E., Cuppels, D.A., Lane, L.C., & Vidaver, A.K) (1980) J. Virol. **33**, 769-773.

123) Buck, K.W. (1975) Nucleic Acids Res. **2**, 1889-1902.

124) Ratti, G. & Buck, K.W. (1978) Nucleic Acids Res. **5**, 3843-3854.

125) Branch, A.D. & Robertson, H.D. (1984) Science **223**, 450-455.

126) Buzayan, J.M., Gerlach, W.L., Bruening, G., Keese, P., & Gould, A.R) (1986) Virology **151**, 186-199.

127) Gerlach, W.L., Buzayan, J.M., Schneider, I.R., & Bruening, G. (1986) Virology **151**, 172-185.

128) Robertson, H., Rosen, D.L., & Branch, A.D. (1985) Virology **142**, 441-447.

129) Kiberstis, P.A., Haseloff, J., & Zimmern, D. (1985) EMBO J. **4**, 817-827.

130) Hutchins, C.J., Keese, P., Visvader, J.E., Rathjen, P.D., McInnes, J.L., & Symons, R.H. (1985) Plant Mol. Biol. **4**, 293-304.

131) Collmer, C.W., Hadidi, A., & Kapen, J.M. (1985) Proc. Natl. Acad. Sci. USA **82**, 3110-3114.

132) Watson, N., Gurevitz, M., Ford, J., & Apirion, D. (1984) J. Mol. Biol) **172**, 301-323.

133) Price, J.V. & Cech, T.R. (1985) Science **228**, 719-722.

134) Cech, T.R. (1986) Cell **44**, 207-210.

135) Acheson, N.H. (1981) In *The Molecular Biology of Tumor Viruses: Second Edition, Part 2. DNA Tumor Viruses* Tooze, J., ed. Cold Spring Harbor Laboratory, Cold Spring Harbor, New York, 125-204.

136) Pope, J.H. & Rowe, W.D. (1964) J. Exp. Med. **120**, 121-128.

137) Feunteun, J., Sompayrac, L., Fluck, M., & Benjamin, T.L. (1976) Proc) Natl. Acad. Sci. USA **73**, 4169-4173.

138) Dulbecco, R., Hartwell, L., & Vogt, M. (1965) Proc. Natl. Acad. Sci. USA **53**, 403-408.

139) Hiscott, J.B. & Defendi, V. (1979) J. Virol. **30**, 590-599.

140) Hiscott, J.B. & Defendi, V. (1981) J. Virol. **37**, 802-812.

141) Tijan, R. & Robbins, A. (1979) Proc. Natl. Acad. Sci. USA **76**, 610-615.

142) Griffin, J.D., Spangler, G., & Livingston, D.M. (1979) Proc. Natl. Acad) Sci. USA **76**, 2610-2614.

143) McCormick, F. & Harlow, E. (1980) J. Virol. **34**, 213-224.

144) Linzer, D.I.H. & Levine, A.J. (1979) Cell **17**, 43-52.

145) DeLucia, A., Lewton, B., Tijan, R., & Tegtmeyer, P. (1983) J. Virol) **46**, 143-150.

146) Stringer, J.R. (1982) Nature (London) **296**, 363-366.

147) Temen, D., Haines, L., & Livingston, D.M. (1982) J. Mol. Biol) **157**, 473-492.

148) Tijan, R. (1978) Cell **13**, 165-179.

149) Margolskee, R.F. & Nathans, D. (1984) J. Virol. **49**, 386-393.

150) Ryder, K., Silver, S., DeLucia, A.L., Fanning, E., & Tegtmeyer, P. (1986) Cell **44**, 719-725.

151) Zahn, K. & Blattner, F.R. (1985) Nature (London) **317**, 451-453.

152) Murakani, Y., Wobbe, C.R., Weissbach, L., Dean, F.B., & Hurwitz, J. (1986) Proc. Natl. Acad. Sci. USA **83**, 2869-2873.

153) Chou, J., Avila, J., & Martin, R. (1974) J. Virol **14**, 116-128.

154) Stahl, H. & Knippers, R. (1983) J. Virol. **47**, 65-76.

155) Hay, R. & DePamphilis, M. (1982) Cell **28**, 167-179.

156) Tapper, D. & DePamphilis, M. (1978) J. Mol. Biol. **120**, 401-412.

157) DePamphilis, M. & Wassarman, P.M. (1982) In *Organization and Replication of Viral DNA* Kaplan, A.S., ed. CRC Press, Boca Raton, 37-114.

158) Wobbe, C.R., Dean, F., Weissbach, L., & Hurwitz, J. (1985) Proc. Natl) Acad. Sci. USA **82**, 5710-5714.

159) Kaiser, D. (1971) In *The Bacteriophage Lambda* A.D. Hershey, ed.

Cold Spring Harbor Laboratory, Cold Spring Harbor, New York, 195-210.

160) Kornberg, A. (1974) *DNA Replication* W.H. Freeman, San Francisco.

161) Kornberg, A. (1982) *1982 Supplement to DNA Replication* W.H. Freeman, San Francisco.

162) Tessman, E.S. & Tessman, I. (1978) In *The Single-Stranded DNA Phages* Denhardt, D.T., Dressler, D., & Ray, D.S., eds. Cold Spring Harbor Laboratory, Cold Spring Harbor, New York, 9-29.

163) Ben-Porat, T. (1982) In *Organization and Replication of Viral DNA* Kaplan, A.S., ed. CRC Press, Boca Raton, 147-172.

164) Jacob, R.J. & Roizman, B. (1977) J. Virol. **23**, 394-411.

165) Roizman, B. & Batterson, W. (1985) In *Virology* Fields, B.N., ed) Raven Press, New York, 497-526.

166) Stanley, J. (1985) Adv. Virus Res. **30**, 139-177.

167) Harrison, B.D. (1985) Ann. Rev. Phytopathol. **23**, 55-82.

168) Mosig, G. (1983) In *Bacteriophage T4* Matthews, C.K., Kutter, E.M., Mosig, G., & Berget, P.B., eds. American Society for Microbiology, Washington, D.C., 120-130.

169) Murti, K.G., Goorha, R., & Granoff, A. (1985) Adv. Virus Res. **30**, 1-19.

170) Jackson, E.N., Jackson, D.A., & Deans, R.J. (1978) J. Mol. Biol) **118**, 365-388.

171) Goorha, R. (1982) J. Virol. **43**, 519-528.

172) Willis, D.B., Goorha, R., & Granoff, A. (1984) J. Virol. **49**, 86-91.

173) Cairns, J. (1960) Virology **11**, 603-623.

174) Pennington, J.H. & Follett, E.A. (1974) J. Virol. **13**, 488-493.

175) Prescott, D.M., Kates, J., & Kirkpatrick, J.B. (1971) J. Mol. Biol. **59**, 505-508.

176) Panicali, D. & Paoletti, E. (1982) Proc. Natl. Acad. Sci. USA **79**, 4927-4931.

177) Mackett, M., Smith, G.L., & Moss, B. (1982) Proc. Nat. Acad. Sci. USA **79**, 7415-7419.

178) Smith, G.L. & Moss, B. (1983) Gene **25**, 21-28.

179) McFadden, G. & Dales, S. (1982) In *Organization and Replication of Viral DNA* Kaplan, A.S., ed. CRC Press, Boca Raton, 174-190.

180) Moss, B., Winters, E., & Jones, E.V. (1983) In *Proceedings of the UCLA Symposium: Mechanisms of DNA Replication and Recombination* Cozzarelli, N.R., ed. Alan R. Liss, Inc., New York, 449-461.

181) Moyer, R.W. & Graves, R.L. (1981) Cell **27**, 391-401.

182) Cavalier-Smith, T. (1974) Nature (London) **250**, 467-470.

183) Forte, M.A. & Fangman, W.L. (1979) Chromosoma **72**, 131-150.

184) Berns, K.I. & Hauswirth, W.W. (1984) In *The Parvoviruses* Berns, K.I., ed. Plenum Press, New York, 1-31.

185) Hauswirth, W.W. (1984) In *The Parvoviruses* Berns, K.I., ed. Plenum Press, New York, 129-152.

186) Cukor, G., Blacklow, N.R., Hoggan, M.D., & Berns, K.I. (1984) In *The Parvoviruses* Berns, K.I., ed. Plenus Press, New York, pp. 33-66.

187) Laughlin, C.A., Jones, N., & Carter, B.J. (1982) J. Virol. **41**, 868-876.

188) Ostrove, J.M. & Berns, K.I. (1980) Virology **104**, 502-505.

189) Janik, J.E., Huston, M.M., & Rose, J.A. (1981) Proc. Natl. Acad. Sci. USA **78**, 1925-1929.

190) Carter, B., Marcus, C., Laughlin, C., & Ketner, G. (1983) Virology **126**, 505-516.

191) Richardson, W.D. & Westphall, H. (198) Cell **27**, 131-141.

192) Jay, F.T., Laughlin, C.A., & Carter, B.J. (1981) Proc. Natl. Acad. Sci. USA **78**, 2927-2931.

193) Meyers, M.W. & Carter, B.J. (1981) J. Biol. Chem. **256**, 567-570.

194) Handa, H. & Carter, B.J. (1979) J. Biol. Chem. **254**, 6603-6610.

195) Srivastava, A., Lusby, E.W., & Berns, K.I. (1983) J. Virol. **45**, 555-564.

196) Berns, K.I., Samulski, R.J., Srivastava, A., & Muzyczka, N. (1983) In *Proceedings of the UCLA Symposium, Mechanisms of DNA Replication and Recombination* Cozzarelli, N.R., ed. Alan R. Liss, Inc., New York, 353-365.

197) Samulski, R.J., Berns, K.I., Tan, M., & Muzyczka, N. (1982) Proc. Natl. Acad. Sci. USA **78**, 2077-2081.

198) Fife, K.H., Berns, K.I., & Murray, K. (1977) Virology **78**, 475-487.

199) Lusby, E., Bohenzky, R., & Berns, K.I. (1981) J. Virol. **37**, 1083-1086.

200) Rhode, S.L. & Paradiso, P.R. (1983) J. Virol. **45**, 173-184.

201) Astell, C.A., Thompson, M., Chow, M.B., & Ward, D.C. (1983) Cold Spring Harbor Symp. Quant. Biol. **47**, 751-762.

202) Wobbe, C.R. & Mitra, S. (1985) Proc. Natl. Acad. Sci. USA **82**, 8335-8339.

203) Walter, S., Richards, R., & Armentrout, R.W. (1980) Biochim. Biophys. Act) **607**, 420-431.

204) Kolleck, R., Tseng, B.Y., & Goulian, M. (1982) J. Virol. **41**, 982-989.

205) Goulian, M., Kollek, R., Revie, D., Burhans, W., Carton, C., & Tseng, B. (1983) In *Proceedings of the UCLA Symposium: Mechanisms of DNA Replication and Recombination* Cozzarelli, N.R., ed. Alan Liss, Inc., New York, 367-379.

206) Baroudy, B.M., Venkatesan, S., & Moss, B. (1983) Cold Spring Harbor Symp. Quant. Biol. **47**, 723-729.

207) Baroudy, B.M., Venkatesan, S., & Moss, B. (1982) Cell **28**, 315-324.

208) Wittek, R. & Moss, B. (1980) Cell **21**, 277-284.

209) Baroudy, B.M. & Moss, B. (1982) Nucleic Acids Res. **10**, 5673-5679.

210) Pickup, D.J., Bastia, D., Stone, H.O., & Joklik, W.K. (1982) Proc. Natl. Acad. Sci. USA **79**, 7112-7116.

211) Pogo, B.G.T., Berkowitz, E.M., & Dales, S. (1984) Virology **132**, 436-444.

212) Pogo, B.G.T., O'Shea, M., & Freimuth, D. (1981) Virology **108**, 241-248.

213) Esteban, M. & Holowczak, J.A. (1977) Virology **82**, 308-322.

214) Mocarski, E.S. & Roizman, B. (1982) Proc. Natl. Acad. Sci. USA **79**, 5626-5630.

215) Spaete, R.R. & Frenkel, N. (1982) Cell **30**, 295-304.

216) Stow, N.D. (1982) EMBO J. **1**, 863-867.

217) DeLange, A.M. & McFadden, G. (1986) Proc. Natl. Acad. Sci. USA **83**, 614-618.

218) Pogo, B.G.T. (1980) Virology **101**, 520-524.

219) Pogo, B.G.T. (1977) Proc. Natl. Acad. Sci. USA **74**, 1739-1742.

220) Esteban, M. & Holowczak, J.A. (1977) Virology **78**, 57-75.

221) Moyer, R.W. & Graves, R.L. (1981) Cell **27**, 391-401.

222) Wittek, R. (1982) Experientia **38**, 285-297.

223) Holowczak, J. (1982) Curr. Topics Microbiol. Immunol. **97**, 27-79.

224) DeLange, A.M., Futcher, B., Morgan, R., & McFadden, G. (1984) Gene 27, 13-21.

225) Merchlinsky, M. & Moss, B. (1986) Cell **45**, 879-884.

226) Sussenbach, J.S. & van der Vliet, P.C. (1983) Curr. Topics Microbiol. Immunol. **109**, 53-73.

227) Tamanoi, F. & Stillman, B.W. (1983) Curr. Topics Microbiol. Immunol. **109**, 75-87.

228) Friefeld, B.R., Lichy, J.H., Field, J., Gronostajski, R.M., Guggenheimer, R.A., Krevolin, M.D., Nagata, K., Hurwitz, J., & Horwitz, M.S. (1984) Curr. Topics Microbiol. Immunol. **110**, 221-255.

229) Futterer, J. & Winnacker, E.-L. (1984) Curr. Topics Microbiol. Immunol. **111**, 41-64.

230) Salas, M. (1983) Curr. Topics Microbiol. Immunol. **109**, 89-106.

231) Desiderio, S.V. & Kelly, T.J., Jr. (1981) J. Mol. Biol. **145**, 319-337.

232) Challberg, M.D. & Kelly, T.J., Jr. (1981) J. Virol. **38**, 272-277.

233) Ruben, M., Bacchetti, S., & Graham, F. (1983) Nature (London) **301**, 172-174.

234) Graham, F. (1984) EMBO J. **3**, 2917-2922.

235) Tamanoi, F. & Stillman, B.W. (1982) Proc. Natl. Acad. Sci. USA **79**, 2221-2225.

236) DeJong, P.J., Kwant, M.M., van Driel, W., Janz, H.S., & van der Vliet, P.C) (1983) Virology **124**, 45-58.

237) Lally, C., Dorpen, T., Groger, W., Antoine, G., & Winnacker, E.L. (1984) EMB) J. **3**, 333-337.

238) van Bergen, B.G.M., van der Ley, P.A., van Driel, W., van Mansfeld, A.D.M., & van der Vliet, P.C. (1983) Nucleic Acids Res. **11**, 1975-1989.

239) Tamanoi, F. & Stillman, B.W. (1983) Proc. Natl. Acad. Sci. USA **80**, 6446-6450.

240) Rijnders, A.W.M., van Bergen, B.G.M., van der Vliet, P.C., & Sussenbach, J.S) (1983) Nucleic Acids Res. **11**, 8777-8789.

241) Lichy, J.H., Field, J., Horwitz, M.S., & Hurwitz, J. (1982) Proc. Natl. Acad. Sci. USA **79**, 5225-5229.

242) Nagata, K., Guggenheimer, R.A., & Hurwitz, J. (1983) Proc. Natl. Acad. Sci) USA **80**, 4266-4270.

243) Nagata, K., Guggenheimer, R.A., Enomoto, T., Lichy, J.H., & Hurwitz, J. (1982) Proc. Natl. Acad. Sci. USA **79**, 6438-6442.

244) Ikeda, J.-E., Enomoto, T., & Hurwitz, J. (1982) Proc. Natl. Acad. Sci., USA **79**, 2442-2446.

245) Guggenheimer, R.A., Nogata, K., Field, J., Lindenbraun, J., Gronostajsky, R.M., Horwitz, M.S., & Hurwitz, J. (1983) In *Proceedings of the UCLA Symposium: Mechanisms of DNA Replication and Recombination* Cozzarelli, N.R., ed.) Alan R. Liss, Inc., New York, 395-421.

246) Blanco, L. & Salas, M. (1985) Proc. Natl. Acad. Sci. USA **82**, 6404-6408.

247) Garcia, J.A., Penalva, M.A., Blanco, L., & Salas, M. (1984) Proc. Natl. Acad. Sci. USA. **81**, 5374-5378.

248) Weissberg, R.A., Gottesman, S., & Gottesman, M.E. (1977) In *Comprehensive Virology, Vol. 8* Frankel-Conrat, H. & Wagner R.R., eds. Plenum Press, New York, 197-258.

249) Stringer, J.R. (1982) Nature (London) **296**, 363-366.

250) Topp, W.C., Lane, D., & Pollack, R. (1981) In *Molecular Biology of Tumor Viruses: DNA Tumor Viruses 2nd Ed.* Tooze, J., ed. Cold Spring Harbor Laboratory, Cold Spring Harbor, New York, 205-296g.

251) Chias, W. & Rigby, P.W.J. (1981) Proc. Natl. Acad. Sci. USA **78**, 6638-6642.

252) Conrad, S.E., Liu, C.-P., & Botchan, M. (1982) Science **218**, 1223-1225.

253) Sambrook, J., Greene, R., Stringer, J., Mitcheson, T., Hu, S.-L., & Botchan, M. (1979) Cold Spring Harbor Symp. Quant. Biol. **44**, 569-583.

254) Merchlinsky, M.J., Tattersall, P.J., Leary, J.J., Cotmore, S.F., Gardiner, E.M., & Ward, D.C. (1983) J. Virol. **47**, 227-232.

255) Berns, K.I., Chung, A.K.-M., Ostrove, J.M., & Lewis, M. (1982) In *Virus Persistence* Mahy, B.W.J., Mirson, A.C., & Darby, G.K., eds. Cambridge University Press, Cambridge, UK, 249-265.

256) Samulski, R.J., Srivastava, A., Berns, K.I., & Muzyczka, N. (1983) Cell **33**, 135-143.

257) Doerfler, W., Gahlmann, R., Stabel, S., Deuring, R., Lichtenberg, U., Schulz, M., Eick, D., & Leisten, R. (1983) Curr. Topics Microbiol. Immunol. **109**, 193-228.

258) Hanahan, D. & Gluzman, Y. (1984) Mol. Cell. Biol. **4**, 302-309.

259) van Doren, K., Hanahan, D., & Gluzman, Y. (1984) J. Virol. **50**, 606-614.

260) Koch, S., Freytag von Loringhoven, A., Kahmann, R., Hofschneider, P.H., & Koshy, R. (1984) Nucleic Acids Res. **12**, 6871-6886.

261) Dejean, A., Brechot, C., Tiollais, P., & Wain-Hobson, S. (1983) Proc. Natl. Acad. Sci. USA **80**, 2505-2509.

262) Mizusawa, H., Taira, M., Yaginuma, K., Kobayashi, M., Yoshida, E., & Koike, K. (1985) Proc. Natl. Acad. Sci. USA **82**, 208-212.

263) Yaginuma, K., Kobayashi, M., Yoshida, E., & Koike, K. (1985) Proc. Natl. Acad. Sci. USA **82**, 4458-4462.

264) Bernstein, L.B., Mount, S.M., & Weiner, A.M. (1983) Cell **32**, 461-472.

265) Mizuuchi, K. (1983) Cell **35**, 785-794.

266) Higgins, N.P., Moncecchi, D., Howe, M.M., Manlapaz-Ramos, P., & Olivera, B.M) (1983) In *Proceedings of the UCLA Symposium: Mechanisms of DNA Replication and Recombination* Cozzarelli, N.R., ed. Alan R. Liss, Inc., New York, 187-201.

267) Craigie, R. & Mizuuchi, K. (1985) Cell **41**, 867-876.

268) Craigie, R. & Mizuuchi, K. (1986) Cell **45**, 793-800.

269) Craigie, R., Mizuuchi, M., & Mizuuchi, K. (1984) Cell **39**, 387-394.

270) Mizuuchi, K. (1984) Cell **39**, 395-404.

271) Chaconas, G., Gloor, G., & Miller, J.L. (1985) J. Biol. Chem. **260**, 2662-2669.

272) Groenen, M.A.M., Timmers, E., & van de Putte, P. (1985) Proc. Natl. Acad. Sci. USA **82**, 2087-2091.

273) Craigie, R., Arndt-Jovin, D.J., & Mizuuchi, K. (1985) Proc. Natl. Acad. Sci) USA **82**, 7570-7574.

274) Harshey, R.M. & Bakhari, A.I. (1983) J. Mol. Biol. **167**, 427-441.

275) Varmus, H.E. & Swanstrom, R. (1982) In *Molecular Biology of Tumor Viruses: RNA Tumor Viruses* Weiss, R.A., Teich, N., Varmus, H.E., & Coffin, J.M., eds. Cold Spring Harbor Laboratories, Cold Spring Harbor, New York, 369-512.

276) Lowy, D.R. (1985) in *Virology* Fields, B.N., ed. Raven Press, New York, 235-263.

277) Champoux, J.J., Gilboa, E., & Baltimore, D. (1984) J. Virol. **49**, 686-691.

278) Resnick, R., Omer, C.A., & Faras, A.J. (1984) J. Virol. **51**, 813-821.

279) Smith, J.K., Cywinski, A., & Taylor, J.M. (1984) J. Virol. **49**, 200-204.

280) Smith, J.K. & Taylor, J.M. (1984) J. Virol. **52**, 314-319.

281) Taylor, J.M. & Hsu, T.W. (1980) J. Virol. **33**, 531-534.

282) Shimotohno, K. & Temin, H.M. (1980) Proc. Natl. Acad. Sci. USA **77**, 7357-7361.

283) Panganiban, A.T. & Temin, H.M. (1983) Nature (London) **306**, 155-160.

284) Misra, T.P., Grandgenett, D.P., & Parsons, J.T. (1982) J. Virol. **44**, 330-343.

285) Varmus, H.E., Padgett, T., Healsey, S., Simon, G., & Bishop, J.M. (1977) Cel) **11**, 307-309.

286) Miller, R.H. & Robinson, W.S. (1986) Proc. Natl. Acad. Sci. USA **83**, 2531-2535.

287) Toh, H., Hyashida, H., & Miyata, T. (1983) Nature (London) **305**, 827-829.

288) Summers, J. & Mason, W.S. (1982) Cell **29**, 403-415.

289) Pfeiffer, P. & Hohn, T. (1983) Cell **33**, 781-789.

290) Tuttleman, J.S., Pugh, J.C., & Summers, J.W. (1986) J. Virol. **58**, 17-25.

291) Korba, B.E., Wells, F., Tennant, B.C., Yoakum, G.H., Purcell, R.H., & Gerin, J.L. (1986) J. Virol. **58**, 1-8.

292) Molnar-Kimber, K.L., Summers, J., & Mason, W.S. (1984) J. Virol. **51**, 181-191.

293) Mason, W.S., Aldrich, C., Summers, J., & Taylor, J.M. (1982) Proc. Natl. Acad. Sci. USA **79**, 3997-4001.

294) Lien, J.-M., Aldrich, C.E., & Mason, W.S. (1986) J. Virol. **57**, 229-236.

295) Buscher, M., Reiser, W., Will, H., & Schaller, H. (1985) Cell **40**, 717-724.

296) Tiollais, P., Dejean, A., Brechot, C., Michel, M.L., Sonigo, P., & Wain-Hobson, S. (1984) In *Viral Hepatitis and Liver Disease* Vyas, G.N., Dienstag, J.L., & Hoofnagle, J.H., eds. Grune & Stratton, Inc., New York, 49-65.

297) Olszewski, N., Hagen, G., & Guilfoyle, T.J. (1982) Cell **29**, 395-402.

298) Covey, S.N., Lomonossoff, G.P., & Hull, R. (1981) Nucleic Acids Res. **9**, 6735-6747.

299) Volovitch, M., Modjtahedi, M., Yot, P., & Brun, G. (1984) EMBO J. **3**, 309-314.

300) Mazzolini, L., Bonneville, J.M., Volovitch, M., Magazin, M., & Yot, P. (1985) Virology **145**, 293-303.

CHAPTER 4

POSTTRANSCRIPTIONAL PROCESSING (SPLICING AND 3'-END FORMATION) OF NUCLEAR MESSENGER RNA PRECURSORS IN VITRO

WALTER KELLER

Division of Molecular Biology, German Cancer Research Center, Im Neuenheimer Feld 280, D-6900 Heidelberg, FRG

INTRODUCTION

In eukaryotic cells the primary RNA transcripts of nuclear genes coding for proteins undergo a series of posttranscriptional modifications before they are transported to the cytoplasm where they function as messenger RNAs (mRNA). These processing reactions include the addition of a 7-methyl guanosine to the 5' end (capping), the formation of the mature 3' ends (usually accompanied by polyadenylation), and finally, the removal of introns by RNA splicing.

First observed in viral systems, polyadenylation, capping and splicing are presently recognized as a general attribute of eukaryotes. Since viral mRNAs are produced by mechanisms that mimic those of the host cell, it is fitting an proper to review in some detail our present understanding of the biochemistry of these processes. The reader will find a more extensive discussion of the individual strategy evolved by each viral family in the proper sections of this volume.

General Outline

Capping of the 5'-end of messenger RNA precursors (pre-mRNA) takes place immediately after the synthesis of the RNA chain is initiated (Fig. 1). The detailed mechanism of this reaction has been elucidated (for a review see ref. 1) and will not be discussed here any further. In general, the synthesis of a pre-mRNA runs well beyond the site of the mature mRNA end. Thus, the 3'-end arises as the result of

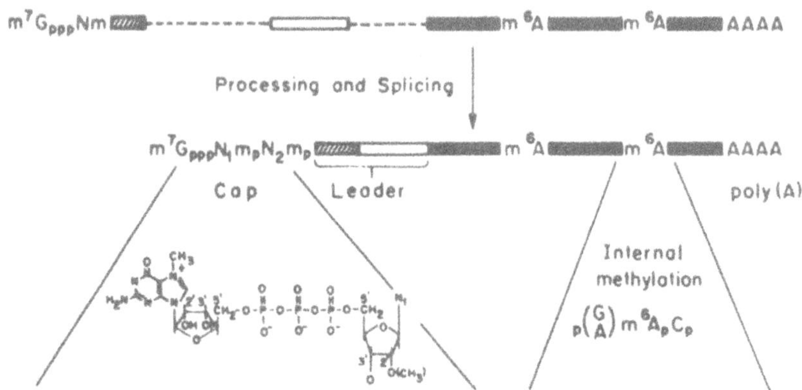

Figure 1. Schematic view of the posttranscriptional processing of pre-mRNAs in eukaryotes (reproduced from ref. 52).

specific cleavage of the pre-mRNA. In most cases this cleavage is followed by the addition of a poly A-tract (reviewed in ref. 2). The results obtained by studies of these reactions *in vitro* will be presented at the end of this chapter.

Finally, intron sequences within the pre-mRNA must be removed by RNA splicing. For any given intron this requires two precise cleavages at its boundaries and the ligation of the adjacent exons. Both 3'-end formation and RNA splicing are complicated processes which require the specific interaction of multiple components. In addition, these reactions can be regulated such that different messenger RNAs can be made from a given precursor molecule (see ref. 2 and 3 for reviews). How such differential gene expression is achieved by alternate RNA processing is one of the important questions in molecular biological research. However, before these problems can be experimentally attacked, the basic mechanisms of the underlying reactions must be worked out. To this purpose, *in vitro* systems with which RNA processing can be studied under defined conditions, are invaluable tools. During the last few years, such in vitro systems have been developed in several laboratories. Below, a summary will be presented of the results obtained by studying splicing and 3'-end formation of simple synthetic pre-mRNAs *in vitro*. For the sake of simplicity, the discussion is restricted to studies on higher eukaryotes. The results obtained with yeast systems will not be included. For further details the reader is referred to more comprehensive recent reviews (4, 5).

A. Intermediates and Products of in vitro *Pre-mRNA Splicing*

The elucidation of the reaction mechanism of pre-mRNA splicing was made possible by the development of *in vitro* systems. Such systems are composed of a synthetic substrate RNA and in most cases an extract from HeLa cell nuclei. The pre-mRNA substrate is usually made by the *in vitro* transcription of a suitably modified plasmid DNA that contains a portion of eukaryotic transcription unit and carries two exons interrupted by an intron. In most cases the plasmids have a phage SP6 promoter fused to the pre-mRNA sequences such that purified SP6 RNA polymerase can be used to synthesize substrate RNA *in vitro* (6). Splicing is carried out by incubating the synthetic pre-mRNA with nuclear extract (7) in the presence of ATP, $MgCl_2$, and KCl as cofactors (6, 8, 9). Analysis of the intermediates and the products of *in vitro* splicing have shown that the reaction proceeds in two discrete steps (see ref. 10 for a review).

First, the pre-mRNA is cleaved at the boundary between the upstream exon and the intron (the 5' splice site). This cleavage is coupled with the covalent joining of the 5' end of the intron to a 2' hydroxyl of an adenosine residue which is located within the intron near the boundary to the down-stream exon. The two products of the first reaction step are therefore the upstream exon ending with a 3' hydroxyl group and a circular RNA with a linear 3' portion (the intron-exon 2 lariat, see Figure 2). In the second step of the reaction, a cleavage occurs at the 3' splice site. This cleavage is accompanied by the covalent joining of the upstream and the downstream exons resulting in spliced RNA as an end product. The other end product is the released intron still in the form of a branched-circle or lariat. Chemically, the two cleavage-ligation steps can be viewed as concerted phosphoester transfers (transesterifications) with the 2' hydroxyl of the branche adenosine and

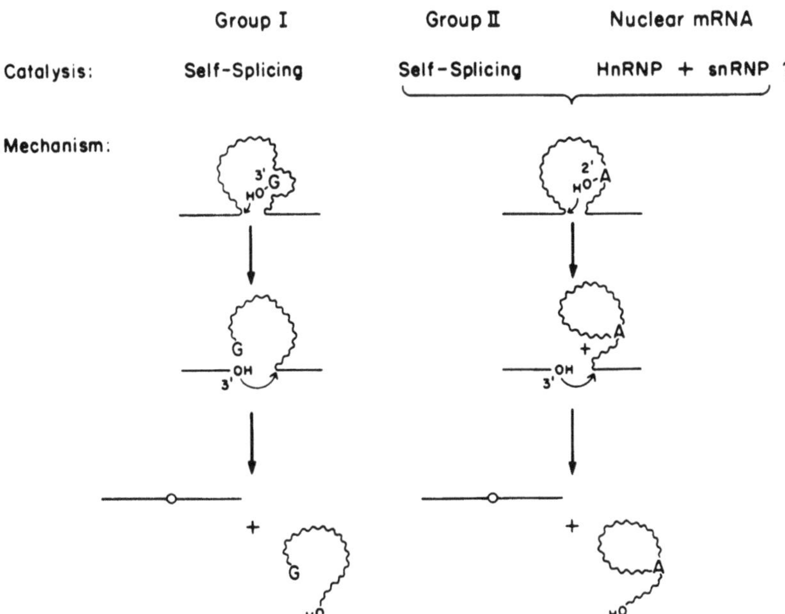

Figure 2. Comparison of the pathways of splicing of self-splicing RNA with that of nuclear mRNA precursors. See text for explanation. Exon sequences are represented by straight lines, introns by a wavy curve. (Reproduced from ref. 13).

the 3' hydroxyl of the cleaved off first exon acting as nucleophilic attacking groups in the reaction steps.

B. Comparison to Other Systems

A comparison of the reaction pathway outlined above for the splicing of nuclear pre-mRNAs with that observed in group I and group II self-splicing RNAs found in ribosomal RNA precursors and mitochondrial ribosomal and messenger RNA precursors is shown in Figure 2. With regard to the structure of the splicing intermediates and end products group II RNAs and nuclear pre-mRNAs appear to use the same mechanism (11, 12). However, the splicing of nuclear pre-mRNAs differs from that of group II RNAs by the fact that it is not self-splicing not self-splicing and that it requires trans-acting factors for the alignment of the splice sites (see below). The mechanism of the self-splicing of group I RNAs is also illustrated in Figure 2. As can be seen, it differs from the lariat-forming pathway in the first of the two reaction steps. Transesterification is initiated here by the 3' hydroxyl group of a guanosine cofactor which becomes covalently attached to the 5' end of the intron as the result of the first cleavage-ligation reaction. The second step of the reaction is analogous to the mechanism operating in group II RNAs and nuclear pre-mRNAs (reviewed in 13 and 14). Despite of these differences the underlying similarity to the other splicing systems is obvious: splicing takes place in two distinct cleavage-ligation steps (phosphoester transfers). In group I RNAs the enzymatic activity with catalyzes the splicing reaction resides within the intron as catalytic RNA (15). This is most likely also the case for self-splicing group II RNAs even though a piece of intron RNA with true catalytic activity has not yet been identified. In the case of nuclear pre-mRNA splicing an RNA catalyst could be provided by the small nuclear snRNPs which are part of the splicing complex in which the reaction takes place (see below). Alternatively, the transesterification reactions may in this case be catalyzed by conventional protein enzymes.

C. The Selection of Splice Sites in Self-splicing RNA: Cis-Alignment

One of the most intriguing questions that comes to mind when considering the high precision of RNA splicing is how distant sites within a large RNA molecule are recognized, brought into proximity and aligned for splicing. In the case of group I self-splicing RNA the answer is provided by comparing the secondary structure of different members of the family. The prototype of this group of RNAs, the Tetrahymena ribosomal RNA precursor contains several stretches of nucleotide sequences which can form intramolecular base-pairings and thereby fold the RNA into a unique secondary and tertiary structure (16, 17). The sequences involved in the complementary base-pairing are conserved among the member of group I RNA (18). The most important feature of this model is the presence of a so-called internal guide sequence that serves to bring the 5' splice site into proximity with the 3' splice site by base-pairing interactions. The resulting structure is shown in Figure 3. The unique secondary structure with its cis-alignment of the splice sites and the built-in segment of RNA with catalytic activity provide the explanation for the capacity of such RNA molecules to carry out the splicing reaction in the presence of a guanosine cofactor and Mg^{2+} ions in an autocatalytic fashion. The mechanism of this reaction is also shown in Figure 3. Presumably, similar principles are involved in the self-splicing of group II RNAs except that the guanosine cofactor is replaced by the 2' hydroxyl group of the branch adenosine which initiates the transesterification reactions by nucleophilic attack of the phosphate at the 5' splice site. As in group I RNAs, the catalytic activity must be provided by an RNA catalyst which most likely is located within group II introns.

D. The Selection of Splice Sites in Nuclear Pre-mRNA: Trans-Alignment

Compared to the situation observed with self-splicing RNAs the mode with which splice sites are selected in nuclear pre-mRNAs is completely different. Because of the enormous variation in the length of the introns found in this class of RNA, a common folding pattern into a unique secondary and tertiary structure is not possible. Short highly conserved sequences have been found only at the extreme 5' and 3' boundaries of the introns of pre-mRNAs (19, 20). (The conserved se-

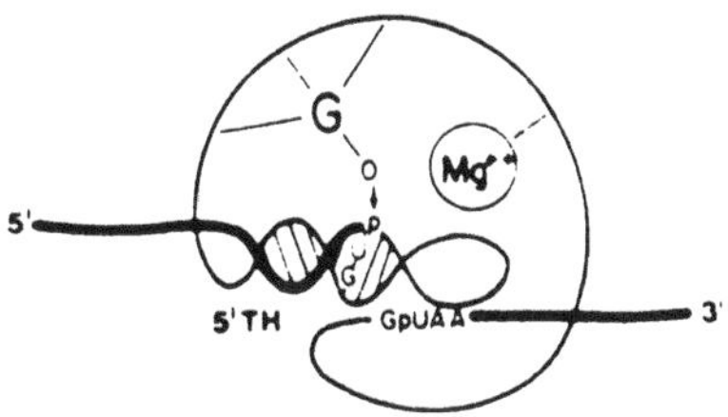

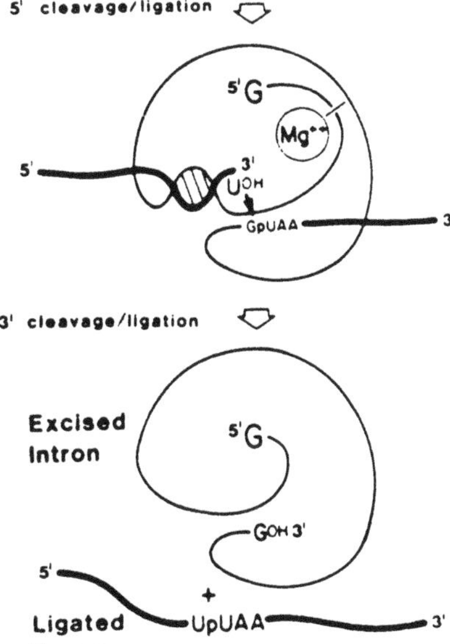

Figure 3. Cis-alignment of splice sites in group I self-splicing RNA. Exon RNA is shown as a thick line, intron RNA as a thin line. 5' TH, 5' splicing target helix, the base-paired region between the 5' splice region and a complementary sequence of the internal guide sequence; G, guanosine cofactor. (The figure is reproduced in modified from ref. 51).

quences are actually not restricted to the introns but extend for a few nucleotides into the adjacent exons). They thus span the exon-intron junctions.

The 5' splice site consensus is:

$$^{C}_{A}\mathrm{AG/}\underline{\mathrm{GU}}\,^{A}_{G}\mathrm{AGU}\ldots$$

where slash indicates the border between exon and intron. The consensus at the 3' splice site reads:

$$\ldots\mathrm{YnX}\underline{\mathrm{AG}}\mathrm{/G}$$

where the Yn represents a sequence containing between four and twelve pyrimidine residues. The importance of these sequences for RNA splicing has been confirmed by studies with naturally occurring mutants as well as with mutated precursor RNAs derived from *in vitro* generated modified transcription units (reviewed in ref. 4 and 5). A third, less well conserved sequence of the type:

$$\mathrm{CU}^{G}_{A}\underline{\mathrm{A}}^{C}_{U}\ldots$$

is found around the branch point adenosine (underlined) which is located in the introns at a distance between ten and fourty nucleotides upstream of the 3' splice site.

Studies performed in a number of laboratories within the last two years have shown that the three conserved sequence elements are the sites of interaction with different U-snRNPs (small nuclear ribonucleoprotein particles) which are trans-acting specificity elements required for splicing. The major types of these particles known as the U1, U2, U4/U6, and U5-snRNPs (21) are involved in the ordered binding to pre-mRNA. They also appear to form contacts with each other which ultimately leads to the assembly of 'splicing complexes' or 'spliceosomes', structures in which the splicing reaction takes place (22, 23). Spliceosomes had first been observed in yeast splicing extracts (24).

The participation of snRNPs in RNA splicing had first been postulated on theoretical grounds for the U1 species on the basis of the finding that the sequence of the 5' portion of U1 snRNA is complementary to the consensus sequence at the 5' splice sites of pre-mRNAs (25, 26). Recognition of 5' splice sites could thus be mediated by a specific base-pairing interaction with U1 snRNP. This hypothesis has been proved to be correct by a number of experimental approaches. *In vitro* splicing could be inhibited by antibodies directed against U1 snRNPs (27). The targeted degradation of the 5' portion of U1 snRNA by complementary oligonucleotides and RNase H in splicing extracts abolishes its splicing activity (28); this indicates that an intact 5' portion of U1 snRNA is essential for splicing. Furthermore, specific binding of U1 snRNPs to the 5' splice site region of synthetic globin pre-mRNA could be demonstrated by immune precipitation of RNase-protected pre-mRNA stretches with anti-U1 snRNP antibodies (29). Finally, the base-pairing model is strongly supported by genetic experiments which showed that the splicing defect due to a mutated 5' splice site could be rescued by a suppressor U1 snRNA with a compensatory mutation (30).

Analogous RNase protection-immune precipitation mapping as well as targeted RNase H degradation of various portions of different U-snRNAs allow the conclusion that all the major U-snRNPs (U1, U2, U4/U6, U5) are required for splicing (31-34). It was shown that U2 snRNPs bind in the intron near the branch point (31) and U5 snRNPs most likely at the 3' splice site (35).

From all these studies a clear correlation emerges: the three short conserved sequence elements (5' splice site, branch point, and 3' splice site) represent regions in the pre-mRNAs for the specific binding of different U-snRNPs. This is illustrated schematically in Figure 4. It is not known at present whether or not the binding of

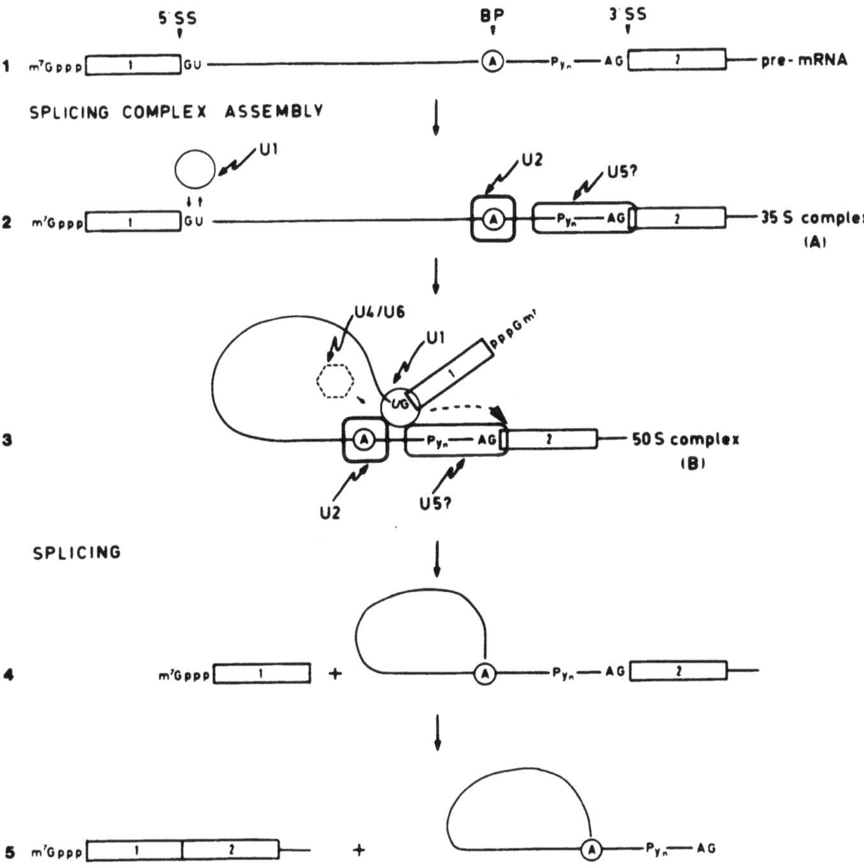

Figure 4. Schematic representation of the interaction of U-snRNPs with conserved intron sequences in nuclear pre-mRNA and the resultant assembly of a splicing complex. Exon RNA is shown as numbered boxes, intron RNA as a line. Level 1: 5' SS, 5' splice site; Bp, branch point; 3' SS, 3' splice site. Level 2: binding of snRNPs (abbreviated U1, U2, U5) to their cognate sites leads to the formation of pre-splicing complexes (35S). Level 3: interaction of the snRNPs with one another and with U4/U6-snRNP converts the pre-splicing complex (35S) to the splicing complex (50S). Levels 4 and 5 show the two steps of the splicing reaction.

U2 snRNPs and U5 snRNP is mediated by base-pairing interactions as documented for U1 snRNP. Short sequences complementary to the branch point consensus can be found within U2 snRNA (36, 37) but experimental evidence in support of base pairing has not yet been obtained. Perhaps, these snRNPs recognize their cognate signals in the pre-mRNA by protein-RNA interaction.

In contrast to the other U-type snRNPs, the U4/U6 snRNPs do not seem to bind directly to the pre-mRNA. However, targeted RNase H hydrolysis on these particles has shown (D. Frendewey and W. Keller, unpublished) that inactivation of the U4/U6 particles leads to an arrest of splicing complex assembly at an intermediate step. It is possible therefore, that the U4/U6 snRNP functions in splicing by interacting with the other snRNP types that are bound to the pre-mRNA and folding the splicing complex into a specific conformation.

Analysis of the time course of splicing complex formation by sucrose gradient sedimentation or native gel electrophoresis has shown that the assembly proceeds in several discrete steps (23, 38). SnRNP binding to the 3' splice site and the branch point leads to a complex of intermediate size with a sedimentation constant of ~35S. With time this pre-splicing complex is converted to a faster sedimenting 50S splicing complex. When the U1 snRNPs or the U4/U6 snRNPs are inactivated by oligonucleotide-targeted RNAse H treatment, the conversion of pre-splicing complex to the splicing complex is blocked. When the 5'-end of U2 snRNA is degraded, no specific complex can be formed at all. After degradation of a different portion of U2 snRNA, represented by an internal loop in the molecule, pre-splicing complexes still assemble, but their conversion to the splicing competent 50S complexes is arrested (D. Frendewey and W. Keller, unpublished results). Analogous experiments with the U5 snRNP have not been possible because the U5 snRNA in the particle is not accessible to oligonucleotide hybridization and can therefore not be degraded by RNAse H. Taken together these results indicate that an intact 5' end of the U2 snRNA is required for the initial binding to the pre-mRNA and that the loop domain of U2 snRNA is important for secondary interactions, perhaps with the U4/U6 and the U1 particles, to form stable splicing complexes. They also show that stable binding interactions first occur at the 3' splice site/branch point region which in turn facilitate the conversion to a competent splicing complex (Figure 4, Level 3). The formation of the 50S spliceosome precedes the first step of the actual splicing reaction. In other words, splicing can only be initiated after all the necessary components have bound to the pre-mRNA and to each other. At present, only the barest outlines of the structure of the splicing complex and its component parts can be perceived. Much future work will be necessary before the precise molecular structure and the interactions involved in the assembly of spliceosomes will be understood.

The snRNPs, although playing the key role in the formation of splicing complexes, are not the only components involved in splicing. Studies with antibodies directed against the so-called hnRNP proteins have shown that these proteins are also required for *in vitro* splicing (39). Whether hnRNPs just provide a scaffold for the pre-mRNA to which other components are then added or whether they have a more specific function in splicing is not known. Biochemical fractionation of nuclear extracts has revealed the existence of additional protein factors needed for *in vitro* splicing (32, 40, 41). Again, the function of these auxiliary splicing factors has not yet been determined. As mentioned earlier, it is not known where the enzymatic activity resides which catalyses the two cleavage-ligaton steps of pre-mRNA splicing. It is possible that, as a modification of the mechanism operating in the lariat-forming self-splicing RNAs, the catalytic activity in nuclear pre-mRNA splicing is provided by the U-snRNAs in the spliceosome. Alternatively, catalysis may be carried out by a transesterification activity associated with one of the auxiliary splicing factors mentioned above. Clearly, this is only one of many questions that are still open in this field and need to be answered by future studies.

1. 3'-PROCESSING

Besides RNA splicing, the formation of the mature 3' ends is the other major RNA processing reaction occurring on nuclear pre-mRNAs. Two classes of precursors are distinguished, one for the generation of histone mRNAs which are not polyadenylated and one for all other pre-mRNAs where cleavage is followed by the addition of a poly A tract to the 3' end of the RNA (reviewed in ref. 2). Both types of 3' processing can be reproduced in vitro with synthetic RNA substrates and nuclear extracts from HeLa cells (42-44). In the case of histone pre-mRNA processing it has been shown by using the Xenopus oocyte injection system that a minor class of ribonucleoprotein particles, the U7 snRNPs, are involved (45, 46). There are two conserved sequence elements near the cleavage site in different histone pre-mRNAs and these sequences are complementary to sequences of the sea urchin U7 snRNA (46, 47). Recognition of the cleavage site is mediated by the binding of U7 snRNPs via base pairing interactions as has recently been demonstrated directly (48). Partially purified snRNPs from HeLa cells can carry out the 3' processing of mouse histone pre-mRNAs efficiently in vitro (44). At the time of writing this review, work is in progress to identify and characterize the human U7 snRNP in these extracts.

The nature of the components responsible for the cleavage and polyadenylation reactions of poly A+ pre-mRNAs is not known. As mentioned above, the processing reaction can be mimicked with high efficiency in vitro with nuclear extracts from HeLa cells. The reaction requires the presence of the highly conserved AAUAAA hexamer sequence as well as sequences downstream of the cleavage site (49). The cleavage step can be uncoupled from polyadenylation by the addition of EDTA or non-hydrolysable ATP analogues (42). Correctly cleaved pre-mRNA can be polyadenylated when re-introduced in the *in vitro* system (43, 49). The *in vitro* processing reaction has been reported to be inhibited by antibodies which specifically react with U1 snRNPs (42). Antisera which react with all known U-type snRNPs can be used to specifically immuneprecipitate an oligonucleotide containing the AAUAAA signal from an *in vitro* reaction (50) indicating that a snRNP associates with this sequence. Since the major U-snRNPs can be separated from the 3' processing activity by biochemical fractionation of HeLa nuclear extracts (G. Cristofori, personal communication) it is improbable that U1-snRNPs are involved. This issue must be resolved by further purification of the responsible components.

2. OUTLOOK

From what was said above, it is clear that the application of in vitro systems for the analysis of RNA processing reactions has been invaluable and productive and will continue to be important in the future. Eventually, all the components and factors involved in splicing and 3' processing will be characterized. This will mainly require the continued fractionation of the components of cell free extracts. Valuable contributions will certainly also come from the work with yeast systems, in which the biochemical approach can be combined with genetic methods. In due time, all the genes for snRNAs, snRNP-proteins, and other factors will be cloned and sequenced. A more challenging area will be the elucidation of the structure of spliceosomes.

Considering the fact that the detailed structure of ribosomes is still not known despite of almost thirty years of efforts, we can accept that results in this field will be forthcoming more slowly. Another, still almost uncharted terrain is the field of regulated RNA processing. Thought initially only thought to be relevant for viral messenger RNA synthesis, many example of complex cellular transcription units have now been found where alternative 3' processing or differential splicing or both lead to the generation of different mRNAs for different proteins from a common pre-mRNA (reviewed in ref. 3). The reproduction of alternate RNA processing in suitible *in vitro* systems will be a difficult but highly rewarding future enterprise.

3. REFERENCES

1) Moss, B. (1983). 5' terminal cap structures of eukaryotic and viral mRNAs. In: *Processing of RNA*, D. Apirion, ed. CRC Press Inc., Boca Raton, Florida.

2) Birnstiel, M.L., Busslinger, M., and Strub, K. (1985). Transcription termination and 3' processing: the end is in site! Cell **41**, 349-359.

3) Leff, S.E., Rosenfeld, M.G. and Evans, R.M. (1986). Complex transcriptional units: diversity in gene expression by alternative RNA processing. Ann. Rev. Biochem. **56**, 1091-1117.

4) Green, M.R. (1986). Pre-mRNA splicing. Ann. Rev. Genet. (in press).

5) Padgett, R.A., Grabowski, P.J., Konarska, M.M., Seiler, S.R., and Sharp, P.A. (1986). Splicing of messenger RNA precursors. Ann. Rev. Biochem. **55**, 1119-1150.

6) Krainer, A.R., Maniatis, T., Ruskin, B. and Green, M.R. (1984). Normal and mutant human B-globin pre-mRNAs are faithfully and efficiently spliced *in vitro*. Cell **36**, 993-1005.

7) Dignam, J.D., Lebovitz, R.M. and Roeder R.G. (1983). Accurate transcription initiation by RNA polymerase II in a soluble extract from isolated mammalian nuclei. Nucl. Acids Res. **11**, 1475-1489.

8) Hernandez, N. and Keller, W. (1983). Splicing of in vitro synthesized messenger RNA precursors in HeLa cell extracts. Cell **35**, 89-99.

9) Hardy, S.F., Grawboski, P.J., Padgett, R.A. and Sharp, P.A. (1984). Cofactor requirements for splicing of purified messenger RNA precursors. Nature (London) **308-**, 375-377.

10) Keller, W. (1984). The RNA lariat: a new ring to the splicing of mRNA precursors. Cell **39**, 423-425.

11) Peebles, C.L., Perlman, P.S., Mecklenburg, K.L., Petrillo, M.L., Tabor, J.H., Jarell, K.A., and Cheng, H.L. (1986). A self-splicing RNA excises an intron lariat. Cell **44**, 213-223.

12) Van der Veen, R., Arnberg, A.C., van der Horst, G., Bonen, L., Tabak, H.F. and Grivell, L.A. (1986). Excised group II introns in yeast mitochondria are lariats and can be formed by self-splicing *in vitro*. Cell **44**, 225-234.

13) Cech, T.R. (1986). The generality of self-splicing RNA: relationship to nuclear mRNA splicing. Cell **44**, 207-210.

14) Cech, T.R. and Bass, B.L. (1986). Biological catalysis by RNA. Ann. Rev. Biochem. **55**, 599-629.

15) Zaug, A.J., and Chech, T.R. (1986). The intervening sequence RNA of Tetrahymena is an enzyme. Science **231**, 470-475.

16) Waring, R.B., Scazzochio, C., Brown, T.A. and Davies, R.W. (1983). Close relationship between certain nuclear and mitochondrial introns. Implications for the mechanism of RNA splicing. J. Mol. Biol. **167**, 595-605.

17) Waring, R.B. and Davies, R.W. (1984). Assessment of a model for intron RNA secondary structure relevant to RNA self-splicing: a review. Gene **28**, 277-291.

18) Davies, R.W., Waring, R.B., Ray, J.A., Brown, T.A. and Scazzochio, C. (1982). Making ends meet: a model for RNA splicing in fungal mitochondria. Nature (London) **300**, 719-724.

19) Breathnach, R., and Chambon, P. (1981). Organization and expression of eucaryotic split genes coding for proteins. Ann. Rev. Biochem. **50**, 349-383.

20) Mount, S.M. (1982). A catalogue of splice junction sequences. Nucl. Acids Res. **10**, 459-472.

21) Busch, H., Reddy, R. Rohtblum, L. and Choi, Y.C. (1981). SsnRNAs, snRNPs, and RNA processing. Ann. Rev. Biochem. **51**, 617-654.

22) Grawboski, P.J., Seiler, S.R. and Sharp, P.A. (1985). A multi-component complex is involved in the splicing of messenger RNA precursors. Cell **42**, 345-353.

23) Frendewey, D. and Keller, W. (1985). Stepwise assembly of a pre-mRNA splicing complex requires U-snRNPs and specific intron sequences. Cell **42**, 355-367.

24) Brody, E. and Abelson, J. (1985). The "spliceosome": yeast pre-messenger RNA associates with a 40S complex in a splicing-dependent reaction. Science **228**, 963-966.

25) Lerner, M.R., Boyle, J.A, Mount, S.M., Wolin, S.L. and Steitz, J.A. (1980). Are snRNPs involved in splicing? Nature (London) **283**, 220-224.

26) Rogers, J. and Wall, R. (1980). A mechanism for RNA splicing. Proc. Nat. Acad. Sci. USA **77**, 1877-1879.

27) Padgett, R.A., Mount, S.M., Steitz, J.A. and Sharp, P.A. (1983). Splicing of messenger RNA precursors is inhibited by antisera to small nuclear ribonucleoprotein. Cell **35**, 101-107.

28) Kramer, A., Keller, W., Appel, B. and Luhrmann, R. (1984). The 5' terminus of the RNA moiety of U1 small nuclear rebonucleoprotein particles is required for the splicing of messenger RNA precursors. Cell **38**, 299-307.

29) Mount, S.M., Pettersson, I., Hinterberger, M., Karmas, A., and Steitz, J.A. (1983). The U1 small nuclear RNA-protein complex selectively binds a 5' splice site *in vitro*. Cell **33**, 509-518.

30) Zhuang, Y. and Weiner, A.M. (1986). A compensatory base change in U1 snRNA suppresses a 5' splice site mutation. Cell **46**, 827-835.

31) Black, D.L., Chabot, B. and Steitz, J.A. (1985). U2 as well as U1 small nuclear ribonucleoproteins are involved in premessenger RNA splicing. **42**, 737-750.

32) Krainer, A.R. and Maniatis, T. (1985). Multiple factors including the small nuclear ribonucleoproteins U1 and U2 are necessary for pre-mRNA splicing in vitro. Cell **42**, 725-736.

33) Black, D.L. and Steitz, J.A. (1986). Pre m-RNA splicing *in vitro* requires intact U4/U6 small nuclear ribonucleoproteins. Cell **46**, 697-704.

34) Berget, S.M. and Robberson, B.L. (1986). U1, U2 and U4/U6 small nuclear ribonucleoproteins are required for *in vitro* splicing but not polyadenylation. Cell **46**, 691-696.

35) Chabot, B., Black, D.L., LeMaster, D.M. and Steitz, J.A. (1985). The 3' splice site of pre-messenger RNA is recognized by a small nuclear ribonucleoprotein. Science **230**, 1344-1349.

36) Keller, E.B. and Noon, W.A. (1984). Intron splicing: a conserved internal signal in introns of animal pre-mRNAs. Proc.Natl.Acad. Sci.USa **81**, 7417-7420.

37) Keller E.B. and Noon, W.A. (1985). Intron splicing: a conserved internal signal in introns of *Drosophila* pre-mRNAs. Nucl. Acids Res. **13**, 4971-4981.

38) Konarska, M.M. and Sharp, P.A. (1986). Electrophoretic separation of complexes involved in the splicing of precursors to mRNAs. Cell **46**, 845-855.

39) Choi, Y.D., Grabowski, P.J., Sharp, P.A. and Dreyfuss, G. (1986). Heterogeneous nuclear ribonucleoproteins: role in RNA splicing. Science **231**, 1534-1539.

40) Furneaux, H.M., Perkins, K.K., Freyer, G.A., Arenas, J. and Hurwitw, J. (1985). Isolation and characterization of two fractions from HeLa cells required for mRNA splicing *in vitro*. Proc. Nat. Acad. Sci. USA **82**, 4351-4355.

41) Kramer, A. and Keller, W. (1985). Purification of a protein required for the splicing of pre-mRNA and its separation from the lariat debranching enzyme. EMBO J. **4**, 3571-3581.

42) Moore, C.L., and Sharp, P.A. (1985). Accurate cleavage and polyadenylation of exogenous RNA substrate. Cell **41**, 845-855.

43) Moore, C.L., Skolnik-David, H. and Sharp, P.A. (1986). Analysis of RNA cleavage at the adenovirus-2 L3 polyadenylation site. EMBO J. **5**, 1929-1938.

44) Gick, O., Kramer, A., Keller, W. and Birnstiel, M.L. (1986). Generation of histone mRNA 3' ends by endonucleolytic cleavage of the pre-mRNA in a snRNP-dependent *in vitro* reaction. EMBO J. **5**, 1319-1326.

45) Galli, G., Hofstetter, H., Stunnenberg, H.G. and Birnstiel, M.L. (1983). Biochemical complementation with RNA in the Xenopus oocyte: a small RNA is required for the generation of 3' histone mRNA termini. Cell **34**, 823-828.

46) Strub, K., Galli, G., Busslinger, M. and Birnstiel, M.L. (1984). The cDNA sequences of the sea urchin U7 small nuclear RNA suggest specific contacts between histone mRNA precursor and U7 RNA during RNA processing. EMBO J.3, 2801-2807.

47) Strub, K., and Birnstiel, M.L. (1986). Genetic complementation in the Xenopus oocyte: Co-expression of sea urchin histone and U7 restores 3' processing of H3 pre-mRNA in the oocyte. EMBO J. **5**, 1675-1682.

48) Schaufele, F., Gilmartin, G.M., Bannwarth, W. and Birnstiel, M.L. (1986). Compensatory mutations suggests that base-pairing with a small nuclear RNA is required to form the 3' end of H3 messenger RNA. Nature (London) **323**, 777-781.

49) Zarkower, D., Stephenson, P., Sheets, M. and Wickens, M. (1986). The AAUAA sequence is required both for cleavage and for polyadenylation of SV40 pre-mRNA *in vitro*. Mol. Cell. Biol. **6**, 2317-2323.

50) Hashimoto, C. and Steitz, J.A. (1986). A small nuclear ribonucleoprotein associates with the AAUAAA polyadenylation signal *in vitro*. Cell **45**, 581-591.

51) Waring, R.B., Towner, P., Minter, S.J. and Davies, R.W. (1986). Splice-site selection by a self-splicing RNA of Tetrahymena. Nature (London) **321**, 133-139.

52) Revel. M., and Groner, Y. (1978). Posttranscriptional modifications of mRNAs and their role in translation. Ann. Rev. Biochem., **47**, 1079-1126.

CHAPTER 5

REGULATION OF TRANSLATION OF VIRAL mRNAs

MARIE-DOMINIQUE MORCH, ROSAURA P.C. VALLE AND ANNE-LISE HAENNI

Institut Jacques Monod[*], 2, Place Jussieu (Tour 43), 75251 Paris Cedex 05, France

Abbreviations

Viruses and corresponding family: AMV = alfalfa mosaic virus (ilarvirus); BBV = black beetle virus (nodavirus); BLV = bovine leukemia virus (retrovirus); BMV = brome mosaic virus (bromovirus); BNYVV = beet necrotic yellow vein virus (furovirus); CaMV = cauliflower mosaic virus (caulimovirus); CPMV = cowpea mosaic virus (comovirus); CPV = cytoplasmic polyhedrosis virus (reovirus); EMCV = encephalomyocarditis virus (picornavirus); FeLV = feline leukemia virus (retrovirus); FMDV = foot-and-mouth disease virus (picornavirus); HSV = Herpes simplex virus (herpesvirus); HTLV = human T-cell leukemia virus (retrovirus); IPNV = infectious pancreatic necrosis virus (binarvirus); Mo-MuLV = Moloney murine leukemia virus (retrovirus); Ra-MuLV = Rauscher murine leukemia virus (retrovirus); Mo-MuSV = Moloney murine sarcoma virus (retrovirus); RSV = Rous sarcoma virus (retrovirus); SSHV = snowshoe hare virus (bunyavirus); STNV = satellite of tobacco necrosis virus (tobacco necrosis virus family); SFV = Semliki forest virus (alphavirus); SV40 = simian virus 40 (papovavirus); TMV = tobacco ringspot virus (nepovirus); TVMV = tobacco mosaic virus (tobamovirus); TNV = tobacco necrosis virus (nepovirus); TYMV = turnip yellow mosaic virus (tymovirus); VSV = vesicular stomatitis virus (rhabdovirus).

Others: ds = double-stranded; VPg = virus protein genomic; kDa = kilodalton; kb = kilobase; ORF = open reading frame; NT = nucleotide; RER = rough endoplamic reticulum.

---

[*]The Institut Jacques Monod is an "Institut Mixte CNRS/Université Paris 7".

INTRODUCTION

Translation constitutes a crucial step in virus development, since it leads to the production of the structural and nonstructural proteins required by the virus. Because of their relative simplicity, viral genomes, and in particular viral RNA genomes, have served as model systems to investigate the steps involved in protein synthesis and have successfully helped to unravel the various strategies of translation known to occur on mRNA templates.

During the last decade, studies of virus gene expression have been performed *in vitro* and *in vivo* and, in many cases, have led to a proposal of genome organization for these viruses. With the advent of recombinant DNA technology, a new phase in the study of virus expression has emerged. The determination of the nucleotide sequence of viral genomes and hence of potential coding sequences has confirmed and in some cases even preceded translation studies. Examples outlined below demonstrate however, that it is important to verify experimentally the translation strategies proposed on the basis of genome sequence. The techniques of molecular cloning and *in vitro* transcription and the use of expression vectors have also helped to confirm the strategies proposed on the basis of translation studies as they made it possible to investigate the expression of individual viral genes or to define the nature of regulatory signals within these genes.

The aim of this chapter is to review our present knowledge on the regulation of translation of viral mRNAs. After an outline of the strategies of expression utilized by viruses, this chapter focuses on the various steps in translation that are subject to regulation: initiation, elongation and termination of protein synthesis. A subsequent section reviews post-translational proteolytic maturation as an additional strategy allowing the production of functional viral proteins.

The reader is referred to other recent reviews for further basic information on the translation mechanisms in eukaryotes (1-10).

1. STRATEGIES OF EXPRESSION

Because of the relatively small size of viral genomes, the genetic organization of viruses is extraordinarily compact. Genes are at times separated by short intercistronic sequences, in many cases they are butt-jointed to each other, and in others they even overlap. Viruses compensate for this limited amount of information by utilizing the translation machinery of the cell for their own benefit; in extreme cases, this leads to complete shut-off of host protein synthesis. In addition, viruses have evolved a number of highly sophisticated strategies to produce the virus-specific structural and nonstructural proteins required for their propagation.

As with eukaryotic cellular mRNAs, viral mRNAs of eukaryotic origin are generally considered functionally monocistronic, even though they are in many cases structurally polygenic, i.e., they contain information for the synthesis of more than one protein. Although exceptions to this rule do exist (see below), in most viral mRNAs only the 5'-proximal gene can be expressed, the other genes being silent. To overcome the problem of internal genes, several strategies have been elaborated by viruses to produce their different proteins.

Three main categories or groups of strategies can be discerned based on the coding properties of the mRNAs:

Group I: 1 AUG → 1 protein:

- I-1. multipartite genome
- I-2. subgenomic RNAs
- I-3. spliced mRNAs
- I-4. ambisense RNAs

Group II: 2 AUGs → 2 proteins:

- II-1. polycistronic mRNA
- II-2. internal in-phase initiation within a gene
- II-3. gene overlap in distinct reading frames

Group III: 1 AUG → 2 proteins:

- III-1. readthrough or suppression of a termination codon
- III-2. arrest of elongation independent of termination codons
- III-3. frameshift during translation
- III-4. post-translational cleavage

These strategies are briefly outlined below and are schematized in Fig. 1.

A. Group I (see Fig. 1 A)

I-1. Multipartite genome

In most of the examples described to date, each RNA genome behaves as a monocistronic mRNA. The multiple RNAs together make up the genome and are either encapsidated in the same particle or, as in the case of some plant viruses, in distinct particles.

I-2. Subgenomic RNAs

In addition to genomic RNAs, many viruses also produce one or more subgenomic RNA species that derive from the genome and are not required for infectivity; they allow expression of cistrons that occupy internal positions in the genome. Many RNA viruses resort to this strategy for the production of their coat protein. The subgenomic RNA may or may not be encapsidated, depending on the virus.

I-3. Spliced mRNAs

In this strategy, multiple mRNAs are produced from a single template by differential splicing. This can result in the production of a protein that utilizes noncontiguous and possibly out of frame information from the template. This strategy is not discussed further here, since it is developed in various other chapters in this book.

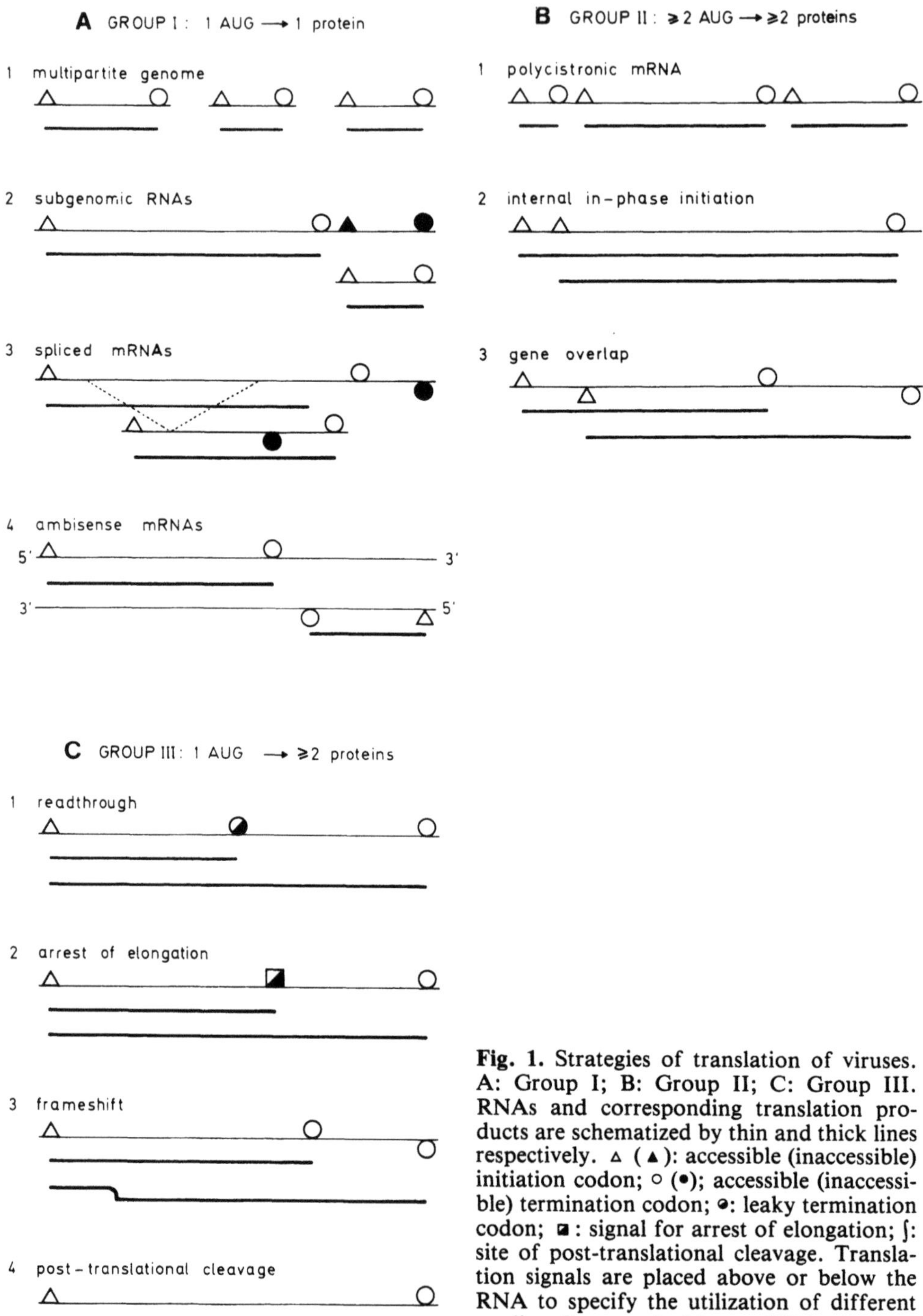

Fig. 1. Strategies of translation of viruses. A: Group I; B: Group II; C: Group III. RNAs and corresponding translation products are schematized by thin and thick lines respectively. △ (▲): accessible (inaccessible) initiation codon; ○ (●); accessible (inaccessible) termination codon; ◕: leaky termination codon; ◪ : signal for arrest of elongation; ʃ: site of post-translational cleavage. Translation signals are placed above or below the RNA to specify the utilization of different reading frames. Ambisense RNA: the two mRNAs are complementary.

I-4. Ambisense RNAs

Here, part of the virus-sense (genomic) RNA contains the gene for one protein, whereas another part within the virus-complementary mRNA codes for another protein. This strategy is not discussed further, since it is developed by D.H.L. Bishop in chapter 14 of this boook.

B. Group II (see Fig. 1 B)

II-1. Polycistronic mRNA

This strategy entails the expression of internal genes in a polycistronic mRNA, analogous to the situation encountered in prokaryotes. The two proteins are initiated on different AUGs, the information for the two proteins occupying distinct regions on the mRNA.

II-2. Internal in-phase initiation within a gene

This leads to the synthesis of two proteins that are initiated on two distinct AUGs in the same reading frame on the mRNA. These proteins are identical except for the fact that the longer one contains additional amino acids at its N-terminus.

II-3. Gene overlap in distinct reading frames

Two distinct proteins are produced from one mRNA template. Here the same nucleotide sequence is read but in two different reading frames.

C. Group III (see Fig. 1 C)

III-1. Readthrough or suppression of a termination codon

This strategy leads to the production of two proteins that differ in size, the longer protein containing additional amino acids at its C-terminus. This results from the elongation of the shorter protein beyond the termination codon, elongation is mediated by a suppressor tRNA.

III-2. Arrest of elongation independent of termination codons

As in the preceding strategy, the two proteins that are synthesized are identical over the total length of the shorter protein. Arrest of elongation occurs at the level of as yet unidentified signals that are different from termination codons.

III-3. Frameshift during translation

A change in reading frame can occur during translation. The two proteins that result from this strategy are identical from their N-termini to the frameshift point and differ in the remainder of their sequences.

III-4. Post-translational cleavage

A polyprotein is first synthesized and then cleaved to produce mature functional proteins.

It is important to distinguish between strategies of *translation* and strategies of *expression*. The former only deals with the translation of a give mRNA species, whereas the latter includes in addition, the different modes of production of the various mRNAs required for viral gene expression. From the above description, it is clear that Group I correspond to various strategies of *expression*: the translation mechanism employed is basically the same, since only one AUG is used as initiator for the synthesis of just one protein. However, since the aim of this review is to discuss the regulation of *translation* of viral mRNAs, the mechanisms involved in the *production* of these different mRNAs will not be further developed here (for reviews, see ref. 10 and other Chapters in this book).

On the other hand, the strategies of Groups II and III correspond to different strategies of translation, since a choice must be made by the translation system between 2 or more initiator AUGs (Group II), or between different possible signals modulating elongation of protein synthesis or post-translational processing pathways (Group III).

In this chapter, the strategies corresponding to Groups I-1 and I-2, and to Group II are covered in section III (*Regulation at the Level of Initiation*), those corresponding to Group III-1, III-2 and III-3 in section IV (*Regulation at the Level of Elongation and Termination*), and those corresponding to Group III-4 in section V (*Post-translational Processing of Viral Polypeptides*).

As will become apparent from the examples discussed below, certain viruses resort to more than one strategy to produce all the proteins required for their propagation.

2. REGULATION AT THE LEVEL OF INITIATION

With the exception of mitochondrial mRNAs, most eukaryotic mRNAs contain an untranslated nucleotide sequence termed *leader sequence* prior to the initiatior AUG codon. The mechanism whereby the 5' leader sequence can affect initiation and dictate the relative efficiency of translation of a given mRNA depends on several factors. Among them are the structure of the 5' terminus of the mRNA and its recognition by specific host factors, the secondary structure and/or the length of the leader sequence, and the sequence surrounding the AUG initiation codon.

A. Nature of the 5' Terminus of Viral RNAs and Role of this Terminus in Initiation Efficiency

Structures encountered: Cap, VPg or ppX

Basically, three types of 5' termini are encountered among eukaryotic viruses.

As in the case of eukaryotic cytoplasmic mRNAs, a large number of viral mRNAs bear a 5'-terminal cap structure (chapter 4, Fig. 1) known to facilitate translation (reviewed in ref. 1, 2, 7, 8).

The capping base is universally m^7G, except for the 26S mRNA of Sindbis virus (an alphavirus) which contains low levels of $m_2^{2,7}G$ and $m_3^{2,2,7}G$ in addition to m^7G (12). Among RNA viruses, only the sense or coding ((+) polarity) strand of the RNA is capped; the antisense strand of the genome of VSV and of influenza virus (an orthomyxovirus), and the antisense strand of the ds RNA genome of reovirus (the type-member of the reovirus family) and of CPV generally carry a 5' di-or triphosphate (discussed in ref. 2).

In certain plant viruses (the como-, nepo-, luteo-, poty-, and sobemoviruses) and animal viruses (the picorna- and calici-viruses), the cap structure in the virion RNA is replaced by a virus-coded protein, VPg. This protein is covalently linked to the 5' pUp in picornavirus RNA (polio-, mengo-, EMC-, FMD virus , ref. 13,14, 244) and CPMV RNA (15,16), and to the 5' pAp in TVMV RNA (17). The linkage, a phosphodiester bond, involves the phenolic (0^4) hydroxyl group of a tyrosine in position 3 in poliovirus (18, 19), or the hydroxyl group of the N-terminal serine residue in CPMV (M. Jaegle, J. Wellink and R. Goldbach, personal communication).

In the case of poliovirus, early experiments had shown that the polyribosome-associated poliovirus RNA is devoid of VPg, suggesting that VPg is removed from the RNA at the time of translation (20-22). An "unlinking enzyme" was purified (23, 24) from uninfected HeLa cells that cleaved the phosphodiester bond between the tyrosine and the 5' terminal pUp of poliovirus RNA.

The observation that VPg is not involved in translation (244) has favored the hypothesis that this polypeptide may operate at the level of RNA synthesis.

The RNA of TNV as well as that of its satellite, STNV, are the only viral RNA genomes known to bear the unmodified sequence ppA at their 5' end (25-27). Yet, they are fully functional in protein synthesis (discussed in ref. 1). To explain the apparent paradox of efficient translation in spite of the absence of a cap structure, it has been suggested (28) that interaction with proteins or a specific conformation of the 5' terminus may protect uncapped mRNAs from exonuclease degradation and allow their translation.

Cap structure and initiation efficiency

The existence of various different 5' ends in viral RNAs prompted several groups to perform to experiments in which the 5' terminus of a given viral RNA has been modified in order to investigate the role of the 5' terminus in *in vitro* translation. From these experiments it appears that removal of the cap structure *from a normally capped mRNA* has a negative effect on the messenger capacity of the RNA: binding to the 40S ribosomal subunit, and the rate and level of translation are decreased compared to the corresponding capped mRNA. The extent of this effect depends on the RNA tested, on the conditions of translation and on the cell-free extract used (29-31).

The availability of STNV RNA which has an uncapped 5' terminus, has prompted a series of experiments in which a 5' cap structure was *added in vitro* to the RNA. Such a modification did not enhance the rate or extent of initiation of protein synthesis of the RNA when compared to untreated STNV RNA (32). Several lines of evidence indicate that elements other than the cap structure such as secondary and/or tertiary folding may account for the efficiency of translation of naturally uncapped mRNAs.

The genomic RNA of picornaviruses provides a first example in this regard, as *internal* (rather than 5' terminal) sequences are recognized by and bind to ribosomes and the eukaryotic initiation factor 2 (eIF-2) at initiation of translation (245-247). The existence of sequences possessing high affinity for initiation factors may obviate the need of the additional stabilization imparted by the cap (247). In fact, competition binding experiments had shown that the affinity of the uncapped Mengovirus RNA for eIF-2 is 35 fold higher than that of the canonical globin mRNA (248).

Preventing the capping of a normally capped mRNA diminishes its stability during *in vitro* translation. This has been demonstrated by comparing the stability of the following mRNAs:

1) capped *vs. in vitro*-synthesized uncapped (ppG-containing) reovirus mRNA translated in *Xenopus laevis* oocytes or in cell-free extracts (28),
2) capped VSV mRNAs *vs.* undermethylated (GpppA-and m^7GpppA-containing) VSV mRNA produced *in vivo* in the presence of cycloleucine (methionine analog which inhibits methylation). Translation of both kinds of messengers was performed *in vivo* (33), and
3) capped versus *in vitro*-decapped CPV (30) or TMV (30,34) RNA.

In all cases, the capped mRNA appeared to be more stable than the un-capped counterpart. It should, however, be borne in mind that the presence of a cap structure does not assure preferential translation of the corresponding mRNA. Indeed, the uncapped STNV RNA, although a relatively inefficient mRNA when added alone to a wheat germ translation system, can outcompete other mRNAs such as globin mRNA and even the subgenomic AMV RNA 4, a normally efficient mRNA, when these mRNAs are all present at saturating levels (37).

Cap binding protein (CBP) and cap recognition

The results outlined above imply that specific structural features of the 5'-terminal cap of mRNAs are required by certain elements of the translation machinery during initiation. This has prompted the search for the cellular components involved in recognition. Using m^7GDP-Sepharose affinity chromatography, cap-binding proteins that differentially stimulated translation of capped mRNAs *in vitro* have been isolated (38, 39); among these, a 24 kDa polypeptide (also known as cap binding protein, eIF-4E or CBPI) was purified from high salt-wash fractions of reticulocyte lysates (38-40). This protein (or possibly a contamination of this protein) appears to specifically stimulate translation of capped Sindbis virus and reovirus mRNAs but has little effect on the VPg-bearing mRNA of EMCV and on the unmodified mRNA of STNV. Added CBPI stimulates translation of the latter RNA after addition of a cap at its 5' end (38).

CBPI cross-links to the oxidized 5' cap structure of mRNAs such as reovirus mRNA 841) and specifically binds capped oligonucleotides reversibly (39). In addition, polypeptides of 28 kDa, 50 kDa and 80 kDa form similar cross-links in the presence of ATP and Mg^{2+} (42, 43). More recently, it was observed that CBPI can be found as part of a high (75 to 10S) molecular weight complex known as CBPII or eIF-4F. This complex is a multimeric protein of 300 kDa. It contains, in addition to CBPI, at least two other subunits, eIF-4A (50 kDa) and a 200-220 kDa (220 kDa) polypeptide (44), and possibly also additional subunits (discussed in ref. 1) CBPII as well as CBPI stimulate Sindbis virus mRNA translation in HeLa cell extracts (54). CBPII in conjunction with eIF-4A and eIF-4B appear to be required for the ATP-dependent binding of capped mRNAs to the 40S pre-initiation complex (46, 47).

Virus-induced shut-off based on cap recognition

In HeLa cells infected with poliovirus, initiation of translation of capped cellular mRNA (48) and of VSV mRNA (49) is inhibited (reviewed in ref. 50,51). Addition of CBPII to poliovirus-infected HeLa cell extracts can restore the ability of these extracts to support the translation of capped VSV mRNA (31). In cells infected with poliovirus, fragmentation of the largest 220 kDa polypeptide present in CBPII is concomitant with inhibition of host protein synthesis and with the appearance of the 220 kDa-related polypeptides of 100 kDa and 130 kDa (52).

The inhibition of capped mRNA translation as a result of the inactivation of the 220 kDa polyprotein contained in CBPII described above for poliovirus-infected HeLa cells may also occur in other picornavirus-infected cells such as Human rhinovirus 14/HeLa cells (58). On the other hand, results have been reported suggesting that a different mechanism may be operative in EMC- (or Mengovirus-) infected HeLa cells (59, 60).

Interestingly, a mechanism of shut off involving the "masking" of the cap-synthesizing machinery of reovirus at late stages of infection (with the consequent production of *uncapped* reovirus mRNAs), and the concomitant inactivation of the cap-recognition machinery of the host cell has been proposed. The reader is referred to chapter 15 and to ref. 251.

Other viruses such as vaccinia virus and the dsDNA-containing frog virus III (an iridovirus) that induce shut-off of host protein synthesis, appear to lack a mechanism for cleaving the 220 kDa polypeptide (mentioned in ref. 59).

Secondary structure and length of leader

The precise role of CBP and of the 5' cap structure in initiation of translation is not fully understood, as exemplified by the fact that viral mRNAs normally not possessing a cap structure are properly and efficiently translated and do not require the CBP. However, because "scanning" of the 5' leader by the 40S ribosomal subunit complex may be the general mechanism of eukaryotic polypeptide chain initiation (reviewed in ref. 3 and see below), it seems likely that at least some relevant features regulating initiation of translation reside in the 5' untranslated leader sequence. It has been considered that CBPII and ATP (required for CBPII cross-linking to oxidized mRNAs, ref. 61) could operate by destabilizing the secondary structure of capped mRNAs, thereby facilitating binding of the mRNA to the 40S pre-initiation complex (43, 46, 61, 62), and that the degree of secondary structure would be correlated with the extent to which mRNA binding is dependent on the cap, ATP and CBPII.

In vitro experiments have demonstrated that CBPII reverses the inhibition of translation conferred to capped mRNAs by the presence of high salt (K^+) concentrations. Since high salt concentrations stabilize the secondary structure of mRNAs, this is in line with a model in which CBPII possibly in conjunction with ATP, participates in melting mRNA sequences thereby facilitating ribosome binding (63). Using reovirus it has been reported (64) that eIF-4A, a component of CBPII, causes a structural change in the viral mRNA as evidenced by increased nuclease sensitivity of the RNA, an observation suggesting that this factor melts or unwinds the RNA structure in an ATP-dependent reaction (64). VPg-bearing viral mRNAs such as CPMV are less dependent on CBPII, or ATP (63, 65), as are mRNAs with little secondary structure at their 5' end such as AMV RNA 4 (63, 66), and inosine-substituted reovirus mRNA (67, 68 and 69 and references therein).

These results are borne out by experiments in which the secondary structure of the leader sequence has been increased by insertional mutagenesis: such modifications performed in the thymidine kinase gene of HSV-1 lead to a decreased translation efficiency of the resulting mRNA *in vivo* and *in vitro* (70).

Finally, the length of the 5' leader sequence of the subgenomic RNA coding for the coat protein of a given plant virus is generally shorter than that of the genomic RNA coding for the nonstructural proteins. A short leader sequence is probably less amenable to secondary structure formation, and this might favor translation of an mRNA bearing such a sequence (see below). Indeed, *in vitro* studies suggest that of two murine c-myc transcripts that contain 5' noncoding regions of 448 and 83 nucleotides, the shorter transcript is ten times more efficient in translation than the longer transcript (72); this could be a consequence of a potentially stronger secondary structure in the leader sequence of the longer transcript.

Competition between viral mRNAs

In several studies the translation ability of different mRNAs has been compared, in an attempt to correlate the differences in translation efficiency with structural features of the leader sequence (37). These results are however difficult to assess since the mRNAs considered differ not only in their 5' noncoding region, but also in their coding region and 3' noncoding regions, and long-range interactions between the 3' and 5' noncoding regions are known to occur in certain viral RNAs (73).

Nevertheless, there is clearly a hierarchy of competitive translational ability among different mRNAs. For instance, the subgenomic RNA coding for the coat protein of many RNA viruses outcompetes the genomic RNA in *in vitro* translation assays. This has been clearly demonstrated in the case of the TYMV subgenomic RNA (74), AMV RNA 4 (37, 75), BMV RNA 4 (76, 77) and BBV RNA 2 (78). Competition probably also occurs *in vivo* (79, 80); this would favor the production of coat protein which is needed in large excess over the nonstructural proteins to secure proper encapsidation of the newly produced viral RNA. Similarly, in mixed infections with a constant amount of TNV and increasing amounts of STNV in the inoculum, there is a considerable decrease in TNV production (81). The mechanism underlying competition is as yet poorly understood. Under conditions of competition, the level of expression of mRNAs *in vivo* reflects the intrinsic translational efficiencies of these mRNAs (82). It is believed that mRNAs must compete for a limiting component (presumed to be an initiation factor), part of the initiation machinery prior to binding to ribosomes. This component would act as a discriminatory factor due to its different affinity for particular mRNAs (83). Intrinsic features of the 5' leader, such as primary and secondary structure or length of the leader as discussed above, could dictate the affinity of this component for one mRNA as opposed to another (83).

B. Choice of Initiator AUG

The observation that eukaryotic ribosomes generally initiate translation at the level of the 5'-proximal AUG of cellular mRNAs with the production of a single protein (reviewed in ref. 6) first led to the proposal of the "scanning" model (84). This model states that the 40S ribosomal subunit, together with the appropriate initiation factors, first binds at or near the 5' end of the mRNA, then migrates along the mRNA until it encounters the first AUG triplet; at this point, it is joined by the 60S subunit and formation of the first peptide bond can occur. There are many exceptions to this model, particularly among viral mRNAs (64, 85-88 etc., see also chapter 8 in this book). In addition, certain viral mRNAs direct the synthesis of more than one protein (see below), and these must therefore be regarded as polycistronic mRNAs.

To account for these exceptions, the following adaptations have been made to the originally proposed "scanning" model:

i) nucleotides adjacent to the AUG codon can modulate the strength of translation initiation at that codon (89). In particular, an AUG codon flanked in positions -3 and $+4$ by a purine residue is preferred for initiation, the sequence ANNAUGG being the strongest (discussed in ref. 88);
ii) the leader sequence in the mRNA could modulate translation initiation by providing recognition sites for discriminatory factors involved in translation of selected mRNAs (69, 70, 83, 90); and
iii) short peptides can be synthesized from upstream AUGs followed by reinitiation at downstream AUGs (see below).

Polycistronic viral mRNAs

Examples of such naturally occurring polycistronic mRNAs among eukaryotic viruses are presented in Table 1A. The name of the virus that possesses a polycistronic mRNA, the proteins translated from the RNA template and the reference of the data presented are indicated. Two situations should be distinguished: *bona fide* polycistronic mRNAs and mRNAs coding for "agnoproteins".

Table 1. - List viruses resorting to the strategies of Group II. **A**: Polycistronic viral mRNAs; **B**: internal in-frame initiation; **C**: gene overlap

Strategy	Virus	mRNA	Proteins involved	Ref.
A. Polycistronic mRNAs				
1. *Bona fide* polycistronic mRNA	CaMV	35S RNA[1]	ORF VI/III; ORF IV/V	p.c.
2. mRNA coding for agnoprotein	SV40	late 16S RNA	agnoprotein	91,92
	RSV	genomic RNA	LP1	93
	CaMV	35S RNA[1]	agnoprotein	94,95
B. Internal in-frame initiation	CPMV	RNA M	105K/95K	96
	HSV-1	TK mRNA	VI43/VI39/VI38	97,98
	FMDV	genomic RNA	P20a/P16[2]	99
	Sendai virus	P/C mRNA unit	P/C	100,101
	Rotavirus	RNA segment 9	37K/53K	102
	FeLV	genomic RNA	$gPr80^{gag}/Pr65^{gag}$	103
C. Gene overlap	Adenovirus d12 and Ad5	E1b mRNA (2.2kb)	21K/55K	104
	SSHV	cRNA to S RNA	N/NS_S	105
	Reovirus	s1 RNA	haemagglutinin/14K	106,107
	Influenza B virus	RNA segment 6	NB/neuraminidase	108
	Sendai virus	P/C mRNA unit	C proteins	100

1: Indicated by the use of *in vitro* transcripts. 2: The proteins are originally part of the polyprotein. p.c. = personal communication (K. Gordon, P. Pfeiffer, J. Fütterer, & T. Hohn); TK = thymidine kinase; cRNA = complementary RNA.

Bona fide polycistronic mRNAs

One example of a *bona fide* polycistronic mRNA has been reported: The CaMV genome contains 8 ORFs. DNA sequence studies have demonstrated that these ORFs are generally closely packed and that in certain cases there is even some overlap between the termination of one ORF and initiation of the next. Most of these are thought to be translated from the 35S or "genomic" RNA, which would thus be a polycistronic mRNA. A "relacy race" model has been proposed to explain this linked translation. According to this model, after reaching the termination codon of one ORF, the ribosome would be capable of reinitiating protein synthesis provided an initiation codon is located in the vicinity (94, 109). Since however, the 35S RNA is poorly translated *in vitro*, constructs corresponding to pairs of ORFs have been transcribed and the RNA assayed in *in vitro* systems. The conclusions suggest that reinitiation can indeed contribute to the appearance of the two corresponding proteins (K. Gordon, P. Pfeiffer, J. Futterer and T. Hohn, personal communication).

Interestingly enough, most *bona fide* polycistronic strategies originally proposed for eukaryotic viruses have since been annuled by further analyses and sequence data, as in the case of flaviviruses (see Section V) or in the case of the fish pathogen, INPV (110, 111). However, this does not preclude the possibility that an mRNA can behave as a polycistronic mRNA as shown recently by the discovery of "agnogenes".

mRNAs encoding "agnogenes"

The term "agnogene" was first coined as a result of observations made with the late 16S mRNA of SV40 that codes for the structural protein VP1. The presence in the genome of a short ORF upstream of the VP1 ORF, has led to the search for a protein that could correspond to this ORF. Indeed, a protein composed of 61 amino acids appears late in the lytic cycle; it has a relatively short half-life of 2h. This protein has been termed agnoprotein, and the corresponding ORF, the agnogene (91, 112). The major 16S mRNA is a likely template for the synthesis of the agnoprotein, since the corresponding coding sequence is found in its leader; it is separated by 45 nucleotides from the VP1 initiator AUG (92). Mutants with deletions in the agnogene regione are viable, but have a reduced growth-rate (113).

The biological significance of the agnogene and its requirement for the expression of the VPI gene on the 16S mRNA are unknown. However, a role for the agnoprotein in the regulation of SV40 gene expression by an attenuator mechanism has been proposed (114) and is discussed in by Y. Aloni in chapter 19 of this book. In any event, it is remarkable that several examples of such "agnogenes" have appeared in recent reports. They are all characterized by the small size of their ORF, their location upstream of the main ORF, and the relatively short hal-life of their product, the agnoprotein.

A similar situation recently described is that of RSV. Its RNA contains a putative untranslated leader sequence, the main polyprotein precursor $Pr76^{gag}$ being initiated at the fourth AUG codon at position 380-382. The three upstream AUGs could have initiation activities according to the modified "scanning" model (see above). Indeed, the first AUG occupying positions 41-43, directs the synthesis *in vitro* of a heptapeptide, Leader Peptide-1 (LP1), that is very unstable (93). Because of the highly conserved position and size of the LP1 ORF among several avian retroviral RNAs, this peptide may play a role in the viral life-cycle.

An agnogene function has been proposed for ORFVII of CaMV (95). As mentioned above, it appears likely that most CaMV ORFs are translated from a *bona fide* polycistronic mRNA. In such a model, the short ORFVII would be the first translation product to be synthesized, followed 60 nucleotides downstream by ORFI. Deletions in ORFVII and in the intercistronic region between ORFVII and ORFI do not affect viral infectivity. However to maintain infectivity, ORFVII must preserve its initiation codon followed in-phase by a termination codon that should map before the initiation codon of ORFI. Furthermore, the location of the ORFVII initiation codon governs infectivity: its AUG codon must be located close to the reverse transcription primer-binding site (94). Mutation of the ORFVII initiation codon results in delayed appearance of disease symptoms (95). These results suggest that this start codon is important in viral expression. However, the protein corresponding to ORFVII has so far not been detected and the question remains open as to whether synthesis of the agnoprotein is required to modulate virus expression.

Experiments have been performed in which an oligonucleotide has been inserted into the leader sequence of an mRNA. Such experiments should shed some light on the manner in which agnoproteins might affect translation. Initiation of translation at the "authentic" AUG of the mRNAs of Hepatitis B virus (a hepadnavirus) surface antigen (115) or of HSV-1 thymidine kinase (116) is decreased (but not abolished) by the insertion upstream of oligonucleotides containing AUG codons. This decrease can be partially overcome if a favorable upstream AUG is followed in phase by a termination codon in the upstream region; in this case internal initiation of the "authentic" AUG can occur, the extent of reinitiation probably depending partly on the relative position of the termination codon versus the reinitiating AUG (115).

Further experiments will be required to establish the precise role of agnogenes, and in particular whether a precise nucleotide sequence in the agnogene or the corresponding translation product favors virus development.

Internal in-frame initiation within a gene

A few cases of internal initiation have been described to date (Table 1B). In the cases where they are based on *in vitro* experiments, the relevance of this mechanism for virus multiplication remains unclear as long as this strategy is not supported by *in vivo* experiments.

The RNA of the middle component (RNA M) of the bipartite RNA virus CPMV directs the synthesis in various *in vitro* translation systems of two colinear polyproteins of 105 kDa and 95 kDa (96). These two proteins are initiated on two AUG codons in the same reading frame. When RNA M is injected into *X. laevis* oocytes, both the 105 kDa and the 95 kDa proteins are produced (117). *In vitro* as *in vivo* systems, the latter protein is synthesized in larger amounts than the former; this might be explained by the fact that the first AUG only partially matches the consensus sequence for initiator codons (see above), whereas the second AUG matches this consensus sequence perfectly (117, 118).

Another case is that of the thymidine kinase mRNA of HSV-1. Translation of this RNA *in vitro* and *in vivo* has provided evidence that two downstream initiation codons are used to produce the V139 and V138 proteins in addition to the first AUG that yields V143 (97, 98). The flanking sequences of the first two AUG codons initiating V143 and V139 are not favorable, whereas those of the AUG codon initiating V138 are favorable (97) according to the modified "scanning" model (89). Thus, some ribosomes may by-pass the first and even the second AUG to initiate translation at the third AUG triplet. It remains to be determined however, whether V139 and V138 have thymidine kinase activity.

The synthesis of the polyprotein of the picornavirus FMDV begins at two in-frame initiator AUG codons at positions 805-807 and 889-891, located upstream of the region coding for the capsid proteins. Upon maturation of the polyprotein, two "leader" peptides, P20a (24 kDa) and P16 (21 kDa) are liberated from the capsid precursor. The presence of these two peptides has been observed by *in vitro* translation of FMDV RNA followed by immunoprecipitation with an antiserum produced against a bacterially synthesized polypeptide corresponding to the putative P16 coding region (99).

In vitro translation of polio- and Mengo-virus RNA had been reported previously to yield *two* initiation peptides (245, 246, 249, 250. The reader is also referred to chapter 8 in this book).

The Sendai virus (a paramyxovirus) genome is composed of a single 15kb RNA of (–) polarity. The information of the virus is expressed via transcription of the genome, yielding several (monocistronic) mRNAs. The second transcriptional unit, the P/C mRNA, is known to code for the structural P protein (see below), but also for a pair of nonstructural C proteins that are initiated on two in-frame AUGs. The two C proteins are observed both *in vivo* and *in vitro* (100, 101).

The utilisation of two in-phase initiator AUGs has been observed *in vitro* and *in vivo* in rotaviruses. The two proteins (37 kDa and 53 kDa) synthesized from genome segment 9 are the structural glycoproteins located in the outer shell of the virus (102).

Finally, two forms of the *gag* precursor of FeLV are detected in cells infected with this virus (119, 120). Nucleotide sequencing of the *gag* gene of FeLV has revealed an ORF (nucleotides 345 to 566) within the leader sequence that is in-frame with the *gag* gene sequence (103). This has led to the proposal that two in-frame initiator AUGs may be responsible for the production of the two forms of the *gag* precursor: AUG 567-569 could specify initiation of the glycosylated *gag* precursor $gPr80^{gag}$, whilst AUG 567-569 would specify initiation of the non glycosylated *gag* precursor $Pr65^{gag}$ (103).

Gene overlap in distinct reading frames

Several cases of gene overlap have been described to date. The major ones are listed in Table 1C. The information for the second protein to be initiated is either entirely contained within the region coding for the protein initiated on the 5' proximal AUG, or it extends beyond the termination point of the first protein.

A single mRNA of 2.2 kb, E1B mRNA, coding for two proteins in overlapping reading frames, has been described in the adenoviruses (104). The proteins involved are the 21 kDa and 55 kDa tumor antigens. The shorter protein initiates at the 5'-proximal AUG, whereas the other initiates at a second AUG in a different reading frame.

The genome of Snowshoe hare virus (SSHV, of the bunyavirus group) is composed of three (–) strand RNA fragments. The RNA complementary to the small RNA (S RNA) has been reported to contain two overlapping reading frames. The larger one corresponds to the viral nucleoprotein N coded by this RNA *in vitro*, and the smaller one to the nonstructural protein NSs induced by the virus in infected cells. Sequence data of the smaller RNA species indicate that the information for protein NSs is contained within the coding region of protein N, but in a different reading frame (105).

The s1 mRNA of human reovirus codes for the viral hemagglutinin and for a nonstructural 14 kDa polypeptide that is initiated on the second AUG in the mRNA (106). These two AUGs are in different reading frames, the second AUG presenting the stronger ribosome binding site (88). The 14 kDa polypeptide is formed in virus-infected mouse L cells, and is synthesized *in vitro* (106, 107).

The RNA corresponding to segment 6 of the influenza virus B genome codes for two proteins *in vitro*. Protein NB (11 kDa), which is also observed *in vivo*, is initiated at the first AUG codon on the mRNA; it overlaps the AUG initiation codon for neuraminidase (~50 kDa) (106). The function of protein NB is unknown.

The P/C mRNA of Sendai virus produced from the (-) RNA genome directs the synthesis of the structural protein P which uses the first AUG of the mRNA in one reading frame, whereas the proteins C (see above) utilize the second and third AUGs in another shorter reading frame (100). The proteins C are thought to be involved at some stage of viral replication.

C. Virus-encoded Regulation Factors

In certain cases, the nature of the factor that controls initiation has been identified. This signal can be a protein as in Rous Sarcoma virus (RSV), or a nucleic acid as in adenovirus.

RSV and its protein p12

The *gag* gene of the 35S RNA of RSV encodes the internal structural proteins of the virus. It is translated into the 76 kDa polyprotein precursor $Pr76^{gag}$ which is processed to the following four proteins: NH_2-p19-p27-p15-COOH (121). The AUG at position 380-382 is the initiator AUG of the polyprotein. As mentioned above, three AUGs located in the leader sequence are closely followed by termination codons in the same reading frame (122). The first AUG is at position 41-43 and it is contained within a strong ribosome binding site that spans nucleotides 9 to 53 (123).

In the virion, the p12 protein is associated with the RNA in a ratio of 2000 to 2500 protein molecules per 70S RNA. The presence of p12 on the RNA does not hinder proviral DNA synthesis in the cell, but it does have a dramatic effect on RNA translation. The following model has been elaborated to explain how a decrease in protein synthesis occurs (124).

RSV RNA is strongly structured. In particular, its 5' untranslated region appears to possess extensive secondary structure, such that the cap, the initiator AUG of the *gag* gene and the strong ribosome binding site lie in close proximity to one another. Ribosomes binding to the 5' region of the mRNA could thus be in a favorable position to "jump" to the AUG of the *gag* gene. The presence of p12 in large amounts unfolds the viral RNA as estimated by the accessibility of the RNA to RNase digestion. It is thus conceivable that in unfolding the leader, p12 causes disruption of the secondary structure of the ribosome binding site, thereby preventing initiation of translation at the "authentic" AUG. Accumulation of p12 on the RNA would thus negatively control RNA translation and would favor the packaging process. This model should not doubt be reconsidered in view of the recent identification of the agnoprotein, LP1, encoded upstream of the "authentic" AUG (see above).

Adenovirus and its VA RNAs

The adenovirus genome contains genes for two virus-associated (VA) RNAs, VAI and VAII, 160 nucleotides in length. They are transcribed by RNA polymerase III from closely-spaced genes on the adenovirus chromosome and accumulate to high concentrations in adenovirus-infected cells once DNA replication has begun. The VA RNAs have been sequenced (125) and are reported to exibit extensive secondary structure (126).

Recently, evidence has accumulated that late in infection, VAI is required for efficient translation of viral and host cell mRNAs. This was first shown by the use of an adenovirus mutant (dl331) which fails to produce VAI (127). Cellular and viral protein synthesis are markedly depressed in cells infected with this mutant, particularly late in infection (128). Mutants lacking VAII do not produce the same translational deficiency, whereas double mutants affect protein synthesis the most severly (129).

In such mutant-infected cells, the target of the lesion resides in initiation of translation, which is strongly reduced. This results from the activation of a protein kinase that phosphorylates the a subunit of the initiation factor eIF-2 and thereby inactivates this factor (128, 130). Indeed, phosphorylated eIF-2 a forms a tight complex with eIF-2B which is present in limiting amounts in the cells, is no longer available for the recycling of eIF-2-GDP. As a consequence of this series of events, initiation of protein synthesis is blocked (reviewed in ref. 131, 132).

The protein kinase responsible for the phosphorylation of eIF-2a is the dsRNA-activated inhibitor (also known as DAI) of protein synthesis (133). This enzyme is present in low levels in many cell types. Its activation *in vitro* requires dsRNA and ATP. It is one of several enzymes involved in the establishment by interferon of an antiviral state.

In cells infected with the adenovirus mutant d1331, the kinase can be activated in a manner similar to that produced by dsRNA. Several lines of evidence suggest that virus-induced dsRNA, resulting from symmetrical transcription of the adenovirus genome is responsible for this activation in the absence of VAI RNA in this mutant.

One model designed to explain how VA1 acts to inhibit protein kinase activation suggests that, in view of its highly base-paired structure, VAI competes with dsRNA for the protein kinase (134). By binding to the kinase, the large amounts of VAI present late in infection would inhibit binding of multiple copies of kinase to dsRNA. This in turn would prevent kinase molecules from phosphorylating each other, leading to the prevention of their inhibitory effect on translation.

Another adenovirus mutant, mutant 709, produces a VAI in which a stretch of 11 bp has been disrupted by 6 mutations in the RNA. The fact that this mutant fails to inhibit activation of the kinase supports the notion that the duplex structured of VAI is important for its function (135).

The observation that VAI can be found associated with polyribosomes (136) and that it can hybridize with adenovirus-specific poly(A)$^+$ mRNA (137) has led to the proposal of another model to explain how VAI RNA can promote initiation of translation late in infection (Mariman, E.C.M., and Van Venrooij, W.J., personal communication). This model rests on sequence data of VAI RNA, of the 3' region of 18S rRNA and of the common tripartite leader sequence of mRNAs derived from the major late transcription unit of adenovirus. In view of potential stable hybrids that can be formed between VAI and the 3' region of 18S rRNA and between VAI and the mRNA leader sequences, it has been proposed that VAI, by binding to the mRNA, could bring the messenger to the ribosome via its interaction with the 18S rRNA. The validity of this interesting model awaits experimental confirmation. It also might need to be adapted to explain how VA RNAs exert their stimulatory effect on protein synthesis programmed not only by the adenovirus mRNAs with 5' tripartite leaders, but also by cellular and other viral mRNAs.

The models proposed above for the action of VAI will probably also have to be adjusted in view of the following results. Experiments in which cells are infected by the adenovirus dl331-mutant and by influenza virus still permit translation of influenza virus mRNAs. Consequently, influenza virus can overcome the block in translation initiation imposed by adenovirus even in the absence of VAI. It is conceivable that an influenza virus-specific product, possibly a small RNA similar to VAI RNA, allows influenza virus but neither adenovirus nor host-cell protein synthesis to proceed (138).

Although the mechanism whereby VAI RNA exerts its effect on protein synthesis is not entirely elucidated, the adenovirus VA RNAs constitute the first example of translational regulation by a small RNA whose mode of action has been extensively studied at the molecular level. It appears likely that the presence of the VA RNAs enables adenovirus to maintain inapparent infection in its host for months or even years, since it counteracts the establishment of the antiviral state.

3. REGULATION AT THE LEVEL OF ELONGATION AND TERMINATION

Regulation of translation of viral mRNAs at the level of elongation and termination has been reported in animal and plant RNA viruses with a (-) RNA genome. Three mechanisms have been described or proposed that are used by such viruses to synthesize more than one polypeptide per cistron and to modulate the level of gene expression. These strategies are schematized in Fig. 1 and include: readthrough of termination codons (Group III-1), arrest of elongation independent of termination codons (Group III-2), and shift in the reading frame (Group III-3). This section deals with each of these strategies and examines a few examples in detail; it then discussed the components of the translational machinery and the features in the mRNA that are involved or may play an important role in such regulation.

A. Strategies of Regulation at the Level of Elongation and Termination among Plant and Animal Viral mRNAs: A Description

Natural suppression

The most widely used mechanism of regulation of translation at the level of elongation and termination among animal and plant viral mRNAs is, no doubt, suppression of termination codons. Termination codons are normally recognized by release factors. However, in some cases, tRNAs called "suppressor tRNAs", may insert an amino acid at the level of the stop codon site. Elongation then proceeds until the ribosome encounters the next in phase termination codon. In this manner a longer protein called a "readthrough protein" is synthesized.

When suppression occurs in a wild-type cell and is performed by a normal tRNA, it is referred to as "natural suppression" (as opposed to suppression by mutated tRNAs, a phenomenon that has been extensively studied in yeast and prokaryotes (139-141). Many viruses have taken advantage of the flexibility of the decoding process illustrated, for instance, by the existence of normal cellular tRNAs capable of misreading to synthesize from the same gene two proteins that, although structurally related, may perform distinct and essential functions. Such a subtle mechanism permits regulation of the level at which the two proteins are produced. Readthrough proteins are synthesized in low amounts, and indeed, in the case where a function for a readthrough proteins was clearly demonstrated (142-145) it corresponded to an enzymatic activity required in catalytic amounts.

Historically, the first observation of readthrough of termination codons in eukaryotic viral mRNAs was reported in 1978 on TMV. The genomic RNA of this virus has a (+) polarity and encodes two high molecular weight nonstructural proteins. These proteins (126 kDa and 183 kDa) are observed both *in vitro* and *in vivo*. They are initiated at the same site on the mRNA; thus they share the same N-terminal sequence. Moreover, the yield of the 183 kDa protein can be increased at the expense of the 126 kDa protein upon addition of yeast amber suppressor tRNA to a reticulocyte cell-free translation system programmed by TMV RNA (146). The existence of an amber codon (UAG) at the end of the 126 kDa ORF was later confirmed by the nucleotide sequence of TMV RNA (147). Taken together these data suggest that a natural suppression mechanism is operative *in vivo* (ad discussed below). In fact, recent experiments show that deletion of the UAG codon abolishes infectivity thereby indicating that regulation at the level of termination is required for TMV multiplication (148, 149).

Since 1978, suppression of termination codons as a mechanism of regulating gene expression has been proposed for an ever growing number of plant and animal RNA viruses, on the basis of *in vitro* translation studies or upon analysis of the nucleotide sequence of viral genomes. Table 2 lists the viral mRNAs for which a readthrough mechanisms has been observed *in vitro* and/or *in vivo*, the nature of proteins, of the termination codons involved, and the nucleotide sequence of those viral genomes for which such information is available. It is interesting to observe that all natural cases of suppression described so far among eukaryotic viruses involve UAG and UGA codons.

Among animal viral mRNAs, a particularly well documented example of natural suppression is the one concerning the expression of the *pol* gene in certain retroviruses. The retrovirus genome consists of three genes, *gag pol* and *env*, encoding the core proteins, reverse transcriptase and envelope proteins, respectively. Like most retroviruses Mo-MuLV expresses its *pol* gene by synthesizing a *gag-pol* polyprotein that is subsequently cleaved to mature components. Mo-MuLV virion RNA can be translated *in vitro* to yield both *gag* and *gag-pol* precursors, and the amount of the latter is increased at the expense of the former by addition of yeast

Table 2. – List of animal and plant viral mRNAs with a suppressible termination codon

Virus		Readthrough *in vitro*	*in vivo*	Nucleotide sequence	Termination codon	Size of proteins	Nature of proteins	Ref.
Moloney Murine Leukemia Virus	Mo-MuLV	+	+	+	**U A G**	Pr76 gag/Pr180 gag-pol	s/ns	142-144
Rauscher Murine Leukemia Virus	Ra-MuLV	+	+	N.D.	**U A G**	Pr80 gag/Pr200 gag-pol	s/ns	145
Sindbis Virus		+	+	+	**U G A**	p230/p270	ns/ns	150,151
Middleburg Virus		+	N.D.	+	**U G A**	p230/p270	ns/ns	150
Tobacco Mosaic Virus	TMV	+	+	+	**U A G**	126K/183K	ns/ns	146,147,152
Carnation Mottle Virus	CarMV	+	+	+	**U A G**	p30/p77/p100	ns/ns	153,154
Beet Necrotic Yellow Vein Virus	BNYVV	+	N.D.	+	**U A G**	21K/75K	s/ns	155,156
Turnip Yellow Mosaic Virus	TYMV	+	N.D.	+[1]	**U A G**	195K/210K	ns/ns	157
Tobacco Rattle Virus	TRV	+	+	+	**U G A**	140K/170K	ns/ns	158
Lucerne Transient Streak Virus	LTSV	+	N.D.	N.D.	N.D.	p78/p100	ns/ns	159
Southern Bean Mosaic Virus	SBMV	+	N.D.	N.D.	N.D.	75K/105K	ns/ns	160
Soil-Borne Wheat Mosaic Virus	SBWMV	+	N.D.	N.D.	N.D.	19.7K/28K/90K	s/ns/ns	161

1: partial sequence data are available. N.D. = not determined; ns = nonstructural protein; s = structural protein.

amber suppressor tRNA (142). Nucleotide sequencing of Mo-MuLV RNA reveales that the *gag* and *pol* genes are in the same reading frame, only separated by an amber codon (143). That *in vivo* suppression of a termination codon by a cellular tRNA is required for *pol* gene expression has now been unambigously shown (144). The protease cleaving the *gag* precursor into the mature core proteins has been purified, and partial amino acid sequence analyses indicate that the first four amino acids of the protease overlap with the 3' end of the *gag* gene. The fifth amino acid of the protease, glutamine, is inserted in response to the amber codon. Furthermore, the C-terminal amino acid is adjacent to the N-terminus of the reverse transcriptase, consequently the protease is almost entirely encoded by the *gag-pol* junction.

Interestingly, even if Ra-MuLV (145) seem to behave similarly to Mo-MuLV, other retroviruses utilize other strategies to express their *pol* gene. *In vitro* translation performed with RSV RNA and yeast suppressor tRNA yields a protein only a slightly longer than the *gag* precursor (162). Indeed, nucleotide sequencing ot the RSV genome reveals that the *gag* gene is terminated by an amber codon but that the N-terminus of the reverse transcriptase maps 20 nucleotides downstream from the *gag* gene in a different reading frame (122). Translational suppression can therefore not explain the synthesis of the *gag-pol* precursor for RSV. Two different mechanisms have been postulated: either frameshift suppression, or a splicing event producing a minor species of mRNA in which the *gag* ORF is fused to the *pol* ORF with removal of the stop codon. The former of the two mechanisms has recently been proven to regulate the synthesis of RSV reverse transcriptase (163 and see below). HTLV I (164) and II (165), BLV (166) and FeLV (103) show similar gene organization to that of RSV and thus could follow the same mode of expression.

Arrest of elongation independent of termination codons

Besides natural suppression, mechanisms of regulation at the level of elongation exist that lead to the differential synthesis of two distinct but related proteins. These mechanisms that result in arrest of elongation independent of termination codons are poorly understood at the molecular level. It is important to draw attention to the fact that in *in vitro* translation experiments various artefacts such as RNA or protein degradation or any deficiency in the translation machinery can lead to incomplete products (often designated "early quitters"). The comparison of the products synthesized in different cell-free systems and the observation of the *in vivo* occurence of these products are mandatory to establish their biological significance.

TYMV

This virus accomodates its genetic information into a single RNA species of (+) polarity. The organization of this genomic RNA, as deduced from *in vitro* translation studies (157), is represented in Fig. 2. In various cell-free translation systems (167,168) TYMV RNA directs the synthesis, in about equal amounts, of two high molecular weight nonstructural proteins of 150 kDa and 195 kDa that are initiated on the same AUG; they share the same N-terminus (169) and have common amino acid sequences (167).

Initially, various hypotheses were formulated to explain how these overlapping proteins are generated during mRNA translation. Production of the 150 kDa protein by proteolytic cleavage of the 195 kDa was excluded on the basis of kinetic experiment (170), and the possibility of a readthrough mechanism for the synthesis of the 195 kDa was discarded by the use of yeast amber, ochre or opal suppressor tRNAs (157). However, the two high molecular weight proteins are synthesized in three different cell-free systems (reticulocyte lysate, wheat germ and ascites (168)), and preliminary observations suggest that they are also produced in TYMV-infected protoplasts (171). Thus, arrest of elongation independent of termination codons could be a strategy employed for the expression of TYMV-encoded nonstructural proteins. Hypotheses such as frameshift, or control of elongation based on codon usage and tRNA-tRNA interactions (see below) have been forwarded.

Nucleotide sequencing and computer assisted analysis of the genomic RNA of TYMV should ultimately lead to the elucidation of the mechanism involved. In addition, amino acid sequence analysis of the C-terminus of the 150 kDa protein should pinpoint, at the codon level, the site where regulation of elongation occurs.

Alfalfa Mosaic Virus (AMV)

Intriguing findings with AMV RNA 1 have led to suggest that this RNA constitutes another example of regulation of protein synthesis at the level of elongation.

RNA1 is the largest of the three genomic (+) stranded RNAs of AMV. It exibits the peculiarity of directing the synthesis of different products during *in vitro* translation depending on its concentration in the cell-free extract. In a reticulocyte lysate or in a wheat germ extract supplemented with wheat germ tRNA and at low mRNA concentrations mainly two smaller products of 58 kDa and 62 kDa are syn-

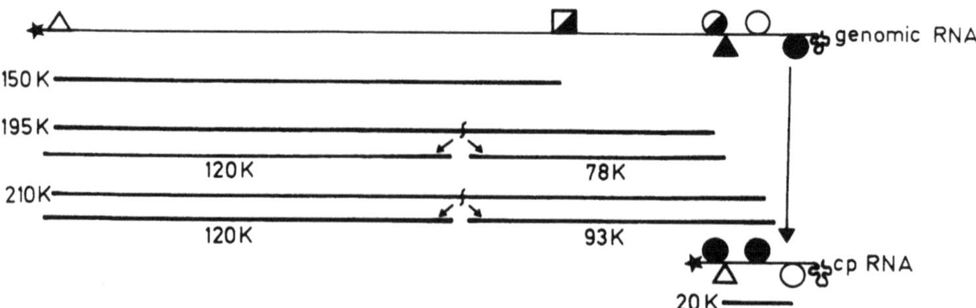

Fig. 2. Genetic map of TYMV RNA. The molecular weight of the products of proteolytic processing is indicated below each fragment. ★: cap structure; : tRNA-like region; other symbols as in Fig. 1.

thesized. These smaller polypeptides share common peptides with the 115 kDa protein and are encoded by the 5' half of RNA1 (172). The nucleotide sequence of RNA 1 (173) reveals no internal termination codons, indicating that a mechanism other than natural suppression must be responsible for the incomplete translation of AMV RNA1.

At high RNA1 concentrations, the full-length product can be formed when additional wheat germ tRNA and/or glutamine are supplied to the translation mixture (172, 174). It has been suggested that premature termination is the result of a decrease in elongation rate at the site corresponding to the C-terminus of the two shorter products and that this decrease is due to a limited availability of tRNAs at these sites. Indeed, a specific glutamine-accepting tRNA from tobacco and wheat germ promotes the full-length translation of RNA1 (174). The present data would support the hypothesis that incomplete translation is due to a limiting concentration of wheat germ Gln-tRNA in the system. However, the demonstration that this mechanism is functional during the life cycle of AMV still awaits the *in vivo* detection of the two shorter polypeptides. So far, of the RNA1 products only the 115 kDa proteins has been detected with the use of antibodies raised against a synthetic peptide derived from the C-terminal part of this protein (175).

Encephalomyocarditis Virus (EMCV)

Finally the case of EMCV RNA translation should be mentioned. The long ORF of EMCV RNA is translated into a polyprotein that undergoes a series of processing events leading to the production of the mature virion proteins. The middle and 3' portions of the genome encode two nonstructural primary products F and C (2C and 3 in the new nomenclature, ref. 176), respectively. The appearance of F in *in vitro* translation studies is somehow delayed. This phenomenon, which is not observed with poliovirus RNA, has been interpreted by some authors (177) in terms of a pause in the translation of the viral RNA. This would be due to the presence of a cluster of codons in structured region in the RNA. Such a proposal was strengthened by preliminary evidence indicating that addition of exogenous tRNAs to the cell-free translation system increased synthesis of the F precursor. The significance of the translational lag is not clear. If pausing is followed by release from the template and recycling of the ribosomes, this mechanism could permit overproduction of capsid proteins, possibly late in infection when viral RNA concentration may be saturating (thus favoring pausing) and when large amounts of capsid proteins are required to form the virus particles. More recently, a detailed kinetic analysis of the *in vitro* synthesis and processing of the EMCV products has been performed (178). This analysis, based both on synchronisation of initiation of translation and calibration of the elongation rate, indicates that there is no discontinuity in the rate of the amino acid incorporation and that translation proceeds at an even pace along the entire genome with no significant pause. The reason for the delayed appearance of F would be that it is excised from the polyprotein only after translation of the region immediately following F. This region encodes the proteolytic activity responsible for this excision.

Frameshift during translation

Frameshift during translation has been described and extensively demonstrated in prokaryotes, where frameshift suppressor tRNAs have been isolated that restore the proper reading frame of translation in frameshift mutants (179).

Just as in natural suppression, where normal tRNAs can read termination codons, normal tRNAs in natural frameshifting are thought to read certain codons in a non-triplet manner. Reading at low efficiency of two nucleotides (frameshift - 1) or four nucleotides (frameshift + 1) leads, by changing the reading frame, to the production in different amounts of two partially related and overlapping proteins.

In eukaryotes, the first well documented example of a frameshift mechanism has been recently described in RSV gene expression (163). In this genome the *pol* ORF is -1 with respect to that of *gag* and the *gag* and *pol* ORF overlap by 58 nucleotides before the *gag* amber termination codon. To discriminate between the two possible mechanisms proposed for *gag-pol* precursor synthesis in RSV, namely splicing and frameshift, SP6-*in vitro* transcripts have been obtained from cloned RSV cDNA. These transcripts begin upstream of the *gag* initiation site and terminate within the *pol* reading frame far downstream from the *gag* termination codon. Upon *in vitro* translation in a reticulocyte lysate, the *gag-pol* RNA transcript produces the expected $Pr76^{gag}$ protein and a larger polypeptide that can be immunoprecipitated both with anti-reverse transcriptase and anti-$p19^{gag}$ serum. The size of the longer polypeptide is in agreement with that of a *gag-pol* fusion protein synthesized from the cloned RSV fragment. These results demonstrated that a - 1 frameshift occurs *in vitro* and suggest that this mechanism could be functional *in vivo*.

Frameshifting has also been proposed for the expression of the dicistronic AMV RNA3. *In vitro* and *in vivo* studies have led to the detection in low amounts of a long polypeptide that could correspond to the translation of the two ORF of RNA3, as indicated by the size of this polypeptide and by immunoprecipitation studies (180, 181). In view of the RNA3 sequence (182), this can only be explained in terms of a frameshift mechanism. However, direct confirmation of such a strategy during translation will have to await sequencing of this long polypeptide.

B. Molecular Bases of Translational Regulation at the Level of Elongation and Termination

The molecular bases of the mechanisms of regulation described so far are poorly understood, and regulation must primarily be considered as the result of the concomitant effect of more than one factor, including components of the translational machinery and features of the mRNA itself. Termination of translation is determined by the recognition of a termination codon by a release factor. Therefore, when dealing with regulation at the level of termination, it is important to view this phenomenon in the light of a competition between the release factor and the suppressor tRNA. However, the bases of the interaction between the termination codon and the release factor are still obscure in eukaryotes and thus will not be discussed further.

Table 3. - Recognition between natural amber suppressor tRNAs and amber termination codons present in different viral mRNAs[1]

	$tRNA^{Tyr}$	$tRNA^{Tyr}$	$tRNA^{Gln}$
SOURCE	*D.melanogaster* Tobacco leaves Wheat germ Wheat leaves	*D.melanogaster* Wheat germ	Mouse [2]
ANTICODON 3'→5'	A ψ G	A ψ Q	not identified
TERMINATION CODON 5'→3'	U A G	–	U A G
COGNATE CODON 5'→3'	U A C U A U	U A C U A U	C A A C A G
VIRAL mRNAs CONCERNED	TMV BNYVV TYMV	-	Mo-MuLV

1: In all cases the two cognate codons for each tRNA species have been included for information; preferential recognition, if any, of one or another codon by the isoacceptor tRNA is not considered here. 2: The glutamine-inserting tRNA involved in suppression of the Mo-MuLV amber codon has not been isolated.

Natural suppression of viral mRNAs seems to involve misreading of termination codons by normal cytoplasmic aminoacyl-tRNAs. Table 3 summarizes the present available data on eukaryotic natural suppressor tRNAs involved in suppression of amber termination codons in viral mRNAs. Two major $tRNA^{Tyr}$ species bearing the same 5' GΨA 3' anticodon and differing only by one base pair and one partial modification have been isolated from non infected tobacco leaves. They act as amber suppressor *in vitro*, recognizing the UAG stop codon of both TMV (152) and BNYVV RNA (155). This suggests that these normal $tRNAs^{Tyr}_{G\Psi A}$ promote read-through in the infected plants *in vivo*.

In contrast, the two tyrosine isoacceptors tRNAs isolated from *Drosophila melanogaster* (183) or wheat germ (184) differ by the presence or absence of the highly modified Q base in the anticodon. Only the $tRNA^{Tyr}$ lacking the Q base is able to overcome the amber codon of TMV. This has been demonstrated *in vivo* upon injection in *X. laevis* oocytes (183) and *in vitro* by translation in a reticulocyte system (152) or in a wheat germ extract (184).

These studies demonstrate that the molecular basis for suppression of the TMV amber codon is the presence of G instead of Q in the wobble position. They also show that the presence of a modification, namely Q, restricts recognition of $tRNA^{Tyr}$ to its cognate codon. This result supports the general theory that modifies bases at position 34 (wobble) and 37 (3' adjacent to the anticodon) of tRNAs modulate codon-anticodon recognition (185), thus providing a means for regulating tRNA selection on the ribosome.

The molecular mechanism, however, by which a 5' GΨA 3' anticodon actually interacts with a 5' UAG 3' codon remains obscure. Such an unorthodox recognition can be explained neither on the basis of the "wobble" hypothesis (186) nor on the basis of the "two out of three" reading mechanism (187). Thus other possibilities such as G.G interaction with the G base of the anticodon adopting a *syn* conformation (188), should be considered (184).

The elucidation of the three-dimensional structure of the Q nucleoside located in the "wobble" position of tRNA reveals that the bulky side chain (a cyclopentenediol group) of this hypermodified nucleoside extends out of the anticodon loop, where it does not interfere with codon-anticodon interactions (189). Therefore it cannot directly prevent QG pairing, but rather would prevent by steric hindrance the Q nucleoside from assuming a conformation such as to base pair with the G of the UAG codon.

The molecular mechanism of amber suppression in Mo-MuLV is no better understood, since if it occurs via misreading by a normal cellular Gln-tRNA, recognition would have to involve a "wobble" type pairing at the 3' instead of the 5' position of the anticodon (Table 3).

It appears that recognition of termination codons in natural suppression, and decoding in general, is not only a matter of hydrogen bonding between antiparallel trinucleotides but also that the conformation of the tRNA plays a role in codon recognition. A classical example is that of the *Escherichia coli* opal suppressor (190) that derived from $tRNA^{Trp}$. The suppressor bears the same anticodon as $tRNA^{Trp}$ but has a G in place of an A at position 24 in the base-paired region of the D arm. It is thus altered "distantly" to the anticodon loop.

Evidence for the involvement of parameters other than codon-anticodon interaction in the regulation of translation at the level of elongation and termination comes from the analysis of translation of TYMV and AMV.

The concept that an arrest in elongation is dependent on the presence of a codon requiring ***a limiting isoacceptor tRNA species*** is a major one. There is evidence based on cellular nucleotide sequences available from a wide variety of sources (bacteria, yeast, *Bombyx mori*, etc.) that there are characteristic codon preferences associated with different genomes, and that in a given organism the distribution of aminoacyl-tRNA isoacceptor species reflects to some degree the codon usage of this organism (191-194). Therefore it is clear that the balance between the codon frequencies in the mRNA and the distribution of the corresponding isoacceptor species is a prerequisite for optimal speed and accuracy of translation (195). A corollary is that rare codons translated by minor tRNA species can be used to regulate the rate of translation. In other words, ribosomes would not move regularly along the mRNA but would "pause" at "hungry codons" (196) corresponding to rare aminoacyl-tRNA species. This pause in elongation would encourage premature termination due to the release of peptidyl-tRNAs from the ribosomes at these sites.

This general concept of regulation of translation by limiting isoacceptor tRNA species, initially developed for cellular mRNAs, may apply to viral mRNA translation as well. Viral codon usage may reflect that of the cellular host thus allowing efficient translation; this could be the molecular basis for host-range specificity. Alternatively, differences in codon usage between the virus and its host could serve regulation of viral protein synthesis at the level of elongation. Hence, such a mechanism may at least in part account for the *in vitro* synthesis of the AMV RNA1 58 kDa and 62 kDa proteins.

If a given distribution of isoacceptor tRNA species corresponds to a given organism, the pattern of translation should be modified by changing the *in vitro* translation system or the source of tRNA supplied. This is remarkably illustrated in the case of AMV where, in contrast to what is observed in the presence of wheat germ tRNA (see above), no differences in the translation pattern are seen between low and high mRNA1 concentrations when yeast or calf liver tRNAs are supplied to the *in vitro* translation system. With yeast or calf liver tRNAs, essentially only the full-length 115 kDa protein is produced (174). This result could be explained either by differences in abundance of a tRNA species in yeast and calf liver tRNAs compared to wheat germ tRNAs or by differences in charging efficiency in the reticulocyte lysate.

In TYMV the control of elongation at the end of the 150 kDa protein was recently correlated to a tRNA species. *In vitro* translation conditions were designed such that the 150 kDa protein was the major high molecular weight protein synthesized. These *in vitro* studies have revealed that a tRNA acceptor of serine isolated from beef liver is capable of by-passing the site of premature termination at the end of the 150 kDa protein gene to yield the 195 kDa protein (our unpublished results). However it is not a minor species among beef liver tRNAs, an observation arguing against the assumption that premature termination in TYMV results from the limited availability of a given aminoacyl-tRNA.

The other major parameter implicated in translational regulation that will be discussed here is codon context. Translation of a codon during polypeptide synthesis may be influenced by the nucleotides surrounding the translated codon. Evidence for the effect of codon context on aminoacyl-tRNA selection comes from studies on nonsense and missense suppression in prokaryotes.

A mutation in *Salmonella typhimurium* affecting the 3' adjacent base of the amber codon in the *his* D gene of the *his* operon has been characterized (197). This mutation increases the efficiency of a glutamine-inserting amber suppressor tRNA, indicating that sequences outside the triplet codon influence tRNA-mRNA interaction. Observations have since accumulated pointing to the effect of bases both at the 5' and 3' side of a given translated codon (197-199).

The molecular mechanism determining context effect is unclear. Two hypotheses have been proposed:

i) the "swollen codon" hypothesis suggests a four base paired codon-anticodon interaction involving the universal U_{33} of the tRNA anticodon and the 3' adjacent base of the codon. If the latter is A or G, one would expect the tRNA-mRNA interaction to be strengthened. Alternatively,

ii) the "tRNA-tRNA interaction" hypothesis suggests that interaction occurs between the peptidyl-tRNA located on the P site of the ribosome and the coming aminoacyl-tRNA located on the A site. Although it is not clear how tRNAs simultaneously present on the ribosome interact, it is reasonable to think that codon-anticodon interaction at the A site may be stabilized by a favorable tRNA-tRNA interaction.

These hypothetical mechanisms can be extended to natural suppression and thus used to explain termination codon recognition by normal cellular tRNAs. In particular, if the efficiency and rate of peptide bond formation depend on optimal fitting on the ribosome of aminoacyl-tRNA and peptidyl-tRNA, ultimately an unfavorable codon-anticodon interaction (such as that between a stop codon and a natural suppressor tRNA) may be compensated by a favorable tRNA-tRNA interaction.

Interestingly, among the viruses that resort to natural suppression as a mode of gene expression, the leaky amber codon of TMV (147), BNYVV (156) and TYMV (our unpublished results) are flanked on either side by a CAA glutamine triplet. For all three mRNAs readthrough occurs in the presence of the same amber suppressor tRNA isolated from tobacco (152, 155 and our unpublished results). Comparison of suppression efficiency in various viral mRNAs presenting different contexts around the amber codon, will help in establishing the significance of codon context in natural suppression.

Finally, if the tRNA-tRNA interaction hypothesis applies to decoding in general, codon context effects appear as yet another regulatory device in translation and may in addition to aminoacyl-tRNA availability account for the regulation at the level of elongation described for TYMV and EMCV RNA.

The data presented so far point out that viral regulation at the level of elongation and termination are very complex mechanisms and represent the outcome of the superimposed contributions of various factors: tRNAs that carry out the decoding process through their anticodon, itself modulated by base modification and tertiary conformation, codon usage and tRNA availability, competition between aminoacyl-tRNA isoacceptor species and between suppressor tRNAs and release factor, and codon context as well as other features that remain to be discerned in the mRNA.

4. POST-TRANSLATIONAL PROCESSING OF VIRAL POLYPEPTIDES

A. Definition and Overview of Virus Specific Proteolytic Processing

Definition

Post-translational modifications are frequently required to produce mature and functionally active viral proteins. They include and may even be the combination of various additions such as phosphorylation, glycosylation, sulphatation, acylation such as acetylation or myristylation (200), as well as proteolytic cleavage of a signal peptide or proteolytic fragmentation of a precursor polypeptide into functional proteins. This section will concentrate on the proteolytic fragmentation since this mechanism is a translation strategy used by many viruses to produce more than one protein from a single mRNA (see Fig. 1, Group III-4) and because it seems to involve virus-specific enzyme and/or signals, whereas the other modifications likely use cellular pathways. However this strategy has to be distinguished from another type of proteolytic processing referre to here as "maturation" cleavage in which a precursor protein is cleaved into fragments (generally two) that remain associated (for example by disulphide bonds) and/or performs a similar function. Fragmentation of the hemagglutinin of orthomyxoviruses (influenza virus (201)), of the fusion protein of paramyxoviruses (Sendai virus (202)), of the precursor to structural proteins VI and VIII of adenovirus (203) and of one of the capsid precursors of picornaviruses (VPO in poliovirus (204, 205)) are examples of such a "maturation" cleavage and will not be considered here, since they are dealt with extensively in the proper sections of this volume. Only precursor polypeptides that upon cleavage yield functionally different proteins will be discussed here; they have been termed "polyproteins". They have been detected by means of translation of viral mRNAs in various *in vitro* systems, and/or by analysis of the proteins formed in infected cells in the presence of amino acids analogs or protease inhibitors such as N-ethyl-maleimide (NEM) or Zn^{2+}. Under normal conditions they undergo rapid and successive cleavage steps some of which can occur before completion of translation.

Overview of virus-specific proteolytic processing

Table 4 summarizes the data reported up to now in the domain of proteolytic processing of viral polypeptides. For each virus group the most studied member chosen as example, the nature of the genome, the size of the mRNA and of the polyprotein, and a short description of the cleavage products are indicated. Reports (if any) about the protease(s) involved in the cleavage(s), in particular evidence for a virus-coded protease, are also mentioned. It has become apparent that both plant and animal viruses can follow this strategy of protein synthesis and that structural as well as nonstructural viral proteins can be produced through proteolytic processing.

This strategy is mainly represented among viruses whose genome is made of RNA with a (+) polarity. The cases of picornaviruses, of the flaviviruses and of the potyviruses are remarkable in the sense that all of their proteins derive from the processing of a unique precursor polyprotein. Virus-encoded proteases have been characterized and even identified in many cases, they are underlined in Table 4. In the following pages, viral proteolytic processing, protease(s) involved and cleavage specificities will be illustrated by the description of several examples and the possible implications of this strategy in the regulation of viral gene expression will be discussed in some detail.

Table 4. – Overview of viral proteolytic processing

Virus group and example[1]	RNA	Polyprotein[2]	Final cleavage products[3]	Protease(s)	Ref.
picornavirus: **poliovirus**	genomic RNA	247K	**VP4/VP2/VP3/VP1/** 2A/2B/2C/3A/VPg/ 3C/3D	2A and 3C + "maturation"[4]	86,206, 207
retrovirus: **RSV**	35S mRNA	Pr76 *gag* Pr180 *gag-pol*	**p19/p10/p27/p12/p15** **gag**/ RT	**p15** + host?	208, 209
	28S mRNA	Pr95 *env*	**gp85/gp37**	+ "maturation"	
alphavirus: **SFV**	42S mRNA	270K	P1/P2/P3/P4	?	210-
	26S mRNA	130K	**C/E3/E2**/6K/**E1**	+ "maturation"	212
flavivirus: **yellow fever virus**	genomic RNA	340K	**C/M/E**/NS1/NS2/NS3/ NS4/NS5	? + "maturation"	213
birnavirus[5]: **IPNV**	segmentA	108K	**VP54/VP31**/ICP29	?	111
comovirus[6]: **CPMV**	B RNA	200K	32K/58K/VPg/110K(24K/87K)[7]	24K	15,16,
	M RNA	95K(105K)[8]	48K(58K)[8]/**VP37/VP23**	and 32K	214,215
tymovirus: **TYMV**	genomic RNA	195K(210K)[9]	120K/78K(93K)[9]	autocatalytic?	216
potyvirus: **TVMV**	genomic RNA	340K	28K/HC/42K/CI/NIa/NIb/ **C**	?	17
nepovirus[6]: **TobRV**	RNA 1	225K	?	?	217,
	RNA 2	116K	**C + 40K + 23K**	+ RNA1-encoded	218

Abbreviations: VP = virion protein, RT = reverse transcriptase, C = capsid protein, NS = non structural protein, ICP = infected cell polypeptide, HC = helper component, CI = cylindrical inclusion, NI = nuclear inclusion, these and all other protein denominations are according to the literature referred to here.

1: Viruses whose sequence is known have been preferentially chosen, other cases from the same group are referred to in the corresponding reference. Except for 5. and 6. all virus groups have a monopartite "+" RNA genome. 2: Molecular weight of the polyprotein is calculated from sequence data. Polyproteins have not always been directly detected in translation studies. 3: Final cleavage products separated by / are written according to their genetic order; when the order is still unknown the cleavage products are separated by +. Structural polypeptides are in thicker characters. Underlined are the viral proteases. 4: "maturation" refers to the "maturation" cleavage defined at the beginning of this section. 5: Birnaviruses have a tripartite double-stranded RNA genome. 6: Comovirus and nepovirus have a bipartite single-stranded "+" RNA genome. 7: Two cleavage pathways are observed. 8: 95K and 105K polyproteins are initiated on two in-frame AUGs on CPMV M RNA (see section III). 48K and 58K are the corresponding N-terminal cleavage products. 9: 210K is the readthrough product of the termination codon of the 195K gene (see section IV). 93K is the C-terminal cleavage product of the 210K.

B. Proteolytic Processing in Various Virus Families

Poliovirus and Cowpea Mosaic Virus

Extensive studies have been carried out to establish the pattern of complete proteolytic processing of the proteins encoded by picornaviruses (chapter 8, and ref. 206, 207) on one hand, and by CPMV on the other one (219). Poliovirus and CPMV infect organisms that are very distantly related among eukaryotes. Their genomes differ from each other in that the former is a unique (+) RNA molecule whereas the latter consists of two (+) RNA molecules, M and B RNAs. In all other respects however poliovirus and CPMV are remarkably similar: in their genome structure, organization and expression (220, 221). For example the genomic RNAs are linked to a VPg protein at their 5' ends and are polyadenylated at their 3' ends. Each RNA codes *in vitro* and *in vivo* for a unique polypeptide that is subsequently cleaved as indicated in Table 5.

The intermediates of cleavage have not been indicated: the processing map is thus simplified, stressing only the position of the cleavage sites and the genetic order of the final products. However, the characterization of the cleavage intermediates has been of great value to determine the pathway of processing, i.e. the order of the successive cleavage steps (207, 222).

The gene order of the viral proteins is very similar in poliovirus and CPMV if one juxtaposes the M and B RNAs of the latter. The nonstructural protein sequences derived from the RNA sequences have been compared (220, 221) and significant homology was found between the 2C, 3C and 3D (pol) proteins of poliovirus, and the 58 kDa, 24 kDa and 87 kDa proteins coded by the B RNA of CPMV respectively. The functions of the 2C and 58 kDa proteins are still unknown. The 3D (pol) polypeptide has polymerase activity involved in RNA

Table 5. - Summary of the cleavage products found in poliovirus and in CPMV

Virus	mRNA	polyprotein	final products
poliovirus	genomic RNA	247K	VP0[1]/VP3/VP1/2A/2B/2C/3A/VPg/ 3C/3D(pol) <--structural--><------nonstructural--------->
CPMV	M RNA	95K(105K)[2]	48K(58K)/VP37/VP23 <--ns--> <-structural->
	B RNA	200K	32K/58K/VPg/110K(24K/87K)[3] <--------nonstructural------->

Underlined are the virus-encoded proteases. ns = nonstructural. 1: VP0 undergoes a "maturation" cleavage producing the capsid proteins VP4 and VP2. This cleavage occurs in the capsid probably only after the viral RNA has been encapsidated [204]. 2: The 95K and 105K are initiated on two in-frame initiation codons on CPMV M RNA (see section III). The 48K and 58K are the corresponding N-terminal cleavage products. 3: Two different cleavage pathways are observed.

replication of poliovirus (223) and a corresponding role has been proposed for the 110 kDa protein of CPMV which contains the 87 kDa region (224). The first indication that the 3C protein of poliovirus could be a protease came from the comparison of this protein with a corresponding p22 polypeptide (now called 3C in the new nomenclature (176)) coded by another picornavirus, EMCV, which maps on a similar region of the genome. This p22 protein is responsible at least partially for the proteolytic maturation of the EMCV polyprotein (225). It is interesting to remark that in comparing the situation between poliovirus and CPMV, the sequence homology between the 3C and the 24 kDa proteins has stimulated the search for a protease activity in the latter polypeptide.

On the contrary it has long been known that the 32 kDa protein coded by the CPMV B RNA is a protease required in the processing of the CPMV structural proteins coded by the M RNA (214). Recently experiments aimed at looking for protease activity in the corresponding 2A protein of poliovirus have been reported (229). Expression in *E. coli* and inhibition with antibodies directed against the 2A protein have shown that this protein is indeed capable of cleaving specific Tyr-Gly bonds in the viral polyprotein. A second proteolytic activity mapping in an homologous position in the genome of EMC or FMD virus has not yet been found (207).

All the proteases tested to date in picornaviruses as well as in CPMV are inhibited by thiol-complexing reagents (207, 230, 231). Sequence data and site-

Table 6. – Cleavage specificities of poliovirus and CPMV proteases

	Protease	Amino acid pair	Location and polypeptides produced	
Poliovirus	2A	Tyr-Gly	VP1 / 2A (3C' / 3D' : alternative cleavage)[1]	
	3C	Gln-Gly	VP0 / VP3 , VP3 / VP1 2A/2B, 2B/2C, 2C/3A 3A/VPg, VPg/3C, 3C/3D(pol)	
CPMV	32K	Gln-Met	48K(58K)/VP37 VPg / 24K or VPg/ 110K	<--M RNA <--B RNA <--B RNA
	24K	Gln-Gly	VP37 / VP23 24K / 87K	<--M RNA <--B RNA
	24K[2]	Gln-Ser	32K / 58K 58K / VPg	<--B RNA <--B RNA

1: The second Tyr-Gly cleavage site that yields 3C' and 3D' is occasionally used. The possible significance of this alternative cleavage is discussed. 2: The different specificities of the 24K protease could rely on the fact that it acts as a precursor polyprotein (see text).

directed mutagenesis have confirmed that at least the 3C protease of poliovirus and the corresponding 24 kDa CPMV protease could be cysteine proteases although they do not share the consensus sequence observed in all members of the cysteine protease family examined far (207).

Table 6 summarizes the specificities of each viral protease activity with regard to the cleavage sites. Underlined is the most striking homologous situation encountered in the two viruses. The corresponding proteolytic activities (3C and 24 kDa) are capable of cleaving the same Gln-Gly bonds at two homologous positions: between two structural proteins on one hand and between the protease itself and the polymerase (part of it in CPMV) on the other hand.

Homologies of genome sequences and organizations between a plant and an animal virus have led to important progress in identifying the proteases involved in the processing of their proteins. The significance of this homology and whether it suggests that a common ancestor has led to both viruses remains a matter of speculation and has been discussed elsewhere (221, 233).

Turnip Yellow Mosaic Virus

The genetic map of TYMV has been presented in section 4 (see above and Fig. 2). Translation studies in various *in vitro* systems have revealed that the 195 kDa protein undergoes a proteolytic cleavage that yields a 120 kDa N-terminal fragment and a 78 kDa C-terminal fragment. The 210 kDa readthrough protein can also be cleaved at the same site as the 195 kDa protein, giving rise to an elongated (~93 kDa) C-terminal fragment (170). The nature of the cleavage site involved in this processing is not yet known. However the proteolytic activity involved in this cleavage is known to be inhibited by amino acid analogs such as canavanine and N-tosyl-L-lysyl-chloroketone (TLCK) (170). Surprisingly the 150 kDa protein which contains the cleavage region is not processed and remains stable during prolonged incubations in the cell-free extracts. Since it is rather unlikely that the same specific proteolytic activity would be present in such different *in vitro* systems are reticulocyte lysate, ascites and wheat germ extracts (168), a more likely possibility is that this activity is encoded by the virus itself. Kinetic studies and dilution experiments have shown that cleavage only occurs when the synthesis of the 195 kDa protein has been completed and that it is at least partially insensitive to dilution (216). These results have supported the hypothesis that the 195 kDa protein could be capable of self-cleavage and that its proteolytic activity could be encoded in the C-terminal region of this protein. Recent reports on the TYMV-specific RNA replicase suggest that one of the subunits of this enzyme is a virus-encoded 115 kDa polypeptide that could correspond to the 120 kDa N-terminal fragment produced by cleavage of the 195 kDa protein (234).

Antibodies have been raised against the 115 kDa subunit and have allowed the detection *in vivo* in TYMV-infected Chinese cabbage protoplasts of a 115 kDa polypeptide as well as of a larger protein that possibly correspond to the uncleaved 150 kDa protein (171). This is the first indication that processing of the TYMV-encoded 195 kDa protein also occurs *in vivo*.

Retroviruses

Retroviruses provide another well documented example in which viral proteases have been identified. In avian retroviruses such as RSV, the 35S virion mRNA is translated into two overlapping polyproteins $Pr76^{gag}$ and $Pr180^{gag\text{-}pol}$ (see section 4). The precursor polypeptide $Pr76^{gag}$ is cleaved to yield the 4 internal viral structural proteins plus the p10 protein that map in the order p19/p10/p27/p12/p15 from the N-terminus to the C-terminus. The p15 protein has been purified and identified as a thiol protease responsible for the processing of $Pr76^{gag}$ (209). Thus in RSV, the protease is part of a structural protein. It is also able to process the *gag-pol* precursor to generate the reverse transcriptase but not the precursor $Pr95^{env}$ for the envelope proteins that is translated from a subgenomic 28S mRNA (209). This latter precursor undergoes a "maturation" cleavage (defined above) in the microsomes after removal of its signal peptide and glycosylation. The cleavage sites of $Pr76^{gag}$ have been identified by amino acid analyses of the products. They are all different and none of them involves basic residues. Comparison of the C- and N-termini of the cleavage products with the corresponding RNA sequence has revealed that several amino acids may be removed at the cleavage site (208): for example there is a gap of 9 amino acids between the C-terminus of p27 and the N-terminus of p12. How these 9 amino acids are removed is unknown. Apart from a second endopeptidase event, the possibility exists that a (host-encoded) exopeptidase activity is associated with the p15 protease to produce the final products of processing. It remains also to be determined how the p15 protease itself is released from the *gag* precursor.

Interestingly and in spite of a rather similar genome organization to that of avian retroviruses, murine retroviruses do not use one of their major structural proteins as a protease for cleavage of the *gag* precursor. In the case of Mo-MuLV a virus-associated protease has been purified that cleaves the $Pr65^{gag}$ into p15/p12/p30/p10 from the N- to the C-terminus (144). Amino acid analysis at the N-terminus of the protease and comparison to the sequence of Mo-MuLV RNA have revealed that this protease maps exactly between the *gag* and the *pol* genes and that its synthesis results from the natural suppression of the amber termination codon of the *gag* gene (see section 4). It is not yet known how this protease is released from the *gag-pol* precursor and whether it can process this precursor. As in the case of avian retroviruses a cysteine residue seems to be involved in the active site of the protease. Sequence comparison suggest that a homologous protease could also be encoded between the sequences corresponding to the *gag* and *pol* proteins in at least three other mammalian retroviruses (144). This is a particularly sophisticated situation, since the processing of the precursor to produce the structural proteins depends on the synthesis via natural suppression of a longer polyprotein that itselfs needs to undergo a processing step to form an active protease.

C. Discussion

Implications of the strategy of proteolytic processing

It is interesting to note that so far, apart from the IPNV, all known viruses using polyprotein processing are (+) RNA viruses. Further investigations will reveal whether or not this is a significant observation. On the other hand some (+) RNA viruses such as TMV do not appear to involve any proteolytic processing for the production of their proteins.

An immediate consequence of the proteolytic cleavage strategy is that in the absence of any regulatory mechanism all cleavage products are produced in equimolar amounts. This may be an advantage for the synthesis of structural proteins when these are required in equal amounts for encapsidation, but this may become a waste of energy for nonstructural proteins such as the polymerase which are theoretically required in lower quantities than the structural proteins.

Three mechanisms can be envisaged to allow a certain regulation of the amount of proteins issued from cleavage:

A) In the first one, regulation at the level of elongation and/or at the level of termination can cause the C-terminal proteins to be produced in lower quantities as illustrated by the *gag-pol* precursor of retroviruses (see section 4).

B) Alternative cleavage may lead to the production in regulated amounts of different cleavage products. This situation has been reported in poliovirus (Table 6) where cleavage of the 3C-3D precursor at the level of a Tyr-Gly amino acid pair can occur, yielding two products: 3C', an elongated 3C protein, and 3D', a truncated 3D polymerase (229). However it is not known whether 3C' or 3D' possess any biological activity. Similarly two cleavage pathways have been described for the 200 kDa polyprotein of CPMV (see Tables 4 and 5), only one of which liberated the 110 kDa protein involved in the CPMV replicase. If proteins resulting from alternative cleavage would possess distinct functions from the standard cleavage products, this mechanism could enlarge the potentiality of expressing the genetic information of the virus. If alternative cleavage does not produce additional functional proteins, it could be a means for the virus of eliminating unnecessary amounts of certain polypeptides, as already suggested (235).

C) Degradation of excess polypeptide in the cell is probably responsible for the removal of unemployed proteins. It would be interesting to evaluate the cost of such a process for the cell. Whether it can play a role in the pathogenesis of the infected cell remains pure speculation. In most cases proteolytic processing is a very specific event: there is a clear preference for viral (rather than host) proteins and, cleavage occurs at very precise sites. The bases of this specificity remain poorly understood with respect to both the cleavage sites and the protease involved. In poliovirus for instance, only 8 of 13 possible Gln-Gly sites are processed efficiently, a ninth site being cleaved only poorly. Sequence comparison at the level of these Gln-Gly bonds does not reveal striking constraints except for the residue at position -4 before the cleavage site. In poliovirus as well as in other picornaviruses this position is always occupied by a non polar residue belonging to a subset of amino acids that is specific for a given picornavirus (207). Similarly an alanine residue is found at position -4 in 5 out of the 6 cleavage sites involved in the processing of the polyproteins encoded by CPMV M and B RNAs.

In the case of many picornaviruses a proline residue and/or non structured sequences are found on either side of the cleavage points of the polyprotein. This argues in favor of the importance of the flexibility of the peptide chain in the vicinity of the cleavage sites and is supported by the three-dimensional structures of the protomers in poliovirus, rhinovirus and mengovirus (205, 236, 237). Flexibility indeed is required for fitting the substrate in the active site of the protease (M. Rossmann, personal communication). In addition the progressive folding of the polypeptide chain in the course of synthesis can be advantageous for succession of consecutive processing steps such as for the structural proteins of alphaviruses (discussed below). Finally, the nature of the cleavage sites as well as three dimensional conformation requirements probably also account for the fact that viral proteases do not seem to cause extensive proteolysis of cellular proteins (56).

Several recent reports point to the involvement of virus-coded proteases in viral proteolytic processing.This correlates well with the specificity of the virus-coded cleavage process. This specificity is especially apparent when this protease is able to act autocatalytically i.e. to cleave itself from a precursor molecule. However this has not always been demonstrated and host proteases may in some cases also be involved.

Processing as a regulating step

Proteolytic processing of viral polyproteins can be considered not only as a strategy but also as a target for regulation.

In alphaviruses the proteolytic cleavage required for the production of the structural proteins is also directly involved in the distribution of these proteins in the cell. This mechanism has been particularly well studied in both Sindbis virus and SFV. A single ORF present on the 26S subgenomic RNA of SFV directs the synthesis of a polyprotein precursor for four different structural proteins. From the N- to the C-terminus, these are the capsid protein C and the three envelope proteins E3, E2 and E1 (see Fig. 6. of chapter 9 of this book). E3 and E2 are released as the precursor p62 that later undergoes "maturation" cleavage. A 6 kDa nonstructural polypeptide is also encoded between E2 and E1 (211). The capsid protein must remain in the cytoplasm to interact with the genomic RNA in forming the nucleocapsid whereas the envelope proteins have to migrate into the RER to be glycosylated and assembled as a viral envelope (equimolar amounts of all these proteins are found in the viral particle) (239). Translation of the 26S RNA begins on free polysomes and correct compartmentalization of the proteins is triggered by the first cleavage step that releases the capsid protein very rapidly after its translation. This protein is thus liberated in the cytoplasm where it can immediately bind to viral RNA. Simultaneously a very hydrophobic amino acid sequence is exposed by this cleavage at the N-terminus of the p62 polypeptide that directs the polysome to bind to the microsomal membrane where the p62-E1 polyprotein can be translocated. As elongation further proceeds the envelope proteins signals are cleaved (238). A good indication that the first cleavage step plays a role in the compartmentalization of the structural proteins E3 and E2 of SFV also comes from observations using a thermosensitive mutant, SFV ts-3 (239). At the non permissive temperature cleavage between C and p62 is inhibited whereas cleavage between p62 and E1 still occurs. In such a situation the fusion protein C-p62 remains in the cytoplasm whereas the E1 protein is recovered in the RER as in the wild type. This means that the El proteins possesses its own signal peptide (possibly contained in the 6 kDa polypeptide (240)), that allows it to migrate into the RER, but also that the absence of cleavage between C and p62 has hindered translocation of p62 to the RER.

Cleavage and host-range pathogenicity

Proteolytic processing can also be related to virus pathogenicity especially when it involves host-proteases. In this case the ability of a virus to infect one cell type is determined by the presence in this cell type of the protease required for correct processing. One example of such a situation is found among retroviruses. RSV, an avian retrovirus, cannot form virus particles in several mammalian cell lines and this has been related to the inability of these cell lines to process the $Pr76^{gag}$ precursor of the structural proteins (241). Although it is known that the protease responsible for this maturation is actually encoded in the precursor (the p15 protein, see above) a host protease is probably first required to liberate the p15 protease from the precursor. This host protease would be absent from the mammalian cell lines thereby preventing virus multiplication in these cells (241).

Cleavage inhibitors and resistance to virus

Finally proteolytic processing can be viewed as a potential target in antiviral research. Comparison of amino acid sequences have shown that several virus-coded proteases form a new class of cysteine proteases. More extensive studies of their active sites as well as of their substrate specificities could lead to the possibility of producing very specific synthetic inhibitors (such as "suicide-substrate" (242)) that would block the processing of the viral polypeptides and thus the replication cycle of the virus without damaging the host organism. The interest of such specific protease inhibitors is documented by a natural situation: the seedlings of the Arlington cowpea line fail to support replication of CPMV and they have been shown to contain a protease inhibitor capable of inhibiting the processing of the CPMV polyprotein in an *in vitro* assay (243).

Thus inhibitors designed on the basis of the cleavage specificity involved in maturation of viral polyproteins may become a novel approach to block replication of viruses utilizing this strategy.

5. CONCLUSIONS

Because of the small size of their genomes and because they can be obtained in relatively large amounts, eukaryotic viruses have over the years served as very valuable model systems for the study of the mechanisms regulating translation. Certain features of translation such as readthrough were first described in plant viral RNAs and only later observed in a cellular mRNA. Other features of translation that are the prerogative of eukaryotic systems, for instance the cap structure, spliced mRNAs and in-frame initiation first observed in eukaryotic viruses, have since been described among cellular mRNAs (for a review, see ref. 10).

The elegant studies aimed at unravelling the mechanism of codon-anticodon interaction constitute a most interesting example of the successful utilization of viral RNAs as a model system. This is strikingly examplified by studies on suppression. The examples of readthrough among eukaryotic viruses have entailed the participation of major isoacceptor tRNA species that behave as natural suppressors. Interaction of these tRNAs was originally thought to be restricted to their cognate codons. However evidence is accumulating that these tRNAs (provided they possess the appropriate anticodon) are capable of interacting with a termination codon, albeit by way of an "unorthodox" mechanism of codon-anticodon interaction.

Thus the study of translation of eukaryotic viral mRNAs has helped to identify certain cellular mechanisms and to demonstrate the flexibility of the cellular translational machinery. Moreover these studies have revealed features unique to eukaryotic viruses such as VPg, as well as features originally thought to be restricted to the prokaryotic kingdom such as polycistronic mRNAs, frameshift and mRNAs with an uncapped 5' end. These features have so far not been observed among eukaryotic cellular mRNAs.

Viral RNAs have exploited the flexibility of the translation machinery more fully than cellular mRNAs; this no doubt reflects the restricted coding capacity of viruses, and one can only marvel at the diversity and at times highly sophisticated strategies evolved by viruses for the synthesis of their proteins. A corollary of this observation is that viruses are not the simple microrganism they were originally thought to be. The vast array of strategies they resort to enable them to produce a larger number of proteins than have been anticipated based on the coding capacity of their genome, i.e., their structural and at least part of their replicating enzyme complex. For example, it has in recent years become clear that probably most of the proteases employed for the post-translational cleavage of viral proteins are encoded by the viral genome. Likewise, the genome of certain viruses codes for enzymes that are present in the cell such as those required for the elaboration of the cap structure. This implies that viruses can display a relative autonomy from the point of view of the post transcriptional and/or post-translational events required in their replication cycle. The balance between this autonomy and the inevitable dependence of viruses on cellular mechanisms for their genetic expression may reflect the closely associated evolution of these microrganisms and their hosts.

Acknowledgements

We are thankful to F. Chapeville for his constant encouragements during the course of this work. We are grateful to the many participants of the following 1986 meetings: the Seventh John Innes Symposium on "Virus Replication and Genome Interactions" (Norwich, England), the N.A.T.O. Summer School on "The Molecular Basis of Viral Replication" (Maratea, Italy), and the E.M.B.O. Workshop on "Mechanisms of Protein Synthesis in Eukaryotes" (Patras, Greece) for their careful reading of the manuscript and for their helpful suggestions. We are indebted to all those who kindly provided unpublished information and have authorized us to quote their data. We thank A. Legoki for his comments and suggestions concerning the manuscript. R.V. is recipient of a fellowship from the "Ministere de la Recherche et de la Technologie". This work was partly supported by a grant from the "Ecole Pratique des Hautes Etudes".

6. REFERENCES

1) Filipowicz, W. (1978), FEBS Lett. **96**, 1-11.

2) Banerjee, A.K. (1980), Microbiol.Rev. **44**, 175-205.

3) *Expression of Eukaryotic Viral and Cellular Genes* (1981), R.F. Pettersson, L. Kaariainen, H. Soderlund, and N. Oker-Blom, eds., Academic Press, London pp. 1-324.

4) *Interaction of Translational and Transcriptional Controls in the Regulation of Gene Expression* (1982), M. Grunberg-Manago and B. Safer, eds. Elsevier Biomedical, New York, pp. 1-524.

5) Kozak, M. (1983), Microbiol. Rev. **47**, 1-45.

6) Kozak, M. (1984), Nucl.Acids Res. **12**, 857-872.

7) Moldave, K. (1985), Amm.Rev.Biochem. **54**, 1109-1149.

8) Rhoads, R.E. (1985), Prog.Mol.Subcell.Biol. **8**, 104-155.

9) Pain, V.M. (1985), Biochem. J. **235**, 625-637.

10) Kozak, M. (1986) Adv.Virus Res. **31**, 229-290.

11) *Viral Messenger RNA Transcription: Processing, Splicing and Molecular Structure* (1985), Y. Becker, ed., Martinus Nijhoff, Boston, pp. 1-383.

12) HsuChen, C.C., and Dubin, D.T. (1967), Nature (London) **264**, 190-191.

13) Lee, Y.N., Nomoto, A., Detjen, B.M. and Wimmer, E. (1977), Proc.Natl.Acad.Sci USA **74**, 59-63.

14) Wimmer, E. (1982), Cell **28**, 199-201.

15) Van Wezenbeek, P., Verver, J., Harmsen, J., Vos, P., and Van Kammen, A. (1983), EMBO J. **2**, 941-946.

16) Lomonossoff, G.P., and Shanks, M. (1983). EMBO J. **2**, 2253-2258.

17) Domler, L.L., Franklin, K.M., Shahabuddin, M., Hellmann, G.M., Overmeyer, J.H., Hiermatch, S.T., Siaw, M.F.E., Lomonossoff, G.P., Shaw, J.G. and Rhoads, R.E. (1986), Nucl. Acids. Res. **14**, 5417-5430.

18) Ambros, V. and Baltimore, D. (1978). J. Biol. Chem. **253**, 5263-5266.

19) Rothberg, P.G., Harris, T.J.R., Nomoto, A. and Wimmer, E. (1978), Proc.Natl.Acad.Sci USA **75**, 4868-4872.

20) Fernandez-Munoz, R. and Darnell, J.E. (1976), J. Virol. **18**, 719-726.

21) Hewlett, M.J., Rose, J.K. and Baltimore, D. (1976). Proc. Natl. Acad. Sci. USA **73**, 327-330.

22) Nomoto, A., Lee, Y.F. and Wimmer, E. (1976), Proc. Natl. Acad. Sci. USA **73**, 375-380.

23) Ambros, V., Pettersson, R.F. and Baltimore, D. (1978) Cell **15**, 1439-1446.

24) Ambros, V. and Baltimore, D. (1980), J. Biol. Chem. **255**, 6739-6744.

25) Wimmer, E. and Reichman, M.E. (1968), Science **160**, 1452-1454.

26) Wimmer, E., Chang, A.Y., Clark, J.M. and Reichman, M.E. (1968), J. Mol. Biol. **38**, 59-73.

27) Lesnaw, J.A. and Reichman, M.E. (1970), Proc. Natl. Acad. Sci. USA **66**, 140-145.

28) Furuichi, Y., LaFiandra, A. and Shatkin, A.J. (1977) Nature (London) **266**, 235-239.

29) Lodish, H.F, and Rose, J.K. (1977), J. Biol. Chem. **252**, 1181-1188.

30) Wodnar-Filipowicz, A., Szczesna, E., Zan-Kowalczewska, M., Muthukrishnan, S., Szybiak, U., Legocki, A. and Filipowicz, W. (1978). Eur. J. Biochem. **92**, 69-80.

31) Muthukrishnan, S., Moss, B., Cooper, J.A. and Maxwell, E. (1978). J. Biol. Chem. **253**, 1710-1715.

32) Smith, R.E. and Clark, J.M. (1979). Biochemistry **18**, 1366-1371.

33) Moyer, S.A. (1981). Virology **112**, 157-168.

34) Shimotohno, K., Kodama, Y., Hashimoto, J. and Miura, K. (1977). Proc. Natl. Acad. Sci. USA **74**, 2734-2738.

35) Muthukrishnan, S., Morgan, M., Banerjee, A.K., and Shatkin, A.J. (1976). Biochemistry **15**, 5761-5768.

36) Rose, J.K. (1975). J. Biol. Chem. **250**, 8098-8104.

37) Herson, D., Schmidt, A., Seal, S.N., Marcus, A., and Van Vloten-Doting, L. (1979). J. Biol. Chem. **254**, 8245-8249.

38) Sonenberg, N., Rupprecht, K.M., Hecht, S.M. and Shatkin, A.J. (1979) Proc.Natl. Acad. Sci. USA **76**, 4345-4349.

39) Hellman, G.M., Chu, L.Y. and Rhoads, R.E. (1982). J. Biol. Chem. **257**, 4056-4062.

40) Webb, N.R., Chari, R.V.J., DePillis, G., Kozarich, J.W. and Rhoads R.E. (1984). Biochemistry **23**, 177-181.

41) Sonenberg, N., Morgan, M.A., Merrick, W.C. and Shatkin, A.J. (1978). Proc.Natl.Acad.Sci.USA **75**, 4843-4847.

42) Lee, K.A.W. and Sonenberg, N. (1982). Proc.Natl.Acad.Sci.USA **79**, 3447-3451.

43) Sonenberg, N. and Lee, K.A.W. (1982) see ref. 4, pp. 373-388.

44) Tahara, S.M., Morgan, M.A. and Shatkin, A.J. (1981). J. Biol. Chem. **256**, 7691-7694.

45) Shatkin, A.J., Darzynkiewicw, E., Nakashima, K., Sonenberg, N. and Tahara, S.M. (1981), see ref. 3, pp. 173-187.

46) Tahara, S.M., Morgan, M.A., Grifo, J.A., Merrick W.C. and Shatkin, A.J. (1982) see ref. 4, pp. 359-372.

47) Edery, I., Humbelin, M., Davreau, A., Lee, K.A.W., Milburn, S., Hershey, J.W.B., Trachsel, H. and Sonnenberg, N. (1983). J. Biol. Chem **258**, 11398-11403.

48) Leibowitz, R. and Penman, S. (1971). J. Virol. **8**, 661-668.

49) Doyle, S. and Holland J. (1972). J. Virol. **9**, 22-28.

50) Ehrenfeld, E. and Lund, H. (1977). Virology **80**, 297-308.

51) Ehrenfeld, E. (1982). Cell **28**, 435-436.

52) Etchison, D., Milburn, C.D., Edery, I., Sonenberg, N. and Hershey, J.W.B. (1982). J. Biol. Chem. **257**, 14806-14810.

53) Sonenberg, N. and Trachsel, H. (1982). Curr. Top. Cell. Regul. **21**, 65-88.

54) Hansen, J. and Ehrenfeld, E. (1981). J. Virol. **38**, 438-445.

55) Hansen, J., Etchinson, D., Hershey, J.W.B. and Ehrenfeld, E. (1982). J. Virol. **42**, 200-207.

56) Lloyd, R.E., Toyoda, H., Etchinson, D., Wimmer, E. and Ehrenfeld, E. (1986). Virology **150**, 299-303.

57) Sarnow, P., Bernstein, H.D. and Baltimore, D. (1986). Proc.Natl.Acad. Sci.USA **83**, 571-575.

58) Etchinson, D. and Fout, S. (1985). J. Virol. **54**, 634-638.

59) Mosenkis, J., Daniels-McQueen, S., Janovec, S., Duncan, R., Hershey, J.W.B., Grifo, J.A., Merrick, W.C. and Thach, R.E. (1985). J. Virol. **54**, 643-645.

60) Jen, G., Detjen, B.M. and Thach, R.E. (1980). J. Virol. **35**, 150-156.

61) Sonenberg, N. (1981). Nucl. Acids. Res. **9**, 1643-1656.

62) Grifo, J.A., Tahara, S.M., Leis, J.P., Morgan, M.A., Shatkin, A.J. and Merrick, W.C. (1982). J. Biol. Chem. **257**, 5246-5252.

63) Edery, I., Lee, K.A.W. and Sonenberg, N. (1984). Biochemistry **23**, 2456-2462.

64) Ray, B.K., Lawson, T.C., Kramer, J.C., Cladaras, M.H., Grifo, J.A., Abramson, R.D., Merrick, W.C. and Thach, R.E. (1985). J. Biol. Chem. **260**, 7651-7658.

65) Jackson, R.J. (1982). in*Protein Biosynthesis in Eukaryotes*, R. Perez-Bercoff, ed., Plenum Press, New York, pp. 363-418.

66) Gehrke, L., Auron, P.E., Quigley, G.J., Rich, A. and Sonenberg, N. (1983). Biochemistry **22**, 5157-5164.

67) Morgan, M.A. and Shatkin, A.J. (1980). Biochemistry **19**, 5960-5966.

68) Kozak, M. (1980) Cell **19**, 79-90.

69) Lee, K.A.W., Guertin, D. and Sonenberg N. (1983). J. Biol. Chem. **258**, 707-710.

70) Pelletier, J. and Sonenberg, N. (1985). Cell **40**, 515-526.

71) Svitkin, Y.V., Maslova, S.V. and Agol, V.I. (1985). Virology **147**, 243-252.

72) Darveau, A., Pelletier, J. and Sonenberg, N. (1985). Proc. Natl. Acad. Sci. USA **82**, 2315-2319.

73) Florentz, C., Briand, J.P. and Giege, R. (1984). FEBS Lett. **176**, 295-300.

74) Benincourt, C. and Haenni, A.L. (1978). Biochem. Biophys. Res. Commun. **84**, 831-839.

75) Godefroy-Colburn, T., Thivent, C. and Pinck, L. (1985). Eur J. Biochem. **147**, 541-548.

76) Zagorski, W. (1978). Eur J. Biochem. **86**, 465-472.

77) Pyne, J.W. and Hall, T.C. (1979). Intervirology **11**, 23-29.

78) Friesen, P.E. and Rueckert, R.R. (1984). J. Virol. **49**, 116-124.

79) Hunter, T.R., Hunt, T. Knowland, J. and Zimmern, D. (1976) Nature (London) . **260**, 759-764.

80) Benincourt, C., Haenni, A.L. and Chapeville, F. (1978). in *Third International Symposium on Ribosomes and Nuclei Acid Metabolism*, J. Zelinka, and J. Balan, eds., Publishing House of the Slovak Academy of Sciences, Bratislava, pp. 95-102.

81) Kassanis, B. (1968) Adv. Virus. Res. **13**, 147-180.

82) Walden, W.E., Godefroy-Colburn, T. and Thach, R.E. (1981). J. Biol. Chem. **256**, 11739-11746.

83) Ray, B.K., Brendler, T.G., Adya, S., Daniels-McQueen, S., Miller, J.K., Hershey, J.W.B., Grifo, J.A., Merrick, W.C. and Thach, R. (1983). Proc. Natl. Acad. Sci. USA **80**, 663-667.

84) Kozak, M. (1981) Top. Microbiol. Immunol. **93**, 81-123.

85) Pinck, M., Fritsch, C., Ravelonandro, M., Thivent, C. and Pinck, L. (1981). Nucl. Acids Res. **9**, 1087-1100.

86) Kitamura, N., Semler, B.L., Rothberg, P.G., Larsen, G.R., Adler, C.J., Dorner, A.J., Emini, E.A., Hanecak, R., Lee, J.J., Van der Werf, S., Anderson, C.W. and Wimmer, E. (1981). Nature (London) **291**, 547-553.

87) Racaniello, V.R. and Baltimore, D. (1981). Proc. Natl. Acad. Sci. USA **78**, 4887-4891.

88) Kozak, M. (1982). J. Mol. Biol. **156**, 807-820.

89) Kozak, M. (1981). Nucl. Acids Res. **9**, 5233-5252.

90) Darlix, J.L., Zucker, M. and Spahr, P.F. (1982). Nucl. Acids Res. **10**, 5183-5186.

91) Jay, G., Nomura, S., Anderson, C.W., Khoury, G. (1981). Nature (London) **291**, 346-349.

92) Ghosh, P.K., Reddy, V.B., Swinscoe, J., Choudary, P.V., Lebowitz, P. and Weissmann, S.M. (1978). J. Biol. Chem. **253**, 3643-3647.

93) Hackett, P.B., Petersen, R.B., Hensel, C.H., Albericio, F., Gunderson, S.I., Palmemberg, A. and Barany, G. (1986). J. Mol. Biol. **190**, 45-57.

94) Sieg, K. and Gronenborn, B. (1982). in*NATO Advanced Study Institute/FEBS Advanced Course on Structure and Function of Plant Genomes* Porto Portese, Italy. Abstract p. 154.

95) Dixon, L., Jiricny, J. and Hohn, T. (1986). Gene **41**, 225-231.

96) Franssen, H., Goldbach, R., Broekhuijsen, M., Moerman, M. and Van Kammen, A. (1982). J. Virol. **41**, 8-17.

97) Preston, C.M. and MacGeoch (1981). J. Virol. **38**, 593-605.

98) Marsden, H.S., Haarr, L. and Preston, C.M. (1983). J. Virol. **46**, 434-445.

99) Beck, E., Fross, S., Strebel, K., Cattaneo, R. and Feil, G. (1983). Nucl. Acids. Res. **11**, 7873-7885.

100) Giorgi, C., Blumenberg, B. and Kolakovsky, D. (1983). Cell **35**, 829-836.

101) Curran, J.A., Richardson, C. and Kolakovsky, D. (1986). J. Virol. **57**, 684-687.

102) Chan, W.K., Penaranda, M.E., Crawford, S.E. and Estes, M.K. (1986). Virology **151**, 243-252.

103) Laprevotte, I., Hampe, A., Sherr, C.J. and Galibert, F. (1984). J. Virol. **50**, 884-894.

104) Bos, J.L., Polder, L.J., Bernards, R., Schrier, P.I., Van den Elsen, P.J., Van der Erb, A.J. and Van Ormondt, H. (1981). Cell **27**, 121-131.

105) Bishop, D.H.L., Gould, K.G., Akashi, H. and Clerx-Vab Haaster, C. (1982). Nucl. Acids Res. **10**, 3703-3713.

106) Enst, H. and Shatkin, A.J. (1985) Proc. Natl. Acad. Sci. USA **82**, 48-52.

107) Sarkar, G., Pelletier, J., Bassel-Duby, R., Jayasuriya, A., Fields, B.N. and Sonenberg, N. (1985). J. Virol. **54**, 720-725.

108) Shaw, M.W., Choppin, P.W. and Lamb R.A. (1983). Proc. Natl. Acad. Sci. USA **80**, 4879-4883.

109) Dixon, L.K. and Hohn, T. (1984). EMBO J. **3**, 2731-2736.

110) Mertens, P.P.C. and Dobos, P. (1982). Nature (London) **297**, 243-246.

111) Duncan, R. and Dobos, P. (1986). Nucl. Acids Res. **14**, 5934.

112) Celma, M.L., Dhar, R., Pan, J. and Weissman, S.M. (1977). Nucl. Acids Res. **4**, 2549-2559.

113) Barkan, A. and Mertz, J.E. (1981). J. Virol. **37**, 730-737.

114) Hay, N., Skolnik-David, H. and Aloni, Y. (1982). Cell **29**, 183-193.

115) Liu, C.C., Simonsen, C.C. and Levinson, A.D. (1984). Nature (London) **309**, 82-85.

116) Bandyopadhyay, P.K. and Temin, H.N. (1984) Mol. Cell. Biol. **4**, 743-748.

117) Vos, P., Verver, J., Van Wezenbeek, P., Van Kammen, A. and Goldbach, R. (1984). EMBO J. **3**, 3049-3053.

118) Huez, G., Bruck, C., Van Vloten-Doting, L., Goldbach, R. and Verduin, B. (1983). Eur. J. Biochem. **130**, 205-209.

119) Neil, J.C., Smart, J.E., Hayman, M.J. and Jarrett, O. (1980). Virology **105**, 250-253.

120) Sherr, C.J., Donner, L., Fedele, L.A., Turek, L., Evens, J. and Ruscetti, S.K. (1980). in*Feline Leukemia Virus*, W.D. Hardy, M. Essex, and A.J. McClelland, eds., Elsevier/North-Holland Biomedical Press, Amsterdam, pp. 293-307.

121) Shealy, D.J., Mosser, A.G., and Rueckert, R.R. (1980). J. Virol. **34**, 431-437.

122) Schwartz, D.E., Tizard, R. and Gilbert, W. (1983)., Cell **32**, 853-869.

123) Darlix, J.L., Spahr, P.F., Bromley,P.A. and Jaton, J.C. (1979). J. Virol. **29**, 597-611.

124) Darlix, J.L., Meric, C., and Spahr, P.F. (1985). see ref. 10, pp. 355-372.

125) Akusjarvi, G., Mathews, M., Andersson, P., Vennstrom, B. and Pettersson, U. (1980). Proc. Natl. Acad. Sci. USA **77**, 2424-2428.

126) Monstein, B. and Philipson, L. (1981). Nucl. Acids Res. **9**, 4239-4250.

127) Thimmappaya, B., Weinberger, C., Schneider, R.J. and Shenk, T. (1982). Cell **31**, 543-551.

128) Reichel, P.A., Merrickm W.C., Siekierka, J. and Mathews, M.B. (1985). Nature (London) **313**, 196-200.

129) Bhat, R.A. and Thimmappaya, B. (1984). Nucl. Acids Res. **12**, 7377-7388.

130) Siekierka, J., Mariano, T.M., Reichel, P.A. and Mathews, M.B. (1985). Proc. Natl. Acad. Sci. USA **82**, 1959-1963.

131) Jagus, R., Anderson, W.F. and Safer, B. (1981). Prog. Nucl. Acid. Res. Mol. Biol. **25**, 127-185.

132) Proud, C.G. (1986). Trends Biochem. Sci. **11**, 73-77.

133) O'Malley, R.P., Mariano, T.M., Siekierka, J. and Mathews, B. (1986). Cell **44**, 391-400.

134) Kitajewski, J., Schneider, R.J., Safer, B., Munemitsu, S.M., Samuel, C.E., Thissappaya, B. and Shenk, T. (1986). Cell **45**, 195-200.

135) Bhat, R.A., Domer, P.H. and Thimmappaya, B. (1985). Mol. Cell. Biol. **5**, 187-196.

136) Schneider, R.J., Weinberger, C. and Shenk, T. (1984). Cell **37**, 291-298.

137) Mathews, M.B. (1980). Nature (London) **285**, 575-577.

138) Katze, M.G., Chen, Y.T. and Krug, R.M. (1984). Cell **37**, 484-489.

139) *Nonsense mutations and tRNA suppressors* (1979). Celis, D.G. and Smith, J.D., eds., Academic Press, New York, London, San Francisco, pp. 1-340.

140) *Transfer RNA: Biological Aspects* (1980). Soll, D.G., Abelson, J. and Schimmel, P.R., eds. Cold Spring Harbor Laboratory, Cold Spring Harbor, New York, pp. 1-569.

141) Celis, J.E. and Piper, P.W. (1981) Trends Biochem. Sci. **6**, 177-179.

142) Philipson, L., Andersson, P., Olshevsky, U., Weinberg, R. and Baltimore, D. (1978). Cell **13**, 189-199.

143) Shinnick, T.M., Lerner, R.A. and Sutcliffe, J.G. (1981).Nature (London) **293**, 543-548.

144) Yoshinaka, Y., Katoh, I., Copeland, T.D. and Oroszlan, S. (1985). Proc. Natl. Acad. Sci. USA **82**, 1618-1622.

145) Murphy, E.C.Jr., Wills, N. and Arlinghaus, R.B. (1980). J. Virol. **34**, 464-473.

146) Pelham, H.R.B. (1978). Nature (London) **272**, 469-471.

147) Goelet, P., Lomonossoff, G.P., Butler, P.J.G., Akam, M.E., Gait, M.J. and Karn, J. (1982). Proc. Natl. Acad. Sci. USA **79**, 5818-5822.

148) Meshi, T., Ishikaza, M., Motoyoshi, F., Semba, K. and Okada, Y. (1986). Proc. Natl. Acad. Sci. USA **83**, 5043-5047.

149) Ishikawa, M., Meshi, T., Motoyoshi, F., Takamatsu, N. and Okada, Y. (1986) submitted for publication.

150) Strauss, E.G., Rice, C.M. and Strauss, J.H. (1983). Proc. Natl. Acad. Sci. USA **80**, 5271-5275.

151) Strauss, E.G., Rice, C.M. and Strauss, J.H. (1984) Virology **133**, 92-110.

152) Beier, H., Barciszewska, M., Krupp, G., Mitnacht, R. and Gross, H.J. (1984). EMBO J. **3**, 351-356.

153) Harbison, S.A., Davies, J.W. and Wilson, T.M.A. (1985). J. Gen. Virol. **66**, 2597-2604.

154) Guilley, J., Carrington, J.C., Balazs, E., Jonard, G., Richards, K. and Morris, T.J. (1985). Nucl. Acids. Res. **13**, 6663-6677.

155) Ziegler, V., Richards, K., Guilley, H., Jonard, G. and Putz, C. (1985). J. Gen. Virol. **66**, 2079-2087.

156) Bouzoubaa, S., Ziegler, V., Beck, D., Guilley, H., Richards, K. and Jonard, G. (1986). J. Gen. Virol. **67**, 1689-1700.

157) Morch, M.D., Drugeon, G. and Benicourt, C. (1982). Virology **119**, 193-198.

158) Pelham, H.R.B. (1979). Virology **97**, 256-265.

159) Morris-Krsinich, B.A.M. and Forster, R.L.S. (1983). Virology **128**, 176-185.

160) Salerno-Rife, T., Rutgers, T. and Kaesberg, P. (1980). J. Virol. **34**, 51-58.

161) Hsu, Y.H. and Brakke, M.K. (1985). Virology **143**, 272-279.

162) Weiss, S.R., Hackett, P.B., Oppermann, H., Ullrich, A., Levintow, L. and Bishop, J.M. (1978). Cell **15**, 607-614.

163) Jacks, T. and Varmus, H.E. (1985). Science **230**, 1237-1242.

164) Seiki, M., Hattori, S., Hirayama, Y. and Yoshida, M. (1983). Proc. Natl. Acad. Sci. USA **80**, 3618-3622.

165) Shimotohno, K., Takahashi, Y., Shimizu, N., Gojobori, T., Golde, D.W., Chen, I.S.Y., Miwa, M. and Sugimura, T. (1985). Proc.Natl. Acad. Sci.USA **82**, 3101-3105.

166) Rice, N.T., Stephens, R.M. Burny, A. and Gilden, R.V. (1985). Virology **142**, 357-377.

167) Benicourt, C., Pere, J.P., and Haenni, A.L. (1978). FEBS Lett. **86**, 268-272.

168) Zagorski, W., Morch, M.D. and Haenni, A.L. (1983). Biochimie **65**, 127-133.

169) Benicourt, C. and Haenni A.L. (1978). Biochem. Biophys. Res. Commun. **84**, 831-839.

170) Morch, M.D. and Benicourt, C. (1980). J. Virol. **34**, 85-94.

171) Candresse, T., Batisti, M., Renaudin, J., Mouches, C. and Bove, J.M. (1986). submitted for publication.

172) Van Tol, R.G.L. and Van Vloten-Doting, L. (1979). Eur. J. Biochem. **93**, 461-468.

173) Cornelissen, B.J.C., Brederode, F.T., Moormann, R.J.M. and Bol, J.F. (1983). Nucl. Acids Res. **11**, 1253-1265.

174) Lindhout, P. (1985). Ph.D. Thesis, University of Leiden, The Netherlands.

175) Berna, A., Briand, J.P., Stussi-Garaud, C. and Godefroy-Colburn, T. (1986). J. Gen. Virol. **67**, 1135-1147.

176) Rueckert, R.R. and Wimmer, E. (1984). J. Virol. **50**, 957-959.

177) Shih, D.S., Shih, C.T., Zimmern, D., Rueckert, R.R. and Kaesberg, P. (1979). J. Virol. **30**, 472-480.

178) Jackson, R.J. (1986). Virology **149**, 114-127.

179) Roth, J.R. (1981). Cell **24**, 601-612.

180) Rutgers, A.S. (1977). Ph.D. Thesis, State University of Leiden, The Netherlands.

181) Joshi, S., Neeleman, L., Pleij, C.W.A., Haenni, A.L., Chapeville, F., Bosch, L., and Van Vloten-Doting, L. (1984). Virology **139**, 231-242.

182) Barker, R.F., Jarvis, N.P., Thompson, D.V., Loesch-Fries, L.S. and Hall, T.C. (1983). Nucl. Acids. Res. **11**, 2881-2891.

183) Bienz, M. and Kubli, E. (1981). Nature (London) **294**, 188-190.

184) Beier, H., Barciszewska, M. and Sickinger, H.D. (1984). EMBO J. **3**, 1091-1096.

185) Nishimura, S. (1979). in *Transfer RNA: Structure, Properties and Recognition*, P. Schimmel, D. Soll and J.N. Abelson, eds. Cold Spring Harbor, pp. 59-79.

186) Crick, F.H.C. (1966). J. Mol. Biol. **19**, 548-555.

187) Lagerkwist, U. (1978). Proc. Natl. Acad. Sci. USA **75**, 1759-1762.

188) Topal, M.D. and Fresco, J.R. (1976) Nature (London) **263**, 289-293.

189) Yokohama, S., Miyazawa, T., Iitaka, Y., Yamaizumi, Z., Kasai, K. and Nishimura, S. (1979). Nature (London) **282**, 107-109.

190) Hirsh, D. (1971). J. Mol. Biol. **58**, 439-458.

191) Chavancy, G., Chevallier, A., Fournier, A. and Garel, J.P. (1979). Biochimie **61**, 71-78.

192) Ikemura, T. (1981). J. Mol. Biol. **151**, 389-409.

193) Ikemura, T. (1982). J. Mol. Biol. **158**, 573-597.

194) Gouy, M. and Gautier, C. (1982). Nucl. Acids Res. **10**, 7055-7074.

195) Varenne, S., Buc, J., Lloubes, R. and Lazdunski, C. (1984). J. Mol. Biol. **180**, 549-576.

196) Goldman, E. (1982). J. Mol. Biol. **158**, 619-636.

197) Bossi, L. and Roth, J.R. (1980). Nature (London) **286**, 123-126.

198) Miller, J.H. and Albertini, A.M. (1983). J. Mol. Biol. **164**, 59-71.

199) Bossi, L. (1983). J. Mol. Biol. **164**, 73-87.

200) Post-translational modifications, Parts A and B (1980), *Methods in Enzymology,* **106** and **107**, Wold, F. and Moldave, K., eds., Academic Press, Orlando, USA.

201) Garten, W., Bosch, F.X., Linder, D., Rott, R. and Klenk, H.D. (1981). Virology **115**, 361-374.

202) Gething, M.J., White, J.M. and Waterfield, M.D. (1978). Proc. Natl. Acad. Sci. USA **75**, 2737-2740.

203) Bhatti, A.R. and Weber, J. (1979). J. Biol. Chem. **254**, 12265-12268.

204) Putnak, J.R. and Phillips, B.A. (1981). Microb. Reviews **45**, 287-315.

205) Hogle, J.M., Chow, M. and Filman, D.J. (1985). Science **229**, 1358-1365.

206) Pallansch, M.A., Kew, O.M., Semler, B.L., Omiliankowski, D.R., Anderson, C.W., Wimmer, E and Rueckert, R.R. (1984). J. Virol. **49**, 873-880.

207) Nicklin, M.H.J., Toyoda, H., Murray, M.G. and Wimmer, E. (1986). Biotechnology **43**, 33-42.

208) Schwartz, D.E., Tizard, R. and Gilbert, W. (1983). Cell **32**, 853-869.

209) Moelling, K., Scott, A., Dittmar, K.E.J. and Owada, M. (1980). J. Virol. **33**, 680-688.

210) Takkinen, K. (1986). Nucl. Acids Res. **14**, 5667-5682.

211) Garoff, H., Frischaud, A.M., Simons, K., Lehrach, H and Delius, H. (1980). Nature (London) **288**, 236-241.

212) Garoff, H., Frischaud, A.M., Simons, K., Lehrach, H. and Delius, H. (1980). Proc. Natl. Acad. Sci. USA **77**, 6376-6380.

213) Rice, C.M., Lenches, E.M., Eddy, S.R., Shin, S.J., Sheets, R.L. and Strauss, J.H. (1985). Science **229**, 726-733.

214) Franssen, H., Moerman, M., Rezelman, G. and Goldbach, R. (1984). J. Virol. **50**, 183-190.

215) Verver, J., Goldbach, R., Garcia, J.A. and Vos, P. (1986) submitted for pubblication.

216) Morch, M.D., Zagorski, W. and Haenni, A.L. (1982). Eur. J. Biochem. **127**, 259-265.

217) Forster, R.S.L. and Morris-Krsinich, B.A.M. (1985). Virology **144**, 516-519.

218) Morris-Krsinich, B.A.M. and Hull, R. (1981). Virology **114**, 98-112.

219) Wellink, J., Rezelman, G., Goldbach, R. and Beyreuther, K. (1986). J. Virol. **59**, 50-58.

220) Franssen, H., Leunissen, J., Goldbach, R., Lomonossoff, G.P. and Zimmern, D. (1984). EMBO J. **3**, 855-863.

221) Argos, P., Kramer, G., Nicklin, M.J.H. and Wimmer, E. (1984). Nucl. Acids Res. **12**, 7251-7260.

222) Goldbach, R. and Rezelman, G. (1983). J. Virol. **46**, 614-619.

223) Flanegan, J.B. and Baltimore, D. (1977). Proc. Natl. Acad. Sci. USA **74**, 3677-3680.

224) Dorssers, L., Van der Krol, S., Van der Meer, J., Van Kammen, A. and Zabel, P. (1984). Proc. Natl. Acad. Sci. USA **81**, 1951-1955.

225) Palmenberg, A.C., Pallansch, M.A. and Rueckert, R.R. (1979). J. Virol. **32**, 770-778.

226) Hanecak, R., Semler, B.L., Ariga, H., Anderson, C.W.A. and Wimmer, E. (1984). Cell **37**, 1063-1073.

227) Klump, W., Marquart, O. and Hofscneider, P.H. (1984). Proc. Natl. Acad. Sci. USA **81**, 3351-3355.

228) Franssen, H., Goldbach, R. and Van Kammen, A. (1984). Virus Research **1**, 39-49.

229) Toyoda, H., Nicklin, M.H.J., Murray, M.G., Anderson, C.W., Dunn, J.J., Studier, F.W. and Wimmer, E. (1986). Cell **45**, 761-770.

230) Pelham, H.R.B. (1978). Eur. J. Biochem. **85**, 457-461.

231) Pelham, H.R.B. (1979). Virology **96**, 463-477.

232) Palmenberg, A.C. and Rueckert, R.R. (1982). J. Virol. **41**, 244-249.

233) Goldbach, R.W. (1986). Ann. Rev. Pytopathology **24** in press.

234) Mouches, C., Candresse, T. and Bove, J.M. (1984). Virology **134**, 78-91.

235) McLean, L., Mathews, T.J., and Rueckert, R.R. (1976). J. Virol. **19**, 903-914.

236) Rossman, M.G., Arnold, E., Erickson, J.W., Frankenberger, E.A., Griffith, J.P., Hecht, H.J., Johnson, J.E., Kramer, G., Luo, M., Mosser, A.G., Rueckert, R.R., Sherry, B. and Vriend, G. (1985). Nature (London) **317**, 145-153.

237) Smith, T.J., Kremer, M.J., Luo, M., Vriend, G., Arnold, E., Kramer, G., Rossmann, M.G., McKinlay, M.A., Dianim G.D. and Otto, M.J. (1986). Science **233**, 1286-1293.

238) Garoff, H., Simons, K. and Dobberstein, B. (1978). J. Mol. Biol. **124**, 587-600.

239) Kaariainen, L., Hashimoto, K., Kalkinnen, N., Keranen, S., Lehtovaara, P., Ranki, M., Saraste, J., Sawicki, D., Sawicki, S., Ulmanen, I. and Vaananen, P. (1981) see ref. 3, pp. 189-203.

240) Welch, W.J. and Sefton, B.M. (1980) J. Virol. **33**, 230-237.

241) Vogt, V.M., Bruckenstein, D.A. and Bell, A.P. (1982). J. Virol. **44**, 725-730.

242) Walsh, C.T. (1984). Ann. Rev. Biochem. **53**, 493-535.

243) Sanderson, J.L., Bruening, G. and Russel, M.L. (1985). in *Cellular and Molecular Biology of Plant Stress* UCLA Symposia on Molecular and Cell Biology, New Series vol. **22**. Rey, J.L. and Kosuge, T., eds., Alan R. Liss, Inc., New York, 401-412.

244) Perez Bercoff, R., & Gander, M. (1978), FEBS Letters, **96**, 306-312

245) Perez Bercoff, R., & Kaempfer, R. (1982), J. Virol., **41**, 30-41

246) Degener, A. M., Pagnotti, P., Facchini, J., & Perez Bercoff, R. (1983), J.Virol. **45**, 889-894

247) Perez Bercoff, R. (1982). In: *Protein Biosynthesis in Eukaryotes*, R. Perez Bercoff, ed. Plenum Press, New York & London, pp.245-252

248) Kaempfer, R. (1982). In: *Protein Biosynthesis in Eukaryotes*, R. Perez Bercoff, ed. Plenum Press, New York & London, pp. 441-457

249) Celma, M.L., & Ehrenfeld, E. (1975), J.Mol. Biol. **98**, 761-780

250) Ehrenfeld, E. (1979). In: *The Molecular Biology of Picornaviruses*, R. Perez Bercoff, ed. Plenum Press, New York & London, pp.223-238

SECTION III

Interference and interferon

CHAPTER 6

BIOCHEMICAL ASPECTS OF INTERFERON ACTION

JEAN CONTENT

Department of Virology, Institut Pasteur du Brabant, Rue Engeland 642, B- 1180 Bruxelles, Belgium

List of abbreviations

IFN = Interferon; dsRNA= double stranded RNA; 2-5A = series of 2'5' linked oligonucleotides with the general structure $pppA(2'p5'A)_n$ with usually $n>2$; Kd = dissociation constant; GBP = guanylate binding protein; DAI = dsRNA activated inhibitor; kb = kilobase; EMC = encephalomyocarditis virus; VA RNA = virus associated RNA; EBV = Epstein-Barr Virus; SV40 = Simian virus 40; VSV = Vesicular Stomatitis Virus; HSV = Herpes Simplex Virus; eIF-2 = eukaryotic initiation factor 2.

INTRODUCTION

Interferons subtypes are the members of a large family of antiviral substances produced by different animal cells stimulated with viruses or different other inducers (Table 1). Beside their antiviral activity, they behave like pleiotropic hormonal effectors, modifying on the one side the biochemistry of the treated cell at various levels (membrane characteristics, antigen exposure, protein expression, among others), and on the other hand the physiology of the entire organism where they not only behave as part of the immediate defense mechanism against viral infection, but also as potent immunomodulating agents (1, 2) and, in several instances as autocrine regulators implicated in the control of cell growth and differentiation (3, 4).

The various IFNs shown in Table 1 are proteins or glycoproteins of 140-166 amino acids, defined by their antiviral activity, protecting pretreated, metabolically active cells (capable of RNA and protein synthesis) against subsequent infection by various types of viruses. IFNs have usually a defined host range or species specificity (5).

Table 1. - The Diversity and Nomenclature of Human Interferons

Interferon type	HuIFN-α	HuIFN-β	HuIFN-γ
Sub-types	> (α... α15)	one (β_1) or more?	Not found
Cell source	Leukocytes (blood buffy coats)	Foreskin or foetal muscle and skin fibroblasts (diploid)	T lymphocytes from blood buffy coat or spleen
Inducer	Sendai virus, other paramyxoviruses, other viruses	Poly(I).poly(C) or other double-stranded RNAs	Lectins, mitogens, some antigens
Specific activity	2-4 · 10^8U/mg	2-4 · 10^8U/mg	>10^8U/mg
mol. wt	15-20 000	20 000	20 000 and 25 000 + dimers
pH2 resistance	+	+	-
Glycosylation	-	+	+
Cloning and expression	Bacteria, yeast	Bacteria, animal cells	Bacteria, yeast and animal cells
Genome location	α_1... α_{15} Chr 9	β_1... Chr 9	Chr 12
Introns	-	-	3

In the past few years, the genes of a number of different interferons have been cloned, sequenced and expressed in heterologous systems and purified to homogeneity. There exist at least fifteen subtypes of human IFN-α, one IFN-β, and one IFN-γ(6). The genetics of murine IFN production has been developed (7) and most of the human and murine IFN structural genes have been mapped in detail and sequenced (8,9) allowing several groups to study the molecular mechanisms controlling the expression of some IFN genes. Some of the most advanced studies in this field concern the regulation of Human IFN-β gene induction by dsRNA (10, 11, 12, 13) or Human IFN-γ induction by viruses (14). Very recently, Maniatis and coworkers (12, 15) have characterized several regulatory sequence elements and located their binding sites in the HuIFN-β gene promoter region.

Interestingly, while the primary structure of most IFN have been determined and HuIFN-γ was obtained in crystallized form in 1982 (16), X-ray diffraction data are altogether lacking. Accordingly, little is known about the possible structure/function relationships in IFNs. At the time of writing this review, it appears as if the first crystallographic data will be obtained for muIFN-B. IFNs are only visualized in space as calculated on the basis of predictive models (17,18,19).

In human β and α IFN one disulfide bridge is essential and very few carboxy terminal amino acids can be removed by proteolysis without altering the biological activity of the molecule. These data are also corroborated by those derived from the preparation of synthetic peptides reproducing some domains of the IFN molecule (reviewed in 20).

Thanks to their expression in E.coli or animal cells, several recombinant IFN types or subtypes have been available in pure form in the last few years. All the IFN species and some of their recombinant hybrids or mutagenised derivatives carry a number of common biological properties but also have distinct spectra of antiviral activities (21).

Interferon activities and in particular antiviral actions are certainly complex and our understanding of this field is progressing more slowly. In contrast to the early concepts (22, 23) that one unique biochemical or enzymatic change could account for this large spectrum of effects, it is now generally accepted that a "multisite" model could best explain and reinforce these multiple activities. The essential difficulty in this field is to correlate the recently discovered, refined mechanisms biochemical with defined biological functions. Here we discuss only some recent developments of the biochemistry of interferon action, with special emphasis on the structure, regulation and role of the IFN-induced proteins, their contribution to the antiviral action of interferons and their possible implication beyond the field of IFN action. Specific examples of cell-virus system will be considered where IFN-regulated biochemical events are known to be required and/or sufficient for the expression of an antiviral state. Several recent reviews have been published on similar aspects of IFN research and should be consulted for background information (24-33).

1. IFN RECEPTORS

IFN binding with high affinity to specific receptors on the cell surface is the first step and an absolute requirement for further development of its biological activities (34). Considering the low level of structural homology between type I (α or β) and type II (γ)-IFN (35, 36), it would be not surprising if the latter would bind to different cellular receptors. It is clear from a number of studies involving binding of radiolabelled IFN (37), affinity labeling experiments (38, 39), studies on cellular variants devoided of type I receptors but responding normally to type II IFN (40) and blocking by specific antibodies (41, 42) that α and β-IFNs interact with a common cellular binding site whereas IFN-γ is recognized by a distinct kind of receptor. As summarized in Table 2, these two kinds of receptors differ in several characteristics. After binding, both α and γ-IFNs are internalized by receptor mediated endocytosis and degraded, presumably in the lysosomes. Down regulation (a phenomenom by which peptidic hormones modulate the expression of their own receptors) has been observed both for α and γ IFN tissue culture (48, 39) but also, *in vivo*, in lymphocytes from patients treated with interferon (49, 50). Such studies will be very helpfull to measure an early biological response to IFN treatment and to monitor and adjust precisely IFN treatments in clinical trials. It is already clear that artificial internalization of IFN-β is not sufficient for expressing its biological activity since microinjection of murine or HuIFN-β in murine or human cells failed to induce an antiviral state (46, 47). Furthermore, microinjection of anti-IFN antibodies did not prevent the antiviral action of HuIFN-α (51).

Receptors for γ-IFN

The situation for IFN-γ, on the other hand, is quite different. First, HuIFN-γ displays antiviral activity "from the inside" in murine cells where, as a result of

Table 2. – IFN cell surface receptors

	IFN type I (α or β)	IFN type II (γ)	Ref.
Number of receptors/cell	$2\text{-}5 \times 10^3$	$25\text{-}50 \times 10^3$	(34, 43)
Binding affinity (kD)	10^{-10}M	$3\text{-}6 \times 10^{-10}$M	(34, 49)
Size of cross-linked glycoprotein subunit	110 kDa	95 kDa	(39, 43)
Cromosomal location of corresponding gene			
- in human	Chr 21	Chr 6	(34, 44)
- in mouse	Chr 16	?	(45)
Lingand induced, down regulation of receptors	+	+	(34, 39, 43)
Biological activity "from the inside"			
- after microinjection HuIFN-β	–		(46)
muIFN-β	–		(47)
- by intracellular expression HuIFN-γ		+	(52)

deleting its signal sequence, it cannot be secreted and even, if released in the medium it would not recognize heterospecific murine receptors (52). Furthermore, preliminary experiments indicate that microinjected HuIFN-γ is also antiviral when microinjected in murine and human cells (Huez *et al.*, personal communication). If confirmed these observations would indicate that not only the structural characteristics of type I and II IFN receptors but also the nature of their functional interaction with ligands and subsequent steps may differ radically. This approach may thus help understanding events occuring beyond the binding of IFN to its receptors.

Binding Affinity

Different purified IFN-α subtypes vary considerably in their capacity to induce antiviral, anticellular or enzymatic activities in a given cell line (53, 54, 56). These differences probably reflect the differences in affinity of the different IFN species for the type I IFN receptor (32, 34). However, some of these differences might also reflect an heterogeneity in the type I receptor population itself (53, 55). An heterogeneity of IFN-γ receptors has been described recently (57) and it has been suggested that the IFN-γ molecule can be "dissected", split into several functional epitopes responsible for the induction of immunoregulatory functions (amino terminal epitope) or antiviral activity (carboxyterminal epitope).

Interestingly, differential responses of various cell types to IFN-γ could then be related to the nature of the IFN-γ receptor they contain (58). Hannigan *et al.* (59) showed recently that the expression of the receptor for IFN α differs according to the state of the cell: in actively proliferating cells, the binding of IFN-α is heterogenous, reflecting the existence of receptors of different affinities and perhaps a negative cooperativity in the interaction of IFN and its receptors. In contrast, non proliferating cells exhibit an homogenous, non cooperative IFN-α binding. The same authors (60) compared T98G human glioblastoma cells in exponential growth or, in growth arrested state where the number of IFN-α receptor per cell dropped seven fold. Looking at the rate of transcription of two IFN induced genes, they concluded that this rate of transcription is closely linked to the IFN-α surface receptor occupancy, and could thus be entirely mediated and controlled by a transmembrane signal. The same conclusion has been reached by Friedman *et al.* (61) when studying the rate of transcription of 7 IFN induced genes in nuclei from T98G cells. The nature of these signals is still unknown. Neither cAMP, cGMP (62) nor cytoplasmic alkalinization, Ca^{++} flux or phosphoinositide turnover could be implied in this transfer (63). Internalization of IFN-α does not appear to be required for the transcriptional activation of IFN-induced genes since increasing the amount intracellular IFN-α by inhibiting its intracellular breakdown did not modify the rate of transcription of two IFN-induced genes. On the contrary, it may be seen in these studies that internalization and degradation of surface bound IFN-α closely precedes and perhaps explains, the down regulation of the transcription of one of the IFN-induced genes.

2. INTERFERON INDUCED GENES AND PROTEINS

Besides the two dsRNA dependent enzymes reviewed in the following section, a number of proteins are induced by different kind of interferons in various cell types. The spectrum of proteins and kinetics of their induction by α and β or γ-IFN are quite different (64). This may explain the observed synergism between type I and type II interferons, although a certain number of polypeptides are induced by type II and type I IFN (64). It remains however to be determined whether the observed differences are the mere consequence of triggering different kind of receptor(s), or else if differences in distal pathways are also implied (see section 5).

A. IFN-induced Proteins

In most cells, IFN induces 15-20 different genes. However, for many of these proteins (mainly characterized by two dimensional gel electrophoresis) we totally ignore their enzymatic activity, function and biological role, if any. Some of these induced proteins are known only by their molecular weight. One interesting example is a human 20 kDa membrane protein whose induction correlates with the cessation of cell growth (65). Other examples include murine proteins of 30 and 89 kDa which are induced and secreted and in the case of the 89 kDa, capable to bind strongly to double stranded RNA (66). Other genes are known because the corresponding cDNA has been cloned by differential screening of a cDNA library prepared from mRNA extracted from IFN-treated cells. By this procedure, Friedman *et al.* (61) have obtained a number of cDNA clones corresponding to IFN induced genes of unknown function (see Table 3: gene 1-8, 9-27 gene 10-Q, 6-16, 6-26).

The 56 kDa and Related Proteins

The 56 kDa protein is one of the most abundantly expressed protein induced by α, β, and sometimes γ-IFN in human cells. Its induction at the protein level has

Table 3. - Interferon induced proteins in human cells

	Protein size (kDa)	dsRNA binding	Enzymatic activity	Antiviral activity	Other role and/or location	
A. 1 *no information of the gene available*						
p 68 Kinase	68	(+?)	kinase	?		
p 48 kinase	48	+	kinase	?		
RNAse L	80	-	2-5A depdt. - Nase	+ or -		
"Mx" human protein	78	-				
Xanthine oxydase						
Guanylate cyclase			GTP→cGMP			
Indoleamine 2-3 deoxygenase						
Nucleoside diphosphate kinase	20					
A.2 *cDNA available*						
2-5A synthetase	42→100	+	2-5A synth.	+ or -	see table 4	
56kDa prot.	56	-	?	?	cytoplasmic	
pIF - 1	42, 58		?	?		
1-8 (multigene family) 9-27 related?						
6-16 (100 related?)	16				secreted	
8-27						
11-25						
MT-IIA (Metallothionein)	7				secreted	
Thymosine β_4 (6-26)	16(→5.25K)					
pIF-γ31	12.4(→10K)				secreted	
GBP	67		Guanylate binding			
15 kDa prot.	15					
20 K	20				associated with growth inhib.	
HLA-A, B, C	44				cell surface	
β2-microglobulin	14				cell surface	
HLA-D R α	34				cell surface	
HLA-D R β	29				cell surface	

Table 3 bis - Interferon induced proteins in murine cells

	Protein size	Activity	Cell location	mRNA size	Gene location	Reference
A.1 *no information on the gene*						
P67		eIF-2 kinase				108
RNase-L		2-5A dep. Nase				135
GBP 5 65		G-binding			Chr 3	225
89K		dsRNA+ + +	secreted			66
30K			secreted			66
A.2 *cloned cDNA's*						
Mx	p75	inh. Flu repl	nuclear	3.5. kb	Chr 16	98, 99
c202	p56K			2kb	Chr 2	77
2-5A synthetase	?	2-A synt	?	1.6 and 4-5kb	?	160

	IFN α, β, γ	Other inducer	RNA size (kb)	cDNA (kb)	Transcription activation	Friedman-Stark module(s)	Chromosomal location	References
	α, β, γ							108
	α, β, γ							108
	α, β							100
	α, β							82, 83
								105
								62
								104
								106
	α, β>γ	steroid PDGF	1.6→3.6	1.4 1.6	 +	−	12	218 and see table 4
	α, β>γ		~2.0	1.7	+	+		71, 74, 75, 214
			2.9		+			74
			0.8			+		61
			1.0					61, 201, 219
	α, γ		1.1					61
	α, γ		0.55-0.65					61
	α, β, γ		0.5					61
								61
	γ>>>α, β		1.5					90
	γ>α, β		3.0					68
			0.7					81
								223,65
	γ>α, β			1.8		+	6	Rev. in 33, 211, 212
	γ>α, β			0.9		+	15	
	γ>>>α, β			1.3		+	6	
	γ>>>α, β							

been described in human fibroblasts by Gupta *et al.* (67), Cheng *et al.* (68) and by Dron *et al.* (69) in lymphoblastoid Daudi cells and, at the mRNA level, by Colonno and Pang (70).

A cDNA corresponding to the 56 kDa protein has been cloned by Chebath *et al.* (71) and we have isolated a full-size cDNA for this 56K protein allowing to express the protein and to determine its complete amino acid sequence (103). Chebath *et al.* (71) and Faltynek *et al.* (72) have studied the expression of the 56K human gene under various conditions of induction and Kusari and Sen (73) have proposed that several pathways are implicated in the induction and regulation of this gene. Larner *et al.* (74, 75) have also studied the transcriptional regulation of this gene (see section 5). A perhaps equivalent murine cDNA called gene e202 has been isolated by Samanta *et al.* (76) from a murine cDNA library. The corresponding gene is located on chromosome 2 (77). There is some clear evidence that the promotor region of this gene can be transcriptionally activated after IFN treatment (78). However, since the cDNA sequence is not yet available, it is not sure that cDNA e202 is the murine equivalent of the 56 kDa protein gene. Nothing is known of the biological activity of the murine or human 56 kDa protein, except for some preliminary assays with antisense RNAs indicating a role of this protein in the antiviral state (79). It is hoped that large scale expression of the protein will permit an evaluation of its cellular localisation and role, if any, in expressing interferon activities.

A similar situation exists for a less studied 15 kDa IFN induced cytoplasmic protein. The protein has been extensively purified (80) and recently a corresponding full size cDNA has been cloned and sequenced (81). The corresponding mRNA of 0.7 kb is induced 20 to 100 fold in IFN-β treated Daudi cells.

Guanylate-binding Proteins

Guanylate binding proteins (GBP) and their mRNAs are also strongly induced after interferon treatment. In human cells the major IFN induced GBP has a molecular weight of 67,000 whereas its murine equivalent has a molecular weight of 65,000. The 67 kDa protein and its mRNA are somewhat better induced by IFN-γ than by β or α-IFN (82). The protein is cytoplasmic, strongly induced after 2-hour treatment with γ-IFN, and can be detected readily in IFN-treated human lymphocytes (83). Its contribution to the expression of IFN activities is totally unknown.

Major Histocompatibility Antigens

IFNs are strong inducers of class I and II major histocompatibility antigen (MHC). This is observed at the protein level (by using appropriate antibodies) or some times more strongly at the RNA level with use of cDNA probes (84, 33). γ-IFN is considerably more efficient in this respect than α or β-IFN and it is remarkable that it enhances MHC antigens at doses that are not sufficient to inhibit viral replication or induce the 2-5A synthetase activity (85).

Rosa *et al.* (86) have suggested that this property of IFN-γ might allow viral antigens to be fully expressed at the cell surface together with HLA antigens, a situation that might considerably favour an antiviral cellular immune response. Some aspects of HLA gene regulation by IFN are also discussed in section 5. It is clear from the vast amount of work accumulated in this field that IFNs modulate and amplify preexisting levels of the MHC class I and class II gene products but in some cases are also real *de novo* inducers of the MHC gene in some cells (e.g. skin fibroblasts, endothelial cells) that are not spontaneously expressing the corresponding antigens (87). It is not yet clear however whether IFNs are the "natural" and only inducers of MHC antigens (84) but it has been suggested that they could be part of the developmental factors responsible for their expression during differentiation (88).

Metallothionein, Thymosin B_4, and β-Thromboglobulin

After cDNA cloning of IFN induced genes some of them were identified on the base of sequence homology to previously known genes. These include: metallothionein II and thymosin β_4 (61, 89) and a protein of 12 kDa specifically induced by IFN-γ (but not by α/β IFN) showing significant amino acid sequence homology to platelet factor 4 and β-thromboglobulin (90). The functional relevance of expressing these genes after interferon treatment to the interferon action is presently unknown. It is possible that, as it is the case for metallothionein, IFN may not be their unique or preferential inducer.

The Mx Gene

The only IFN-induced protein whose genetics (91, for review) structure and function is known in some details is a karyophilic (92), 75 kDa protein (93), the presumed product of the Mx gene, a mouse gene which was shown to confer a specific cellular resistance to the growth of influenza virus when induced by interferon α/β (but not by IFN-γ) (94) preventing specifically transcription (95, 96) and/or translation (97) of influenza mRNAs. Furthermore, the cDNA corresponding to the protein has been cloned, sequenced and expressed constitutively in Mx^- cells and shown to confer them a specific resistance to influenza virus, even in the absence of IFN-treatment (98). The Mx^- phenotype results from a deletion in the Mx gene. The Mx gene has been localized on mouse chromosome 16 (99). Analogs of the Mx protein exist in human (100) and bovine cells where the protein is cytoplasmic (Horisberger, personal communication) and in rats where the protein is nuclear and cytoplasmic (101) and confers also a specific antiviral state against influenza virus (102). It will be of extreme interest to see whether the Mx gene carries other regulatory functions, and to search for other examples of "partial" virus-specific, IFN-mediated antiviral mechanisms.

IFN-induced Enzymatic Activities

A number of enzymatic activities are enhanced in IFN-treated cells. These include a very early guanylate cyclase (clearly not required for the biological activity of IFN (62), an indoleamine 2,3-dioxygenase (104), and a xanthine oxidase (105) and an early induced, 20 kDa nucleoside-diphosphate kinase (106). Their biological significance is presently unknown and requires further evaluation.

3. dsRNA DEPENDENT IFN-INDUCED ENZYMATIC ACTIVITIES

The Interferon-induced Protein Kinase Pathway

The essential features of the system are described in Fig. 1. However, with the recent use of monoclonal antibodies (107) it has been possible to characterize a 110 kDa kinase complex containing a p68 and a p48 subunit, both induced by IFN. In the presence of dsRNA, p48 phosphorylates p68, the latter becoming capable of phosphorylating exogenous substrates such as histones or the α-subunit of the eukaryotic initiation factor 2 (eIF-2), even in the absence of dsRNA (108). Evidence has been obtained that the kinase pathway is functional in interferon-treated intact reovirus infected cells (109,110), and that the α subunit of eIF2 is phosphorylated in EMC virus infected IFN-treated cells at 9-12 hrs post infection

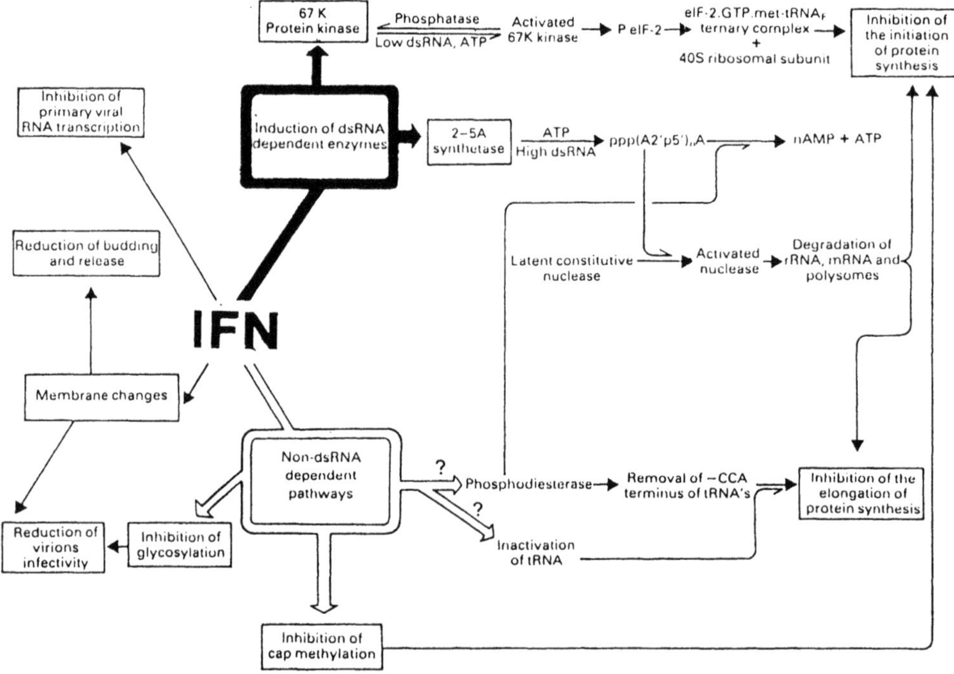

Figure 1. The various pathways proposed as possible, non-mutually exclusive mechanisms of antiviral action of interferon. On the left side of the diagram the mechanisms not entirely understood have been represented very schematicly. The emphasis is on the existence of various mechanisms showing the control or lack of control of translation and in the first case the presence or absence of dsRNA. Question marks indicate controversial or not yet demonstrated pathways.

(111). However, there is no direct proof that the kinase pathway can confer by itself (in the absence of IFN treatment) an antiviral effect nor that this effect would be related to the phosphorylation of e IF-2 *in vivo*.

It is interesting that the same eIF2 kinase, also called double-stranded RNA activated inhibitor (DAI) is activated during adenovirus infection (112) probably as a result of the synthesis of viral dsRNA. This would result in an arrest of polypeptide chain initiation but the kinase activation by dsRNA is normally prevented by VA RNA I, a small RNA of 160 nucleotides with partial double-stranded structure that tends to accumulate to high intracellular concentrations ($\approx$100 µg/ml) late in infection (see chapter 20 of this volume). VA RNA is supposed to interfere with the activation of the kinase (113-115) Adenovirus VA RNA I exerts thus through this new pathway a direct control on the rate of protein synthesis. Vaccinia virus infection elicites the production of a factor inhibiting the interferon induced dsRNA dependent protein kinase in a similar way (116, 117, 118). Influenza virus, by another unknown mechanism, can also counteract the DAI related inhibition of protein synthesis (119). The Herpes-like Epstein-Barr virus encloses two small RNAs with a structure similar to VA RNAs which could play a similar role (120). *In vitro* Mengovirus RNA can inhibit the activation of the DAI by dsRNA (121). It is not yet clear if these newly discovered mechanisms represent compensating defense pathways developed by viruses to escape the IFN-mediate inhibition (113, 122) or, if animal cells and in particular virus-infected cells, utilize currently some of these mechanisms, including those implying small RNA's for translational control (112, 113).

B. The 2-5A Synthetase Pathway

Enzymatic Components of the 2-5A System

This system (see Fig. 1) has been extensively studied in the past few years (24, 28, 30, 124,125). 2-5A synthetases now considered as a family or group of enzymes are strongly (10-100 fold) induced by IFN. In the presence of dsRNA, 2-5A synthetases convert ATP to a series of unusual oligonucleotides with the general structure ppp5'A(2'p5'A)$_{n1}$ (n > 2) referred to as 2-5A (126-129).

These oligonucleotides specifically activate a latent endoribonuclease, RNase L, that is responsible for the cleavage of rRNA and mRNA (130) in intact cell systems (131). In its active form, the nuclease is a dimer of 80 kDa subunits, and this activation as well as 2-5A binding are reversible (26). Other 2-5A binding proteins of unknown activity, have been found in murine cells (132). Although initially considered as constitutive and invariable, intracellular levels of the 2-5A dependent endonuclease may increase in growth arrested cells (133, 134), IFN treated cells or differentiating cells (135). HLA and 2-5A synthetase (88, 136) as well as RNase L modulation, represent three different indications that endogenous, autocrine, IFNs may play a role at some stages of differentiation.

The production of 2-5A has been demonstrated in several IFN-treated virus-infected cell systems: in IFN-treated encephalomyocarditis (EMC) virus-infected cells, both the triphosphate (2-5A) and its core were detected by HPLC, radioimmuno- and radiobinding assays, and the appearance of these products correlated with the establishment of the antiviral state (131, 137). That RNase L is activated under these conditions was attested by the degradation of rRNA and EMC virus RNA (138, 139). Furthermore the synthesis of 2-5A occurs in IFN-protected HeLa cells infected with reovirus and is accompanied by reovirus mRNA degradation (140). A more complex situation may occur in IFN-treated vaccinia virus-infected cells (where controversial results have been reported with regard to the role of the 2-5A system (141-144), and in Semliki Forest virus-infected cells (28).

Genetics of the System

Despite numerous attempts (27), very few cell variants have been obtained which are deficient in one of the enzymes implicated in the 2-5A pathway. One example is a NIH 3T3 receptor positive murine fibroblast subclone which is deficient in the induction of 2-5A synthetase and resistant towards the antiviral effects of IFN against mengovirus and vesicular stomatitis virus (145). The failure to detect 2-5A dependent endonuclease (RNase L) activity in some cell clones (146, 147) indicate that further assessment is required before these cell clones could be accepted as true RNase L-deficient mutants.

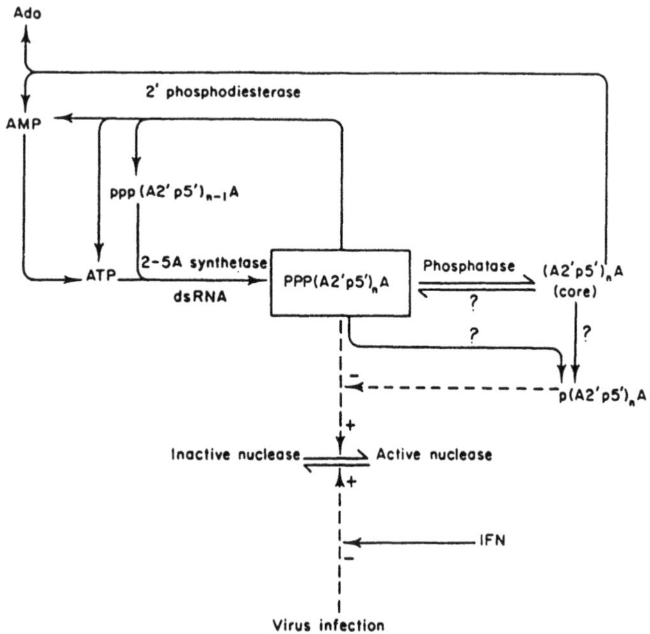

Figure 2. The metabolism of 2-5A as deduced from cell-free experiments; question marks indicate hypothetical pathways. Ado means adenosine.

Related Enzymic Activities

At least two additional enzymes are involved in the 2-5A pathway (Fig. 2)

i) a *phosphodiesterase*, which specifically cleaves (2'-5') phosphodiester linkages starting at the 3'-terminal; the induction of this enzyme IFN was demonstrated only in murine L cells (reviewed in 148) and its role in intact cells is not yet understood (149); and

ii) one or several *phosphatases* which are able to dephosphorylate 2-5A to its core; the role of this enzyme(s) in intact cells is not clear either (27). The presence of the phosphodiesterase and phosphatases may account for the rapid degradation of 2-5A in intact cells and cell extracts. Therefore, several analogs of 2-5A have been synthesized either chemically or enzymatically with the aim to increase their metabolic stability towards degradative enzymes (see 150-154).

The Multiple Forms of 2-5A Synthetases

Both murine (132), porcine (155), avian (156,157) and human cells (158) contain several molecular species of 2-5A synthetase. Table 4 summarizes some characteristics of the murine and human 2-5A synthetases. Some of the structural differences of human 2-5A synthetases result from a cell or tissue specific differential splicing of the transcripts of a common gene (159, 160, 161, 162). Neither the regulation of this process nor the cellular localisation and specific function of the various forms of 2-5A synthetases are yet fully understood. Full size cDNA clones corresponding to the small 42 kDa 2-5A synthetase have been sequenced and expressed *in vitro* (163) or in Cos-7 cells (164).

Biological Role of the 2-5A Pathway in the Antiviral Activities of IFN

Direct proof that 2-5A may suffice to exert an antiviral effect has been obtained recently by microinjecting 2-5A (47) or stabilized 2-5A analogs in intact cells (154, 167), therefore confirming the results of earlier experiments in which 2-5A oligonucleotides were introduced into permeabilized cells (165, 166, 168) However, in some virus-cell systems, the contribution of the 2-5A-mediated activation of RNase L to the antiviral activity of IFN could not be ascertained (55, 147, 169) and numerous reports have been published describing situations where virus-infected IFN-treated cells failed to show any evidence of 2-5A production and/or rRNA or viral RNA degradation. This is the case for vesicular stomatitis virus-infected cells (170, 171), poliovirus-infected cells (172, 173), reovirus-infected cells (174) and SV40 virus infected cells (169). To clarify the issue, a 2-5A analog that inhibits specifically the activation of RNase L by 2-5A was introduced in IFN-treated, EMC infected cells (175). This resulted in a partial inhibition of the IFN effect on virus development indicating that in this case, although the 2-5A system cannot inhibit picorna virus by itself (167, 176), it may contribute, to some extent, depending probably on the virus/host system cell considered, to an antiviral effect. Also Goren *et al.* (54) found a similar 2-5A synthetase-inducing activity for three dif-

Table 4. – Murine and human 2-5 A synthetases

			MURINE 2-5A SYNTHETASES				
			mRNA's	Proteins	Localisations	IFN induction	S/G2↑
			< 5 kb ↓				
			4 kb	100*kDa (132)	cytoplasmic	+ + +	+ + (216)
			1.7 kb	30-40 kDa (132)	nuclear	+ + +	
			HUMAN 2-5A SYNTHETASES				
HeLa	WISH	DAUDI	mRNA's (161, 162)	Proteins (224)	Localisations	IFN induction	
(+)			(4 kb)	Δ 100*kDa	cytoplasmic (RWF) low activation dsRNA (158)	+	
+	+		3.6 kb	Δ 69 kDa		+	
+			(2.7 kb)	Δ 67 kDa		+	
+		+	1.8 kb**(46K)† →	46 kDa		+	
+	+		1.6 kb (42K)† →	42 kDa		+	
				~ 30 kDa	cytoplasmic (s100) high activation dsRNA (158)	+	

† coding capacity deduced from the complete sequence of cDNA.
Δ no mRNA assigned so far.
** human exon 8 (specific for 1.8 kb RNA) hybridizes only to murine >5 and 4 kb mRNA's.
* see 161, 162 and 163 for review.

ferent IFN- subtypes that differ at least 300 times in their antiviral (or anticellular) potency. It is remarkable that several classes of animal viruses have developped various means of counteracting the 2-5A system. In EMC-infected cells the endoribonuclease L cannot be activated by 2-5A (Fig. 2). The mechanisms of this inhibition and of the restoration of the endonuclease reactivity by IFN-treatment are not known (176). It fits well with the observation that a picorna virus like mengovirus is very insensitive to inhibition by microinjected 2-5A (167). Infection of IFN-treated cells with SV40 (169), Herpes Simplex Virus (177), vaccinia virus (144, 178) or the *in vitro* incubation of reovirion cores in a cell-free extract from IFN-treated cells (179) generate atypical 2-5A nucleotides perhaps degradation products that, in all cases, inhibit the activation of the 2-5A dependent endoribonuclease by authentic 2-5A. The structure of these nucleotides has not been completely characterized. In the case of vaccinia virus infection, two additional virus mediated inhibitors have been described, that also inhibit the 2-5A synthesis in the early phase of infection through degradation of ATP (by an ATPase) or 2-5A (by a phosphatase) (180).

The real significance of these inhibitory mechanisms is still unknown, though by analogy with the inhibitors of the eIF-2 kinase, it is tempting to speculate that they constitute defense mechanism(s) evolved by viruses relatively insensitive (e.g: vaccinia) to escape the action of IFN (180). Alternatively, they can be viewed as important "negative loops" controlling the inhibitory effect of 2-5A dependent mechanisms, with the specific purpose of maintaining them localized in time and space in order to insure their discriminative action as suggested by Baglioni and coworkers.

4. DO WE UNDERSTAND THE BIOCHEMICAL MECHANISMS OF ANTIVIRAL ACTIVITY OF IFN ?

A number of operative pathway have been described in IFN-treated cells. What is their relevance (if any) to the establishment of a specific state? There is no simple answer for the moment. It is clear that IFN can inhibit viral replication at several levels: penetration, primary transcription, translation (here several possibilities are available, see Fig. 1 and Table I in ref. 4), or assembly, maturation and release of virus particles (reviewed in 26, 27, 29, 31, 148, 181-185). We should bear in mind, however, that these observations are relevant to studies performed in tissue culture where IFN acts as a multi-target inhibitor. An additional antiviral activity that broadens considerably the effect of IFN is provided by the very potent immunomodulating effect of IFN, that can provide excellent protection *in vivo* against viruses such as vaccinia which are very resistant to its action in tissue culture conditions (1, 2, 186).

The assortment of specific pathways that are implicated in these antiviral actions depends on the challenge virus and the conditions of infection, the type of the host cells, and the type or subtypes of IFN used (32). Thus, depending on the experimental circumstances, one single virus type may be inhibited at various levels by IFN. This is the case for instance for SV40, inhibited by IFN at the level of translation (187) or primary transcription (188), for HSV inhibited at the level of translation (189) or virus maturation (190) for VSV inhibited at the level of protein glycosylation (191), maturation (192), inhibition of viral mRNA translation perhaps related to a defect in the methylation of their cap structure (170) or as a result of the protein kinase activation (193), or following inhibition of primary transcription of the viral genome (reviewed in 31). Reovirus was considered by Baglioni *et al.* (194) to be inhibited by a discriminative activation of the 2-5A system, whereas no activation of the 2-5A system was observed by Munoz & Carrasco in IFN-treated, reovirus infected cells (195). More detailed and systematic reviews on different viruses will be found elsewhere (29,31). These pathways, although not exclusive and variable from one experimental system to the other, may also co-exist and complement each other. The only known partial, virus-specific, IFN-induced, antiviral state is the one corresponding to the expression of the Mx gene.

Interestingly, it is also the only case where the expression of a single protein suffices to mediate this specific antiviral state (this, of course, does not mean that influenza viruses cannot be inhibited through several other additional IFN-mediated pathways). It is, *however,* possible that in the future some assortments of mechanisms or pathways will be assimilated to other virus-type specific, partial antiviral states.

The picture emerging from this review should make it obvious that at the present time no unqualified, unambiguous answer can be given to the original question.

Genetical and biochemical tools (27) however exist now to approach these questions *in specific situations.*

Vaccinia virus

Numerous studies on the vaccinia virus response to IFN have been published (for reviews see 118, 141, 143, 196, 197), the general outcome appears to be that while the sensitivity to IFN varies considerably among different vaccinia virus strains (196), it also depends on the culture conditions (197). In some cases *ts* mutations of vaccinia virus seem to affect viral functions required for the sensitivity or resistance to IFN antiviral effect. This approach is currently being developed (198).

Stable 2-5A oligonucleotide analogs

In the case of the well known 2-5A system, 2-5A oligonucleotide analogs inhibiting the activation of nuclease L have been described and used to evaluate the contribution of the 2-5A system in the antiviral effect of IFN towards EMC (175). Chemical modifications that enhance their stability (154) and penetration in intact cells (199) may become very usefull for these approaches.

DNA-dependent Kinase

It is possible, although more complex, to inhibit the kinase pathway by superinfecting cells with several viruses inhibiting the activation of *the DNA dependent* kinase vaccinia (118, 193); influenza (119, 200). But the most selective approach in this direction may be the infection of cells with adenovirus 5 (or 2) in parallel with an infection with its corresponding dl 331 mutant. In cells infected with wild type adenovirus, VAI RNA (known to be responsible for preventing the activation of the IFN-induced kinase by dsRNA) is made, whereas in dl 331 infected cells VAI RNA is not expressed, the kinase is activated and the α-subunit of eIF-2 is phosphorylated (for review: 112, 122; and section 3 above).

Antibodies (68, 107, 162), cDNA probes (Table 3), antisense DNAs or RNAs (79), full-size cDNA or cosmids allowing expression in heterologous cells (98, 164, 201) of the genes corresponding to various IFN-induced proteins are — or will become — increasingly available in the near future, allowing a systematic evaluation of their role.

Concluding a recent paper, Rice *et al.* (111) summarized the present views of many groups working in this field:

"Attempts by a number of groups to correlate sensitivity to interferon for different viruses with the presence or absence of the kinase or one or other of the enzymes of the 2-5A system, although highly suggestive in some cases, have been generally inconclusive. In no case has it been possible to establish rigorously the absence of enzyme activity in intact cells. (...) Resolution of the question of the importance of the kinase and 2-5A systems in the antiviral action of IFN against different viruses will, therefore, have to await future analysis with either rigorously characterized mutants in their function or specific inhibitors of their activation".

5. REGULATION OF THE EXPRESSION OF THE IFN-INDUCED GENES

Over the past few years considerable progress has been made in this area since several cDNA clones have been available and used to probe the expression of various genes induced by IFN (Table 3). Various functional classifications have been proposed for these genes on the basis of:

a) *their kinetics of mRNA accumulation* showing either a transient peak (e.g: gene 11-25, 2-5A synthetase, metallothionein) or being maintained for several days, in the presence of IFN (e.g: gene 6-16). The amplitude of induction varies between 2-3 fold to several hundred folds (202).

b) *the effect of cycloheximide on the induction pattern* allows to distinguish those genes whose induction is a primary response, independent of protein synthesis (1-8, 9-27, 6-16, 8-27, 11-25, 2-5A synthetase, MT II, HLA class I) (61, 72, 202) and those genes requiring protein synthesis for their induction (e.g. the 56K gene in HeLa cells) (73).

c) *the nature of the inducer required*: i) in some instance IFN is not the only in ducer as it is the case for metallothionein (also induced by Zn^{++} or steroid hormones) or 2-5A synthetase (also induced by steroid hormones 203-205) or PDGF (206). ii) other genes are specifically induced by IFN. Some of them are indifferently stimulated by γ, α, or β IFN (e.g: the 1-8 gene family or MT II) although some times with a different kinetics (202). Other genes are *better* induced by type I IFN

(2-5A synthetase) or *exclusively* induced by type I IFN: the Mx gene (94), the murine C202 gene (207), the human 6-16 gene (201, 202), pIF-1 (74), 15kDa protein gene (81). β-IFN, on the other hand appears *better* to induced the 67 kDa GBP (68, 82), HLA class I (208), and *very predominantly* to induce platelet factor 4 (90), HLA class II (209).

d) at least two groups of genes may be *activated through different pathways* in human cells as indicated by their independent loss in interferon-resistant Daudi cell (69, 89).

Little is known about the mechanisms regulating the expression of these various genes or their induction. For all the genes listed in Table 3, induction has been shown by *in vitro* transcription assays, to correspond to transcriptional activation (74, 90, 210, 207) occuring sometimes within 5-10 minutes after IFN-treatment and reaching a maximal rate corresponding to a 10-50 fold increase within 0.5-3 hr after initiating IFN treatment (74, 61, 60, 211). Transcriptional activation is strictly dependent on the transfer of an unknown signal resulting from the binding of type I IFN to extracellular specific cell surface receptors (60, 61).

Some of the IFN-induced genes like gene 1-8 and the 56K protein gene are also controlled by post-transcriptional mechanisms and their accumulation is increased by inhibiting protein synthesis at appropriate time (61, 73).

Larner *et al.* (75) have also shown that the transcription of this same 56K protein gene is negatively controlled by an additional protein responsible for the desentization of the transcription machinery to IFN.

The Friedman-Stark Consensus Sequence

Gene promotor sequence of the cognate genes are certainly an important element in the complex regulatory circuits observed. This has only been demonstrated directly by functional studies in the case of the HLA class I gene promotor where a region of 670 bp upstream the mRNA cap site suffices to confer IFN inducibility in heterologous cells (212) and for the murine gene C202 (78). Promotor gene sequences, however, are available at this time for the human 2-5A gene, 1-8 gene, gene C202 and for several HLA class I gene (213) as well as for the metallothionein gene (214) and for the human 56K protein gene (Wathelet *et al,* in preparation). This has led to sequence comparisons and to the discovery of a conserved motive named the Friedman-Stark consensus sequence TTCN$^{G}_{C}$NACCT CNGCAGTTTCTC$^{G}_{T}$TCT-CT, present in several of the known gene promotors: gene 1-8, gene C202 in the coding region, HLA genes, MT II (211), and in the human 56K gene promotor (Wathelet *et al.*, in preparation). No functional studies have been performed that could clarify the role of this consensus sequence. It was recently shown however, to act necessarily in concert with an enhancer sequence (217). It should be noted that its presence is not universal since, e.g., the human 2-5A synthetase gene does not contain this module. It is thus possible that the Friedman-Stark sequence allows the modulation or amplification of the expression of IFN induced genes but may not be the regulatory element requested for IFN inducibility. Several experimental evidence indicate already that the sequence(s) required for IFN inducibility may be strongly conserved in different animal species, since human genes are induced by IFN and expressed when introduced in murine cells with their promotor (212, 201). This justifies further sequence studies and the comparison of various IFN induced gene promotors.

Further studies probably including the fusion of Friedman-Stark sequence to other promotors or systematic deletions of one or several gene promotors and the use of techniques of genomic sequencing (215) may help to reach a refined understanding of the structural features of IFN-regulated gene promotors such as the location of other regulatory modules responsible for the inducibility and binding site(s) for regulatory proteins.

A good model for such studies is provided by the elegant work of Maniatis and his coworkers on the regulation of the HuIFN-β gene (12, 15).

6. IFN, 2-5A AND THE CONTROL OF CELL GROWTH

Recent observations indicate that the expression of 2-5A synthetase may vary during cell growth (222) or cell cycle in synchronized cells (216). Zullo *et al.* (206) have proposed that 2-5A synthetase(s) and endogenous β-related autocrine IFN (220), which are both induced by PDGF and expressed in differentiating cells (136), could be considered as negative growth factors. Other explanations of the cytostatic effect of IFN have been envisaged (3, 4), but their discussion would fall out of the scope of this review.

In a similar vein, the relationship between IFN and expression of oncogenes have been reviewed recently (221).

It appears that IFNs, 2-5A synthetases (and presumably other IFN modulated genes as well) may have a much wider physiological role than initially anticipated on the basis of their antiviral potential.

Acknowledgements

I am very gratefull to G. Huez, P. Defilippi and M. Wathelet for critical comments and discussion. I thank doctors G. Jonak, L. Mallucci, M. Knight, Y. Cheng, M. Esteban, L. Carrasco, B. Lebleu, G. Sen, P. Staeheli, O. Haller, A. Kimchi, T. Maniatis, R. Silverman and M.I. Johnston for kindly sending preprint of their work. I thank J. Herinckx for secretarial help and typing the manuscript and C. Content for help in preparing the bibliography.

The author's work described in this review was supported by grant 2.9006.79 from Fund for Joint Basic Research (Belgium).

7. REFERENCES

1) Kirchner H. (1986) In: *Antiviral Research*, 6, Elsevier Science ed. (Biomedical division), 1-17.

2) De Maeyer-Guignard, J. and De Maeyer, E. (1985) in *Interferon 6*, I. Gresser, ed. Academic Press (London), 69-91.

3) Sreevalsan, T. (1984) in *Interferon 3*. Mechanism of production and action. R.M. Friedman, ed. Elsevier (Amsterda,, N.Y., Oxford), 343-387.

4) Clemens, M.J. and Mc Nurlan M. (1985). Biochem.J., **226**, 345-360.

5) Stewart, W. II (1979) *The Interferon System*. Springer Verlag, Berlin and New York.

6) Dijkmans R and Billiau A. (1985). in *Interferons-Their impact on Biology and Medicine*. J. Taylor-Papadimitriou, ed. Oxford University, Press, Oxford, New York, Toronto, 1-18.

7) De Maeyer E and De Maeyer-Guignard J. (1979). in *Interferon 1* I. Gresser, ed. Academic Press New York, 75-100.

8) Revel M (1983) in *Interferon 5*, I. Gresser, ed. Academic Press, London, 205-239.

9) Weissmann C. and Weber, H. (1986). Progress in Nucl. Acid Res. and Mol. Biol. (in press).

10) Enoch, T., Zinn K, and Maniatis T. (1986). Mol. Cell. Biol., **6**, 801-810.

11) Zinn, K., Di Maio D. and Maniatis, T. (1983). Cell, **34**, 865-879.

12) Goodbourn, S., Burstein H. and Maniatis T. (1986). Cell, **45**, 601-610.

13) Fujita, T., Ohno S., Yasumitsou, H. and Taniguchi, T. (1985) Cell, **41**, 489-496.

14) Ryals, J., Dierks P, Ragg, H. and Weissmann, C. (1985) Cell, **41**, 497-507.

15) Zinn, K. and Maniatis, T. (1986). Cell, **45**, 611-618.

16) Miller DL, Hsiang-Fu Kung, Pestka S. (1982). Science, **215**, 689-690.

17) Sternberg, MJE and Cohen, FE (1982). Int. J. Biol. Macromol., **4**, 137-144.

18) Zav'yalov V.P. and Denesyuk, A.I. (1982). Immunol. Lett., **4**, 7-14.

19) Hayes, T.G. (1980) Biochem Biophys.Res. Comm., **2**, 872-879.

20) Content J. and Derynck, R. (1983). La Recherche, **141**, 323-325.

21) Kingsman, S.M. and Kingsman A.J. (1983) in:*Interferon: from Molecular Biology to Clinical Applications*. D.C. Burke and A.G. Morris, eds. Cambridge University Press (Cambridge), 211-254.

22) Marcus, P.I. and Salb, J.M. (1966). Virology **30**, 502-516.

23) Joklik, W.K. and Merigan, T.C. (1966) Proc.Natl. Acad. Sci. USA, 56, 558-565.

24) Lengyel, P. (1981). Methods Enzymol., **79**, 135-148.

25) Lengyel, P. (1982). in *Interferon 3*. I. Gresser ed., Academic Press, London, 78-79.

26) Lengyel, P. (1982) Ann.Rev.Biochem., **51**, 251-282.

27) Lebleu, B. and Content J. (1982) in *Interferon 4*, I. Gresser, ed. Academic Press, London. 47-94.

28) McMahon, M. and Kerr, I.M. (1983) in *Interferons: from Molecular to Biology to Clinical Applications*, D.C. Burke and A.G.Morris, eds., Cambridge University Press (Cambridge), 89-108.

29) Sen, G.C. (1984). Pharmacol.Ther., **24**, 235-257.

30) Johnston, M.I. and Torrence, P.F. (1984) in *Interferon 3* - Mechanisms of production and action. R. Friedman ed., Elsevier North Holland, Amsterdam, 189-298.

31) Galabru, J. and Hovanessian, A. (1984). Bull. Inst. Pasteur, 82, 283-334.

32) Williams, B.R.G. and Fish, E.N. (1985) in *Interferon, their impact in Biology and Medicine*, J. Taylor-Papadimitriou, ed. Oxford University Press, Oxford, New York, Toronto, 40-60.

33) Revel, M. and Chebath, J. (1986). TIBS, **11**, 166-170.

34) Aguet, M. and Mogensen, K.E. (1983) in *Interferon 5*, I. Gresser, ed., Academic Press, London, 1-22.

35) Epstein, L.B. (1981). Nature (London), **295**, 453-454.

36) Derynck, R. (1983) in *Interferon 5*, I. Gresser, ed., Academic Press, London, 181-203.

37) Branca, A.A. and Baglioni, C. (1981). Nature (London), **294**, 768-770.

38) Joshi, A.R., Sarkar, F.H. and Gupta, S.L. (1982). J. Biol. Chem., **257**, 13884-13887.

39) Littman, S.J., Faltynek, C.R. and Baglioni, C. (1985) J.Biol.Chem. **260**, 1191-1195.

40) Ankel, H., Krishnamurti, C., Besancon F, Stefanos S. and Falcoff, E. (1980). Proc.Natl.Acad. Sci. USA, **77**, 2928.

41) Raziuddin, A., Sarkar, F.H., Dutkowski, R., Shulman, L, Ruddle, F.H. and Gupta, S.L. (1984) Proc.Natl.Acad.Sci. USA, **81**, 5504-5508.

42) Shulman, L.M., Kamarek, M.E., Slate, D.L., Ruddle, F.K., Branka, A.W., Baglioni, C., Maxwell, B.L., Gutterman, J., Anderson, P., Nagler, C. and Vilcek, J. (1984). Virology, **137**, 422-427.

43) Sarkar, F.H. and Gupta, S.L. (1984). Proc. Natl. Acad. Sci. USA, **81**, 5160-5164.

44) Rashidbaigi, A., Langer, J.A., Jung, V., Jones, C., Morse, H.G., Tischfield, J.A., Trill, J.J., Hsiang-Fu King and Pestka, S, (1986). Proc. Natl. Acad. Sci. USA, **83**, 384-388.

45) Cox, D.R., Epstein, L.B. and Epstein, C.J. (1980). Proc. Natl. Acad. Sci. USA, **77**, 2168-2172.

46) Huez, G., Silhol, M. and Lebleu, B. (1983). Biochem. Biophys. Res. Comm., **110**, 155-160.

47) Higashi, Y. and Sokawa, T. (1982). J. Biochem., **91**, 2021-2028.

48) Branca, A.A. and Baglioni, C. (1982). J. Biol. Chem., **257**, 13197-13200.

49) Lau, A.S., Hannigan, G.E., Freedman, M.H. and Williams BRG (1986). J. Clin. Invest., **77**, 1632-1638.

50) Billard, C., Sigaux, F., Castaigne, S., Valensi F., Flandrin, G., Degos, L., Falcoff, E. and Aguet, M. (1986). Blood, **67**, 821-826.

51) Arnheiter H. and Zoon, K.C. (1984). J. Virol., **52**, 284-287.

52) Sanceau, J., Lewis, J.A., Sondermeyer, P., Beranger, F., Falcoff, R. and Vaquero, C. (1986). Biochem. Biophys. Res. Comm., **135**, 894-901.

53) Rehberg, E., Kelder, B., Hoal E.G. and Petska, L. (1982). J. Biol. Chem., **257**, 11497-11502.

54) Goren, T., Kapitkovsky, A., Kimchi, A. and Rubinstein, M. (1983). Virology, **130**, 273-280.

55) Vandernbussche, P., Divizia, M., Verhaegen-Lewalle, M., Fuse, A., Kuwata, T., De Clerq, E. and Content, J. (1981) Virology, **111**, 11-22.

56) Ortaldo, J.R., Mantovani, A., Hobbs, D., Rubinstein, M., Petska, S. and Herberman, R.B. (1983). Int. J. Cancer, **31**, 285-289.

57) Yonehara, S., Yonehara, M., Fukunaga, R. and Nagata, S. in *Abstracts from the 1985 TNO-ISIR meeting on the Interferon System*, Clearwater Beach USA, 13-18 october 1985.

58) Orchansky, P., Rubinstein, M. and Fischer, D.G. (1986). J. Immunol., **136**, 169-173.

59) Hannigan, G.E., Lau, A.S. and Williams B.R.G. (1986). Eur. J. Biochem., **157**, 187-193.

60) Hannigan, G. and Williams, B.R.G. (1986). Embo J., 5, 1607-1613.

61) Friedman, R.L., Manly, S.P., Mc Mahon, M, Kerr, I.M. and Stark, G.R. (1984). Cell, **38**, 745-755.

62) Tovey, M.G. (1982). in: *Interferon 4*, I. Gresser ed., Academic Press London, 23-46.

63) Mills, G.B., Hannigan, G., Steward, D., Mellors, A., Williams, B. and Gelfand, E.W. (1985). in *The 2-5A System: Molecular and clinical aspects of the Interferon-regulated Pathway*. B.R.G. Williams and R.H. Silverman, eds. Alan R. Liss, New York, 357-367.

64) Weil, J., Epstein, L.B., Sedmak, JJ, Sabran, J.L. and Grossberg, S.E. (1983). Nature (London), **301**, 437-439.

65) Knight, E. Jr., Fahey, D. and Blomstrom, D.C. (1985). J. Interferon. Res., **5**, 305-313.

66) Tominaga, S., Tominaga, K. and Lengyel, P. (1985). J. Biol. Chem., **260**, 16406, 16410.

67) Gupta, S.L., Rubin, B.Y. and Holmes, S.L. (1979). Proc. Natl. Acad. Sci. USA, **7**, 4817-4821.

68) Cheng, Y.S.E., Becker manley, M.F., Nguyen, T.D., De Grado, W.F. and Jonak, G.J. J. Interferon Res. (submitted).

69) Dron, M, Tovey, M.G. and Eid, P. (1985). J. Gen. Virol., **66**, 787-795.

70) Colonno, R.J. and Pang, R.H.L. (1982). J. Biol. Chem., **257**, 9234-9237.

71) Chebath, J., Merlin, G., Metz, R., Benech, P. and Revel, M. (1983). Nucl. Acid. Res., **11**, 1213-1226.

72) Faltynek, C.R., Mc Candless, S., Chebath, J. and Baglioni, C. (1985). Virology, **144**, 173-180.

73) Kusari, J. and Sen, G.C. (1986). Mol. and Cell. Biol., **6**, 2062-2067.

74) Larner, A.C., Jonak, G.C., Cheng, Y.S.E., Korant, B., Knight, E. and Darnell, J.E. Jr. (1984). Proc. Natl. Acad. Sci., **81**, 6733-6737.

75) Larner, A.C., Chaudhuri, A. and Darnell, J.E. Jr. (1986). J. Biol. Chem., **261**, 453-459.

76) Samanta, H., Dougherty, J.P., Brawner, M.E., Schmidt, H. and Lengyel, P. (1982) in: UCLAS Symposia on Molecular and Cellular Biology. T.C. Merigan and R.B. Friedman, eds., **25**, 59-72.

77) Samanta, H., Pravtcheva, D.D., Ruddle, F.H. and Lengyel, P. (1984). J. Interferon Res., **4**, 295-300.

78) Lengyel, P., Samanta, D., Engel, D., Chao, H., Tominaga, S., Garcia-Blanco, M. and Thakur, A. in: Abstracts from the 1985 TNO-ISIR meeting, Clearwater Beach, USA, 13-18 october 1985, p. 14.

79) Moutschen, S., Wathelet, M., Cravador, A., Defilippi, P., Huez, G. and Content, J. in *The 2-5A System*, B.R.G. Williams and R.H. Silverman, Eds., Alan R. Liss, New York, Vol. **202**, 183-192 (1985).

80) Korant, B.D., Blomstrom, D.C., Jonak, G.J. and Knight, E. Jr. (1984). J. Biol. Chem., **259**, 14835-14839.

81) Blomstrom, D.C., Fahey, D., Kutny, R., Korant, B.D. and Knight, E. Jr. (1986). J. Biol. Chem. (in press).

82) Cheng, Y.S.E., Colonno, R.J. and Yin, F.H. (1983). J. Biol. Chem, **258**, 7746-7750.

83) Cheng, Y.S.E., Becker-Manley, M.F., Chow, T.P. and Horan, C. (1985). J. Biol. Chem., **260**, 15834-15839.

84) Rosa, F. and Fellous, M. (1984). Immunol. Today, **5**, n. 9, 261-262.

85) Fellous, M., Nir, U., Wallach, D., Merlin, G., Rubinstein, M. and Revel, M. (1982). Proc. Natl. Acad. Sci. USA, **79**, 3082-3086.

86) Rosa, F., Hatat, D., Abadie, A., Wallach, D., Revel, M. and Fellous, M. (1983). Embo J., **2**, 1585-1589.

87) Collins, T., Korman, A.J., Wake, C.T., Boss, J.M., Kappes, D.J., Fiers, W., Ault, K.A., Gimbrone, M.A., Strominger, J.L. and Pober, J.S. (1984). Proc. Natl. Acad. Sci. USA, **81**, 4917-4921.

88) Yarden, A., Shure, Gottlieb, H., Chebath, J., Revel, M. and Kimchi, A. (1984). Embo J. **3**, 969-973.

89) Mc Mahon, M., Stark, G.R. and Kerr, I.M. (1986). J. Virol., **57**, 362-366.

90) Luster, A.D., Unkeless, J.C., Ravetch, J.V. (1985). Nature (London), **315**, 672-676.

91) Haller, O. (1981). Current topics Microbiol., Immunol, **92**, 25-52.

92) Dreiding, P., Staeheli, P. and Haller, O. (1985). Virology, **140**, 192-196.

93) Horisberger, M.A., Staeheli, P. and Haller, O. (1983). Proc. Natl. Acad. Sci. USA, **80**, 1910-1914.

94) Staeheli, P., Horisberger, M.A. and Haller, O. (1984). Virology, 132, 456-461.

95) Krug, R.M., Shaw, M., Broni, B., Shapiro, G. and Haller, O. (1985). J. Virol., **56**, 201-206.

96) Ransohoff, R.M., Maroney, P.A., Nayak, D.P., Chambers, T.M. and Nilsen, T.W. (1985). J. Virol., **56**, 1049-1052.

97) Meyer, T. and Horisberger, M. (1984). J. Virol., **49**, 709-716.

98) Staeheli, P., Haller, O., Boll, W., Lindenmann, J. and Weismann, C. (1986). Cell, **44**, 147-158.

99) Staeheli, P., Pravtcheva, D., Lundin, L.G., Acklin, M., Ruddle, F., Lindenmann, J. and Haller, O. (1986). J. Virol., **58**, 967-969.

100) Staeheli, P. and Haller, O. (1985). Mol. and Cell. Biol., **5**, 2150-2153.

101) Staeheli, P. (1985). in *The Biology of the Interferon System*. H. Schellekens and Stewart, eds., Elsevier Biochemical Press

102) Meier, E., Arnheiter, H. and Haller, O. (1986). Abstract from the 1986 TNO-ISIR Meeting, Espoo, Finland 7-12 september 1986.

103) Wathelet, M., Moutschen, S., Defilippi, P., Cravador, A., Collet, M., Huez, G. and Content, J. (1986). Eur. J. Biochem., **155**, 11-17.

104) Yoshida, R., Imanischi, J., Oku, T., Kishida, T. and Hayashi, O. (1981). Proc. Natl. Acad. Sci. USA, **78**, 129-132.

105) Ghezzi, P., Bianchi, M., Mantovani, A., Spreafico, F. and Salmona, M. (1984). Biochem. Biophys. Res. Comm., **119**, 144-149.

106) Ohtsuki, K., Yokoyama, M., Ikeuchi, T., Ishida, N., Sugita, K. and Sathoh, K.S. (1986). FEBS Lett., **196**, 145-150.

107) Laurent, A.G., Krust, B., Galabru, J., Svab, J. and Hovanessian, A.G. (1985). Proc. Natl. Acad. Sci. USA, **82**, 4341-4345.

108) Galabru, J. and Hovanessian, A.G. (1985). Cell, **43**, 685-694.

109) Gupta, S.L., Holmes, S.L. and Mehra, L.L. (1982). Virology, **120**, 495-499.

110) Nilsen, T.W., Maroney, P.A. and Baglioni, C. (1982). J. Biol. Chem., **257**, 14593-14596.

111) Rice, A.P., Duncan, R., Hershey, W.B. and Kerr, I.M. (1985). J. Virol., **54**, 894-896.

112) O'Malley, R.P., Mariano, T.M., Siekierka, J. and Mathews, M.B. (1986). Cell, **44**, 391-400.

113) Schneider, R.J., Safer, B., Munemitsu, S.M., Samuel, Ch.E. and Shenk, T. (1985). Proc. Natl. Acad. Sci., **82**, 4321-4325.

114) Reichel, P.A., Merrick, W.C., Siekierka, J. and Mathews, M.B. (1985). Nature (London), 313, 196-200.

115) Siekierka, J., Mariano, T.M., Reichel, P.A. and Mathews, M.B. (1985). Proc. Natl. Acad. Sci USA, 82, 1959-1963.

116) Rice, A.P. and Kerr, I.M. (1984). J. Virol., **50**, 229-236.

117) Whitaker-Dowling, P. and Youngner, J.S. (1984). Virology, **137**, 171-181.

118) Paez, E. and Esteban, M. (1984). Virology, **134**, 12-28.

119) Katze, M.G., Detjen, B.M., Safer, B. and Krug, R. (1986). Mol. and Cell. Biol., 6, 1741-1750.

120) Bhat, R.A. and Thimmappaya, B. (1985). J. Virol., **56**, 750-756.

121) Rosen, H., Knoller, S. and Kaempfer, R. (1981). Biochemistry, **20**, 3011-3020.

122) Kitajewski, J., Schneider, R.J., Safer, B., Munemitsu, S.M., Samuel, C.E., Thimmappaya, B. and Shenk, T. (1986). Cell, **45**, 195-200.

123) Winkler, M.M., Lashbrook, C., Hershey, J.W.B., Mukherjee, A.K. and Sarkar, S. (1983). J. Biol. Chem., **258**, 15141-15145.

124) Ball, A. (1982). in *The Enzyme*,**15**, Boyer, D. ed., Academic Press, New York, 281-313.

125) Baglioni, C. and Nilsen, T.W. (1983) in: *Interferon 5*, I. Gresser, ed., Academic Press (London), 23-40.

126) Kerr, I.M. and Brown, R.E. (1978). Proc. Natl. Acad. Sci., **75**, 256-260.

127) Ball, A. and White, C.N. (1978) Proc. Natl. Acad. Sci USA, **75**, 1167-1171.

128) Slattery, E., Ghosh, N., Samanta, H. and Lengyel, P. (1979). Proc. Natl. Acad. Sci. USA, **76**, 4768-4778.

129) Baglioni, C., Benvin, S., Maroney, P.A., Minks, M.A., Nilsen, T.W. and West, D.K. (1980). Ann. N.Y. Acad. Sci., **350**, 497-509.

130) Clemens, M.J. and Williams, D.R.G. (1978). Cell, **13**, 565-572.

131) Williams, B.R.G., Golgher, R.R., Brown, R.E. and Gilbert, C.S. (1979). Nature (London), **282**, 582-586.

132) St-Laurent, G., Yoshie, O., Floyd-Smith, G., Samanta, H., Sehgal, P.B. and Lengyel, P. (1983). Cell, **33**, 95-102.

133) Jacobsen, H., Czarniecki, C.W., Krause, D., Friedman, R.M. and Silverman, R.H. (1983). Virology, **125**, 496-501.

134) Jacobsen, H., Krause, D., Friedman, R.M. and Silverman, R.H. (1983). Proc. Natl. Acad. Sci. USA, **80**, 4954-4958.

135) Krause, D., Silverman, R.H., Jacobsen, H., Leisy, S.A., Dieffenbach, C.W., Friedman, R.M. (1985). Eur. J. Biochem, **146**, 611-618.

136) Resnitzky, D., Yarden, A., Zipori, D. and Kimchi, A. (1986). Cell, **46**, 31-40.

137) Knight, M., Cayley, P.J., Silverman, R.H., Wreschner, D.H., Gilbert, C.S., Brown, R.E. and Kerr, I.M. (1980). Nature (London), **288**, 189-192.

138) Silverman, R.H., Cayley, P.J., Knight, M., Gilbert, C.S. and Kerr, I.M. (1982). Eur. J. Biochem., **124**, 131-138.

139) Baglioni, C., D'Alessandro, S.B., Nilsen, T.W., Den Hartog, J.A.J., Crea, R. and Van Boom, T.H. (1982). J. Biol. Chem., **256**, 3253-3257.

140) Nilsen, T.W., Moroney, P.A. and Baglioni, C. (1982). J. Virol., **42**, 1039-1045.

141) Esteban, M., Benavente, J. and Paez, E. (1984). Virology, **134**, 40-51.

142) Goswani, B.B. and Sharma, O.K. (1984). J.Biol.Chem., **259**, 1371-1374.

143) Paez, E. and Esteban, M. (1984b) Virology, **134**, 29-39.

144) Rice, A.P., Roberts, W.K. and Kerr, I.M. (1984). J. Virol., **50**, 220-228.

145) Salzberg, S., Wreschner, D.H., Oberman, F., Panet, A., Bakhanashvili, M. (1983). Mol. Cell. Biol., **3**, 1579-1765.

146) Epstein, D.A., Czarniecki, C.W., Jacobsen, H., Friedman, R.M., Panet, A. (1981). Eur. J. Biochem., **118**, 9-15.

147) Sen, G.C. and Herz, R.E. (1983). J. Virol., **45**, 1017-1027.

148) Revel, M. (1979). in *Interferon 1*, I. Gresser, ed., Academic Press London, 101-163.

149) Verhaegen-Lewalle, M. and Content, J. (1982). Eur. J. Biochem., **126**, 639-643.

150) Doetsch, P., Wu, Jm, Sawada, Y. and Suhadolnik, R.J. (1981). Nature (London), **291**, 355-358.

151) Epstein, D.A., Marsh, Y.V., Schryver, B.B., Larsen, M.L., Barnett, J.W., Verheuden, J.P.H., Prisbe, E.J. (1982). J. Biol. Chem., 13390-13397.

152) Imai, J., Johnston, M.I. and Torrence, P.F. (1982). J. Biol. Chem., **257**, 12739-12745.

153) Haugh, C.M., Cayley, J.P., Serafinowska, H.T., Norman, D.G., Reese, C.B. and Kerr, I.M. (1983) Eur. J. Biochem., **132**, 77-84.

154) Bayard, B., Bisbale, C., Silhol, M., Cnockaert, J., Lebleu, B. and Huez, G. (1984). Eur. J. Biochem., **42**, 291-298.

155) Shimizu, N., Sokawa, Y. and Sokawam J.)1984). J. Biochem., **95**, 1827-1830.

156) Sokawa, J., Shimizu, N. and Sokawa, Y. (1984). J. Biochem., **96**, 215-222.

157) Sokawa, J. and Sokawa, Y. (1986). J. Biochem., **99**, 119-124.

158) Ilson, D.H., Torrence, P.F. and Vilcek, J. (1986). J. Interferon Res., **6**, 5-12.

159) Benech, Ph., Merlin, G., Revel, M. and Chebath, J. (1986). Nucl. Acid. Res., **13**, 1267-1281.

160) Saunders, M., Gewert, D., Castelino, M., Rutherford, M., Flenniken, A., Willard, H., Williams, B.R.G. (1985). in *The 2-5A system*, Vol. 202, B.R.G. Williams and R.H. Silverman, eds. Alan R. Liss, New York, 163-174.

161) Benech, P., Mory, Y., Revel, M. and Chebath, J. (1985). Embo J., **4**, 2249-2256.

162) Chebath, J., Benech, P., Mory, Y., Federman, P., Berissi, H., Gesang, O., Forman, J., Danovitch, S., Lehrer, R., Aloni, N. and Revel, M. (1985). in *The 2-5A system* vol. 202, B.R.G. Williams and R.H. Silverman, eds., Alan R. Liss, New York, 149-161.

163) Wathelet, M., Mouthschen, S., Cravador, A., Dewit, L., Defilippi, P., Huez, G. and Content, J. (1986). FEBS Lett., **196**, 113-120.

164) Shiojiri, S., Fukunaga, R., Ichii, Y. and Sokawa, Y. (1986). J. Biochem., **99**, 1455-1464.

165) Williams, B.R.G., Golgher, R.R. and Kerr, I.M. (1979). Febs. Lett., **105**, 47-52.

166) Hovanessian, A.G. and Wood, J. (1980). Virology, **101**, 80-90.

167) Defilippi, P., Huez, G., Verhaegen-Lewalle, M., De Clerq, E., Imai, J., Torrence, P. and Content J. (1986). Febs. Lett., **198**, 326-332.

168) Alarcon, B., Bugany, H. and Carrasco, L. (1984). J. Virol., **52**, 183-187.

169) Hersch, C.L., Brown, R.E., Roberts, W,L,, Swyryd, E.A., Kerr, I.M. and Stark, G.R. (1984). J. Biol. Chem., **259**, 1731-1737.

170) De Ferra, F. and Baglioni, C. (1981) Virology, **112**, 426, 435.

171) Masters, P.S. and Samuel, C.E. (1983). J. Biol. Chem., **258**, 12026-12033.

172) Munoz, A. and Carrasco, L. (1983). Eur. J. Biochem., **137**, 623-629.

173) Munoz, A., Harvey, R. and Carrasco, L. (1983). FEBS Lett., **160**, 87-92.

174) Miyamoto, N.g., Jacobs, B.L. and Samuel, C.E. (1983). J. Biol. Chem., **258**, 15232-15237.

175) Watling, D., Serafinowska, H.T., Reese, C.B. and Kerr, I.M. (1985). Embo, J., **4**, 431-436.

176) Cayley, P.J., Knight, M. and Kerr, I.M. (1982). Biochem. Biophys. Res. Comm., **104**, 376-382.

177) Cayley, P.J., Davies, J.A., Mc Cullagh, K.G. and Kerr, I.M. (1984). Eur. J. Biochem., **143**, 165-174.

178) Rice, P.A., Kerr, S.M., Roberts, W.K., Brown, R.E. and Kerr, I.M. (1985). J. Virol., **56**, 1041-1044.

179) Williams, G.J., De Benedetti A. and Baglioni, C. (1986). Virology, **151**, 233-242.

180) Esteban, M. and Paez, E. (1985). in *The 2-5A system*, vol. 202, B.R.G. Williams and R.H. Silverman, eds. Alan R. LISS, New York, 25-34.

181) Sen, G.C. (1981). in *Progress in Nucleic Acid Research*, J.N. Davidson and W.E. Cohn, eds., 27, Academic Press, New York and London, 105-156.

182) Friedman, R.M. (1977). Bacteriol. Rev., **41**, 543-567.

183) Gordon, J. and Minks, M.A. (1981). Microb. Rev., **45**, 244-266.

184) Content, J. (1984). in *Interferon 1*. A. Billiau, ed., Elsevier Amsterdam, 125-138.

185) Jacobsen, H. (1986). Arzneim-Forsch./Drug. Res., **36**, 512-516.

186) Schellekens, H., Weimar, W., Cantell, K., Stitzt, L. (1979). Nature (London), 278, 742.

187) Yacobson, E., Revel, M. and Winocour, E. (1977). Virology, 80, 225-228.

188) Brennan, M.B. and Stark, G.R. (1983). Cell, **33**, 811-816.

189) Chatterjee, S., Hunter, E. and Whitley, P. (1985). J. Virol., **56**, 419-425.

190) Munoz, A. and Carrasco, L. (1984). J. Gen. Virol., **65**, 1069-1078.

191) Maheshwari, R.K., Friedman, R.M. (1980). Virology, **101**, 399-407.

192) Maheshwari, R.K., Demsey, A.E., Mohanty, S.B. and Friedman, R.M. (1980). Proc. Natl. Acad. Sci. USA, **77**, 2284-2287.

193) Whitaker-Dowling, P.A., Wilcox, D.K., Windell, C.C.C. and Youngner, J.S. (1983). Proc. Natl. Acad. Sci. USA, **80**, 1083-1086.

194) Baglioni, C., de Benedetti, A., Williams, G.J. (1984). J. Virol., **52**, 865-871.

195) Munoz, A. and Carrasco, L. (1984). J. Gen. Virol., **65**, 377-379.

196) Esteban, M., Paez, E. (1985). Prog. Med. Virol., **32**, 159-176.

197) Esteban, M., Benavente, J. and Paez, E. (1986). J. Gen. Virol., **67**, 801-808.

198) Pacha, R.F. and Condit, R.C. (1985). J. Virol., **56**, 395-403.

199) Bayard, B., Bisbal, C., and Lebleu, B. (1986). Biochem., **25**, 3730-3736.

200) Katze, M.G., Cheng, Y.T. and Krug, R.M. (1984). Cell, **37**, 483-490.

201) Kelly, J.M., Porter, A.C.G., Chernajovsky, Y., Gilbert, C.S., Stark, G.R. and Kerr, I.M. (1986). Embo J., **5**, 1601-1606.

202) Kelli, J.M., Gilbert, C.S., Stark, G.R. and Kerr, I.M. (1985) Eur. J. Biochem, **153**, 367-371.

203) Oikarinen, J. (1982). Biochem. Biophys. Res. Comm., **105**, 876-881.

204) Krishnan, I. and Baglioni, C. (1980) Proc. Natl. Acad. Sci. USA, **77**, 6506-6510.

205) Watson, G.H. and Baker, K.L. (1985). in: *The 2-5A system*, Vol. 202, B.R.G. Williams and R.H. Silverman, eds. Alan R. Liss, New York, 369-375.

206) Zullo, J.N., Cochran, B.H., Huang, A.S. and Stiles, C.D. (1985). Cell, 43, 793-800.

207) Engel, D.A., Samanta, H., Brawner, M.E. and Lengyel, P. (1985). Virology, **142**, 389-397.

208) Wallach, D. (1983). Eur. J. Biochem., **13**, 794-798.

209) Pober, J.S., Gimbrone, M.A.Jr., Cotran, R.S., Reiss, C.S., Burakoff, S.J., Fiers, W. and Ault, K.A. (1983). J. Exp. Med., **157**, 1339-1353.

210) Staeheli, P., Haller, O. and Sutcliffe, J.G. (1986). Abstracts from the 1986 TNO-ISIR Meeting. J. Int. Res., 6, suppl. 1, II **48**, p. 85.

211) Friedman, R.L. and Stark, G.R. (1985). Nature (London), **314**, 637-639.

212) Yoshiem O., Schmidt, H., Reddy, E.S.P., Weissman, S. and Lengyel, P. (1982). J. Biol. Chem., **257**, 13619-13172.

213) Malissen, M., Malissen, B. and Jordan, B.R. (1982). Proc. Natl. Acad. Sci. USA, **79**, 893-897.

214) Karin, M. and Richards, R.I. (1982). Nature (London), **299**, 797-802.

215) Church, G.M. and Gilbert, W. (1984). Proc. Natl. Acad. Sci. USA, **81**, 1991-1995.

216) Mallucci, M., Chebath, J., Revel, M., Wells, V. and Benech, P. (1982). in *The Interferon System*, Serono Symposia 24, Dianwani F. and Rossi, G.B. eds. Raven Press, 165-170.

217) Israel, A., Kimura, A., Fournier, A., Fellous, M. and Kourilsky, P. (1986). Nature (London), **322**, 743-746.

218) Williams, B.R.G., Saunders, M. and Williard, H.F. (1986). Somatic Cell and Mol. Genetics, **12**, 403-408.

219) Burrone, O.R. and Milstein, C. (1982). EMBO J., **1**, 345-349.

220) Gresser, I., Vignaux, F., Belardelli, F., Tovey, M.G. and Maunoury, M.T. (1985). J. Virol., **53**, 221-227.

221) Jonak, G.J. and Knight, E. Jr. (1986). in: *Interferon 7.*, I. Gresser, ed.

222) Etienne-Smekens, M., Vandenbussche, P., Content, J. and Dumont, J.E. (1983). Proc. Natl. Acad. Sci. USA, **80**, 4609-4613.

223) Knight, E. Jr., Fahey, D., Kutny, R. and Blomstrom, D.C. (1986). J. Interferon Res., 6, s. **1**, 26.

224) Chebath, J., Benech, P. and Revel, M.J. of Interferon Res., 6, s. **1**, 21.

225) Staeheli, P., Prochazka, M., Steigmeier, P.A. and Haller, O. (1984). Virology, **137**, 135-142.

CHAPTER 7

THE ROLE OF DEFECTIVE INTERFERING (DI) PARTICLES IN VIRAL INFECTION

ALICE S. HUANG

Division of Infectious Diseases, Children's Hospital & Department of Microbiology and Molecular Genetics Harvard Medical School Boston MA 02115, U.S.A.

A special class of viral deletion mutants was first detected biologically about 35 years ago and characterized physically about 20 years ago. They are formed by almost every virus group and given the name defective interfering (DI) particles. DI particles have the following properties:

i) they are derived from infectious standard virus;
ii) they contain the same structural proteins as standard virus;
iii) they contain nucleic acid deletions and/or rearrangements;
iv) they amplify themselves within the viral population, generally, but not necessarily always, at the expense of the standard virus (interference); and
v) they require helper standard virus for their complete replication.

These DI particles are of considerable interest because they are thought to be natural modulators of viral infections. It is hoped that by understanding their role in pathogenesis that a new strategy for antivirals based on genetic engineering can be developed, especially for those lethal viral diseases where conventional vaccines fail to protect. Moreover, because DI particles retain two essential functions, that of encapsidation and nucleic acid replication, studying their nucleic acid sequences tells the molecular virologist a great deal concerning the regulation of viral morphogenesis and the binding of polymerases. Since the molecular mechanisms involved in the generation of DI particles are extensively discussed by Schubert and Emerson (chapter 11) and by Kolakofsky and Roux (chapter 12) elsewhere in this volume, this review will concentrate principally on the role that DI particles may play in the control of viral infections.

Despite the identification and physical characterization of defective interfering (DI) particles among many groups of animal viruses (reviewed in ref. 1-3), little is known about the role that these mutants play in natural viral disease. In cell cultures they inhibit homologous viral replication, sometimes resulting in the establishment and maintenance of persistent infections. However, *in vivo* protection experiments have been less clear. Inoculation of purified viral preparations lacking DI particles into animals leads to their generation and replication, but attempts to reproducibly demonstrate an ameliorating effect on disease symptoms or subsequent mortality require an unnaturally large inoculum of DI particles. Dose-response studies with lower amounts of DI particles relative to standard infectious virus produced unpredictable results. Furthermore, with animal studies there are additional complications induced by nonspecific host defense mechanisms, interferon and specific immune responses against the viruses.

Moreover, virologists have assumed for some time that natural disease results from exposure to one or very few infectious virus particles and that the replication of viruses under these circumstances results in such low titer that cells in the animal infected with more than one virus are unlikely to be found. Several lines of evidence, however, appear to contradict this assumption. Recent *in vivo* studies demonstrate interaction among viruses in the inoculum. As viral spread occurs, continued interaction among the inoculated members goes on throughout the disease process. Moreover, with the recognized persistence of many viruses within a host, cells may be dually or even multiply infected with virus particles during the course of the host's lifetime.

Our own recent studies have shown that if mice were infected with a mixture of standard infectious vesicular stomatitis virus (VSV) and its DI particles, a dynamic continuous interaction occurs between the two populations for several days (4, 5). The outcome, in terms of mortality, can be predicted from the composition of the initial inoculum only if a cyclic pattern of production of infectious virus and DI particles occur in these animals. In cell cultures this pattern is known to occur with a reproducible pattern (6).

If the host has no previous experience with the virus, theoretically, alternate cyclic patterns may be set up in the host depending on the amount of total virus particles inoculated and the relative ratios of standard virus to DI particles. For example, if more DI particles are present, the initial synthesis of infectious virus would be lower, if less DI particles are present, the initial synthesis of infectious virus may be increased. Several days after the inoculation the amount of infectious progeny as well as the total antigenic load of viral particles may be quite different from what was initially inoculated. At this time, the interaction of mounting host defenses against the viral population may be the determining factor for the outcome of disease. Therefore, it may not be the initial inoculum that the host is directly interacting with, but rather with a concentration of viral particles that is produced within the host at some later time after inoculation. Such cycling patterns are, of course, more likely to occur when the virus spreads from the site of inoculation into some interior organ. Under these circumstances, the general ability to demonstrate a consistent dosage effect of DI particles on disease is probably due to our lack of knowledge of the particular cyclic pattern that is being introduced.

On the other hand, highly localized disease, especially those at the inoculation site, would be expected to have an outcome that is more predictable. With localized disease, immediate replication of the inoculum at the site might determine the outcome without the delay and recruitment of uninfected cells. Similarly, an immunologically primed host may confine the spread of the virus and prevent subsequent cycling of viral production. Along these lines, recent studies with young seronegative calves, inoculated intradermally with VSV and its DI particles, have shown a good dose-response effect of DI particles on the inhibition of localized tongue lesions (Yilma and Huang, unpublished observations).

Further support for the role of VSV DI particles in natural disease comes from analysis of two virulent isolates obtained from the epizootic of vesicular stomatitis among cattle and horses in the Western United States in 1982-83. These isolates generate and amplify DI particles more slowly than laboratory strains and appear to be more resistant to pre-existing DI particles, particularly heterotypic ones (7). Further analysis of isolates from different locations and isolates obtained later during the epizootic would confirm whether DI particles played a role in altering the severity of local outbreaks and/or were responsible for the eventual self-limited nature of the epizootic. Parenthetically, this natural disease in cattle is particularly suited for examining the role of DI particles because protection does not appear to correlate with the immune status of the host.

Such studies suggest that persistent infections initiated by a mixture of DI particles and standard virus are likely to lead to more virulent, or serotypically altered variants which exacerbate the infection and produce overt disease. Therefore, isolations of virus from chronic infections, such as subacute sclerosing panencephalitis and progressive multifocal leukoencephalopathy, are not necessarily expected to contain DI particles, but should contain standard variants with altered responses to interference by DI particles.

If direct viral isolates from patients are heterogeneous and if intermittent shedding of viral particles occur during the disease process, many implications come to mind other than the effects on pathogenesis. In particular, the implications for viral diagnosis and for viral transmission will be discussed here.

Much of viral diagnosis depends on the amplification of infectious, cytopathic viruses in cell cultures. The success rate of such isolations during a clinically well recognized epidemic is usually around 30-50% at best. Rather than a lack of sensitivity, this may be due to periods of viral production which results in patient specimens that contain viral soluble antigens or noninfectious particles. Therefore, diagnostic techniques based on the detection of antigen or nucleic acids should be used as well. However, even with these selective techniques, it is unlikely that a positive diagnosis would be made each time, particularly during the periods of a cycle when all viral products are minimally produced. Sequential specimen collection from patients may overcome such problems.

In relation to viral transmission, the cyclic pattern of viral synthesis may be reflected in the quality and quantity of the shed virus from an infected individual. During the course of illness the shed population of viruses from an individual may be very virulent, whereas later in time, from the same individual, the viral population may be avirulent but highly immunogenic. Transmission of heterogeneous and qualitatively different populations of the same virus from one individual to several others may explain the high degree of variability in disease transmission and resultant symptoms. Therefore, to control virus spread, it would be important to determine when an infected individual is transmitting highly virulent populations versus avirulent ones.

Parenthetically, anecdotal practices support what in all likelihood may be the use of DI particles for vaccination without disease.

This first case involves canine distemper where dogs recently recovering from the disease are introduced into kennels exposing other dogs to the virus: It is thought that dogs who have survived the infection shed virus in a form that will protect unexposed dogs against the overt disease. In a second example, my Chinese colleagues mention the traditional practice of rubbing of brain material from rabid dogs into the site of bites from rabid animals. Presumably, DI particles from such preparations help to prevent the onset of rabies in the human host. Obviously, much more characterization of viral populations produced in different species and organ sites will be necessary before such intriguing presumptions can be accepted as true. But these hypotheses are testable. Sequential data obtained from human rotavirus infections (8) are good examples of the necessary approach to be taken. As the role that DI particles play in natural disease becomes clearer, safe methods to harness their interfering capability should become possible.

Acknowledgements

The author's work described here was supported by a research grant from the National Institute of Health AI 16625/20896. I am grateful for revealing conversations with Dr. David R. Cave and Dr. Bernard Fields and to Barbara Connolly for computerized text editing.

REFERENCES

1) Huang A.S. and Baltimore D. (1977). Defective interfering animal viruses. In: *Comprehensive Virology*,**10**, 73-116.
2) Perrault J. (1981). Origin and replication of defective interfering particles. Curr. Topics. Microbiol. Immunol. **93**, 151-207.
3) Holland J.J., Kennedy SIT, Semler B.L., Jones C.L., Roux L. and Grabau E.A. (1980). Defective interfering RNA viruses and the host-cell response. In: *Comprehensive Virology*, H. Fraenkel-Conrat & R.R. Wagner, eds., **16**, 137-192.
4) Cave D.R., Hagen F.S., Palma E.L. and Huang A.S. (1984). Detection of the RNA of vesicular stomatitis virus and its defective interfering particles in individual mouse brains. J. Virol.**50**, 86-91.
5) Cave D.R., Hendrickson F.M. and Huang A.S. (1985). Defective interfering virus particles modulate virulence. J. Virol. **55**, 366-373.
6) Palma E.L. and Huang A.S. (1984). Cyclic production of vesicular stomatitis virus caused by defective interfering particles. J. Infect. Dis. **129**, 402-410.
7) Huang A.S., Wu T.Y., Yilma T. and Lanman G. (1986). Characterization of virulent isolates of vesicular stomatitis virus caused by defective interfering particles. Microb. Pathogen. **1**, 205-215.
8) Pedley S., Hundley F., Chrystie I., McCrae M.A. and Desselberger U. (1984). The genomes of rotaviruses isolated from chronically infected immunodeficient children. J. Gen. Virol. **65**, 1141-1150.

SECTION IV

The strategies of replication

CHAPTER 8

PICORNAVIRUSES AT THE MOLECULAR LEVEL

R. PEREZ BERCOFF
The Institute for Virology, University of Rome, Rome, Italy

OVERVIEW

Defined as the smallest RNA containing animal viruses, *Picornaviruses* (an acronym derived from **P**olio, **I**nsensitive to ether, **C**oxsackie, **O**rphan, **R**hino, **RNA** genome, that also stands for *pico* = small, *RNA* viruses, ref.1) include the etiological agents of a large number of human and animal diseases such as hepatitis A, poliomyelitis, common cold, foot-and-mouth disease (FAMD) of cloven-foot animals, encephalomyocarditis (EMC), among others.

Over the past two decades, systematic vaccination campaigns have virtually eradicated epidemic poliomyelitis from the Northern Hemisphere, though in developing countries a very high number of fatal cases are officially reported every year and very likely far more go completely unreported.

Human Hepatitis A, a highly transmissible disease, is still endemic in vast areas of the world — including many European countries — where over 80% of the population between 20 and 40 years bear serological evidence of past infection.

If these two examples (and the list can be extended) illustrate the relevance of this group of pathogens in the medical field, the interest of the molecular virologist for picornaviruses has not diminished since the early days when Enders and collaborators (2) succeeded in growing poliovirus in tissue culture. Almost 40 years later, the complete *sequence of the genome* of several picornaviruses (including polio and hepatitis A) have been determined, and our knowledge of the *structure and organization* of the viral particle goes down to the atomic level, at least for three members of the group. We have also gained a substantial understanding of the mechanism of translation and replication of picornaviruses. The purpose of this review (which cannot be either complete nor exhaustive) is to summarize what we do know (and what we would like to know) about the structure and the expression of the picornavirus genome.

Classification

Although the naked, icosahedrical virions (Fig. 1) are remarkably similar in size (27-30 nm), morphology, structure, and organization, they differ substantially in buoyant density, pH-stability, sedimentation behavior, and thermo-sensitivity (3, 4), differences that provide the basis for the classification into four groups:

i) the acid-stable ENTEROVIRUSES (polio, coxsackie, echo, hepatitis A);
ii) RHINOVIRUSES, (common cold);
iii) APHTOVIRUSES (seven serotypes of the world-wide spread FAMD), and
iv) CARDIOVIRUSES, the agent of different varieties of murine, swine (and human ?) meningo-encephalo-myocarditis (EMC, Mengo, Columbia SK, Theiler's disease, GD VII, Maus Eberfeld, etc)

Table 1 summarizes the properties of the four genera.

1. THE VIRION

A. The Protein Shell

A single-stranded, linear RNA molecule of messenger-like polarity is surrounded by a protein shell of 60 identical subunits (6S "protomers") arranged in groups of 5 (14S "pentamers") (3, 5). Each protomer is composed of four structural peptides called VP1, VP2, VP3, and VP4 in decreasing order of size. Before the adoption of a common nomenclature (ref. 6) at the 3rd. Meeting of the European Study Group on Molecular Biology of Picornaviruses (Urbino, Italy, 1983),

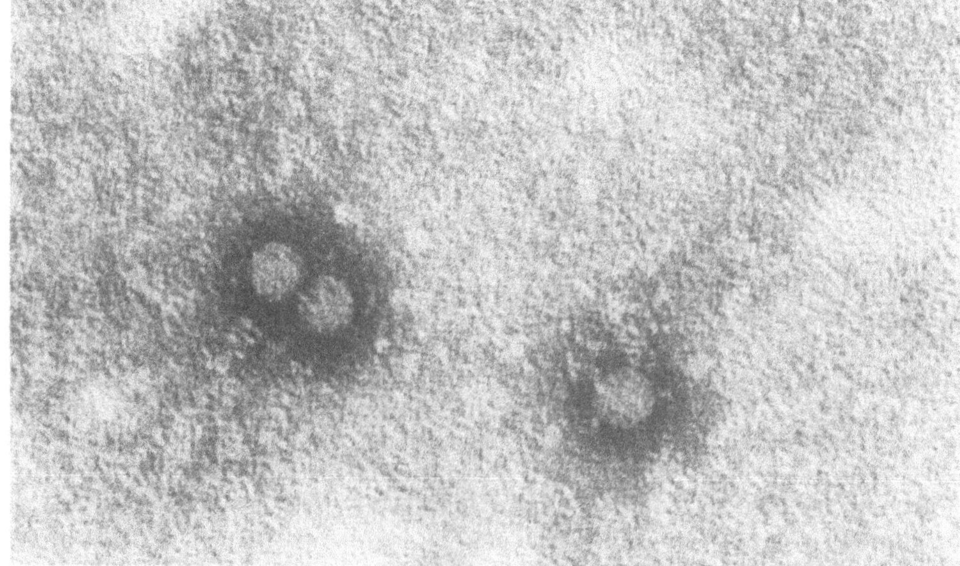

Figure 1. Electronmicroscopy of Human Hepatitis A virus (Courtesy of F. Ruggeri, Ist. Sup. Sanita', Rome, Italy)

Table 1. - Picornaviruses

Enterovirus	Hepatitis A Polio (3 serotypes) Coxsackie A (23) Coxsackie B (6) Echo (31) Enteroviruses of mice, swine, cattle Enterovirus 70 (conjunctivitis virus)	Sedimentation coefficient ~155 S Buoyant density (CsCl) ~1.34 g/ml Virions stable at pH 3-10 Empty capsids produced *in vivo*
Cardiovirus	EMC Mengo Columbia-SK MM Theiler's disease GD VII (Serologically very closely related)	Sedimentation coefficient ~155 S Buoyant density ~1.34 g/ml Virions labile 5<pH<7 in the presence of 0.1 M Cl^- or Br^- No empty capsids *in vivo*
Rhinovirus	More than 120 serotypes	Sedimentation coefficient ~155 S Buoyant density ~1.40 g/ml Virions labile pH<5 Empty capsids produced *in vivo*
Aphtovirus	Foot-and-Mouth Disease Virus (7 serotypes)	Sedimentation coefficient ~145 S Buoyant density ~1.34 g/ml Virions labile pH<6.5 Empty capsids produced *in vivo*

the corresponding peptides of *cardioviruses* were designated α, β, γ, and δ. The molecular weights of the structural proteins range from 35 through 5 kDa, with slight variations of these values among the different genera.

Topological relationships among structural proteins

Early work focusing on the antigenic properties of several enteroviruses revealed that *empty capsids* and *infectious virions* differed to a substantial extent in their antigenic composition: antibodies raised against the former failed very often to neutralize the latter. These observations provided the bases for a distinction between "D"-type and "C"-type antigenicity, proper of infectious particles and empty capsids, respectively. Upon heating, mild denaturing treatment, or attachment to followed by elution from susceptible cells, infectious enterovirions lose their infectivity and change their antigenic properties ("A"-type antigenicity). "A"-type particles lack VP4 (the smallest structural polypeptide) are susceptible to protease degradation, and are not infectious, although they carry intact RNA (7). Pretreatment of the cells with sulphydryl-containing compounds (which renders the cell refractory to virus attachment) prevents the formation of "A" particles (7). Interestingly, FAMD virus does not generate "A" particles upon interaction with the cell receptor (8, 9). These changes were correctly interpreted as a result of conformational modifications of the surface of the virion.

The most rewarding approach to elucidate the topological relationships among the structural peptides in the capsid was the systematic analysis of the substructures generated by chemical or physical disruption of the picornavirus particle (3, 10-14).

Chemical labeling of the peptides accessible at the surface of the virion (15, 16), and chemical or U.V. light induced cross-linking studies (17, 18) provided a consistent picture of the virion that recent X-ray crystallographic analysis have confirmed and extended to the atomic level (see chapter 2 of this volume, and ref. 3, 19, 20). Briefly,

(a) the structure of VP1, VP2, and VP3 is similar to one another, and basicly similar to the structural proteins of the plant virus SBMV, TBSV: a topologically identical "core" of 8-stranded anti-parallel β sheets, plus two flanking helices and additions at the N- and C- end (see Fig. 2 chapter 2)

b) the N-terminal of VP1, VP2, and VP3 fold beneath the barrels, and lie in close contact with the genomic RNA;

c) VP4 presents a more extended configuration (only 2-strand β sheets), appears as a mere N-terminal extension of VP2, and lies below of and in close contact with VP1 and VP3;

d) the uppermost 3 loops of VP1, and 2 loops of VP2 and VP3 are exposed at the outer surface of the virion, the former at the summit of the vertices;

e) the *external surface* of the virion presents "peaks" and "valleys", notably a 25 Å-deep, 12 Å-wide "canyon" or moat around the 12 vertices of the icosahedron;

f) the *inner surface* of the protein shell is formed by the bottom surface of the β barrels of VP1, VP2, and VP3, the whole of VP4, plus the N-terminals of VP1, VP2, and VP3, which form a network linking the protomers and pentamers;

g) VP4 appears to lie at the bottom of the moat, in a site inaccessible to either external proteases or antibodies.

The tertiary folding of the structural proteins holds a striking similarity with the conformation of plant viruses (SBMV, TBSV): the subunits A, B, and C of SBMV (of identical amino acid sequence but arranged in different geometrical environment) find their functional equivalent in VP1, VP3, and VP2, respectively. A comparison between VP2 and the C subunit makes the homology all the more evident (Fig. 2, chapter 2).

Two clearly defined areas have been identified in the C subunit of SBMV: amino acids 64-260 constitute the *shell* domain organized in anti-parallel β barrels (Fig. 2, chapter 2), while the first 63 residues are associated with the RNA. Interestingly, amino acids 38-63 are highly ordered, and form the βA arm.

In picornaviruses, only VP2 presents a similar organization, and through its βA arm appears to play a role comparable to that of the C subunit in holding the architecture of the virion: the N-end of the βA arm interacts at the 3-fold axis of symmetry with two other VP2, while the C-end of the arm interacts with the equivalent area of another VP2 at the 2-fold axis. The transition from native to "A" particle with loss of VP4 indicates important re-arrangements in the architecture of the virion.

B. The RNA Component

The 7.5-8.5 kb long genomic RNA of picornaviruses is one of the most efficient messenger that can be translated in eukaryotic systems. It contains a poly(A) tail of 16-50 residues at the 3'-end, two extra-cistronic terminal regions of unequal length flanking both ends of a large open reading frame that extends practically over the entire molecule (6, 21).

The complete sequence of several picornavirus RNAs (and partial sequences of some isolates) have been determined. The list includes poliovirus and the attenuated Sabin strains (22-24, 42-44), Human Hepatitis A (four different isolates, ref. 25-29), coxsackie B (29, 30), rhinoviruses (31-33), FAMD (34-38), echovirus (39), EMC (40), and Theiler's disease virus (41).

A 3' untranslated region of variable length (47 to 126 nucleotides) extends from the major open reading frame (ORF) up to the terminal poly(A) tract (16-80 nucleotides).

The 5' extra-cistronic sequences, 0.7-1.2 kb long, contain several short ORF with potential coding capacity for proteins of <10 kDa, but for the time being no peptide assignable to this region has been identified. A string of eleven pyrimidines interrupted just once by a purine appear to precede the initiation AUG codon (41). Secondary structure predictions have shown that the sequences 5' distal to the major initiation site can be arranged to form stable base-pairing. Whether these predicted hairpin structures do actually form or are disrupted by tertiary folding or upon interaction with proteins, is still an open question.

The RNA of *cardio-* and *aphto-*viruses (with the exception of Theiler's disease virus, ref. 41) contains a poly(C) tract about 80-150-nucleotide long very close to (but not at) the 5'-end (46-50). The function (if any) of the poly(C) tract is still obscure.

A comparison of the nucleotide sequence of several picornavirus genomes indicates that *rhinoviruses* use codons ending in A or U more frequently (67%) that in G or C (33%), while the opposite appears to hold true for *aphtoviruses*. Entero- and cardioviruses have no particular preference in codon usage (45).

The 5'-end of picornavirus RNA is exceptional among eukaryotic mRNAs for the lack of a terminal "cap". Instead, the terminal nucleotide (pUp) is covalently linked via a tyrosine-O^4-phosphodiester bond to a small virus-coded protein, VPg. Removal of VPg does not affect the exceptional translation ability of picornavirus RNA (51). Translation and sequencing studies have provided evidence of a remarkable conservation of the genome structure and organization throughout the entire group.

2. THE REPLICATION CYCLE

A. Adsorption, Penetration and Uncoating

Picornaviruses penetrate the host cell by receptor-mediated endocytosis (52, 53): following recognition of and interaction with proper cellular receptors, the viral particles are internalized from coated-pits into coated vesicles, which in turn fuse with intracellular vesicles and bring the virion to the endosome. Uncoating is a pH-dependent process that appear to take place in the lysosomes. Consequently, NH_4Cl, chloroquine, or ionophores able to raise the pH of the endosomes or lysosomes prevent the uncoating, and block infectivity. The entire process of penetration and uncoating is completed in 20-30 minutes (53). Less than 5% of the input virions does actually uncoat (54).

Cellular receptors play a crucial role in determining the viral tropism and consequently their pathogenic potential. Receptors are an integral part of the cell membrane, under the genetic control of the host. Hybrid human/mouse cells express the receptors for poliovirus until they segregate human chromosome 19 (52). Transfection of mouse tk^-, $aprt^-$ L cells with human DNA resulted in the appearance of receptors for poliovirus in the mouse cells (60).

In HeLa cells there are about 3,000 receptor sites /cell for poliovirus, ~500,000 for coxsackie B, and between 10^4-10^6 receptors for rhinoviruses (56). Treatment of the host cell with trypsin inactivates the receptors for polio or rhinoviruses; those for coxsackie, echo and cardioviruses are susceptible to chymotrypsin, and removal of sialic acid residues with neuraminidase prevents the attachment of cardioviruses, bovine enterovirus (BEV), and equine rhinovirus (52). The inactivation is reversible unless cellular protein synthesis is inhibited (55).

The receptors for coxsackie B viruses have been isolated and characterized: they are glycoproteins, containing α-mannosyl like residues, constitutively present in the cell membrane (57-59).

Attachment of picornaviruses to the host cell involves a close interaction of cell receptors with VP4 (the smallest and most internal structural protein, deeply buried at the bottom of the "canyon"), and a less strong one with VP1, suggesting that the process may involve multiple polypeptides acting cooperatively.

B. Shut Off of Host Cell Synthesis

Picornaviruses are highly cytolytic viruses that can induce a strong inhibition of all cellular macromolecular synthesis. Translation of viral RNA is a prerequisite for inhibition to occur (61). The shut off of cell protein synthesis seems to be the primary event, upon which inhibition of cellular RNA and DNA ensues.

Following an initial lag (whose length depends on the viral strain, multiplicity of infection, cell line, physiological state of the cell, among other conditions), the cellular polysomes disaggregate, and the overall rate of protein synthesis declines (61). During the second half of the infectious cycle, virtually all proteins synthesized in picornavirus-infected cells are virus-coded. Capped messenger RNAs are not degraded, but cannot be translated (62), suggesting that a most efficient mechanism of discrimination between cellular and viral messengers is operative.

The intensity of the inhibition depends again on the MOI, the virus strain and the cell line. Early reports by Plagelmann and Swim underlined the role of the host cell in the establishment of the shut-off and subsequent development of cytopathic effect (CPE): the same strain of mengovirus that caused a strong inhibition of the host cell RNA and protein synthesis in Novikoff hepatoma cells, did not interfere noticeably with the host cell macromolecular synthesis when it replicated in a nutritional derivative of the same cells (63, 64). More recently, a fast-growing strain of human hepatitis A virus has been isolated that induces a strong cytopathic effect in Frp/3 cells (rhesus), but is unable to shut-off Vero cells (cercopithecus), although it replicates with comparable efficiency in both cell lines (25).

Ribosomes (or crude lysates) of picornavirus-infected cells fail to support the translation of "capped" messenger RNAs of cellular or viral origin. Addition of either ribosomal wash or fractionated initiation factors from un-infected cells relieves the inhibition (65, 66), suggesting that a ribosome-associated factor is the main target hit in picornavirus-infected cells.

Analysis of the ribosome-associated proteins from poliovirus-infected HeLa cells led to the discovery of cap-binding proteins (CBP), required for the recognition of the 5'-terminal "cap", and whose inactivation results in a distinctive translational advantage of the uncapped polio RNA over any "capped" cellular messenger (67, 68). The role of the CBPs in initiation of translation is discussed in chapter 5.

The biochemistry of the virus-induced inactivation of the CBP is still unclear: early work suggested that the inhibition of HeLa cell protein synthesis following poliovirus infection correlated with the proteolytic degradation of a 220 kDa peptide associated with (or part of) the initiation factor eIF-3/CBP complex (62). A recent report, however, indicates that eIF-3 is still active after poliovirus infection has abolished the cap-binding activity (70). Inactivation of CBP has also been observed in human rhinovirus-infected HeLa cells (71).

Double-infection experiments have demonstrated that picornaviruses do not shut-off each other, suggesting that the inhibitory mechanism discriminates very efficiently against cellular mRNAs (72). While the inactivation of the cap-binding complex is the most plausible explanation of the selective inhibition of the host protein synthesis in *polio* and *rhino*virus infected cultures, direct experimental evidence proving that *cardio* and *aphto*viruses do operate a similar mechanism is altogether lacking.

The crude ribosomal fraction of Mengovirus infected L-929 cells appears to contain an inhibitor of *in vitro* protein synthesis (73). The inhibitor can be removed by high salt washing, and it has been recently characterized: it is a protein kinase that phosphorylates *in vitro* the α subunit of the initiation factor eIF-2 at sites apparently different from those phosphorylated by either the double-stranded RNA activated kinase, or the haem-controlled inhibitor constititively present in reticulocyte lysates (74). It remains to be seen: a) whether a similar virus-activated enzyme can be detected in cells infected by picornaviruses outside the cardiovirus group, and b) whether the Mengo-induced kinase can phosphorylate (and thereby inactivate) the cap-binding complex

C. Translation of Picornavirus RNA

Initiation

Translation of most eukaryotic mRNA appears to require the recognition of the 5' end of the mRNA molecule by ribosomes and associated initiation factors (see chapter 5). The presence of a capping group at the 5' terminus of most eukaryotic mRNAs facilitates initiation complex formation (75, 76).

The 5' end of picornavirus RNA is exceptional among mRNAs translated in animal cells for the lack of a 5' terminal cap; also this structure is not found in the viral polysomes (77). Yet the genomic RNA of picornaviruses serves directly as mRNA and can be translated *in vivo* or *in vitro*, with the RNA of *cardioviruses* ranking among the most efficient mRNA species so far studied in eukaryotic systems (78-82). The molecular basis of such an efficiency is still unknown. Clearly, in the absence of a 5' terminal cap other mechanisms of recognition should be operative to secure the translation ability of picornavirus RNA.

RNase T_1 fingerprint mapping of the sequences of Mengovirus RNA protected by ribosomes at initiation of translation provided the first experimental evidence that, in contrast to eukaryotic mRNAs, translation of the un-capped picornavirus RNA initiates well *inside* the molecule, in areas that are *internal* rather than 5' terminal. Moreover, the ribosome-protected sequences originated in at least two widely separated domains (83).

The sequences of Mengovirus RNA recognized by eIF-2 virtually overlapped with the ribosome-binding site. The specificity of the eIF-2 binding to *internal* sequences of Mengovirus RNA indicated that a free 5' end was not required for this interaction (83).

The extra-cistronic regions

The cap-independent mechanism of initiation of translation used by picornaviruses involves internal (instead of terminal) sequences, and this raises two specific questions.

The first one concerns the *translatability* of the long extra-cistronic region intervening between the 5' end and the authentic initiation site. As already mentioned, these sequences contain several short ORFs that upon translation would generate so far undetected low molecular weight peptides. In view of the growing list of eukaryotic mRNAs encoding "agnoproteins" it would not be too surprising if the sequences preceding the major ORF were indeed translated into short-lived "agnopeptides", a possibility that can be tested by readily designable experiments.

If they are not (as the presently available evidence suggests) it seems legitimate to raise the question of whether the long sequence 5' distal to the initiation AUG codon acts as a functional *leader*, i.e: whether ribosomes translate the uncapped picornavirus RNA by a mechanism involving the initial attachment of the initiation complex at the 5' end followed by migration to the initiation AUG codon as proposed in the original "scanning model" (84). Several lines of evidence argue against this mechanism. First, migration of the 40S ribosomal subunits from the 5' end to the initiation AUG codon apparently requires ATP hydrolysis (84, 85), yet in the case of EMC virus RNA initiation complex formation can occur in the absence of ATP or in the presence of non-hydrolysable ATP analogs (86). Moreover, *in vitro* translation of FAMD virus RNA deprived of the entire extra-cistronic region (and consequently of the legitimate 5' end) generates a set of peptides indistinguishable from the one obtained upon translation of the intact RNA (87).

Two initiation sites

All the available evidence, therefore, suggests that "picornaviruses may have evolved (or conserved during evolution) a most efficient mechanism of initiation that can by-pass the need of either a 5'-end or a cap structure" (83, 88). Such a cap-independent mechanism of initiation, obviously, is compatible with *initiation at more than one site*. Analysis of the peptides released upon tryptic digestion of the products of *in vitro* translation of polio (89-91) or mengovirus (92) RNA revealed that this was indeed the case: two N-termini were readily resolved, a finding supporting the notion of two initiation events. Independently of the cell lysate or the conditions of translation used, the relative proportion of the mengo-coded initiation peptides remained unvaried (92). Also pre-translation did not modify the pattern of the sequences of mengo RNA protected by ribosomes, suggesting that no regulatory mechanism was operative during the *in vitro* translation (83). These findings have been recently confirmed and extended: *in vivo* translation of FAMD virus in BHK-21 cells has been reported to initiate at two sites (bases 805-807 and 889-891, 28 codons apart in the same reading frame) that are used with equal frequency during the *in vitro* (but not *in vivo*) translation (93).

Proteolytic processing

Twelve to fifteen peptides were identified in picornavirus infected cells (94). Their combined molecular weights exceeded largely the total coding capacity of the viral genome (5, 95). The apparent paradox was solved when pulse-chase labeling experiments, kinetics studies, and peptide fingerprint analysis showed that the observed intracellular peptides were precursor/products in a chain of cascade cleavages from a single, large polyprotein (96-98). Pactamycin mapping appeared to provide additional support to this model (95, 99).

Proteolytic processing of the polyprotein generates a series of intermediate products following a well preserved pattern. This is summarized in Fig. 2, where the peptides are identified according to the nomenclature described in ref. 6. The scheme (also known as L-4-3-4 after the number of final peptides generated from each primary cleavage product) can be summarized as follows:

a) primary cleavage of the polyprotein generate 3 large peptides: P1, P2, and P3, which in *cardio-* and *aphto-* viruses are preceded by a short Leader (L) protein. Two carboxy co-linear leader peptides have been described in FAMD virus (Lab and Lb, 20 and 16 kDa, respectively) as a result of the initiation of translation at two different sites (see above and ref. 93).

The first cleavage (P1/P2) occurs in the nascent chain, and is probably catalysed by peptide 2A (M.W. 15,000). In *cardio-* and *aphto-*viruses this first cleavage appears to take place within P2 instead.

b) subsequent processing of P1 produces the four *structural* proteins VP1, VP2, VP3, and VP4;

c) cleavage of P3 originates the viral RNA polymerase (3D), a protease responsible for all but the first and last cleavage of the polyprotein (3C), the RNA-linked VPg (3B), and its N-terminal extension (3A) (ref. 100);

d) out of the 3 peptides originated from P2, the function(s) of 2B have not yet been defined, 2A is another protease (conceivably responsible for the first cleavage, i.e: between P1 and 2A it self). 2C appears to be involved in the development of the guanidine-resistant character (101);

e) the cysteine-type protease 3C (M.W. 20,000) seems to cleave specifically at Glu/Gly pairs. In the polyprotein these cleavage sites are surrounded by extensive flanking regions rich in helix-breaking amino acids (notably, histidine and threonine);

f) the final maturation step (cleavage of VP4/VP2 out of peptide 1AB) is an *auto-catalytic* event, triggered by the vicinity of the viral RNA acting as a proton acceptor (see chapter 2, and ref. 3).

The viral proteases (2A and 3C) cleave them selves out of the polyprotein while still in the nascent chain. The genome of FAMD virus differs slightly from this scheme in as much as it encodes: a) two leader peptides and a much shorter protein 2A (35-37, 102), and b) three in-tandem VPgs (103).

D. Replication of Picornavirus RNA

Newly synthesized viral RNA can be detected about 1 hour after infection: following an initial phase of synthesis at exponential rate (2-3 hours), viral RNA accumulates at linear rate during 3-4 hours, and then tends to plateau.

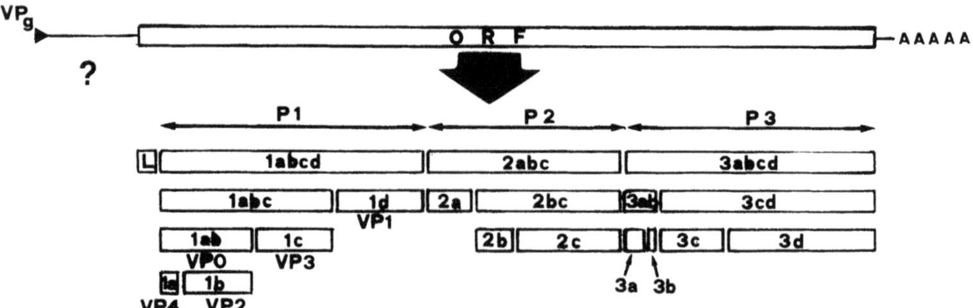

Figure 2. Proteolytic cleavage of picornavirus polyprotein. Peptides are identified according to the nomenclature adopted at the 3rd. Meeting of the European Study Group on Molecular Biology of Picornaviruses (Urbino, Italy, 1983) (ref. 6)

Three viral RNA species can be extracted from picornavirus infected cells: (a) single-stranded RNA which is identical with virion RNA; (b) the Replicative Form (RF), a very stable double-stranded molecule formed by an intact RNA chain, hydrogen-bond to a full-length complementary strand; and (c) the Replicative Intermediate (RI), a complex molecule with a d-s core and single-stranded nascent chains hydrogen-bond to the core at the growing point (for a more extensive description, see ref. 104).

Single-stranded virion-like RNA is by far the most abundant RNA species. Negative-strand RNA is never found *free*, but in the d-s RF or in the core of RI. This asymmetry of transcription suggests the existence of a self-regulating mechanism that keep balance of the (+) and (−) strand production.

Every viral RNA molecule, be it (+), (−) strand contain VPg attached to the 5'-terminal pUp (105-107).

The actual site of RNA synthesis is a membrane-bound structure, the Replication Complex (RC) (108), where RI RNA can be found in tight association with the enzymic complex. Isolated RC can continue the synthesis of viral RNA *in vitro*, but they proved unable to re-initiate RNA synthesis: in the absence of protein synthesis, there is no initiation of RNA synthesis, suggesting that a continuous supply of a viral protein may be required to secure initiation.

Viral replicase, VPg, and host factor

The virus-coded RNA polymerase has been purified from crude preparations of RC (110, 111). The solubilized enzyme can transcribe *in vitro* from different RNA templates, provided the reaction is properly primed (112-114): oligo(U) has been used in reactions where poly(A)$^+$ RNA served as template. It was later observed that a protein from uninfected cells (termed "host factor") could obviate the need of an oligo(U) primer (112, 114). The host factor appears to be a terminal uridyl transferase (TuT), able to add uridylate residues at the 3' end of virion RNA (115, 116). Presumably, the short U-tract so formed is able to anneal back to the poly(A) stretch and to form a transient hairpin structure that the RNA polymerase may use as a starter. The involvement of VPg (or its precursor, peptide 3ab) in RNA replication is borne out by several observations, though its precise role is still unclear: antibodies raised against VPg inhibit the replicase reaction primed by host factor, but are ineffective against the oligo(U) primed transcription (116, 117). Moreover, anti-VPg antibodies precipitate VPg-RNA complexes only in host factor — but not oligo(U) — primed reactions, suggesting that VPg (or its precursor) enter the complex *after* priming.

3. GENETIC RECOMBINATION

Recombination between two genetically distinct picornavirus variants was observed more than 20 years ago (118). Several recent reports have described "cross-over"-like recombinants of picornaviruses, whose genomes carried defined portions of the RNAs of the parental strains (119-125). In at least two cases, conclusive evidence of this peculiar event has been provided by sequencing the parental and the recombinant RNAs (125, 126). While the mechanism of the "cross-over" is still far from clear, several remarkable features emerged from the analysis of the recombination site:

i) the nucleotide sequence was consistent with the protein analysis of the recombinants;
ii) the number of nucleotides involved in the "cross-over" site varied from a minimum of 2 to a maximum of 32;
iii) there was no common or recognisable "consensus" pattern as in the sequences involved in RNA splicing; and
iv) recombination does not require large regions of perfect homology at the site of the "cross-over" site (126).

4. MOLECULAR CLONING

The genome of representatives of all groups of picornaviruses (including four independent isolates of Hepatitis A, virulent and attenuated poliovirus) have ben cloned into recombinant plasmids.

A. Infectious cDNA

Recombinant plasmids containing a complete representation of the genome of poliovirus I have been constructed, and used to transfect HeLa cells. Such recombinants proved infectious, i.e: able to express the viral genome, and to generate progeny virions (127). Insertion of SV-40 transcription and replication signals apparently enhanced the production of virus following transfection of *COS-1* cells (128). However, transfection of *HeLa* cells with a similar construct carrying a full-length cDNA copy of human rhinovirus 14 genome failed to produce progeny virions, although the RNA transcribed *in vitro* from the recombinant plasmid was infectious (129). A plausible explanation of such different behaviour came from a series of elegant experiments by K. Kean and collaborators, who showed that SV-40 regulatory signals were of little effect on the infectivity of the recombinant plasmid in the absence of T- antigen (which is constitutively expressed in COS-1 cells), and plasmid replication (130).

In vitro transcription of cloned cDNA yields RNA representations of both strands of the recombinant plasmids, i.e: (+) or (−) strand RNA, containing additional sequences at the ends. RNA transcripts of *virion* polarity that were obtained by either direct transcription of the plasmid cDNA (129, 131), or after secondary transcription of the (−) strand RNA (132) were infectious. The additional sequences present in those transcripts appear to reduce their infectivity, suggesting a defined functional role of the native 5' end (129, 131).

Chimeras

The possibility of generating progeny virions upon transfection with the recombinant plasmid, or following infection with *in vitro* produced RNA paves the way to virtually unlimited genomic manipulations by site-directed mutagenesis or insertion of entire fragments of the genome of one variant into another one, as a first step to investigate viral functions or to transfer desired characters into the progeny. Both approaches proved successful: insertion of a 8-bp fragment into the 3' extra-cistronic region of poliovirus cDNA (133), or the transfer of a 405-nucleotide piece from the 5' extra-cistronic region of coxsackie B3 RNA into polio 1 cDNA (134), generated *ts* polio mutants. Conceivably, the construction of such chimeras may prove also a valuable tool in the design and construction of new vaccine strains (135).

B. Conservation, Homology and Divergency

Comparison of the predicted amino acid sequence of the polyprotein of several picornaviruses revealed that the peptides derived from P3 (involved mainly in replication) are the most conserved ones. Predictably, the reverse holds true for P1 and the capsid proteins where selective pressure favors mutation.

The closest relationships were found between Rhino- and Entero-viruses, and then between Aphto- and Cardio-viruses. Hepatitis A holds a place somehow apart. Similar computer assisted analysis of their sequences proved that although the Theiler's murine encephalomyelitis virus lacks a poly(C) tract, it is closely related to the cardioviruses EMC and mengo (41).

Besides the obvious taxonomical interest of these studies, a comparison of the sequences of virulent and attenuated vaccine viruses may shed some light into the molecular grounds of the attenuation process. In the case of poliovirus type 1 and the corresponding attenuated Sabin strain (LSc,2ab) 51 point mutations were observed, scattered all over the genome (34). Only 21 out of those 51 mutations resulted in amino acid substitutions (135). Ten point mutations (and three amino acid changes) separate the parental neurovirulent poliovirus type 3 (Leon/37) from the corresponding attenuated Leon 12 a,b vaccine (135). None of the substitutions observed in the transition from wild type to attenuated vaccine in poliovirus type 1 have a similar counterpart at the corresponding position of poliovirus type 3, although the is over 90% homology between the genomes of these two serotypes. Surprisingly, the reversion to neurovirulence proved to be consistently associated with a back mutation (U back to C) at position 472 of poliovirus type 3, i.e: in the middle of the 5' extra-cistronic (silent ?) region of the genome (136). The role (if any) of this point mutation in the reversion to wild type remains to be clarified.

5. OPEN QUESTIONS

"In the beginning, it was poliovirus..."

Over the past few years we have witnessed profound changes in fundamental concepts and ideas, and the generally accepted view of picornaviruses as "the simplest animal viruses" is undergoing overdue revision. Such unorthodox views as the *internal* initiation of translation, and the existence of two active, legitimate initiation sites constitute now integral part of the common wisdom in the field.

Though, the list of open, challenging questions has kept (fortunately !) growing at faster a pace than our ability to provide answers.

We would like to know, for instance, the role of the unusually long extracistronic regions preceding and following the major ORF: are they really devoid of attributes related to gene expression, or do they encode so far undetected short-lived "agno-peptides"?

We would like to know whether VPg plus its N-extension 3a (and possibly 2c) are integral part of the polymerase complex, and what is the role that each of these peptides play in the process of RNA replication.

Similarly, the function(s) of peptide 2b remains to be clarified.

We would like to see whether all picornaviruses make use of a similar mechanism to shut off the host cell macromolecular synthesis, and if not we would have to explain how is it so that they do not shut off each other.

We would have to determine to what extent can we artificially exchange portions of the genomes of different picornaviruses, and what are the structural limits beyond which viable chimeras cannot be constructed.

The list can be extended, and for the time being we can only make plausible guesses, but there is reason to hope that the efforts to unravel these questions will provide further vigour to research in this field.

Acknowledgements

The author's work described in this review was supported with grants from the Italian National Research Council (CNR), within the "Progetto Finalizzato Ingegneria Genetica" and with funds from the Istituto Pasteur/Fondazione Cenci Bolognetti.

6. REFERENCES

1) Melnick, J. L. (1983), Intervirology, **20**,61-100

2) Enders, J.F., Weller, T.H., & Robbins, F.C. (1949), Science, **109**, 85

3) Scraba, D.G. (1979) The Picornavirion:Structure and Assembly. In: *The Molecular Biology of Picornaviruses*, pp.1-24, R. Perez Bercoff, ed. Plenum Publishing Co., N.York & London

4) Rueckert, R. R. (1976), in *Comprehensive Virology*, pp. 131-213, H. Fraenkel-Conrat & R.R. Wagner, eds., Plenum Publishing Co, New York.

5) Rueckert, R. R., Mathews, T. J., Kew, O.M., Pallansch, M., McLean, C., & Ominianowski, D. (1979) Synthesis and Processing of Picornavirus Polyprotein. In: *The Molecular Biology of Picornaviruses*, pp.113-126, R. Perez Bercoff, ed., Plenum Publishing Co., N. York & London.

6) Rueckert, R.R. & Wimmer, E. (1984) J. Virology, **50**, 957-959.

7) McGeady, M.L., & Crowell, R.L. (1981), J. Gen. Virol., **55**, 42-55.

8) Cavanagh, D., Rowlands, D.J., & Brown, F. (1978), J. Gen. Virol. **41**, 255-264.

9) Baxt, B., & Bachrach, H.L. (1980), Virology, **104**, 42-55.

10) Dunker, A.K., & Rueckert, R.R. (1971), J. Mol. Biol. **58**, 217.

11) Dunker, A.K. (1979), Virology, **97**, 141.

12) Hordern, J.S., Leonard. J.D., & Scraba, D.S. (1979), Virology, **97**,131.

13) Brown, F., & Cartwright, B., (1961), Nature (London), **192**, 1163.

14) Talbot, P., & Brown, F., (1972), J. Gen. Virol. **15**, 163.

15) Benecke, T.W., Habermehl, K.O., Diefenthal, W., & Buckholz, M. (1977), J. Gen. Virol., **34**, 387-390.

16) Lonberg-Holm, K., & Butterworth, B.E. (1976), Virology, **71**207-216.

17) Hordern, J.S., Leonard, J.D., & Scraba, D.S. (1979), Virology, **97**,131-140.

18) Wetz, K., & Habermehl, K.O. (1982), J. Gen. Virol. **59**, 387-401.

19) Hogle, J.M., Chow, M., & Filman, D.J. (1985), Science, **229**, 1358-1365.

20) Rossmann, M.G., Arnold, E., Erickson, J.W., Frankenberger, E.A., Griffith, J.P., Hecht, H.J., Johnson, J.E., Kamer, G., Luo, M., Mosser, A.G., Rueckert, R.R., Sheng, B., & Vriend, G. (1985), Nature (London), **317**, 145-153.

21) Rueckert, R.R. (1985) Picornavirus and Their Replication. In: *Virology*,pp.705-738. B. Fields, D.M. Knipe, R.M. Chanock, J.L. Melnick, B. Roizman, and R.E. Shope, eds. Raven Press, N. York.

22) Racaniello, V.R., & Baltimore, D. (1981), Proc. Natl. Acad. Sci. USA, **78**, 4887-4891.

23) Kitamura, M., Semler, B., Rothberg, P.G., Larsen, G.R., Adler, C.J., Dorner, A.J., Emini, E.A., Hanecak, R., Lee, J.J., van der Werf, S., Andersen, C.W., & Wimmer, E. (1981), Nature (London), **291**, 547-553.

24) Toyoda, H., Kohara, M., Kataoka, Y., Suganuma, T., Omata, T., Imura, M., & Nomoto, A. (1984), J. Mol. Biol. **174**, 561-585.

25) Venuti, A., Di Russo, C., Del Grosso, N., Patti, A-M., Ruggeri, F., De Stasio, R., Martiniello, M.G., Pagnotti, P., Degener, A.M., Midulla, M., Pana', A., & Perez Bercoff, R. (1985), J. Virol.**56**, 579-588.

26) Najarian, R., Caput, D., Gee, W., Potter, S.J., Renard, A., Merrywheather, J., Van Nest G., & Dina, D. (1985), Proc. Natl. Acad. Sci. USA, **82**, 2627-2631.

27) Baroudy, B.M., Ticehurst, J.R., Miele, T.A., Maizel, J.V., Purcell, R.H., & Feinstone, S.M. (1985), Proc. Natl. Acad. Sci. USA, **82**, 2143-2147.

28) Linemeyer, D.L., Menje, J.G., Martin-Gallardo, A., Hughes, J.V., Young, A., & Mitra, S.W. (1985), J. Virol. **54**, 247-255.

29) Stalhandske, P., Lindberg, M., & Patterson, U. (1984), J. Virol. **51**, 742-746.

30) Tracy, S., Liu, H.L., & Chapman, N.M. (1985), Virus Res. **3**, 263-270.

31) Callaghan, P., Mizutami, S., & Colonno, R.J. (1985), Proc. Natl. Acad. Sci. USA **82**, 732-736.

32) Skern, T., Sommergruber, W., Blaas, D., Gruendler, P., Fraundorfer, F., Pieler, C., Fogy, I., & Kuechler, E., (1985), Nucleic Acids Res. **13**, 2111-2126.

33) Stanway, G., Hughes, P., Mountford, R., Minor, P., & Almond, J.W. (1984), Nucleic Acids Res. **12**, 7859-7875.

34) Boothroyd, J., Harris, T., Rowlands, D., & Lowe, P. (1982), Gene, **17**, 153-161.

35) Carroll, A., Rowlands, D.J., & Clarke, B.E. (1984), Nucleic Acids Res. **12**, 2461-2472.

36) Robertson, B.H., Grubman, M.J., Wendell, G.N., Moore, D.M., Welsh. J.D., Fischer, T., Dowberko, D.J., Yansura, D.G., Small, B., & Kleid, D.G. (1985), J. Virol. **54**, 651-660.

37) Forss, S., Strebel, K., Beck, E., & Schaller, H. (1984), Nucleic Acids Res. **12**, 6587-6601.

38) Newton, S.E., Carroll, A.R., Campbell, R.O., Clarke, B.E., & Rowlands, D.J. (1985), Gene **40**, 331-336.

39) Werner, G., Rosenwirth, B., Bauer, E., Seifel, J-M., Werner, F.J., & Besemer, J. (1986), J. Virol. **57**1084-1093.

40) Palmenberg, G.A., Kirby, E.M., Janda, M.J., Drake, N.L., Duke, G.M., Potratz, K.F., & Collet, M.S., (1984) Nucleic Acids Res. **12**, 2969-2985.

41) Pevear, D.C., Calenoff, M., Rozhon, E., & Lipton, H.L. (1987), J.Virol. in press.

42) La Monica, N., Merian, C., & Racaniello, V. (1986), J. Virol. **57**, 515-525.

43) Stanway, G., Cann, A.J., Hauptmann, R., Hughes, P., Mountford, R.C., Minor, P.D., Schild, G.C., & Almond, J.W. (1983), Nucleic Acids Res. **11**,5629-5643.

44) Stanway, G., Hughes, P.J., Mountford, R.C., Reeve, P., Minor, P.D., Schild, G.C., & Almond, J.W. (1984), Proc. Natl. Acad. Sci. USA, **81**, 5885-5889.

45) Palmenberg, A.C. (1986). In: *The Molecular Biology of Positive Strand Viruses*, D.J. Rowlands, B.H.J. Mahy, & M. Mayo, eds., Academic Press, New York.

46) Fellner, P. (1979), General Organization and Structure of the Picornavirus Genome. In: *The Molecular Biology of Picornaviruses*, pp.25-48, R. Perez Bercoff, ed. Plenum Publishing Co., New York.

47) Perez Bercoff, R., & Gander, M. (1977), Virology, **80**, 426-429.

48) Harris, T.J.R., & Brown, F. (1976), J. Gen. Virol. **33**, 493-501.

49) Black, D.N., Stephenson, P., Rowlands, D.J., & Brown, F. (1979), Nucleic Acids Res. **6**, 2381-2390.

50) Costa Giomi, M.P., Bergman, I.E., Scodeller, E.A., Auge de Mello, P., Gomez, I., & La Torre, J.L., (1984), J. Virol. **51**, 799-805.

51) Perez Bercoff, R., & Gander, M. (1978), FEBS Lett. **96**, 378-386.

52) Crowell, R.L., & Landau, B.J. (1983) Receptors in the Initiation of Picornavirus Infections. In: *Comprehensive Virology*, **18**, pp.1-42, H. Fraenkel-Conrat & R.R. Wagner, eds. Plenum Publishing Co., N.York & London

53) Zeichardt, H., Wetz, K., Willingmann, P., & Habermehl, K.O. 1985), Ξ. Γεν. Ωιρολ. **66**, 483-492.

54) Mandel, B. (1965), Virology **25**, 152.

55) Levitt, N.H., & Crowell, R.L. (1967), J. Virol. **1**, 293.

56) Lonberg-Holm, K., & Phillipson, L. (1974). In: *Monographs in Virology*, **9**, 1-148.

57) Mapoles, J.E., Krah, D.L., & Crowell, R.L. (1985), J. Virol. **55**, 560, 566.

58) Krah, D.L., & Crowell, R.L. (1982), Virology **118**, 148-156.

59) Krah, D.L., & Crowell, R.L. (1985), J. Virol. **53**, 867-870.

60) Mendelsohn, C., Johnson, B., Lionetti, K.A., Nobis, P., Wimmer, E., & Racaniello, V.R. (1986), Proc. Natl. Acad. Sci. USA **83**, 7845-7849.

61) Lucas-Lenard, J. (1979) Inhibition of Cellular Protein Synthesis After Virus Infection. In: *The Molecular Biology of Picornaviruses*, pp.73-100. R. Perez Bercoff, ed., Plenum Publishing Co., N. York & London.

62) Etchinson, D., Milburn, S.C., Ederg, I., Sonenberg, N., & Hershey, J.W.B. (1982), J. Biol. Chem. **257**, 14806-14810.

63) Plagemann, P.G.W. (1968), J. Virol. **2**, 461-473.

64) Plagemann, P.G.W. & Swim, H.E. (1966), J. Bacteriol. **91**, 2317-2326.

65) Rose, J.K., Trachsel, H., Leong, K., & Baltimore, D. (1978), Proc. Natl. Acad. Sci. USA **75**, 2732-2736.

66) Helentjaris, T., & Ehrenfeld, E. (1978), J.Virol. **26**, 510-521.

67) Sonenberg, N., Morgan, M.A., Merrick, W.C., & Shatkin., A.J. (1978), Proc. Natl. Acad. Sci. USA **75**, 4843-4847.

68) Sonenberg, N., Rupprecht, K.M., Hecht, S.M., & Shatkin, A.J. (1979), Proc. Natl. Acad. Sci. USA **76**, 4345-4349.

69) Sonenberg, N., (1979), Nucleic Acids Res. **9**, 1643-1656.

70) Etchinson, D., Hansen, J., Ehrenfeld, E., Edery, I., Sonenberg, N., Milburn, S., & Hershey, J.W.B. (1984), J. Virol. **54**, 832-837.

71) Etchinson, D., & Fout, S. (1985), J. Virol. **54**, 634-638.

72) Alonso, M.A., & Carrasco, L. (1982), J. Gen. Virol. **61**, 15-24.

73) Pensiero, M.M., & Lucas-Lenard, J. (1985), J. Virol. **56**, 161-171.

74) Pani, A., Julian, M., & Lucas-Lenard, J. (1986), J. Virol. **60**,1012-1017.

75) Lodish, H.F., & Rose, J.K. (1977), J. Biol. Chem. **252**,1181-1188.

76) Shatkin, A.J. (1976), Cell **9**, 645-653.

77) Petterson, R.F., Flanegan, J.B., Rose, J.K., & Baltimore, D. (1977), Nature (London) **268**, 270-272.

78) Egberts, E., Hackett, P.B., & Traub, P. (1977), Hoppe-Seyler's Z. Physiol. Chem. **358**, 463-474.

79) Hackett, P.B., Egberts, E., & Traub, P. (1978), Eur. J. Biochem. **83**, 341-352.

80) Hackett, P.B., Egberts, E., & Traub, P. (1978), Eur. J. Biochem. **83**, 353-361.

81) Ehrenfeld, E. (1979), *In vitro* Translation of Picornavirus RNA. In: *The Molecular Biology of Picornaviruses*, pp. 223-238, R.Perez Bercoff, ed., Plenum Publishing Co. New York.

82) Lawrence, C., & Thach, R.E. (1974), J. Virol. **14**, 598-610.

83) Perez Bercoff, R., & Kaempfer, R. (1982), J. Virol. **41**, 30-41.

84) Kozak, M. (1978), Cell **15**, 1109-1123.

85) Kozak, M. (1980), Cell **22**, 459-467.

86) Jackson, R.J. (1982) The Cytoplasmic Control of Protein Synthesis. In: *Protein Biosynthesis in Eukaryotes*, pp.363-418, R. Perez Bercoff, ed., Plenum Publishing Co., New York.

87) Sangar, D.V., Black, D.N., Rowlands, D.J., Harris, T.J.R., & Brown, F. (1980), J. Virol. **33**, 59-68.

88) Perez Bercoff, R. (1982). But is the 5'-End of Messenger RNA *Always* Involved in Initiation ? In: *Protein Biosynthesis in Eukaryotes*, pp.245-252. R. Perez Bercoff, ed. Plenum Publishing Co., New York.

89) Celma, M.L., & Ehrenfeld, E. (1975), J.Mol.Biol. **98**, 761-780.

90) Ehrenfeld, E., & Brown, D. (1981), J. Biol. Chem. **256**, 2656-2661.

91) Jense, H., Knauert, F., & Ehrenfeld, E. (1978), J. Virol. **28**, 387-394.

92) Degener, A.M., Pagnotti, P., Facchini, J., & Perez Bercoff, R. (1983), J. Virol. **45**, 889-894.

93) Clarke, B.E., Sangar, D.V., Burroughs, J.N., Newton, S.E., Carroll, A.R., & Rowlands, D.J. (1985), J. Gen. Virol. **66**, 2615-2626.

94) Summers, D.F., Maizel, J.V., & Darnell, J.E. (1965), Proc. Natl. Acad. Sci. USA **54**, 505-513.

95) Lucas-Lenard, J. (1979) Virus-directed Protein Synthesis. In: *The Molecular Biology of Picornaviruses*, pp. 127-148, R. Perez Bercoff, ed., Plenum Publishing Co., New York.

96) Jacobson, M.F., & Baltimore, D. (1968), Proc. Natl. Acad. Sci. USA **61**, 77-84.

97) Jacobson, M.F., Asso, J., & Baltimore, D. (1970), J. Mol. Biol. **49**, 657-669.

98) Butterworth, B.E., Hall, L., Stoltfuss, C.M., & Rueckert, R.R. (1971), Proc. Natl. Acad. Sci. USA **68**, 3083-3087.

99) Summers, D.F., & Maizel, J.V. (1971), Proc. Natl. Acad. Sci. USA **68**, 2852-2856.

100) Pallansch. M.A., Kew, O.M., Palmenberg, A.C., Golini, F., Wimmer, E., & Rueckert, R.R. (1980), J. Virol. **35**, 414-419.

101) Anderson-Silman, K., Bartal, S., & Tershak, D.R. (1984), J. Virol. **50**, 922-928.

102) Beck, E., Forss, S., Strebel, K., Cattaneo, R., & G-Feil, G. (1983), Nucleic Acids Res. **11**, 7873-7885.

103) Forss, S., & Schaller, H. (1982), Nucleic Acids Res. **12**, 6587-6601.

104) Perez Bercoff, R. (1979) The Mechanism of Replication of Picornavirus RNA. In: *The Molecular Biology of Picornaviruses*, pp. 293-318, R. Perez Bercoff, ed., Plenum Publishing Co., New York.

105) Petterson, R.F., Ambros, V., & Baltimore, D. (1978), J. Virol. **27**, 357-365.

106) Wu, M., Davidson, N., & Wimmer, E. (1978), Nucleic Acids Res. **5**, 4711-4723.

107) Thornton, G.B., Robberson, D.L., & Arlinghaus, R.D. (1981), J. Virol. **39**, 229-237.

108) Girard, M., Baltimore, D., & Darnell, J.E. (1967), J. Mol. Biol. **24**59-74.

109) Girard, M. (1969), J. Virol. **3**, 376-384.

110) Flanegan, J., Petterson, R., Ambros, V., Hewlett, M., & Baltimore, D. (1977), Proc. Natl. Acad. Sci. USA **74**, 961-965.

111) Dasgupta, A., Baron, M.H., & Baltimore, D. (1979), Proc. Natl. Acad. Sci. USA **76**, 2679-2683.

112) Dasgupta, A., Zabel, P., & Baltimore, D. (1980), Cell **19**, 423-429.

113) Baron, M.H., & Baltimore, D. (1982), J. Biol. Chem. **257**, 12351-12358.

114) Baron, M.H., & Baltimore, D. (1982), J. Biol. Chem. **257**, 12359-12366.

115) Andrews, N.C., Levin, D.H., & Baltimore, D. (1985), J. Biol. Chem. **260**, 7628-7635.

116) Andrews, N.C., & Baltimore, D. (1986), Proc. Natl. Acad. Sci. USA **83**, 221-225.

117) Baron, M.H., & Baltimore, D. (1982), Cell **30**, 745-752.

118) Pringle, C.R. (1965), Virology **25**, 48-54.

119) Romanova, L.I., Tolskaya, E.A., Kolesnikova, M.S., & Agol, V.I. (1980), FEBS Lett. **118**, 109-112.

120) McCahon, D., & Slade, W.R. (1981), J.Gen. Virol. **53**, 333-342.

121) Saunders, K., King, A.M.Q., McCahon, D., Newman, J.W.I., Slade, W.R., & Forss, S. (1985), J. Virol. **56**, 921-929.

122) King, A.M.Q., McCahon, D., Saunders, K., Newman, J.W.I., & Slade, W.R. (1985), Virus Res. **3**373-384.

123) McCahon, D., King, A.M.Q., Roe, D.S., Slade, W.R., Newman, J.W.I., & Cleany, A.M. (1985), Virus Res. **3**, 87-100.

124) Agut, H., Bellocq, C., van der Werf, S., & Girard, M. (1984), Virology **139**, 393-402.

125) Agut, H., Kean, K., Bellocq, C., Fichot, O. & Girard, M. (1987), J. Virol. (submitted)

126) Scientific Report 1984/85, Animal Virus Res. Inst., Pirbright, U.K.,pp. 11-12.

127) Racaniello, V.R., & Baltimore, D. (1981), Science, **214**, 916-919.

128) Semler, B., Dorner, A.J., & Wimmer, E. (1984), Nucleic Acids Res. **12**, 5123-5141.

129) Mizutami, S., & Colonno, R.J. (1985), J. Virol. **56**, 628-632.

130) Kean, K., Wychowski, C., Kopecka, H., & Girard, M. (1986), J. Virol. **59**, 490-493.

131) van der Werf, S., Bradley, J., Wimmer, E., Studier, F.W., & Dunn, J.J. (1986), Proc. Natl. Acad. Sci. USA, **83**, 2330-2334.

132) Kaplan, G., Lubinski, J., Dasgupta, A., & Racaniello, V. (1985), Proc. Natl. Acad. Sci. USA **82**, 8424-8428.

133) Sarnow, P., Berstein, H.D., & Baltimore, D. (1986), Proc. Natl. Acad. Sci. USA **83**, 571-575.

134) Semler, B.L., Johnson, V.H., & Tracy, S. (1986), Proc. Natl. Acad. Sci. USA **83**, 1777-1781.

135) Almond, J.W., Stanway, G., Cann, A.J., Westrop, G.D., Evans, D.M.A., Ferguson, M., Minor, P.D., Spitz, M., & Schild, G.C. (1984), Vaccine **2**, 177-184.

136) Evans, D.M.A., Dunn, G., Minor, P.D., Schild. G.C., Cann, A.J., Stanway, G., Almond, J.W., Currey, K., & Maizel, J.V. jr. (1985), Nature (London), **314**,548-550.

CHAPTER 9

THE REPLICATION OF TOGAVIRIDAE AND FLAVIVIRIDAE AT THE MOLECULAR LEVEL

MILTON J. SCHLESINGER

Department of Microbiology and Immunology, Washington University School of Medicine, St. Louis MO 63130, U.S.A.

INTRODUCTION

Until quite recently, the small enveloped viruses that contain single stranded (+) RNA (their genome is infectious and serves as an mRNA) were grouped into the *Togaviridae* family, with two major subgroups noted as alphaviruses and flaviviruses. With the elucidation of the genomic sequence of the Yellow Fever virus and additional knowledge of the structure and replication of the flaviviruses, it became clear that these viruses should be placed into a family distinct from that of the *Togaviridae*. Even so, the viruses in these two families share a number of properties and many lead to similar pathologies upon infection of mammalian hosts. Most of the members of these families were previously noted as *arboviruses* (acronym from "arthropod borne") since a portion of their natural life cycle involves growth in an arthropod vector. There are a very large number of species within these families with genotypically and serologically distinct isolates obtained throughout the world. A thorough description of the epidemiology of these viruses can be found in the Togaviruses: Biology, Structure, Replication (1).

The molecular biology of these viruses advanced rapidly in the early 1980's with the ability to prepare cDNAs of the virus' genome and to define biochemical events in the replication of these viruses in tissue culture cells. The entire genomic sequence has been determined for prototype viruses representative of the *Togaviridae* (Sindbis virus) and the *Flaviviridae* (Yellow Fever virus). In this chapter I present these sequences and describe recent information available about the structure and replication of these prototype viruses. Specific details and references can be found in *The Togaviridae and Flaviviridae* (2).

1. THE VIRION STRUCTURE

A. Togaviridae

The virions consist of a nucleocapsid core surrounded by a lipid bilayer containing glycoproteins which project outwards as spikes from the membrane (3). These spikes consist of heterodimers, noted E1 (439 amino acids) and E2 (423 amino acids) arranged in a trimeric structure (Fig. 1). The heterodimers form an icosahedral surface lattice of triangulation number T = 4 (3, 4).

The core contains a single RNA (11,703 bases, see Fig. 2) and 240 copies of a single capsid protein (264 amino acids) also arranged in icosahedral symmetry with a T = 4 lattice. The radius of the core is about 20-23 nm and the spikes protrude to an outer radius of about 30-32 nm. Both E1 and E2 are embedded in the lipid bilayer and the E2 glycoprotein has an additional 33 amino acids (the carboxy-terminal region of the polypeptide) inside the membrane. The latter are postulated to interact with the capsid during the final assembly of the virion. The trimeric clustering of the glycoprotein heterodimers resembles the influenza virus hemagglutinin trimer and may serve to sequester hydrophobic regions responsible for the fusogenic properties of the glycoprotein. Little is known at present about the packing of the RNA in the core, but spherical plant viruses with similar geometries have their RNA tightly condensed. Most of the RNA secondary structure is retained with segments binding to flexible portions of the amino-terminal region of the capsid polypeptide.

B. Flaviviridae

Much less detail is known about the virions' structure although the overall dimensions are similar to that of the *Togaviridae*. A single glycoprotein, E (493 amino acids), forms the spike but its oligomeric state is not known. A small non-glycosylated protein, M (75 amino acids) is also bound to the envelope lipid. The nucleocapsid core consists of a single strand of RNA (10,862 bases) and multiple copies of a capsid protein (~100 amino acids). The flavivirus nucleocapsid is somewhat smaller than the cores of the *Togaviridae* and, unlike the latter, interacts

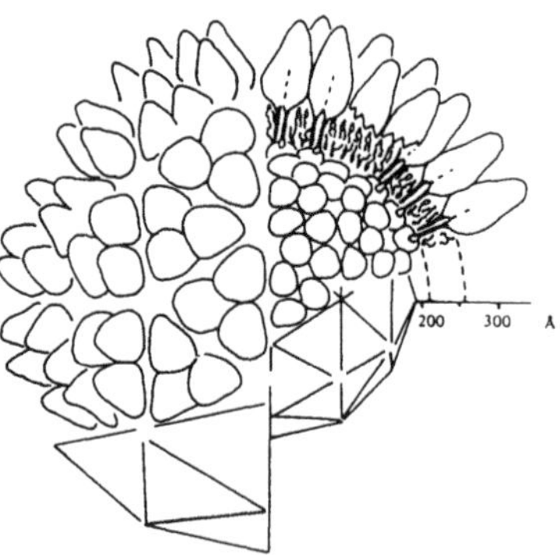

Figure 1. Organization of protein and lipid in Sindbis virus particles and Semliki Forest is basicly the same, although the latter is about 10 A smaller. E1 + E2 heterodimers are represented by the pear-shaped units. When clustered in trimers, they produce the grooved patterns observed in electronmicrographs. The hydrophobic C-termini of E1 and E2 are deeply rooted in the lipid bilayer. Those of E2 have a small internal domain shown here making contact with a site on a nucleocapsid subunit. The bilayer is symbolized in the diagram by an array of lipid molecules between radii 210 and 255 A. The lower part of the drawing depicts the icosahedrical lattices typical of the outer glycoprotein layer and inner core at the surface of the virion.

with the E glycoprotein at the Golgi membrane where assembly of virions occurs (5). The *Togaviridae*, however, assemble almost exclusively at the cell surface membrane.

2. GENOMIC ORGANIZATION

A. *Togaviridae*

Seven genes (eight if the 6 kDa "linker" between E2 and E1 is included) are encoded in the Sindbis virus RNA. The sequence of this RNA has been determined from cDNAs (6) and is presented in Fig. 3. RNA transcribed from this sequence, which had been cloned and inserted into an *E. coli* vector containing an SP6 promoter immediately adjacent to the 5' end of the virus genomic sequence, is capable of producing infectious Sindbis virions when the *in vitro* transcribed RNA is added to chicken embryo fibroblasts (unpublished data of C. Rice, H. Huang, R. Levis, J.H. Strauss and S. Schlesinger).

The genomic RNA has a ^{7}mG 5' cap structure followed by 59 nucleotides before the first AUG codon which serves to initiate translation of the genomic RNA. Sequences for four nonstructural polypeptides, noted as nsP1, nsP2, nsP3 and nsP4, follow (see Fig. 6) with an OPAL termination codon located at the putative terminus of nsP3 and a set of three closely spaced termination codons at the carboxy terminus of nsP4. The latter termination signals lie within a 48 base sequence referred to as the "junction" region. This region is also present as the 5'

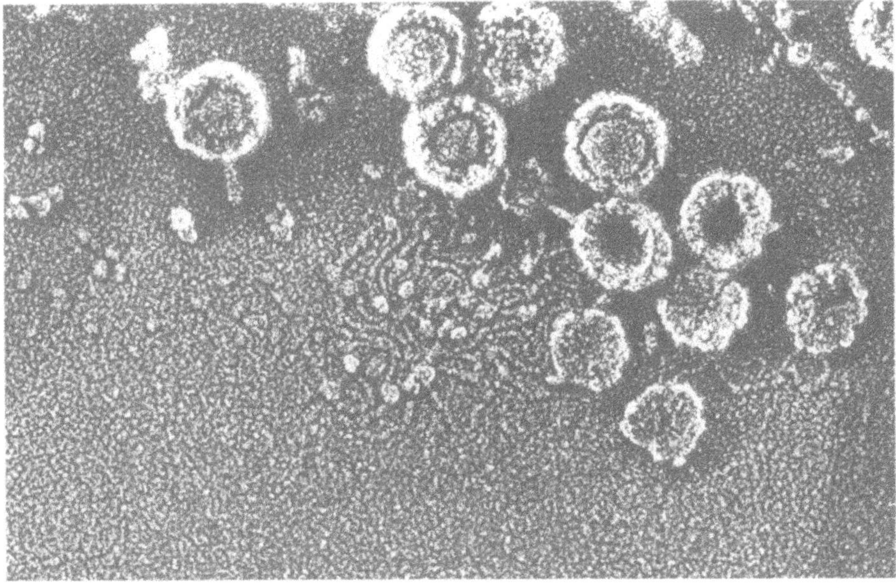

Figure 2. Sindbis virions and empty nucleocapsids with the viral RNA genome spread. Sample prepared by Dr. J. Heuser, Washington University Medical School.

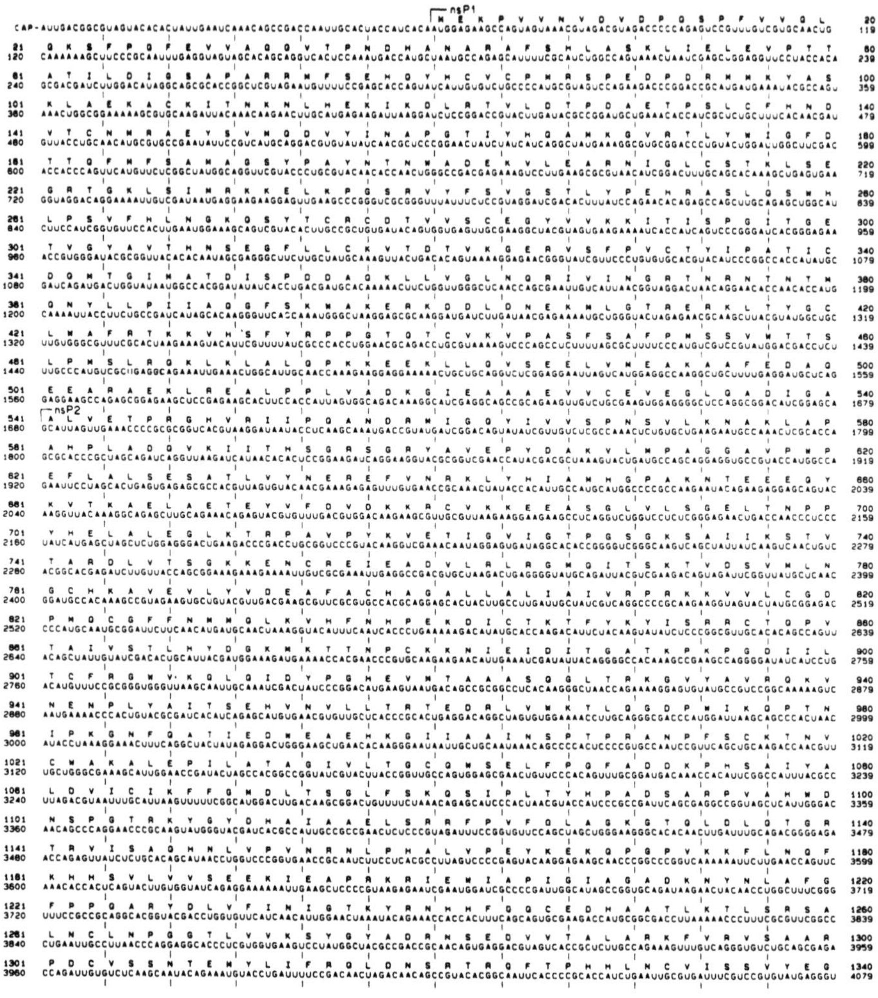

Figure 3. **A.** The nucleotide sequence of Sindbis virus genomic RNA. Reprinted from Strauss E.G. and Strauss, J.H. (6).

untranslated leader for the 26S subgenomic mRNA. The junction sequence is followed by the sequences for the capsid protein and then for the p62/E2 glycoprotein, the 6 kDa linker protein, and the E1 glycoprotein. A 3' untranslated sequence of 322 bases followed by a poly(A) sequence complete the genome structure. Four regions in the Sindbis virus sequence show a very high degree of conservation with other alphaviruses. These are:

i) a 19 nucleotide stretch at the 3' terminus adjacent to the poly(A) tract;
ii) 21 nucleotides in the junction region;
iii) 51 nucleotides beginning ~150 bases from the 5' end of the RNA; and
iv) a conserved "stem-loop" structure at the extreme 5' terminus.

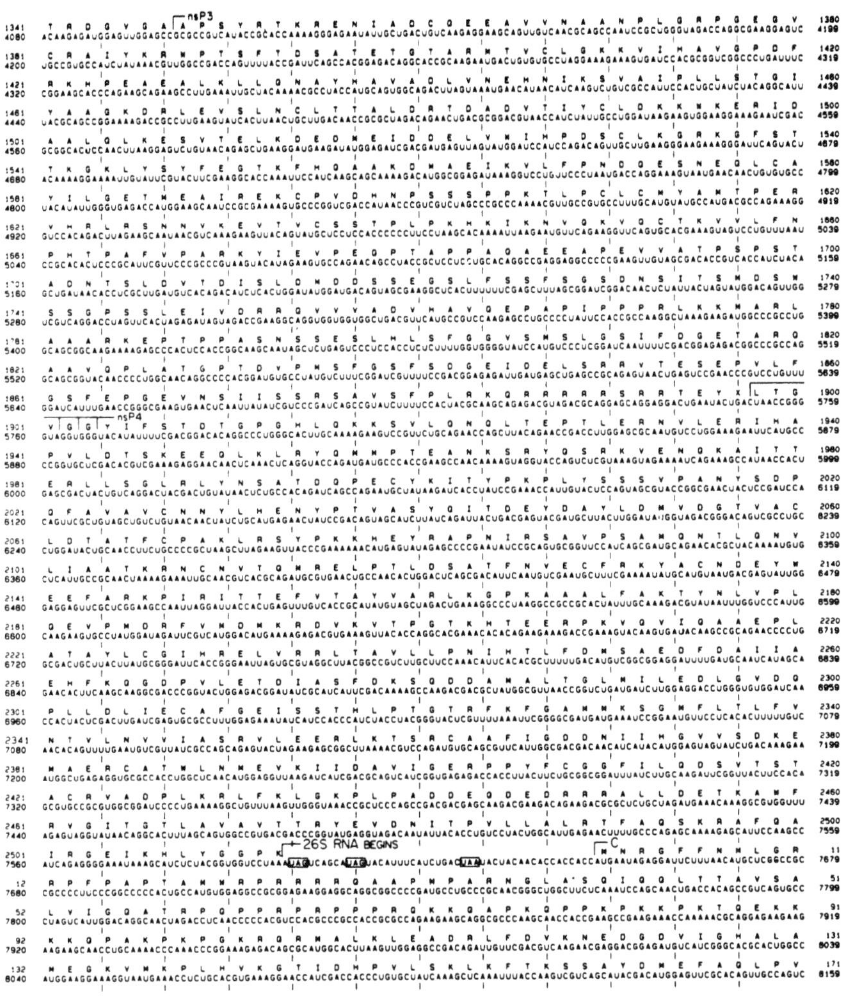

Figure 3. B

B. *Flaviviridae*

The sequence of genomic RNA for yellow fever virus, derived from cDNAs (7), is presented in Fig. 4, but the assignments for some of the protein has not been definitively made.

There is an initiation site for translation located 118 nucleotides from a ^{7}mG capped 5' end of the RNA. In contrast to the *Togaviridae*, the virus structural genes are located at the immediate 5' segment of the RNA in the order capsid-preM/M-E followed by the nonstructural genes in the order Ns1-ns2a-ns2b-Ns3-ns4a-ns4b-Ns5. There is a 3' non translated region of 511 nucleotides and no poly(A) tract.

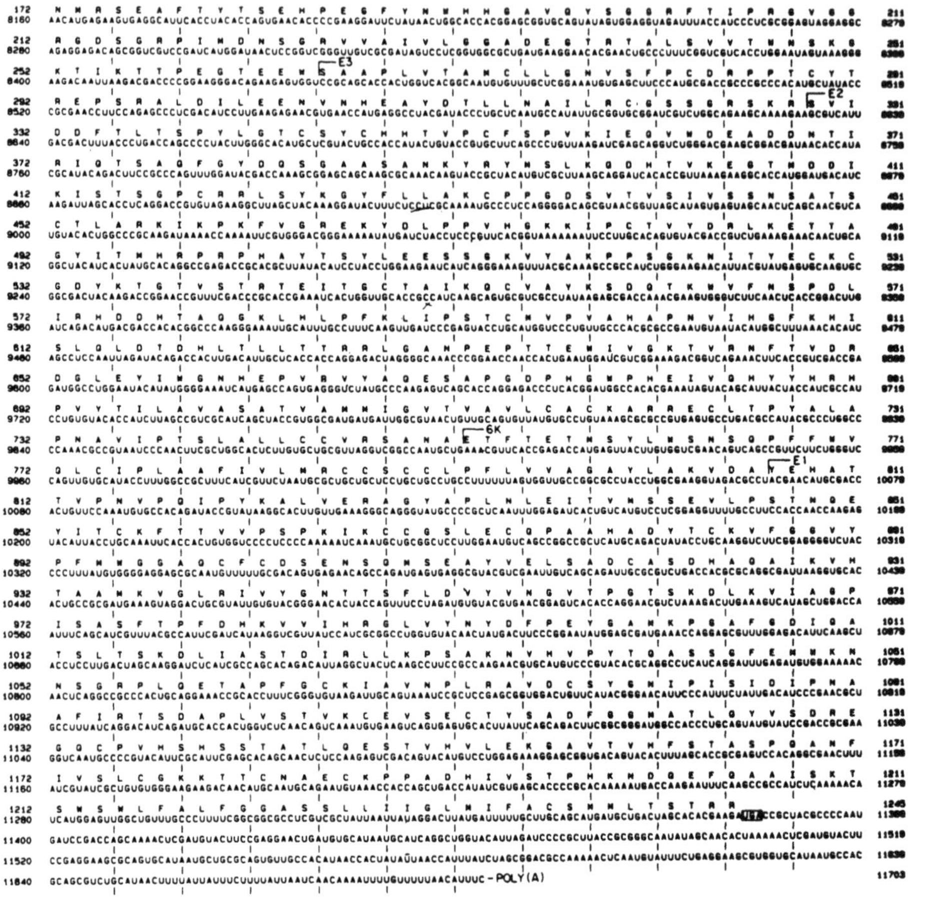

Figure 3. C

There is also no "junction" region and apparently *no subgenomic mRNA species* expressed during replication. There are insufficient data from sequences of other flaviviruses to determine which regions of the genome are conserved.

3. VIRUS REPLICATION

A. Togaviridae

The stages of viral replication described here are primarily those carried out by alphaviruses, specifically in vertebrate cell cultures infected with the prototype viruses, Sindbis and Semliki Forest. The replication cycle can be very rapid with up to 10,000 new virions per cell appearing in the culture fluid by 7 to 8 hours post infection when new virions first appear in the culture fluid.

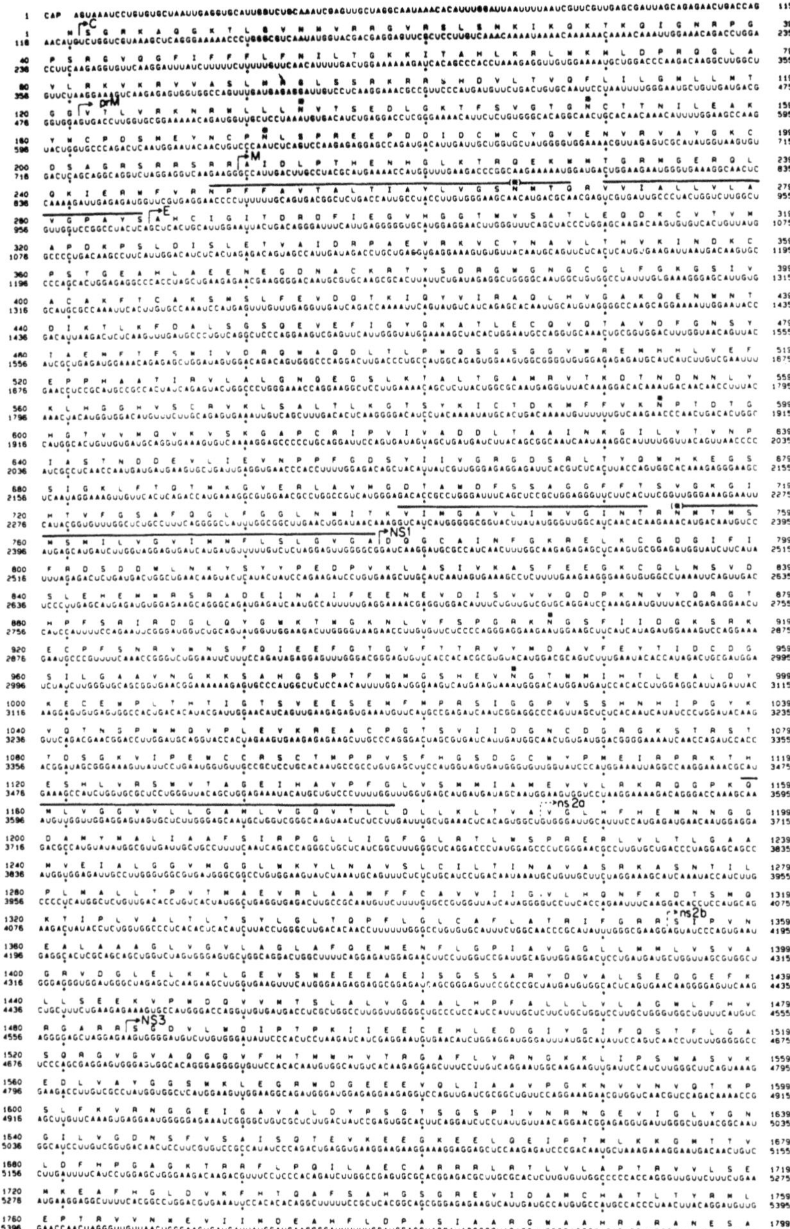

Figure 4. A. The nucleotide sequence of yellow fever virus genomic RNA. Reprinted from Rice, C. et al. (7).

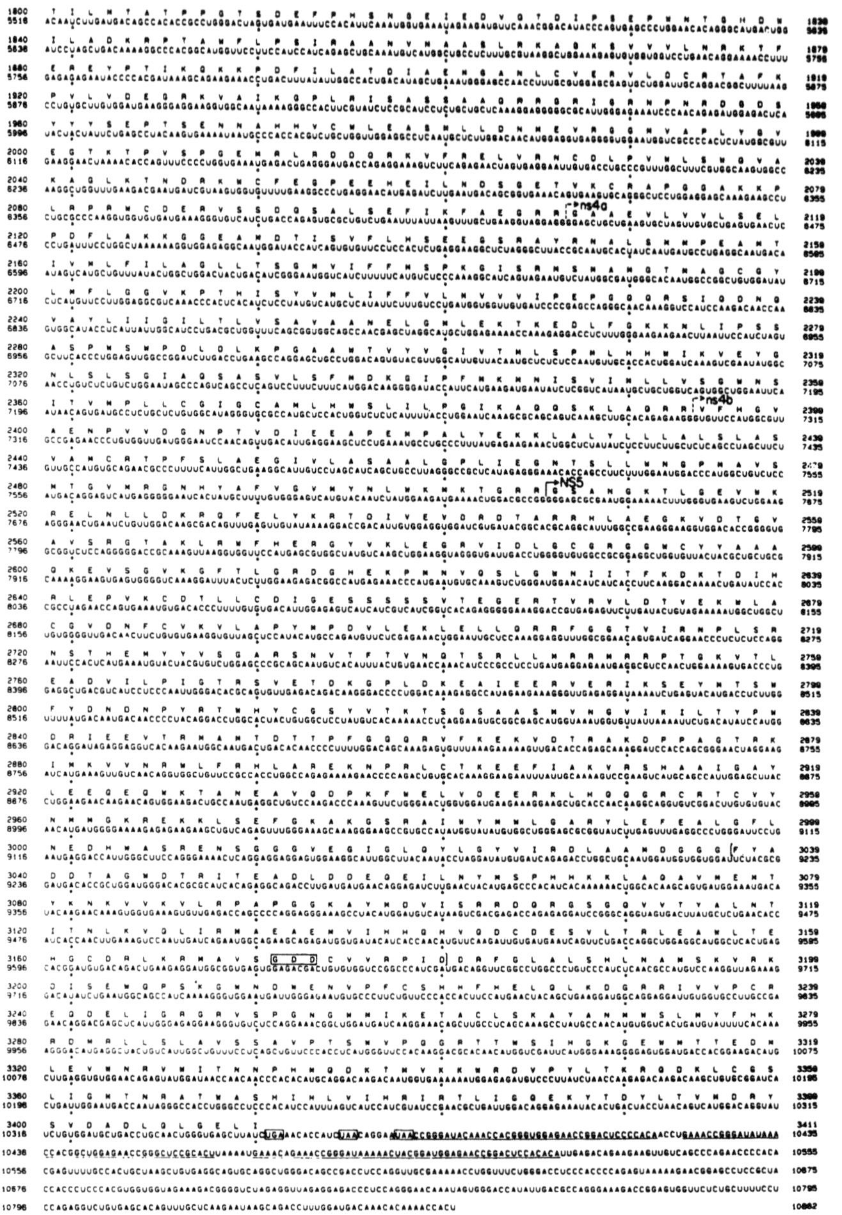

Figure 4. B

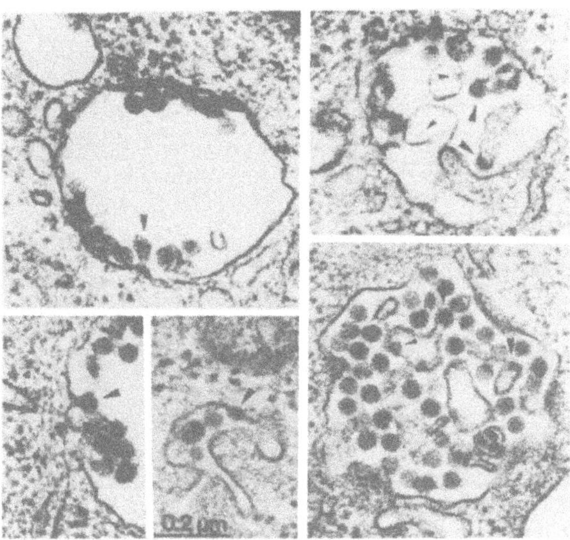

Figure 5. Fusion of Semliki Forest Virus from endosomes. Arrows show fusion of virion with endosomal membranes. Reprinted from Kielian, M. and Helenius, A. (9).

Virus Tropism

These viruses grow in a wide variety of animal tissues, in arthropods and in many different kinds of cell cultures. In nature, they are found primarily in birds and arthropods. Despite considerable effort, investigators have not yet definitively identified a specific cell surface receptor to which an alphavirus binds. There are, however, data suggesting that ionic interactions between virus and host cell are a major determinant for cell tropism. For example, "mutations" leading to changes in the isoelectric points of Sindbis virus E1 and E2 glycoproteins were solely responsible for extending the tropism of the virus from chicken embryo fibroblast cells to mouse lymphocytes (8).

Virus Uptake and Uncoating

Once bound to a cell's surface, alphaviruses utilize the host's receptor-mediated endocytotic pathway for gaining entry to the cell's cytoplasm (9). In this pathway, the virion is first brought to a coated pit where endocytosis occurs. A coated vesicle forms and subsequently is converted to an acidified endoplasmic vesicle. At the lower pH a fusogenic activity present in the E1/E2 glycoprotein is activated. This leads to the fusion of virion lipid membrane with the endosome membrane and the entry of the virus nucleocapsid into the cytoplasm (Fig. 5). The lower pH may also release the RNA from the capsid; however, the molecular details of uncoating are not known.

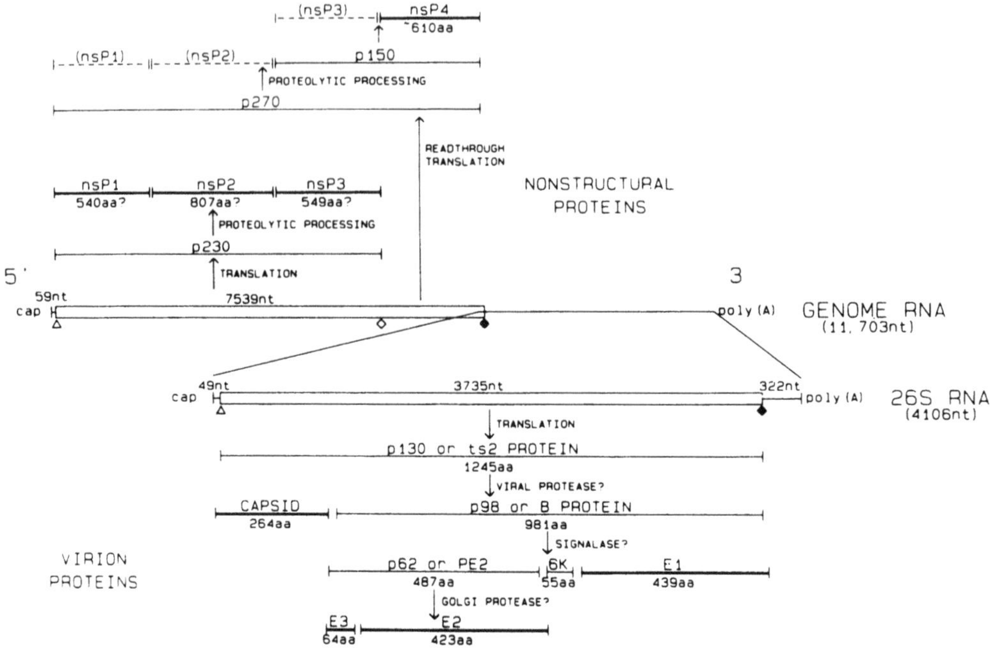

Figure 6. Replication strategy of Sindbis virus. The symbols are: △, initiation codons; ◆, termination codons; ◇, UGA codon. Reprinted from Strauss, E.C. and Strauss, J.H. (6).

Translation of the Virion RNA

Cell ribosomes bind to the genomic RNA at a site near the 5' terminus and translation initiates at a single AUG. A polyprotein consisting of the nsP1 + 2 + 3 sequences is made but there is also read-through at the OPAL codon at the nsp3 terminus to produce a polyprotein of nsP1 + 2 + 3 + 4 (Fig. 6). The latter accounts for about 10% of the total polyproteins made.

Very soon after synthesis, an autoprotease activity that appears to reside in nsP2 cleaves at the nsP2-3 junction to produce nsP1 + 2 and nsP3, as well as nsP1 + 2 and nsp3 + 4. The same autoprotease is believed to cut at nsP1-2, but this cleavage is slower than the nsp2-3 reaction. An even slower cleavage occurs between nsp3 + 4. The proteolytic processing of the polyprotein can be studied both *in vivo* in virus-infected cells (Fig. 7) and *in vitro* using isolated virion genomic RNA or RNAs synthesized from the Sindbis virus cDNA clone (Fig. 8). Studies with truncated virus mRNAs show that synthesis of the nsP2 sequence is necessary to produce the nsP1-2 and 2-3 proteolytic cuts. None of the structural genes are translated from the virion RNA.

Virus RNA Replication and Transcription

Few details are available currently to define the molecular events in the synthesis of negative strand RNA, the transcription of positive strand subgenomic

RNA, the formation of new full length genomic RNA and the regulation of these multiple activities (Fig. 9). How each of the nonstructural proteins functions in these various activities is still unknown. It has been possible, however, to define those regions of the Sindbis virus genome RNA essential for replication and/or packaging by studying properties of deleted forms of RNA from defective-interfering(DI) RNA.

DI-RNAs are themselves naturally-occurring deletions of the virus genomic RNA but they retain those sequences of the normal RNA required for RNA

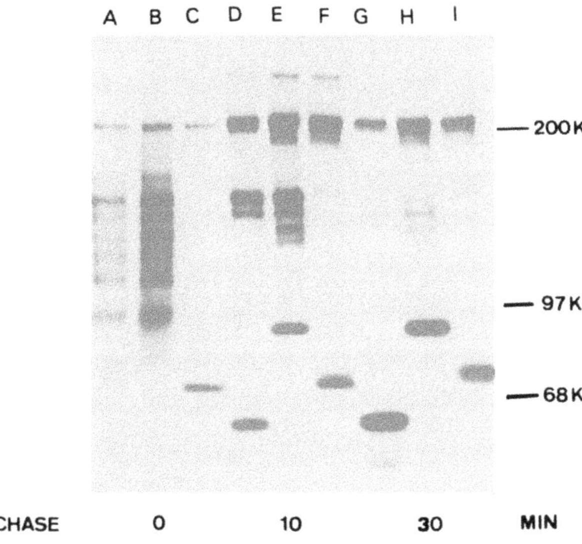

Figure 7. *In vivo* processing of Sindbis virus nonstructural proteins. Chicken embryo fibroblasts (10 cells/dish) were infected with Sindbis virus (MOI = 50). At. 2.5 hr post infection, 5M NaCl was added to a final concentration of 0.25 M for 40 min. Cells were then washed and incubated with MEM lacking methionine for 10 min. followed by a 5 min label with 100 μCi of ^{35}S methionine (1000 μCi/mM). Label was removed and unlabeled methionine (1mM) was added. After 0, 5, 10, 20, 30 and 40 min, media was removed and cells lysed with an SDS/PAGE gel buffer. Samples were diluted 10-fold with a radioimmunoprecipitation buffer and incubated with 4 μl of different rabbit sera containing antibodies against Sindbis virus nsP1,2,3 or 4 (prepared by R. Hardy, California Institute of Technology). Protein A-sepharose was added and precipitates washed 4 times followed by solubilization with gel buffer. Analysis was by SDS/PAGE using a 10% acrylamide gel. Figure is an autofluorograph. Lanes: A,B,C, 0 min chase; D,E,F, 10 min chase; G,H,I, 30 min. chase; samples in lanes B,E,H were precipitated with anti-nsP1; samples in lanes B,E,H, were precipitated with anti-nsP2; samples in lanes C,F,I were precipitated with anti-nsP3. NsP3 but neither nsP1 nor nsP2 was detected in the 5 min. chase. No free nsP4 was seen although a protein equivalent to the nsP3 + 4 sequences was clearly detected by the anti-nsP4 sera (data not shown). Conversion to nsP4 must take longer than 30 min in these infected cell under conditions used for labeling. These experiments and those noted in Fig. 8 have been carried out by M. Ding and M.J. Schlesinger (Washington University Sch. Med.) in collaboration with R. Hardy and J.H. Strauss (Cal.Inst.Tech.).

replication and for packaging into virions. Their accumulation during normal virus replication leads to an interference in the replication of the genomic RNA and a decrease in the infection titer of virus. S. Schlesinger and colleagues (Washington University, St. Louis, MO) have cloned and sequenced a Sindbis virus DI-RNA (Fig. 10, ref. 10) and subsequently engineered the clone to determine which parts of the DI sequence could be removed withouth blocking its ability to interfere with

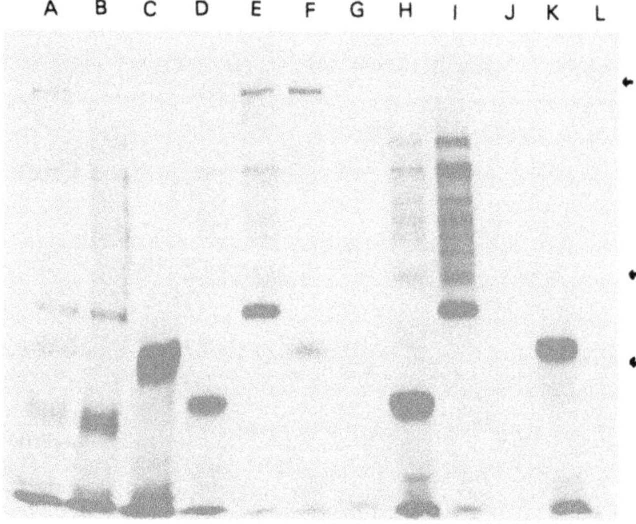

Figure 8. *In vitro* transcription, translation and processing of Sindbis virus RNAs containing the nonstructural protein genes. Virion RNA was isolated by SDS/phenol/chloroform extraction of freshly prepared virions produced by 6 hr infection of chicken embryo fibroblasts (MOI = 100). Purity of this RNA preparation and those produced from *in vitro* transcription were ascertained by agarose gel electrophoresis. Truncated RNAs were prepared by *in vitro* transcription of the TOTO 1000 clone (unpublished data, C. Rice, H. Huang and J.H. Strauss) digested with either *Sac*1, *Stu*1, *Bam* H1 or *Eco* R1. DNAs were isolated by phenol/chloroform extraction, ethanol precipitated and their size confirmed by electrophoresis in 0.6% agarose. *In vitro* transcription with SP6 polymerase and the capping substrate, ^{7}mGppp, was carried out for 60 min at 40° C. The RNA product was extracted and analyzed by agarose gel electrophoresis. These RNAs were added to a rabbit reticulocyte lysate translation reaction mixture (BRL, Bethesda, MD.) but supplemented with tRNA, additional ATP and GTP and ^{35}S methionine. After 30 min incubation at 30° C, non labeled methionine was added and reactions continued for 60 min. Gel loading buffer was added, samples boiled for 3 min, diluted and immunoprecipitates as described in the legend to Fig. 7. Lanes: A, D,E,F,G are proteins translated by mRNA transcribed from *Stu* 1 digested DNA (includes nsP1-4); B,H,I,J are proteins translated on mRNAs transcribed from *Bam* H1 digested DNA (includes nsP1 + nsP2 + 80 aa of nsP3); C,K,L are proteins translated by mRNA transcribed from Eco R1 digested DNA (includes nsP1 + 40 aa of nsP2). A, B, C are reaction products before radio immunoprecipitation; D,H,K were treated with anti-nsP1 sera; E,I,L were treated with anti-nsP2 sera; F,J were treated with anti-nsP3 sera; G was treated with anti-nsP4 sera.

normal virus replication. For the latter, they established a transcription-transfection system which allowed for the synthesis of DI-RNAs from cDNAs and the infection of cells with the DI-RNAs in the presence of normal "helper" Sindbis virus (11). The initial results of these studies have shown that only structures at the 5' end and the 19-nucleotide sequence at the 3' end are needed for DI-RNA function (Fig. 11).

Virus Structural Protein Formation

The genes encoding the structural proteins reside in the 3' terminal segment of the genomic RNA but are only expressed from a subgenomic mRNA (the 26S RNA, Fig. 6) that is transcribed from negative strands of the genomic RNA. Near the 5' end of 26S RNA (which has a ^{7}mG cap) is an AUG which serves as the only site for initiation of translation from this mRNA and all the gene products are the result of proteolytic processing from a polyprotein.

The capsid protein is the first translation product and it is released from the nascent polyprotein by an autoprotease activity residing in the capsid sequences (12, 13).

The second product of translation is the p62/E2 glycoprotein, and a sequence of hydrophobic amino acids at the amino terminus act as a signal for attachment of the mRNA to the endoplasmic reticulum (ER) membrane. Subsequent insertion of the p62/E2 into the ER vesicle leads to glycosylation and acylation of the protein (14). Unlike most polypeptides following this pathway of ER insertion, the p62/E2 signal sequence is not cleaved from the protein during transport into the ER lumen.

Another hydrophobic sequence near the carboxy terminus of p62/E2 serves to stop transfer of the polypeptide into the lumen and ultimately anchors the E2 in the envelope of the virion. An additional 33 amino acids of this glycoprotein appear to remain in the cytoplasm and are presumed to interact with the virus nucleocapsid

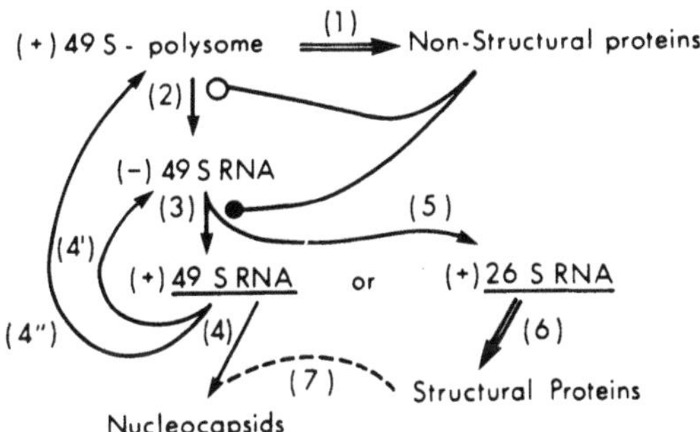

Figure 9. Alternate pathways for RNA transcription and replication. Genome (+) RNA binds to polysomes to forme nonstructural proteins (1). The latter produce (−) RNA (2) which can be transcribed to either (+) 49S RNA (3) or 26S RNA (5). New progeny (+) 49S RNA combine with capsid (4) to form nucleocapsids, make (−) RNA (4') or bind ribosomes (4"). With permission from Raven Press, New York.

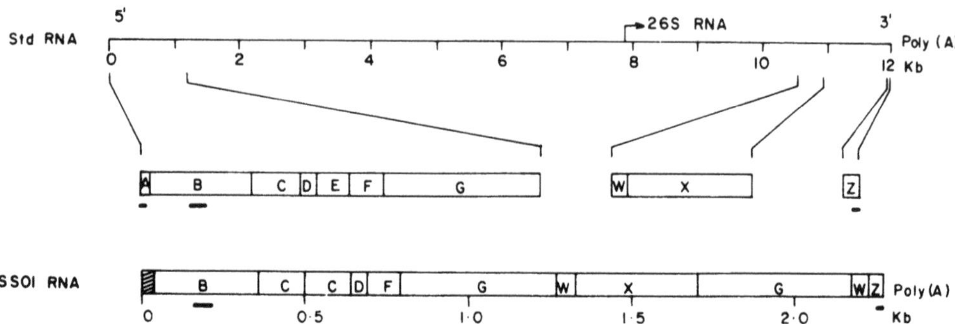

Figure 10. A comparison between the deduced sequences of the standard Sindbis virus RNA and the SS01 DI RNA (10).

during assembly (see below). A small 6 kDa "linker" polypeptide is translated from RNA sequences lying between the p62/E2 and E1 genes (Fig. 6). It is very hydrophobic and serves as an additional signal sequences for insertion of E1 into the ER lumen. Two proteolytic cleavages, postulated to occur by the cellular "signalase" in the ER membrane, release the 6 kDa from the p62/E2 and E1. Insertion of E1 into the ER is also accompanied by glycosylation and acylation. E1 has a sequence of hydrophobic amino acids at its carboxy terminus which anchors this protein to the lipid bilayer.

Post Translational Events and Assembly of Virus

The p62/E2 and E1 glycoproteins are transported as a non-covalent, weakly interacting heterodimer through the cellular organelles that are normally used for transporting protein to the cell surface. It takes about 30 minutes for the proteins to reach the cell surface from the ER. During this time, both proteins undergo modifications in primary structure and in conformation. The glycosyl residues (three on p62/E2 and two on E1) are extensively altered and both proteins are acylated with palmitic acid close to their membrane domains. Acylation probably occurs in late ER.

Glycosyl groups begin to be modified in the ER and completion occurs in the trans-Golgi vesicles. The earliest event in processing the oligosaccharide is removal of 3 glucose residues. Blocking this removal with deoxynorjirmycin leads to inhibition of p62 conversion to E2 and virus assembly although transport of the glycoprotein to the cell surface is not affected. These and other data from experiments utilizing several additional drugs or cell mutants affecting oligosaccharide processing have provided clear evidence that protein-bound glycosyl groups play a major role in the early stages of polypeptide chain folding and are nonessential for the intracellular transport of these proteins to the cell surface (15).

Late in the trans-Golgi, a critical change occurs in conformation of the glycoproteins when p62 is proteolytically converted to E2. The cleavage is at a dibasic amino acid located 64 amino acids from the amino terminus of p62. The enzyme acting on p62 is believed to be a host-specific protease which cleaves at dibasic sites in several secreted "preproteins". As a result of cleavage, stable E2-E1 heterodimers form and subsequently associate to the trimers which are seen on the virion's surface (Fig. 12). The cytoplasmic face of this trimer is postulated to have the E2 cytoplasmic tails which serve as "receptors" for the nucleocapsid. Binding of the latter to E2 initiates budding of virus from the cell surface.

Because of these post translational processing events and passage through the cell's membrane transport apparatus, there is a substantial delay (~30 minutes) between the end of polypeptide chain synthesis and the appearance of glycoprotein in newly released virions. Such a delay is not seen in the appearance of capsid in virions — which occurs almost immediately after its synthesis. No "free" capsid subunits have been detected in cells, although a transient association of capsid with ribosomes has been reported. Capsids associate with newly made genomic RNAs to form nucleocapsids. The latter move to the cell's plasma membrane where newly transported glycoproteins appear. A "nucleation" site forms where glycoproteins and capsid interact and additional glycoproteins move into the region, leading to a budding and release of virions (Fig. 12).

B. Flaviviridae

Although their genome is a (+) RNA of about the same size as that of the *Togaviridae*, viruses in this family express their genes differently from the *Togaviridae* (16). For example, *subgenomic mRNAs have not been detected during virus replication*.

It is likely that all the virus gene products are formed from a single polyprotein whose synthesis initiates at a site near the 5' end of the genomic RNA. Unlike the *Togaviridae*, the structural proteins are synthesized first in the order of capsid, preM/M, envelope protein.

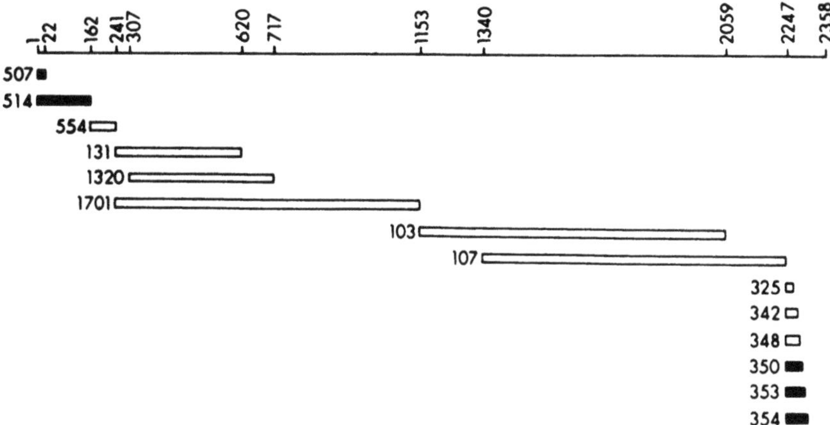

Figure 11. A deletion map of Sindbis DI RNA. Open boxes are regions which are deleted without causing loss of biological activity. Closed boxes are regions which when deleted led to loss of activity (11).

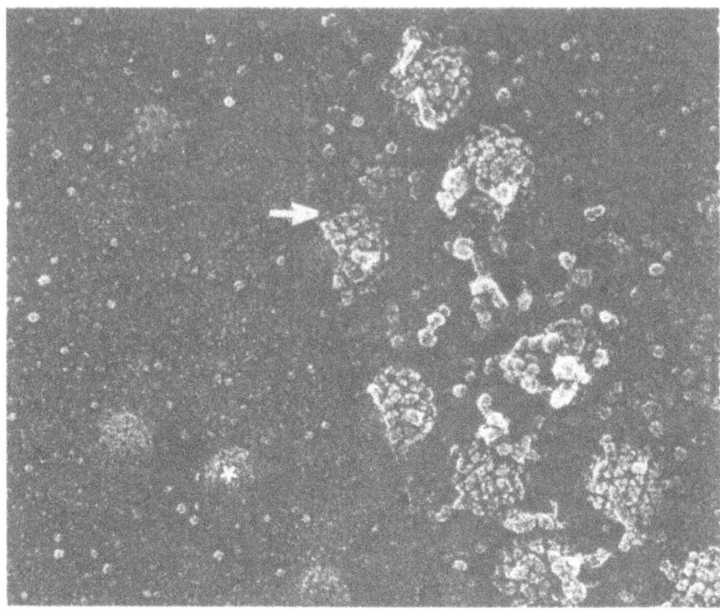

Figure 12. Budding of Sindbis virions from infected chicken embryo fibroblasts. Cells were quick frozen and freeze fractured. Electron micrograph show the fracture across the surface membrane of the cell (white arrow). On the right is the outer surface of the cell with clusters of virus glycoproteins at sites of budding. On the left is the inner leaflet of the surface lipid bilayer with viral nucleocapsids (white star) bound to the inner cytoplasmic side of the membrane. Sample preparation for electron microscopy was carried out by Dr. J. Heuser (Wash.U.Sch.Med., St. Louis, MO).

A hydrophobic sequence in the preM/M is probably the signal sequence for membrane association of the polysome and subsequent insertion of the M and E proteins into the ER membrane and lumen, respectively. Figure 13 shows the current state of knowledge about the organization and processing of the Yellow Fever virus genome and Table 1 lists the polypeptides produced during replication of their virus. Numerous proteolytic cleavages are needed to produce the individual virion structural and nonstructural proteins. Undoubtedly some of these proteolytic activities will be encoded in the virus gene products and function as autoproteases while others (such as a "signalase") will be contributed by the host cell.

An analysis of the proposed cleavage sites for the individual proteins shows that M, ns2b, Ns3, ns4a and 4b, and Ns5 have a dibasic sequence at the carboxy side of the cleavage (7). A cellular enzyme with specificity for these sequences is normally found in a Golgi compartment. It seems unlikely that this cellular enzyme would be involved in processing virus nonstructural proteins, and more information will have to be forthcoming before we can sort out the various proteolytic events in flavivirus replication.

Table 1. – Flavivirus Polypeptides

Protein[a]					
Proposed nomenclature	Old nomenclature	M_r	$M_{pred.}$	Glycosylated	Comments
	Structural region				
C	V2 ($NV1^1_2$)	13,000-16,000	13,432	No	Nucleocapsid protein
prM	(NV2) ($NV2^1_2$)	19,000-23,000	18,813	Yes	Precursor to M
M	V1	8,000-8,500	8.526	No	Virion envelope protein
E	V3	51,000-60,000	53,712	Both forms[d]	Major virion envelope protein
	Nonstructural region				
NS1	NV3	44,000-49,000	45,869	Yes	Soluble complement-fixing antigen
ns2a	($NV2^1_2$) (NV2)	16,000-21,000	18,086	No	Hydrophobic; function unknown
ns2b	($NV1^1_2$)	12,000-15,000	13,823	No	Hydrophobic; function unknown
NS3	NV4	67,000-76,000	69,319	No	Replicase component
ns4a	(NVX) ($NV2^1_2$)	24,000-32,000	31,196	No	Hydrophobic; function unknown
ns4b	(NV1)	10,000-11,000	12,159	No	Hydrophobic; function unknown
NS5	NV5	91,000-98,000	104,079	No	Replicase component

As with the *Togaviridae*, virtually nothing is known about RNA replication, but replication complexes are found associated with nuclear membrane fractions (16). Replicative intermediates and replicative forms of viral RNA have been detected, indicating that negative strand RNA synthesis and positive RNA transcription occur via a mechanism analogous to those for other single-strand (+) RNA viruses.

Encapsidation of new genomic RNA to form nucleocapsids is probably also analogous to similar events in *Togaviridae* replication. Final assembly of new virions, however, occurs on intracellular membranes. This process is probably distinct from that of the *Togaviridae* since there is evidence that the preM/M conversion accompanies maturation but not assembly.

Unlike infection of cells with alphaviruses, flavivirus infection does not lead to a significant inhibition of host cell protein synthesis. In addition, the virus growth cycle is slower and the yields of virus per cell much lower that found in alphavirus infection.

4. GENETIC ENGINEERING OF TOGAVIRUSES

Genomic sequences from Semliki Forest and Sindbis viruses have been copied into cDNAs and inserted into expression vectors such that the virus-coded proteins could be expressed in cells in the absence of a complete virus replication cycle. The most extensive studies have been carried out by Garoff and colleagues with sequences from the structural genes of Semliki Forest virus (17-19). Table 2 summarizes the results of experiments in which genetically-engineered mutations were created in various regions of the virus glycoprotein genes with accompanying alterations in the expression and function of these proteins. Changes in the cytoplasmic and transmembrane domains of the glycoprotein showed varying effects on intracellular transport and post translational processing of the glycoprotein. The role of the hydrophobic 6 kDa linker polypeptide as a signal sequence for the E1 protein was established by these methods.

The structural genes for Sindbis virus have been inserted in a yeast vector with expression of the virus proteins under regulation of a galactose promoter (Fig. 14). When grown on galactose, yeast cells carrying the vector made up to 3% of their

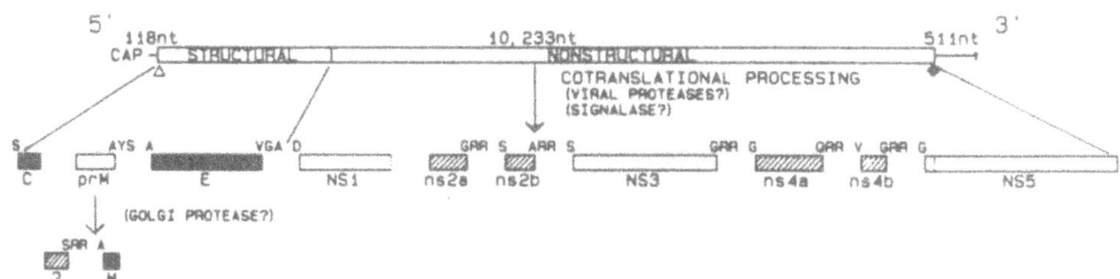

Figure 13. Organization and processing of proteins encoded by yellow fever virus (7).

Table 2. - Site-specific Mutations in Semliki Forest Virus

Mutation	Effect
1. Deletions of 16 to 31 aa from the carboxy terminus of P62/E2. Each deletion has an additional 8 new amino acids added	1. P62/E2 in all cases is transported to the cell surface at rates similar to the wild type
2. Change in the basic charge cluster at the cytoplasmic side of the P62/E2 transmembrane sequence from Arg-Ser-Lys to Gly-Ser-Glu or Gly-Ser-Met	2. The orientation of the P62/E2 in the membrane is unperturbed but the protein is less stably anchored in the membrane
3. Insertion of Arg into the sequence Gly-Asp-Ser-Gly at the putative active site of capsid autoprotease activity	3. No cleavage of capsid from P62/E2 occurs, and a polyprotein is made.
4. Deletions of varying amounts of the 6 kDa peptide located at the 5' side of E1	4. E1 translocated into the ER lumen occurs until only 9 amino acids remain in the 6 kDa

protein as virus capsid and glycoproteins p62 and E1 (Fig. 15, ref. 20). Much of the capsid formed aggregates of 30S particles (D. Wen, M. Ding and M.J. Schlesinger, unpublished data) while most of the glycoprotein formed disulfide-linked oligomers containing high mannose oligosaccharides (21). Very little of the glycoprotein was transported to the yeast surface membrane and low amounts of E2 were formed.

I noted earlier that a cDNA for the entire genome of Sindbis virus has been cloned into a plasmid immediately downstream from the bacteriophage SP6 promoter. Transcription of the cDNA yields an mRNA which can be translated in vitro to produce authentic virus nonstructural proteins (Fig. 8). Mutagenesis of this cDNA at many "selected" sites should produce mutants whose phenotypes will reveal much about how these viruses replicate.

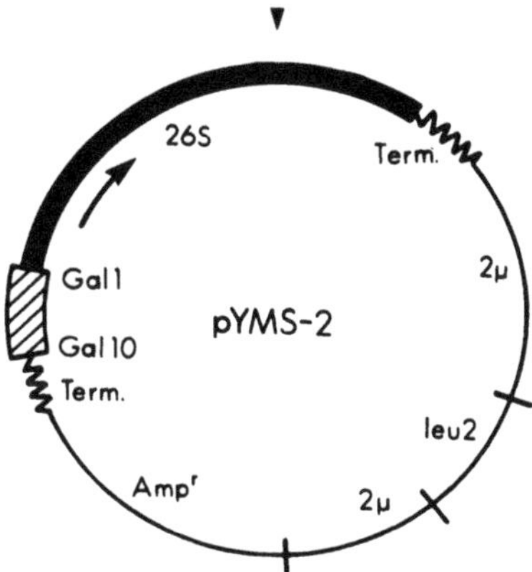

Figure 14. Plasmid constructed for expression of Sindbis virus structural genes in S. cerevisiae (20).

5. UNSOLVED PROBLEMS

The wealth of data accumulated over the past few years has revealed much of the molecular details of the virus genome including the sequences of virus-coded proteins. We also know how these viruses enter mammalian cells and how viral glycoproteins function during virus attachment and uptake. Many of the molecular events involved in the post translational modifications of virus glycoproteins have been elucidated. But despite considerable effort, we still know very little about how viral RNAs replicate and what controls viral RNA transcription. These latter represent probably the greatest gaps in our knowledge of the molecular basis of virus replication.

But detailed mechanisms of other events in the virus infection cycle are also lacking. These include the uncoating of the nucleocapsid, proteolytic processing of polyproteins, packaging of RNA into new nucleocapsids and membrane glycoproteins. The next few years should see many of these problems solved as the amazing tools of recombinant DNA technology are applied to these viruses.

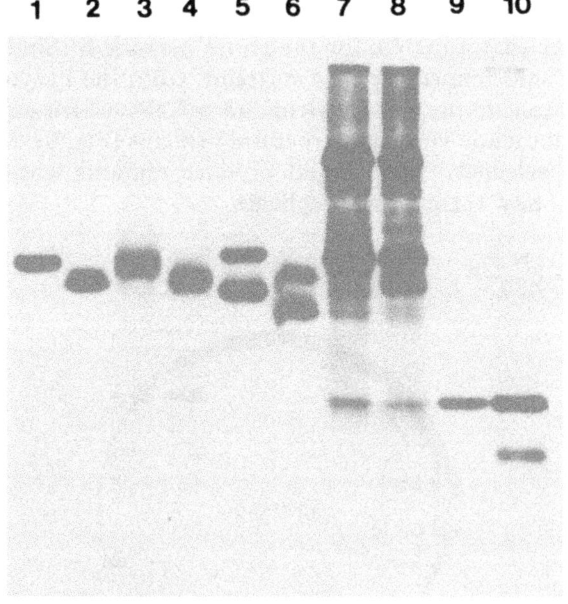

Figure 15. Expression of Sindbis virus capsid and glycoproteins in S. cerevisiae. Lane: 1,2,5,6,9 - lysates of chicken embryo fibroblasts infected with Sindbis virus; 3,4,7,8,10 - extracts of yeast grown on galactose. Anti E1 antibodies were used for 1,2,3,4; anti E2 antibodies were used for 5,6,7,8; anti Capsid antibodies were used for 9,10. Samples in lanes 2,4,6 and 8 were treated with endo F. See ref. 20 for detail.

6. REFERENCES

1) Schlesinger, R.W. ed. (1980), *The Togaviruses-Biology, Structure, Replication* Academic Press, New York

2) Schlesinger, S. and Schlesinger, M.J., eds. (1986)., The Togaviridae and Flaviviridae. In: *Comprehensive Virology*, H. Fraenkel-Conrat and R.R. Wagner, eds., Plenum Pub. Corp., New York

3) Harrison, S.C. (1986). In: "*The Togaviridae and Flaviviridae*, S. Schlesinger and M.J. Schlesinger, eds., Plenum Pub. Corp., New York, 21-37.

4) Vogel, R.H., Provencher, S.W., von Bonsdorff, C.H., Adrian, M. and Dubochet, J. (1986). Nature (London) 320, 533-535.

5) Dubois-Dalcq, M., Holmes, K.V. and Rentier, B. (1985), *Assembly of Enveloped RNA Viruses*, Springer-Verlag, Wien New York, 136-147.

6) Strauss, E.G. and Strauss, J.H. (1986), In: *The Togaviridae and Flaviviridae*, S. Schlesinger and M.J. Schlesinger, eds., Plenum Pub. Corp., New York, 35-90.

7) Rice, C.M., Strauss, E.G. and Strauss, J.H. (1986). In: *The Togaviridae and Flaviviridae*, S. Schlesinger and M.J. Schlesinger, eds., Plenum Pub. Corp., New York, 279-327.

8) Symington, J. and Schlesinger, M.J. (1978). Arch. Virol. **58**, 127-136.

9) Kielian, M. and Helenius, A. (1986). In: *The Togaviridae and Flaviviridae*, S. Schlesinger and M.J. Schlesinger, eds., Plenum Pub. Corp., New York, 91-120.

10) Schlesinger, S. and Weiss, B.G. (1986). In: *The Togaviridae and Flaviviridae*, S. Schlesinger and M.J. Schlesinger, eds., Plenum Pub. Corp., New York, 91-120.

11) Levis, R., Weiss, B.G., Tsiang, M., Huang, H. and Schlesinger, S. (1986), Cell **44**, 137-145.

12) Aliperti, G. and Schlesinger, M.J. (1978), Virology **90**, 366-369.

13) Hahn, C.S., Strauss, E.G. and Strauss, J.H. (1985). Proc. Natl. Acad. Sci. USA **82**, 4648-4652.

14) Schlesinger, M.J. and Schlesinger, S. (1986). In: *The Togaviridae and Flaviviridae*, S. Schlesinger and M.J. Schlesinger, eds., Plenum Pub. Corp., 121-148.

15) Schlesinger, S., Malfer, C. and Schlesinger, M.J. (1985). J. Biol. Chem. **259**, 7597-7601.

16) Brinton, M.A. (1986). In: *The Togaviridae and Flaviviridae*, S. Schlesinger and M.J. Schlesinger, eds., Plenum Pub. Corp., New York, 327-374.

17) Garoff, H., Kondor-Koch, C., Petterson, R. and Burke, B. (1983). J. Cell. Biol. **97**, 652-658.

18) Cutler, D.F., Melancon, P. and Garoff, H. (1986). J. Cell Biol. 102, 902-910.

19) Garoff, H., Melancon, P., Cutler, D., Roman, L., Hare, J., Zerial, M. and Huylebroeck, D. (1986). In: *Positive Strand Viruses*, Brinton, M. and Rueckert, R. R. eds., Alan Liss, New York, in press.

20) Wen, D. and Schlesinger, M.J. (1986). Proc. Natl. Acad. Sci. USA 83, 3639-3643.

21) Wen, D., Ding, M. and Schlesinger, M.J. (1986) Virology 153, in press.

CHAPTER 10

THE MOLECULAR BIOLOGY OF CORONAVIRUSES

BRIAN W.J. MAHY
The Animal Virus Research Institute, Pirbright, U.K.

INTRODUCTION

The family Coronaviridae consists of eleven viruses which infect vertebrates and cause a range of diseases mostly affecting the respiratory or the gastro-intestinal tract (1). The name *coronavirus* was coined by Tyrrel *et al.* (2) on the basis of the characteristic *corona* of widely spaced bulbous surface projections, 12-24 nm long, seen in electron micrographs of these viruses (Figure 1). More recently, it has become clear that members of the *Coronaviridae* share a number of molecular biological features which constitute a unique replication strategy and set these viruses apart from other virus groups. These features may underlie the fascinating biological properties of certain murine coronaviruses which provide one of the best available models for human demyelinating neurological diseases (3).

1. BIOLOGICAL PROPERTIES

Coronaviruses affect several species, including chicken, cattle, pigs, dogs, cats, mice, rats and man (Table 1). Although acute respiratory or enteric diseases are common features of coronavirus infections, there are several examples of chronic persistent infections, and these may involve the central nervous system. Mouse hepatitis virus (MHV) exists in four strains, one of which (MHV-4, JHM strain) can induce chronic or recurrent central nervous system demyelination as well as encephalomyelitis in both mice and rats (3-5). The exact outcome of infection depends upon the age and strain of the rodent host as well as the dose and route of inoculation. There is evidence that pathogenicity of JHM virus is genetically controlled, since virus mutants causing demyelination rather than encephalitis can be selected as *ts* mutants (6) or using monoclonal antibodies (7). Perhaps the most interesting rodent model is the infection of 10-15 day old rats with a mutant of JHM

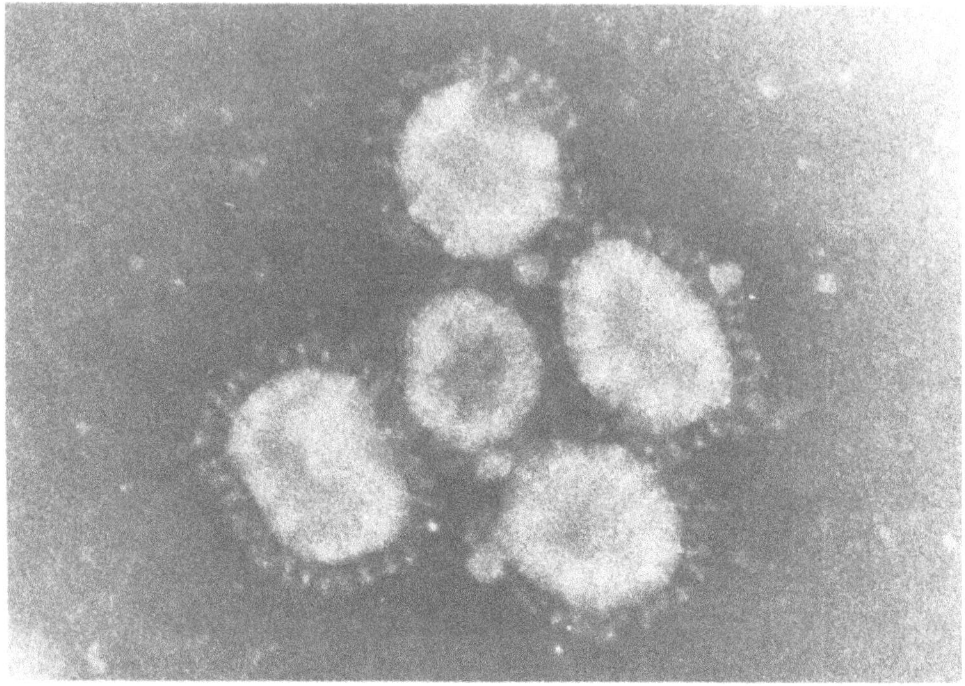

Figure 1. Virions of avian infectious bronchitis virus (electron micrograph courtesy of Dr. Cavanagh and J.K.A. Cook).

virus (*ts* 43), which results in a high proportion of persistently infected animals that develop a subacute relapsing-remitting demyelinating disease with close parallels to multiple sclerosis in man (5). Although the isolation of two human coronaviruses, which may have originated from the brain tissue of two multiple sclerosis cases, was reported some years ago (8), the observation has not been confirmed and is not supported by serological surveys (9). Moreover, owing to the circumstances of their isolation, the viruses in question may be of murine rather than human origin (10). However, the JHM strain of mouse hepatitis virus provides an important virus model for multiple sclerosis (11), despite the fact that clinical epidemiological evidence is more supportive of a paramyxovirus model (12).

In general, human coronavirus infections are found to cause only acute respiratory disease. The avian virus causing acute infectious bronchitis is of considerable economic importance, but may also infect non-respiratory tissues such as lymph node or kidney, to cause long-term persistent infectious even in the presence of high serum antibody wich result eventually in a fatal disease (13,14).

A different type of persistence, involving cells of the reticuloendothelial system, is found in cats with feline infectious peritonitis virus infection. This virus primarily infects macrophages, where a persistent infection becomes established and immune complexes of virus and IgG subsequently develop and become deposited in the kidney, the severity of the disease being greatest in cats with high antibody titres (15).

The ability of several coronaviruses to establish persistent infections in their hosts can also be studied, using *in vitro* model systems, where it has proved relatively easy to obtain persistently infected cell cultures, particularly in cells of neural origin (11). In a detailed study, Baybutt *et al.* (16) established a persistent infection of Sac(-) cells with JHM virus. The cultures released high levels of virus for more than 100 passages, but the cells were not killed, as in normal JHM infection. It was concluded that a variant JHM virus (JHM-Pi) of reduced cytotoxicity was responsible for the persistent infection (16). Clearly, the molecular events which lead to the production of such a virus variant require further study; comparison of variant and wild-type replication could shed light on how persistent infections become established in the central nervous system of the infected host.

2. CORONAVIRUS STRUCTURE: GENOME RNA AND POLYPEPTIDES

The coronavirus genome is a large, non-segmented, single-stranded RNA molecule which reportedly varies in length in different viruses between 15 kb and 27 kb (5×10^6 to 8×10^8 molecular weight). Recent estimates for the best-studied coronaviruses put the lengths of both avian coronavirus (IBV) and of mouse hepatitis virus RNA at 27 kb (7, S.G. Siddell, personal communication) and of porcine transmissible gastroenteritis virus (TGEV) RNA at 23.6 kb (18). Most

Table 1. - Coronaviruses

Antigenic group	Name	Natural Host	Principal Disease
I	Avian infectious bronchitis (IBV - Many serotypes)	Chicken	Respiratory, Lymphatics, Kidney
II	Turkey coronavirus (TCV)	Turkey	Respiratory, Enteritis
III	Human coronavirus (HCV-229E)	Human	Respiratory
	Transmissible gastroenteritis Virus (TGEV)	Pig	Enteritis
	Feline infectious peritonitis (FIPV)	Cat	Peritonitis, Granulomas in many organs
	Feline enteric coronavirus (FEC)	Cat	Enteritis
	Canine coronavirus	Dog	Enteritis
IV	Mouse hepatitis virus (MHV - Many serotypes)	Mouse	Hepatitis, Enteritis, Encephalomyelitis
	Rabbit coronavirus (RbCv)	Rabbit	Enteritis
	*Bovine coronavirus (BCV)	Cow	Enteritis
	*Haemagglutinating encephalomyelitis virus (HEV)	Pig	Vomiting and wasting
	*Human coronavirus (HCV-OC43)	Human	Respiratory

* These viruses all haemagglutinate.

molecular biological studies have been carried out either on IBV or on MHV, and these form the basis for this review.

The genome RNA is polyadenylated at the 3' end, contains a 7-methyl guanosine cap structure at the 5' end, and is infectious (1). The virion RNA is therefore of positive polarity, and can function directly as mRNA without the requirement for virion enzymes or other proteins.

The genes encoding virion structural polypeptides are located within the 3' half of the genome, and this has facilitated sequence analysis and, hence, predicted their primary structure. For both IBV and MHV, sequences of the major structural proteins have now been determined after molecular cloning of cDNA synthesized from the 3' poly A tail hybridized to oligo dT as the initial primer.

Within the virion, the genome RNA is associated with a phosphoprotein of 50K to 60K molecular weight, termed the nucleocapsid (N) protein, to form a ribonucleoprotein (RNP). This structure is surrounded in the virion by a lipoprotein envelope which consists of host cell-derived lipids and two major virus-specified glycoproteins. One of these is an integral transmembrane glycoprotein which has a variable molecular weight, depending upon the degree of glycosylation, of 23K to 34K (20). The other envelope glycoprotein projects from the virion to form the spikes which make up the *corona* structure seen in electron micrographs of the virus. The nomenclature in current use for these two envelope proteins is confusing: the smaller transmembrane glycoprotein is termed M (matrix) in IBV, but E_1 (envelope 1) in MHV; the larger glycoprotein is termed S (spike) in IBV, but E_2 (envelope 2) in MHV. For reasons which have been discussed previously (10), the terms M and S will be used in this review. A diagrammatic representation of a coronavirus particle is shown in figure 2.

A third envelope glycoprotein, the haemagglutinin (H) protein, has been described in haemagglutinin mammalian coronaviruses, including bovine enteric coronavirus, porcine haemagglutinating encephalomyelitis virus, and human coronavirus OC 43 (21, 22).

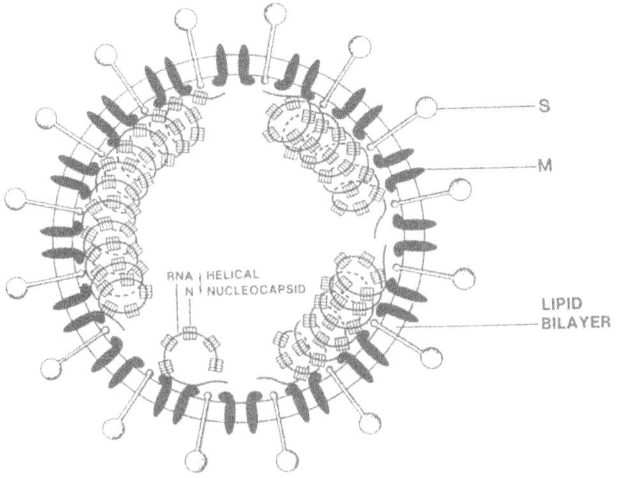

Figure 2. Model of a coronavirus. (Adapted from Holmes, K.V. (87)).

A. The N Polypeptide

The primary structure of the N polypeptide has been determined for several coronaviruses, since this protein is encoded in the genome adjacent to the 3' terminus, which is polyadenylated. In MHV, the nucleotide sequence of this gene revealed a single open reading frame encoding a protein of 50,000 molecular weight which is rich in basic residues (23, 24). The N protein is phosphorylated on serine residues, and a protein kinase activity has been detected in virions of MHV strain JHM which may be responsible for this phosphorylation (25).

Nucleotide sequences of the N protein genes of two strains of IBV (Beaudette and M41) have also been determined and compared with the MHV sequences (26). The predicted molecular weight of the IBV N protein is 45,000 and the two IBV strains differ mainly in that the 3' non-coding region of the Beaudette contains a 184 nucleotide segment not present in M41. Considerable homology was found between the N protein sequences of IBV and MHV, including the positions of clusters of serine residues where phosphorylation occurs (26).

B. The M Polypeptide

The matrix polypeptide is an integral membrane protein which is extremely hydrophobic and has unusual features when compared with most other viral glycoproteins, including the coronavirus S glycoprotein. In MHV and in bovine coronavirus, the carbohydrate moiety lacks mannose and is O-linked to serine and threonine residues (27-29), but in IBV it is mannose-rich and N-glycosidically linked (20, 30, 31).

An elegant series of *in vitro* experiments by Rottier *et al.* (28) showed that integration of M polypeptide into membranes occurs without cleavage of an N-terminal signal peptide sequence, although it requires a signal-recognition particle (32). These features may be important in the accumulation of M polypeptide in the endoplasmic reticulum, where coronavirus budding occurs (rather than at the plasma membrane as with most other enveloped viruses).

With both MHV (28) and IBV (33), studies using proteases have shown that only small portions of M protein at the amino- and carboxy termini are susceptible to digestion and hence lie outside the membrane. The extremely hydrophobic nature of the M protein has been confirmed from its predicted sequence for both IBV and MHV. Although the nucleotide sequences of the M protein gene show little homology between the avian and murine viruses, the amino-acid sequences can be aligned to 27% homology (29), this being concentrated in the amino-terminal half of the protein. From theoretical analyses, a topological model has been suggested according to which the M protein is anchored in the lipid bilayer by three successive membrane-spanning helices present in the amino-terminal half, whereas the carboxy-terminal half is associated with the membrane surface (29). Recently, the complete sequence of the M gene of MHV strain IHM has been obtained; the amino-acid sequence shows only 7 conservative changes as compared to the A59 sequences, and all the potential O-glycosylation sites are conserved (29a).

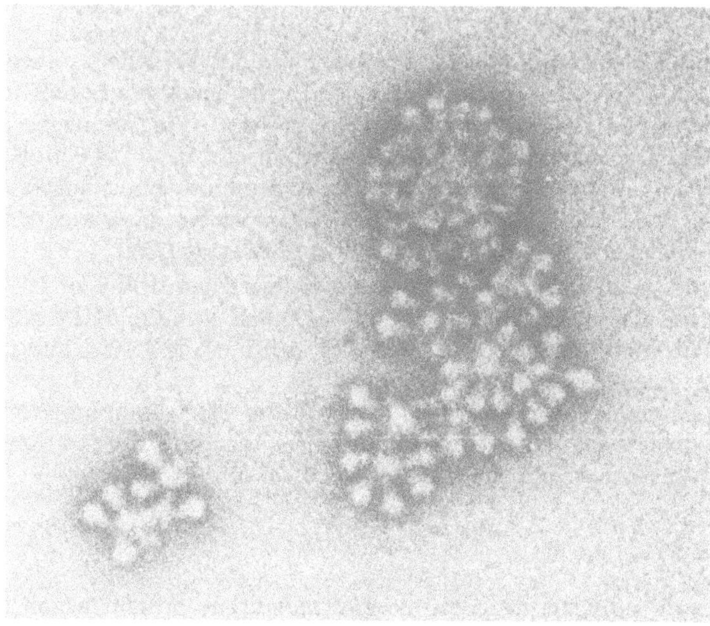

Figure 3. Purified spikes of avian infectious bronchitis virus prepared as described in reference 35 (electron micrograph courtesy of D. Cavanagh).

C. *The S Polypeptide*

The S polypeptide projects from the virion surface and can be seen in the electron microscope (Figure 3) as bulbous or "tear-drop" shaped in purified preparations (34, 35). Only a small anchor region of the protein is embedded in the envelope lipid bilayer, and the peplomer portion is responsible for attachment to cells as well as mediating cell fusion, since both activities are inhibited by monoclonal antibodies specific for S (36, 37). Clearly the S protein appears analogous to the F protein of paramyxoviruses or the hemagglutinin (HA) protein of influenza virus, and this suggests that proteolytic cleavage of S may be necessary for full biological activity.

Recent structural studies on the S protein of the avian coronavirus, IBV (Beaudette strain), largely support this analogy. The sequence of the S gene predicts a large polypeptide of 127,000 molecular weight (38), which is close to the estimate of 125,000 obtained for S after removal of carbohydrate residues by endoglycosidase H (20). The sequence also predicts a hydrophobic signal sequence which is not present in the mature protein, 28 potential N-glycosylation sites, and a membrane *anchor* region of 44 non-polar amino-acids close to the carboxy-terminus (38).

Stern and Sefton (39) first demonstrated that the S protein of IBV is a precursor to two smaller polypeptides, S1 and S2, which are derived by proteolytic cleavage. The mature spike protein is an oligomer comprising two or possibly three copies of each of S1 and S_2, anchored in the virus membrane by S_2 (35). It is possible to remove the S_1 portion of the spike selectively using urea, and this abolished infectivity and haemagglutination but not attachment to cells (40). Monoclonal antibodies binding to S_1 alone neutralise infectivity and haemagglutination (37). An arginine-rich sequence has been found at the cleavage site between S_1 and S_2; this consists of 5 residues which after removal by carboxypeptidase activity would leave an S_1 component of 514 residues and an S_2 component of 625 residues (41).

There is evidence from monoclonal antibody binding studies that the S protein of murine viruses undergoes antigenic variation, in marked contrast to the M and N proteins (42). Recently, the sequence of the S protein gene was determined for MHV strain JHM (S.G. Siddell, personal communication) and for a second strain of IBV (M41). The M41 sequence revealed 50 amino-acid differences from the Beaudette strain, and these were clustered in two regions which may be "hot spots" where epitope variations occur naturally (43). The MHV sequence was quite different from the IBV sequences, although DIAGON plots show considerable similarity in the S_2, but not the S_1 region of the polypeptide. Overall, the S polypeptide of MHV has a molecular weight in the unglycosylated form of 137,000, with 21 potential glycosylation sites. The proteolytic cleavage site between S_1 and S_2 has the basic sequence Arg-Arg-Ala-Arg-Arg.

D. Other Proteins

Although most coronaviruses have three major virion structural proteins, N, M and S, a fourth protein (H) has been reported on the haemagglutinating mammalian coronaviruses, including porcine encephalomyelitis virus, bovine enteric coronavirus, and human coronavirus OC43. This glycoprotein has a molecular weight of 130,000 and consists of two disulphide-linked subunits each of 65,000 molecular weight (21, 22). The putative gene encoding this H protein has not been described; a surface glycoprotein of 65,000 molecular weight distinct from S has also been found in MHV strain JHM (44), although haemagglutination has not been described with this virus.

In addition to these major virion proteins, coronavirus-infected cells contain virus-coded polypeptides of 14,000 and 30,000 molecular weight (1, 44-48). A polypeptide of 14,000 molecular weight has occasionally been reported as a minor virion component, but the 30,000 molecular weight proteins seems to be nonstructural.

Now that extensive nucleotide sequences are available for both the IBV and MHV genomes, various open reading frames have been described which predict further virus-specified polypeptides not yet identified in infected cells (49-52). The current position regarding the various gene products of IBV and MHV is summarized in Table 2.

3. REPLICATION

Coronaviruses attach to their host cells by the peplomer S protein, but virus from which the S_1 subunit has been selectively removed is still able to attach (40).

Table 2. – Coding assignments of coronavirus RNAs

IBV		RNA	MHV	
mRNA	Product*	Length (Kb)	mRNA	Product
A	Nucleoprotein	2.0	7	Nucleoprotein
B	(B_1 - 9.5K) (B_2 - 7.5K)	2.5	6	Matrix
M	?	3.2	5	(5_1 - 12K) (5_2 - 10K)
C	Matrix	3.6	4	Nonstructural 14K
D	(D_1 - 6.7K) (D_2 - 7.4K) (D_3 - 12.4K**)	4.5		
E	Spike	7.8	3	Spike
		10.3	2	Nonstructural 30K
F	(Polymerase)	27	1	(Polymerase)

* Products in brackets are predicted from the sequence but have not yet identified in infected cells.

** A membrane associated protein corresponding to the product of the D3 open reading frame has been identified in infected cells (S.C. Inglis, personal communication).

Such virus is non-infective, however, so that entry requires S_1; it is likely, though not yet demonstrated, that receptor-mediated endocytosis is involved in virus entry. Coronavirus RNA alone is infectious, so no virion enzymes are involved.

Replication does not require a functional host cell nucleus (53) and occurs in the cytoplasm even in the presence of inhibitors of DNA-dependent RNA synthesis such as α-amanitin or actinomycin D (1, 46, 47, 54). The virus multiplication cycle lasts about 10 hours and begins with the synthesis, from the virion RNA template, of a genome-length negative strand RNA (55, 56). This negative strand RNA functions in replicative-intermediate structures as the template for the synthesis of a family of positive-strand RNAs; the RNA-dependent RNA polymerases involved in these synthetic events appear to be membrane-bound and to require continuous protein synthesis, but have not been extensively characterized (54-57).

A. Virus-specific RNA in Infected Cells

In coronavirus-infected cells, multiple virus-specific RNA species can be detected by pulse-labelling with ^{3}H-uridine in the presence of actinomycin D. In addition to full-length genome-sized RNA, six smaller, subgenomic RNAs are synthesized in cells infected with MHV, and five in cells infected with IBV. All the intracellular RNAs of both avian and murine coronaviruses are capped and polyadenylated, and can be found in association with polysomes. They also share common sequences, as originally demonstrated by RNase T_1-resistant oligonucleotide mapping studies on RNAs from IBV-infected cells by Stern and Kennedy (58, 59) and subsequently shown for MHV-infected cell RNAs by several groups (60-62). Each RNA contains the sequences present in the next smaller RNA plus a unique 5' terminal sequence, so that the RNAs form a 3' co-terminal

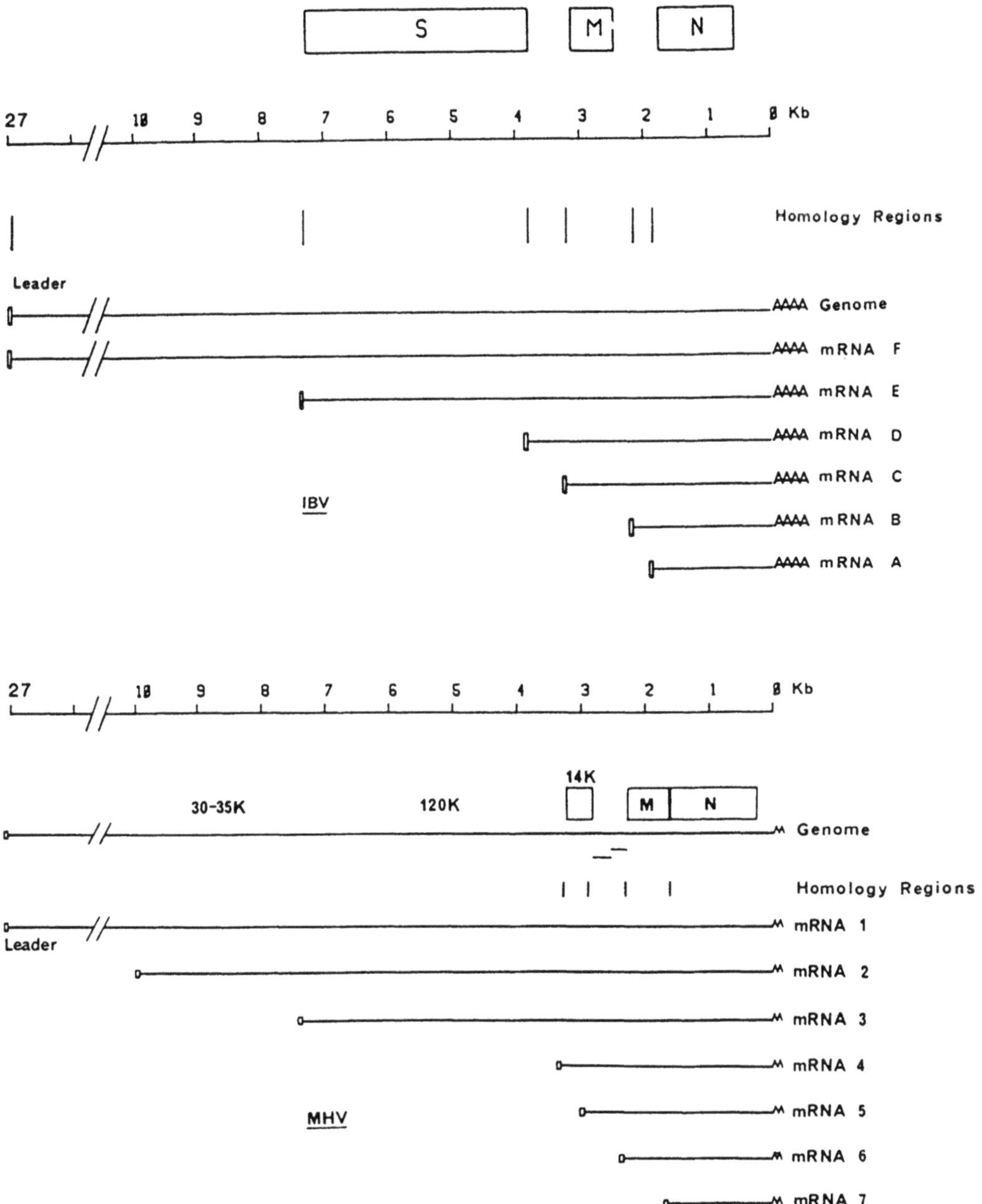

Figure 4. Diagrammatic representation of the "nested set" of coronavirus mRNAs, adapted from Siddell (91).

overlapping 'nested set' (Figure 4). The coding assignment of each RNA species has been determined for both avian and murine coronavirus-infected cells by partial purification of the RNA followed by *in vitro* translation (1, 46, 63, 64). Only the unique sequences at the 5' end are translated, so that each mRNA specifies a single polypeptide even though it may include coding sequences for several polypeptides.

Numerous sequencing studies of the coronavirus genome have established that the order of the genes specifying the major structural polypeptides is 3' N, M, S, 5'. The genome also contains putative genes for non-structural proteins, and it is likely that the very large gene(s) towards the 5' end must specify the RNA polymerase(s). These coding assignments are summarized in Table 2.

B. Mechanism of RNA Synthesis

The first event in coronavirus-infected cells must be the translation of the 5' region of genome RNA to yield an RNA-dependent RNA polymerase. Attempts to reproduce this event by *in vitro* translation of virion-derived genome RNA have met with little success, although Leibowitz *et al* (65) reported the translation of MHV virion RNA into several related high molecular weight polypeptides of around 200,000 molecular weight. An "early" RNA-dependent RNA polymerase activity has been detected in MHV-infected cells one hour post-infection, followed by a second peak of polymerase activity at 6 hours, the latter being coincident with the bulk of virus specific RNA synthesis (66). The early RNA polymerase activity synthesises an RNA copy (negative-strand) of the genome RNA (56), only genome-length negative strand RNA can be detected in infected cells (55) and no subgenomic negative strand RNAs have been reported. From the foregoing, it can be concluded that the multiple subgenomic plus strand RNAs are synthesised from a full-length negative-strand RNA template.

An important experimental finding concerning the mechanism of subgenomic RNA synthesis was made by Jacobs *et al* (67), who showed that the UV-target sized of MHV-induced subgenomic RNAs corresponded to their physical sizes. A similar result was obtained when IBV-induced subgenomic RNAs were analysed by this method (68), and it was concluded that coronavirus subgenomic RNAs are synthesized independently and not derived by processing from a larger precursor molecule. However, in apparent contradiction to this idea, it was subsequently found that all the subgenomic RNAs share a common 5' sequence (leader sequence), which is 72 nucleotides long in MHV- and 60 nucleotides long in IBV-induced RNAs (69-72). The leader sequences are apparently derived from the 5' end of genome RNA but not by processing a full-length precursor RNA. They appear instead to act as primers for the transcription of individual mRNAs at specific internal initiation sites on the negative strand template RNA (73-74). Small virus RNA species containing leader sequences have been detected in MHV-infected cells, and a *ts* mutant of MHV was isolated which synthesises only small leader RNAs, not mRNAs, at the non-permissive temperature (75). Finally, it was demonstrated by mixed infection of cells with two strains of MHV that leader RNA sequences can reassort. The strains used were recombinant viruses containing either JHM strain sequences at the 5' end and A59 'body' sequences, or the other way around. The leader sequences were exchanged between the mRNAs of the two co-infecting viruses at high frequency (76).

If the leader RNAs act as primers, they should contain sequences which are complementary to sequences at the initiation site for mRNA synthesis on the negative strand template RNA. Regions of homology between the 3' end of the genomic leader sequence and sequences at the intergenic boundaries have been found in MHV-infected cells (51, 52, 77) and in IBV-infected cells (72). In all the major IBV RNA species the sequence CUUAACAA is present, and this could base-pair with complementary sequences at the intergenic boundaries on the negative strand template RNA. Related sequences containing the trinucleotide AAC are present in the MHV leader. However, the presence of this sequence on free leader RNA has not been formally proven. Alternative hypotheses are that this common intergenic sequence is a recognition site for the RNA polymerase rather than a primer-binding site, or serves both functions. Further evidence is required to decide between these and other possibilities (78).

C. Recombination

In 1986 Lai *et al* (79) reported the isolation of a recombinant coronavirus following mixed infection of DBT cells with two strains (A59 and JHM) of MHV. They used *ts* mutants of each strain, and derived a recombinant (B1) from plaques formed at the non-permissive temperature. From oligonucleotide fingerprint analysis of the B1 virus genome, it was clear that, although most of the genome was derived from A59, some 3 kilobases at the 5' end were derived from JHM virus, so that a cross-over event must have occurred within the 5'-proximal gene. Moreover, analysis of the subgenomic RNA species induced in B1 virus-infected cells showed that they all contained "hybrid" sequences, where the coding sequences of A59 virus were joined to leader sequences of JHM. Subsequently, it was found by co-infecting cells with *ts* mutants of A59 virus and *wt* JHM virus that recombinants could be obtained at high frequency (80).

Analysis of further recombinants (81) revealed some where more than one cross-over site was present, and some where there was a cross-over site within the leader sequence itself. By crossing a *ts* mutant of A59 virus, which synthesises only leader RNA at the restrictive temperature (75), with *wt* JHM virus, recombinants were obtained with up to three cross-over sites clustered close to the 5' end of the genome RNA. These sites correspond to the termination points of three leader-containing RNA species, less than full length (47,50 and 57 nucleotides respectively), which were detected in A59 virus-infected cells (81).

The high recombination frequency of murine coronavirus is remarkable. Lai *et al* (80, 81) have interpreted this as evidence for discontinuous, non-processive RNA species which can dissociate from and reassociate with the negative strand template RNA. Such a model would also provide an explanation for the generation of defective-interfering coronavirus RNAs which contain deletions within the genome. However, high frequency recombination at multiple sites is also observed with another positive strand virus, foot-and-mouth disease virus (see chapter 8 of this volume and ref. 82), and five cross-over sequences have been sequenced in these recombinants (83). The number of common nucleotides varied from 2 to 32 at the cross-over sites, and there were no common features or recognisably unique sites such as those involved in RNA splicing (84). It will be important to sequence cross-over sites within the coronavirus recombinats to see how general is the recombination event.

D. Virus Assembly and Morphogenesis

One of the early steps in coronavirus assembly is the formation of helical nucleocapsids by interaction of newly synthesised genome RNA with molecules of N protein. Aggregates of nucleocapsids have been visualized in infected cells late in the infectious cycle (85). The nucleocapsid forms a complex with the M protein *in vitro* (86) and it is likely that *in vivo* this complex formation occurs at the maturation site where budding occurs from intracellular membranes in the rough endoplasmic reticulum (RER). Although the S protein may accumulate at the plasma membrane, virions never bud there, and it seems that the M protein alone determines the site of budding. Virions are released into the lumen of the RER and migrate through the Golgi apparatus, where they are transported into smooth-walled vesicles. Release from the cell presumably involves fusion of these vesicles with the plasma membrane (87).

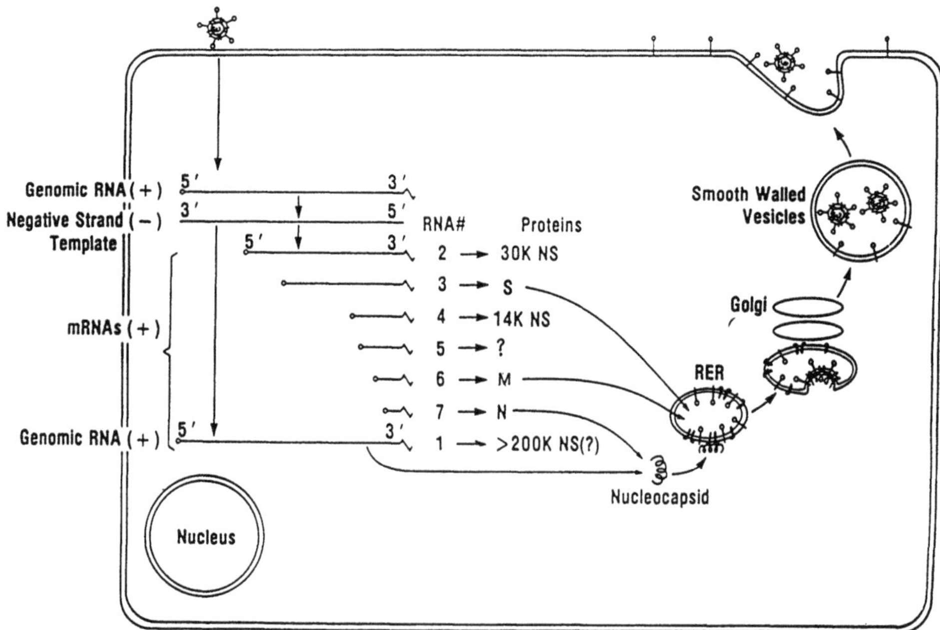

Figure 5. Steps in the replication and assembly of coronaviruses. Adapted from Holmes K.V. (87).

Studies using inhibitors support the importance of the M protein in determining the budding process. In MHV-infected cells, treatment with tunycamycin blocks S protein synthesis but allows synthesis of glycosylated M, since O-linked glycosylation is resistant to the drug. In such cells, virions lacking spikes still form in the RER and the Golgi apparatus (88). The production of IBV (47) and MHV (89) was also inhibited by monensin, an ionophore which inhibits transport of secretory glycoproteins at the level of the Golgi complex (90). In MHV-infected cells, virions accumulated in large amounts within the RER in the presence of monensin, which also blocked the O-linked glycosylation of the M polypeptide (89). An outline of the steps in replication and assembly of coronaviruses is given in Figure 5.

4. CONCLUDING REMARKS

When first discovered, the coronaviruses were thought to be negative-strand RNA viruses akin to influenza or paramyxoviruses on the basis of their general morphology including the helical structure of the ribonucleoprotein. It is now clear that they form a unique group of positive strand viruses with unusual molecular biological properties. At the time of writing, the complete nucleotide sequence of one coronavirus, IBV, is virtually completed (M.E.G. Boursnell, personal communication). This shows the enormous information content of the coronavirus genome, which is the largest infectious RNA molecule yet described. Unravelling the way in which this information is expressed in the infected cell is a fascinating challenge for the future.

Acknowledgements

I wish to thank Mike Boursnell, Dave Cavanagh, Steve Inglis, Michael Lai, and Stuart Siddell for valuable discussion, information and preprints on coronavirology. I also thank Kathryn Holmes for permission to reproduce Figures 2 and 5 in adultered form, and Dave Cavanagh for Figures 1 and 3. Finally, I am grateful to my secretary, Nicki Snook, for her excellent rendition of my manuscript.

5. REFERENCES

1) Siddell, S., Wege, H. and ter Meulen, V. (1983). J. gen. Virol. **64**, 761-776.

2) Tyrrell, D.A.J., Almeida, J.D., Berry, D.M., Cunningham, C.H., Hamre, D., Hofstad, M.S., Mallucci, L. and McIntosh, K. (1968). Nature (London) **220**, 650.

3) Wege, H., Siddell, S. and ter Meulen, V. (1982). Current Topics in Microbiol. Immunol. **99**, 165-200.

4) Knobler, R.L., Tunison, L.A., Lampert, P.W. and Oldstone, M.B.A. (1982). Amer. J. Path. **109**, 157-168.

5) Wege, H., Watanabe, R. and ter Meulen, V. (1984). J. Neuro-immunol. **6**, 325-336.

6) Knobler, R.L., Lampert, P.W. and Oldstone, M.B.A. (1982). Nature (London) **298**, 279-280.

7) Fleming, J.O., Trousdale, M.D., El-Zaatari, F.A.K., Stohlman, S.A. and Weiner, L.P. (1986). J. Virol. **58**, 869-875.

8) Burks, J.S., Devald, B.L., Jankovsky, L.D. and Gerdes, J.C. (1980). Science **209**, 933-934.

9) Leinikki, P.D., Holmes, K.V., Shekarchi, I., Iivanainen, M., Madden, D. and Sever, J.L. (1981). Adv. Exp. Med. Biol. **142**, 323-326.

10) Mahy, B.W.J. (1981). Adv. Exp. Med. Biol. **142**, 261-270.

11) Mahy, B.W.J. (1983). In: *Viruses and Demyelinating Diseases*, ed. C.A. Mims, Academic Press, pp. 81-87.

12) Larner, A.J. (1986). J. Roy. Soc. Med. **79**, 412-417.

13) Hofstad, M.S. (1978). In: *Diseases of Poultry*, ed. M.S. Hofstad, B.W. Calnek, C.F. Helmboldt, W.M. Reid and H.W. Yorke. Iowa State University Press, Ames, pp. 487-504.

14) Alexander, D.J., Gough, R.E. and Pattison, M. (1978). Res. Vet. Sci. **24**, 228-233.

15) Jacobse-Geels, H.E., Daha, M.R. and Horzinek, M.C. (1982). Amer. J. Vet. Res. **43**, 666-670.

16) Baybutt, H.M., Wege, H., Carter, M.J. and ter Meulen, V. (1984). J. gen. Virol. **65**, 915-924.

17) Brown, T.D.K., Boursnell, M.E.G. Binns, M.M. and Tomley, F.M. (1986) In: *Molecular Organisation of Positive Strand Viruses*, Rowlands, D.J., Mahy, B.W.J., and Mayo, M.A. eds, Academic Press (in press).

18) Jacobs, L., van der Zeijst, B.A.M. Horzinek, M.C. (1986). J. Virol. **57**, 1010-1015.

19) Lai, M.M.C. and Stohlman, S.A. (1981). J. Virol. **38**, 661-670.

20) Cavanagh, D. (1983). J. gen. Virol. **64**, 1187-1191.

21) King, B., Potts, B.J. and Briaan, D.A. (1985). Virus Res. **2**, 53-59.

22) Hogue, B.G. and Brian, D.A. (1986). Virus Res. **5**, 131-144.

23) Armstrong, J., Smeekens, S. and Rottier, P. (1983). Nucleic Acid Res. **11**, 883-891.

24) Skinner, M.A.S. and Siddell, S.G. (1983). Nucleic Acid Res. **11**, 5045-5054.

25) Siddell, S.G., Barthel, A. and ter Meulen, V. (1981). J. gen. Virol. **52**, 235-243.

26) Boursnell, M.E.G., Binns, M.M., Foulds, I.J. and Brown, T.D.K. (1985). J. gen. Virol. **66**, 573-580.

27) Niemann, H. and Klenk, H.D. (1981). J. mol. Biol. **153**, 993-1010.

28) Rottier, P., Brandenburg, D., Armstrong, J., van der Zeijst, B. and Warren, G. (1984). Proc. Natl. Acad. Sci. **81**, 1421-1425.

29) Rottier, P.J.M., Welling, G.W., Welling-Webster, S., Niesters, H.G.M., Lenstra, J.A. and van der Zeijst, B.A.M. (1986). Biochemistry **25**, 1335-1339.

29a) Pfleiderer, M., Skinner, M.A. and Siddell, S.G. (1986). Nucleic Acids Res., in press.

30) Stern, D.F. and Sefton, B.M. (1982). J. Virol. **44**, 804-812.

31) Boursnell, M.E.G., Brown, T.D.K. and Binns, M.M. (1984). Virus Res. **1**, 303-313.

32) Rottier, P., Armstrong, J. and Meyer, D.I. (1985). J. biol. Chem. **260**, 4648-4652.

33) Cavanagh, D., Davis, P.J. and Pappin, D.J.C. (1986). Virus Res. **4**, 145-156.

34) Davies, H.A. and MacNaughton, M.R. (1979). Arch. Virol. **59**, 25-33.

35) Cavanagh, D. (1983). J. gen. Virol. **64**, 2577-2583.

36) Collins, A.R., Knobler, R.L., Powell, H. and Buchmeier, M.J. (1982). Virology **119**, 358-371.

37) Mockett, A.P.A., Cavanagh, D. and Brown, T.D.K. (1984). J. gen. Virol. **65**, 2281-2286.

38) Binns, M.M., Boursnell, M.E.G., Cavanagh, D., Pappin, D.J.C. and Brown, T.D.K. (1985). J. gen. Virol. **66**, 719-726.

39) Stern, D.F. and Sefton, B.M. (1982). J. Virol. **44**, 794-803.

40) Cavanagh, D. and Davis, P.J. (1986). J. gen. Virol. **67**, 1443-1448.

41) Cavanagh, D., Davis, P.J., Pappin, D.J.C., Binns, M.M., Boursnell, M.E.G. and Brown, T.D.K. (1986). Virus Res. **4**, 133-143.

42) Talbot, P.J. and Buchmeier, M.J. (1985). Virus Res. **2**, 317-328.

43) Niesters, H.G.M., Lenstra, J.A., Spaan, W.J.M., Zijderveld, A.J., Bleumink-Pluym, N.M.C., Hong, F., van Scharrenburg, G.J.M., Horzinek, M.C. and van der Zeijst, B.A.M. (1986). Virus Res. **5**, 253-264.

44) Siddell, S.G. (1982). J. gen. Virol. **62**, 259-269.

45) Siddell, S.G., Wege, H. and ter Meulen, V. (1982). Curr. Topics Microbiol. Immunol. **99**, 131-163.

46) Sturman, L.S. and Holmes, K.V. (1983). Advances in Virus Res. **28**, 35-111.

47) Alonso-Caplen, F.V., Matsuoka, Y., Wilcox, G. and Compans, R.W. (1984). Virus Res. **1**, 153-167.

48) Bond, C.W., Anderson, K. and Leibowitz, J.L. (1984). Arch. Virol. **80**, 333-347.

49) Boursnell, M.E.G. and Brown, T.D.K. (1984). Gene **29**, 87-92.

50) Boursnell, M.E.G., Binns, M.M. and Brown, T.D.K. (1985). J. gen. Virol. **66**, 2253-2258.

51) Skinner, M.A., Ebner, D. and Siddel, S.G. (1985). J. gen. Virol. **66**, 581-592.

52) Skinner, M.A. and Siddell, S.G. (1985). J. gen. Virol. **66**, 593-596.

53) Wilhelmsen, K.C., Leibowitz, J.Z., Bond, C.W. and Robb, J.A. (1981). Virology **110**, 225-230.

54) Mahy, B.W.J., Siddell, S.G., Wege, H. and ter Meulen, V. (1983). J. gen. Virol. **64**, 103-111.

55) Lai, M.M.C., Patton, C.D. and Stohlman, S.A. (1982). J. Virol. **44**, 487-492.

56) Brayton, P.R., Stohlman, S.A. and Lai, M.M.C. (1984). Virology **133**, 197-201.

57) Sawicki, S.G. and Sawicki, D.L. (1986). J. Virol. **57**, 328-334.

58) Stern, D.F. and Kennedy, S.I.T. (1980). J. Virol. **34**, 665-674.

59) Stern, D.F. and Kennedy, S.I.T. (1980). J. Virol. **36**, 440-449.

60) Lai, M.M.C., Brayton, P.R., Armen, R.C., Patton, C.D., Pugh, C. and Stohlman, S.A. (1981). J. Virol. **39**, 823-824.

61) Leibowitz, J.L., Wilhelmsen, K.C. and Bond, C.W. (1981). Virology **114**, 39-41.

62) Spaan, W.J.M., Rottier, P.J.M., Horzinek, M.C. and van der Zeijst, B.A.M. (1982). J. Virol. **42**, 432-439.

63) Siddell, S.G. (1983). J. gen. Virol. **64**, 113-125.

64) Stern, D.F. and Sefton, B.M. (1984). J. Virol. **50**, 22-29.

65) Leibowitz, J.L., Weiss, S.R., Paavola, E. and Bond, C.W. (1982). J. Virol. **43**, 905-913.

66) Brayton, P.R., Lai, M.M.C., Patton, C.D. and Stohlman, S.A. (1982). J. Virol. **42**, 847-853.

67) Jacobs, L., Spaan, W.J.M., Horzinek, M.C. and van der Zeijst, B.A.M. (1981). J. Virol. **39**, 401-406.

68) Stern, D.F. and Sefton, B.M. (1982). J. Virol. **42**, 755-759.

69) Lai, M.M.C., Patton, C.D., Baric, R.S. and Stohlman, S.A. (1983). J. Virol. **46**, 1027-1033.

70) Lai, M.M.C., Baric, R.S., Brayton, P.R. and Stohlman, S.A. (1984). Proc. Natl. Acad. Sci. **81**, 3626-3630.

71) Brown, T.D.K., Boursnell, M.E.G. and Binns, M.M. (1984). J. gen. Virol. **65**, 1437-1442.

72) Brown, T.D.K., Boursnell, M.E.G., Binns, M.M. and Tomley, F.M. (1986). J. gen. Virol. **67**, 221-228.

73) Spaan, W., Delius, H., Skinner, M., Armstrong, J., Rottier, P., Smeekens, S., van der Zeijst, B. and Siddell, S.G. (1983). EMBO J. **2**, 1839-1844.

74) Brown, T.D.K. and Boursnell, M.E.G. (1984). Virus Res. **1**, 15-24.

75) Baric, R.S., Stohlman, S.A., Razavi, M.K. and Lai, M.M.C. (1985). Virus Res. **3**, 19-33.

76) Makino, S., Stohlman, S.A. and Lai, M.M.C. (1986). Proc. Natl. Acad. Sci. **83**, 4204-4208.

77) Buldzilowicz, C.J., Wilczynski, S.P. and Weiss, S.R. (1985). J. Virol. **53**, 834-840.

78) Mahy, B.W.J. (1984). Adv. Exptl. biol. Med. **173**, 1-10.

79) Lai, M.M.C., Baric, R.S., Makino, S., Keck, J.G., Egbert, J., Leibowitz, J.L. and Stohlman, S.A. (1985). J. Virol. **56**, 449-456.

80) Makino, S., Keck, J.G., Stohlman, S.A. and Lai, M.M.C. (1986). J. Virol. **57**, 729-737.

81) Lai, M.M.C., Makino, S., Baric, R.S., Soe, L., Shieh, C.K., Keck, J.G. and Stohlman, S.A. (1986). In: *Positive strand RNA Viruses*, UCLA Symposium on Molecular and Cellular Biology. M.A. Brinton and R.R. Rueckert, eds. in press.

82) King, A.M.Q., McCahon, D., Saunders, K., Newman, J.W.I. and Slade, W.R. (1985). Virus Res. **3**, 373-384.

83) King, A.M.Q., Ortlepp, S.A., Newman, J. and McCahon, D. (1986). In: *Molecular Organisation of Positive Strand Viruses*, Rowlands, D.J., Mahy, B.W.J., Mayo, M.A., ed. Academic Press, in press.

84) Cech, T.R. (1983). Cell **34**, 713-716.

85) Dubois-Dalcq, M.E., Doller, E.W., Haspel, M.V. and Holmes, K.V. (1982). Virology **119**, 317-331.

86) Sturman, L.S., Holmes, K.V. and Behnke, J. (1980). J. Virol. **33**, 449-462.

87) Holmes, K.V. (1985). **In** 'Virology', edited B.N. Fields*et al.*Raven Press, 1331-1343.

88) Holmes, K.V., Doller, E.W. and Behnke, J.N. (1981). Adv. Expt. Med. Biol. **142**, 133-142.

89) Niemann, H., Boschek, B., Evans, D., Rosing, M., Tamura, I. and Klenk, H.D. (1982). EMBO J. **1**, 1499-1504.

90) Tartakoff, A.M. (1983). Cell **32**, 1026-1028.

91) Siddell, S. (1986). In: *Molecular Organization of Positive Strand Viruses* Rowlands, D.J., Mahy, B.W.J. and Mayo, M.A., eds., Academic Press, in press.

CHAPTER 11

MOLECULAR BASIS OF RHABDOVIRUS REPLICATION

SUZANNE U. EMERSON AND MANFRED SCHUBERT
University of Virginia, Charlottesville VA 22908, and National Institutes of Health, Bethesda MA 20892, U.S.A.

INTRODUCTION

A virus was originally classified as a rhabdovirus if it displayed a distinctive elongated morphology (1). Rhabdo means rod-like and all rhabdoviruses have a bacilliform or conical appearance. Over 60 different viruses currently comprise the rhabdovirus group and some members are listed in Table 1 (2). Many of these viruses are transmitted to their plant or vertebrate hosts by insects and, therefore, must be able to successfully infect and replicate in both invertebrate and vertebrate or invertebrate and plant cells. However, transmission by an arthropod vector is not a mandatory requirement for classification as a rhabdovirus since rabies virus is most commonly spread by direct animal contact via the bite of an infected vertebrate. Many viruses in the rhabdovirus group are of interest because as pathogens of livestock (vesicular stomatitis), fish (spring viremia of carp), or plants (lettuce necrotic yellows), they have a significant economic impact. Others, such as rabies virus are feared because they are virulent human pathogens. In humans (and most animals), once a rabies virus infection is established and symptoms appear, there is no effective treatment and death invariably follows due to destruction of the central nervous system. Aside from the obvious importance of rhabdoviruses as pathogens, certain members of the group have generated considerable interest as models for viral replication, glycoprotein biosynthesis, and membrane structure. Vesicular stomatitis virus (VSV) has been particularly useful in this regard because it can be easily propagated to high titers in tissue culture. The virus has an astounding host range and replicates in tissue culture cells from such diverse species as mammals, fish, and insects. The virion is relatively stable and easily purified, and although VSV can infect humans, unlike rabies virus infections, the resulting disease is not serious and resembles a bad case of flu. For these reasons, VSV has

Table 1. - Examples of Rhabdoviruses

Vertebrate or Invertebrate Host	Plant and Invertebrate Host
Vesicular stomatitis	Lettuce necrotic yellows
Rabies	Strawberry crinkle
Kern Canyon	Maize mosaic
Lagos bat	Potato yellow dwarf
Infectious hematopoietic necrosis	Rice transitory yellowing
Oregon sockeye disease	Sugar beet leaf curl
Spring viremia of carp	Eggplant mottled dwarf
Sigma	Wheat chlorotic streak

been extensively studied. Much of our knowledge of rhabdovirus replication is derived from studies of VSV and it is considered the prototype rhabdovirus.

Classification of a virus as a rhabdovirus solely on the basis of morphology is obviously tentative. Many of the viruses in the group have not yet been analyzed extensively enough to assure their proper classification. However, morphology has proved surprisingly reliable since numerous viruses have proved to have the rhabdovirus characteristics as defined by VSV. Of primary importance is the genome character and polarity. The genome of rhabdoviruses consists of a single strand of RNA (appr. 4-5 X 10^6 MW) which is complementary to the viral messenger RNAs (1). The viruses are, therefore, negative-strand viruses. The non-segmented nature of the genome and its negative-sense mandate the viral replication strategy.

Historically, VSV is important because until 1970, virus particles in general were thought to be too simple to contain enzymes. However, while attempting to discover why the deproteinized genome of VSV was not infectious while that of a positive-stranded virus was, Baltimore *et al.* (3) made the exciting discovery that highly purified VS virions could synthesize RNA *in vitro* and that this RNA was complementary to the viral genome. This demonstration of a polymerase in the virion was crucial to our understanding of the rhabdovirus replication cycle. Rhabdovirus genomic RNA is not infectious because it is negative-sense and viral proteins can not be synthesized until the genomic RNA is transcribed into mRNA. Host cell enzymes cannot carry out this required transcription. Therefore, all rhabdoviruses must package a virally encoded RNA-dependent RNA polymerase in the virion in order to initiate the replication cycle. The discovery led to an extensive study of VSV because it indicated that purified VSV virions readily provided template and polymerase in a form that was amenable to analysis *in vitro*.

1. THE VIRION

Rhabdoviruses are not pleomorphic and each virus type produces particles of uniform dimensions and structure (1). For instance, infectious VSV is bullet-shaped with one planar end and one curved end and each rod is approximately 180 nm long and 65 nm wide. In general, the other animal rhabdoviruses have similar morphology and dimensions while the plant rhabdoviruses are often longer and curved at both ends. Rhabdoviruses also produce deletion mutants or defective particles which retain the morphology and width of the parent but are shorter in length (see Section 4.E).

Electron micrographs of partially disrupted virions indicate the virion contains a long helical nucleocapsid coiled in a regular manner to give the virion its characteristic shape (Fig. 1). The condensed nucleocapsid is enclosed within a lipid envelope which is studded with small projections or spikes. Polyacrylamide gel analysis of purified VS virions revealed five viral proteins. Labeling experiments with radioactive precursors demonstrated that one protein was a *glycoprotein* while another was a *phosphoprotein*. Proteolytic digestion of intact virions removed the spikes, resulted in the selective degradation of the glycoprotein, and greatly decreased infectivity of the virus. Additionally, antibodies to the glycoprotein were shown to neutralize viral infectivity. Therefore, the glycoprotein, which is the only

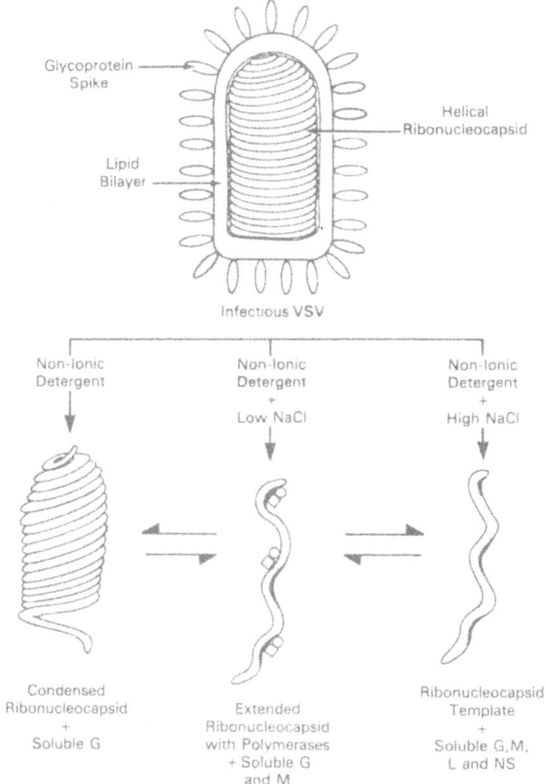

Figure 1. Schematic diagram of vesicular stomatitis virions depicting controlled disruption by non-ionic detergent and NaCl. The G protein comprising the spike is released from the particle when the lipid is dispersed with non-ionic detergents such as Triton X-100. The M (matrix) protein located internal to the envelope is responsible for condensation and is released from the ribonucleocapsid complex as the NaCl concentration is raised. The resulting extended ribonucleocapsid consists of the N protein-genomic RNA (template) and the polymerase (L and NS). The L and NS are dissociated from the template and each other with 0.72 M NaCl. The NaCl mediated dissociations are reversible.

protein on the surface of the virion, is believed to mediate viral attachment to host cells.

Nonionic detergents in the presence of low salt disperse the virion envelope and release the glycoprotein and a second protein, the matrix or M protein (4). The M protein is the most abundant protein in the virion and is thought to form a matrix layer just internal to the lipid bilayer.

The remaining three viral proteins remain with the RNA as a ribonucleocapsid (RNP). When these three proteins are present, the RNP is infectious, albeit at low efficiency, and is capable of *in vitro* transcription (4). Removal of any one of the three RNP proteins destroys infectivity and transcription capacity.

The major RNP protein is the nucleocapsid (N) protein. Approximately 1300 molecules of N protein are tightly arranged along the genomic RNA (5). The N protein apparently never detaches from the RNA and renders it resistant to nuclease digestion. This RNA-N protein complex is the template for both transcription and replication.

The remaining two RNP proteins, L and NS, are present in 50 and 470 copies (5), respectively, per genome and together constitute the viral RNA-dependent RNA polymerase (6). The L (large) protein contains 2109 amino acids and most likely is the catalytic unit of the polymerase (7). The much smaller NS protein (265 amino acids) is the phosphoprotein (8). Its exact function is not yet defined, but it is required for L protein attachment to the template and may interact with cytoplasmic N protein to prevent N protein aggregation (9, 10).

Another protein representing the carboxy terminal region of NS protein has recently been detected in both virions and infected cells (11). Whether this polypeptide has any function is not known. Table 2 summarizes important characteristics and functions of the five structural VSV proteins.

Rhabdoviruses, therefore, are relatively simple viruses which consist of a single strand of RNA, a lipid envelope and five or six proteins. Yet, the prototype virus, VSV, is able to infect and replicate efficiently in virtually any cell. In spite of its simplicity, VSV is a highly cytopathic virus which shuts off host DNA, RNA, and protein synthesis. Thus, it appears that rhabdoviruses are relatively independent viruses which only rely on the host cell to supply universal and basic functions such as translation facilities.

It remains to be discovered how VSV can carry out all of the required processes with such a small repertoire of proteins. Elucidation of the exact structure-function relationships of the viral components promises to provide an example of a highly efficient biological complex.

2. GENOME ORGANIZATION

Rhabdovirus genomes consist of a single stranded RNA of negative-sense (anti-message sense) polarity which is completely encapsidated by the viral nucleocapsid protein N. The genomic organization of rhabdoviruses as well as of paramyxoviruses share striking similarities which point to a common strategy in their gene expression and replication (15). The prototype rhabdovirus genome of VSV contains five genes (Fig. 2). The gene order (3'-N-NS-M-G-L-5') is highly con-

Table 2. – VSV mRNAs and Proteins

Protein	mRNA Length (15)	Predicted Amino Acids (15)	Molecules per Virion (5)	Characteristics	Function
N	1,333	422	1,300	Self-aggregates	Encapsidates genomic RNAs Anti-terminator of replication (12)
NS	821	265	470	Heterogeneously phosphorylated	Promotes L binding to template (9) Polymerase component (6) Prevents N self-aggregation? (10)
M	838	229	1,800	Basic Interacts with lipids, G and N	Condenses nucleocapsids (13) Clusters G in plasma membrane? (14)
G	1,672	511 (precursor) 495 (mature)	1,200	Transmembrane glycoprotein	Cell attachment (1) Endosome fusion (15) Budding (1)
L	6,380	2,109	50	Genetically and physically labile	Catalytic unit of polymerase Post-transcriptional modification of mRNAs?

served among rhabdoviruses. The nucleocapsid protein needed in large amounts to completely encapsidate the genomic and anti-genomic RNAs is always encoded by the first gene while the catalytic polymerase protein L is always encoded by the last gene. As discussed below, this gene order together with the sequential and attenuated mode of transcription ensures that the five genes are expressed in a preprogrammed manner based on this genomic organization.

The entire nucleotide sequence of the VSV genome has been determined (15). The 5' terminal nucleotide carries a triphosphate, suggesting an initiation event during replication. The 3' terminal nucleotide carries a 3' hydroxyl group. The rabies virus genome has also been sequenced except for the large L gene. It differs from VSV in that a small, possibly rudimentary gene was detected between the G and L genes of rabies virus. This short gene contains a message start and a polyadenylation site; however, all reading frames carry multiple translational stop codons. Interestingly, a sixth gene which encodes a small 12K nonstructural protein (NV) was tentatively mapped between the G and L genes of the rhabdovirus hematopoietic necrosis virus (16). Its function is unknown. Future sequence analysis may reveal whether this gene is related to the rabies virus rudimentary gene. Since VSV does not carry this sixth gene,its function may be nonessential or may be carried out by the other viral or host proteins.

The rhabdovirus genome is extremely compact (Fig. 2). From a total of 11161 nucleotides of the VSV (Indiana serotype) genome, 11091 nucleotides are transcribed and a total of 10608 nucleotides encode the five proteins. Thus, the untranslated regions of the messenger RNAs are short. The extracistronic regions consist of the 50 nucleotide 3' terminal *leader* region (**l**), which is transcribed into a 47 nucleotide plus-sense leader RNA (17), and the 59 nucleotide untranscribed 5' terminal **trailer** region (**t**) (18). The untranscribed short intercistronic (**i**) regions consist of only two nucleotides $^{G}/_{C}A$ (Fig. 3). The intercistronic regions are flanked directly upstream by the highly conserved polyadenylation site $AUACU_7$ and directly downstream by

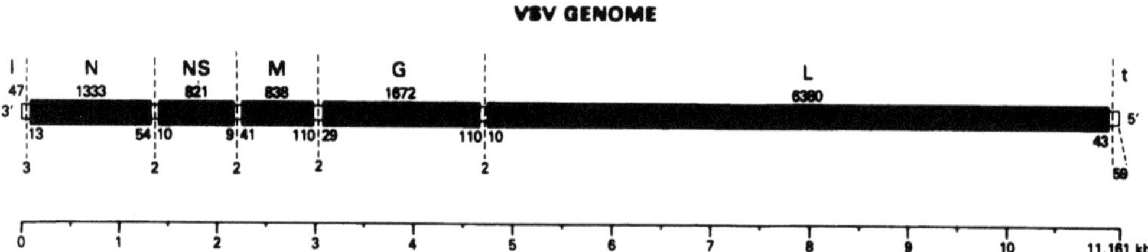

Figure 2. Map of the VSV genome. The VSV genome consists of 11161 nucleotides which are sequentially transcribed into a short 47 nucleotide plus-sense leader RNA and five monocistronic messenger RNAs encoding the structural proteins N (nucleocapsid), NS (phosphoprotein, originally thought to be absent in the virion), M (matrix protein), G (glycoprotein) and L (large polymerase protein). The coding regions of each gene are marked by black bars. The sizes of the untranslated regions of the corresponding mRNAs are indicated. The two extracistronic regions are the 3' terminal leader region (l) and the 5' terminal, nontranscribed trailer region (t). The nontranscribed intercistronic regions are two nucleotides long. There are three nontranscribed nucleotides between the leader region and the cap site of the N gene.

the message start sequence (cap site) UUGUC of the next gene. During *replication*, these regulatory sequences are ignored (see below). The 3' terminus of the genomic RNA contains the origin of replication. A complementary sequence is present at the genomic 5' end which allows the replication of the anti-genome. Rhabdoviruses transcribe and replicate in the cytoplasm of the host cell and, in contrast to orthomyxoviruses (influenza virus), there is no nuclear involvement and messenger RNAs are not spliced.

To expand the genetic information, paramyxovirus messenger RNAs encoding the phosphoprotein P also encode a C protein which is translated from an overlapping reading frame by using a different start codon during translation. Both translational start codons are found close to the 5' terminus of the messenger RNA. With VSV, a 7K protein is translated *in vitro* from the NS messenger RNA (11). It is translated in the same reading frame as the NS protein itself and contains the carboxy terminal amino acids of NS. The start codon for the 7K protein translation is, therefore, closer to the 3' rather than the 5' end of the messenger RNA. This protein has also been detected *in vivo* and the strategy of expanding genetic information by encoding two proteins in one gene has to be considered as a possibility for the phosphoprotein gene of rhabdoviruses.

Like most other negative-strand viruses, the rhabdovirus genome presents itself as a highly compact unit which is efficiently expressed and replicated in an autoregulated mode.

3. MOLECULAR BASIS OF VIRAL REPLICATION

A. Overview of the Cycle

Analysis of the prototype, VSV, indicates rhabdovirus replication is an efficient process. Progeny virions can be detected as soon as two to three hours after infection of BHK cells and virus production continues for up to 20 hours or until

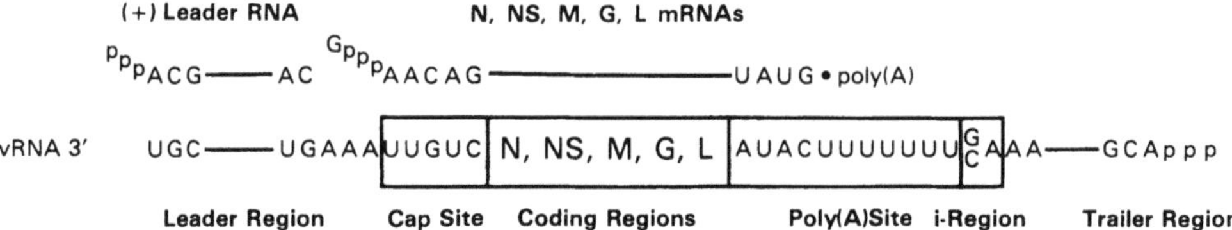

Figure 3. Highly conserved VSV transcription sequences. Each VSV gene starts with a 3' terminal cap site which contains five conserved nucleotides and terminates with a conserved poly (A) site containing seven uridine residues which are repeatedly transcribed into, on the average, 200 nucleotide long poly (A) tails. The poly (A) site is followed by the intercistronic region (i) which consists of the conserved sequence GA at 3 out of 4 junctions. The 3' terminal and 5' terminal extracistronic leader region and trailer region, respectively, are indicated. The leader region is transcribed into a 47 nucleotide plus-sense leader RNA. The trailer region is not transcribed.

the cell eventually dies. As many as 10,000 virus particles can be released from an infected cell.

Penetration

Rhabdovirus attachment to target cells is mediated by the viral glycoprotein (1). Tunicamycin studies suggest that in at least one strain of VSV, the carbohydrate residues are not essential for virion receptor activity (19). The host cell receptor has not yet been conclusively identified for any rhabdovirus. In the case of VSV, many virus particles can attach to a cell, and it is difficult to demonstrate saturation of receptor sites. However, in general only about 10% of the virus in an inoculum results in a productive infection so it is not clear whether all attachments to the cell surface represent specific receptor recognition. Since VSV can infect cells from very divergent species, the cell receptor must be a common component of virtually all cell plasma membranes.

Saturable binding of VSV to Vero cells is inhibited by phosphatidylserine so this common membrane lipid may constitute all or part of the VSV receptor (20). Competitive binding studies with α-bungarotoxin or d-tubocurarine suggest rabies virus utilizes the acetylcholine receptor for high affinity binding to neural cells (21). Unlike the paramyxoviruses, the rhabdoviruses do not induce membrane fusion at neutral pH and viral entry into cells is achieved via endocytosis (22).

Electron micrographs demonstrate that VS virions accumulate on the cell surface in clathrin coated pits. After endocytosis of these regions, the virions are deposited in secondary endosomes which maintain an acidic pH. Both rabies virus and VSV exhibit membrane fusion capacity at acidic pH: for VSV, the glycoprotein acts as a fusion protein (23). Therefore, it is believed that the acidic environment of the endosome induces glycoprotein mediated fusion of the virion envelope with the endosome membrane such that the RNP is uncoated and is expelled into the cytoplasm.

Rhabdovirus replication occurs totally in the cytoplasm since VSV can grow in enucleated cells. Therefore, as soon as the RNP is released from the envelope and enters the cytoplasm, the biosynthetic portion of the replication cycle begins.

Primary Transcription

Primary *transcription* is the obligatory first biosynthetic event in rhabdovirus infections. Removal of the virion envelope enables the endogenous nucleocapsid-associated viral RNA polymerase to initiate transcription. Leader RNA and messenger RNAs are transcribed sequentially and in a polar manner (24, 25). As a result of this transcription strategy, the molar abundance of each mRNA is regulated by the location of its gene. The abundance of an RNA decreases as the distance of the gene from the 3' end increases. Consequently, the N mRNA is transcribed most frequently and L mRNA least frequently. Since the viral mRNAs are translated with similar efficiencies, there is also a corresponding differential accumulation of viral proteins (26). This strategy provides the virus with optimal amounts of each protein.

Translation of the Glycoprotein mRNA

Each viral mRNA is capped, methylated, and polyadenylated so it resembles host mRNAs and direct and efficient translation on ribosomes occurs. The glycoprotein mRNA is translated on membrane bound polyribosomes (27).

The first 16 amino acids at the amino terminus of G protein serve as a signal sequence to direct the G protein through the membrane of the rough endoplasmic reticulum (rER). As the nascent G polypeptide is elongated it is extruded into the lumen of the rER where cellular enzymes add high-mannose core sugars to two acceptor asparagine residues. The carbohydrate chains are further processed in the rER and in the Golgi apparatus to yield complex oligosaccharide chains. Since the host enzymes synthesize the oligosaccharides, the sugar content of the G protein varies somewhat according to the host. The major function of the sugars appears to be transport related and, generally, non-glycosylated G protein aggregates and does not appear on the cell surface. The G protein of the Indiana, but not the New Jersey serotype of VSV, also has palmitate esterified to the carboxy terminal region. The palmitate, which is also added in the Golgi apparatus, serves no known function.

The modified G protein lacking the terminal signal sequence is transported from the rER through the Golgi apparatus to be inserted in the cell plasma membrane. Here it is oriented with the carboxy terminus internal to the membrane and 95% of the polypeptide, including the carboydrate, external (28). The glycoprotein is anchored in the membrane by a stretch of 20 hydrophobic amino acids proximal to the carboxy terminus. Synthesis, processing, and transport of the glycoprotein requires at least 15 minutes. New glycoprotein continues to be inserted into the membrane during the course of the infection and accumulates there in preparation for viral maturation.

Posttranslational Processing of Viral Proteins

The remaining viral proteins are translated on nonmembrane bound polysomes and accumulate in cytoplasmic pools. Neither the L nor the N protein is believed to undergo processing or modification. The M protein is sometimes phosphorylated to a low extent. The NS protein, on the other hand, is extensively phosphorylated at serine and threonine residues so that a heterogeneous population of NS molecules is generated (29). The soluble NS protein in the cytoplasm of BHK cells has a lower level of phosphorylation than that isolated from mature virions, but the functional significance of these differences is unknown. Many of the phosphorylations are provided by cell enzymes and there is a report that the L protein also phosphorylates NS protein (30). Although there is abundant speculation as to whether phosphorylation affects NS function, there is no unambiguous data.

The NS Protein

Since the gene order determines the amount of each viral protein synthesized and the NS gene is second in order, large amounts of NS protein are synthesized. Yet the NS protein is a minor component of the virion. In contrast, to the other viral proteins which accumulate to only a limited extent in cytoplasmic pools, the NS protein enters a large and constantly expanding cytoplasmic pool. Pulse-chase studies indicate about 75% of the NS protein enters this pool and never interacts with nucleocapsids (31). Since in general, small viruses have evolved to function with utmost efficiency, this result suggests the cytoplasmic pool of NS protein performs some function.

Although recent *in vitro* assays indicate newly synthesized N protein is sufficient to drive replication (32, 33), *ts* mutant studies demonstrated NS protein was also required for *in vivo* replication.

Reconciliation of these observations is achieved by postulating that the NS protein in the cytoplasmic pool functions to keep N protein from self-aggregating prior to nucleocapsid assembly. If this is proven, then the NS protein has a different role in replication than in transcription.

Viral RNA Replication

Accumulation of sufficient amounts of newly synthesized viral proteins triggers viral RNA replication (Fig. 4). In the first step of replication, an exact full length complement of the genomic RNA is synthesized from a template previously

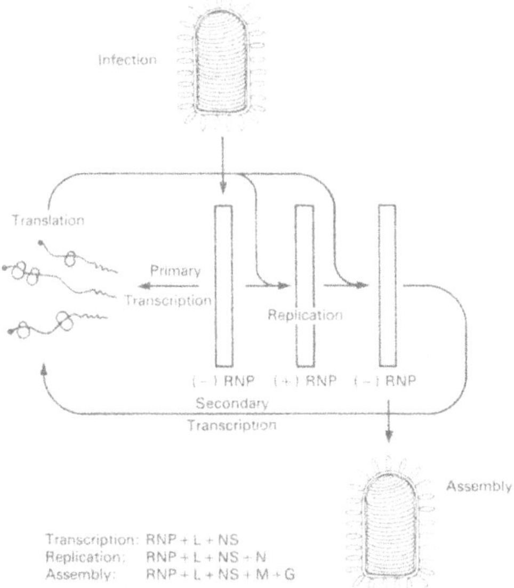

Figure 4. VSV transcription and replication. The minus-sense RNP of the virion is transcribed by the packaged RNA-dependent RNA polymerase (L + NS) into a 47 nucleotide plus-sense leader RNA and five monocistronic messenger RNAs (primary transcription) which are translated into the five viral proteins. When sufficient nucleocapsid protein N has accumulated, the genomic RNP is replicated and encapsidated into plus-sense genomic RNPs. These in turn serve as a template for the synthesis of minus-strand RNPs. Newly synthesized minus-strand RNPs are transcribed into high amounts of mRNAs (secondary transcription). Increased synthesis of M and G finally results in the selective packaging of minus-sense RNPs during assembly into mature virions.

involved in primary transcription. The N protein has a central role in this process. As the genomic plus-strand is being synthesized, it is concomitantly encapsidated with N protein (34). The L and NS proteins also associate with the nascent nucleocapsid complex so that at completion, the plus-strand RNP is ready to function in the second step of replication, the synthesis of new minus-strand RNPs. The nascent minus-strands are similarly encapsidated with N protein and associated with L and NS proteins. Synthesis of replicative RNA is regulated such that minus-strands are preferentially made.

The new minus-strand RNPs can enter a number of pathways. In the secondary transcription pathway, the new RNPs amplify the infection by providing additional leader and mRNAs (Fig. 4). Alternatively, if the soluble pool of nucleocapsid associated proteins is high enough, they can remain in the replication pathway. Finally, these RNPs can undergo morphogenesis into mature virions.

Viral Maturation

Viral maturation and virion egress occur simultaneously as condensed RNPs bud through regions of the plasma membrane modified by insertion of G protein. Although the outline of rhabdovirus assembly is known, virtually nothing is known about the details. While studies with *ts* mutants demonstrate that M protein is absolutely required for assembly, they do not provide information as to why M is required.

It has long been assumed that M protein acts as a cement to hold the envelope and RNP together. The conclusion that M protein in virions contacts both G and N proteins is based on detection of both N-M and G-M dimers after cross-linking proteins with chemical reagents (35).

However, since G protein is in the plasma membrane when the RNP is in the cytoplasm, it seems reasonable that M must interact initially with one or the other during morphogenesis. There are data suggesting M interacts first with RNPs and other data suggesting interaction with G is first. Nucleocapsids must condense prior to budding and M protein may accomplish this condensation. Certainly *in vitro*, M protein affects RNP structure (13).

Detergent extraction of the lipids and G protein from virions produces residual structures which contain M protein and resemble condensed nucleocapsids (Fig. 1). Removal of the M protein in addition to G protein and lipids, yields highly extended RNPs. Readdition of M protein to extended RNPs causes condensation. *In vivo*, M protein has not been detected on RNPs by biochemical means. However, immune electron micrographs of infected cells show extended nucleocapsids with regions of coiling associated with M protein (36). Strangely enough, the M protein could not be detected by immunostaining of fully condensed coils apparently ready to bud. One explanation for the lack of immunostaining is that M is located on the inside of tight coils. In that case, it is not obvious how it also interacts with G protein. *In vitro*, M protein interferes with transcription and theoretically M protein could bind to transcribing RNPs, cause release of nascent RNA chains, and thus prepare the RNP for migration to the membrane (37).

The question of when M protein interacts with RNPs is relevant for a number of reasons. Currently, there is no way to distinguish this pathway from the alternative which is that the RNP migrates to the membrane prior to its first contact with M protein. It is unknown what signals RNP migration to the plasma membrane, but it is possible that M protein may be involved. It is also important to determine whether M initiates condensation at a specific site on the RNP. This point is of interest because although both plus- and minus- strand RNPs are present in the cytoplasm, only minus-strands are incorporated into virions. If M protein recognizes a specific nucleation site for condensation, it could be the factor responsible for the negative-strand designation of the rhabdoviruses.

Other experiments suggest M protein orchestrates budding by causing clustering of G protein in patches prior to RNP arrival at the membrane. Because G protein spans the membrane and its carboxy terminus extends into the cytoplasm, there is a region of G protein that is potentially available for M protein interaction. Fluorescent photobleaching experiments demonstrate that the majority of G protein on the cell surface is highly mobile throughout the infection. However, there are *ts* mutations affecting M protein which result in a decreased surface mobility of G protein (14). The conclusion was drawn that soluble M protein interacts with G protein to immobilize it and form patches of modified membrane adequate for budding. It is probably at the stage of G protein clustering that cellular proteins are somehow excluded from these regions of the lipid bilayer. As a result, when budding does occur, the viral envelope contains only traces of host proteins.

The convergence of the M protein, the RNP, and the G protein at the plasma membrane results in virus particles budding from the cell surface. The driving force for this process is undefined. Once budding has occurred, the replication cycle is completed and the progeny virions are ready to initiate a new cycle of cell attachment and infection.

B. Transcription

The virions of all negative-strand RNA viruses contain an RNA-dependent RNA-polymerase which carries out both viral transcription and replication. Removal of the VSV lipid envelope by treatment with nonionic detergents activates the viral endogenous RNA polymerase *in vitro* (Fig. 1) (3). In the presence of ribonucleotide triphosphates, the genomic RNP template is transcribed (starting at the precise 3' end of the genome) into a short 47 nucleotide plus-sense leader RNA and five monocistronic messenger RNAs.

It is commonly accepted that the leader RNA and the messenger RNAs are sequentially transcribed following the gene order. Sequential transcription of the genes as well as the gene order itself were first established by ultraviolet (UV) light inactivation kinetics: transcription of genes proximal to the 3' end of the genome is (independent of their size) more UV resistant than the transcription of genes further downstream (24, 25). In fact, the UV target size of a transcribed gene depends on its position in the genome and is the sum of the target sizes of the preceeding genes (*polar effect*).

The UV inactivation studies were confirmed by kinetics on the sequential appearance of the messenger RNAs *in vitro* and *in vivo* (38). The kinetics studies were possible primarily because of the relatively slow elongation rate of the polymerase which incorporates approximately 3.7 nucleotides per second *in vitro*. At this rate, it would take about 65 minutes to transcribe the entire VSV genome. Interestingly, the elongation rate across gene junctions is reduced to about 1.0 nucleotides per second, suggesting that the polymerase may pause at gene junctions. In addition, at each gene junction there is a decrease of about 31% in the molar amount of mRNA synthesized.

Thus, the N mRNA is the most abundant and L mRNA is the least abundant of the mRNAs. This attenuation as well as the sequential mode of transcription are important features which are the basis for the highly efficent and autoregulated mode of transcription.

Single *vs* Multiple Initiation Events

Although transcription occurs sequentially, it is not clear whether the polymerase enters the template only at a single site at the 3' terminus of the genome or at multiple internal sites, i.e., at the gene starts.

If there are multiple entry sites for the polymerase, then transcription of a gene must depend on (and must follow) the transcription of the upstream gene in order to account for the sequential mode of transcription. In the absence of UTP and GTP, polymerase which is bound to the template can synthesize short leader starts (pppAC) and message starts (pppAACA) (Fig. 3) (39, 40). These internal message starts are independent of trancription of the upstream gene. When the polymerase is dissociated from the template and is allowed to re-enter, only leader starts (pppAC) are synthesized (39).

Message starts which should have occurred if the multiple entry model was correct were not observed but the data are consistent with a single entry site for the polymerase at the 3' end of the genome. The data also suggest that the polymerase can not translocate or freewheel along the template in the absence of UTP and GTP but may require simultaneous RNA synthesis. This observation could also explain the short length of the intercistronic regions which are only two nucleotides long.

C. The Polymerase Complex and Its Multiple Functions

Dissociation and reconstitution of the VSV polymerase complex *in vitro* demonstrated that besides the ribonucleoprotein (RNP) template, the viral NS and L proteins are required for transcription (6). Neither L nor NS alone is transcriptionally active. Unencapsidated VSV RNA or foreign RNA does not serve as a template (4). The polymerase complex, therefore, consists of the RNP and the L and NS proteins in an as yet unknown molar ratio. NS alone, but not L alone can bind to the RNP (9). Its binding site for polymerase entry has been mapped by footprinting to reside within positions 16 and 30 at the 3' end of the minus-strand genome and within positions 17 and 37 at the 3' end of the plus-strand genome (se-

Table 3. - Multiple Functions of Viral Polymerase (L & NS protein)

Function	Supporting Data
Recognition of signal sequences	Transcripts are initiated and terminated at highly conserved sites
Elogation	Ts mutants of L and NS defective in transcription and replication Both L and NS required for *in vitro* transcription
Capping	Different pathway than that used by known host enzyme
Methylation	Host range mutants with decreased methylation *in vivo* and *in vitro*
Polyadenylation	L mutant that synthesizes abnormal poly A Rare viral m RNAs covalently linked by poly A tracts

quence A and B, respectively, Fig. 5) (41, 42). One of the functions of the small 30K NS protein is that it promotes binding of the large 241K L protein which is believed to contain the catalytic sites of the complex.

Posttranscriptional Processing of Viral mRNAs

In the presence of the methyl donor S-adenosylmethionine, messenger RNAs synthesized by detergent disrupted virions *in vitro* are capped, methylated and polyadenylated (43). Therefore, all enzymatic functions necessary for the post-transcriptional modifications of the mRNAs must be packaged in the virion. It is uncertain, however, whether all of these functions are carried out by viral proteins or whether packaged host proteins are involved. Attempts to either cap, methylate or polyadenylate exogeneous RNAs during an *in vitro* transcription reaction have

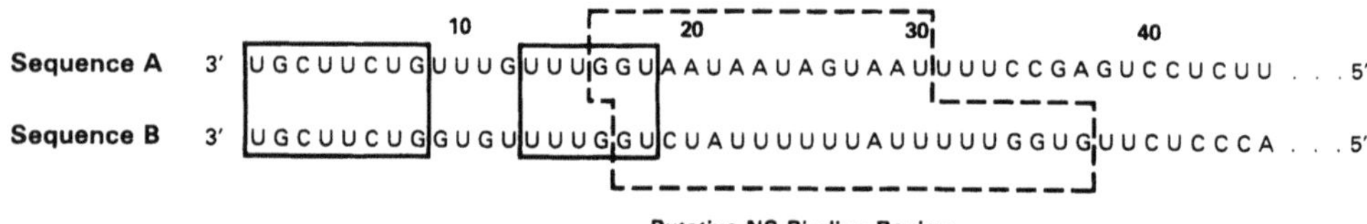

Figure 5. Nucleotide sequences of 3' termini of VSV genomic and anti-genomic RNAs. Sequence A is present at the 3' end of the minus-strand genomic RNA. Sequence B is present at the 3' end of the anti-genome as well as most DI particle minus-strand RNAs. The complementary sequence B' is found at the 5' termini of all VSV and DI particle RNAs analyzed to date. The terminal 18 bases contain two blocks of identical sequences. The putative sites where NS protein binds to the RNP are also boxed. Note that the nucleotide sequences in these regions differ between A and B. This difference may cause the more efficient replication of minus-strand RNPs over plus-strand RNAs.

not been successful and this result raises the possibility that only nascent transcripts are accepted as substrates. It appears that these modification functions are closely linked and are carried out by the polymerase complex itself. Since the virus transcribes and replicates exclusively in the cytoplasm of the host cell, it may rely on its own post-transcriptional modification system to synthesize mRNAs which can be efficiently translated.

"Cap" Formation

The mechanism of cap formation differs from that of the host cell or other viruses which code for their own guanylyltransferase activity (44). During capping of VSV mRNAs *in vitro* the endogenous guanylyltransferase adds pppG to the 5' terminal pppAACAG... mRNA start to yield a cap structure GpppAACAG...

In this cap structure, only the phosphate in the α position of the message is retained. This is in contrast to the cellular guanylyltransferase which conserves both the α and β phosphates of the mRNA start. Although an alternate mechanism of cap formation like that in the host cell can be detected under conditions when the α and β phosphates of the cap donor GTP is blocked by an imido residue, the preferred mechanism seems to differ from that of the host cell, suggesting that the guanylyltransferase may be a viral protein.

There are two types of methylation of the cap core structure *in vitro* which give rise to a methylated cap ^{7}mGpppAm... The guanosine residue is methylated in the 7 position. The ribose of the first nucleotide of the transcript is methylated in the 2'0 position (see Fig. 1, chapter 4 of this volume). Analysis of VSV host range mutants revealed that some of these mutants hypomethylate the cap structure *in vitro* as well as *in vivo* (45, 46). Hypomethylation is undoubtedly caused by a mutation in the viral genome. Further studies are necessary to identify the viral protein which is affected by the mutation and to rule out the alternative possibility that binding of cellular methylases to the polymerase complex could have been impaired by the viral mutation.

Polyadenylation

The first indication that the addition of, on the average, 200 nucleotide long poly (A) tails at the 3' end of mRNAs is carried out by the viral polymerase was derived from nucleotide sequence analyses of VSV genomic RNA (47). As described earlier, each gene terminates with a highly conserved stretch of 7 uridine residues (Fig. 3). It was proposed that polyadenylation is carried out by repeated transcription of these uridine residues. Further support that the viral polymerase itself polyadenylates the mRNAs came from the detection of polycistronic transcripts found *in vitro* and *in vivo* (48). At a frequency of about 3%, polycistronic transcripts of two or more adjacent genes are detected. Most importantly, these transcripts are covalently linked by poly (A). Nucleotide sequence analyses of these transcripts showed that the viral polymerase can, without termination, polyadenylate a mRNA and resume faithful transcription through the two nucleotide intercistronic region into the next gene. More recently, a VSV mutant was identified which synthesizes extremely long poly (A) tails of about 1000 nucleotides in length. By dissociation and reconstitution of the polymerase complex using various combinations of the RNP template, L, and NS proteins of wild-type and mutant virus, the mutation was clearly shown to reside in the L protein (49). All of these data taken together demonstrate that polyadenylation results from reiterative transcription of the short uridine stretch at the end of each gene. It is a transcriptional event carried out by the viral polymerase and differs from the post-transcriptional polyadenylation of cellular mRNAs. This mechanism of polyadenylation is common among most negative-strand RNA viruses.

In summary, the polymerase appears as a multifunctional enzyme complex which in addition to polymerization carries out polyadenylation and possibly methylation and capping (Table 3). Earlier calculations suggested that the functional VSV polymerase complex consisted of one L protein and one NS protein subunit (50). However, recent accurate molecular weights of the proteins which were deduced from nucleotide sequences suggest that the ratio of L to NS is closer to 1:2. It is uncertain whether the polymerase complex contains more than one L subunit. The observed intragroup complementation between L gene mutants could indicate that the complex contains more than one L subunit (51).

D. Replication

The viral genes are, as described earlier, sequentially transcribed by what appears to be a stop-start mechanism. Therefore, in addition to the multiple enzymatic functions it performs, the polymerase when in the transcription mode must also recognize the highly conserved nucleotide sequences at the beginning and end of each gene. During the replication mode when a full length genomic or antigenomic sense RNA is synthesized, these regulatory sequences are ignored. How this switch from transcription to replication is regulated is not completely understood on the molecular level. *In vitro* replication has shown that the availability of free nucleocapsid protein N allows replication to occur (32, 33). N protein preferentially binds to leader RNA *in vitro* and the leader RNA can be found encapsidated in the infected cell (12, 52). It is suggested that the N protein starts encapsidating leader RNA and functions as an anti-terminator of transcription at the leader/N gene junction as well as at the intercistronic regions (53). If N protein is the anti-terminator then post-transcriptional encapsidation of the transcript must somehow control the elongation of the nascent transcript. The precise mechanism by which this anti-termination occurs is being studied.

E. Generation of Defective Interfering Particles

Aberrant replication events can lead to the generation of defective interfering (DI) particles. Most DI particles are deletion mutants with up to 80% of the parental genome deleted (54). Propagation of DI particles is, therefore, dependent on co-infection with parental virus. Repeated passage of the virus at high multiplicity favors DI particle selection and mixtures of DI particles can readily be detected. Unlike with pleomorphic viruses such as measles virus, the size of rhabdovirus DI particles reflects the size of the packaged nucleocapsid. Since rhabdovirus DI particles can easily be purified because of their smaller size, their genomic structures have been extensively characterized.

DI particle genomes often have a selective advantage over parental virus genomes because their genomes are replicated more efficiently. In addition, most DI particles compete for viral gene products at the expense of the parental virus. This competition results in autointerference with a drastically reduced total yield of parental virus as well as DI particles. Autointerference is usually most efficient with homologous parental virus; however, heterotypic autointerference between DI particles and wild-type virus of a different serotype has also been described. The precise mechanisms of homotypic and heterotypic autointerference are unknown, but the structures of various DI particle genomes raise some possibilities. These structures also give insight into how DI particle genomes may be generated (55). Fig. 6 shows some representative VSV DI particle genome structures. Most of the DI particle genomes described to date have the N, NS, M, G and most of the L gene deleted.

There is a simple deletion DI particle which encodes the N, NS, M and G proteins but has a large deletion in the L gene. Unlike the other DI particles, this DI particle heterotypically interferes with wild-type virus of a different serotype probably through the mixing of proteins from the two serotypes. The other DI par-

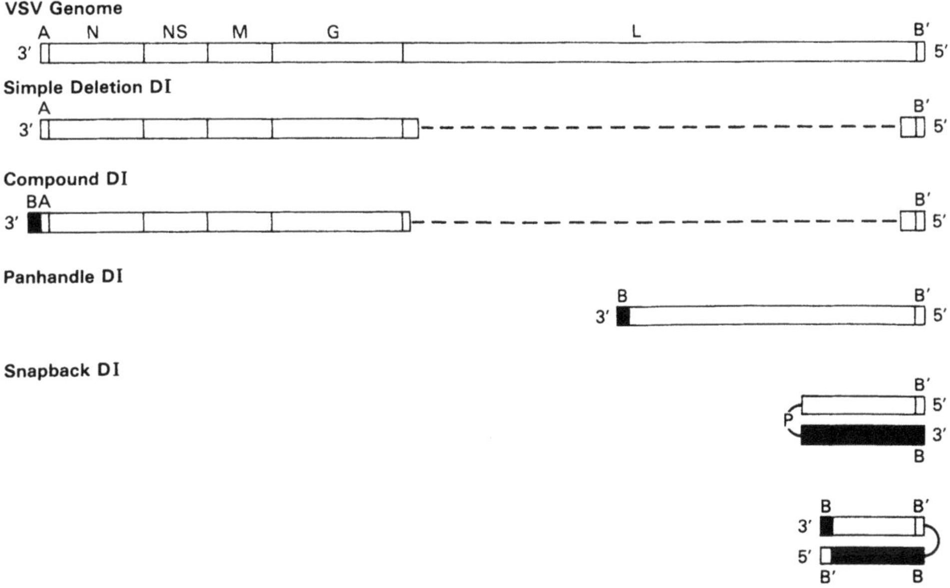

Figure 6. Structures of representative defective interfering (DI) particle RNAs. Regions of the standard VSV genome which are conserved are indicated by open boxes. Complementary sequences are marked by black boxes. Most DI RNAs carry the efficient 3' terminal origin of replication (B) and its complement (B') at their 5' terminus. Snapback DI RNAs consist of complementary RNAs which are covalently linked. Removal of the nucleocapsid protein allows immediate annealing into a hairpin structure.

ticles, such as the *panhandle* type, the *snapback* type, and probably the *compound* type, do not heterotypically autointerfere, most likely because they do not encode a viral protein product. They can, however, to some extent be propagated with heterologous parental virus since they contain the conserved 3' terminal replicase binding sites and encapsidation sites which can be recognized by heterologous polymerase and N protein, respectively.

All DI particle genomes contain, like the parental genome, competent replicase binding sites at the 3' termini of both the minus-sense and plus-sense genomic RNAs (A or B, Fig. 5). The terminal sequences A and B are not only identical for 8 bases between plus- and minus-strands, but also between different VSV serotypes (56). Starting at position 19, including the putative binding site for the NS proteins, these terminal sequences differ greatly. The efficiency of NS and, subsequently, L binding to the template may determine the frequency of replication.

The minus-strand VSV genome is four times more abundant in infected cells than the plus-strand genome (57), suggesting that the origin of replication B is more efficient than A. Most DI particle RNAs, however, contain the B sequence at the 3' end of the minus-sense RNA and an exact complement B' at the 5' terminus. DI particle genomes which have complementary ends are found intracellularly in a 1:1 ratio. The efficiency of replication is, therefore, dependent on terminal nucleotide sequences and the selection of DI particle genomes with the complementary B and B' sequences may reflect an advantage during their replication.

In vitro transcription of DI particle genomes yields a short 46 nucleotide minus-sense leader RNA, also referred to as DI product RNA (58). It is transcribed from the precise 3' end (B) of the DI genomic RNA (59). As with the plus-sense leader RNA which is transcribed from the 3' end (A) of parental genomic RNA, anti-termination by viral N protein gives rise to complete genomic RNA. It should be pointed out, that the preferred synthesis of the minus-strand VSV RNP during replication, could also be a result of a more efficient anti-termination at this site rather than by more efficient replicase binding.

Although the overall structures of DI particle genomes vary greatly, their generation can be explained by a common mechanism (55). This mechanism seems to involve premature termination of replication. With the replicase still attached to the nascent RNA which may in part be encapsidated already, replication is resumed at a different site on the same or a different RNP template (copy choice). A simple deletion DI particle genome results when replication is resumed on the same template thus conserving the terminal replicase binding sites of the parental genome. When the nascent RNP is used as a template, genomic structures like the snapback type DI genome or the panhandle type DI genome are generated with complementary terminal replicase binding sites (B and B', Fig. 6). The shortest complementary region detected to date is 45 nucleotides and many panhandle type DI particle genomes contain complementary regions which are very close to this size. The compound DI particle RNA was most likely generated in two steps. These steps may have included a simple deletion followed by the addition of a short B' sequence at the precise 3' end of the plus-sense DI RNA.

Copy choice of replication appears to lead to a virtually limitless shuffle of genomic sequences without recombination events via breakage and reunion of the

RNA strands. The polymerase may switch templates frequently; however, only RNAs with an origin of replication at the 3' end and a complement thereof at the 5' end are replicated. A replicative advantage over parental virus leads to the accumulation of defective and interfering particles.

F. Viral Mutations

The appearance of DI particles in a rhabdovirus population is the most dramatic result of the inaccuracy of the viral RNA polymerase. DI particles can easily be detected by simple physical separation. Single nucleotide changes, however, are more difficult to identify, but they can also have dramatic effects. Spontaneous temperature sensitive mutants at levels as high as 2% have been detected in purified VSV populations (60). Misincorporations of nucleotides by the viral polymerase range in the order of 1 misincorporation per 104 bases (61). Since the VSV genome is 11kb and there is no proofreading mechanism, this high inaccuracy by the polymerase suggests that every single VSV genomic RNA may differ in at least one nucleotide. The polymerase gene L alone comprises more than 50% of the entire genome; therefore, most of these base changes can be expected within this gene. Base changes in the polymerase gene L as well as in the NS and N genes may in turn affect the multiple functions of the polymerase complex itself as well as its accuracy. It is tempting to speculate whether an erosion in the accuracy of the polymerase may contribute to the generation of DI particles or the high frequency of replication mutants often observed during persistent viral infections (62).

Nucleotide sequence analysis of the polymerase gene L revealed that genomic cDNA clones of plaque purified virus contained many nucleotide differences (7). Two different cDNA clones were sequenced which had a single base deletion in the exact same position while two other cDNA clones derived from the same region did not. This suggests that a plaque purified VSV population may contain lethal L gene mutants which can only be propagated with the help of parental virus. It can be expected that these lethal mutants interfere with the replication of parental virus to the same extent as the simple deletion DI particle described earlier or temperature sensitive L gene mutants at the restrictive temperature. Therefore, a plaque purified wild-type virus population may already contain defective and possibly interfering particles which are indistinguishable in size from parental virus. Lethal polymerase mutants may in part explain why most VSV particles can infect and kill a cell, however, only one fifth to one tenth of these particles yield progeny particles. Cell killing is independent of viral replication, but requires transcription (63). Transcription of lethal polymerase mutants could still be carried out by the packaged wild-type polymerase of the helper virus, but replication and secondary transcription would not be efficient because of the lethal mutation in the newly synthesized polymerase. Consequently, the cell is killed but no virus is released.

The virus population consists of a constantly changing mixture of variants. The rates at which mutations accumulate may differ among the genes. Comparisons of the deduced amino acid sequences for the N, NS, M and G proteins of two serotypes of VSV demonstrate that the sequence homology between proteins of the two serotypes varies from about 32-68% (64). It appears that amino acid changes are tolerated more readily with some proteins, (i.e., NS) while other proteins are more conserved. The fact that a high number of mutations can accumulate in the NS gene, despite the fact that NS protein interacts with multiple components, suggests that tertiary rather than primary structures may be important for its functionality (65).

4. RELEVANT PROBLEMS IN RHABDOVIROLOGY

RNA virus genomes are less accessible to genetic manipulation than DNA virus genomes. Extremely high mutation rates and pseudoreversion within the same or a different gene make it extremely difficult to obtain meaningful data on the interactions of the viral proteins and genomic RNA through classical genetics. The cDNA cloning and sequencing of mRNA and genomic RNA together with the expression of the genes using eucaryotic expression vectors has now made it possible to specifically alter single viral gene products. The ultimate goal is to have the genome itself accessible to site specific mutagenesis. Transcription of DNA into functional infectious genomic RNA has been demonstrated with a number of plus-strand RNA viruses. With negative-strand RNA viruses, however, the RNA must be encapsidated for propagation. The proper encapsidation of RNA transcribed from recombinant DNA is the key reaction to bring negative-strand RNA viruses to the same level of experimentation now possible with DNA and plus-strand RNA viruses.

There are many open questions which can be addressed after this technology is established. How do negative-strand viruses exclude plus-strand genomic RNA from virus particles? Questions concerning the cytopathogenicity of the virus can be addressed with respect to lytic infection and viral persistency. Genomic regulatory sequences and protein interactions can be studied on a molecular level. The generation of recombinant DI particles may shed light on the mechanism of autointerference. Recombinant DI particles may serve as safe live vaccines. These recombinant DI particles could potentially be used as autonomous expression vectors utilizing the packaged RNA polymerase, and the study of the multiple functions of the polymerase may give insight into essential viral specific mechanisms involved in transcription and replication. which could be a target for anti-viral therapy. The cloning and functional expression of rhabdovirus genes opens new avenues for the study of the molecular mechanisms of rhabdovirus replication and virus/host interactions.

5. REFERENCES

1) Wagner, R.R. (1975) In: *Comprehensive Virology*, H.Fraenkel-Conrat & R.R. Wagner, eds., Plenum Publishing Corp., New York, 1-93.

2) Knudson, D.L. (1973), J. Gen. Virol. **20**, 105-130.

3) Baltimore, D., Huang, A.S. & Stampfer, M. (1970), Proc. Natl. Acad. Sci. USA **66**, 572-576.

4) Emerson, S.U. & Wagner, R.R. (1972), J. Virol. **10**, 297-309.

5) Thomas, D., Newcomb, W.W., Brown, J.C., Wall, J.S., Hainfield, J.F., Trus, B.L., & Alasdair, S.C. (1985), J. Virol. **54**, 598-607.

6) Emerson, S.U. & Yu, Y.-H. (1975), J. Virol. **15**, 1348-1356.

7) Schubert, M., Harmison, G.G., & Meier, E. (1984), J. Virol. **51**, 505-514.

8) Gallione, C.J., Greene, J.R., Iverson, L.E., & Rose, J.K. (1981), J. Virol. **39**, 529-535.

9) Mellon, M.G. & Emerson, S.U. (1978), J. Virol. **27**, 560-567.

10) Bell, J.C., Brown, E.G., Takayesu, D., & Prevec, L. (1984), Virology **132**, 229-238.

11) Herman, R.C. (1986), J. Virol. **58**, 797-804.

12) Blumberg, B.M., Giorgi, C., & Kolakofsky, D. (1983), Cell **32**, 559-567.

13) Newcomb, W. & Brown, J. (1981), J. Virol. **39**, 295-299.

14) Reidler, J.A,. Keller, P.M., Elson, E.L. & Lenard, J. (1981), Biochemistry **20**, 1345-1349.

15) Rose, J. & Schubert, M. (1987) In: *The Rhabdoviruses*, R.R. Wagner, ed., Plenum Publishing Corp., New York, in press.

16) Kurath, G. Ahern, K.G.. Pearson, G.D., & Leong, J.C.(1985), J. Virol. 53, 469-476.

17) Colonno, R.J. & Banerjee, A.K. (1976), Cell **8**, 197-204.

18) Schubert, M. & Lazzarini, R.A. (1981), J. Virol. **38**, 256-262.

19) Gibson, R., Leavitt, R. Kornfeld, S., & Schlesinger, S. (1978), Cell **13**, 671-679.

20) Schlegel, R., Tralka, T.S., Willingham, M.C. & Pastan, I. (1983), Cell 32, 639-646.

21) Lentz, T.L., Smith, A.L., Crick, J. & Tignor, G.H. (1981), Science **215**, 182-184.

22) Matlin, K.S., Reggio, H., Helinius, A. & Simons, K. (1982), J. Mol. Biol. 156, 609-631.

23) Florkiewicz, R. & Rose, J. (1984), Science **225**, 721-723.

24) Abraham, G. & Banerjee, A.K. (1976), Proc. Natl. Acad. Sci.USA **73**, 1504-1508.

25) Ball, L.A. & White, C.N. (1976), Proc. Natl. Acad. Sci. USA **73**, 442-446.

26) Villarreal, L.P., Briendl, M. & Holland, J.J. (1976), Biochemistry **15**, 1663-1667.

27) Hubbard, S.C. & Ivatt, R.J. (1981), Ann. Rev. Biochem. **50**,555.

28) Rose, J.K. & Gallione, C. (1981), J. Virol. **39**, 519-528.

29) Hsu, C.-H. & Kingsbury, D.W. (1982), J. Virol. **42**, 342-345.

30) Sanchez, A., De, B.P., & Banerjee, A.K. (1985), J. Gen. Virol. **66**, 1025-1036.

31) Hsu, C.-H., Kingsbury, D.W., & Murti, K.G. (1979), J. Virol. **32**, 304-313.

32) Patton, J.T., Davis, N.L., & Wertz, G.W. (1984), J. Virol.49, 303-309.

33) Peluso, R.W. & Moyer, S.A. (1984) In: *Nonsegmented Negative Strand Viruses*, D.H.L. Bishop & R.W. Compans, eds., Academic Press, San Francisco, 153-160.

34) Hill, V. & Summers, D. (1982), Virology **123**, 407-419.

35) Dubovi, E.J. & Wagner, R.R. (1977), J. Virol. **22**, 500-509.

36) Odenwald, W.F., Arnheiter, H., Dubois-Dalcq, M., & Lazzarini, R.A. (1986), J. Virol. **57**, 922-932.

37) Pinney, D.F. & Emerson, S.U. (1982), J. Virol. **42**, 897-904.

38) Iverson, L.E. & Rose, J.K. (1981), Cell **23**, 477-484.

39) Emerson, S.U. (1982), Cell **31**, 635-642.

40) Chanda, P.K. & Banerjee, A.K. (1981), J. Virol. **39**, 93-103.

41) Keene, J.D., Thornton, B.J., & Emerson, S.U. (1981), Proc. Natl. Acad. Sci. USA **78**, 6191-6195.

42) Isaac, C.L. & Keene, J.D. (1982), J. Virol. **43**, 241-249.

43) Banerjee, A.K., Abraham, G., & Colonno, R.J. (1977), J. Gen. Virol. **34**, 1-8.

44) Banerjee, A.K. (1980), Microbiol. Rev. **44**, 175-205.

45) Horikami, S.M. & Moyer, S.A. (1982), Proc. Natl. Acad. Sci. USA **79**, 7694-7698.

46) Horikami, S.M., de Ferra, F., & Moyer, S.A. (1984), Virology **138**, 1-15.

47) Schubert, M., Keene, J.D., Herman, R.C., & Lazzarini, R.A. (1980), J. Virol. **34**, 550-559.

48) Herman, R.C., Adler, S., Lazzarini, R.A., Colonno, R.J., Banerjee, A.K., & Westphal, H. (1978), Cell **15**, 587-596.

49) Hunt, D.M., Smith, E.F., & Buckley, D.W. (1984), J. Virol. **52**, 515-521.

50) Naito, S. & Ishihama, A. (1976), J. Biol. Chem. **251**, 4307-4314.

51) Belle-Isle, H.D. & Emerson, S.U. (1982), J. Virol. **43**, 37-40.

52) Blumberg, B.M. & Kolakofsky, D. (1981), J. Virol. **40**, 568-576.

53) Blumberg, B.M., Leppert, M., & Kolakofsky, D. (1981), Cell **23**, 837-845.

54) Perrault, J. (1981), Curr. Topics Microbiol. Immunol. **93**, 151-207.

55) Lazzarini, R.A., Keene, J.D., & Schubert, M. (1981), Cell **26**, 145-154.

56) Giorgi, C., Blumberg, B., & Kolakofsky, D. (1983), J. Virol. **46**, 125-130.

57) Simonsen, C.C., Batt-Humphries, S., & Summers, D.F. (1979), J. Virol. 31, 124-132.

58) Emerson, S.U., Dierks, P.M., & Parsons, J.T. (1977), J. Virol. **23**, 708-716.

59) Schubert, M., Keene, J.D., Lazzarini, R.A., & Emerson, S.U. (1978), Cell **15**, 103-112.

60) Pringle, C.R. (1982), Arch. Virol. **72**, 1-34.

61) Steinhauer, D.A. & Holland, J.J. (1986), J. Virol. **57**, 219-228.

62) Preble, O.T. & Youngner, J.S. (1972), J. Virol. **9**, 200-206.

63) Marcus, P.I. & Sekellick, M.J. (1975), Virology **63**, 176-190.

64) Gill, D.S. & Banerjee, A.K. (1986), Virology **150**, 308-312.

65) Gill, D.S. & Banerjee, A.K. (1985), J. Virol. **55**, 60-66.

CHAPTER 12

THE MOLECULAR BIOLOGY OF PARAMYXOVIRUSES

DANIEL KOLAKOFSKY AND LAURENT ROUX
Department of Microbiology, University of Geneva Medical School, 9 Ave. du Champel, 1211-Geneva, Switzerland

1. THE PARAMYXOVIRUS FAMILY

PARAMYXOVIRUSES

Parainfluenza	*Morbillivirus*	*Pneumovirus*
mumps	measles	respiratory syncitial
human parainfluenza (HPV)1-4	canine disease (CDV)	
	rinderpest	
Newcastle disease (NDV)		
Simian virus 5 (SV5)		
Sendai		

A. Classification

The paramyxoviruses were originally classified together with the influenza viruses as myxoviruses because of their ability to attach to neuraminic acid containing receptors on erythrocytes (hemagglutination) and their ability to elute themselves due to a neuraminidase activity. The further findings that unlike influenza viruses, paramyxoviruses contain a non-segmented genome and that they replicate in the presence of actinomycin D, led to their reclassification as a separate family.

The present family includes the parainfluenza, morbilli, and pneumo-virus genera. All the paramyxoviruses are morphologically similar by electronmicroscopy examination; their most distinctive feature being that they are polymorphic in shape, and often in size as well. The size distribution of the virions is accompanied by polyploidy, and in some virus-host systems, virions with multiple nucleocapsids (containing antigenomes as well as genomes), represent the vast majority of the particles. The pneumoviruses differ from the parainfluenza and morbilliviruses in having a helical nucleocapsid of slightly smaller diameter and a genome containing 10 rather than 6 trans-criptional units. The genome organization of parainfluenza and morbilli-viruses are extremely similar, that of pneumoviruses are basically similar but also show clear differences.

Morbilliviruses differ from parainfluenza viruses in not having neuraminidase. This latter distinction may, however, turn out to be less significant than at first appears. Influenza C viruses were also though not to contain neuraminidase, until it was discovered that they do contain neuraminidase, but for a different form of neuraminic acid (1). The neuraminidase or receptor-destroying activity commonly found in myxoviruses and paramyxoviruses is though to be a consequence of the fact that these viruses use relatively simple host cell receptors that are present on the host cell surface (as well as in serum and respiratory tract secretions) in enormous numbers. The neuraminidase would then help in the liberation of the virus from cells, and its spread as well. Assuming that the entire paramyxovirus family uses high density cellular receptors, they would all then be expected to contain a receptor destroying activity.

The measles, canine disease virus (CDV) and rinderpest morbilliviruses have long been known to be antigenically related and probably represent human, canine and bovine associated forms of basically the same virus. In a similar vein, mumps, human parainfluenza virus 3 (HPV3) and Sendai virus (HPV1 ?) form one group within the parainfluenza genera as judged by shared surface antigens, whereas SV5 (originally thought to be a simian virus) and HPV2 another. SV5 is also referred to as canine parainfluenza virus 2 (SV5 is most prevalent in dogs and it or a closely related virus is responsible for the kennel cough well known to some) and Sendai virus is also referred to as murine parainfluenza virus 1 since this virus is endemic in mice (and a constant threat to laboratory mouse colonies). However, many of the laboratory strains of Sendai virus now used were originally isolated from young children with severe respiratory disease during epidemics in Japan (and passed only in embryonated chicken eggs). These viruses and the mouse virus are indistinguishable by RNA:RNA hybridization (unpublished), and are probably the same or extremely close. More recently, aminoacid sequence comparisons of two strains of Sendai virus and HPV3 have shown that these viruses are also closely related, with about the same amount of homology as that found between the Indiana and New Jersey serotypes of vesicular stomatitis virus (VSV).

2. THE VIRION

The virion (Fig. 1) contains a lipid envelope (obtained by budding from the host cell plasma membrane) with projection or spikes on the outside surface. These spikes are composed individually of the F and HN gene products, which are both transmembrane glycoproteins. Inside the lipid envelope lies the viral (–) genome in the form of a helical nucleo-capside, structurally similar to tobacco mosaic virus (TMV) but with important differences (Fig. 2). Unlike TMV which is a rigid rod, the Sendai nucleocapsid appears as a flexible coil apparently composed of discs whose interaction can easily be broken. Heggeness *et al.* (2) have shown that the nucleocapsid structure varies with ionic strength; at low salt the nucleocapsid is flexible, but becomes increasingly rigid as the salt is increased. The interactions of the disc subunits thus depends, at least in part, on ionic interactions. Furthermore, while TMV contains a (+) RNA and must therefore be disassembled for translation, the (−) parainfluenza virus genome nucleocapsid need never be disassembled since the viral polymerase transcribes the nucleocapsid structure and not free RNA.

The paramyxovirus nucleocapsid has many interesting properties. By mass, like TMV, it is composed of 96% protein and 4% RNA, hence its relatively low buoyant density in CsCl (1.31 gm/ml). The genomic RNA in this structure is bound tightly to the NP protein, from which it can be dissociated only by agents such as SDS or high concentrations of guanidine. This core nucleocapsid is also resistant to the high salt conditions of CsCl density gradients, a property not found among cellular RNA:protein complexes. More loosely associated with the nucleocapsid are the phosphoprotein P and the large protein L, which together are though to con-

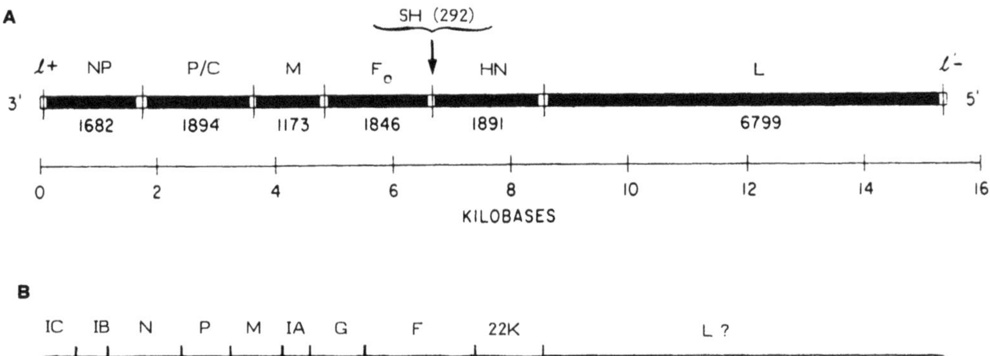

Figure 1. Genetic map of Paramyxoviruses.

The top line (A) shows the Sendai virus genome as (−) RNA. The dark bars show the translated regions of each gene bounded by the untranslated regions shown as open bars. The vertical lines represent the gene boundaries, i.e: the intergenic trinucleotides (see Table I). The length of each mRNA exclusive of its poly A tail is shown underneath. The position and length of the SH gene of SV5 is indicated above the line. Although only Sendai virus has been sequenced entirely to date, this map is considered to also represent the other parainfluenza and morbilliviruses.

The bottom line (B) shows the RSV viral genome with the vertical line representing the gene boundaries as above. The RSV genome has been sequenced only up to the L gene so that beyond this point the map is still tentative. The RSV genes have been drawn to scale. Note that only the RSV F Gene is similar in size to its Sendai virus counterpart.

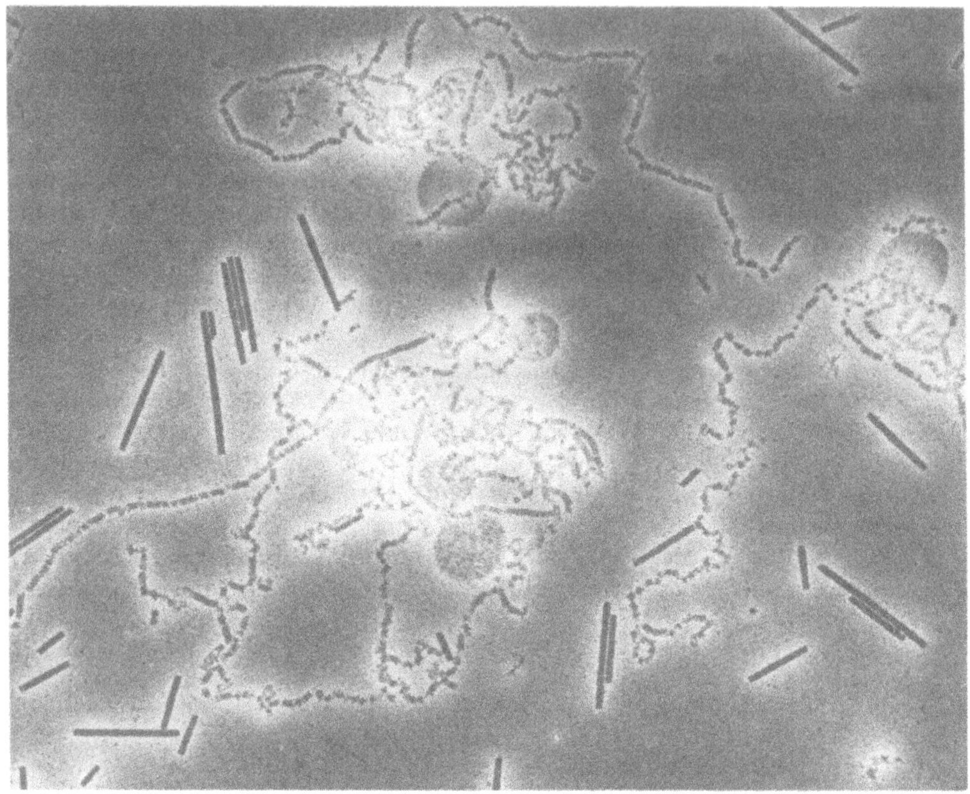

Figure 2. Electron micrograph of Sendai virus

A typical virion, stained with 2% phosphotungstic acid, is shown. The viral envelope has been broken allowing the stain to enter the virion and allowing visualization of the internal nucleocapsid, which is partially spilling out of the envelope. The surface projections on the outside of the membrane are also visible.

stitute the polymerase activity which transcribes the (–) genome RNA-NP protein template during both mRNA and antigenome synthesis. The (–) genome is replicated via a full-lenght complementary strand, the antigenome, also found only in nucleocapsid structures.

How the viral polymerase deciphers the bases within this structure is not clear, but some information about the topology of the RNA and protein in this structure can be deduced. All the RNA within nucleocapsids composed only of the NP protein, including both ends of the chains, is resistant to RNase A attack regardless of salt concentration (3), consistent with a uniform interaction of NP protein and RNA limiting access of the endonuclease to the RNA. The RNA within VSV nucleocapsids (which are structurally very similar to parainfluenza nucleocapsids), however, is as susceptible to attack by small molecules, such as those used in the Maxam-Gilbert sequencing reactions, as is the free RNA (4). Since RNase A must simultaneously recognize both the phosphodiester linkage and the pyrimidine base, one possibility is that the NP protein contacts only the ribose-phosphate backbone in a stable fashion, leaving the bases free to interact with the polymerase directly, or after the local displacement of the NP protein from the bases, a function that may be due to the viral P protein.

It is important to note that the only sub-structure of the virus which is infectious is the (-) genome nucleocapsid containing the P and L proteins. The (+) antigenome, even as naked RNA, is unlikely to be infectious for many reasons (e.g: a polycistronic mRNA without a 5' cap group). It is this structure which contains not only the genetic information but also the enzymatic activity required to express this information as mRNA and ultimately to lead to the production of infectious progeny. To a first approximation, the virus can be divided into its envelope and its nucleocapsid core. The envelope serves the function of getting the nucleocapsid into the cell to initiate the replicative cycle, and then out of the cell in a form capable of reinfecting other cells. The nucleocapsid core is responsible for the intracellular viral replication.

One virion protein, the matrix protein M, cannot be neatly separated according to the above division. When purified virions are treated with non-ionic detergents under low ionic strength and then analyzed by centrifugation, the M protein is found to sediment with the nucleocapsid whereas the F and HN proteins are found in the soluble supernatant. However when the extraction is performed under high ionic strength, the M protein is solubilized as well (5, 6). If the high salt supernatant is dialyzed against low salt buffers, the M protein precipitates and can be easily purified from the soluble glycoproteins by low speed centrifugation. When this precipitate was examined in the electronmicroscope, it was found to be composed of a paracrystaline array of hexagonal periodicity (7). A paracrystaline array of identical periodicity has also been detected in the inner leaflet of the infected cell plasma membrane bilayer by freeze-fracture electronmicroscopy studies (8). The most plausible explanation for these results is that the M protein is amphiphilic and is associated with both the inside surface of the envelope (which can be disrupted by non-ionic detergents) and the nucleocapsid (which can be disrupted by high salt) (9). This protein is thus thought to play a crucial role in virus maturation, by mediating the association of progeny nucleocapsids with the plasma membrane modified by insertion of the viral glycoproteins. It should be noted however, that this hypothesis has received little direct experimental support to date.

The Sendai, HPV3, CDV and measles genomes also code for a non-structural protein (i.e: a protein found in infected cells but not in mature virions) which is expressed from an overlapping reading frame of the P mRNA (10-14). In the case of Sendai virus, this protein, the C protein, is approximately 5 times as abundant as the P protein in infected cells (15). The C protein has been co-localized to the nucleocapsids in measles virus infected cells (11). A nucleocapsid association would argue for a role in RNA synthesis, and its relative absence from mature virions could argue for a role in genome replication. Interestingly, other parainfluenza viruses such as NDV, SV5 and mumps also express NS protein(s) from the P mRNA, but here they derive from the same reading frame as the P protein, since they share tryptic peptides with their respective P proteins.

A. The Viral Genome

The genome of paramyxoviruses, like rhabdoviruses, is a single segment of single-stranded (s-s) RNA of negative polarity (as defined by translation, see chapter 1 of this volume). The Sendai viral genome whose entire sequence has been recently determined and is used as the prototype here, is 15,383 nucleotide long and contains 6 mRNA transcription units, the NP, P/C, M, F, HN and L mRNAs, and the (+) leader RNA (9, 10, 16-21). The viral mRNAs are both capped and polyadenylated, whereas the short leader RNA is unmodified at either end. Repeated at each junction between mRNAs is the sequence TAAGAAAAA CTT AGGGTNAAAG (written 5' to 3' as (+) DNA). Each mRNA is thought to terminate on the A_5 stretch of the first block, with polyadenylation presumably occuring by the polymerase stuttering to create the poly A tail. Each succeeding

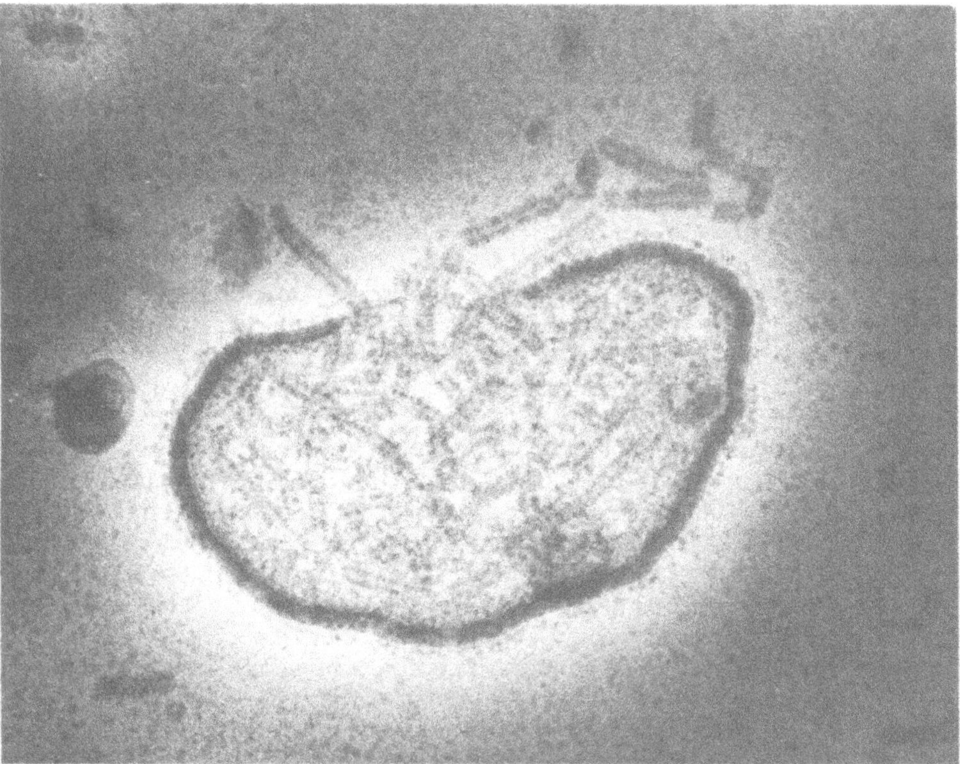

Figure 3. Electron micrograph of Sendai virus nucleocapsids.

Sendai virus nucleocapsids, which were isolated by equilibrium centrifugation on CsCl density gradients and thus contain only the NP protein in addition to the RNA, were stained with uranyl acetate. For comparison, tobacco mosaic virus has been included which appear as rigid rods, the longest of which are 0.2 µm in length. Note that the Sendai nucleocapsid discs can also be seen lying on their side and that, like the TMV discs, they appear to have a hollow center.

mRNA starts on the A of the AGGG sequence. At the (+) leader-NP junction, the TAAGAAAAA is missing, consistent with the unmodified 3' end of the (+) leader RNA. At the right hand end of the viral genome, i.e: at the L gene/(−) leader RNA junction, both the termination polyadenylation signal and the CTT are present, but only the A of the 10 nucleotide mRNA start signal is found, suggesting that the terminal 56 nucleotides are not transcribed into (+) RNA. The presence of the intergenic CTT just following the L gene suggests that these pyrimidines may also contribute to the termination-polyadenylation signal. Thus of the 15,383 nucleotides of the (−) genome, only the five mRNA intercistronic junctions of 3 bases each, and the terminal 56 nucleotides beyond the end of the L gene, are not copied during transcription.

Whereas the TAAGAAAAA termination-polyadenylation signal is invariant in the Sendai genome, the 10 nucleotide mRNA start signal does contain minor variations between mRNAs. The F mRNA contains an A rather than a T at position 5, the L mRNA contains a T instead of an A at position 9, and position 6 appears to have little conservation. In addition, the CTT intergenic sequence is replaced by a CCC at the HN-L junction and a TTT at the (+) leader/NP junction. The reasons for these minor variations are unclear, but may reflect variations in "promoter" strength. In the case of the non-segmented paramyxovirus genome where the polymerase is thought to initiate at each cistron only after having transcribed the preceding gene (the single entry promoter model, see below), promoter strength would reflect the probability of reinitiation after termination of the preceding gene.

All the junctional sequences of HPV3 except that at the L/(−) leader junction have also been determined recently (13, 14) and are compared to their Sendai virus counterparts in Table I. Since these viruses are known to be closely related in their amino acid sequences, the fact that their junctional sequences are similarly organized and highly conserved is not surprising. However, considering this, the differences that do exist here are particularly notheworthy. The start sequences of HPV3 are almost identical to those of Sendai virus and can be written as the consensus sequence AGGANNAAAG versus the Sendai virus AGGGTNAAAG. This start sequence is again preceded by the 3 nucleotide intergenic sequence CTT in all cases, except notably at the (+) leader/NP junction. The stop signal (i.e: termination-polyadenylation) for the first two mRNA junctions, ie NP/P and P/M, are again $TAAGA_5$, but from here on the departures from the consensus sequence are indeed intriguing. The M gene stop signal is now $TTATA_6$ (which is also used by at least some of the measles virus genes), and the HN gene stop signal is $ATATA_6$. Thus, in HPV3, unlike Sendai virus, neither the TAAG nor the A_5 part of the termination-polyadenylation signal are inviolate.

Less is known about the junctions of other parainfluenza and morbilliviruses, but again comparison of the limited information is of interest. Measles virus appears to contain similarly organized junctions, including the intergenic CTT (22). However, for SV5, the intergenic sequences are conserved neither in sequence nor in length (23, 24). An extreme case of the non-conservation of the intergenic sequences is that of the pneumovirus RSV, where these sequences vary from 1 to 52 nucleotides in length (25). It will be of interest to determine whether these dif-

Table 1. - Sequences at the gene boundaries of Sendai virus (top line) and HPV3 (bottom line)

Junction	Stop seq.	Intergenic seq.	Start seq.
(+)1/NP	ATACAGGAT	TTT	AGGGCAAG
	.C. . . CTTG	.AC	. . . A . T . .
NP/P	TAAGAAAAA	CTT	AGGGTGAAAG
	. . . .	. . .	. . . A . T . . . .
P/M	TAAGAAAAA	CTT	AGGGTGAAAG
	. . . .	. . .	. . . A . T . . . .
M/F	TAAGAAAAA	CTT	AGGGATAAAG
	AGATAATC	. . .	. . . ACA . . . .
F/HN	TAAGAAAAA	CTT	AGGGTGAAAG
	.T . T. . . . A	. . .	. . . AGT . . . .
HN/L	TAAGAAAAA	CCC	AGGGTGAAAG
	AT . T. . . . A	. TN	. . . AGC . . A.
L/(-)1	TAAGAAAAA	CTT	ACTAGAAGAC

Underlined bases in the Sendai sequences indicate departures from the consensus sequence. Homologies between the HPV3 and Sendai sequences are indicated by dots.

ferences between the sequences at the various paramyxoviruses gene junctions can be related to promoter strength.

Within the parainfluenza virus genus, the entire genetic map of Sendai virus and HPV3 have been determined and are identical (their NP, F and HN proteins are also highly conserved). In the case of SV5, a seventh mRNA has been located between the F and HN genes, called SH (coding for a small (44 amino acids) hydrophobic protein which is tiny in comparison (23). To date, this situation is unique among the parainfluenza viruses (it also does not appear to be the case for NDV), but considering the antigenic similarity of SV5 and HPV2, it will be of interest to determine with HPV2 also expresses this function via a separate mRNA.

In the case of the morbilliviruses, their genetic map is organized similar to that of the parainfluenza viruses, also without a separate transcription unit between F and H (Blumberg and Dowling, personal communication). The only noticeable difference here to date is that the 3' and 5' ends of the M and F mRNAs respectively contain relatively long "untranslated regions" (the M 3' "untranslated region" contains AUGs followed by respectable ORFs, but their putative translation products have not been detected). The pneumovirus genetic map is by far the most distant from the other paramyxoviruses (26, 27). It contains NP, P, M, F, G (the receptor binding equivalent of HN) and L genes in the usual order except that F and G are inverted, but it contains in addition four separate transcription units, called 1a, 1b, 1c and 22k. The 1c and 1b genes precede the NP gene at the extreme left of the map, 1a (which is also small and hydrophobic) is found between M and G, and the 22k gene lies between F and L (see Fig. 3). The function of these genes, like that of the parainfluenza viruses and morbillivirus C gene, is unclear.

3. THE REPLICATIVE CYCLE

Virions attach to carbohydrate containing receptors with terminal neuraminic acid via the hemagglutinin of the HN protein. This attachment takes place without energy (at 4° C) and is often studied by the agglutination of erythrocytes by virions or purified HN protein as "rosettes". Following attachment, the fusion protein F is responsible for fusion of the viral envelope to the cytoplasmic membrane with the consequent penetration of the nucleocapsid core into the cytoplasm, (which requires elevated temperature). This reaction has been studied both as hemolysis (since the viral envelope, especially on storage, contain holes, when virions fuse to erythrocyte, the hemoglobin leaks out) or by the fusion of artificial liposomes containing various protein combinations. These studies showed that the fusion protein alone was sufficient for fusion if some way other than the HN protein were found to attach membranes together (e.g: plant lectins) (28).

The fusion protein of all paramyxoviruses, like their myxovirus counterpart (the HA protein) is a transmembrane protein (anchored at its C-terminus) with a cleaved N-terminal signal peptide, and which is made as an inactive precursor Fo (19, 29). Activation of Fo takes place by proteolysis of an arg-phe bond during transport to the cell surface, by a host cell trypsin-like protease. The activated protein, called F12, is held together by a single disulfide bond. When Sendai virus is grown in cell lines in culture (as opposed to embryonated chicken eggs), the vast majority of the F protein is *not* cleaved and the progeny are not infectious. Such Fo containing virions can however be activated from the outside by mild trypsin treatment, and Sendai virus can be plaqued in culture by including trypsin in the agar overlay (30, 31). Because of this unusual situation, Scheid and Choppin (32) have isolated variant viruses able to plaque in the presence of other proteases such as chymotrypsin and elastase. As expected, these variants contain mutations near the cleavage-activation site (Hsu and Choppin, personal communication).

Cleaved paramyxovirus F proteins, unlike the influenza HA protein (or the rhabdovirus or alphavirus fusion proteins) are active at neutral pH and do not require the low pH environment of endosomes for activity. Paramyxoviruses are thus not endocytosed and one consequence of their fusion directly with the plasma membrane is that the virion envelope proteins are left on the infected cell surface during penetration (33). Upon cleavage-activation, the Sendai F protein undergoes conformational changes, judged both by an increase in alpha-helical content, and an increased ability to bind non-ionic detergents (34). This latter change, probably similar to the irreversible conformational change induced in HA1-HA2 protein of influenza virus by low pH, is most likely indicative of the unmasking or the creation of hydrophobic domains capable of interaction with the lipid bilayer.

The paramyxovirus HN proteins, like the influenza neuraminidase, is anchored in the membrane near its N-terminus, which is not cleaved from the mature protein like a classical signal sequence (20). Comparison of the amino acid sequences following the new N terminus created on cleavage activation of the F proteins of Sendai virus and NDV with that of the influenza C virus HA2 protein showed statistically significant homology (19, 35). Comparison of the Sendai HN sequence with the known hemagglutinin and neuraminidase domains of their respective influenza virus counterparts has also suggested that the H domain of the HN protein is located at the C-terminus of the molecule, and the N domain in the center. These studies suggest that paramyxo and influenza viruses shared a common ancestor during evolution. Furthermore, alignment of the neuraminidase, hemagglutinin and fusion domains of these viruses suggests that the hypothetical ancestor may have had a single surface glycoprotein containing all three functions, and membrane anchored at both ends (20). More recent work from Portner's laboratory mapping amino acid changes which allow viruses to escape from neutralyzing monoclonal antibodies which inhibit hemagglutinin activity have found changes covering virtually the entire chain (personal communication). These results clearly show that the H function is not restricted to the C terminus of the molecule, and fits nicely with the observation that disruption of the secondary structure (by SDS and β-mercapto-ethanol)) of the Sendai HN protein destroys all the available native antigenic sites (36).

A. Primary Transcription and Translation

Primary transcription is the process of RNA synthesis which takes place using the infecting genome as template after penetration of the nucleocapsid core, and *before genome amplification occurs in the cytoplasm.*

The fate of the viral M protein during penetration is not known, but it may well, like the surface glycoproteins, also remain in the host cell membrane, thus leaving the nucleocapsid free for transcription. Indirect evidence suggests that the parainfluenza virus M protein, like its rhabdovirus counterpart, acts to inhibit transcription, which may be one of its functions during virus maturation.

Operationally, primary transcription is defined as that which takes place in cells infected in the presence of inhibitors of protein synthesis (which is required for genome amplification). Alternatively, primary transcription can be examined by using purified virions whose envelope has been permeabilized with non-ionic detergents, since all the activities required for the synthesis of active mRNAs (capping, methylation and polyadenylation) are contained in the nucleocapsid core. The virion polymerase reaction *in vitro* is stimulated by the addition of polyanions (e.g: poly-L-glutamic acid) and phosphate rather than Cl or acetate, presumably to weaken the M protein-nucleocapsid interactions.

As far as is known, all the viral mRNAs are produced during primary transcription which is identical to secondary transcription in all respects except for the small amounts of mRNAs produced due to the limiting amounts of the infecting templates. The polymerase is thought to start at the precise 3' end of the (-) genome templates and to sequentially synthesize the (+) leader RNA and the succeeding mRNAs. This conclusion follows from U/V inactivation kinetics studies in which the target size for downstream transcription unit plus the upstream units as well (37, 38). According to this model, the (-) genome template contains a single entry promoter, at its 3' end.

The precise mechanism by which the individual mRNAs are produced from the polycistronic template is unclear. Using ATP and GTP labelled in the β-phosphate, the leader RNA is labelled with β-GTP but not ATP (Blumberg and Kolakofsky, unpublished). Thus, as is the case for VSV transcription, only the α-phosphate of the adenosine residue at the beginning of each mRNA is retained in the cap group. The Sendai virus capping reaction thus differs from cellular capping reaction (or reovirus and vaccinia virus reactions) where both the α and β phosphates of the first residue are retained in the cap. The fact that the middle phosphate of the cap group of Sendai mRNAs comes from the capping GTP makes it impossible to decide whether the first A residue of each mRNA originated as a triphosphate (in which case each mRNA must have been independently initiated) or could have been generated by a processing endonuclease which left a 5' phosphate which was subsequently capped. Gupta and Kingsbury (39) have pointed out that since the intercistronic CTT is not found in mature mRNAs, this would argue against a processing nuclease, since this enzyme would have to remove the CTT. However, considering that this putative activity is not well described in other systems, this does not appear to be an insurmontable obstacle.

Using Northern blot analysis, several groups have shown that dicistronic mRNAs are found at low frequency in infected cells, with a poly A region at the 3' end, but not at the junction between the cistrons (40, 41). These distronic mRNAs have been useful in determining the gene order, but they do not speak directly to the mechanism of individual mRNA generation. These distronic mRNAs would be expected to be functional, but only for the upstream cistron. The most straightforward explanation of these dicistronic mRNAs, however, is that they arise by the failure of the viral polymerase to recognize a termination signal.

Most recently, sequence work on the gene junctions of RSV and HPV3 have also provided some evidence on the nature of the transcription process. In the case of HPV3, the inviolate TAAGAAAAA found as a termination signal in Sendai virus is also found at the NP/P and P/M junction in HPV3, but that at the M-F junction contains an 8 nucleotide insertion between the G and the A, and presumably as a consequence of this, the M-F dicistronic mRNA is found at greatly elevated levels (13, 14). In the case of RSV, a consensus mRNA start signal for the mRNA is not found just downstream of the end of the 22K mRNA, but 64 nucleotide upstream, within the 3' non-translated region of the 22K mRNA. Because of this unusual situation, production of the L mRNA is dependent on the polymerase not recognizing the stop signal at the end of the 22K gene and indeed, the major transcript from the L-mRNA start site is a 64 nt mRNA with a poly A tail longer than the mRNA body; the true L-mRNA being the minor species (Collins, personal communication). In both of these situations, it is difficult, if not impossible, to reconcile these events with a mechanism by which the mRNAs are produced by a processing endonuclease rather than a stop-start mechanism. With this latest evidence in hand, a processing nuclease mechanism would seem to be excluded, leaving the stop-start model as the only option.

B. Genome Replication (Amplification)

Translation of the primary transcripts allows the accumulation of viral proteins which leads to genome amplification via a full-length (+) antigenome nucleocapsid. Antigenome synthesis is thought to be initiated identically to mRNA transcription, except that the viral polymerase does not stop and restart at each junction, but reads through continuously to produce a full-length complement of the (-) genome. Since antigenomes, like genomes, are found only as nucleocapsids, it is logical to assume that encapsidation and synthesis are coupled. One current view is that it is the assembly of the NP proteins on the nascent chain which is responsible for causing the polymerase to disregard the junctional termination signals and to transcribe the genome in an uninterrupted fashion.

According to this model which was determined in large part for VSV genome replication, the switch between (-) genome transcription and antigenome synthesis is controlled by the intracellular concentration of unassembled NP protein. Under limiting NP, the (-) genome is transcribed to produce mRNA, which then leads to the accumulation of unassembled NP. When NP reaches a critical concentration, it would act by presumably binding to the nascent (+) leader chain, causing antigenome synthesis whose product would be an assembled nucleocapsid, thus depleting the pool of unassembled NP. The initiation of one nucleocapsid chain would commit approximately 2000 NP molecules for final assembly. The switch is then part of a self regulatory system in which NP controls both is own synthesis and utilization (42).

The antigenome nucleocapsid would then serve as a template for genome synthesis in an analogous manner, except that the antigenome template would presumably contain only a single internal termination signal, that for the putative (-) leader RNA. It should be noted, however, that a parainfluenza (-) leader RNA has not been demonstrated, either in infected cells, or as the product of a virion polymerase reaction containing suitable templates (i.e: antigenomes or snapback DI RNAs). In addition, neither the (+) or (-) leader RNAs have been shown to exist intracellularly assembled into nucleocapsids, which is part of the essential evidence that they would contain the site for the initiation of nucleocapsid assembly, as has been demonstrated for VSV (see 42). Without the demonstration of a unique (-) leader RNA, there is no way of knowing whether the polymerase initiates at the 3' end of the antigenome template independently, as is thought to be the case for VSV. For example, it is not impossible that a non-structural protein, such as the C protein, is required for the polymerase to initiate on the antigenome but not the genome template. Thus, until further work is carried out, a coherent scheme for parainfluenza virus genome replication must remain sketchy.

C. Genetic Stability of Paramyxoviruses and Their Evolution

Work with RNA phages and VSV have shown that the viral polymerase products contain point mutation at a frequency of around 3×10^{-4} (see 43). This high mutation rate is considered to be due to the intrinsic error rate of the polymerase and the absence of repair mechanisms which act at the level of DNA. Since there is no reason to believe that the parainfluenza virus polymerases are any less error prone than other RNA polymerases, one would expect similar mutation frequencies here (44), which would lead to rapid divergence of sequences in the absence of selective pressure to maintain function. In the case of the surface glycoproteins containing the neutralizing epitopes, mutations which allow escape from immune surveillance could then be selected for, similar to the situation described for influenza viruses.

Interestingly, when the amino acid sequences of the various paramyxovirus proteins were compared by the statistical method of Needleman and Wunsch, larger than expected homologies were detected, especially in the receptor binding protein (Table 2). This situation contrasts strongly with the influenza viruses, where the HA proteins undergoes the greatest divergence and is responsible for the antigenic diversity of the various strains. The antigenic stability of paramyxoviruses has, however, long been known. For example, there is only one known serotype for each of these viruses (except for RSV which contains two). Furthermore, the Edmonston vaccine strain of measles virus is effective in all developed countries, i.e: over a period of more than 20 years the measles neutralizing epitopes have not been able to diverge to escape immune surveillance, as their influenza virus counterparts have clearly done. The reason for this remarkable stability in the antigenic structure of the paramyxovirus receptor binding protein is unclear. The fact that paramyxoviruses contain non-segmented rather than segmented genomes does not appear to offer any obvious explanation, and the possibility that the paramyxo-

Table 2. - Relationship of paramyxovirus proteins with their counterparts in other paramyxoviruses and also myxoviruses

Standard Deviation of HPV3 HN vs Significance[1]		Standard Deviation of HPV3 NP vs Significance[1]	
SV HN	72.7	SV NP	59.8
SV5 HN	20.4	MV NP	19.7
MV H	4.6	CDV NP	9.5
RSV G	-0.9	RSV N	1.8
HPV3 F vs		*HPV3 P vs*	
SV F	37.5	SV	30.3
SV5 F	15.7	MV	4.5
RSV F	8.5	CDV	2.8
		RSV	1.4
Sendai F vs		*HPV3 C vs*	
Flu A	1.9	SV	25.2
B	3.6	MV	2.9
C	5.0	CDV	3.3
RSV	7.2		
SV5	10.0		
Flu A vs			
Flu B	16.8		
Flu C	4.2		

[1] Each alignment score is expressed as the number of standard deviations of significance of the direct comparison versus that of the average of 100 randomized permutations of one of the two sequences, using the mutational data matrix of 1978 (55). For proteins that are thought to serve similar functions, scores of 3 and above are considered indicative of an authentic relationship (56).
TheHPV3 comparison are from Spriggs and Collins (14). The others are from Blumberg *et al.* (19) and Kolakofsky *et al.* (57).

virus neutralizing epitopes are all part of the receptor binding site and therefore cannot vary without loss of function, for example, seems unlikely. The ramarkable stability of the paramyxovirus surface proteins thus deserves further investigation.

We have previously suggested that paramyxoviruses and myxoviruses shared a common ancestor during evolution, and that the rearrangement of the genetic information from segmented to non-segmented form (or vice versa) could be accomplished by the "jumping polymerase" thought to function during the creation of Defective Interfering (DI) genomes (19). With our increasing knowledge of the genomic structure of the various paramyxoviruses, genetic rearrangements within the non-segmented paramyxovirus family also appear to have taken place during evolution. For example, as previously noted, SV5 contains an extra transcription unit (SH) between the F and HN genes, whereas the other parainfluenza viruses do not, and the F and HN proteins are among the most conserved of the parainfluenza virus genes. Similarly, in RSV, the F and G genes are inversed in order, and 4 other transcription units have been introduced. As shown in Table II, since all the paramyxoviruses can be shown to have evolved from a common ancestor, at least as far as their F genes are concerned, genetic rearrangements must also have occured during paramyxovirus evolution. It thus seems likely that the genetic rearrangements seen to take place during DI genome generation, i.e: the jumping polymerase model (45), would also have operated throughout evolution and have been responsible for the genetic rearrangements of non segmented genomes. Further, since the jumping polymerase model allows for the genetic equivalent of gene conversion, this mechanism could also be responsible for recombination between non-defective genomes which could then offer a way of suppressing the unwanted effects of point mutations. In this way, the jumping polymerase could also contribute to the stability of the paramyxovirus non-segmented genome.

4. THE GENERATION AND BIOLOGY OF DEFECTIVE INTERFERING PARTICLES

When Sendai virus is repeatedly passaged at high moi, defective interfering (DI) particles arise, as is the case for all RNA viruses examined to date (46). These particles contain DI genomes which are deletion mutants of the non-defective (ND) or standard virus genome and Sendai DI genomes in which more than 90% of the nondefective (ND) genome has been deleted are not uncommon (47). In these deletion mutants, of the copy back variety, the right end of the viral genome has been retained (see Fig. 4) whereas in general the left end has not, but has been replaced by a duplication of the right end, as an inverted repeat (45). When Sendai DI genome nucleocapsids are deproteinized, the complementary ends of both the (+) and (−) strands anneal in an concentration independent manner. The length the ds-RNA stem thus formed, which is not conserved in length between different DI genomes (but is generally 100-200 bp long) is easily visualized as a panhandle on the circular DI RNAs viewed in the electron microscope. Subsequent mapping experiments showed that these DI genomes were derived from the right end of the viral genome for their entire length minus that of the inverted terminal repeat. The absence of extensive inverted repeats at the termini of the ND genome suggested that these DI genomes must have been generated by the jumping of the viral polymerase during the copying of the (+) antigenome template from this template to its nascent (-) genome chain. The position of this jump or cross-over to the nascent strand then determines the length of inverted terminal repeat (45).

These DI genomes are clearly defective as they contain only the C-terminal portion of the L gene (Fig. 4), and they also interfere with the replication of the ND helper genome. This interference phenomenon probably results from two effects. Since they do not contain the left end of the ND genome they are missing the *cis*-acting signals for transcription and DI genomes do not make any transcripts, except for the putative (-) leader RNA.

Secondly, since ND (-) genomes predominate over (+) antigenomes throughout the replicative cycle intracellularly and are the respective templates for each other, the right end of the ND genome clearly contains stronger *cis*-acting sequences for genome replication than does the left end, and DI genomes have

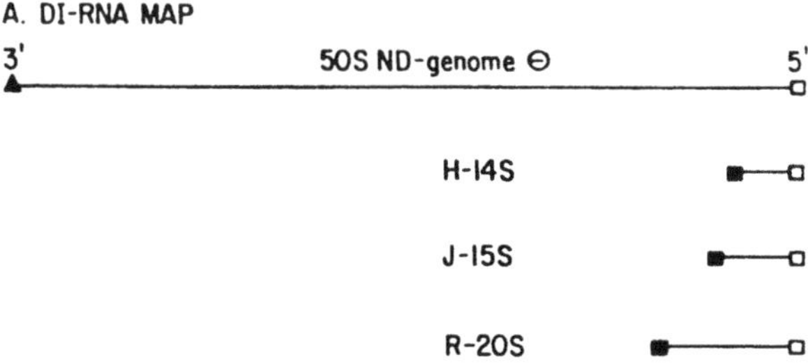

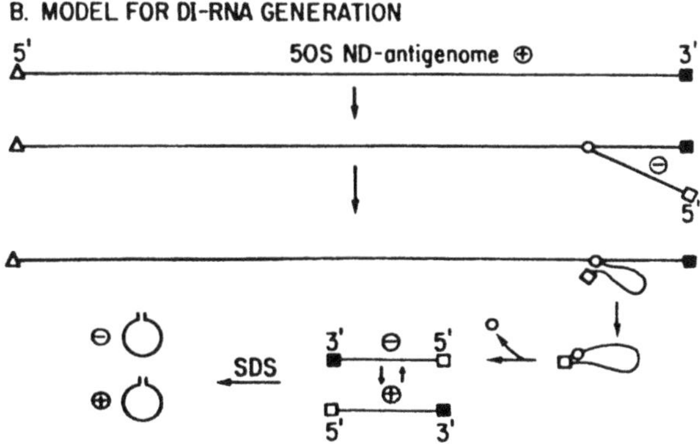

Figure 4. Model for DI-RNA generation. **A:** The DI-RNAs are approximately in scale relative to the non-defective (ND) RNA, 15 kb long. Open squares indicate the 5' terminal sequence of the ND genome, and the filled squares represent their complement. Filled and open triangles represent the 3' end of the ND genome and its complementary sequences, respectively.
B: Closed triangles and squares represent the recognition sites for the viral RNA replicase (open circles) Reproduced from ref. 45.

duplicated these stronger *cis*-acting sequences at both ends. One consequence of this is that (+) and (-) DI genomes, unlike their ND counterparts, are found intracellularly in equimolar amounts.

Another consequence is that under limiting amounts of viral proteins required for genome replication, (L, P and NP at least), the DI genomes will out-compete the ND genomes and be preferentially amplified, leading to ND genome interference.

Another important biological property of DI particles is their ability to convert a cytolytic Sendai virus infection to a non cytopathic persistent infection. Sendai DI particles offer a unique opportunity to study this latter biological property biochemically, since unlike the VSV system where the addition of DI particles to the infecting ND stock allows the survival of a extremely rare cell in a persistently infected manner, addition of Sendai DI particles to an infection allows virtually all the cells to survive, again as a persistent infection (48). This Sendai system has therefore been recently studied in an attempt to determine what features of the infection are responsible for the survival of the entire cell population and its subsequent establishment as persistent infection.

The most likely model for cell survival due to the presence of DI genomes was thought to be the attenuation of the ND viral infection due to interference, i.e: since DI genome out-compete ND genomes for limiting viral components, less ND genomes is amplified and consequently less viral mRNAs and proteins are made, and that it is this lower level of viral macromolecules intracellularly which allows the cell to survive. However, early work on this system showed that this model was grossly oversimplified, or even incorrect. When, for example, BHK cells are infected only with ND virus, cytopathic effect (CPE) becomes pronounced at 12-18 hours post-infection and cell death follows swiftly (all cells are dead by 48 hours after infection). When DI particles are included in the infecting stock, cytopathic effect first appears at about the same time, but rather than increase in severity, the cytopathic effect actually diminishes with time, such that the culture appears "normal" by 48 hr pi. When the level of viral macromolecules intracellularly was measured, the mixed viral infection in many cases was found to contain more rather than less nucleocapsids than the standard infection, especially at 24-48 hr pi when the standard virus infected cells are all in the process of dying whereas the mixed virus infected cells are recovering their normal morphology (49). In these experiments, the severity of the cytopathic effects and the level of viral macromolecular synthesis as reflected in the accumulation of nucleocapsids could clearly *not* be correlated.

Since the level of viral macromolecular synthesis in itself did not correlate with cell death, other criteria were examined. One difference between cells acutely infected with standard virus alone and long-term persistently infected cells was found to be the amount of HN protein present at the plasma membrane, either relative to that of the F protein also on the membrane, or to the viral core proteins intracellularly (50-52). As might be expected for cells which would have to avoid immune surveillance, persistently-infected cells contained little if any surface HN. Interestingly, when cells are infected with mixed virus and compared to standard virus infected cells, in the mixed virus infection surface HN expression was again down regulated, and this change could be seen to be taking place as early as 40 hours post-infection, when cell survival was being determined. To date, the only change upon addition of DI particles to the standard virus stock used for infection that can be correlated with cell survival is the expression of HN on the infected cell surface. However, cell death cannot be simply associated with the amount of surface HN, since infection with a *ts* mutant in which HN surface expression is non existant gives both virus particles without detectable HN and also leads to cell death, both at the non-permissive temperature (53).

How the presence of DI genomes intracellularly affects the surface expression of HN is indeed intriguing because it was so unexpected. According to our current understanding of virus assembly, the HN protein is modified in its transport through the RER and Golgi and then is carried to the plasma membrane in smooth vesicles which bud from the trans-Golgi apparatus. The patches of the cell surfaces from which the virion buds are then modified not only by the F and HN surface proteins, but also contain a paracrystalline array of M protein on the inner surface, and nucleocapsids which have associated with the cytoplasmic tails of the HN and F proteins, or the layer of M protein, or both. Moreover, not all viral nucleocapsids appear to be equally efficient in the budding process. For example, (-) ND genome nucleocapsids are always more efficiently assembled into virus particles than (+) ND antigenome nucleocapsids (54), and DI genome nucleocapsids of the copy-back variety are assembled less efficiently than ND nucleocapsids (Tuffereau and Roux, unpublished). The reasons for the preferential assembly of (-) genomes over the others is not clear, but only the (-) ND genomes of this group contain the *cis*-acting sequences required for mRNA synthesis at their 3' ends and can thus be distinguished at least by this criteria.

The preferential assembly of (-) ND genome nucleocapsids may offer an insight in the effect of DI genomes on virion assembly in general and on HN surface expression in particular. It is clear that (-) ND genome nucleocapsids have an assembly advantage, and this advantage is presumably due to the fact that only these nucleocapsids associate "correctly" with either the cytoplasmic tails of the surface glycoproteins or the assembled M proteins, or both. Assuming for the sake of discussion that efficient budding takes place only when (-) ND genome nucleocapsids make their contacts, it is easy to explain why the other nucleocapsids are less efficiently assembled, and further why the presence of large amounts of DI genomes intracellularly inhibit virion maturation, i.e: by competing for sites which are then blocked to (-) ND genome nucleocapsids.

However, it is not clear from this model how the association of the paracrystalline array of M takes place with the modified plasma membrane. Electronmicroscopy studies have shown that the M protein modifies the inside surface of the plasma membrane only where the F and HN proteins are already present (6, 8), presumably due to initial interactions between M and the cytoplasmic tails of the F and/or HN proteins. The self association of the M protein as a layer below the membrane probably serves to keep the surface glycoproteins localized in the membrane and allows the correct association of the nucleocapsids with the membrane for budding to occur. However, it is not impossible that this association of M with the F and HN proteins take place not at the plasma membrane but on the smooth vesicles which carry the surface proteins from the Golgi to the plasma membrane. Nucleocapsids might possibly then also associate with the modified smooth vesicles such that upon integration into the plasma membrane the fully modified patch of plasma membrane would already be formed. Although such associations of M and nucleocapsids with smooth vesicles is entirely hypothetical, it might explain how the presence of DI genomes intracellularly could after the surface expression of HN. If the stable surface expression of HN were to depend in some way on an interaction between the (-) ND genome nucleocapsid and the M/HN complex at the level of smooth vesicles, then the presence of large amounts of DI genomes intracellularly could again compete for these sites (to which it would presumably have a lower affinity) and block them to (-) ND genome nucleocapsids. Models such as these are indeed without experimental support so far, but only in such ways can a coherent scheme been drawn to explain why DI genomes can affect surface expression of HN.

5. REFERENCES

1) Herrler, G., R. Rott and H.-D. Klenk. 1985. Neuraminic acid is involved in the binding of influenza C virus to erythrocytes. Virology **141**, 144-147.

2) Heggeness, M.H., A. Scheid and P.W. Choppin. 1980. Conformation of the helical nucleocapsids of paramyxoviruses and vesicular stomatitis virus: Reversible coiling and uncoiling induced by changes in salt concentration. Proc. Natl. Acad. Sci. USA **77**, 2631-2635.

3) Lynch, S. and D. Kolakofsky. 1978. Ends of the RNA within Sendai virus defective interfering nucleocapsids are not free. J. Virol. **28**, 584-589.

4) Keene, J.D., B.J. Thornton and S.U. Emerson. 1981. Sequence-specific contacts between the RNA polymerase of vesicular stomatitis virus and the leader RNA gene. Proc. Natl. Acad. Sci. USA **78**, 6191-6195.

5) McSharry, J.J., R.W. Compans, H. Lackland and P.W. Choppin. 1975. Isolation and characterization of the nonglycosylated membrane protein and a nucleocapsid complex from the paramyxovirus SV5. Virology **67**, 365-374.

6) Bachi, T. 1980. Intramembrane structural differentiation in Sendai virus maturation. Virology **106**, 41-49.

7) Heggeness, M.H., P.R. Smith and P.W. Choppin. 1982. In vitro assembly of the nonglycosylated membrane protein (M) of Sendai virus. Proc. Natl. Acad. Sci. USA **79**, 6232-6236.

8) Buechi, M. and T. Bachi. 1982. Microscopy of internal structures of Sendai virus associated with the cytoplasmic surface of host membranes. Virology **120**, 349-359.

9) Blumberg, B.M., K. Rose, M.G. Simona, L. Roux, C. Giorgi and D. Kolakofsky. 1984. Analysis of the Sendai virus M gene and protein. J. Virol. **52**, 656-663.

10) Giorgi, C., B.M. Blumberg and D. Kolakofsky. 1983. Sendai virus contains overlapping genes expressed from a single mRNA. Cell **35**, 829-836.

11) Bellini, W.J., G. Englund, S. Rozenblatt, H. Arnheiter and C.D. Richardson. 1985. Measles virus P gene codes for two proteins. J. Virol. **53**, 908-919.

12) Barrett, T., S.B. Shrimpton and S.E.H. Russel. 1985. Nucleotide sequence of the entire protein coding region of canine distemper virus polymerase-associated (P) protein mRNA. Virus Research **3**, 367-372.

13) Spriggs, M.K. and P.L. Collins. 1986. Human parainfluenza virus type 3: Messenger RNAs, polypeptide coding assignments, intergenic sequences, and genetic map. J. Virol., in press.

14) Spriggs, M.K. and P.L. Collins. 1986. Sequence analysis of the P and C protein genes of human parainfluenza virus type 3: Patterns of amino acid homology among paramyxoviral proteins. J. Gen. Virol., in press.

15) Curran, J.A., C. Richardson and D. Kolakofsky. 1986. Ribosomal initiation at alternate AUGs on the Sendai virus P/C mRNA. J. Virol. **57**, 684-687.

16) Shioda, T., Y. Hydaka, T. Kanda, H. Shibuta, A. Nomoto and K. Iwasaki. 1983. Sequence of 3,687 nucleotides from the 3' end of Sendai virus genome RNA and the predicted amino acid sequences of viral NP, P and C proteins. Nucleic Acids Res. **11**, 7317-7330.

17) Shioda, T., K. Iwasaki and H. Shibuta. 1986. Determination of the complete nucleotide sequence of the Sendai virus genome RNA and the predicted amino acid sequences of the F, HN and L proteins. Nucleic Acids Res. **14**, 1545-1563.

18) Morgan, E.M., G.G. Re and D.W. Kingsbury. 1984. Complete sequence of the Sendai virus NP gene from a cloned insert. Virology **135**, 279-287.

19) Blumberg, B.M., C. Giorgi, K. Rose and D. Kolakofsky. 1985. Sequence determination of the Sendai virus fusion protein gene. J. Gen. Virol. **66**, 317-331.

20) Blumberg, B., C. Giorgi, L. Roux, R. Raju, P. Dowling, A. Chollet and D. Kolakofsky. 1985. Sequence determination of the Sendai virus HN gene and its comparison to the influenza virus glycoproteins. Cell **41**, 269-278.

21) Morgan, E.M. and K.M. Rakestraw. 1986. Sequence of the Sendai virus **L** gene: Open reading frames upstream of the main coding region suggest that the gene may be polycistronic. Virology, in press.

22) Richardson, C.D., A. Berkovich, S. Rozenblatt and W.J. Bellini. 1985. Use of antibodies directed against synthetic peptides for identifying cDNA clones, establishing reading frames, and deducing the gene order of measles virus. J. Virol. **54**, 186-193.

23) Hiebert, S.W., R.G. Paterson and R.A. Lamb. 1985. Identification and predicted sequence of a previously unrecognized small hydrophobic protein, SH, of the paramyxovirus simian virus 5. J. Virol. **55**, 744-751.

24) Paterson, R.G., T.J.R. Harris and R.A. Lamb. 1984. Fusion protein of the paramyxovirus simian virus 5: Nucleotide sequence of mRNA predicts a highly hydrophobic glycoprotein. Proc. Natl. Acad. Sci. USA **81**, 6706-6710.

25) Collins, P.L., L.E. Dickens, A. Buckler-White, R.A. Olmsted, M.K. Spriggs, E. Camargo and K.V.W. Coelingh. 1986. Nucleotide sequences for the gene junctions of human respiratory syncytial virus reveal distinctive features of intergenic structure and gene order. Proc. Natl. Acad. Sci. USA. **83**, 4594-4598.

26) Collins, P.L. and G.W. Wertz. 1983. cDNA cloning and transcriptional mapping of nine polyadenylated RNAs encoded by the genome of human respiratory syncytial virus. Proc. Natl. Acad. Sci. USA **80**, 3208-3212.

27) Collins, P.L., Y.T. Huang and G.W. Wertz. 1984. Identification of a tenth mRNA of respiratory syncytial virus and assignment of polypeptides to the 10 viral genes. J. Virol. **49**, 572-578.

28) Hsu, M.-C., A. Scheid and P.W. Choppin. 1979. Reconstitution of membranes with individual paramyxovirus glycoproteins and phospholipid in cholate solution. Virology **95**, 476-491.

29) Choppin, P.W. and Compans, R.W. 1975. Reproduction of paramyxoviruses. In: *Comprehensive Virology*, **4**,pp. 95-178. Ed. by Frankel-Conrat and Wagner. Plenum Press, NY.

30) Homma, M. and M. Ohuchi. 1973. Trypsin action on the growth of Sendai virus in tissue culture cells. III. Structural difference of Sendai viruses grown in eggs and tissue culture cells. J. Virol. **12**, 1457-1465.

31) Scheid, A. and P.W. Choppin. 1974. Identification of biological activities of paramyxovirus glycoproteins. Activation of cell fusion, hemolysis, and infectivity by proteolytic cleavage of an inactive precursor protein of Sendai virus. Virology **57**, 475-490.

32) Scheid, A. and P.W. Choppin. 1976. Protease activation mutants of Sendai virus: activation of biological properties by specific proteases. Virology **69**, 265-277.

33) Fan, D.P. and B.M. Sefton. 1978. The entry into host cells of Sindbis virus, vesicular stomatitis virus and Sendai virus. Cell **15**, 985-992.

34) Hsu, M.-C., A. Scheid and P.W. Choppin. 1981. Activation of the Sendai virus fusion protein (F) involves a conformational change with exposure of a new hydrophobic region. J. Biol. Chem. **256**, 3557-3563.

35) Chambers, P., N.S. Millar and P.T. Emerson. 1986. Nucleotide sequence of the gene encoding the fusion glycoprotein of Newcastle disease virus. J. Gen. Virol., in press.

36) Mottet, G., A. Portner and L. Roux. 1986. Drastic immunoreactivity changes between the immature and mature forms of the Sendai virus HN and F_0 glycoproteins. J. Virol. **59**, 132-141.

37) Glazier, K., R. Raghow and D.W. Kingsbury. 1977. Regulation of Sendai virus transcription: Evidence for a single promoter *in vivo*. J. Virol. **21**, 863-871.

38) Collins, P.L., L.E. Hightower and L.A. Ball. 1980. Transcriptional map for Newcastle disease virus. J. Virol. **35**, 682-693.

39) Gupta, K.C. and D.W. Kingsbury. 1984. Complete sequences of the intergenic and mRNA start signals in the Sendai virus genome: Homologies with the genome of vesicular stomatitis virus. Nucleic Acids Res. **12**, 3829-3841.

40) Wilde, A. and T. Morrison. 1984. Structural and functional characterization of Newcastle disease virus polycistronic RNA species. J. Virol. **51**, 71-76.

41) Gupta, K.C. and D.W. Kingsbury. 1985. Polytranscripts of Sendai virus do not contain intervening polyadenylate sequences. Virology **141**, 102-109.

42) Kolakofsky, D. and B.M. Blumberg. 1982. A model for the control of non-segmented negative strand virus genome replication. In: *Virus Persistence*, B.W.J. Mahy, A.C. Minson, and G.K. Darby, eds, Cambridge University Press, pp. 203-213.

43) Holland, J., K. Spindler, F. Horodyski, E. Grabau, S. Nichol and S. VandePol. 1982. Rapid evolution of RNA genomes. Science **215**, 1577-1585.

44) Portner, A., R.G. Webster and W.J. Bean. 1980. Similar frequencies of antigenic variants in Sendai, vesicular stomatitis, and influenza A viruses. Virology **104**, 235-238.

45) Leppert, M., L. Kort and D. Kolakofsky. 1977. Further characterization of Sendai virus DI-RNAs: A model for their generation. Cell **12**, 539-552.

46) Holland, J., Kennedy, S.I.T., Semler, B.L., Jones, C.L., Roux, L. and Grabau E. 1980. Defective interfering RNA viruses and the host cell response. In: *Comprehensive Virology*,**16**, Fraenkel-Conrat and Wagner eds., Plenum Press, New York & London, pp. 137-192

47) Kolakofsky, D. Isolation and characterization of Sendai virus DI-RNAs. 1976. Cell **8**, 547-555.

48) Roux, L. and Holland, J.J. 1979. Role of defective interfering particles of Sendai virus in persistent infections. Virology **93**, 91-103.

49) Roux, L. and Waldvogel, F. 1981. Establishment of Sendai virus persistent infection: biochemical analysis of the early phase of a standard plus defective interfering virus infection of BHK cells. Virology **112**, 400-410.

50) Roux, L. and Waldvogel, F. 1983. Defective interfering particles of Sendai virus modulate HN expression at the surface of infected BHK cells. Virology **130**, 91-104.

51) Roux, L., Beffy, P. and Portner, A. 1984. Restriction of cell surface expression of Sendai virus hemagglutinin-neuraminidase glycoprotein correlates with its higher instability in persistently and standard plus defective interfering virus infected BHK-21 cells. Virology **138**, 118-128.

52) Roux, L., Beffy, P. and Portner, A. 1985. Three variations in the cell surface expression of the haemagglutinin-neuraminidase glycoprotein of Sendai virus. J. Gen. Virol. **66**, 987-1000.

53) Tuffereau, C., Portner, A. and Roux, L. 1985. The role of haemagglutinin-neuraminidase glycoprotein cell surface expression in the survival of Sendai virus-infected BHK-21 cells. J. Gen. Virol. **66**, 2313-2318.

54) Kolakofsky, D. and Bruschi, A. 1975. Antigenomes in Sendai virions and Sendai virus-infected cells. Virology **66**, 185-191.

55) Dayhoff, M. 1978. Atlas of protein sequence and structure. Vol. 5, Suppl. 3. National Biomedical Research Foundation, Washington, D.C.

56) Barker, W.C. and M.D. Dayhoff. 1982. Viral src gene products are related to the catalytic chains of mammalian cAMP-dependent protein kinase. Proc. Natl. Acad. Sci. USA **79**, 2836-2839.

57) Kolakofsky, D., J. Curran, B.M. Blumberg, C. Giorgi, R.A. Lamb, R.G. Paterson, S.W. Hiebert, S. Venkatesan and N. Elango. 1986. Sequence comparison of paramyxovirus surface glycoproteins. In: *The Biology of Negative Strand RNA Viruses*, B.W.J. Mahy & D. Kolakofsky, eds., Elsevier.

CHAPTER 13

INFLUENZA VIRUSES: GENOME STRUCTURE, TRANSCRIPTION AND REPLICATION OF VIRAL RNA

PETER PALESE AND DEBORAH A. BUONAUGURIO
Mount Sinai School of Medicine, One Gustave Levy Plaza, New York NY 10029, U.S.A.

INTRODUCTION

The influenza viruses, or *orthomyxoviridae*, are enveloped animal viruses containing a segmented RNA genome. They are classified into three types, A, B and C, based on their immunologically distinct nucleoprotein (NP) and matrix (M) protein antigens. Influenza A viruses are further grouped into antigenic subtypes according to their specific hemagglutinin (HA) and neuraminidase (NA) surface proteins. The first human influenza A virus was isolated in 1933. In addition to man, influenza A viruses have been isolated from avian species and a wide variety of mammals including horses, pigs, and seals. The only known reservoir for influenza B viruses is man and up until recent times, influenza C viruses were only found in the human population.

Influenza A viruses are intensely studied due to their association with periodic pandemic and epidemic outbreaks of disease in man. The problem of prevention of widespread disease is a reflection of the unique ability of influenza A viruses to undergo rapid changes in antigenic character. Variability is not restricted to the major surface antigens, hemagglutinin and neuraminidase, of the virus, but also occurs in genes coding for nonstructural proteins as well as proteins comprising the core of the virus particle. As soon as an immune response can be generated to an antigenic variant, the virus can mutate into a "new" variant, rendering the antibodies raised to the previous challenge ineffective. In this way, the influenza virus can evade the host immune response, thus making it quite difficult to develop an effective vaccine against the virus. Today's attenuated strains used to vaccinate individuals may not provide protection against tomorrow's newly emerging epidemic variants.

Influenza B viruses are associated with epidemic disease and do undergo antigenic variability, which appears not to be as extreme or frequent as that found in influenza A viruses. Influenza C viruses only cause mild respiratory infections in man and exhibit few antigenic differences.

1. STRUCTURE OF THE INFLUENZA A PARTICLE

Intact influenza virus particles contain approximately 1% RNA, 70% protein, 6% carbohydrate, and 23% lipid. The virions have a spherical shape when visualized in the electron microscope using negative staining techniques and their morphologically distinct spikes represent the hemagglutinin and neuraminidase proteins (1, 2).

The hemagglutinin is involved in the binding of the virus to cellular sialic acid containing receptors and mediates virus entry into cells by a low pH-induced membrane fusion event in endosomal vesicles (3-5). The neuraminidase hydrolyzes sialic acid from the hemagglutinin receptors on the cell surface, thereby aiding in the release of the progeny virions from the cell membrane (6, 7). Influenza viruses derive their lipid envelope by budding at the plasma membrane of the infected cell.

Directly beneath the viral lipid bilayer is an electron-dense layer composed of the most abundant virion protein, M1. It is believed that the M1 contributes to the structural integrity of the virion and also plays a role in the maturation of the virus at the plasma membrane prior to budding (8). Disruption of the outer coat of the virus, reveals the internal ribonucleoprotein (RNP) structure which contains the eight single-stranded RNA segments of the viral genome in close association with four proteins. The nucleoprotein is the predominant polypeptide of the nucleocapsid. The other minor constituents of the RNP complex are the polymerase (P) proteins, PB1, PB2, and PA. The three P proteins form a complex which has RNA-dependent transcriptase activity required for initiating viral RNA transcription in the infected cell (9, 10). A schematic diagram of the structure of the virion is shown in figure 1.

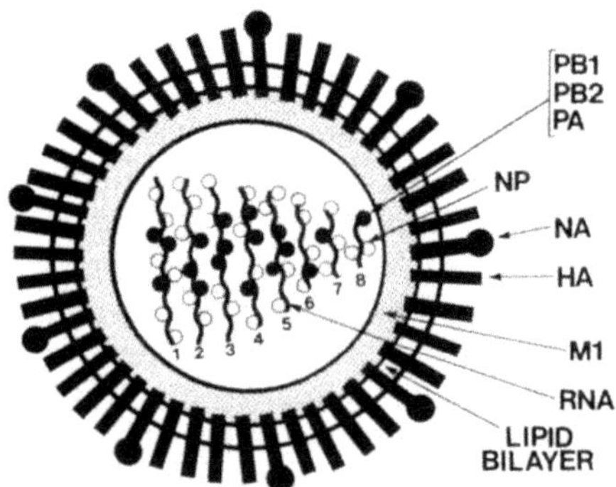

Figure 1. Schematic diagram of the structure of an influenza virus particle. Proteins PB1, PB2, PA, and NP comprise the virion core and are associated with the genome RNAs. M1 is in close contact with the lipid bilayer. The hemagglutinin and neuraminidase spikes are surface glycoproteins.

2. STRUCTURE OF THE RNA GENOME

Influenza A viruses possess a segmented genome of eight single-stranded RNAs of negative polarity (11). The genetic information of the virus codes for at least ten polypeptides: seven proteins that have both structural and enzymatic functions are found in the virion (PB1, PB2, PA, hemagglutinin, neuraminidase, nucleoprotein, and M1) and three nonstructural proteins (NS1, NS2, and M2) are synthesized only in infected cells (for a review see ref. 12). The complete genome of the influenza A virus has been mapped (Fig. 2) and recombinant DNA technology has made possible the determination of the complete nucleotide sequence of the entire genome of the influenza virus A/PR/8/34 (12). Over recent years, many additional nucleotide sequences of influenza virus RNA segments have been elucidated which have extended our knowledge of the molecular biology of the virus. Furthermore, considerable information on the structure of influenza virus proteins has been generated, highlighted by the determination of the 3-dimensional structure at approximately 3Å resolution of the hemagglutinin (13) and neuraminidase (7) proteins.

A. RNA Segments 1, 2 and 3: Polymerase Proteins PB1, PB2 and PA

The proteins of greatest molecular weight of the virus are the three polymerase proteins, PB1, PB2 and PA (apparent MW 94,000-82,000). The precise functions of the polymerase proteins in transcription and replication of viral RNAs have not been completely elucidated. Studies with temperature-sensitive (ts) mutants defective in RNA synthesis have suggested that PB1 and PB2 are involved in messenger RNA (mRNA) synthesis and PA, possibly, in virion RNA (vRNA) production (14-16). Detailed experiments using an *in vitro* transcription system have indicated that the role of PB2 is in binding to the cap structure of the host cell mRNA primer (17-21) and that of PB1 is in initiating transcription (9, 17, 22) (Table 1).

Table 1. - Influenza A virus RNAs and proteins[a]

Segment	Length[b] (Nucleotides)	Encoded Protein	Amino Acid[c] Length (Nascent Protein)	Approx. No. Molecules per Virion
1	2,341	PB2	759	30-60
2	2,341	PB1	757	30-60
3	2,233	PA	716	30-60
4	1,778	HA	566	500
5	1,565	NP	498	1,000
6	1,413	NA	454	100
7	1,027	M1	252	3,000
		M2	97	
8	890	NS1	230	
		NS2	121	
	13,588			

[a] Adapted from (12).
[b] Values for A/PR/8/34 strain.
[c] Deduced from nucleotide sequence.

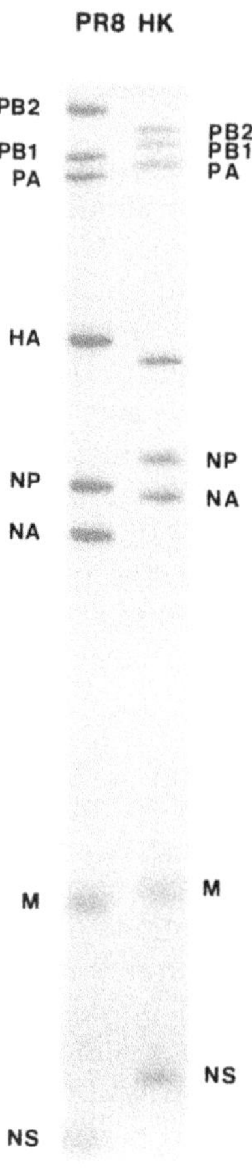

Figure 2. Genetic map of influenza A/PR/8/34 (PR8) and A/HongKong/8/68 (HK). Purified RNAs of the two influenza viruses were separated on a polyacrylamide gel. Specific gene products were assigned to each RNA segment following RNA and protein analysis of recombinants derived from the two parent viruses. PB1, PB2 and PA are the large genes coding for the polymerase proteins. HA, NA and NP indicate the genes coding for the hemagglutinin, neuraminidase and nucleoprotein, respectively. M (matrix protein) and NS (nonstructural protein) genes code for two proteins each: the M1 and M2 and the NS1 and NS2 proteins, respectively. Based on Palese, 1977 (11).

B. RNA Segment 4: Hemagglutinin Protein (HA)

The hemagglutinin protein is encoded by RNA segment 4 of the virus. Segment 4 of influenza A/Aichi/68(H3) is 1765 nucleotides in length, whereas the corresponding gene of influenza B is 1882 nucleotides. In the former, two short untranslated stretches of 29 and 35 bases precede and follow an open reading frame of 1698 bases.

The nascent HA polypeptide contains an N-terminal hydrophobic signal sequence (15-17 amino acids long) that is proteolytically removed after translocation across the endoplasmic reticulum membrane (Fig. 3). This signal sequence is required for glycosylation and for the transport of the hemagglutinin to the cell surface (23, 24).

The HA molecule is further processed by a trypsin or trypsin-like protease to generate the HA1 and HA2 subunits followed by the loss of at least one arginine residue between the chains (25). The removal of the arginine residue(s) suggests that an additional exopeptidase cleavage is involved in hemagglutinin activation. Following the proteolytic cleavage and subsequent removal of residue 329 (Arg), the newly generated C-terminal of HA1 (Thr) is 21 Å apart from the strongly hydrophobic N-end of HA2, suggesting that a substantial re-arrangement of the tertiary folding follws the cleavage.

The N-terminus of the HA2 chain consists of a hydrophobic sequence that is conserved among strains within the same or different subtypes (13, 26). The N-terminus of HA2 has been implicated in the process of virus penetration of the host cell membrane to initiate infection (27).

Recently, the mechanism of hemagglutinin-mediated membrane fusion has been further elucidated through analysis of naturally occurring fusion variants (28) and through expression of mutant hemagglutinins in SV40 virus vectors, which were engineered by oligonucleotide-directed mutagenesis within the fusion peptide coding region of the hemagglutinin cDNA (4). A second hydrophobic sequence (approximately 24 residues) is located near the C-terminus of HA2. This sequence spans the lipid bilayer and anchors the hemagglutinin in the lipid envelope (23, 29). The primary amino acid sequence of the hemagglutinin contains several possible glycosylation signals (Asn-X-Ser/Thr) where carbohydrate can attach to asparagine residues via N-glycosidic linkages and there are also several sites where disulfide bonds may form between cysteine residues within HA1 and HA2 or between the two chains. On the surface of the virion envelope the hemagglutinin protein exists as a trimer (Fig. 4) of non-covalently linked HA1/HA2 monomers (13). The molecule has been shown by 3-dimensional crystallography (13) to consist of an elongated 135 Å long cylinder with:

i) a long fibrous stem extending 76 Å from the membrane, and
ii) a distal globular region.

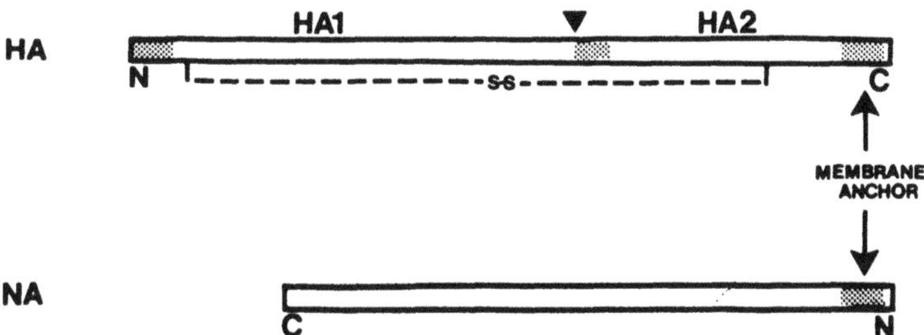

Figure 3. Structure of the hemagglutinin (HA) and neuraminidase (NA) of influenza A virus.
The hemagglutinin precursor protein contains three hydrophobic regions (stippled boxes): 1) signal sequence at the N-terminus of HA1, 2) fusion peptide at the N-terminus of HA2, and 3) membrane anchor at the C-terminus of HA2. The triangle represents the site where the cleavage activation occurs. The two resulting subunits, HA1 and HA2, are linked by a disulfide bridge.
The neuraminidase protein contains a single extended hydrophobic signal sequence at the N-terminus which serves to anchor the polypeptide in the viral membrane.

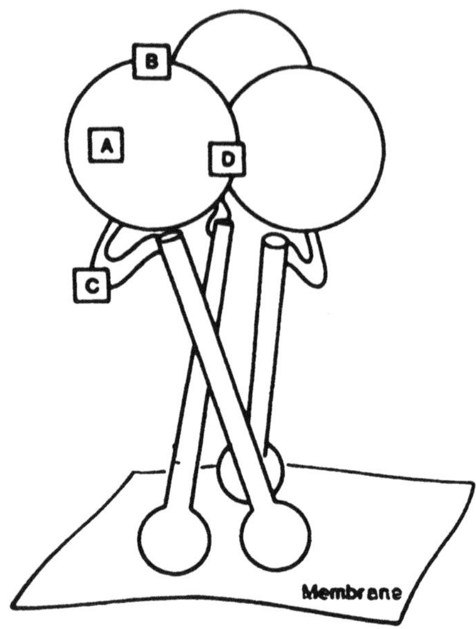

Figure 4. Schematic representation of the three-dimensional structure of the hemagglutinin trimer as it exists on the virion surface.
The hemagglutinin monomers are non-covalently linked and each consists of a globular head domain and an elongated stem region. Four antigenic sites have been mapped to the globular head of the H3 hemagglutinin. These regions, A, B, C, and D, are indicated on one of the hemagglutinin monomers. The diagram is modified from Wilson *et al.* (ref. 13).

At least four antigenic sites (A, B, C and D) have been proposed which are found at the top of the globular region. These sites were derived from an analysis of amino acid changes in naturally occurring antigenic variants and laboratory variants selected with monoclonal antibodies (13, 30, 31).

C. RNA Segment 5: Nucleoprotein (NP)

RNA segment 5 codes for the nucleoprotein which interacts with the viral RNA and the three P proteins to form RNP particles (32). It is believed that the nucleoprotein constitutes the backbone of the RNP complex. Based on the total length of the viral genome and the number of nucleoprotein molecules per virion, it can be estimated that a single nucleoprotein molecule interacts with approximately 20 nucleotides (33). Although the precise function of nucleoprotein is unclear, the analysis of ts nucleoprotein mutants has suggested that NP is involved in viral RNA synthesis (14, 34, 35). In addition, Van Wyke *et al.* (36) have demonstrated that monoclonal antibodies directed against two antigenic domains of the NP can inhibit *in vitro* transcription of viral RNA. Recently, Kato *et al.* (10) have purified an RNA polymerase-RNA complex from influenza virus that lacks NP protein. They have shown that this complex is active in synthesizing viral mRNA *in vitro*. However, the synthesis of full-length complementary RNA (cRNA) transcripts may require the complex containing NP protein.

The NP protein has been shown to be modified post-translationally by phosphorylation and proteolytic cleavage. NP is phosphorylated at a single serine residue per molecule by cellular phosphokinases, but it is not clear what role the phosphate group plays (37, 38). Zhirnov and Bukrinskaya (39) have reported that two forms of phosphorylated NP protein are detected in cells infected with human influenza viruses: one uncleaved 56,000 MW form (NP56) and a cleaved form of 53,000 MW (NP53) derived from the uncleaved product.

The nucleoprotein protein accumulates in the nucleus of virus-infected cells (40) and Davey *et al.* (41) have identified a nuclear accumulation signal sequence (amino acids 327-345) by *in vitro* mutagenesis of NP cDNAs which directed the synthesis of mutant nucleoprotein proteins after injection into Xenopus oocytes.

D. RNA Segment 6: Neuraminidase (NA)

Neuraminidase is an integral membrane glycoprotein of the virus encoded by the sixth longest RNA. The deduced amino acid sequence of the NA protein reveals an N-terminal hydrophobic extended signal sequence which serves to anchor the protein in the virus membrane and is long enough to span the lipid bilayer (Fig. 3) (42). In addition, this sequence is required for translocation of the nascent neuraminidase polypeptide into the rough endoplasmic reticulum, which leads to glycosylation and cell surface expression (43-45). The signal sequence of the neuraminidase is not proteolytically removed following translocation across the membrane (42) and no processing at the C-terminus of the molecule has been observed.

There are a number of possible functions associated with the enzymatic activity of the neuraminidase in virus replication and maturation:

i) release of budding virus particles from the host cell membrane (46, 47);
ii) prevention of self-aggregation of the virus (46);
iii) exposure of the hemagglutinin to proteolytic cleavage, thereby altering viral virulence (48); and
iv) penetration of virus into the host cell via low pH-induced membrane fusion (49).

Antibodies to neuraminidase do not neutralize or prevent virus infection (50), but do act to reduce the extent of pathology in the host by restricting multiple cycles of replication (51).

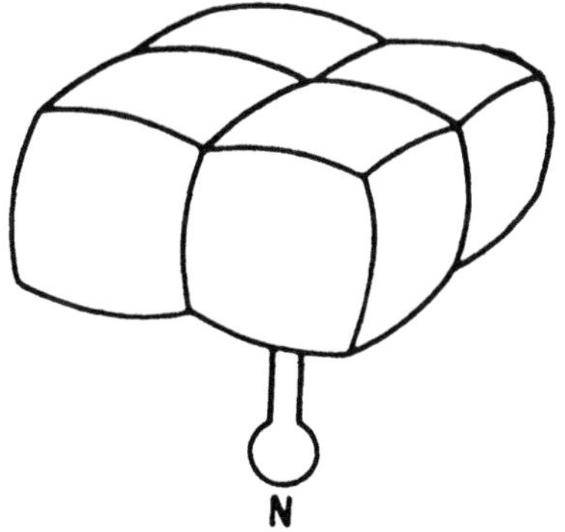

Figure 5. Schematic diagram of the three-dimensional structure of the neuraminidase tetramer.
The tetramer has a box-shaped globular head with four-fold symmetry, and is attached to virus membrane by a slender stalk. The N-terminus of the neuraminidase molecule is the site of attachment to the viral membrane. The diagram is drawn from data by Varghese *et al.* (ref. 7).

X-ray crystallography (7) has shown that the molecule has a box-shaped globular head with four-fold symmetry, connected to the virus membrane by a long slender stalk (Fig. 5). Viewed from above the head, each monomer contains six topologically identical beta-sheets arranged in a propeller formation. The antigenic sites cluster preferentially on the distal surface loops which connect the various strands of the beta-sheets. Also, the catalytic and sialic acid binding sites of the neuraminidase have been localized to the head portion of the mushroom-shaped structure.

E. RNA Segment 7: Matrix Protein (M1); Nonstructural Protein (M2)

This RNA segment codes for two proteins, M1 and M2, and possibly a small peptide M3. Three separate mRNAs derived from RNA 7 are found in virus-infected cells, one of which is a colinear transcript of the genome segment and two are spliced mRNAs probably derived from the colinear transcript (Fig. 6).

Electron microscopy of virions reveals that the M1 forms an electron-dense layer beneath the lipid bilayer. In addition to providing structural stability to the virion envelope, M1 may recognize the viral glycoproteins and form a domain on the inner surface of the plasma membrane which subsequently provides a binding site for the RNP segments during virus assembly (52, 53). The precise mechanism of virus particle formation, however, has not been elucidated.

In addition to the long open reading frame encoding the M1 protein, there is a second open reading frame (+1 frame) at the 5' end of the vRNA which provides the bulk of the coding sequence of the M2 on a spliced mRNA (Fig. 6).

The nonstructural M2 protein is an integral membrane protein, not N-glycosylated, and is expressed abundantly at the cell surface (54, 55). A minimum of 18 N-terminal amino acids of M2 are exposed at the cell surface and the hydrophobic amino acids 25-43 are thought to anchor the protein in the membrane.

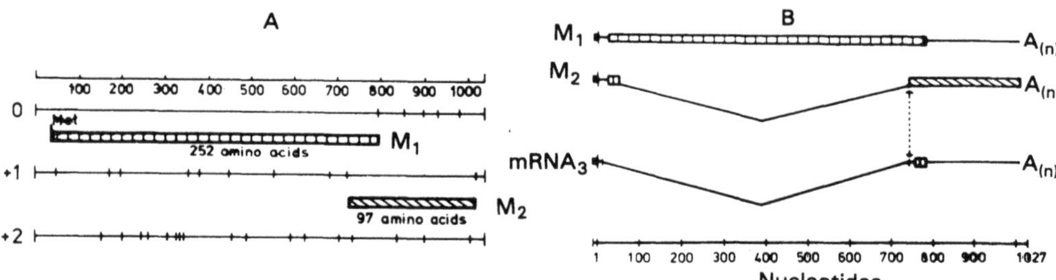

Figure 6. - **A:** Schematic representation of segment 7 of influenza virus RNA. Two long open reading frames encoding M1 and M2 are shown as hatched boxes. Vertical bars mark termination codons in each reading frame.
B: Organization of M1, M2, and M3 mRNAs. The filled boxes at the 5' end of the mRNAs preceding nucleotide 1 of the corresponding genes indicate the capped sequences (13-18 nucleotides) derived from cellular mRNAs (see text and Fig. 9). Thin horizontal lines represent the untranslated 5' and 3' terminal regions of each messenger. V-shaped lines indicate the splices. Note that M1 and M2 mRNAs are translated in different reading frames in the region comprised between nucleotides 740 and 1004. (adapted from ref. 12).

The remainder of the M2 protein, 54 C-terminal amino acids, comprise a long tail on the cytoplasmic side of the membrane. The function of the M2 in the virus life cycle has not been defined. A second spliced mRNA derived from RNA segment 7 has been isolated from virus-infected cells (56, 57) but the existence of a corresponding protein M3 has yet to be proven.

F. RNA Segment 8: Nonstructural proteins (NS1 and NS2)

The smallest RNA segment of the influenza virus genome codes for two polypeptides, NS1 and NS2 (58, 59), that are only found in infected cells (Fig. 7). NS1 is a phosphoprotein with phosphate attached to one or two threonine residues per molecule (60). The NS1 polypeptide (MW 26,000) is synthesized in large amounts early in infection and accumulates in the cell nucleus (40). Within the nucleus, the NS1 is localized predominantly in the nucleoli or the nucleoplasm depending on the time of infection and also the strain of influenza virus (61). The smaller of the two nonstructural proteins, NS2 (MW 14,000), is made late in infection and also has been shown by immunofluorescence analysis to be localized in the cell nucleus (62).

There are two mRNAs corresponding to RNA segment 8 that are found in influenza virus-infected cells (Fig. 7). Nucleotide sequencing of the longer transcript, that codes for NS1, revealed a mRNA of about 860 bases which is colinear with the vRNA segment. Sequence analysis of the NS2 mRNA showed that it contains an interrupted region of 473 nucleotides. The first 56 virus-specific nucleotides at the 5' end of the spliced NS2 mRNA are the same as those found at the 5' end of the NS1 mRNA. This NS2 leader sequence contains the initiation codon for protein synthesis plus information to code for nine amino acids which are common to NS1 and NS2. The remainder of the NS2 mRNA (340 bases) is translated in the + 1 reading frame (Fig. 7) (56).

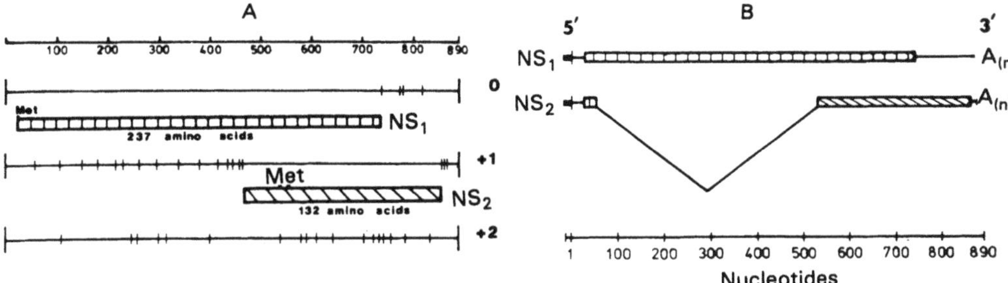

Figure 7. - **A:** Gene organization of segment 8 of influenza virus RNA. Hatched boxes indicate the sequences encoding NS1 and NS2; the short vertical bars mark the location of termination codons in each reading frame.
B: Arrangement of NS1 and NS2 mRNAs: The filled boxes preceding nucleotide 1 at the 5' termini indicate the capped sequences (13-18 nucleotides) derived from cellular mRNAs (see text and Fig. 9). Thin horizontal lines represent the untranslated 5' and 3' terminal regions of each messenger. V-shaped lines indicate the splices. Note that NS1 and NS2 mRNAs are translated in different reading frames in the region comprised between nucleotides 529 and 861. (Adapted from ref. 12).

The functions of NS1 and NS2 in virus replication have not been elucidated. Examination of ts mutants of A/FPV/Rostock/34 defective in RNA segment 8 has revealed a variety of phenotypic characteristics including reduced M1 synthesis, defects in vRNA synthesis, and decreased hemagglutinin production at the nonpermissive temperature (63-65). Major efforts in several laboratories are now aimed at understanding the roles of these viral proteins.

3. TRANSCRIPTION AND REPLICATION OF INFLUENZA A VIRUS RNA

Analysis of influenza virus-infected cell extracts revealed two distinct classes of RNA transcripts which are complementary to the vRNA segments (66-68). The first class of cRNA contains a 3' poly (A) tail and a 5'-terminal 7-methyl guanosine ($m^7GpppNm$) cap structure. There are two species of these capped cRNA transcripts. One consists of colinear copies of the eight vRNA segments lacking a sequence of 17-22 bases present in a conserved region at the 5' termini of the vRNAs. The second species of capped cRNA transcripts also lack these sequences and are spliced transcripts derived from the colinear transcripts complementary to RNAs 7 or 8. The second class of cRNAs found in the infected cell are complete transcripts of the vRNAs, which are uncapped and unpolyadenylated. These complete cRNAs serve as templates for the synthesis of vRNAs which are incorporated into progeny virus particles. The virion-associated transcriptase complex is responsible for the synthesis of both classes of cRNAs in the nucleus of infected cells (69-71) (Fig. 8). In addition to the structural differences between the two types of cRNAs, differences in kinetics of synthesis of individual transcripts and regulation have been identified.

A. Transcriptional Controls

The synthesis of influenza viral mRNAs is controlled during infection with respect to the relative amount of each mRNA synthesized and to the time at which each mRNA is synthesized in greatest amount (67, 68, 72, 73). Immediately after infection, similar amounts of all eight viral mRNAs are detected and their production is independent of viral protein synthesis (primary transcription). This is followed by an early phase of secondary transcription, dependent on viral protein synthesis, during which the production of the nucleoprotein and NS1 mRNA predominates. Subsequently, during the late phase of secondary transcription, the rate of synthesis of NS1 mRNA relative to NP mRNA decreases while that of the M, hemagglutinin, and neuraminidase mRNAs is greatly amplified. At all times except immediately after infection, the rate of synthesis of the three P protein mRNAs remains low relative to that of the other viral mRNAs. The relative rates of synthesis of the various viral mRNAs at different times of infection correlates closely with the relative rates of synthesis of the proteins encoded by these mRNAs, indicating that viral gene expression is regulated mainly at the level of transcription (67, 68, 73-75).

As mentioned above, the mRNAs encoding the M2 and NS2 polypeptides of influenza virus are produced in infected cells through splicing of colinear transcripts from vRNA segments 7 (76) and 8 (56), respectively (figures 6 and 7). It appears that the spliced transcripts are formed through the action of cellular RNA processing enzymes since unspliced and correctly spliced (as *in vivo*) RNAs are expressed from SV40 virus vectors containing a cDNA copy of either the M or NS gene in the absence of other influenza virus products (77, 78). During the course of influenza virus infection, there is an increase in the level of virus-specific spliced mRNAs, most evident for the NS2 mRNA, relative to that of their unspliced precursors (79, 80). It appears that virus-specific products may regulate the production of spliced mRNAs in influenza virus-infected cells.

Unlike the viral mRNAs, the production of full-length transcripts is not regulated. Approximately equimolar amounts of each of the eight cRNA transcripts are synthesized throughout infection (67). Influenza virus infection in the presence of cycloheximide results in the formation of only mRNA (primary transcription), indicating that newly synthesized viral or possibly host proteins are required for the synthesis of the full-length cRNAs (67, 72). The nascent virus protein may act to modify the transcriptase complex in some way so that transcription initiates without a primer and continues past the termination site utilized during viral mRNA synthesis (Fig. 8).

A unique feature of influenza virus replication, not found in other nononcogenic RNA viruses, is the requirement of a functional host nuclear RNA polymerase II (81, 82). This enzyme synthesizes the precursors to cellular mRNAs. It was found that a specific inhibitor of RNA polymerase II, α-amanitin, inhibits influenza virus replication at the level of viral RNA transcription (83).

B. Mechanism of Viral mRNA Synthesis

The mechanism of how influenza viruses synthesize mRNAs has been dissected using *in vitro* transcription systems which, in most respects, reflect the process occurring *in vivo*. The initiation of influenza viral mRNA transcription requires 5' capped primers derived from nascent host cell mRNAs synthesized by RNA polymerase II (Fig. 8 and 9, and ref. 22). The host mRNAs are cleaved by a virus specific endonuclease at a purine residue 10 to 15 nucleotides from the capped 5' end. It is this fragment containing the 5'-terminal m^7GpppNm (cap 1 structure) that serves as a primer and is transferred to the 5' ends of the influenza virus mRNAs during transcription ("cap snatching") (22, 84, 85). Shaw and Lamb (86) have reported that host RNA primers containing a 3'-terminal Py-G-C-A sequence before the presumed endonuclease cleavage site are preferred as primers in influenza virus mRNA synthesis. The viral PB2 protein functions in cap recognition during the endonuclease reaction (17, 19-21). The capped fragments do not have to be hydrogen-bonded to the 3' end of the vRNA template to prime viral transcription (87).

The second step in the reaction is the initiation of transcription via incorporation of a G residue onto the 3' end of the capped primer fragment (Fig. 9), directed by the penultimate C residue at the 3' end of each vRNA template (3' U-C-G-...). It is the PB1 protein that catalyzes the initiation step of transcription (17, 88, 89). After initiation with a G residue, chain elongation proceeds. It is not known which P protein(s) functions in this process, but there is evidence to suggest that PB1 may be involved (9, 89). Elongation is terminated at a tract of U residues (polyadenylation signal) about 17-22 nucleotides from the 5' end of the template and poly(A) addition occurs at the 3' end of the mRNA (90).

Recently, an RNA polymerase-viral RNA complex has been purified from influenza virions by cesium trifluoroacetate centrifugation and phosphocellulose column chromatography (10, 84). The complex, composed of PB1, PB2, PA, and vRNA, has been shown to be catalytically active in viral mRNA synthesis *in vitro*. These data sugest that NP is not required for mRNA synthesizing activity.

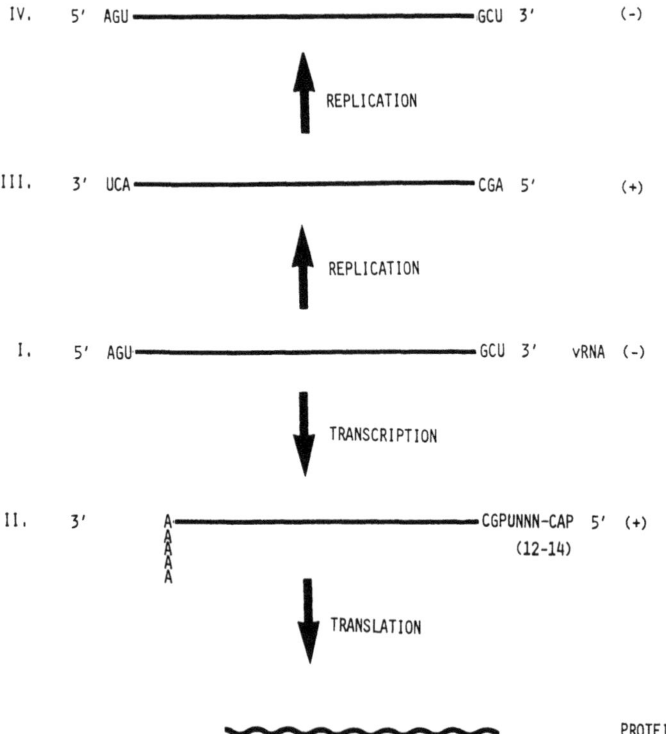

Figure 8. Transcription and replication of influenza virus RNA. Following virus infection, the vRNAs (I) are transcribed into (+) RNAs (II) that are incomplete complementary copies of the template vRNAs. These transcripts (mRNA), containing a 5' cap structure and a 3' poly (A) tail, are translated to produce virus-specific proteins. Protein synthesis is required for the production of unpolyadenylated (+) RNAs (III) that are full-length complementary copies of the vRNA segments. These complete (+) RNA transcripts serve as templates for the synthesis of (-) RNAs (IV) that are incorporated into progeny virus particles.

C. Synthesis of Full-length cRNA Transcripts

The unpolyadenylated cRNAs are complete copies of the vRNA segments and consequently the termination of transcription during mRNA synthesis, which occurs 17-22 bases from the 5' ends of the vRNA templates, is inoperative during the synthesis of these transcripts. In contrast to mRNA synthesis, initiation of full-length transcripts *in vivo* occurs without a primer at the 3'-terminal U of the vRNA templates (71, 91). The synthesis of full-length cRNA requires ongoing viral protein synthesis. Since these transcripts are complete copies of the vRNA segments and lack host-derived primer sequences, they most likely function as the templates for vRNA synthesis. To aid in deciphering the mechanism of full-length cRNA and vRNA production, *in vitro* systems have been established in which full-length

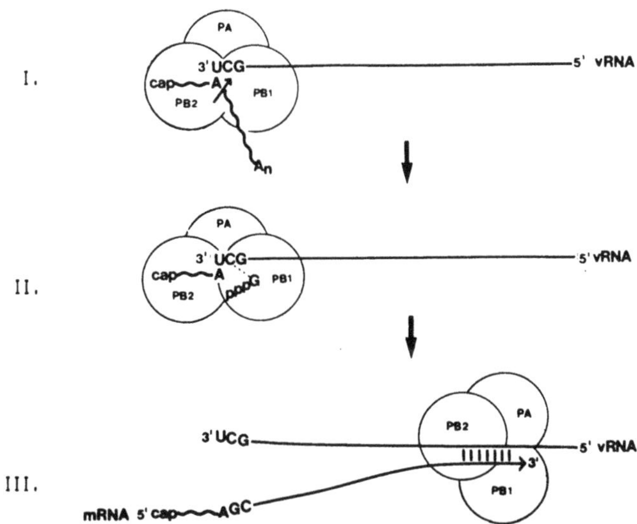

Figure 9. Model showing the steps involved in influenza virus mRNA synthesis.
The PB1, PB2, PA complex is required for transcription of viral mRNAs in the nucleus of the host cell.
I. Capped primer fragments are generated by cleavage (indicated by arrow) of host cell RNA polymerase II transcripts (wavy line), near the 5' ends. Capped fragments (12-14 bases) terminating in **A** residues are preferred primers for influenza mRNA synthesis. The PB2 protein is responsible for cap recognition of the primer by the complex and it may also possess endonuclease activity.
II. Initiation of transcription occurs by addition of a **G** residue onto the 3' end of the capped primer directed by the penultimate **C** residue at the 3' end of the viral RNA template. PB1 is involved in this step of mRNA synthesis.
III. Initiation is followed by elongation of the mRNA chain. It is not clear which of the **P** protein(s) is active in this step, but PB1 is thought to be involved. Termination of transcription and polyadenylation occurs 17-22 bases from the 5' end of the viral RNA template. (Modified from Braam *et al.*, ref. 9).
It is not yet clear whether the P proteins stay together as a complex throughout all the stages of mRNA synthesis. Also, their orientation with respect to one another and relative to the template RNA is not known.

cRNA, in addition to incomplete mRNA, can be synthesized (71, 92). Beaton and Krug (71) have found an anti-termination factor in the cytoplasmic extract of virus-infected cells that enables the transcriptase complex to continue transcription past the site at which mRNA synthesis terminates. In addition, this cytoplasmic extract allows the transcriptase complex to initiate transcription without the capped primers used in viral mRNA synthesis. The identification and purification of the active cytoplasmic factor(s) may help to characterize the switch mechanism from mRNA to full-length cRNA synthesis.

D. Synthesis of vRNA (Replication)

There is little information available concerning the synthesis of vRNA and these data include conflicting reports. The synthesis of minus strand vRNA most likely occurs in the nucleus of infected cells (70). Smith and Hay (93) have shown that the eight vRNA segments are produced in dissimilar amounts and in different relative proportions at various times of infection. Virion RNAs 5 and 8 are preferentially synthesized early in infection, similar to mRNA synthesis of these templates. In contrast, Enami *et al.* (73) have reported that all eight segments of minus strand RNA are produced coordinately at nearly equimolar ratios. Analysis of ts mutants defective in vRNA production has implicated at least two of the P proteins in vRNA synthesis (94), but the precise role of these P proteins and possibly of other viral proteins, such as the NP (14,58) in catalyzing vRNA synthesis is not known. Furthermore, the mechanism has not been elucidated as to how approximately equimolar amounts of the different vRNA segments are packaged into virions and how budding of the virus occurs through the plasma membrane.

E. Effect of Interferon on Influenza Virus Replication

Interferon induces an antiviral state against influenza virus (95) and the mechanism of inhibition is of particular interest because influenza virus is the sole nononcogenic RNA virus which requires host cell nuclear function for replication. Ransohoff et al. (96) have investigated the molecular mechanism whereby interferon induces resistance to influenza virus by monitoring the accumulation of both plus and minus strand viral RNAs in infected MDBK cells treated with human α A interferon. This group found that influenza viral primary transcripts failed to accumulate in interferon-treated cells. The result is consistent with observations made by Krug et al. (97), who have reported that the interferon-induced Mx gene product inhibits influenza viral mRNA synthesis in mouse embryo cells bearing the Mx gene. In mouse cells carrying the dominant influenza resistance allele, Mx^{++}, interferon α/β induces a 75,000 MW protein (Mx) which accumulates in the cell nucleus and is responsible for an enhanced antiviral state specifically against influenza virus (see chapter 6 of this volume and refs. 98-101). There are conflicting reports concerning the stage of viral replication at which the Mx protein exerts its inhibitory effect. In contrast to the results of Krug et al. (97), Meyer and Horisberger (102) suggest that interferon-treated Mx-bearing macrophages exhibit inhibition of influenza virus replication at the level of protein translation. Staeheli and Haller (103) have detected an interferon-induced 80,000 MW protein in human cells which appears to be the human homolog of Mx protein in mouse cells. It will be interesting to explore the significance of the human Mx protein for defense against influenza viruses in man.

4. REFERENCES

1) Laver, W.G. and Valentine, R.C. (1969), Virology **38**, 105-119.
2) Murti, K.G. and Webster, R.G. (1986), Virology **149**, 36-43.
3) White, J., Kartenbeck, J. and Helenius, A. (1982), EMBO J. **1**, 217-222.
4) Gething, M.-J., Doms, R.W., York, D. and White, J. (1986), J. Cell. Biol. **102**, 11-23.
5) Wharton, S.A., Skehel, J.J. and Wiley, D.C. (1986), Virology **149**, 27-35.
6) Palese, P., Tobita, K., Ueda, M. and Compans, R.W. (1974) Virology **61**, 397-410.
7) Varghese, J.N., Laver, W.G. and Colman, P.M. (1983), Nature (London) **303**, 35-40.
8) Gregoriades, A., Christie, T. and Markarian, K. (1984), J. Virol. **49**, 229-235.
9) Braam, J., Ulmanen, I. and Krug, R.M. (1983), Cell **34**, 609-618.
10) Kato, A., Mizumoto, K. and Ishihama, A. (1985), Virus Res. **3**, 115-127.
11) Palese, P. (1977), Cell **10**, 1-10.
12) Lamb, R.A. (1983). In: *Genetics of Influenza Viruses*, P. Palese and D.W. Kingsbury, eds., Springer-Verlag, New York, pp. 21-69.
13) Wilson, I.A., Skehel, J.J. and Wiley, D.C. (1981), Nature (London) **289**, 366-373.
14) Krug, R.M., Ueda, M. and Palese, P. (1975), J. Virol. **16**,790-796.
15) Palese, P., Ritchey, M.B. and Schulman, J.L. (1977), J. Virol. **21**, 1187-1195.
16) Scholtissek, C. and Bowles, A.L. (1975), Virology **67**, 576-587.
17) Ulmanen, I., Broni, B.A. and Krug, R.M. (1981), Proc. Natl. Acad. Sci. USA **78**, 7355-7359.
18) Ulmanen, I., Broni, B.A. and Krug, R.M. (1983), J. Virol. **45**, 27-35.
19) Blaas, D., Patzelt, E., and Kuechler, E. (1982), Virology **116**, 339-348.
20) Blaas, D., Patzelt, E. and Kuechler, E. (1982), Nucl. Acids Res. **10**, 4803-4812.
21) Penn, C.R., Blaas, D., Kuechler, E., and Mahy, B.W.J. (1982), J. gen. Virol. **62**, 177-180.
22) Plotch, S.J., Bouloy, M., Ulmanen, I. and Krug, R.M. (1981), Cell **23**, 847-858.
23) Gething, M.-J. and Sambrook, J. (1982), Nature (London) **300**, 598-603.
24) Sekikawa, K. and Lai, C.-J. (1983), Proc. Natl. Acad. Sci. USA **80**, 3563-3567.
25) Garten, W., Bosch, F., Linder, D., Rott, R. and Klenk, H.-D. (1981), Virology **115**, 361-374.
26) Min Jou, W., Verhoeyen, M., Devos, R., Saman, E., Fang, R., Huylebroeck, D., Fiers, W., Threlfall, G., Barber, C., Carey, N. and Emtage, S. (1980), Cell **19**, 683-696.

27) Daniels, R.S., Downie, J.C., Hay, A.J., Knossow, M., Skehel, J.J., Wang, M.L. and Wiley, D.C. (1985), Cell **40**, 431-439.

28) Doms, R. W., Gething, M.-J., Henneberry, J., White, J. and Helenius, A. (1986), J. Virol. **57**, 603-613.

29) Sveda, M., Markoff, L.J. and Lai, C.-J. (1982). Cell **30**, 649-656.

30) Both, G.W. and Sleigh, M.J. (1981), J. Virol. **39**, 663-672.

31) Yewdell, J.W., Webster, R.G. and Gerhard, W. (1979), Nature (London) **279**, 246-248.

32) Compans, R.W., Content, J. and Deusberg, P.H. (1972), J. Virol. **10**, 795-800.

33) Winter, G. and Fields, S. (1981), Virology **114**, 423-428.

34) Ritchey, M.B. and Palese, P. (1977), J. Virol. **21**, 1196-1204.

35) Thierry, F. and Danos, O. (1982), Nucl. Acids Res. **10**, 2925-2937.

36) Van Wyke, K.L., Bean, W.J., Jr., and Webster, R.G. (1981), J. Virol. **39**, 313-317.

37) Petri, T. and Dimmock, N.J. (1981), J. gen. Virol. **57**, 185-190.

38) Kistner, O., Muller, H., Becht, H. and Scholtissek, C. (1985), J. gen. Virol. **66**, 465-472.

39) Zhirnov, O. and Bukrinskaya, A.G. (1984), J. gen. Virol. **65**, 1127-1134.

40) Briedis, D.J., Conti, G., Munn, E.A. and Mahy, B.W.J. (1981), Virology **111**, 154-164.

41) Davey, J., Dimmock, N.J., and Colman, A. (1985), Cell **40**, 667-675.

42) Blok, J., Air, G.M., Laver, W.G., Ward, C.W., Lilley, G.G., Woods, E.F., Roxburgh, C.M. and Inglis, A.S. (1982), Virology **119**, 109-121.

43) Bos, T.J., Davis, A.R. and Nayak, D.P. (1984), Proc. Natl. Acad. Sci. USA **81**, 2327-2331.

44) Markoff, L., Lin, B.-C., Sveda, M.M. and Lai, C.-J. (1984), Mol. Cell. Biol. **4**, 8-16.

45) Jones, L.V., Compans, R.W., Davis, A.R., Bos, T.J. and Nayak, D.P. (1985), Mol. Cell. Biol. **5**, 2181-2189.

46) Palese, P. and Schulman, J.L. (1974), Virology **57**, 227-237.

47) Webster, R.G., Brown, L.E. and Laver, W.G. (1984), Virology **135**, 30-42.

48) Schulman, J.L. and Palese, P. (1977), J. Virol. **24**, 170-176.

49) Huang, R.T.C., Dietsch, E. and Rott, R. (1985), J. gen. Virol. **66**, 295-301.

50) Kilbourne, E.D., Laver, W.G., Schulman, J.L. and Webster, R.G. (1968), J. Virol. **2**, 281-288.

51) Schulman, J.L., Khakpour, M. and Kilbourne, E.D. (1968), J. Virol. **2**, 778-786.

52) Choppin, P.W., Compans, R.W., Scheid, A., McSharry, J.J. and Lazarowitz, S.G. (1972). In: *Membrane Research*, C.F. Fox, ed., Academic Press, New York, pp. 163-179.

53) Lohmeyer, J., Talens, L.T. and Klenk, H.-D. (1979), J. gen. Virol. **42**, 73-88.

54) Lamb, R.A., Zebedee, S.L. and Richardson, C.D. (1985), Cell **40**, 627-633.

55) Zebedee, S., Richardson, C.D. and Lamb, R.A. (1985), J. Virol. **56**, 502-511.

56) Lamb, R.A. and Lai, C.-J. (1980), Cell **21**, 475-485.

57) Inglis, S.C. and Brown, C.M. (1981), Nucl. Acids Res. **9**, 2727-2740.

58) Ritchey, M.B., Palese, P. and Schulman, J.L. (1976), J. Virol. **20**, 307-313.

59) Lamb, R.A. and Choppin, P.W. (1979), Proc. Natl. Acad. Sci. USA **76**, 4908-4912.

60) Privalsky, M.L. and Penhoet, E.E. (1981), J. Biol. Chem. **256**, 5368-5376.

61) Young, J.F., Desselberger, U., Palese, P., Ferguson, B., Shatzman, A.R. and Rosenberg, M. (1983), Proc. Natl. Acad. Sci. USA **80**, 6105-6109.

62) Greenspan, D., Krystal, M., Nakada, S., Arnheiter, H., Lyles, D.S. and Palese, P. (1985), J. Virol. **54**, 833-843.

63) Robertson, J.S., Robertson, E., Roditi, I., Almond, J.W., and Inglis, S.C. (1983), Virology **126**, 391-394.

64) Wolstenholme, A.J., Barrett, T., Nichol, S.T., and Mahy, B.W.J. (1980), J. Virol. **35**, 1-7.

65) Koennecke, I., Boschek, C.B. and Scholtissek, C. (1981), Virology **110**, 16-25.

66) Skehel, J.J. and Hay, A.J. (1978), J. gen. Virol. **39**, 1-8.

67) Hay, A.J., Lomniczi, B., Bellamy, A.R. and Skehel, J.J. (1977), Virology **83**, 337-355.

68) McCauley, J.W. and Mahy, B.W.J. (1983), Biochem, J. **211**, 281-294.

69) Herz, C., Stavnezer, E. and Krug, R.M. (1981), Cell **26**, 391-400.

70) Jackson, D.A., Caton, A.J., McCready, S.J. and Cook, P.R. (1982), Nature (London) **295**, 366-368.

71) Beaton, A.R. and Krug, R.M. (1984), Proc. Natl. Acad. Sci. USA **81**, 4682-4686.

72) Barrett, T., Wolstenholme, A. and Mahy, B.W.J. (1979), Virology **98**, 211-225.

73) Enami, M., Fukuda, R. and Ishihama, A. (1985), Virology **142**, 68-77.

74) Lamb, R.A. and Choppin, P.W. (1976), Virology **74**, 504-519.

75) Inglis, S.C. and Mahy, B.W.J. (1979), Virology **95**, 154-164.

76) Lamb, R.A., Lai, C.-J., and Choppin, P.W. (1981), Proc. Natl. Acad. Sci. USA **78**, 4170-4174.

77) Lamb, R.A. and Lai, C.-J. (1982), Virology **123**, 237-256.

78) Lamb, R.A. and Lai, C.-J. (1984), Virology **135**, 139-147.

79) Inglis, S.C. and Brown, C.M. (1984), J. gen. Virol. **65**, 153-164.

80) Smith, D.B. and Inglis, S.C. (1985), EMBO J. **4**, 2313-2319.

81) Lamb, R.A. and Choppin, P.W. (1977), J. Virol. **23**, 816-819.

82) Spooner, L.L.R. and Barry, R.D. (1977), Nature (London) **268**, 650-652.

83) Mark, G.E., Taylor, J.M., Broni, B.A. and Krug, R.M. (1979), J. Virol. **29**, 744-752.

84) Kawakami, K. and Ishihama, A. (1983), J. Biochem. **93**, 989-996.

85) Kawakami, K., Mizumoto, K. and Ishihama, A. (1983), Nucl. Acids Res. **11**, 3637-3649.

86) Shaw, M.W. and Lamb, R.A. (1984), Virus Res. **1**, 455-467.

87) Krug, R.M., Broni, B.A., LaFiandra, A.J., Morgan, M.A. and Shatkin, A.J. (1980), Proc. Natl. Acad. Sci. USA **77**, 5874-5878.

88) Horisberger, M.A. (1982), Virology **120**, 279-286.

89) Romanos, M.A. and Hay, A.J. (1984), Virology **132**, 110-117.

90) Robertson, J.S., Schubert, M. and Lazzarini, R.A. (1981), J. Virol. **38**, 157-163.

91) Hay, A.J., Skehel, J.J. and McCauley, J. (1982), Virology **116**, 517-522.

92) Del Rio, L., Martinez, C., Domingo, E. and Ortin, J. (1985), EMBO J. **4**, 243-247.

93) Smith, G.L. and Hay, A.J. (1982), Virology **118**, 96-108.

94) Mahy, B.W.J., Barrett, T., Nichol, S.T., Penn, C.R. and Wolstenholme, A.J. (1981). In: *The Replication of Negative Strand Viruses*, D.H.L. Bishop and R.W. Compans, eds., Elsevier, New York, pp. 379-387.

95) Horisberger, M.A., and de Staritzky, K. (1985), FEMS Microbiol. Lett. **29**, 207-210.

96) Ransohoff, R.M., Maroney, P.A., Nayak, D.P., Chambers, T.M. and Nilsen, T.W. (1985), J. Virol. **56**, 1049-1052.

97) Krug, R.M., Shaw, M., Broni, B.A., Shapiro, G. and Haller, O. (1985), J. Virol. **56**, 201-206.

98) Horisberger, M.A., Staeheli, P. and Haller, O. (1983), Proc. Natl. Acad. Sci. USA **80**, 1910-1914.

99) Horisberger, M.A. and Hochkeppel, H.K. (1985), J. Biol. Chem. **260**, 1730-1733.

100) Dreiding, P., Staeheli, P. and Haller, O. (1985), Virology **140**, 192-196.

101) Staeheli, P., Haller, O., Boll, W., Lindenmann, J. and Weissmann, C. (1986), Cell **44**, 147-158.

102) Meyer, T. and Horisberger, M.A. (1984), J. Virol. **49**, 709-716.

103) Staeheli, P. and Haller, O. (1985), Mol. Cell. Biol. **5**, 2150-2153.

CHAPTER 14

THE MOLECULAR BIOLOGY OF ARENAVIRUSES

DAVID H.L. BISHOP
NERC Institute of Virology, Oxford, U.K

INTRODUCTION

Five families of viruses are recognized as negative-stranded RNA viruses. These are the Arenaviridae, Bunyaviridae, Orthomyxoviridae, Paramyxoviridae and Rhabdoviridae. All have a lipid envelope, an external fringe of glycoprotein and internal components consisting of one or more species of single-stranded RNA in addition to structural proteins and RNA polymerase components. The RNA species of negative-stranded viruses are not infectious *per se* since the infection process of each of these viruses requires that the virion polymerase transcribes the viral RNA into complementary mRNA species before the infection can proceed. Therefore removal of viral protein from the RNA of these viruses eliminates the required enzymes and renders the RNA non-infectious.

Two families of negative-stranded viruses have single species of genomic RNA (Rhabdoviridae, Paramyxoviridae), the others have either seven, or eight (Orthomyxoviridae), or three (Bunyaviridae), or two viral RNA species (Arenaviridae). In this article the evidence will be reviewed that show that arenaviruses have some proteins coded in sub-genomic, *viral sense* mRNA species and other proteins coded in sub-genomic, *viral-complementary* mRNA sequences (i.e., the viruses have genomes with an ambisense coding strategy). This unique feature is discussed in relation to the implication it has on the arenavirus intracellular infection process and how such a coding arrangement may have evolved.

1. MOLECULAR ATTRIBUTES OF THE ARENAVIRIDAE

A. The Members of the Arenaviridae

Arenaviruses are grouped into the Old World species (lymphocytic choriomeningitis (LCM), Lassa, Mobala, Mopeia) and New World species (Amapari, Flexal, Junin, Latino, Machupo, Parana, Pichinde, Tacaribe, Tamiami), although LCM virus (the prototype of the family) has been found in Africa, the Americas, Europe and Asia. An alternate designation for the New World arenaviruses that is based on serological considerations is the Tacaribe complex. Recent information indicates that at the nucleotide and protein sequence level members of both groups are closely related. All of the viruses have been isolated from rodents. The exception is Tacaribe virus which was isolated from fruit-eating bats. Several of the viruses (LCM, Junin, Lassa and Machupo viruses) have been recovered from naturally acquired human infections.

In terms of human disease, Lassa virus is the etiologic agent of Lassa fever in West Africa, Junin virus causes Argentine hemorrhagic fever, and Machupo virus causes Bolivian hemorrhagic fever. LCM virus also infects humans. It usually produces a mild, influenza-like, infection that on occasion may lead to aseptic meningitis. Antibodies to these four viruses and to Flexal virus have been detected in human sera. Laboratory acquired infections to LCM, Lassa, Junin, Machupo, Flexal, Pichinde and Tacaribe viruses have been reported.

Reviews concerning the biological aspects of arenavirus replication in animals and the immunological responses to arenavirus infections in relation to the host species, infection route and virus type are available from other sources (1,2). Other than providing contextual information, this review primarily concerns the genetic and biochemical information that is available on arenaviruses with regard to the RNA coding strategy. A comprehensive earlier review on the structural features of arenaviruses, including information on the biophysical and antigenic properties, purification and replication processes has been published by Pedersen (2). The extensive literature on the biology of LCM and an able discussion of the subject matter can be found in the book edited by Lehman-Grube (3).

B. Arenavirus Structural Components

Electron microscopic analyses of intact arenaviruses, as well as thin sections of arenavirus preparations and infected tissues, have shown that the viruses usually contain ribosomes, and that virions are frequently spherical but often pleomorphic with sizes ranging from 50-300 nm, although their average size is 110-130 nm (1-3). A schematic arenavirus particle is shown in Fig. 1. Embedded in the lipid envelope of the virus are 5-10 nm long club-shaped surface projections of glycoprotein that appear to have a hollow axis. The projections either consist of two protein species with distinct sizes that are present in essentially equal numbers (for Pichinde, LCM, Lassa, Mopeia and Machupo viruses: G1, size: 50-72 kDa; G2, size: 34-41 kDa), or one size class of protein (for Junin, Tacaribe and Tamiami viruses: G, size: 35-44 kDa). Estimates of the order of 400 molecules of each glycoprotein species have been reported for Pichinde virus preparations. Probably the single size class of glycoprotein for Junin, Tacaribe and Tamiami viruses means that they two similarly sized proteins. The surface projections can be removed from virus particles by protease digestion leaving a spikeless particle that exhibits reduced infectivity. The G1 and G2 proteins of Pichinde and LCM viruses each have distinct amino acid sequences. They have been shown to be derived from a common glycosylated precursor (GPC, size: ca. 80 kDa). The nonglycosylated form of the

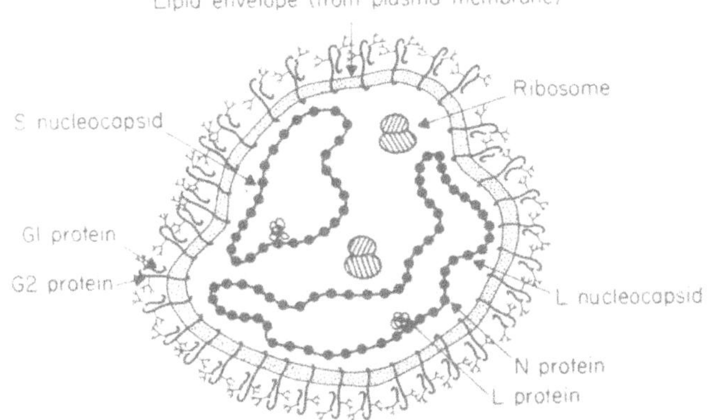

Figure 1. A schematic arenavirus particle.

LCM and Pichinde virus GPC primary gene products have been deduced from DNA sequence analyses to have sizes of 56-57 kDa (4, 5).

The internal components of arenaviruses include two other viral-coded proteins. There are minor quantities of a large protein L (size: 180-200 kDa) that is believed to be a transcriptase/replicase component, and large quantities of the nucleocapsid protein, N (size: 63-72 kDa). The latter constitutes some 60-70% of the total viral protein corresponding to some 1500 molecules of N protein per virion. It has been shown that the N protein is responsible for the antigenic cross-reactivity among the Tacaribe complex of arenaviruses. It is closely associated with the two viral RNA species, the complexes corresponding to the two viral nucleocapsids. The extended forms of the nucleocapsids are long and convoluted, 3-5 nm in diameter and without any obvious helical symmetry. Almost no information is available on the viral L protein, other than its distinct size and tryptic peptide profile. Other proteins that have described in arenavirus preparations may be alternative or derived forms of the major structural proteins. Their functions (if any) are not known. Like other negative polarity RNA viruses, RNA polymerase activities have been identified in extracts of arenavirus preparations. The L protein is a candidate for the viral RNA polymerase.

The genetic information of arenaviruses is resident in two species of RNA, designated small, S (size: 1.1×10^6 Da) and large, L (size: 2.2×10^6 Da). For Pichinde and LCM viruses it has been shown that the L and S RNA species have different sequences (as evidenced by fingerprint analyses) although their 3' ends are homologous in sequence. From cloning and sequence analyses it has been found that the 5' terminal sequences of the S segment of the Pichinde and LCM RNA are complementary to the 3' terminal sequences for approximately 20 residues (4,5). As discussed below, the L RNA is believed to code for the L protein, the S RNA for the GPC and N proteins. Whether either RNA codes for non-structural proteins is not known.

In addition to the protein, RNA and envelope components of arenaviruses, a variety of other macromolecules have been identified in arenavirus preparations.

These include host ribosomes, various RNA species and enzymes. Reviews of this subject are available elsewhere (1-3). The presence of ribosomes is a characteristic feature of arenaviruses that sets them apart from other families of viruses. With regard to the question of the origins of the ribosomes and the other minor components, it should be borne in mind that arenaviruses are pleomorphic and form virus particles at the surfaces of cells. It is perhaps not surprising that virus morphogenesis may result in the acquisition of cellular components, including bulky host ribosomes if the processes of viral morphogenesis do not include ways for their exclusion. Other pleomorphic viruses (e.g., paramyxoviruses) do not exhibit this property, possibly because they have another internal structural protein, the matrix protein. Whether the arenavirus ribosomes have any function in the processes involved in initiating an infection is not known. Leung and Rawls (6) have shown that Pichinde virus grown in cells with thermolabile ribosomes yielded viruses that were able to productively infect other cells at temperatures that were non-permissive for the ribosomes carried in the virion. It appears therefore that ribosomes are not essential for the infectivity of arenavirus particles. This does not mean however that competent ribosomes incorporated into arenavirus particles do not function upon gaining entry to a permissive cell (e.g., they may become involved in the translation of an associated mRNA or participate in the *de novo* initiation of mRNA translation). "Northern" analyses have shown that among the minor RNA species that can be identified in extracts of purified virus preparations are the two S mRNA species (unpublished data). If such species are associated with the virion ribosomes, then they may be involved with the continued or *de novo* synthesis of viral proteins after virus penetration.

C. The Infection Cycle

In the rodent species vertical transmission appears to be frequent: transuterine, transovarian as well as postpartum involving milk, saliva, or urine routes. Veneral transmission may also be involved in intraspecies infections. Interspecies transmission (e.g., to man) is thought to be caused by the acquisition of virus throught contamination, either due to rodent infestations of dwellings (e.g., Lassa), or through encounters in the field (e.g., Junin), or the laboratory. For transuterine, or neonatally acquired virus infections, rodents usually become persistently infected and develop viremia and viruria, secreting virus throughout their life. An attribute of such infections is the hypoimmune response and coexistence of circulating antibody and virus. Depending on the host and virus species, experimental infections of adult rodents are either inapparent (e.g., Tacaribe virus), or may be lethal (e.g., LCM virus). In neonates, experimental infection with Tacaribe virus is lethal, for LCM viruses a persistent infection is frequently induced. LCM virus infection of young rodents induces a chronic immune complex disease involving virus infection of lymphocytes and an overall immunodepression and possible autoimmune effects. Extensive reviews of this subject are available elsewhere (7). The abilities of arenaviruses to elicit short-term acute or long-term persistent infections have been studied both from the viewpoint of the virus strain, the host species and *in vitro* culture (3).

In cell culture, arenaviruses can productively infect a variety of cell lines. Some studies have indicated that the host cell nucleus is required to obtain a productive infection and also that arenaviruses are inhibited by actinomycin D or alpha-amanitin. However the intracellular processes in the replication cycle and their relationships to cellular functions are not understood in detail (2). In broad terms, the viral glycoproteins are involved in the adsorption, penetration and uncoating process, although how these results are achieved is not known. Following uncoating it is assumed that the viral polymerase synthesizes mRNA (presumably, the S segment codes for N and GPC, whereas the L segment encodes the homologous mRNA species). As described below, RNA replication must proceed before the S coded GPC mRNA can be synthesized. With the availability of the newly synthesized gene products and replicated viral RNA, viral morphogenesis takes place at the cell surface. The molecular steps involved of these processes have yet to be defined.

D. The Genetic Attributes of Arenaviruses

Intraspecies virus recombination involving the reassortment of the two viral RNA species has been demonstrated using temperature-sensitive (*ts*) mutants of two strains of Pichinde virus. Reassortment has been documented using wild-type or *ts* mutants of LCM virus strains (8, e.g., the WE and ARM varieties). No reassortment has yet been detected between representatives of Pichinde and LCM viruses. Analyses of recombinant arenaviruses have shown that the S RNA species codes for the N protein and GPC (and therefore the two viral glycoproteins, G1 and G2) and that the L RNA codes for the L protein. Using LCM virus strains that exhibit different phenotypes, it has been shown that the S RNA codes for functions that result in growth hormone induced disease and other virulence markers.

Genetically diploid LCM reassortants have been detected among the viruses obtained from crosses of complementing *ts* mutants representing the S RNA (i.e., N and GPC mutants). The diploid viruses readily segregate *ts* mutants upon passage, in agreement with the postulate that they contain complementing S RNA species representing both *ts* genotypes. The results of crosses that provided evidence for diploid virus formation (8) are exemplified in Fig. 2. Whether the propensity to produce polyploid viruses has a biological consequence is not known. Conceivably, it could confer genetic stability in nature by acting against the biological cloning of variants.

E. The Ambisense Coding Arrangement of the S RNA Species

In order to determine how arenaviruses code for gene products, the S RNA species of Pichinde virus and that of a viscerotropic strain of LCM virus (LCM-WE) have been cloned into DNA and sequenced (4, 5). Analyses of the 1.1×10^6 Da sequences of each S RNA have, like the earlier genetic and biochemical studies, confirmed that two gene products are coded by the arenavirus S RNA. One (the 62-63 kDa N protein) is coded in a viral-complementary sequence corresponding to the 3' half of the viral RNA. The other (the 56-57 kDa primary gene product corresponding to GPC) is coded in the 5' half of the viral RNA in a viral-sense sequence. Comparison of the gene products of Pichinde and LCM viruses indicates that the N proteins of the two viruses exhibit 51% direct amino acid sequence homology; the GPC primary gene products have only 39% sequence homology.

Using the appropriate single-stranded S DNA probes, it has been shown that extracts of Pichinde virus infected cells contain two sub-genomic S RNA species in addition to the full-length viral and viral-complementary S RNA species (4). The subgenomic RNA species are each approximately half the size of the viral S RNA. They correspond to a viral-complementary N mRNA (as demonstrated by immune precipitation of N protein from *in vitro* translation products of the mRNA) and a viral-sense subgenomic species that is deduced to be the GPC mRNA species. The inability to bind Pichinde N mRNA to oligo-dT cellulose columns (4) indicates that the N mRNA species lacks 3' polyadenylated sequences. Tryptic peptide analyses have shown that for Pichinde virus the protein order in the GPC precursor is G1 (amino half) then G2 (carboxy half).

In view of the observations that have been made concerning the coding arrangement, the arenavirus S RNA is described as having an *ambisense* strategy, to denote the fact that both viral and viral complementary sequences are used to make

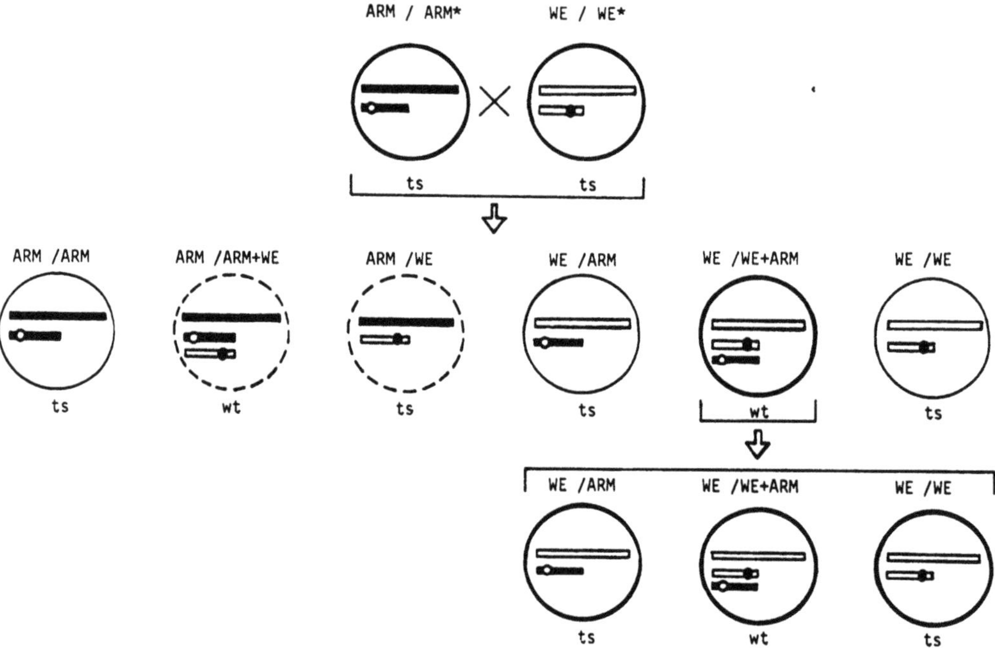

Figure 2. Crosses of LCM *ts* mutants that generate diploid reassortants. Small (S) RNA *ts* mutants of ARM and WE strains of LCM virus (having mutations in the glycoprotein, ARM left circle, or nucleoprotein, WE right circle) can be used to coinfect cells and produce reassortants. Two of the derived genotypes that are diploid with respect to the S RNA exhibit a wild-type phenotype. They segregate *ts* mutants upon subsequent passage.

gene products. Only partial sequence information has been reported for the L RNA of the WE strain of LCM virus (ca. 1000 nucleotides from the 3' end). The limited data that have been obtained suggest that there is a gene product coded in the L RNA viral-complementary sequence (presumably the L protein). "Northern" analyses have so far only identified a single L mRNA species of approximately the same size as the viral L RNA (unpublished data). It may be, therefore, that the arenavirus L RNA has a simple negative strand coding arrangement. Alternatively it may have an ambisense arrangement. Until the complete L RNA has been cloned and sequenced and the clones used to identify all the L mRNA species, the answer will not be known.

One implication of the arenavirus S RNA coding arrangement is that the GPC subgenomic mRNA species and the viral glycoproteins cannot be made in infected cell until after viral RNA replication has commenced and a replicative intermediate, full-length, viral-complementary, RNA is produced that can function as a template for GPC mRNA synthesis (Fig. 3). This is unlike the organization of the negative-stranded rhabdoviruses and paramyxoviruses which have been shown to synthesize all their (viral-complementary) mRNA species in a consecutive manner from the viral RNA. It is also unlike the negative-stranded, segmented genome, orthomyxoviruses that also only code for proteins in their viral-complementary sequences. An advantage of the arenavirus S RNA strategy is that it allows the syntheses of the two S coded mRNA species be regulated independently so that different quantities of each can be made. Another feature is that the GPC mRNA (and protein) species are not synthesized until the time that they are required (i.e., after the onset of RNA replication and before initiation of the processes of viral morphogenesis).

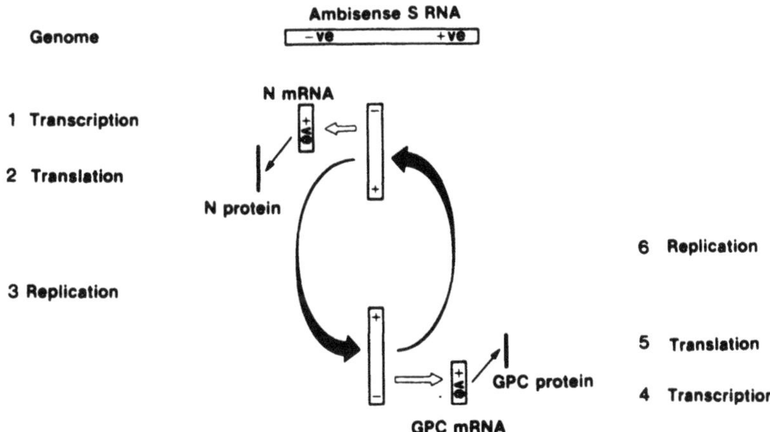

Figure 3. Coding, transcription, translation and replication strategy of the ambisense S RNA species of arenaviruses.

It is quite possible that GPC mRNA synthesis may be curtailed by competition with viral RNA synthesis if, for instance, the availability of N protein regulates the process. Whether this occurs and contributes to the establishment (and maintenance) of persistently infected cèlls *in vivo* or *in vitro* remains to be determined. Conceivably the curtailment of glycoprotein synthesis would prevent virus morphogenesis (or at least the synthesis of virus paricles coated with the homologous viral protein), but may allow viral RNA replication to proceed. It may also render the infected cell sublimal to effective recognition by the host immune procedures. While these observations are speculative, the procedures and tools for the analysis of such hypotheses are now available.

The intergenic region of the Pichinde and LCM viral S RNA species has a unique feature, that of an inverted complementary sequence that may be arranged into a hairpin configuration (Fig. 4). Although there is no information on the trascription initiation and termination processes of mRNA synthesis from the viral S (or L) RNA species, in unpublished experiments using oligonucleotides representing viral and viral-complementary sequences of the intergenic region of Pichinde S RNA, we have found that transcription of both N and GPC mRNA species terminates near the top of the 18-21 base-pair intergenic hairpin. How transcription termination in the intergenic region is effected is not known. In agreement with the absence of binding to oligodT-cellulose, there are no polyuridylate tracts in the intergenic region of the viral S RNA (or viral-complementary S RNA) that could serve as templates for polyadenylation of the 3' ends of the S mRNA species.

F. How an Ambisense Genome May Arise?

Concerning the question of how arenaviruses with an ambisense coding strategy may have arisen, unless one invokes an origin from a DNA source in which the arrangement of proteins coded on opposite strands of nucleic acid has been maintained, the simplest explanation is that a chimeric RNA was derived at some stage of arenavirus evolution. Such a chimeric RNA could have been formed during the processes of RNA replication and represent a consolidation of genetic information (i.e., a virus with three RNAs each coding for a single gene product giving rise to an arenavirus with a consolidated genome and, subsequently, only two RNAs by the exclusion of the redundant third RNA species). Such consolidation of genetic information could occur by a viral replicase copying the coding strand of one RNA species and, instead of terminating, continuing RNA synthesis on the non-coding strand of another RNA. This would result in a chimeric RNA molecule composed of two genes coded on opposite strands of the RNA. Together with the subsequent loss of the redundant RNA species, a virus would be generated in which all the original genetic information had been conserved. Of course, the reverse situation may have occurred, i.e., the formation of a virus with three RNA species from a virus having two, one of which originally had an ambisense coding arrangement. Whatever the origins, the ambisense coding arrangements that have been observed for arenaviruses open yet another dimension to the way in which viruses replicate in cells.

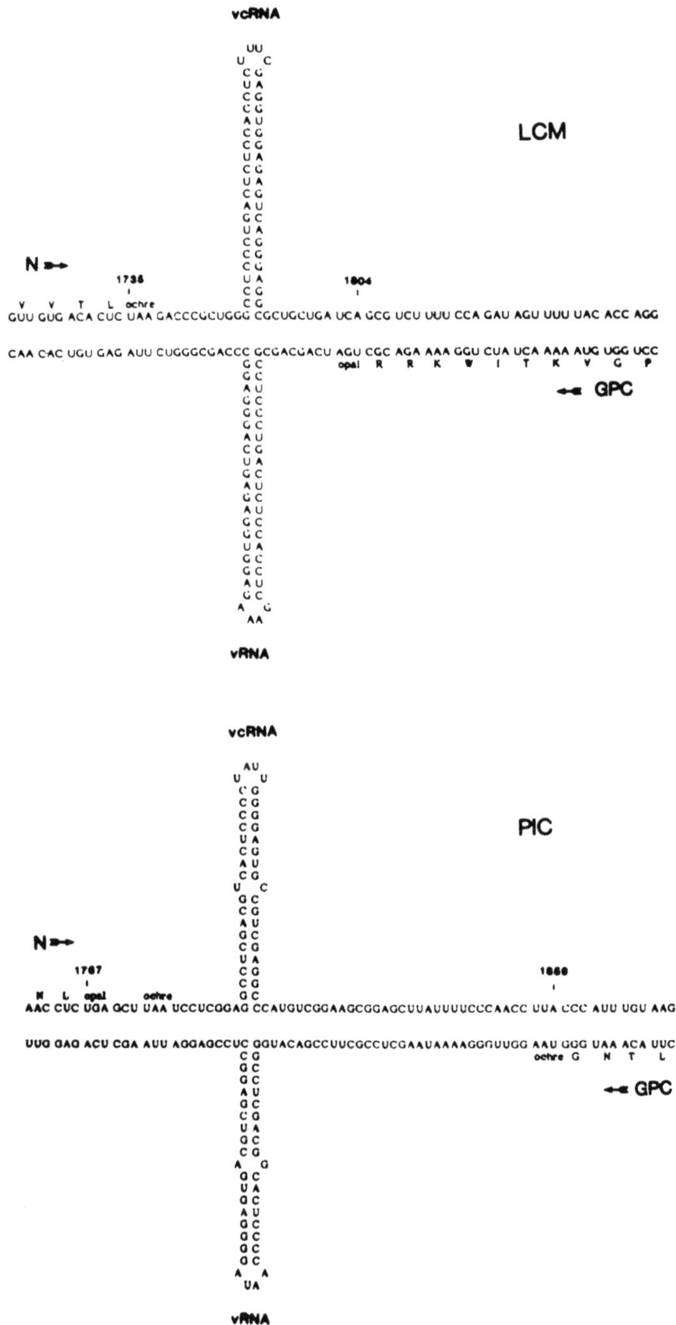

Figure 4. The intergenic region of the S RNA of Pichinde and LCM arenaviruses.

2. REFERENCES

1) Rawls, W.E. and Leung, W.C. (1979). Arenaviruses. In: *Comprehensive Virology*, (H. Fraenkel-Conrat and R.R. Wagner, eds.), Vol. 14, pp. 157-192 Plenum Press, New York.

2) Pedersen, I.R. (1979). Structural components and replication of arenaviruses. *Advs. Virus Res.*,**24**, 277-330.

3) Lehmann-Grube, F., (1973) ed. *Lymphocytic choriomeningitis virus and other arenaviruses* , Springer-Verlag, Berlin and New York. 4) Auperin, D.D., Romanowski, V., Galinski, M. and Bishop, D.H.L. (1984). Novel coding strategy of arenaviruses-An ambisense viral S RNA". J. Virol. **52**, 897-904.

5) Romanowski, V., Matsuura, Y. and Bishop, D.H.L. (1985). Complete sequence of the S RNA of lymphocytic choriomeningitis virus (WE strain) compared to that of Pichinde virus. Virus Res., **3**, 101-114.

6) Leung, W.C. and Rawls, W.E. (1977). Virion associated ribosomes are not required for the replication of Pichinde virus. Virology, **81**, 174-176.

7) Johnson, K.M. (1985). Arenaviruses. In: **Virology**, B.N. Fields, *et al.* eds., pp. 1033-1053. Raven Press, New York. 8) Romanowski, V. and Bishop, D.H.L. (1983). The formation of arenaviruses that are genetically diploid. Virology, **126**, 87-95.

CHAPTER 15

THE REOVIRUS FAMILY AT THE MOLECULAR LEVEL

MALCOLM A. McCRAE

Department of Biological Sciences, University of Warwick, Coventry, CV 4 7AL, England

INTRODUCTION

In the late 1950's a wide range of new viruses were being isolated from the respiratory and enteric tracts of humans, This work was fueled in the main by the studies on poliomyelitis and amongst the new isolates was a virus designated ECHO 10 which was originally placed in the picornavirus family. However with a particle size of approximately 70 nm this virus was far larger than any other number of the *picornaviridae* and primarily as a result of this it was designated as the prototype of a new virus group which Sabin (1) proposed should be called Respiratory Enteric Orphan or REO viruses. This virus was the original serotype 1 of the mammalian reoviruses.

Additional work particularly by Gomatos and his collaborators (2) in the early 1960's quickly established further differences between reoviruses and other virus groups. A crucial observation, which first suggested the true nature of the viral genome, was the finding of cytoplasmic inclusions in infected cells that gave a pale green fluorescence on staining with acridine orange which is characteristic of double stranded nucleic acid. In the ten years succeeding these original studies on the mammalian reoviruses, numerous other viruses which also contain double-stranded RNA as their genome were isolated from a wide range of species across both the animal and plant kingdoms. The majority of these viruses have now been classified into a single virus family, the *Reoviridae*, which is at present composed of six genera (Table 1).

The only genus that contains members of major medical importance is the *Rotavirus* genus. In the last ten years rotaviruses have become recognnized as the predominant etiological agents of acute viral gastroenteritis in human infants (3). As such they are responsible for an estimated 2-5 million human deaths per annum,

Table 1. - The Reoviridae

Genus	Host Range	No of genomic Segments	Examples
Orthoreovirus	Mammals and Birds	10	3 serotypes of mammalian reovirus 5 serotypes of avian reovirus
Orbivirus	Mammals and Insects	10 and 12	Bluetongue African Horsesickness Colorado Tick Fever
Rotavirus	Mammals and Birds	11	Bovine rotavirus Sa11
Cypovirus	Insects	10	Cytoplasmic polyhedrosis virus
Phytoreovirus	Plants and Insects	12	Wound tumour virus Rice dwarf virus
Fijivirus	Plants and Insects	10	Fiji disease virus Maize rough dwarf virus

probably more than all the other human infections combined (4). These deaths are confined almost exclusively to children under the age of five and result from the rapid and severe body dehydration caused by the diarrhoea, consequently they are localised to areas of the third world where primary health care facilities are poor. However even in countries which have well organised and effective primary health care rotaviruses are still the cause of significant infant morbidity.

The orbivirus Colorado Tick Fever virus is also of some medical importance in areas where its transmitting ticks are found. It causes an acute febrile illness in man (5) although this is only rarely fatal.

In veterinary medicine the rotaviruses are again of major importance. Their worldwide geographical distribution and occurence in all the major species of domestic livestock means that they cause significant economic losses to agriculture globally (6). A number of orbiviruses, particularly bluetongue virus and african horsesickness virus are also of considerable veterinary importance. Bluetongue virus causes a degenerative and sometimes fatal disease of sheep and is of major economic importance in the southern half of Africa and around the Caribbean basin (7). African horsesickness virus causes an acute febrile illnes in horses, mules and donkeys that is often fatal (8), and consequently it is of some importance particularly to the bloodstock industry in areas where it occurs.

Individual members of the three *reoviridae* genera infecting insects or plants are of some agricultural importance. Thus the cytoplasmic virus (cypovirus) of *Bombyx mori* is responsible for considerable economic loss to the Japanese sericulture industry (9). Fiji Disease Virus (10) and rice ragged stunt virus (11) causes quite severe losses in their respective hosts of sugarcane and rice.

The members of the sixth genus, the *Orthoreoviruses* are of limited pathological importance. Some avian reovirus isolates have been associated with respiratory and enteric illness (12) and they have also been isolated from birds with tenosynovitis/arthritis (13). The mammalian reoviruses are clearly "viruses in search of a disease" since despite their widespread occurence in man and their detailed biochemical characterisation, with the exception of a possible association with the rare neonatal and pediatric liver disease biliary atresia, (14) they have not been associated with any infection of medical or veterinary importance.

A. General Characteristics of Reoviridae

The criterion normally taken as being synonimous with the *reoviridae* namely the possession of a genome composed of d-s RNA is in fact not unique to this virus family. A variety of other infectious agents including infectious pancreatic necrosis (IPN) virus of fish (15), infectious bursal disease virus (IBDV) of chickens and turkeys (16), and the phage Φ6 (17) and yeast killer factor particles all contain d-s RNA. However in these cases the number of distinct segments making up the viral genome is in all cases three or less. In the *reoviridae* there are a minimum of 10 discrete segments of d-s RNA making up the genome (Table 1), These RNAs range in size from approximately 500 base pairs up to a maximum of about 4000 base pairs with a total genomic molecular weight range from about 12-15 $\times$ 10^6. With one well characterised exception (18) all genomic segments of the *reoviridae* analysed to date are monocistronic.

Virion symmetry

The virion in the *reoviridae* is in all cases iscohedral or quasi-spherical and about 70-80 nm in diameter with three major types of structural pattern having been discerned.

In the first, which is exemplified by the *orthoreoviruses* and *phytoreoviruses* there are two very distinct capsid shells. Each shell has its own regular capsomeric structure although defining the exact geometrical parameters of the outer shell has proved difficult due to the sharing of protein sub-units between capsomers. By contrast the inner capsid or core has twelve well defined and prominent projections or spikes which appear to be situated on the vertices of a regular icosahedron (19).

In the second particle pattern seen in the *orbiviruses* and *rotaviruses*, there are again two concentric capsid shells but in this case the outer one is smoother and seemingly more detached from the inner shell. The underlying inner capsid or core again appears to be a regular icosahedron.

The third and final pattern is that seen in the *cypoviruses* and *Fijiviruses*. Here the particles seem to lack the outer shell completely, resembling the orthoreovirus core structure with 12 large projections or spikes. However given the susceptibility of the outer shell (when present) to removal by a wide variety of chemical and physical agents then some caution is required in concluding that at no stage is there a second outer capsid shell in this third type of particle.

Enzymic component

All members of this virus family share two additional characteristics. First, since uninfected cells do not contain enzymes that can transcribe d-s RNA into mRNA then such enzymes must be carried by the infecting virion, and consequently the extracted genomic RNA of these viruses is non-infectious. In fact *reoviridae* particles carry both the transcriptive enzymes for synthesising the m-RNA chains (20, 21) and the enzymes necessary for elaborating the 5' terminal methylated cap structure found on mRNA, the phenomenon of 5' capping of eucaryotic mRNAs having been first described for cytoplasmic polyhedrosis virus (22). An unusual feature of the mRNAs of the virus family is that they represent one of the few well characterised eukaryote mRNAs in which *there is no 3' polyadenylation*.

The final diagnostic feature of this virus family is that its members do not completely uncoat the parental viral genome following infection. When it is present the outer virion shell is removed following infection, but the parental input viral genomes remain enclosed within the inner capsid or core structure throughout the infectious cycle.

B. General Properties of Mammalian Reoviruses: The Model System

The mammalian reoviruses of which there are three distinct serotypes are all able to grow to high titre in a wide range of mammalian cell lines (23). This simple technical advantage coupled with their apparent lack of pathogenicity in man has meant that these viruses and in particular the Dearing strain of reovirus serotype 3 have become the molecular model for the whole *reoviridae* family. Thus many of the detailed molecular mechanisms involved in *reoviridae* replication have to-date only been studied in reovirus type 3 and the assumption made that with the basic mechanisms discerned will operate in the other genera of the family. In the absence of additional detailed information this is a reasonable approach which also simplifies the task of describing the replication cycle by allowing one to focus on type 3 mammalian reovirus as the model. However such an approach does have obvious disadvantages and hopefully these will ensure that work will continue in the future on the type members of the other five genera to see to what extent the detailed mechanisms unravelled for reovirus type 3 can be applied to the rest of this quite diverse virus family.

1. THE REOVIRUS VIRIONS

A. Morphology

The reovirus particle has two very distinct capsid shells, each is composed exclusively of protein arranged into discrete morphological sub-units or capsomers (24) with no lipid containing envelope being present. Accurate measurement gives diameters for the inner and outer capsid shells of 76 nm and 52 nm respectively with the diameter of the central cavity being 38 nm (19). These electron microscope values are somewhat smaller than the "hydrodynamic" values of 96 and 70 nm obtained for the outer and inner shell sizes by measuring diffusion co-efficients of intact virions and cores (25). This indicates that considerable shrinkage of the particles must occur during the fixation and dehydration procedures used in electron microscopic preparation.

Despite the obvious presence of capsomers in both the outer and inner capsid shells, to-date, it has not proved it possible to definitively determine their exact spatial relationship.

The first results were interpreted as showing that the outer shell was composed of 80 hexagonal and 12 pentagonal hollow columnar or prismatic capsomers arranged according to a 5: 3: 2 icosahedral symmetry (26).

An alternative interpretation of essentially the same data produced an outer shell with 92 surface holes, 12 of them surrounded by 5 structural subunits and the rest by six to give a sub-unit total of 180 (27).

Later data (19) indicated that the outer capsid shell appeared to have 20 peripheral capsomers and not the 18 demanded by either of the earlier interpretations. This later study (19) showed the capsomers as either cylinders or truncated pyramids, 9 nm long and separated from each other by 1 nm. This picture was not consistent with work done on particles stored in water (28) which supported the second (27) of the earlier interpretations.

Palmer and Martin (29) re-examined the virus using negative staining and rotational image enhancement. This indicated that the capsid was composed of large capsomers, 18 nm in diameter. These capsomers were mainly hexagonal units composed of six wedge-shaped sub-units and these in turn were made up of three smaller sub-units with the architecture of a T = 9 icosahedron (29). These workers also noted that the number of peripheral capsomers discernable, depended on the orientation of the particle, if viewed along the five fold axis of symmetry there were 20 large peripheral capsomers whereas when seen along the two or three fold axis there were 36 smaller periheral sub-units. The symmetrical axis of viewing also influenced whether a five-fold or six-fold clustering of morphological units was seen (29). Thus our understanding of the exact symmetry of the outer capsid shell is still somewhat confused, due mainly to the sharing of morphological sub-units which is a prominent and apparently unique feature of the *reoviridae*.

The spikes

The removal of the outer capsid shell which can be achieved using a variety of physical and chemical treatments reveals the very stable core structure whose most obvious morphological feature is the 12 projections or spikes. These spikes which are arranged on the twelve 5-fold vertices of an icosahedron, are about 10 nm in diameter. They have a central channel which is about 5 nm wide and they project about 5.5 nm beyond the surface of the core (19).

There is serological evidence which suggests that these spikes actually penetrate through the outer capsid shell, when it is present, to the surface of the intact virion. Thus antibodies to the λ2 protein which makes up the spikes (30) can react with intact virions to both neutralise their infectivity and also block virus directed haemagglutination of red blood cells (31). The rest of the core shell is composed of morphological sub-units of 4 nm in diameter, that is smaller than those of the outer shell (19). The precise symmetry of these sub-units has not been analysed to-date.

Top component

A by-product of the growth of reovirus in tissue culture is the release from infected cells of empty virions containing no nucleic acid. The yield of these particles, termed top component because of their lower buoyant density, is generally less than 10% of that of intact virions although it can vary quite widely. The morphology of these particles appears identical to that of normal virus particles except that they are penetrated by phosphotungstic acid stain (32).

B. Protein Constitution

The protein composition of the reovirus particle was first examined in the late 1960's on polyacrylamide gels using the phosphate continuous buffer system (32, 33). This revealed seven structural polypeptides divided into three clones termed λ, μ and σ encoded by the L (large), M (medium) and S (small) size classes of gene respectively (Table 2). Early work also established that the major capsid polypeptide, μ1c was derived by proteolytic cleavage of an amino terminal fragment from the primary gene product μ1 (34). Later work (35) using the higher resolution discontinuous buffer system (36) allowed the identification of two additional minor structural proteins λ3 and μ2 which had co-migrated with other structural proteins on the earlier gels.

Of the 9 structural polypeptides clearly identified to-date only three have been definitively localised to the outer of the two capsid shells namely, μ1c, σ1 and σ3 (Table 2).

The location of the minor structural protein μ2 is less clear, it is normally listed as a component of the virus core, based on the fact that it can still be detected in virion core preparations. However, when virus cores are prepared by the standard proteolytic digestion method (37) approximately 80% of the μ2 present in whole virions is lost (McCrae, unpublished observation) making it difficult to be confident as to its precise location within the virion. The remaining five structural proteins λ1-λ3, μ1 and σ2 are all clearly constitutents of the virion core.

Proteolytic cleavage

The virion proteins undergo a number of post-translational modifications whose significance is not clear. The best characterised of these is the proteolytic cleavage of approximately eight thousand daltons of protein from the amino terminus of the primary gene product μ1 to generate the most abundant virion protein μ1c (34). The significance of the fact that both precursor and product are constituents of the virion is obscure. In the infected cell approximately 80% of the μ1/μ1c is complexed to the other major outer shell protein σ3 (38) and that in this complex about 95% of the μ1 has been cleaved to μ1c which agrees with the ratios of the two proteins in purified virions.

From this it is generally assumed that the apparent encapsidation of μ1 simply reflects an inefficiency in the μ1 to μ1c cleavage reaction, although at present there

Table 2. – Summary of Available Information on Reovirus Proteins

Protein Species	Encoding gene	Approximate Molecular Wt.	Approximate Percentage of Total virion protein	Location within Virion	Post-translational Modification
Structural					
λ1	L3	155,000	15	Core	No
λ2	L2	140,000	11	Core	Yes, proteolytic cleavage
λ3	L1	135,000	<2	Core	No
μ1	M2	76,000	2	Core	Yes, cleavage to μlc
μ1c	M2	68,000	35	Outer Shell	Yes, glycosylation? phosporylation? & ADP ribosylation
μ2	M1	70,000	<2	Core/Outer Shell	No
σ1	S1	42,000	1	Outer Shell	No
σ2	S2	38,000	7	Core	No
σ3	S4	34,000	28	Outer Shell	Yes ADP ribosylation
Non-Structural					
μNS	M3	75,000	—	—	Yes, proteolytic cleavage
σNS	S3	36,000	—	—	No
p14	S1	14,000	—	—	No

is no data to exclude completely separate functional roles for these two proteins within the virion.

Post-translational modifications

Early work indicated that μ1c was phosphorylated (39) at one or more serine residues and that 2-5% of the molecules were glycosylated (40) with a tri- or tetra-saccharide linked O-glycosidically to a serine or threonine residue. However, the interpretation of these observations has come into question with the finding the μ1c is both adenylated by the addition of a 7-11 residue chain of AMP and ADP-ribosylated (41, 42). The functional significance of these post-translational additions whatever they are finally resolved to be has not been studied to-date.

The λ2 protein is also subjected to post-translational processing with a small percentage of it (<10%) being proteolytically cleaved to remove approximately 15,000 daltons worth of protein converting it into λ2C (38). This cleavage product has only been clearly detected in infected cell extracts and it remains an open question as to whether or not it is also a virion structural protein.

C. Nucleic Acid Composition

The genome of the virus is enclosed within the inner of the two capsid shells and consists of ten discrete segments of double stranded (d-s) RNA falling into 3 size classes, L, M and S. The size classes have molecular weights in the range 2.4-2.6 $\times$ 10^6 for the three L-species, 1.2-1.4 $\times$ 10^6 for the three M's and 0.6-0.8 $\times$ 10^6 for the four S RNAs (43), giving them lengths of about 3700-3900, 2100-2300 and 1100-1400 base pairs respectively (44) and an overall genome size of approximately 15 $\times$ 10^6 or about 23000 base pairs. The RNA segments are therefore between 250 nm and 1000 nm in length compared to a core diameter of approximately 50 nm. Consequently it would be expected that the genome is densely packed within the core and low angle X-ray diffraction studies indicate that this is the case with the RNA being in a well-ordered para-crystalline packaging format, with adjacent RNA helices lying next to each other (45).

The composite RNA complement

Particle/p.f.u. ratios approaching unity can be obtained for reovirus preparations which strongly suggest that each virus particle carries the complete complement of 10 RNA segments. This interpretation would be in agreement with the fact that extraction of the nucleic acid from a virus preparation yields equimolar amounts of the ten RNA segments. However, this does pose the question, what mechanism is used to ensure that each virion receives the correct complement of RNA molecule ?

Two electron microscopic studies (46, 47) have provided evidence that this problem may be solved by physically linking the RNA segments either through RNA-protein or direct RNA-RNA interaction before insertion into the particle, since RNA molecules larger than the maximum segment length were seen at low frequency on extracting virion RNA. By contrast the biochemical evidence overwhelmingly suggests that within the virion the RNA segments are physically separate entities. Thus Millward and Graham (48) showed that even before extraction there were twenty 3' termini in the RNA within each virion and Banerjee and Shatkin (49) found that all ten d-s RNA segments in the virion have ppGp at one of their 5' termini. A non-linked arrangement for the genomic RNAs would also be more consistent with the results of *in vitro* transcription assays. These show that the frequency of transcription of the individual genes is inversely proportional to their size and this is most readily accounted for by postulating that each gene is transcribed independently of all others.

The single-stranded oligonucleotides

In addition to the ten protein coding genes, reovirus particles carry about 25% of their total RNA content in the form of short (2-20 nucleotide) single stranded oligonucleotides (50, 51). These oligonucleotides fall broadly into two classes and since they have a triphosphate grouping at their 5' ends then they are almost certainly not breakdown products. Approximately one third of the total is composed solely of adenine residues range from 2-20 nucleotides in length. The remainder form an ordered series (52, 53) ranging from 2-9 nucleotides in length. About 10% of this second group are guanylated at their 5' termini (54) but not methylated. Since the sequence of this nested set of oligonucleotides closely resembles that found at the 5' end of viral mRNAs it is assumed that they represent abortive transcription products which become sealed into the virion during the last stages of virion maturation. Immediately following their recognition several hypotheses were put forward giving them a role in the infectious process, however Carter *et al.*, (55) were able to show that under special conditions, cores which lack them are still infectious.

D. Virion-associated Enzymes

Reoviruses are members of Class III viruses according to the classification scheme proposed by Baltimore (56) in which viruses are divided into one of six groups or classes according to the nature of their genetic material and what is done to it in order to convert it into mRNA which is viewed as the pivotal molecule in virus replication (see chapter 1 of this volume).

Obviously in the case of reoviruses the first event following infection must be the transcription of the genomic d-s RNA to generate plus sense single-stranded RNA copies which are then further modified by capping and methylation to yield functional mRNAs.

The discovery by Kates and McAuslan (57) of a DNA dependent RNA polymerase within purified vaccina virions stimulated the search for synthetic reaction enzymes in other virus systems. Two groups (58) working independently discovered that reovirions also carry an enzyme capable of transcribing the viral genome into mRNA. Shortly following this nucleoside triphosphate phosphohydrolase (59, 60) was demonstrated in purified virions, and sometime later Furuichi and Shatkin and their collaborators were able to demonstrate that the virus particle carries the ***guanylyl-transferase*** and ***methylase*** activities (61, 62) that elaborate the 5' terminal cap structure found on eukaryote mRNAs. These activities were all demonstrated initially in virus core preparations and they operate in a concerted fashion to generate the 10 capped species of mRNA which are extruded through the 12 surface spikes on cores (63).

Location of the enzymes

All of the virion associated enzymic activities are lost when either whole virions or cores are disrupted in attempts to purified the enzymatically active proteins and therefore the nature and exact location of the active centres of the different enzymes have not been resolved. The current working hypothesis is that the enzymes are components of the core shell and that the d-s RNAs move past the catalytic sites of these enzymes located at the base of the spikes such that the d-s RNA template remains within the core which the growing transcript is extruded through the hollow centre of the core spikes which are composed predominantly of λ2. Consistent with such a hypothesis is the observation that when cores transcribe d-s RNA in the presence of labelled pyridoxyl phosphate, which is hypothesised to react with the reactive centre of the transcriptase enzymes, then both λ1 and λ2 become labelled (64) and the four capping enzymes are inhibited. An alternative location for the catalytic sites of the transcriptase on λ3 is suggested by the work of Drayna and Fields (65) who used inter-typic recombinants to show that the pH optimum of the transcriptase is specified by gene L1 which encodes λ3 (see Table 2).

This relatively simple picture in which the transcription complex was wholly core associated has been complicated by two sets of data. First Yamakawa *et al.*, (66) showed that whole virions also possess some RNA synthetic capacity which in contrast to that seen in cores, synthesises di, tri and tetranucleotide oligonucleotides which may be capped and methylated (67) and not complete mRNA molecules. Secondly Furuichi and Shatkin (68) showed that cores but not whole virus particles are able to cap and methylate exogenously added mRNA molecules. At present nothing is known about the changes in structure that are induced during virus core formation, allowing them to produce complete transcripts as opposed to just oligonucleotides. Neither is the significance of the exogenous template capping experiments to the location of these activities within the core clear.

2. BIOLOGICAL FUNCTIONS OF REOVIRUS-CODED PEPTIDES

A. Structural Proteins

To-date the most rewarding approach in attempting to define the functions of the various proteins in the overall process of viral replication and pathogenesis has been the genetic one taken by Fields and his collaborators. This general area has been the subject of a number of excellent reviews in recent years (69-71) and will therefore not be discussed in great detail here, however the basic approach adopted will be outlined and the results achieved summarised.

Gene shuffling and the genetic approach

The mammalian reoviruses can be divided into three distinct serotypes and while the majority of biochemical studies have focused on the Dearing isolate of serotype 3, isolates from serotypes 1 (Laing) and 2 (Jones) have been adapted to routine growth in tissue culture. These three isolates can be easily differentiated by virtue of differences in the mobilities of their genomic RNA segments on polyacrylamide gels (72). They are also able to form inter-typic recombinants at high frequency in which the parental origin of individual genomic segments can be defined by analysing genome profile of the recombinant on polyacrylamide gels. The high frequency with which recombinants are formed when a cell is co-infected with two genetically distinct viruses is due to recombination being the result of genome segment reassortment or "gene shuffling" rather than true inter-molecular recombination. This high level of recombination makes it logistically feasible to isolate inter-typic recombinants of any desired genotype.

The biological properties of the types viruses isolates from each of the three serotypes show a number of clear differences particularly between the Laing (type 1) and Dearing (type 3) isolates. Using inter-typic reassortants between these two serotypes it has been possible to localise particular biological phenotypes to the possession of a single viral gene from one or other of the two parental viruses and hence the protein that it encodes.

To-date the type of work has only produced definitive data for the three outer capsid shell proteins, μ1c, σ1 and σ3. It has however allowed specific biological functions to be ascribed to each of these proteins both in the process of extra-cellular pathogenesis, that is the steps involved in the passage of the infecting virion from outside its host to inside the target cell for replication and intracellular pathogenesis, the processes by which the virus subverts the metabolic machinery of the target cell and achieves its own replication. A summary of the information currently available for these three proteins is presented in Table III.

Function assignement

In the case of the function given to σ1 it should be remembered that the experimental results actually ascribed functions to the protein(s) encoded by a particular genomic segment and in the case of the S1 gene, in addition to σ1 it also encodes the small non-structural protein p14 or σ1S (18, 73). Therefore it remains a formal possibility that some of the functions currently ascribed to σ1, e.g: stimulation of the specific cytotoxic T-cell response are in fact localised on this small protein.

For the minor capsid protein σ1, which appears to have a number of important biological functions some additional information is available from domain mapping using monoclonal antibodies. Thus using a number monoclonals raised

Table 3. – Brief summary of the functions of the virion outer capsid proteins in the process of viral pathogenesis

Gene	Protein	Functions in extracellular pathogenesis	Functions in intracellular pathogenesis
M2	$\mu 1 \rightarrow \mu 1c$	1. Determines the sensitivity of the virion to protease digestion. 2. Generation of Suppressor T-lymphocytes following peroral inoculation i.e. – induction of immunologic tolerance. 3. Controls the capacity of the virus to grow in intestinal tissue.	1. Regulates virus growth in neuronal tissue and hence neurovirulence within a serotype.
S1	$\sigma 1$ (structural) and p14 or $\sigma 1S$ (non structural)	1. Controls cell and tissue tropism. 2. Interacts with cellular receptors. 3. Determines the specificity of the humoral and cell mediated immune responses.	1. Controls the inhibition of DNA synthesis.
S4	$\sigma 3$	1. Role in establishing persistent infections.	1. Controls the inhibition of and protein synthesis. 2. Binding to ds RNA.

N.B.: In the case of the S1 gene where two distinct protein products are encoded by the same gene, the experimental data does not exclude the possibility that some of the functions listed are carried out by p14 (see text).

against this protein three distinct domains were recognised, one is involved in type specific neutralisation of the virus by antibodies, a second in the haemagglutinating ability of the virus and the third does not at present have a function (74). Monoclonal antibodies have also been valuable in confirming the role of this protein both in recognition of cell-surface receptors and in determining cell tropism in the brain of infected mice (75).

λ proteins

The two λ proteins, λ2 and λ3 have also been given biological functions using the combined intertypic recombinant and monoclonal antibody isolation approach. Thus monoclonal antibodies against λ2 are able to interact with reovirus particles to both precipitate and give group specific neutralisation (31) showing that this virus core component must project through the outer capsid layer in the intact double shelled virion. This would be consistent with early biochemical evidence indicating that λ2 makes up the spike projections on the virion core (76). Genetic studies have shown that the L2 gene product (λ2) controls the ability of the virus to generate deletion mutants of the defective interfering type upon serial passage (77), stocks containing this type of mutant are less lytic and more able to initiate persistent infections. Therefore although it is the S4 gene product (σ3) that appears to control the initiation of persistent infections (see Table 3 and 78) it is the functioning of λ2 that determines whether a virus stock carries the requisite mutation.

In the case of λ3 genetic analyses have indicated that the protein may carry the catalytic site of the virion associated transcription since it controls the pH optimum of the enzyme.

The functions of the other proteins namely, λ1, μ2 and σ2 remains obscure. λ1 and σ2 are both virion core proteins and there is some evidence that λ1 carries the guanylyl transferase activity necessary in mRNA capping (79). However the function(s) of the other virion proteins and the various minor cleavage products that are produced from them is not known at present.

B. Non-structural Proteins

Reovirus encodes three non-structural proteins, μNS, σNS, and σ1S (p14). Two of these, namely μNS and σNS are the sole primary gene products of their respective genes M3 and S3 (35, 80) whereas the small protein variously termed p14 and σ1S is encoded by a second short open reading frame within the S1 gene (18). Little is known about the biological function(s) of any of these proteins at present.

μNS is made in large amounts in the infected cell and approximately 50% of it exists as a cleavage product μNSc from which about 5,000 daltons of the 75,000 daltons of protein have been removed (38). Some recent evidence (cited in 81) suggest that this protein complex may have a fundamental role in the early stages of virus morphogenesis since monoclonals antibodies directed against it react with the earliest immature virus particles that can be isolated from infected cells.

σNS is characterised by exibiting a high affinity for single stranded RNA, which has been used to purify it (82) and means that in infected cells it exists in complexes with viral and cellular RNAs. It can be isolated from infected cells in

13-19S particles which possessed a poly C dependent RNA polymerase activity (83). These particles were able to bind s-s RNA and this evidence (84, 85) together with recent monoclonal antibody studies (cited in 81) suggest that this protein may be involved in assembling together the correct complement of 10 species of viral mRNA that forms a first step in viral morphogenesis.

The small protein p14 or σ1S has only recently been defined and there has been insufficient time to carry out definitive studies on its function. However it is very basic and could therefore interact with nucleic acids and since S1 gene products are known to control the rate at which reovirus infection inhibits host DNA synthesis (86), then it is possible that σ1S is involved in this.

3. REOVIRUS REPLICATION CYCLE

A. Virus Adsorption

Virus adsorbs to cells through a specific interaction between σ1, the virion cell attachment protein, and receptors on the cell surface (87). The interaction occurs with approximately equal efficiency 4° and 37° C, whereas virus penetration is blocked at 4°C and therefore adsorbing virus at 4° C gives an excellent method for generating a synchronised infection. A wide variety of cells possess receptors which have affinity for all three of the reovirus, e.g: mouse L-cells. However in some cases (e.g: mouse and human ciliated ependymal cells) there are only receptors for one serotype (serotype 1), and hence only serotype 1 can cause ependymitis in humans and hydrocephalus in experimental mice (88). These differences in cellular affinities of various reovirus strains both provide at least a partial explanation for their differences in tissue tropism and suggests that each virus serotype will have its own unique receptor.

Considerable progress in the isolation and characterisation of the cellular receptor for serotype 3 reovirus has been made by exploiting anti-idiotype antibodies. Thus anti-receptor antibodies have been isolated (89) by generating anti-idiotype antibodies against an anti-σ1 monoclonal that blocked cell-surface binding. These anti-receptor reagents have been used to isolate a 67-68K glycoprotein from a range of cells (90, 91) which appears to show some structural relationship to the β-adrenergic receptor present on a wide variety of cells (92). Further studies are however required to confirm the relationship and unravel the molecular details of the interaction of the virus with its receptor.

B. Penetration and Uncoating

Two routes of cellular penetration have been observed for infective reovirus particles. Following their specific interaction with cell surface receptors whole virions appear to penetrate the cytoplasmic membrane by an active but non-specific phagocytic process termed viropexis in which the virions first enter phagocytic vesicles, these then move to the interior of the cell and fuse with lysosomes by 60 mins post infection (93). The alternative route of penetration is that observed for intermediate sub-viral particles or ISVP's. This virus particle can be produced *in vitro* by digestion of intact virions with chymotrypsin in the presence of sodium ions (94); it differs from intact virions in that the outer capsid proteins σ3 and σ1 are removed and μ1c has undergone cleavage to a specific product. ISVPs can be converted to cores with a concomitant activation of the virion associated transcriptase by exposure to potassium ions (94).

These ISVPs are able to directly penetrate the cytoplasmic membrane, bypassing the lysosomal pathway followed by intact virions, and undergo the transcriptase activation step in the potassium rich environment of the cytoplasm (95). Although this alternative route of entry may play a role in inter-cellular spread of virus, since particles similar to ISVPs may be produced late in the infectious cycle, it is probably not the normal pathway for virus entry.

Within the lysosome the conversion of the intact virus into the parental subviral particle or SVP occurs. This process, which is temperature dependent (93), involves the hydrolytic cleavage of the proteins of the outer capsid shell to yield a 57 nm particle which is transcriptionally active and which retains at least some of the μ1c component of the outer capsid in the form of a 65 kDa cleavage product σ (96, 97).

Despite the removal of 40-50% of the virion protein the inner capsid shell of SVP's remains intact and the genome remains completely resistent to ribonucleases even after prolonged intra-cellular incubation. This suggests that the parental SVP, which is morphologically identical to the viral cores that can be produced *in vitro* by protease digestion is the final product of uncoating. There are however some unresolved problems with this suggestion: cell fractionation and electron-microscopic studies (93) indicate that the parental SVP's remain within the lysosomes throughout the infectious cycle, but a lysosomal localisation for SVP's would not be conducive with their role in primary transcription to produce the viral mRNAs.

The two-step model

Attempts to explain this apparent paradox have been made by reference to the mechanism of virus core production observed *in vitro*. In a two-step process virions are first converted to transcriptionally inactive ISVP's by proteolysis in the presence of sodium ions and then further proteolysis in the presence of potassium ions both removes the residual fragments of μ1c to give virus cores and activates the transcriptase (95). Transposing this *in vitro* situation to that seen *in vivo*, the parental SVP would be the equivalent of the ISVP, which is capable of directly penetrating cellular membranes; if it too were able to penetrate the lysosomal membrane it would enter the potassium rich cytoplasmic environment to complete the uncoating process and begin transcribing viral mRNA. It this lysosomal escape on the SVP were relatively slow and inefficient it would be consistent with both the electron microscopic and cell fractionation data. It would also explain the curious lag in the onset of mRNA synthesis, since although uncoating begins 20-30 minutes post infection viral mRNA is not detectable until 2 hours post-infection (98). Despite the plausibility of the two step model for *in vivo* uncoating it clearly does not completely explain the mechanics of virus uncoating and transcriptase activation. For example it provides no insight into why ISVPs are transcriptional inactive whereas parental SVP's possess an activated transcriptase, nor why the transcriptase activated *in vivo* initially transcribes only four genomic segments (L1, M3, S3 and S4) but when activated *in vitro* is immediately able to transcribe all ten genomic RNA species (99).

C. *Viral Transcription*

The presence of a highly active RNA dependent RNA polymerase (transcriptase) (20, 21) and associated 5' terminal capping and methylating activities (61, 62) within the infecting virion, means that viral mRNA production can commence without the need to induce either a new transcriptase or modify host cellular enzymes.

In vivo synthesis of viral mRNA is most easily studied in cells that have been pre-treated with low levels of actinomycin D to partially inhibit cellular DNA dependent transcription, making the RNA templated viral mRNA the bulk of newly synthesised RNA (100, 101). Temporal studies on *in vivo* transcription allow its division into two stages, early and late.

Early and late transcription

Early mRNAs are transcribed from the parental SVP's, are first detected at about 2 hours after infection and reach a maximum at between 6 and 8 hrs post infection (98). By definition these early transcripts are synthesised on the input genomes and serve a dual function, being used as mRNA for viral protein synthesis and also being assembled into progeny SVP's within which they act as template for minus strand synthesis (see later) and hence progeny genome production.

The late mRNA, which represents the bulk of the total viral mRNA produced during the infectious cycle, is transcribed from these progeny SVP's and its synthesis begins at about 4-6 hrs post infection and reaches a maximum at 12 hrs post infection.

One group (102-103) have shown an apparent structural difference between early and late mRNA. In these experiments early mRNA was found to possess the normal capped and methylated 5' terminus consistent with the finding that the 5' end of the plus strand of progeny genomes is capped and methylated (104). By contrast late mRNA was found to be uncapped as the result of an apparent masking of the guanyltransferase and methyltransferase activities in progeny SVP's (102). These interesting observations have been used in conjunction with additional work on the translational preferences at different times post infection (105-106), to provide an explanation for why only early mRNA is apparently used for minus strand synthesis but they require independent confirmation.

Pre-early and early transcription

In addition to the above division in the transcription process, studies using both metabolic inhibitors and defective virions have provided for qualitative regulation within the early mRNA class. Thus following early pulse labelling experiments (99) which indicated that some mRNAs may be preferentially transcribed at the earliest times post infection, studies using the protein synthesis inhibitor cycloheximide found that a *pre-early* phase of early transcription existed in which only four of the viral genes, L1, M3, S3 and S4 were transcribed (99).

Lau *et al* (107) provided supporting evidence for the pre-early pattern of transcription by taking advantage of the fact that inhibition of protein synthesis by cycloheximide is reversible. They infected cells in the presence of cycloheximide and allowed pre-early transcripts to accumulate for 17.5 hrs, cycloheximide was then removed at the same time the transcriptional inhibitor cordycepin was added to block the formation of new viral transcripts (107). Labelling of the viral proteins synthesised following removal of the cycloheximide revealed that the products of the four genes listed above, namely λ3, μNS, σNS and σ3 were the only viral proteins made. If cordycepin was not added at the time of cycloheximide removal then within three to four hours the pattern of protein synthesis was the same as that in uninfected cells, i.e: all ten genes were expressed (107). Therefore the picture that has emerged of the early phase of transcription from these studies is one in which initially transcription is restricted to only four viral genes (L1, M3, S3 and S4), and then *in a mechanism which requires cellular protein synthesis* there is a switch to transcription of all ten viral genes.

Studies (108) using defective virions which lack the L1 gene (λ3) and are therefore unable to replicate their genome have shown that replication is not a prerequisite for the early transcriptional switch to operate (108). Thus at early times post infection (5 hrs) there was preferential transcription of the M3, S3 and S4 genes whereas at later times (10 hrs) all nine of the remaining genes were transcribed.

A similar type of result was achieved using class C *ts* mutants, these are defective in replication at the non-permissive temperature but nevertheless show the early transcriptional switch from restricted to complete genomic transcription when grown under non-permissive conditions (109).

Pre-early/early transition

At present there is little information as to the nature of the early transcriptional switch. Two working models have been proposed, the first supposes that the initial *in vivo* uncoated particle can only transcribe four genomic segments and that one of the protein products from these transcripts four genomic segments and that one of the protein products from these transcripts then causes some change in the parental SVP allowing it to begin transcribing the remaining six genomic species.

The second model involves a cellular repressor that inhibits transcription from six of the genomic species but is inactivated when combined with the protein product from one of the pre-early genes (10). The observation that isolated particles uncoated *in vivo* are able to transcribe all ten genes *in vitro* (111) is most easily accommodated by the second model but much more work is be needed before a clear and unequivocal picture will emerge.

Regulation of RNA synthesis

In addition to the qualitative regulation of the transcription pattern some quantitative regulation has been noted (34). Thus in the period 2-8 hrs post infection approximately 20 times as much mRNA is synthesised for the S3 and S4 genes than for any of the L genes. Since the s mRNAs are in the region of one quarter to one third of the size of the l mRNAs then they must be either initiated more often and/or transcribed at a faster rate than the l mRNAs.

Pulse labelling experiments done on viral cores synthesising mRNA *in vitro*, which produce the same relative amounts of mRNAs suggests that some of the disparity is due to differences in rates of transcription on the different templates (112). The molecular basis for this apparent preferential utilisation of the small class of genomic RNAs during transcription of all ten templates at later times in infection is not known.

D. Virus Replication

The unusual fully non-conservative mechanism of reovirus progeny genome production was suggested by the observation that the parental input genome is never completely uncoated; in fact at late times post infection the parental particles appear to be re-assembled into whole virions (113). Thus since both strands of the parental genome are retained within the infecting particle then the (+) sense transcripts made by these particles must have dual function acting both as mRNA for viral protein synthesis and as the template for the synthesis of new minus strands. A number of experimental approaches were taken to verify this duality of (+) sense transcripts function and hence mechanism of genome replication but the most definitive of these centred on the demonstration that the complementary strands of the progeny genomes were formed asynchronously (114).

In this study infected cells were pulse labelled with ^{3}H-uridine in the mid-logarithmic phase of virus replication, total d-s RNA extracted, denatured, and hybridised to an excess of unlabelled viral mRNA synthesised *in vitro*. This procedure allowed the distribution of ^{3}H-uridine label between the (+) and (-) strands of the *in vivo* synthesised RNA to be analysed (114). The logic of the experiment argued that if genome replication involved a semi-conservative mechanism similar to that for d-s DNA replication then both progeny genome strands should be equally labelled. Hybridisation of the denatured d-s RNA to an excess of unlabelled mRNA should displace most of the labelled plus strands from the re-formed hybrid leaving approximately 50% of the label retained in d-s RNA. However if the (+) strand did indeed serve as template for (-) strand synthesis, then *pulse* labelling should lead to the incorporation of radioactivity predominantly into the (-) strand of any d-s RNA. Under these circumstances hybridisation of the denatured d-s RNA to an excess of unlabelled mRNA should lead to the formation of hybrids in which 100% of the label was retained in the hybrid. When the experiments were performed then this latter result was obtained indicating that genome replication is fully non-conservative with neither strand of the parental genome being retained in progeny genomes.

Location of RNA synthesis

Studies on the location of (-) strand synthesis indicated that replicase activity was sequestered into a provirion which also contained the plus strand RNA template (115). This in turn led to the concept that the progeny d-s RNA was actually formed within the nascent core of developing progeny virions.

The initial characterisation of these provirions done by analysing the distribution of replicase activity in the large particle fraction of cells extracted with non-ionic detergents showed them to the heterogenous in size sedimenting between 300 and 600S on sucrose gradients (116). A more detailed analysis of these particles on sucrose gradients showed that their sedimentation coefficients ranged between 180 and 600S with two peaks of activity at 280 and 550S (117). When the d-s RNA products synthesised by the various gradient fractions were studied then those with the lowest S value made predominantly d-s RNA of the small (S) size class. Particles of intermediate S value made increased amounts of middle size of M class d-s RNA and those sedimenting most rapidly synthesised predominantly large-size or L class d-s RNA (117). These results imply a sequential pattern of d-s RNA synthesis (S → M → L) through which the initial provirion passes as it progresses towards maturation.

Direct examination of the infected cells by light and electron microscopy using a wide variety of techniques (118-120) have shown that virus replication is localised to large cytoplasmic inclusions or factories which first appear in the perinuclear region at 6-8 hrs post infection. Electron microscopy of these inclusion bodies indicates that there is an association between microtubules and both mature and immature virions but this cannot be obligatory for virus replication since the virus grows to normal yields in cells treated with cytoskeletal poisons such as colchicine (121).

Finally the ability of reovirus to replicate in enucleated cells (122) indicates that host cell nuclear functions are not required for viral replication.

E. Translation of Reovirus mRNAs

In analysing the translation of reovirus mRNAs in infected cells, two levels of regulation are evident:

a) control of the level of translation of the different viral mRNAs, and
b) regulation of viral *versus* host mRNA translation.

Intrinsic translation ability of reovirus mRNAs

Dealing with the first of these, it was evident from the early quantitative studies on the levels of the various viral mRNAs and proteins found in infected cells that there was a profound regulation of viral gene expression at the translational level (123). The apparent differences in translatability of the various mRNAs was also observed in *in vitro* translation systems derived from a wide variety of cell types (124-126). From this it was clear that the relative efficiencies with which the individual viral mRNAs were translated was an inherent property of the mRNAs themselves.

However early attempts to investigate the mechanistics of this translational control were thwarted by the lack of a complete set of RNA-protein assignments. This problem was overcome by the development of a procedure that allowed the (+) sense strand of genomic d-s RNA to be translated in *in vitro* systems (35). Since the individual species of genomic RNA can easily be purified in bulk, then it was possible to do the coding assignments.

These biochemically determined coding assignments (see Table II and IV) were independently confirmed for the μ and σ polypeptides using a genetic approach (80). The completion of the RNA-protein coding assignment allowed estimates of the translation efficiencies with which the primary gene products are made to be generated (see Table IV). The only gene for which a full set of data is not available is S1 since the mRNA for this gene (S1) is now known to encode two polypeptides σ1 and p14 (σ1S) (18) and as yet the relative levels of the smaller non-structural protein (p14) made *in vivo* have not been determined.

However ribosome protection studies carried out *in vitro* (127) suggest that it will be synthesised to a higher level than the other protein product (σ1) of this mRNA and therefore the ratio value given in Table IV for this gene is a minimum one. Despite this shortfall in the data it is nevertheless apparent that there is at least a 20 fold variation between the most and least efficiently translated messengers, with one more recent analysis estimating that this value may be greater than 50-fold (128).

The simplest level at which these variations in translational efficiency might be effected is at the level of ribosome binding during the initiation process and this stimulated a large number of studies to elucidate the 5' terminal sequences of the different reovirus mRNAs (129-134). Unfortunately there are considerable differences between the sequences actually obtained by the different groups which probably reflects the technical difficulties encountered in direct RNA sequence analysis. Nevertheless comparative analysis of the sequences by the different

Table 4. - Gene Coding Assignments and Relative Expression Efficiences for Reovirus type 3 (Dearing)

Gene	m-RNA	Protein encoded	Relative *In vivo* Translation Frequency	Relative *In vivo* Translation Frequency	Transcription /Translation
L1	12	λ3	0.05	0.03	0.6
L2	11	λ2	0.05	0.15	3.0
L3	13	λ1	0.05	0.1	2.0
M1	m1	μ2	0.15	0.03	0.2
M2	m3	μ1/μ1c	0.3	1.0	3.3
M3	m2	μNS	0.5	0.5	1.0
S1	s1	σ1	0.5	0.05	0.1
		p14 (σ1S)	0.5	?	?
S2	s2	σ2	0.5	0.2	0.4
S3	s3	σNS	1.0	0.3	0.3
S4	s4	σ3	1.0	0.7	0.7

groups have been done but these have not revealed any features of primary or secondary structure that can be correlated with translational efficiency. Even Kozak's rules (135) on eukaryote initiation sites do not hold true for all of the reovirus mRNAs, since although it contains a "strong" initiation codon pattern the μ1 mRNA is translated very poorly.

The availability of cDNA clones of the various reovirus genes (136) should however facilitate site direct mutagenesis studies aimed at defining the RNA sequences involved in regulating the translational efficiency of these mRNAs. Thach and his collaborators have carried out studies which strongly support the hypothesis that translational regulation is effected at the level of initiation (137). These *in vitro* studies measured the ability of globin mRNA to compete with the translation of the different reovirus mRNAs. The results were interpreted as indicating that the different reovirus mRNAs have intrinsically different initiation rates under competitive conditions and these differences are controlled by their relative affinities for a mRNA discriminatory factor that binds before 40S ribosomal subunits become complexed (137). The only difficulty with this model, for which the same group have produced supporting data from *in vivo* studies (138), is that the hierarchy of mRNA efficiencies established in these studies is different from that obtained by direct measurement (Table IV). In contrast to many other virus systems in reovirus with the exception of the preferential expression of the four pre-early genes immediately following infection, there appears to be no kinetic regulation of viral gene expression over the time course of the infectious cycle.

Translation of host *vs.* reovirus mRNAs

In addition to the translational regulation of viral gene expression there is also a regulation of host versus viral protein synthesis. In many cells, reovirus does not give a rapid inhibition of host protein synthesis but rather a gradual shift from host cell to viral such that by 10 hours post infection viral proteins represent the majority of the translation products (34).

Two models have been put forward to account for this. The first (137, 138) is a passive one which proposes that as the level of viral mRNA increases during infection its higher affinity for rate a limiting message-discriminatory pre-initiation factor results in viral mRNA becoming progressively more translated until the protein synthesis machinery is swamped with viral mRNA.

The second model put forward by Millward and his collaborators is based on their observations that the progeny SVP's which account for the bulk of viral mRNA (90%) synthesised in infected cells, have inactive or masked 5' capping activities and therefore synthesise uncapped mRNA (102). This uncapped mRNA cannot be translated *in vitro* by extracts from *un-infected* cells but is efficiently translated by extracts from infected cells (106). They therefore proposed that reovirus infection induces a gradual shift in the cap dependence of the translational apparatus to one in which the uncapped viral mRNAs are preferentially translated compared to the capped host cell-mRNAs are preferentially translated compared to the capped host cell mRNA. Despite the attraction of this second model, which has been used to explain additional facets of the viral replication cycle, other works were unable to confirm this transition in cap dependence of translation (139) and in fact further work is required before either of the above or any other model can be substantiated.

F. Virion Morphogenesis and Maturation

RNA segment selection

The detailed mechanics of reovirus assembly remains as a major unsolved problem of virus replication, in particular how does the virus ensure that each virion receives one copy of each of the ten genomic species making up the genome?

The low particle/p.f.u ratios that can be obtained for reovirus indicate that the equimolar RNA segment distribution obtained when genomic RNA is extracted from virions is not due to a random or quasi-random distribution of particular genomic segments within the virion population. However all the biochemical structural and genetic evidence strongly suggests that the parental (+) sense transcripts which are assembled into provirions are not covalently linked to each other at any stage. Since the random probability of forming a single particle containing one copy of each of the ten single strand RNA species has been estimated at one in ten billion (140) then a very precise mechanism must exist to ensure that it occurs. Although the established mechanism of genome replication dictates that the segment selection mechanism must operate on single strand plus some RNA transcripts details of how it works have not progressed significantly beyond the speculation stage. The main problem is that the earliest immature particles are both complex and unstable and to-date it has not been possible to isolate and characterise them. There is a presumption that the σNS polypeptide is in some way intimately involved in the segment selection process because of its affinity for single stranded RNA (82, 84, 85) but there is nothing in these interactions as they have been investigated so far that would account for the exquisite specificity of the assembly process.

The intermediate structures

Once the initial segment selecting has been completed and the first provirion formed it seems likely that there is a progression through the various replication competent sub-viral particles (117) discussed earlier towards the final double shelled virion. Two of these particles have been characterised in a little more detail (141). One with an S value of 400 which was first detected at 4 hrs post infection, before mature virions are seen, and which declined at later times post infection contained all the viral core polypeptides and σNS.

The second sedimented at 600S, contained all of the virion capsid polypeptide but still retained a small amount of σNS. Whole virions have a S-value of 630 and the current working hypothesis is that the above particles represent sequential late intermediates in the maturation process. Nothing is known concerning the mechanism of capsid protein condensation other than that it appears to be independent of d-s RNA formation. This conclusion can be reached since empty capsids (top component) which appear equivalent to whole virions in both their polypeptides and morphology are formed as by-product during normal infections (32).

4. CONCLUDING REMARKS

From the above description it is clear that the bulk of our current knowledge of replication events within the *reoviridae* is derived from the study of the mammalian reoviruses. Despite their apparent lack of major disease causing potential they have offered sufficient technical advantages to cause molecular studies to focus on them. The continuing technical advances particularly in recombinant DNA technology together with the need to provide means of combatting some members of this virus family such as rotaviruses and orbiviruses which are major medical and/or veterinary pathogens is ensuring that increasing attention is being paid to these viruses. It will be interesting to see to what extent the broad details of the replication strategy established for the reoviruses are maintained within the other genera of the family. Perhaps detailed molecular analysis of viruses from the other five genera will provide a speedier route to understanding at the molecular level the two major remaining questions of the replication process: How is the qualitative regulation of individual virus mRNA translation achieved? And what is the mechanism which ensures that each progeny virion receives one copy of each genomic segment?

Acknowledgements

I am grateful to Carol Trimnell for typing this article. The author is a Lister Institute Research Fellow and work in his laboratory is also supported by grants from the MRC, AFRC, EEC, W.H.O. and Wellcome Trust.

5.REFERENCES

1) Sabin, A.B. (1959). Science **130**, 1387-1389.
2) Gomatos, P.J., Tamm, I., Dales, S., and Franklin, R.M. (1962). Virology **17**, 441-454.
3) Du Pont, H.L. (1984). J. Infect. Dis. **149**, 663-666.
4) Agarwal, A. (1979). Nature (London) **278**, 389.
5) Florio, L., Miller, M.S. and Mugrage, E.R. (1950). J. Immunol. **64**, 265-272.
6) Flewett, T.H. and Woode, G.N. (1978). Arch. Virol. **57**, 1-25.
7) Robertson, A. (1976). *Handbook on Animal Diseases in the Tropics*, 3rd ed. published by P. Burgess and San Abingdon, Oxfordshire.
8) Henning, M.W. (1956). *Animal Diseases in South Africa*, 3rd ed., Central News Agency, South Africa.
9) Aruga, H. (1971). *The Cytoplasmic Polyhedrosis Virus of the Silkworm*. University of Tokio Press, Tokio.
10) Teakle, D.S. and Steindl, D.R.L. (1969). Virology **37**, 139-145.
11) Ling, K.C., Tiongco, E.R. and Aguiero, V.M. (1978). Plant. Dis. Rep. **62**, 701.

12) Peshmukh, D.R., Sayed, H.I. and Pomeroy, D.S. (1968). Avian Dis. **13**, 243-251.

13) Van der Heide, L. (1977). Avian Path. **6**, 271-284.

14) Morecki, R., Glaser, J.H., Cho, S., Balistreri, W.F. and Horwitz, M.S. (1982). New Engl. J. Med. **307**, 481-484.

15) Dobos, P., Hill, B.J., Hallet, R., Kells, D.T.C., Becht, H. and Teninges, D. (1979). J. Virol. **32**, 593-605.

16) Nick, H., Cursiefen, D. and Becht, H. (1976). J. Virol. **18**, 227-234.

17) Van Elten, J., Lane, L., Gonzales, C., Partridge, J. and Vidauer, A. (1976). J. Virol. **18**, 652-658.

18) Ernst, H. and Shatkin, A.J. (1985). Proc. Natl. Acad. Sci. USA **82**, 48-52.

19) Luftig, R.B., Kilham, S., Hay, A.J., Zweerink, H.J. and Joklik, W.K. (1972). Virology **48**, 170-181.

20) Skehel, J.J. and Joklik, W.K. (1969). Virology **39**, 822-831.

21) Shatkin, A.J. and Sipe, J.D. (1968). Proc. Natl. Acad. Sci. USA **61**, 1462-1469.

22) Furuichi, Y. (1974). Nucleic Acids Res. **1**, 809-820.

23) McCrae, M.A. (1985). *Virology. A Practical Approach*, B.W.J. Mahy, ed. IRL Press, Oxford and Washington, p. 151-168.

24) Jordon, L.E. and Mayor, H.D. (1962). Virology **17**, 597-599.

25) Harvey, J.D., Farrell, J.A. and Bellamy, A.R. (1974). Virology **62**, 154-160.

26) Vasquez, C. and Tournier, P. (1962). Virology **17**, 502-510.

27) Vasquez, C. and Tournier, P. (1964). Virology **24**, 128-130.

28) Amano, Y., Katagiri, S., Ishida, N. and Watanabe, Y. (1971). J. Virol. **8**, 805-808.

29) Palmer, E.L. and Martin, M.L. (1977). Virology **76**, 109-113.

30) Ralph, S.J., Harvey, J.D. and Bellamy, A.R. (1980). J. Virol. **36**, 894-896.

31) Hayes, E.C., Lee, P.W.K., Miller, S.E. and Joklik, W.K. (1981). Virology **108**, 147-155.

32) Smith, R.E., Zweerink, H.J. and Joklik, W.K. (1969). Virology **39**, 791-810.

33) Loh, P.C. and Shatkin, A.J. (1968). J. Virol. **2**, 1353-1359.

34) Zweerink, H.J. and Joklik, W.K. (1970). Virology **41**, 501-518.

35) Mccrae, M.A. and Joklik, W.K. (1978). Virology **89**, 578-593.

36) Laemmli, U.K. (1970). Nature (London), **227**, 680-685.

37) Joklik, W.K. (1972). Virology **49**, 700-715.

38) Lee, P.W.K., Hayes, E.C. and Joklik, W.K. (1981). Virology **108**, 134-146.

39) Krystal, G., Winn, P., Millward, S. and Sakuma, S. (1975). Virology **64**, 505-512.

40) Krystal, G., Perrault, J. and Graham, A.F. (1976). Virology **72**, 308-321.

41) Carter, C.A. (1979). Proc. Natl. Acad. Sci. USA, **76**, 3087-3091.

42) Carter, C.A., Lin, B.Y. and Metley, M. (1980). J. Biol. Chem., **255**, 6479-6485.

43) Shatkin, A.J., Sipe, J.D. and Loh, P.C. (1968). J. Virol., **2**, 968-991.

44) Cashdollar, L.W., Esparza, J., Hudson, G.R., Chemeol, R., Lee, P.W.K. and Joklik, W.K. (1982). Proc. Natl. Acad. Sci. USA **79**, 7644-7648.

45) Harvey, J.D., Bellamy, A.R., Earnshaw, W.C. and Schutt, C. (1981). Virology **112**, 240-249.

46) Granboulan, N. and Niveleau, A. (1967). J. Microsc., **6**, 23-34.

47) Kavenoff, R., Talcove, D. and Mudd, J.A. (1975). Proc. Natl. Acad. Sci. USA, **72**, 4317-4321.

48) Millward, S. and Graham, A.F. (1970). Proc. Natl. Acad. Sci. USA **65**, 422-426.

49) Barnerjee, A.K. and Shatkin, A.J. (1971). J. Mol. Biol., **61**, 643-653.

50) Shatkin, A.J. and Sipe, J.D. (1968). Proc. Natl. Acad. Sci. USA **59**, 246-251.

51) Bellamy, A.R. and Joklik, W.K. (1967). Proc. Natl. Acad. Sci. USA **58**, 1389-1395.

52) Nichols, J.L., Bellamy, A.R. and Joklik, W.K. (1972). Virology **49**, 562-572.

53) Bellamy, A.R., Hole, L.V. and Baguley, B.C. (1970). Virology **42**, 415-420.

54) Carter, C.A. (1977). Virology **80**, 249-259.

55) Carter, C.A., Stoltzfus, C.M., Banerjee, A.K. and Shatkin, A.J. (1974). J. Virol., **13**, 1331-1337.

56) Baltimore, D. (1971). Bacteriol. Rev. **35**, 235-241.

57) Kates, J. and McAuslan, B.R. (1967). Proc. Natl. Acad. Sci. USA, **57**, 314-320.

58) Borsa, J. and Graham, A.F. (1968). Biochem. Biophys. Res. Comm. **33**, 895-901.

59) Kapuler, A.M., Mendelsohn, N., Klett, H. and Acs, G. (1970). Nature (London); **225**, 1209-1213.

60) Borsa, J., Grover, J. and Chapman, J.D. (1970). J. Virol. **6**, 295-302.

61) Shatkin, A.J. (1974). Proc. Natl. Acad. Sci. USA **71**, 3204-3207.

62) Furuichi, Y., Morgan, M., Muthukrishnan, S. and Shatkin, A.J. (1975). Proc. Natl. Acad. Sci. USA **72**, 362-366.

63) Gillies, S., Bullivant, S. and Bellamy, A.R. (1971). Science **174**, 694.

64) Morgan, E.M. and Kingsbury, D.W. (1981). Biochemistry **19**, 484.

65) Dragna, D. and Fields, B.N. (1982). J. Virol. **41**, 110.

66) Yamakawa, M., Furuichi, Y. and Shatkin, A.J. (1982). Virology **118**, 157.

67) Carter, C.A. (1978). Virology **88**, 222.

68) Furuichi, Y. and Shatkin, A.J. (1977). Virology **77**, 566.

69) Fields, B.N. and Greene, M.I. (1982). Nature (London); **300**, 19-23.

70) Sharpe, A.H. and Fields, B.N. (1985). N. Eng. J. Med. **312**, 486-497.

71) Fields, B.N. (1975). Trends in Genetics **1**, 284-287.

72) Ramig, R.F., Cross, R.K. and Fields, B.N. (1977). J. Virol. **22**, 726-733.

73) Jacobs, B.L. and Samuel, C.E. (1985). Virology **143**, 63-74.

74) Burstin, S.J., Spriggs, D.R. and Fields, B.N. (1982). Virology **117**, 146-155.

75) Weiner, H.L., Drayna, D., Averill, D.R. Jnr., Fields, B.N. (1977). Proc. Natl. Acad. Sci. USA **74**, 5744-5748.

76) White, C.K. and Zweerink, H.J. (1976). Virology **70**, 171-180.

77) Brown, E.G., Nibert, M.L. and Fields, B.N. (1983). In: *Double-stranded RNA viruses* R.W. Compans and D.H.L. Bishop, eds. Elsevier p. 275-287.

78) Ahmed, R. and Fields, B.N. (1982). Cell **28**, 605-612.

79) Zarbl, H. and Millward, S. (1973). In: *The Reoviridae* W.K. Joklik ed. Plenum Press, New York and London, p. 107-197.

80) Mustoe, T.A., Ramig, R.F., Sharpe, A.H. and Fields, B.N. (1978). Virology **89**, 594-604.

81) Joklik, W.K. (1985). Ann. Rev. Genet. **19**, 537-575.

82) Huismans, H. and Joklin, W.K. (1976). Virology **70**, 411-424.

83) Gomatos, P.J., Stamatos, N.M. and Sarkar, N.H. (1980). J. Virol. **36**, 556-565.

84) Gomatos, P.J., Prakash, O., Stamatos, N.M. (1981). J. Virol. **39**, 115-124.

85) Stamatos, N.M. and Gomatos, P.J. (1982). Proc. Natl. Acad. Sci. USA **70**, 3457-3461.

86) Sharpe, A.H. and Fields, B.N. (1981). J. Virol. **38**, 389-392.

87) Lee, P.W.K., Hayes, E.C. and Joklik, W.K. (1981). Virology **108**, 156-163.

88) Tardieu, M. and Weiner, H.L. (1982). Science **215**, 419-421.

89) Ertl, H.C.J., Greene, M.I., Noseworthly, J.H., Fields, B.N. and Neporn, J.T. (1982). Proc. Natl. Acad. Sci. USA **79**, 7479-7483.

90) Lee, P.W.K. Personal Communication.

91) Co, M.S., Gaulton, G.N., Fields, B.N. and Greene, M.I. (1985). Proc. Natl. Acad. Sci. USA **82**, 1494-1498.

92) Co, M.S., Gaulton, G.N., Tominaga, A., Homcy, C.J., Fields, B.N. and Greene, M.I. (1985). Proc. Natl. Acad. Sci. USA **82**, 5315-5318.

93) Silverstein, S.C. and Dales, S. (1968). J. Cell. Biol. **36**, 197-230.

94) Borsa, J., Copps, T.P., Sargent, M.D., Long, D.G. and Chapman J.D. (1973). J. Virol. **11**, 552-564.

95) Borsa, J., Morash, B.D., Sargent, M.D., Copps, T.P., Lievaart, P.A. and Szekely, J.G. (1979). J. Gen. Virol. **45**, 161-170.

96) Chang, C.T. and Zweerink, H.J. (1971). Virology **46**, 544-555.

97) Silverstein, S.C., Astell, C., Christman, J., Klett, H. and Aes, G. (1972). Virology **47**, 797-806.

98) Joklik, W.K. (1974). In: *Comprehensive Virology* H. Fraenkel-Conrat and R.R. Wagner, eds. Plenum Publishing Co, New York and London, **2**, p. 231-334.

99) Monoyama, M., Millward, S. and Graham, A.F. (1974). Nucleic Acids Res. **1**, 373-385.

100) Shatkin, A.J. (1965). Biochem. Biophys. Res. Comm. **19**, 506-510.

101) Kudo, H. and Graham, A.F. (1965). J. Bacteriol. **90**, 936-945.

102) Skup, D. and Millward, S. (1980). J. Virol. **34**, 490-496.

103) Zarbl, H., Skup, D. and Millward, S. (1980). J. Virol. **34**, 497-505.

104) Furuichi, Y., Muthukrishnan, S. and Shatkin, A.J. (1975). Proc. Natl. Acad. Sci. USA **72**, 742-745.

105) Skup, D. and Millward, S. (1980) Proc. Natl. Acad. Sci. USA **77**, 152-156.

106) Skup, D., Zarbl, H. and Millward, S. (1981). J. Mol. Biol. **151**, 35-55.

107) Lau, R.Y., Van Alystyne, D., Berckmans, R. and Graham, A.F. (1975). J. Virol. **16**, 470-478.

108) Spandidos, D.A., Krystal, G. and Graham, A.F. (1976). J. Virol. **18**, 7-19.

109) Spandidos, D.A. and Graham, A.F. (1975). J. Virol. **16**, 1444-1452.

110) Zarbl, H. and Millward, S. (1983). In: *The Reoviridae*, W.K. Joklik ed. Plenum Press, New York and London, p. 107-197.

111) Silverstein, S.C., Astell, C., Levin, D.H., Schonberg, M. and Acs, G. (1972). Virology **47**, 797-806.

112) Banerjee, A.K. and Shatkin, A.J. (1970). J. Virol. **6**, 1-11.

113) Silverstein, S.C., Schonberg, M., Levin, D.H. and Acs, G. (1970). Proc. Natl. Acad. Sci. USA **67**, 275-281.

114) Schonberg, M., Silverstein, S.C., Levin, D.H. and Acs, G. (1971). Proc. Natl. Acad. Sci. USA **68**, 505-508.

115) Acs, G., Klett, H., Schonberg, M., Christman, J., Levin, D.H. and Silverstein, S.C. (1971). J. Virol. **8**, 684-689.

116) Zweerink, H.J., Ito, Y. and Matsuhisa, T. (1972). Virology **50**, 349-358.

117) Zweerink, H.J. (1974). Nature (London), **47**, 313-315.

118) Rhini, J.S., Jordan, L.E. and Mayor, H.D. (1962). Virology **17**, 342-355.

119) Silverstein, S.C. and Schur, P.H. (1970). Virology **41**, 564-566.

120) Dales, S., Gomatos, P.J. and Hsu, K.C. (1965). Virology **25**, 193-211.

121) Spendlove, R.S., Lennette, E.H. and John, A.C. (1963). J. Immunol. **90**, 554-560.

122) Follett, E.A.C., Pringle, C.R. and Pennington, T.H. (1975). J. Gen. Virol. **26**, 183-196.

123) Fields, B.N., Laskov, R. and Scharff, M.D. (1972). Virology **50**, 209-215.

124) McDowell, M.J., Joklik, W.K., Villa-Komoroff, L. and Lodish, H.F. (1972). Proc. Natl. Acad. Sci. USA **69**, 2649-2653.

125) Both, G.W., Lavi, S. and Shatkin, A.J. (1975). Cell **4**, 173-180.

126) Levin, K.H. and Samuel, C.E. (1980). Virology **106**, 1-13.

127) Kozak, M. (1982). J. Mol. Biol. **156**, 807-820.

128) Gaillard, R.K. Jr. and Joklik, WK. (1985) Virology **147**, 336-348.

129) Nichols, J.L., Hay, A.J. and Joklik, W.K. (1972). Nature New Biol. **235**, 105-107.

130) Kozak, M. (1977). Nature (London), **269**, 390-394.

131) Kozak, M. and Shatkin, A.J. (1977). J. Mol. Biol. **112**, 75-96.

132) Kozak, M. (1982). J. Virol. **42**, 467-473.

133) McCrae, M.A. (1981) J. Gen. Virol. **55**, 393-403.

134) Antczak, J.B., Chmelo, R., Pickup, D.J. and Joklik, W.K. (1982). Virology **121**, 307-319.

135) Kozak, M. (1981). Nucleic Acids Res. **9**, 5233-5252.

136) Cashdollar, L.W., Chmelo, R., Esparza, J., Hudson, G.R. and Joklik, W.K. (1984). Virology **133**, 191-196.

137) Brendler, T., Godefroy-Colburn, T., Cartill, R.D. and Thach, R. (1981). J. Biol. Chem. **256**, 11747-11754.

138) Walden, W.E., Godefroy-Colburn, T. and Thach, R.E. (1981). J. Biol. Chem. **256**, 11739-11746.

139) Detjen, B.M., Walden, W.E. and Thach, R.E. (1982). J. Biol. Chem. **257**, 9855-9860.

140) Silverstein, S.C., Christman, J.K. and Acs, G. (1976). Ann. Rev. Biochem. **45**, 375-408.

141) Zweerink, H.J., Morgan, E.M. and Skyler, J.S. (1976). Virology **73**, 442-453.

CHAPTER 16

THE MOLECULAR BIOLOGY OF RETROVIRUSES

GIOVANNI BATTISTA ROSSI, SIMONETTA PULCIANI AND MAURIZIO FEDERICO

Laboratory of Virology, Istituto Superiore di Sanita', Rome, Italy

INTRODUCTION

The discovery of the first RNA-containing tumor virus dates back to the beginning of this century, but only with the discovery of the viral enzyme involved in their replication cycle, i.e. the reverse transcriptase, it has been possible to point out the peculiarities of this group which is unique among all known animal viruses.

In 1908-1909 Ellerman and Bang correlated avian leukemia, and other related diseases of the blood-forming tissues, with virus infection (1). Likewise, Peyton Rous also correlated the onset of spontaneous chicken sarcomas with viral infection (2). Since then great efforts have been made to unravel the roles of viruses in the onset and progression of neoplasia in animals as well as in humans and even in plants. In 1936 Bittner isolated a virus related to the onset of mammary tumors in female mice of some strains (3). Several other viruses were thereafter isolated also from many other murine tumors: It's enough to mention Gross leukemia virus in 1951 (4), and Friend Erythroleukemia virus in 1957 (5). The discovery of the viral etiology of neoplasia also in some feral (wild) animal species (i.e. feline) helped to dispel the last doubt, and the virus-cancer relationship was not any longer considered a laboratory artifact (6). In this context, the recent demonstration of the retroviral etiology of some forms of human leukemia and immunodegenerative disorders (7) opened new avenues in cancer research.

Retroviruses are enveloped viruses with a roughly spherical structure (8) and a single-stranded, messenger-like polarity RNA genome present in diploid fashion, i.e. two identical molecules. The chemical composition of all retroviruses is about 60-70% proteins, 30-40% lipids, 2-4% carbohydrate and about 1% RNA. The virion consists of an internal core containing the genome and reverse transcriptase, and an outer membrane (= envelope) that the virus acquires at the host cell membrane during the maturation step. The lipids and glycoproteins play an important role in virus infectivity and host selection. In the viral envelope low-molecular-weight hydrofobic polypeptides are found which anchor the envelope components in the lipid bilayer. The cores contain a major nonglycosylated polypeptide (usually with a molecular weight of 24-30 kDa) and a smaller, but highly basic, polypeptide. Internal to the cores is the RNA genome associated with a highly basic polypeptide and the reverse transcriptase. This enzyme has a crucial role in the life cycle of all retroviruses as it catalyzes the conversion of the single-stranded RNA genomes into duplex DNA strands which are the cornerstones of virus replication. Each virion contains 20-70 copies of reverse transcriptase.

In avian retroviruses the reverse transcriptase is a bimolecular complex composed of the alpha and beta subunits; in murine and primate retroviruses, instead, this enzyme is a single polypeptide chain. The murine mammary tumor virus (MMTV) contains two different species of reverse transcriptase, the biological roles of which are not yet well understood.

Once converted into double-stranded DNA, the viral genome is usually integrated into the host cell DNA, where it directs the production of viral progeny employing, as any other set of host genes, the cell machinery devoted to gene expression. This unique strategy for replication may well explain why most retroviruses are not cytocidal. The unusual possibility to be conserved in cellular genome as any other gene is also related to the retroviruses mode of replication.

When integrated viral genomes, i.e. stably linked to host cell DNA, are poorly or not expressed at all, they are named endogenous proviruses. These can be transmitted as Mendelian elements from parent to offspring (vertical transmission) (9). Retroviruses also spread horizontally as do most infectious agents. Also endogenous retroviruses can be transmitted horizontally, once their expression leads to production of infectious particles. Under those circumstances usually they cannot infect cells of the species they derived from (xenotropic viruses) (10).

Since the retroviruses become integrated into the cellular genome during their replication, they may acquire host genetic sequences. Indeed, the neoplastic properties of acute transforming retroviruses are correlated with the expression of some genes (oncogenes or *v-onc*) transduced by the parental virus from the host genome (*c-onc*) (11).

1. CLASSIFICATION

Retroviruses have been found in vertebrates and in invertebrates as well, and they are usually divided into three major subfamilies: Oncornaviruses, lentiviruses, and spumaviruses (Table 1).

The *oncornaviruses* (oncos = tumor) induce a large variety of neoplasias once they infect the susceptible host, and represent the largest subfamily (12), which will be the main topic of this review.

The *lentiviruses* subfamily includes Visna and Maedi viruses which induce neurological diseases and chronic pneumonia in sheep after a long latent period (13). The viruses of this subfamily derive their denomination, by the unusually protracted incubation period and by the chronicity of the diseases they induce. There is an on-going dispute whether the Human Immunodeficiency Virus (HIV), also named Human T lymphotropic virus (HTLV)-III or Lymphoadenopathy-associated virus (LAV), should be included among the Lentiviruses rather than in the

Table 1. - Classification of retroviruses

Genus	Species
Spumavirus	Foamy viruses of: human primates bovines cats
Lentivirus	Visna virus of sheep Maedi virus
Oncovirus	
— Intracellular A-type particles (found in mice, hamster and guinea pigs)	
— B-type particles	Mouse mammary tumor viruses
— C-type particles	Rous sarcoma virus, other avian transforming viruses: MC29, AMV, etc.
	Murine leukemia viruses: Gross, Friend, Moloney, Rauscher viruses, etc.
	Murine endogenous viruses Rat leukemia virus Feline leukemia viruses Feline sarcoma virus Feline endogenous virus (RD114) Hamster leukemia virus (HLV) Porcine leukemia virus Bovine leukemia virus Primate sarcoma viruses (woolly monkey; gibbon ape) Primate sarcoma-associated virus Primate endogenous viruses Viper virus
— D-type particles	Mason-Pfizer monkey virus (MPMV) Langur virus Squirrel monkey virus

previously mentioned Oncoviruses, which comprise the HTLV family. For practical purposes, HIV viruses have been discussed together with the HTLV family, but this must not be taken as a side-taking position by these reviewers on an issue that still requires more debate.

The *Spumaviruses* have been found in a number of mammalian species. Usually they are not pathogenic and have been isolated mainly as contaminants of primary cell cultures. These viruses derive their name from the characteristic foamy degeneration they induce in cultured cells (14).

2. STRUCTURE OF THE VIRIONS

A. Morphology

Even if all retroviruses have an overall similar composition, electron-microscopy allowed to subdivide them (8) according to their morphology in A-, B-, C-, and D-type particles (Fig. 1).

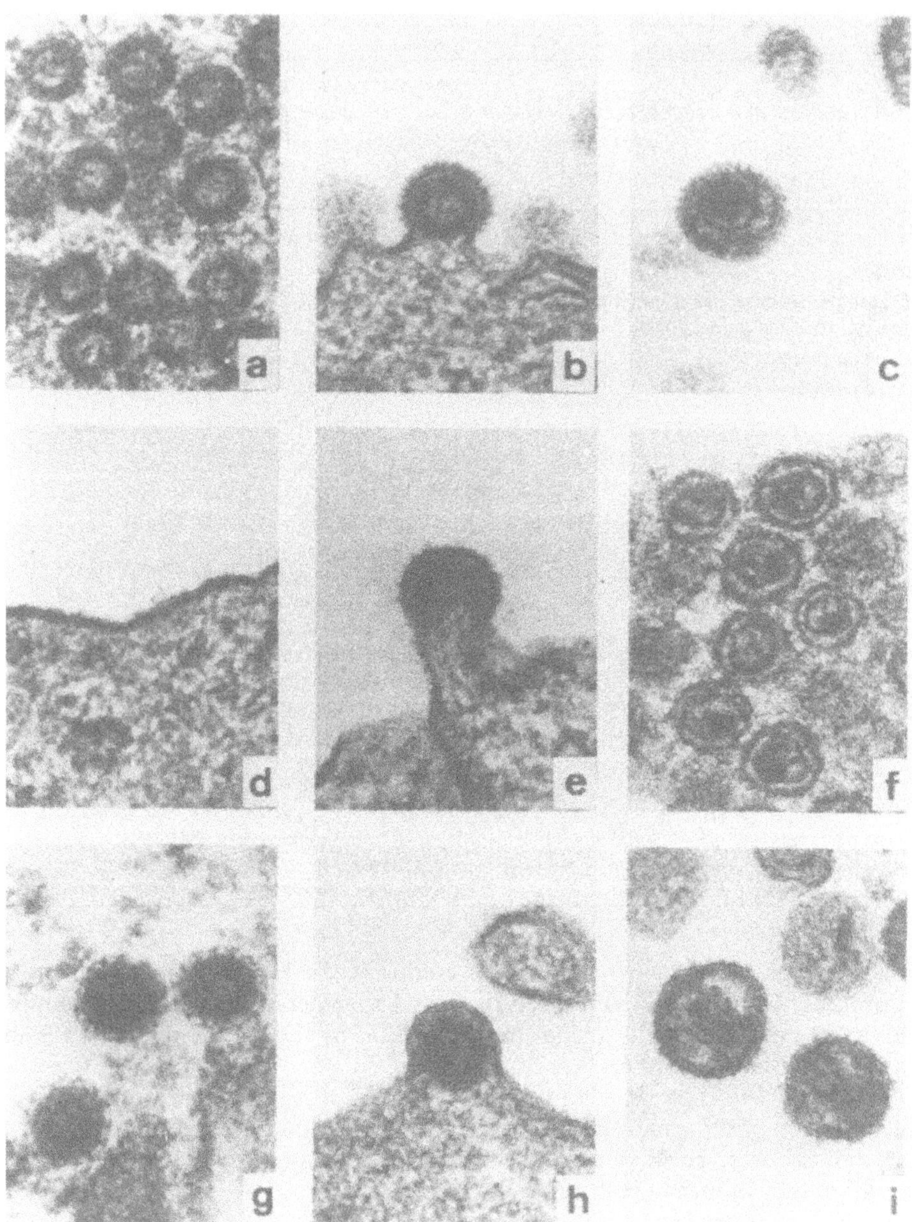

Figure 1. Electron micrographs of Retroviral Particles. **A**-type particles: intracytoplasmic (a), and intracisternal (b). **B**-type particle: budding (c), and extracellular (d). **C**-type particle: budding (e), and extracellular (f). **D**-type particle: budding (g), and extracellular (h).

(Reproduced with permission from Fine, D., & Schochetman, G: *Type D primate retroviruses: A review* Cancer Res. (1978), **38**, 3123-3139, (15)).

A-type particles

These particles are only found intracellularly and do not have any infectivity. A-type particles have an electron-lucent center surrounded by a double shell, generally 60-90 nm in diameter. Intracisternal A-type particles can be found in cells producing C- or D-type retroviruses, but their role in the morphogenesis of retroviruses is not yet well known (15).

B-type particles

B-type particles show morphologically distinguishable changes while they bud through the cellular plasmamembrane. Just before budding, particles with toroidal cores are attached to the plasmamembrane with long glycoproteic spikes protruding through the cell surface. Following budding, the mature forms are composed of electron-dense nucleoids, eccentrically located within the enveloped particles.

C-type particles

These particles can be observed only when budding has started at the plasmamembrane. An electron-dense core is the first viral structure that can be seen. Usually production of spikes on the virion envelope in the same area of the cell surface can be also observed. While budding, the virus particles are surronded by the plasmamembrane and the central core becomes electron-lucent. Two forms of extracellular particles have been described: The immature and the mature forms, with electron-lucent and electron-dense cores, respectively.

D-type particles

Mason-Pfizer monkey virus (MPMV) is the prototype of the group. The intracellular particles are ring-shaped and located near the plasmamembrane. The extracellular forms are the largest retroviruses, 100-120 nm in diameter, containing an eccentrically located electron-dense core. The D-type particles present short surface spikes (16).

B. Protein Components

Structural proteins

The viral proteins are identified by their molecular weight, p30 = a protein of 30.000 daltons (=30 kDa), for instance. Whenever the proteins are phosphorylated, they add a **p** before their coded designation: pp19 = a phosphoprotein of 19.5 kDa. The glycosylated proteins, on the other hand, are specified by a gp: gp37 stands for a glycoprotein of 37 kDa (17).

Five non glycosylated proteins, named pp19, p12, p27, p15 and p10 according to their molecular weight, compose the core of all *avian* C-Type retroviruses (Table 2). The *murine* C-Type viruses, instead, contain four core proteins, i.e. p15, pp12, p30, and p10, whereas the murine B-Type MMTV contains only three core polypeptides, p30, p27 and p14 (Tables 3 and 4) (18).

Table 2. – Structural proteins of avian C-type viruses.

Proteins	Mr kDa	Location	Properties	Glycosylation	Phosphorylation	mRNA size	Function
p19, pp19	19	associated with RNA and between core and envelope	major hydrophobic	no	found phosphorylated (pp19) and not phosphorylated	genomic size	virion RNA binding; association of the precursor with the inner surface of the plasma membrane (?)
p27	27	virion core	moderately hydrophobic	no	no	genomic size	subunits of the core shell
p15	15	between virion core inner envelope	protease	no	no	genomic size	protease involved in cleavage of gag-protein precursors
p12	12	bound to RNA; core	rich in arginine and lysine; highly basic	no	no	genomic size	virion RNA packaging and folding
p10	10	virion membrane (?)	no lysine residues	no	no	genomic size	(?)
p23	23	(?)	containing p19 residues	no	no	genomic size	(?)
p58 (alpha subunit) p92 (beta subunit)	58	internal core		no	no	genomic size	transcription of genomic RNA
p32	32	internal core	from proteolytic digestion of p180 $^{gag\text{-}pol}$ precursor of alfa/beta	no	no	genomic size	endonuclease
gp37	37	envelope	hydrophobic	yes	no	subgenomic size	transmembrane protein spike structure
gp85	85*	envelope	major glycoprotein	yes	no	subgenomic size	virion membrane knob structure associated with gp37

* 15% carbohydrate by weight.

Table 3. – Structural proteins of murine C-type viruses.

Proteins	Mr kDa	Location	Properties	Glycosylation	Phosphorylation	mRNA size	Function
p15	15	between membrane and core	highly hydrophobic	no	no	genomic size	may align $Pr65^{gag}$ during morphogenesis
p30	30	core	moderately hydrophobic	no	no	genomic size	subunits of the core shell
pp12	12	may lie under viral membrane	acidic	no	yes	genomic size	binding to viral RNA, may be involved in regulatory function
p10	10	internal core	very basic	no	no	genomic size	binds to RNA
p70	70-80	virion core		no	no	genomic size	transcription of genomic RNA; DNA-RNA dependent polymerase and RNase-H activities
gp70	70*	envelope	major glycoprotein	yes	no	subgenomic size	subgroup specificity knob structure
p15(E)	22(?)	envelope	hydrophobic	no	no	subgenomic size	anchors gp 70 to membrane spike structures

* 30% carbohydrate by weight.

Table 4. – Structural proteins of murine B-type viruses.

Proteins	Mr kDa	Location	Properties	Glycosylation	Phosphorylation	mRNA size	Function
p10	10	associated with virion envelope	hydrophobic	no	no	genomic size	important in aggregation and assembly
pp21	21	between core and envelope	major phosphoprotein	no	yes	genomic size	structural role (?)
p27/pp27	27	virion core	moderately hydrophobic		may be found phosphorylated (pp27) and not phosphorylated	genomic size	major component of core shell
p14	14	bound to RNA	highly basic	no	no	genomic size	may be involved in viral DNA packaging
p30	30	virion core	related to p14 by aminoacid sequences	no	no	genomic size	minor virion component with unknown function
p8	8	(?)	basic protein	no	no	(?)	minor virion component with unknown function
$p100^{pol}$	100	virion core	dimeric form of the enzyme (?)	no	no	genomic size	RNA-DNA dependent polymerase RNase-H activities
gp36	36*	envelope	very hydrophobic	no	no	subgenomic size	transmembrane protein may anchor gp52 to membrane; spike structure
gp52	52*	envelope		no	no	subgenomic size	knob structure directly associated with gp36

* 10% carbohydrate by weight.

The pp19 polypeptide of avian C-Type viruses is observed in association with both the viral RNA and the lipid envelope. Recent studies have suggested that pp19 may have the important function of blocking RNA processing leading to a greater accumulation of genomic RNA (19). In fact, at least 13 binding sites for pp19 have been found in the genome of the PR-RSV-B strain of the Rous Sarcoma Virus (RSV), two located in the coding region of pp19, the others located at the 5' end of the *gag*, at the 3' end of the *pol* and the *env* genes, respectively. Alternatively, it has been shown that the first 40 amino acids of pp19 have the potential activity to anchor the *gag*-precursor to the inner side of the plasma membrane at the site of virus budding suggesting that pp19 is associated with the envelope glycoprotein gp37 (12, and Pepinsky and Vogt, personal communication). As for the murine retroviruses, p15 for the C-type group, and p10 for the B-Type MMTV, are the equivalent of avian p19. Further, the hydrophobicity of p15 might suggest its role on aligning *gag*-precursor $Pr65^{gag}$ in the cellular membrane during virus morphogenesis. The MMTV p10 is a small hydrophobic polypeptide, which bridges the core and the viral envelope. It is synthesized at the aminoterminal end of *gag*-precursor, and it is thought to tie the precursor to the cell membrane and to help the association of the core with the glycoprotein envelope during budding (20, 21).

The avian p12 is the major protein of the ribonucleoprotein (RNP) complex. p12 binds to RNA at non specific sites by ionic interactions (22). The corresponding proteins are p10 for C-type and p14 for D-type MMTV (23). The p10 polypeptide is a highly basic protein, and it binds the viral genome, approximately 140 molecules each 38S RNA molecule. Actually p10 is purified taking advantage of its affinity for single-stranded or double-stranded nucleic acid. Hence, it has been suggested that their main role resides in the regulation of viral genome expression and virion assembly.

The avian p27 is not found within the RNP, but it has been suggested that it may form the subunits of the core shell in view of the fact that p27 forms higher-order homotypic multimers when chemically cross-linked in virions (24).

The avian p15 may reside between the core and the inner part of the viral envelope; its major function is a proteolic activity which may be responsible of the cleavage of the *gag*-precursor polyproteins into the virion core proteins (see also "Expression of the Viral Genome" in this Chapter) (25).

The p10 protein represents 7% of total virion proteins. The coding sequences for p10 are located between those of p19 and p27. The exact p10 molecular weight is 6,000 daltons, and it exhibits a high content of proline and glycine. p10 is located between the core and the lipid envelope, but its role(s) have not been elucidated yet (26).

In avian virions a p23 polypeptide is found, that is a phosphoprotein and contains the N-terminus of pp19. There are not conclusive data about the role that p23 plays in virion structure; it may be a cleavage intermediate packaged during budding.

In murine C-type viruses p30 forms the icosahedral shell of the viral core, is moderately hydrofobic, and its epitopes provide the antigenic determinants of murine C-type viruses. Modifications in p30 are correlated with FV-1 gene restriction for virus tropism (27).

The MMTV core exhibits, beside the already mentioned p10 and p14, the phosphoproteins pp27 and pp21 and the two minor components p30 and p8. MMTV pp27 is postulated to be the structural protein of the virion core. This polypeptide has strong hydrophobic properties as already described for the avian p27 and the murine p30. The MMTV p27 is also highly immunogenic. pp21 is not highly represented in the virions. Its exact location has not been defined yet. The polypeptide is hydrophobic and it seems unlikely that it may have a membrane function. Since in avian and murine C-type viruses phosphoproteins have been shown to be associated with the viral RNA, a similar function for this MMTV core component has been proposed (28). The MMTV p30 is related to p14 by amino acid sequence and it can be an intermediate in the processing which generates the smallest polypeptide. Alternatively, since p30 contains unique sequences that have not been detected in the other *gag* proteins nor in the *gag*-precursor $Pr77^{gag}$ while they are, instead, present in precursor $Pr110^{gag}$, it has been suggested that p14 and p30 are different proteins with their own specific functions. p8 is a polypeptide rich in arginine and lysine, but lacks methionine. Its function is still unknown.

The reverse transcriptase

This enzyme is encoded by the *pol* gene carried by all replication-competent retroviruses. Each virion contains at least 20 reverse transcriptase molecules which have been purified to homogeneity and characterized. Like all other known DNA polymerase, the reverse transcriptase is primer-dependent (29). Studies performed *in vitro* have shown a high error rate for the avian reverse transcriptase when used to transcribe under certain conditions homopolymeric RNA and DNA templates. This data has suggested that reverse transcription may promote a high mutation rate, although the error rate on viral RNA is not known. Each viral genomic RNA contains five to ten N^6-methyladenines/subunit which the reverse transcriptase should be able to transcribe (30). However, not all modified bases can be transcribed and recent data present evidence that viral DNA synthesis *in vivo* terminates at an N^1-methyladenine.

The reverse transcriptase is the only virion-associated protein that can be readily purified from the virions. This has allowed its structural and biological characterization. The reverse transcriptase of avian retroviruses is composed of two polypeptide subunits designated alpha (58 kDa) and beta (92 kDa). Studies on the aminoacid sequence of alpha and beta subunits have shown that alpha is derived from the proteolitic cleavage of beta subunit. The reverse transcriptase activity is associated with both alpha and beta subunits, but the alpha subunit is itself a functional enzyme. It has been postulated that the biological function of alpha subunit is to stabilize the binding of the enzyme to the template. The murine retroviruses have a monomeric reverse transcriptase of 70-84 kDa. The endogenous virus of mice, MOPC-315, is an exception having a reverse transcriptase composed of two subunits of about 28 kDa and 26 kDa. Mammary Tumor viruses have a monomeric reverse transcriptase of about 85-110 kDa which is unrelated to those of all C-type viruses. Several pieces of evidence suggest that intracellular virions may have bigger reverse transcriptase molecules than those found in the extracellular virions: Presumably, the former molecules are the precursors of the smaller ones. The natural template of reverse transcriptase is the genomic RNA, but a large variety of RNA and DNA can be used as templates. The enzymatic reaction is carried out *only* in presence of a primer; in this respect this RNA-dependent DNA polymerase is like any other known DNA polymerase (See Chapter 3 of this volume and ref. 31). Several species of cellular RNA, including tRNAs, are found in the virions. Interestingly, they are specifically selected among all the available cellular tRNA species.

Besides the ability of synthesizing DNA on RNA templates, the reverse transcriptase has a ribonuclease activity on RNA-DNA hybrids (32). The existence of a ribonuclease (RNase H) capable to degrade the RNA strand of a DNA: RNA hybrid was detected by Housen and Stein in 1970 (33). Later, Moelling et al. (ref. 33) showed that this activity is contained in avian retroviral virions. Moreover, the ribonuclease H could not be dissociated from the viral RNA-directed DNA polymerase. The ribonuclease activity has been always associated with the virions of all retroviruses, strongly supporting the idea that RNase H and DNA-polymerase activities may be intrinsically associated. To date there is not a precise definition of all functions of RNase H, even if there is circumstancial evidence that it may play important roles in viral DNA synthesis. RNase H removes the tRNA from minus-strand DNA once it has been used as primer. Further, several experiments have shown that the reverse transcriptase specifically selects the primer to be packaged. It has also been shown that RNase H activity also consists of the catalysis of viral RNA, once it has been transcribed by the reverse polymerase activity, resulting in formation of oligomeric RNA fragments to be used as primers for plus-strand DNA synthesis.

As already mentioned, the avian retroviruses are endowed with a reverse transcriptase formed by two subunits, but they also contain another polypeptide of 32 kDa (p32) exhibiting a DNA-dependent endonuclease activity. This polypeptide is also coded for by the *pol* gene. Since the endonuclease activity can be found associated also with the alpha/beta complex, it has been proposed that the cleavage of alpha leads to the production of the beta subunit plus p32, and therefore to the generation of the alpha/beta/p32 complex always present in the virions (35).

The reverse transcriptase of murine retroviruses consists of single polypeptides of 70-80 kDa (36). Murine mammary tumor viruses present a reverse transcriptase which employs magnesium ions to successfully carry out its enzymatic reaction. This property has been exploited to distinguish MMTV from the murine C-type viruses that, instead, employ manganese. It is not yet known whether the MMTV reverse transcriptase consists of two subunits or is a single polypeptide of 100 kDa (37).

Envelope proteins

The viral envelope consists of virus coded proteins and host membrane components as well. The *env* gene (Fig. 2, see also Figs. 8 and 9) codes for the viral proteins present in the envelope (38). Their role in the virus replication cycle is to mediate the adsorption to, and penetration of virus into the susceptible host cells. In fact, treatments of intact virions with proteases that remove the surface spikes and knobs, lead to loss of infectivity. Further, the envelope glycoproteins are correlated with host-range, neutralization, interference and subgroup specificities of retroviruses (39).

The avian retroviruses exhibit on the virion surface knobs connected to the virions by spikes. Knobs and spikes are mainly formed by the viral gp85 and gp37, respectively. Treatment of virions with reducing agents releases gp85 but not gp37,

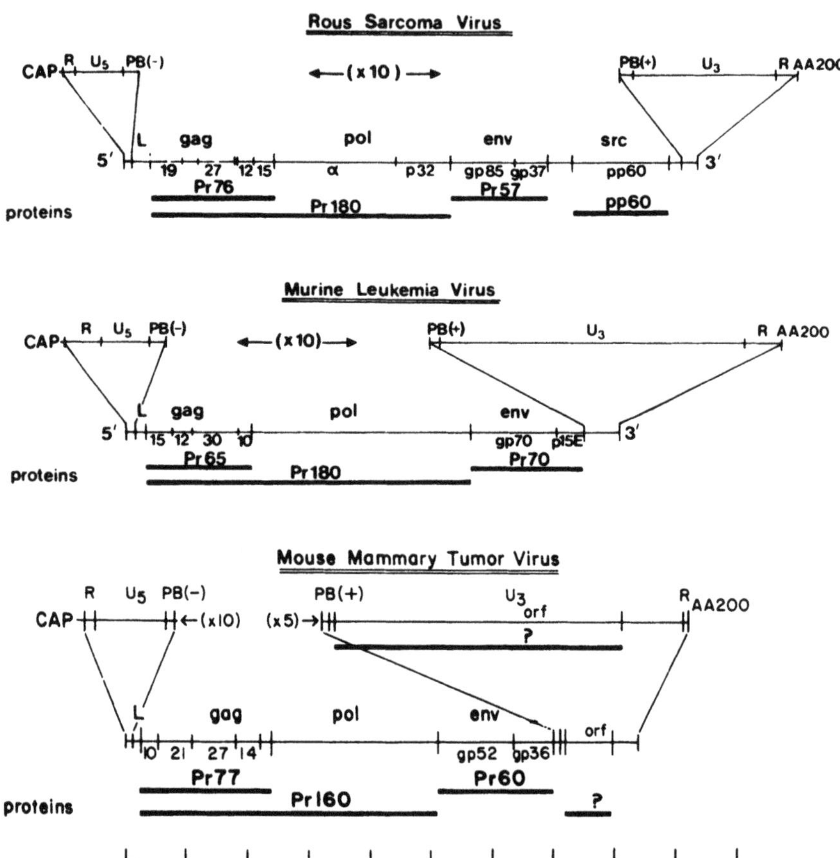

Figure 2. Genomic organization of nondefective Rous Sarcoma Virus, Murine Leukemia Virus and of Murine Mammary Tumor Virus.

5' and 3' terminal regions are shown expanded in the upper lines. Under the genomes, solid bars indicate the precursor proteins, and the approximate molecular weight of the mature peptides is shown.

(Reproduced with permission from Coffin, J: *Structure of the Retroviral Genome* In: *Molecular Biology of Tumor Viruses: RNA Tumor Virus* edited by Weiss, R.A., Teich, N., Varmus, H.E., & Coffin, J.M., Cold Spring Harbor Laboratory, Cold Spring Harbor, NY (1982), 261-368, (41).

which suggests that gp37 is membrane-associated and has the function to anchor gp85 at the virion surface. The carbohydrate content of gp85 and gp37 varies among virus strains and may even depend on the host (38).

The murine C-type retroviruses exhibit two main viral envelope proteins: gp70, a glycoprotein of at least 70 kDa, and p15(E), a protein of 15 kDa. Besides these two major components, a small protein of 12 kDa, p12(E), and a glycoprotein of 45 kDa (gp45) can be found in variable amounts in the viral particles. p12(E) derives from p15(E) by cleavage of carboxyterminal peptides. Also gp45 may represent a degradation product of gp70. p15(E) is membrane-associated and anchors gp70 to the viral surface. p15(E) and gp70 can be covalently linked by disulfide bouds as well as by noncovalents bonds, unlike the avian gp85 and gp37 which are linked only by disulfide-covalent bonds (38).

In the murine B-type MMTV the viral envelope contains two major glycoproteins, gp52 and gp36. Other glycoproteins may be found in the same virus preparation, but they are not virus-coded. Both gp52 and gp36 form MMTV spikes and they accumulate in the host cell membrane at the site where budding will occur (40).

C. The RNA Genome

All known retroviruses have a similar complement of nucleic acids. The retroviral particles contain a double set of single-stranded viral RNA, tRNA as well as ribosomal, messenger RNA and DNA. Retroviruses are the only virus group with a diploid genome, and several hypotheses have been proposed to explain the selective advantage of such an unusual arrangement. The process of DNA synthesis involves a series of jumps and therefore it may require that certain portions of genome be present twice so that the whole coding sequence will be maintained in the viral progeny. Moreover, the presence of two genomes may allow the transfer of the reverse transcriptase from one subunit to another, thus ensuring the repair of any damage occurred in the RNA. How and why the two RNA subunits are held together hasn't been explained yet, though the two subunits are joined by base pairing through an inverted repeat (41, 42).

The dimeric complex of genomic RNA sediments at 60-70S. After heating or denaturing treatments the sedimentation coefficient changes to about 34-38S, that of a molecule of about 8-10kb. The genomic RNA of retroviruses is similar to eukaryotic mRNAs in that it presents a poly(A) sequence of about 200 residues at the 3' end and a typical *capping* group at the 5'end. Since the virion RNA lacks a terminal U tract, the addition of poly(A) must be posttranscriptional (43).

The other nucleic acids found in the retroviral particles include small amounts of 18S and 28S rRNAs, 5S and 7S RNA, and host DNA as well. Since the amounts of these species is small and variable, they are not supposed to play any role in the viral life cycle. Conversely, about 125 molecules of 4S host tRNA can be found in each virion (44). Most of these 4S tRNAs do not cosediment with the genomic RNA, only 20% of it being weakly bound to the viral genome. Among all the different 4S tRNAs there is always one which is tightly associated with the genome as it plays the role of primer for the reverse transcription process (45). These tRNA molecules are associated with the genomic RNA by base pairing of the 3'-terminal 16-19 nucleotides of the primer with the PB(-) primer-binding site, 100-200 nucleotides aways from the 5' extremity of the viral genome (46).

In addition to the sequences coding for viral specific proteins, the retroviral genome contains regulatory sequences involved in viral expression and multiplication (Figs. 2 and 3).

Typically a retrovirus RNA presents, in the 5' to 3' direction, the following arrangement (Fig. 2):

R = REPEAT: A short sequence repeated at both ends of the genome, which plays a relevant role in the reverse transcription process, allowing the transfer of the nascent DNA chains and of the polymerase during the synthesis of proviral double-stranded DNA (47, 48).

U_5 = UNIQUE sequence at 5' end (48).

PB(-) = PRIMER BINDING site of the tRNA primer for negative-strand DNA synthesis (49).

L = LEADER: An untranslated segment which may contain a sequence required for packaging (ca. 250 nucleotides) (50).

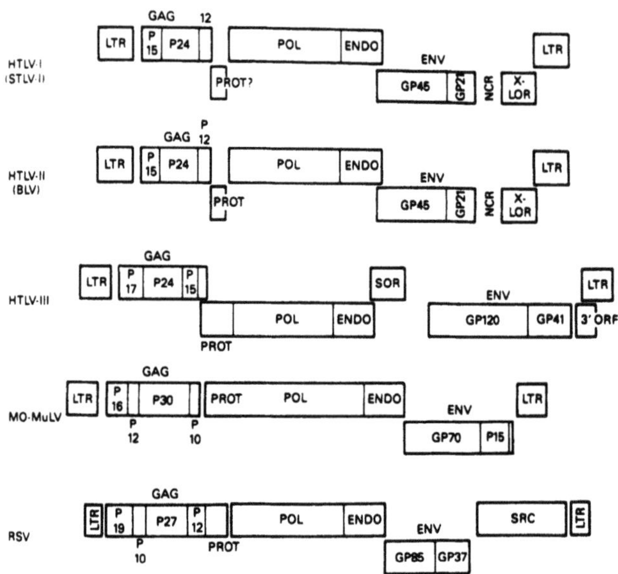

Figure 3. Genetic organization of non-defective Rous Sarcoma Virus (RSV) and Murine Leukemia Virus (Mo-MLV) compared to the family of human retroviruses HTLV. Abbreviations: ENDO = endonuclease; PROT = protease; NCR = noncoding region; X-LOR = long open reading frame in px; Sor = short open reading frame; 3' ORF = 3' open reading frame.

(Reproduced with permission from Wong-Staal, F., & Gallo, R.C: *Human T-lymphotropic retroviruses* Nature (London) (1985), **317**, 395-403, (7)).

gag = GROUP-SPECIFIC ANTIGEN: The region coding for the internal structural proteins (ca. 2 kb) (51). The *gag* genes are pretty much conserved within each group but not between groups. The avian C-type viruses are the only analyzed retroviruses which encode the *gag* protease (p15) as part of the *gag* product. In the murine C-type viruses instead, the protease is encoded as part of *gag-pol* product. The *gag-pol* organization of the HTLV family will discussed later.

pol = POLYMERASE: The region coding for the RNA-dependent DNA polymerase (ca. 3 kb) (30).

env = ENVELOPE: The region encoding virion envelope proteins (ca. 2 kb)(51). The *env* genes of different retroviral groups show a fairly high structural similarity. For instance, the *pol* and *env* genes overlap since the splice acceptor site at the 5' end of *env* falls just before the 3' end of *pol*. The sequences encoding the amino terminus of the *env* primary product lie within the nucleotide sequences coding for the carboxyl terminus of the *pol* product, even if in a different reading frame.

src = Abbreviation for SARCOMA. Present only in RSV. It contains the coding sequences (2 kb) for a phosphoprotein with protein kinase activity, the expression of which is correlated with the RSV transforming properties (52).

PB(+) = PRIMER BINDING site for positive-strand DNA synthesis (53). A purine- rich region.

U_3 = UNIQUE sequence at the 3' end, which lies between the PB(+) site and the R region. The MMTV genome is characterized by a particularly long U_3 region which may contain the coding sequences for other viral proteins (53).

LTR = LONG TERMINAL REPEAT. Once the RNA genome has been converted to double-stranded DNA, the combination of R, U_5, U_3 regions constitute the "long terminal repeats" located at both ends of proviral DNA (Fig. 6). The cloning and sequencing of the LTR of several strains of retroviruses have shown their significant roles as regulatory sequences involved in promotion, initiation and polyadenylation of transcripts of viral and, most interestingly, of eukaryotic genes (54, 55) (Table 5).

onc = Abbreviation for ONCOGENE. The acute transforming retroviruses contain *onc* sequences responsible for the onset and development of the transformation process. These sequences derive from cellular genes (see Section D) and have been integrated in the viral genomes replacing large portions of *gag*, *pol* or *env* genes. RSV is the only exception since its *onc* gene, *src*, is located between the *env* gene and the PB(+) sequences, thus lacking deletions of any gene(s) needed for the replication process (56).

Table 5. – Critical sequences in LTRs in some major species of retroviruses.

	U_3				R	U_5	
Virus	length (bp)	inverted repeat sequence	TATA Box	PolyA signal	length (bp)	Inverted repeat sequence	length (bp)
RSV	230	AATGTAGTCTTATGC...	...TATTTAG...	...AATAAA...	21	...GCAGAAGGCTTCATT	80
ev-1	172	AATGTAGTC...	...TATATAA...	...AATAAA...	21	...GGCTTCATT	80
MoMLV, MSV	371-442	AATGAAAGACCCC...	...AATAAAAG..	...AATAAA...	70	...GGGGTCTTTCATT	75
SNV	369	AATGT...	...TATAAAG...	...AATAAA...	80	...ACATT	100
MMTV	1192	AATGCCGC...	...TATAAAAG..	...AGTAAA...	13	...GCGGCAGC	122
HTLV-I	376	GATGACAATG...	...TATAAAAG..	...AATAAA...	299	...TTAGTACAGT	176

Tat = TRANS ACTIVATING TRANSCRIPTION. The human T-lymphotropic and the Bovine Leukemia viruses (BLV) contain sequences, the *Tat* gene, capable of activating transcription of viral genes and of positively or negatively regulating transcription of some cellular genes. Molecular analysis of HTLV regulation has shown an unusually complex splicing process involved with the maturation of mRNAs for the transacting transcription-regulation (TAT) proteins. For instance, in HTLV-I and HTLV-II (see below), the x-lor gene, located in the pX region, codes for the TAT proteins. Its mRNA is generated by joining three exons using i) a donor splice site located in the R region of the LTR, ii) an acceptor site upstream of the *env* gene, followed by a donor site 190 nucleotides apart, and iii) a second acceptor splice site that is also located at the beginning of x-lor (57) (Fig. 3).

The R Regions

The localization of the tRNA primer near the 5' end of the genome underlines the importance of the 5' and 3' extremities in the replication of the virus. In fact, if a short sequence is repeated at both ends of the genome, the synthesis of the "minus" DNA, initiated close to (but not at) the 5' end of one strand of the dimeric genome, may continue from the 3' end of the other strand by base-pairing of the template with the growing chain (see Fig. 13, Chapter 2 of this volume). The length of this redundant sequence (R) may vary among different virus strains.

Such a mechanism of replication implies that the 3' redundant sequence is not copied into DNA but serves only as a bridge for the minus strong-stop DNA (see below); therefore, any mutant or recombinant genome with different R regions at 3' and 5' ends will exhibit the same redundant sequences after one cycle of replication with the genetic continuity residing only in the R region at the 5' end. In fact, a viral genome bearing two different R regions has never been found. The hexanucleotide AAUAAA is present in the redundant sequence of some viral genomes. These sequences are found even in all eukaryotic viral and cellular mRNAs, possibly as a signal for adding poly(A) tails to mRNAs. The presence of these sequences at both ends raises the question of how the system chooses the correct end to be polyadenylated. In RSV and ALV genomes the AAUAAA is contained in the U_3 region. It has been suggested that the R sequence also protects viral RNA from nuclease digestion, thus favouring its binding to ribosomes (58, 59).

The U_5 region

The U_5 region is represented by noncoding sequences unique to the 5'end of viral genome. This region is located between R and the primer binding site, PB(–). The R-U_5 is known as "strong-stop" region, since the DNA synthesis reaction *in vitro* gives as major product a copy of these genomic sequences only. The strong-stop sequences vary among distantly related retroviruses, even if there are some conserved features such as the 9 nucleotides forming the U_5-PB(–) junction. The role of this sequence and therefore the reasons of this conservation are not well understood as yet. Conversely, the strong-stop sequences are not quite divergent among closely related virus strains. The "strong-stop" regions of HTLV-1, HTLV-2 and BLV exhibit a very long R region, and the sequence AAUAAA is located near the 3' end of U_3 at least 250 bases from the polyadenylation site (60). A hairpin structure has been proposed as a mechanism to pair the sequences AAUAAA to the poly(A) addition site (61). The primer binding site (PB –) in loosely related exogenous viruses consists of highly conserved sequences complementary to the 3' 18 bases of the suitable tRNA. Conversely, in endogenous viruses PB(–) sequences may vary in this region. Germ-line mutation or incorrect tRNA usage have been proposed to explain these changes (62).

The L region

The L (= leader) region is an untraslated segment located between the 3' end of PB(–) and the initiation codon of *gag*. The length of this region varies among different retroviruses. In fact, it is 250 nucleotides long in RSV and 61 nucleotides long in Moloney-Murine Sarcoma Virus (Mo-MSV). The L region has functions of importance in the replication cycle. In fact, almost all known retroviruses show the splicing donor site for subgenomic mRNAs within this region, with the exception of ALV-related viruses that have the splice donor sequence in the *gag* gene (see Fig. 8). For HTLV the splice donor site has not been determined yet, and the L region is only 26 nucleotides long (53, 63). The L region is involved in the dimer linkage formation and it represents a recognition signal for packaging the genomic RNA into virions. It has been also postulated that the L region may play a role in the regulation of the genome splicing, as to mantain the suitable ratio among messenger RNAs and genomic RNAs, besides being a region which allows the selective incorporation of genomic RNA vs mRNA species into virions.

The U_3 region

The U_3 region is the sequence unique to the 3' end of viral RNA and is repeated at both the ends of unintegrated linear proviral DNA. The U_3 region does not seem to encode any viral product. Only the MMTV U_3 region shows a large open reading frame (ORF), but the product that it encodes has not been identified as yet. The MMTV U_3 region is also involved in the inducibility of virus expression by glucocorticoid hormones albeit its molecular mechanism is not yet elucidated (64). This region contains signals for integration, initiation, and, for several virus strains, polyadenylation. The sequences more likely involved in polyadenylation are AAUAAA, and these are present in U_3 avian viruses. MMTV, instead, presents the related sequence AGUAAA. The U_3 sequence also contains the TAT (A or $T)_{3\text{-}4}G$, the Hogness box, that represents the initiation site for RNA transcription. Differences in the U_3 regions have been associated with phenotypic changes (65, 66). The lack of leukomogenic potential of endogenous viruses may be correlated with changes of their U_3 sequences. Some special mention deserves the HTLV U_3 region, which seems the target of viral product *pX* or *Lor*; this will discussed in more details in the proper section of this review.

D. *Oncogenes*

Definition

Retroviruses induce onset and development of neoplasia through the persistence of at least a portion of the viral genome as an integral part of the host chromosomes. Analysis at a molecular level of acute transforming retrovirus genomes has allowed to define sequences (the oncogenes) the expression of which induces and maintains the neoplastic process.

The retroviral oncogenes are transduced from cellular genetic loci, with the only exception of the spleen focus-forming virus (SFFV) of the Friend Leukemia Virus complex, the oncogene of which is a recombinant form of the *env* gene with an endogenous retrovirus. The oncogene of RSV (*v-src*) was the first one to be identified and characterized. At the time of writing this review, more than 20 retroviral oncogenes have been isolated from viruses that infect at least six animal species (Table 6). Oncogenes are designated by a term derived from the name of the virus that bears the gene, without any obvious implication of target-cell specificity or function of the transforming protein. Independent virus isolates may bear related oncogenes; these are often identified by a prefix with the name or abbreviation of the virus strain. The viral oncogenes always bear the prefix **v-** to distinguish them from their cellular counterpart, also named proto-oncogene, which bears the prefix **c-**. *c-onc* sequences have been highly conserved throughout evolution, as they are found in all the species besides that where the corresponding virus had been isolated from. Therefore, it is convenient to identify the source of the cellular sequences by indicating the animal source in brackets, e.g. *c-myc* (chicken), *h-myc* (human) (67).

Historical background

Rous Sarcoma Virus was isolated from extracts of tumor tissues after the original sarcoma had been transplanted several times from one animal to another one. Likewise, murine sarcoma viruses (Harvey, 1964; Moloney, 1966) (68, 69) and the Abelson murine leukemia virus (Abelson and Rabstein, 1970) (70) have been isolated after the passage of leukemia viruses in susceptible hosts. These findings suggested that viruses acquire transforming properties only after being passaged into host animals. Detailed analysis of the *acute* transforming retroviruses revealed that they constitute a mixed population formed by the original replication-competent virus (helper) plus a virus in which portions of structural genes had been replaced by new nucleotide sequences (*v-onc*).

The first transforming gene (and corresponding product) to be identified belonged to RSV, the only retrovirus bearing an onc-sequence, which a) can replicate without the requirement of a helper virus, and b) may induce transformation *in vivo* and *in vitro* (71, 72). These properties proved instrumental in the discovery of viral sequences involved in the induction and mantainance of transformation. The gene, named *v-src*, was shown to code for a phosphoprotein (pp60) of 60,500 dalton, specifically endowed with tyrosin kinase activity (73). DNA fragments bearing *v-src*, but no other viral gene, induce transformation once

Table 6. – Retroviral oncogenes and their products.

Oncogenes	Product	Function/protein feature	Cellular localization of product	Virus prototype	*v-onc* origin
src	$pp60^{src}$	tyrosine kinase	plasma membrane	Rous sarcoma	chicken
yes	$p90^{gag\text{-}yes}$	tyrosine kinase	membrane	Y73 sarcoma	chicken
abl	$p90\text{-}p160^{gag\text{-}abl}$	tirosyne kinase	plasma membrane	Abelson murine sarcoma	mouse
ros	$p68^{gag\text{-}ros}$	tyrosine kinase	membrane	URII sarcoma	chicken
fes	$p85^{gag\text{-}fes}$	tyrosine kinase	plasma membrane	Snyder-Theilen feline sarcoma	cat
fps	$p130^{gag\text{-}fps}$	tyrosine kinase	plasma membrane	Fujinami sarcoma	chicken
fgr	$p70^{gag\text{-}fgr}$	tyrosine kinase	(?)	Gardner-Rasheed feline sarcoma	cat
V-kit	$p80^{gag\text{-}kit}$	tyrosine kinase	(?)	Hardy-Zuckerman 4	cat
erb B	$pg65^{erb\ B}$	truncated EGF receptor	plasma membrane	Avian erythroblastosis	chicken
fms	$gp180^{gag\text{-}fes}$	G-CSF receptor like	membrane virus	Mc Donough feline sarcoma	cat
mos	$p37^{env\text{-}mos}$	homologous to precursor protein for EGF	cytoplasm	Moloney murine sarcoma	mouse
raf	$gp90^{gag\text{-}raf}$	(?)	(?)	3611 murine sarcoma	mouse
ras^{H}	$pp21^{ras}$	GDP/GTP binding	membrane	Harvey sarcoma	rat
ras^{K}	$pp21^{ras}$	GDP/GTP binding	membrane	Kirsten sarcoma	rat
myc	$p100^{gag\text{-}myc}$	DNA binding	nucleous	Avian myelocitomatosis	chicken
myb	$p45^{myb}$	(?)	nucleous	Avian myeloblastosis	chicken
fos	$pp50^{fos}$	(?)	nucleous	FBJ osteosarcoma	mouse
ski	$p110^{gag\text{-}ski\text{-}pol}$	(?)	(?)	Avian SKV 770	chicken
erb A	$p75^{gag\text{-}erb\ A}$	partial homology steroid receptor	cytoplasm	Avian erythroblastosis	chicken
ets	$p135^{gag\text{-}ets\text{-}myb}$	(?)	nucleous	E26	chicken
mil/mht	$p100^{gag\text{-}mht}$	serine kinase	cytoplasm	MH2	chicken
sea	$gp155^{sea}$	tyrosine kinase	(?)	S13	chicken

transfected into normal fibroblasts, and the transformed cells do synthesize pp-60 and induce sarcomas when injected into syngeneic recipients (74). Similar experiments demostrated that the acquisition of new *v-onc* sequences correlated with the oncogenic properties of the viruses. The acquired genetic sequences had been *stolen* from the host genome (67).

The first *onc* sequence isolated from *cellular* DNA was the homologue to v-*src*. This was done by using radioactive complementary DNA ($cDNA_{src}$) that hybridized only with nucleotide sequences encoding *v-src* . These experiments allowed to detect the genetic *src* loci not only in chickens but even in evolutionarily distant species. The same experimental approach was followed to search for cellular genetic sequences related to transformation of all replication-competent viruses. The general outcome seemed to favour the "oncogene hypothesis" proposed by Huebner and Todaro (1969) (75), who had suggested that the oncogenic process was correlated with the activation of cryptic retrovirus genes already resident in cellular genomes. Subsequent molecular cloning and characterization of viral and cellular *onc* sequences revealed that these were of cellular and not viral origin in murine and avian defective acute transforming retroviruses.

Intriguingly, the same cellular sequences present in oncogenic viruses were found in non-transformed cells of distantly related species: *src* related sequences were found in Drosophila, as genes belonging to the family of *ras* oncogenes were found in yeast (76). The fact that these genetic loci have been highly conserved in the course of evolution suggests that they have important physiological roles. *c-onc* sequences have been conventionally named *proto-oncogenes* if isolated from normal cells, and *cellular* oncogenes if present with an altered structure in tumor cells (77).

Cellular *onc* sequences have introns (with the only exception of *c-mos* gene) whereas *v-onc* transforming genes do not. These findings suggested the molecular mechanisms by which retroviruses could pick-up host sequences in their genomes (67). The fact that *v-onc* are intronless indicates that the recombinant process had taken place after transcription and maturation of viral and cellular mRNA. The transduction process begins with the integration of a retroviral genome near the *c-onc*, followed by deletion of cellular and viral sequences which brings the cellular gene under the regulatory LTR viral sequences (see Section "Replication" in this Chapter) (Fig. 4). The cellular gene (still containing introns) is transcribed and its RNA maturated to mRNA by splicing (with loss of all intron sequences), capping and poly-A addition. This new genomic-size mRNA may form a heterodimer structure with the RNA genome of another retrovirus accidentally present in the same cell by the same as yet unknown mechanism that ties together the two strands of any diploid RNA retroviral genome. If this happens, the heterodimer will resemble a diploid retroviral RNA genome, that is packed into retroviral virions and may be reintroduced into susceptible hosts. Once in the *new* host cell, the heterodimer structure is reverse-transcribed; copy-choice during reverse transcription can generate a new retrovirus genome containing the *c-onc* sequences. The insertion in the retroviral genome of the cellular genes usually replaces part of the sequences devoted to viral replication.

The virus-related oncogenes found so far are at least 20. They have been grouped in several families according to their sequence homology(ies) (Table 6).

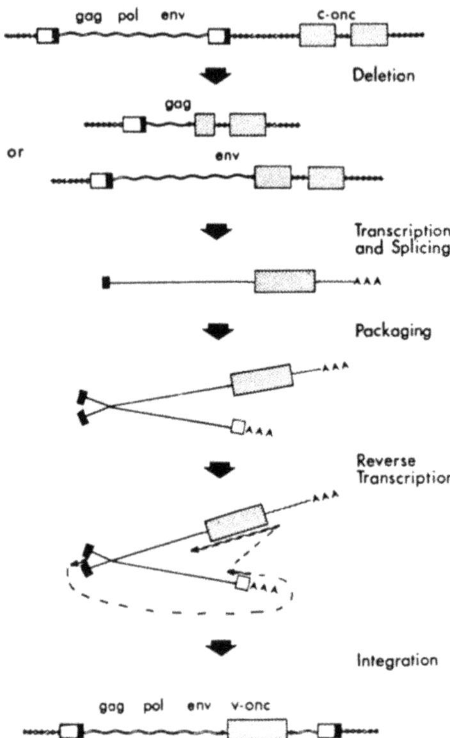

Figure 4. A model for transduction by retroviruses of cellular proto-oncogenes.
First: An intact retroviral provirus integrates next to a cellular proto-oncogene (*c-onc*).
The viral LTRs are illustrated by black and white boxes, and exons of the **c-*onc*** locus by stippled boxes.
Second: The provirus and the *c-onc* come into a single genetic unit by a deletion process, which eliminates portions of the viral genome and the LTR.
Third: A polyadenylated mRNA (solid lines) is generated by transcription of the hybrid unit, and the introns of the *c-onc* are removed by splicing.
Fourth: The recombined RNA and the genome of another retrovirus in the same cell may form a heterodimer, which can be packaged into virions. Following budding from the cell, the packaged heterodimer may infect another permissive cell host.
Fifth: Once in the new host, copy-choice during reverse transcription can generate a new retrovirus genome containing the *c-onc* sequences.
(Reproduced with permission from Bishop, J.M: *Retroviruses and cancer genes* Adv. Cancer Res. (1982), **37**, 1-32, (67)).

The largest family is defined by its homology to the 30kDa carboxyterminal of $pp60^{src}$ where the tyrosine kinase activity resides. This group can be divided in two subgroups: One having protein products with tyrosine-specific kinase activity, and another coding for proteins without clear kinase activity. Oncogenes such as *yes*, *fps*, *fes*, *abl*, *ros* and *fgr* belong to the first group, whereas *mos*, *erb-B*, *fms*, *raf*, and *mht/mil* belong to the other. Among these *v-raf* lacks the site for phosphorylation of tyrosine, present, instead, on *v-erb-B* and *v-fms*. The product of *v-mos* oncogene seems more closely related to serine-specific kinases (78).

Another family is composed by *v-myc*, *v-myb* and the E1A region of the tumorigenic domain in the adenoviral genome. This assortement must be taken with caution, even if these three oncogenes are all located in the cell nucleus and share the capacity to complement the activity of certain *ras* genes in the transformation of embryonic rodent cells (79). Moreover, this grouping represents the first provocative attempt to tie the oncogenes of at least some retroviruses to those of DNA viruses which seem to share common functions.

The *ras* genes constitute a family on their own. They code for proteins immunologically related, and defined p21 according to their molecular weight. To this group belong the oncogenes isolated from murine sarcoma viruses and *N-ras* identified by gene transfer of DNA from human neuroblastoma cells (80).

Besides the above-mentioned families, there are several other oncogenes that cannot be grouped in any defined family such as *v-sis*, the oncogene of Simian Sarcoma Virus (SSV), which codes for a protein related to platelet derived growth factor (PDGF) (81).

Four oncogenes will be discussed in some detail: *src*, *v-sis*, *v-fms*, and *erb*.

Src

V-src appears to encode a single protein: pp60^{src}. Recent studies on mutations that affect *v-src* translation, have revealed an open reading frame within the gene allowing for translation of a 7kDa protein (82). These findings suggest a) that *src* is a gene similar to the multiple genetic domains that direct tumorigenesis by DNA tumor viruses (Adeno and Papova), and b) that more than one biochemical function contributes to neoplastic transformation by *v-src*. So far the search for the 7kDa product of *v-src* has been unsuccessful.

pp60^{src} is synthesized on free polyribosomes, forms a complex with two cellular proteins (50kDa and 89kDa) in the cytoplasm, 5-10 minutes later leaves the complex and reaches the inner surface of the plasmamembrane. pp60^{src} is a protein kinase specific for tyrosine. Moreover, the 30kDa catalytic domain of the protein is related structurally to the cAMP-dependent protein kinase of mammals. However, evidence for a relationship between enzymatic activity and neoplastic transformation is not yet conclusive.

As much as 85-90% of pp60^{src} is located on the cytoplasmic side of the plasmamembrane, clustered in specialized regions thereof known as adhesion plaques. A signal peptide (aminoacids 2-14) at the N-terminus of pp60^{src} appears to anchor the protein to the plasmamembrane (83). Since removal of these amino acids destroy the transforming activity and tumorigenicity of the protein, it follows that pp60^{src} may have neoplastic properties only when bound to the plasmamembrane.

In addition to phosphorylation of tyrosine, pp60^{src} can also catalyze phosphorylation of the lipid phosphatidylinositol, phosphatidylinositol-4-phosphate and diacylglycerol. Similar properties belong also to the oncogene *v-ros*. This newly discovered biological activity of pp60^{src} may give a special clue to figure out the molecular mechanisms involved in triggering cellular transformation, as the turnover of phosphatidylinositol is thought to play an important role in the regulation of the cellular metabolism (81).

V-sis

The cellular sequences acquired by SSV code for a protein of 28 kDa which is closely related to the B chain of PDGF. They are inserted into the *env* gene near its 3' end. This gene is expressed as spliced subgenomic mRNA, which presumably uses the RNA splice acceptor site of *env* (See section "Replication" of this Chapter) (84).

The *v-sis* product is a surface glycoprotein which is synthesized on membrane-bound polysomes. The primary translation product is of 28 kDa and is rapidly dimerized to a 56 kDa form which is then processed proteolytically in a series of steps down to a 24 kDa dimer. Lysates from SSV-transformed cells as well as culture medium of most SSV-transformed cells possess a mitogenic activity for normal fibroblasts which can be partialy neutralized by anti-PDGF antibodies. Therefore, it has been proposed that the mature *v-sis* gene product interacts with PDGF receptors leading to unrestricted growth of SSV-transformed cells in an autocrine fashion. Moreover, since SSV-induced tumors are restricted to cell types which have PDGF receptors, it has been postulated that their presence is necessary for transformation (see Fig. 5).

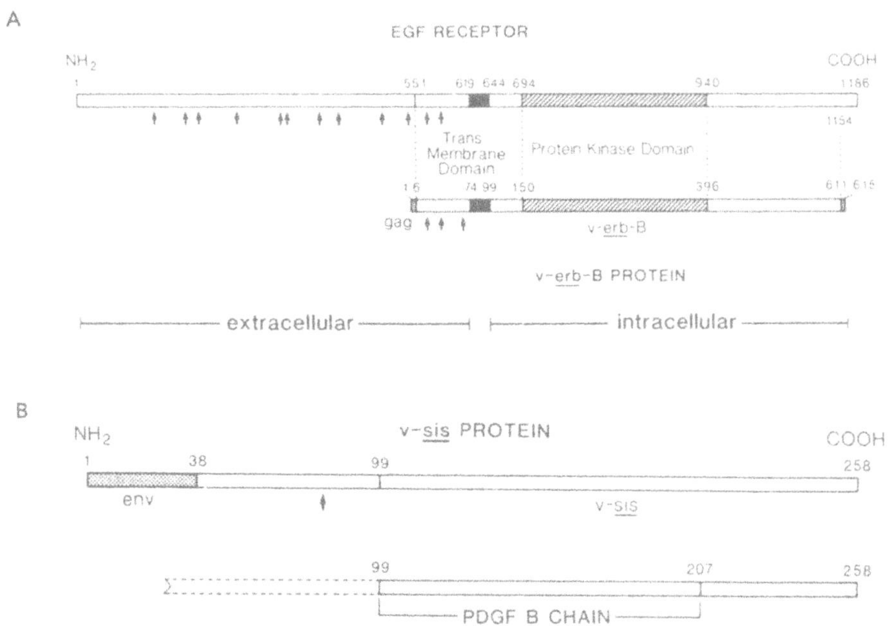

Figure 5. Top panel: The homologous domains of *v-erb B* protein and human EGF receptor have been aligned.

The primary product of human EGF receptor has a 24 aminoacid signal sequence which is removed by cleavege. The *v-erb B* protein is the product of a spliced mRNA which comprises six aminoacids of the viral *gag* gene joined to 609 aminoacids encoded by *v-erb B* sequences. Most likelv also *v-erb B* has a signal sequence which is proteolytically removed.

The bottom panel: The *v-sis* product is compared with the PDGF B chain by superimposing the homologous domains of the two proteins.
The structure of *v-sis* gene is predicted from data sequencing, 38 aminoacids residues derived from *env* gene are probably present on the primary product, and then cleaved. The arrow indicates the potential glycosylation site.

(Reproduced with permission from Hunter, T: *Oncogenes and growth control*, Trends in Bioch. Sciences (1985), **10**, 275-280, (88)).

V-fms

The viral oncogene *v-fms* of the Mc Donough strain of feline sarcoma virus (SM-FeSV) encodes a 140 kDa integral transmembrane glycoprotein ($gp140^{V\text{-}fms}$). The *v-fms* gene product is oriented in the plasma membrane with its glycosylated amino-terminal domain outside the cell and its carboxy-terminal domain in the cytoplasm. The *v-fms* product exhibits an associated tyrosine kinase activity; moreover, the cytoplasmic carboxy-terminal domain of this glycoprotein is closely related to sequences of prototypic tyrosine-specific protein kinases. However, in contrast to proteins encoded by other members of this family, the *v-fms* glycoproteins are not highly phosphorylated *in vivo*.

The *c-fms* proto-oncogene is expressed at relatively high levels in cat spleen, and even the *c-fms* gene product identified in normal cat spleen is a glycoprotein that functions as a substrate *in vitro* for an associated tyrosine kinase. The *c-fms* product has been shown to bind the granulocyte/macrophage-colony-stimulating factors. Therefore, the *c-fms* proto-oncogene and the gene(s) for their receptor(s) appear to be related, if they are not the same gene (85).

erb

Avian erythroblastosis virus (AEV) induces erythroid leukemia and sarcomas in infected chickens, and transforms *in vitro* both erythroid hematopoietic cells and fibroblasts. All the strains of AEV are defective in replication, since portions of structural genes have been replaced with the oncogenes *v-erb A* and *v-erb B*. Their cellular counterparts are located in different loci in the chromosomes. Site-directed mutagenesis demonstrated that only *v-erb B* is required and even sufficient for all the transforming and neoplastic properties of AEV. *v-erb A* may have a helper function on erythroid cell transformation.

The *erb A* gene comprises residues of *gag* gene and *v-erb A* itself with the structure of a "fusion" gene. It is transcribed into a genomic size mRNA, and its product is a protein of 75 kDa ($p75^{gag\text{-}erb\ A}$) which is partly related to carbonic anhydrase.

The primary product of the *erb B* gene, instead, is translated from a subgenomic mRNA, and has an apparent molecular weight of 62 kDa ($p62^{erb\ B}$). This protein is then glycosylated and further phosphorylated to give rise to the final product of molecular weight of 68 kDa ($gp68^{erb\ B}$). *v-erb B* protein is mostly retained in cellular membranes, a small fraction (1-5%) undergoes glycosylation changes, acquires an apparently larger molecular weight (70-72 kDa) and reaches the outer surface of the plasmamembrane. However, there is lack of decisive evidence about the need of the *v-erb B* protein on the cell surface to achieve transformation. The *v-erb B* gene shares extended nucleotide homology with tyrosine-specific protein kinases. Further amino acid sequencing has revealed that the *v-erb B* product is a truncated version of the cell surface receptor for the epidermal growth factor (EGF). This finding opens new insight into the mechanism by which *v-erb B* mediates neoplastic transformation. It has been proposed that the deletion renders the *v-onc* protein constitutively active, without any requirement for stimulation by a growth factor (86) (Fig. 5).

Oncogenes in non-virus-induced tumors

Theories about onset and development of cancer had always postulated the presence of cellular genes with potential transforming activity. The discovery that retroviral oncogenes originate from cellular genes gave strong support to this hypothesis. Moreover, the knowledge of the chromosomal localization of cellular proto-oncogenes, along with what was already known on chromosomal aberrations in tumors, helped to shed some ligth to define the role of those cellular sequences in non-virus-induced neoplasia (Table 7).

Human Burkitt's lymphoma and mouse plasmocytoma are two types of lymphoid tumors that produce immunoglobulin and are characterized by specific chromosomal translocations, which involve a) chromosome 8 and either chromosomes 14, 2 or 22 in man, and b) chromosome 15 and either chromosomes 12 or 6 in the mouse.

These translocations displace the *c-myc* genes on human chromosome 8 or on mouse chromosome 15 in the proximity of an activated immunoglobulin gene(s), thereby activating its expression (87, 88).

Chromosomal translocations which involve *c-onc* loci have been unveiled in many other tumors the description of which at length would fall out the scope of this review (The reader is advised to consult refs. 12, 76, 77). For the sake of completeness, it is enough to mention the translocation present in the Philadelphia chromosome in cases of chronic myeloid leukemia. This involves the *c-abl* gene on chromosome 9 which is fused to sequences on chromosome 22 called *bcr* (89).

Table 7. - Non-viral oncogenes

c-onc	Prototype of tumor where detected	Detected by	References
Blym-1	Burkitt's lymphomas	transfection	80
Tlym-1	Human T-cell lymphomas	transfection	80
Tx-1	Human pre B-cell neoplasm	transfection	80
Tx-2	human myelomas	transfection	80
Tx-3	Human mature T-cell neoplasm	transfection	80
Tx-4	human mammary carcinoma	transfection	80
N-ras	human and murine neoplasm	transfection	80
N-myc	neuroblastoma	amplification	97
L-myc	small cell lung cancer	amplification	98
Neu/Mac	neuroectodermal tumor chemically induced in rats	transfection	91
Trk	human colon carcinoma	transfection	93
Met	MNNG-HOS (in vitro chemically transformed human cells)	transfection	92
Mas	Spontaneously selected during transfection experiments	transfection	99
hst	human stomach cancer	transfection	95
dbl	diffuse B-cell lymphoma	transfection	94
ret	spontaneously selected during transfection experiments	transfection	96

Chromosomal aberration such as homogeneously staining regions (HSR_s) and double minutes (DM_s) have been observed. HSR are elongated regions of chromosomes with abnormal banding patterns. DM_s appear as small paired, usually spherical, chromosomal structures in preparations of metaphase chromosomes. In COLO 320, a colon carcinoma cell line, *c-myc* sequences are amplified in an HSR (90). Analysis of DNA from a series of neuroblastoma cell lines and from one tumor showed the presence of amplified sequences with a limited homology to *v-myc*. These sequences have been named *N-myc*. Analysis of expression of the amplified or translocated *c-onc* sequences has revealed high levels of their mRNAs. This has been tentatively interpreted as an indication that activation of oncogenes leads to overexpression of their products.

Transfection experiments had shown that viral oncogenes could induce cellular transformation once integrated in the host genome. Applying a similar approach to search for a transforming gene in non-virus-induced tumors it has been possible to show the presence of cellular sequences capable to induce transformed phenotype once transfected into competent normal cells. These transforming genes have been detected in animal neoplastic cells as well as in human tumors and tumor cell lines (79). These oncogenes have been shown to be either closely related to retroviral oncogenes, or else cellular sequences never detected in retroviral genomes. Most oncogenes identified by this assay are members of the *ras* gene family. Two of these genes are members of the cellular omologues of the *v-onc* present in Harvey or Balb (*H-ras*), and Kirsten (*K-ras*) Murine Sarcoma Viruses. The third gene is called *N-ras*, a gene recognized as a member of the *ras* group by immunological cross-reaction of its product with antisera raised against products of other *ras* genes and by comparison of the deduced aminoacid sequence of *N-ras* protein with corresponding sequences from other family members (80).

Nucleotide substitution and the ensuing amino acid change in members of the *ras* gene family correlates with the acquired transforming activity. Molecular analysis of transforming activated *ras* genes found in animal as well as in human neoplasias has shown that oncogenic mutations occur only at a restricted number of sites, including codons 12, 13, 59, 61, and 63. The replacement of the glycine residue normally encoded by the twelfth codon of *c-Ha-ra-1* by amino acids with side chains will significantly alter the conformation of the resulting protein.

In neuroectodermal tumors induced in perinatal BDIX rats by single dose of ethylnitrosourea a transforming gene has been detected by the NIH-3T3 focus-forming assay. This gene is designated *neu*, is unrelated to the *ras* gene and codes for a membrane protein of 185 kDa (p185). Comparison of the normal and transforming cloned sequences has allowed to define that a single point mutation is responsible for the activation of the *neu* proto-oncogene. The point mutation results in the substitution of a valine with a glutamic acid in the transmembrane domain of the protein (91).

Treatment of the human osteogenic sarcoma (HOS) cells *in vitro* with N-methyl-N'-nitrosoguanidine (MNNG) activates an oncogene related to the tyrosine kinase gene family and unrelated to any known viral oncogene. This novel oncogene has been named *met* and its activation is related to its translocation from chromosome 7 (q21-31) to a region defined *tpr* (translocated promoter region) located on chromosome 1. In this case oncogene activation resulted from fusion of two chromosomally disparate loci mediated by the direct-acting carcinogen MNNG (92). The activation of the *met* oncogene resembles that already observed in the Philadelphia chromosome translocation in chronic myeloid leukemia, where the *c-abl* gene is fused to *bcr* sequences on chromosome 22.

The *trk* transforming gene isolated from a human colon carcinoma is formed by sequences from both nonmuscle tropomyosin and a novel tyrosine protein kinase (93) (Table 7).

Mechanism(s) of action of oncogenes

Attempts to correlate proto-oncogenes alterations to the uncontrolled proliferation of tumor cells have shown a tight linkage between growth factors and proto-oncogenes. In fact, several oncogene products are similar to growth factors (*c-fms*) or to growth factor receptors (*c-erb B*). Moreover, all available experimental data indicate that transcription of certain proto-oncogenes (*myc* and *fos*) is increased by several growth factors. These results may explain the decrease dependence on growth factors of tumor cells.

Different mechanisms have been proposed to elucidate how oncogenes can confer such independence to the cells. Several years ago Sporn and Todaro (100) proposed that tumor cells can produce increased amounts of growth factors. Once released into the medium and bound to the specific receptors, they lead the cells to an uncontrolled growth (autocrine mechanism). A typical example of this mechanism was demonstrated for the product of the oncogene *sis* which is an altered version of PDGF. Moreover, the finding that a portion of the epidermal growth factor (EGF) receptor has a high homology with the protein specified by *c-erb B* oncogene, suggests an alternative mechanism able to alter the cell growth control. In this case malfunctioning receptors could keep the cell disinformed about the signals from the microenvironment and particularly about the right concentration(s) of growth factor(s) in the medium.

The *ras* gene provides an example of still another oncogene-induced independence from cell growth factors: The p21 proteins coded by the proto-oncogenes belonging to the *ras* family show a specific affinity for guanosintriphosphate (GTP) and guanosindiphosphate (GDP). In fact, the normal p21 protein is able to bind GTP and hydrolyze it to GDP. Because of this ability to bind GTP molecules, a role in the signal transduction from the cellular membrane has been proposed: In fact, this is also a property of the G protein, which is involved in transduction signals from plasmamembrane to the enzyme adenylcyclase. The oncogenic counterparts of p21 proteins, instead, exhibit a reduced hydrolytic activity. Therefore, in cells containing mutated p21 proteins the signals from the plasmamembrane may be altered, presumably misleading the cell into an incorrect proliferation.

Furthermore, since the expression of several proto-oncogenes (e.g. *myc, fos*) can be induced by growth factors suggesting a relevant role of oncogenic proteins in cell multiplication, it follows that constitutive high-level expression of these proto-oncogenes can make the cells independent from the growth factor action and set the cell in a continuous proliferating status (81, 88).

None of the above discussed hypotheses, however, suffice to explain completely how the altered expression of proto-oncogenes may free the cell from its dependence on growth factors, and one can easily anticipate that the current efforts aimed at elucidating the nature of the putative link between regulation of proto-oncogens, on the one hand, and cellular differentiation and proliferation, on the other one, will yield important fruits in a near future.

3. THE REPLICATION CYCLE

In the exogenous infection the replication cycle begins with the adsorption of the virus particle to the cell membrane and its penetration into the host cytoplasm. The adsorption process is not specific as is regulated only by ionic interactions among virions and host surface components; the penetration, instead, is highly specific (39) and both cellular structures and virus envelope glycoproteins play important roles in this process. However, little is known about the entry of the infectious particles inside the cells. There is evidence that virions are taken into the cytoplasm inside vacuoles. The uncoating process is difficult to follow once the virions are penetrated in the cytoplasm, since there is always an excess of noninfectious particles that adsorb to cell membranes as compared to those that will effectively start an infectious cycle. As for these initial steps, the life cycle of retroviruses is comparable to that of all known viruses. Conversely, the reverse transcription of genomic RNA molecules resulting in the synthesis of double-stranded linear DNA (the DNA provirus) makes retroviruses multiplication a process which has opened new frontiers in the understanding the molecular mechanism(s) involved in the evolution and regulation of all living organisms.

A. The Synthesis of the DNA Provirus

This process is carried out by the viral enzyme reverse transcriptase, present in the mature virions and encoded by a viral gene, designated *pol* , that is carried by all replication-competent retroviruses. Interestingly, virions lacking reverse transcriptase activity cannot be complemented by coinfection with replication-competent viruses (101). This appears to indicate that the viral genome can be used as template only by reverse transcriptase molecules present in the same virion.

Virions of the avian retroviruses contain $tRNA^{trp}$ and $tRNA^{met}$, whereas murine retroviruses do contain $tRNA^{pro}$. Not much is known about the molecular mechanisms by which these tRNAs are selected and bind to the viral RNA genome and to the reverse transcriptase, although there is evidence suggesting that the viral enzyme plays a major role in the process (102). An intact tRNA molecule is a crucial requirement to start and to carry out the correct reverse transcription of the template. The tRNA primers do have the function of initiating the synthesis of minus-stranded DNA, but it has also been suggested that they have a role in loosely holding together the 3' ends of the two RNA genomic molecules present in each virion (103).

The binding site in the $tRNA^{pro}$ consists of 19 bases at the 3' terminus, and in the $tRNA^{trp}$ of 16 bases of the same end. Primers bind to a short sequence located at 100-200 bases from the 5' end of viral genome. The localization of this sequence was made possible by studies of the process of reverse transcription of viral genome *in vitro*. During this reaction the first region of the RNA genome to be reverse-transcribed is represented by the sequences between the primer binding site and the 5' end of the genome ("strong-stop" region). The first experimental evidence of such unusual a mechanism was presented in 1976 by Charles Weissmann. The tRNA isolated in the retroviral particles is used as primer *only* for the synthesis of minus-strand DNA. It appears that the RNAse activity of the reverse transcriptase generate the effective primer involved in starting the synthesis of the plus-strand DNA chain, whose sequence is $(2A(2G)_3rA(2G)_5rA)$.

The cloning in prokaryotic vectors of viral genomes by recombinant DNA procedures has made available large amounts of the viral DNA, allowing to define the structural characteristics of the major species of viral DNA within the infected cells. Soon after infection, linear double-stranded molecules of almost the same size of viral RNA genome represent the most abundant species of viral DNA. When analyzed after denaturation, these DNA molecules appear to consist of a full-length minus-strand DNA chain and plus-strand DNA fragments of heterogeneous size. Molecular analysis of these linear duplex DNA molecules has revealed that they terminate with the same sequences, in the same orientation, about 250-1200 bp in size. These sequences have been called *Long Terminal Repeats* (LTR) and are composed (Fig. 6) by the combination of the regions U_3, U_5 and R located at the two ends of viral RNA. The generation of LTR, their structure and the role they play in the integration process will be discussed later (104).

The linear duplex DNA of almost all retroviruses is synthesized in the cytoplasm of infected cells and then transported into the nucleous. The Visna virus is an exception, since its DNA copy seems to be generated in the nucleus.

Transfection experiments have shown that linear duplex DNA molecules are infectious, therefore they represent the viral DNA sequences ready to generate the new progeny. Once the linear duplex DNA molecules reach the nucleus, a minority of them are circularized. There are still many unanswered questions about the circular and the linear DNA forms found in the nuclei of the infected cells. For instance, it has not yet been defined whether the molecules which circularize have some structural peculiarities. As for the linear DNA forms, it's not known whether they can integrate into the host chromosomes or they can be used as templates to synthesize viral messengers or genomic RNAs. Accumulation of free linear DNA in avian cells infected by cytopathic viruses seems correlated with cell death. This has also been postulated for the infectious cycle of HIV, but the evidence appears still scanty.

The circular forms can be found only in the nucleus. The use of restriction enzymes which cut outside the LTR (or only once inside the LTR) has allowed to establish that almost all DNA circles have 1 or 2 LTR. There are also several circles with just a fragment of LTR, and others formed by more then one linear DNA linked head-to-tail.

The initial step of the complex process which will eventually lead to the synthesis and integration of viral DNA into the cellular chromosome and thereafter to the production of progeny virus deserves a more detailed discussion (Fig. 6). The synthesis of viral minus-strand DNA starts by adding deoxynucleotides at the 3' end of the tRNA primer located 100-200 bases from the 5' end. Obviously, the nascent DNA chain runs out of the template shortly after the onset of its synthesis, forcing the replication complex to quit the exhausted template in search for a new

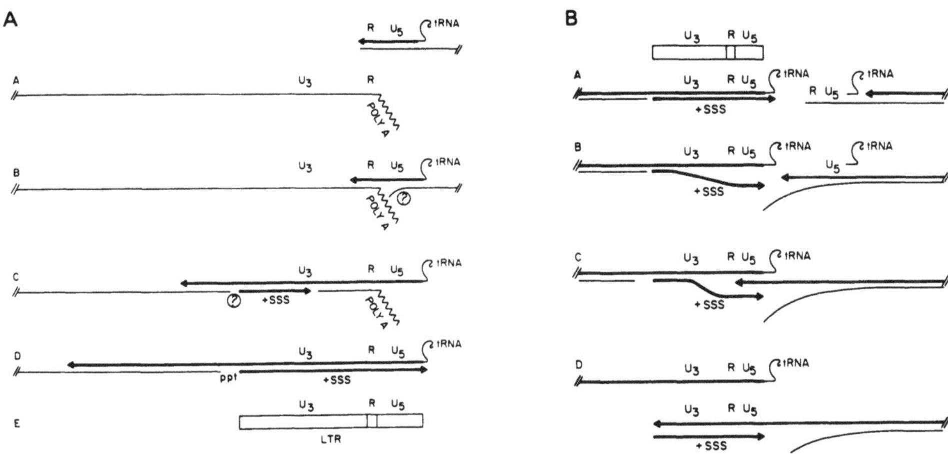

Figure 6. Proviral DNA synthesis and LTR formation.

Left panel:

A: Reverse transcription of the minus strand DNA initiates close to the 5' end of the viral RNA genome. The newly synthesized DNA (thick line) is still associated with its template, and linked to the tRNA primer. The RNA strands are depicted by thin lines.

B: The transcriptional complex jumps to 3' end of one of the two RNA copies of the viral genome. The R sequences present at both ends of virion RNA play a crucial role in this step of the process. The mechanism(s) allowing the base pairing of R to the cDNA (question mark) is not fully understood.

C: The minus DNA strand synthesis proceeds along the new template and the plus strand DNA starts. The question mark indicates that the mechanisms involved in this process are not yet clear.

D: The synthesis of (+) strand strong stop fragment (+SSS) is completed, once the first few bases of tRNA primer have been copied.

E: The LTR which will reside at the 3' end of the completed proviral DNA molecule.

Right panel: The second transfer of DNA synthesis between templates.

A: The complete proviral 3' distal LTR, and the viral minus strand DNA with a complete plus strand strong stop DNA (+SSS) attached to it are shown. The growing minus strand DNA has the primer still attached to it.

B: The growing minus DNA strand displaces the tRNA primer and go through the double-stranded complex between the +SSS and its template.

C: The minus strand DNA displaces the (+SSS) fragment, and can be elongated using this free DNA template.

D: There is shown the completed double stranded structure generated by the process described in panel C.

Bottom: The LTR which will be present at the left end of the unintegrated linear proviral DNA.

(Reproduced with permission from Hughes, S.H *Synthesis, integration, and trascription of the retroviral provirus* Curr. Top. Microbiol. Immunol. (1983), **103**, 23-49).

one. As described previously, genomic RNA molecules exhibit the same sequence R at both ends. The minus DNA chain leaving its exhausted template (see Fig. 13 of Chapter 3 of this volume) can grab the right complementary sequences located at the other terminus, the 3' end of the diploid viral genome, finding therefore a new template on to which its synthesis may resume (53). One can visualize the jump for the 5' end to the 3' end as follows:

i) the sequences extending from the primer-binding site up to the 5' end of the RNA genome are reverse-transcribed into a short piece of cDNA (*minus strong-stop DNA*).
ii) the 3' end of the minus DNA is released from the 5' end of the RNA template. This can be accomplished either by displacement or by digestion of a portion of the 5' end of the viral RNA, i.e. the R sequence. It is not known whether the jump from the 5' end to the 3' end of viral RNA involves the same viral genomic molecule, or else involves the second molecule of the diploid viral genome present in the same virion.
iii) extension of the minus DNA strain on the new template. This process seems to be continuous and relatively slow, but nonetheless the fidelity of the reverse transcriptase in carrying on its activity is much poorer than that observed with other DNA polymerases.

Whether the slow rate of the synthesis of the minus DNA chain and the error-prone activity of the reverse transcriptase are inherent to the structure of the template or are peculiar features of the viral enzyme are still open questions.

With the jump of the minus DNA chain from the 5' to the 3' end one of the LTR present at the extremity of the viral RNA, linear DNA is generated. In order to generate the second LTR the synthesis of the plus-strand DNA chain must start at the 5' boundaries of U_3, proceed to 3' and make another jump to link the complementary sequences present at the two ends of the minus DNA chain. For this event to take place, another crucial step must occur, i.e. the endonucleolytic cleavage of viral RNA genomes, that is apparently carried out by the RNase H activity of the reverse transcriptase. As a consequence of this cleavage, oligomeric RNA fragments are generated that will serve as primers for the synthesis of the plus-strand DNA. Apparently the RNase H activity effects a single endonucleolytic scission either near or at that portion of the tRNA primer which is distal with respect to the 3' end of tRNA that remains either base-paired with plus-strand DNA or linked to the minus-strand DNA. This activity is present in the alpha subunit of avian reverse transcriptase, and diminishes in the presence of a homopolymeric inhibitor of RNAse H.

Recent sequencing studies have demonstrated that the synthesis of the plus-strand DNA chain starts in all retroviral genomes at a specific site, having the sequence AATG or GATG, near to the 5' end of U_3. The plus-strand DNA chain is then synthesized from left to right through the 3' end.

In vitro reactions carried out with purified avian myeloblastosis virus (AMV) DNA polymerase and viral RNA, give rise to a full-length minus-strand and to a plus-strand around 300-400 bp in length. The plus-strand chain is covalently linked to a fragment of 12 ribonucleotides representing the primer derived from RNase

H-digested genomic RNA; this short plus-strand sequence is called "plus-strand strong stop" and is the sequence which makes the second jump looking for its own complementary sequences present on the minus-strand chain, which has already been extended into or beyond the primer-binding site (Fig. 6). Once base-paired, the minus strand would be extended to the point at which the "plus-strand strong-stop" initiates, whereas the synthesis of plus-strand could generate long plus-strands even of genomic length. It has been pointed out, however, that the plus-strand can start also from other sites along the viral genome (105). These sites are those where the synthesis of the minus-strand "pauses". The "pauses" may be inherent to the secondary structure of the RNA template. Analysis of *pol* mutants has shown that some as yet undefined activities of reverse transcriptase may have some role in taking over the template hindrance.

Once the reverse transcription has been completed, the provirus DNA must be trimmed by removing the tRNA via the RNase H activity present in the reverse transcriptase. The linear duplex DNA is transported into the nucleus, where is converted into circular forms. In this process both viral and cellular factors are involved. For instance, under conditions of restrictions by the FV-1 locus, full length linear DNA but not circular DNA can be found. Recently, a quite satisfactory progress in understanding at a molecular level the process of proviral integration in the host genome has been made, favouring the idea that integration occurs without apparent specificity for host sequences, whereas the LTR ends of the viral sequences are those quite precisely used at the integration site. It has been established that a viral nucleoprotein complex rather than naked viral DNA is the effector for integration. Evidence has been provided by Temin and coworkers (106) that the close circles with two tandem LTR are most likely the precursors of the integrated forms. The short inverted repeats present at the ends of the LTR represent the essential signal for virus integration (Fig. 7). These short inverted repeats generate a unique palindromic sequence which is the "att" site. The enzymes involved in the integration process have not been clearly identified as yet (107).

An endonuclease activity has been associated with the products of the *pol* genes of RSV, AMV and MLV, and appears to be encoded in the 3' terminal third of the gene. It has not been yet defined the real role of the viral endonuclease which can be involved either in DNA binding, or in trimming linear DNA. However, analysis of chromosomal DNA at the virus insertion site has shown that the size of nucleotide duplication appears to depend on the infectious virus. Therefore, it appears as the most likely function of the viral endonuclease to cleave host DNA generating a short duplication. Conceivably, the enzyme may recognize the "att" site formed by the palindromic sequence located between the U_5-U_3 regions of proviral circular DNA containing the two LTR. Once integrated, the proviruses are quite stable, and they may undergo amplification (108).

B. Expression of the Viral Genome

Once integrated in the host chromosome, the proviral DNA starts to function as template to generate in most cases the production of viral particles. As every other virus, retroviruses depend mainly upon the host transcriptional and translational apparatus for the expression of their genetic information. Since eukaryotic cells have a different pattern of genetic organization, the retroviruses have adopted several strategies to effectively exploit host resources.

Retroviral genes are transcribed in genomic or subgenomic mRNAs, which are policystronic. Since animal cells cannot translate single gene products from polycistronic mRNAs, retroviruses like many other animal viruses, have developed the mechanism to cleave precursor polypeptides to generate viral proteins.

Most if not all viral RNAs are probably transcribed by the host RNA polymerase II, using as template the integrated proviral DNA. The primary trascription product from the provirus is a full length RNA subunit and represents 0.1 to 1% of total cellular RNA, and about 20% of the polyadenylated RNA. In the life cycle of replication-competent viruses about half of the full length RNA is used to provide the progeny virion RNA, while the other half serves as the mRNA for viral gene products. As discussed previously (see Fig. 2), the usual gene order of transcription of replication-competent retrovirus, from the 5' end to the 3' end of viral RNA, is *gag-pol-env* (or *gag-pol-env-src*, in the case of RSV). The *gag* and *pol* proteins are translated from genomic length mRNAs (Fig. 8), whereas the *env* and *src* genes from subgenomic RNAs. *gag* mRNA species are always more abundant than *pol* or *env* mRNAs. Studies performed to characterize in detail the viral mRNAs have shown that i) the major species of viral mRNAs are transcribed from a single initiation site, and ii) that the subgenomic mRNAs are generated by splicing. The 5' ends of cell mRNAs are identical to the 5' end of virion RNA. The *gag* and *pol* gene products are translated from genomic length mRNAs which differ

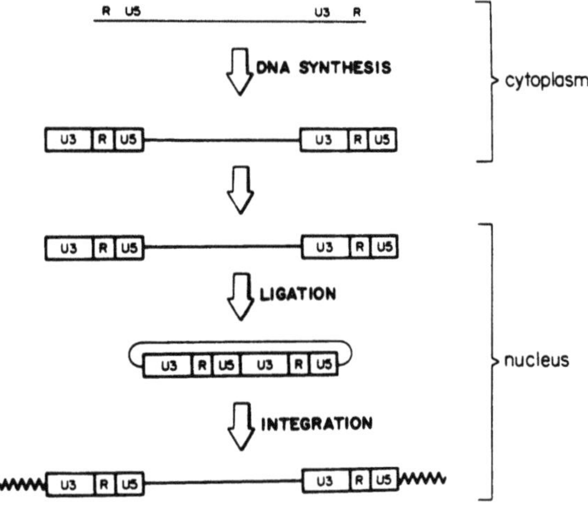

Figure 7. Scheme of the integration of proviral DNA in the host chromosome.
Linear, double strand proviral DNA is synthezised in the cytoplasm and converted into a circular form in the nucleous prior to integration process.

(Reproduced with permission from Panganiban, A.T., & Temin, H.M: *Circles with two tandem LTRs are precursors to integrated retrovirus DNA* Cell (1984), **36**, 673-679, (107)).

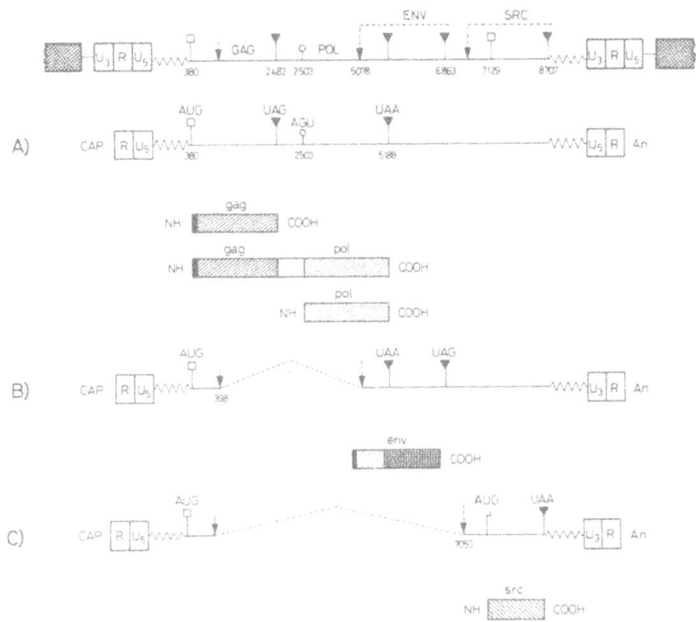

Figure 8. Schematic representation of Rous Sarcoma Virus genome expression from an integrated provirus.

Top line: The coding sequences are shown as straight line, the crosshatched boxes represent host chromosomal DNA, the open boxes are the viral LTR, the zig zag lines indicate the adjacent non coding sequences.

A: Synthesis of *gag* and *pol*: The proviral DNA is transcribed into a genomic size mRNA. The initiation AUG codon is at position 380-383. The amber termination codon UAG (2483-2486) will stop translation after the synthesis of a *gag*. The precursor is represented by a right hatched box, the aminoterminal sequences by a black box.
The *pol* gene product is translated as a polyprotein from the same genomic size mRNA, by a frameshifting mechanism that suppresses the termination after *gag*. The polyprotein is then proteolytically processed, and the peptide coded from the *gag* sequences and from nucleotide 2483 through 2503 are deleted.
The *pol* precursor is represented by a black and white box.

B: The *env* gene products are translated as a polyprotein from a subgenomic mRNA generated by a splice which takes off most of the coding sequences of the *gag* and *pol* genes.
The composite mRNA for *env* contains, therefore, 5' and 3' terminal regulatory sequences (open boxes) and the 18 nucleotides (380-398) from the *gag* mRNA, in addition to the coding sequences of the *pol* gene.
The AUG initiation codon used for *gag* polyproteins is also used for the translation of *env* precursor.
Pol and *env* genes are read on different reading frames. The terminator codon (UAG) for the *env* product is located at position 6863-6866. The *env* product is represented by a crossed line box.

C: The *src* gene product is translated from a spliced subgenomic mRNA.
The acceptor site for this splice is located at nucleotide 7053, and the donor site is the same used for *env* mRNA at position 398. The initiation codon (AUG) for the *src* product is at position 7129 and the terminator codon (UAA) at position 8706.
The *src* product is represented by a left hatched box.

with respect to their reading frames. The *env* gene and in general all the genes located at the 3' end, are translated from subgenomic mRNAs generated by splicing out the region between the LTR at the 5' end and the specific starting point of the gene (53).

Regulatory sequences

Molecular cloning of viral DNA has allowed to define the regulatory sequences of viral transcription within the LTRs. Nucleotide sequence analysis of LTRs cloned from several strains of retroviruses suggests that LTRs provide functions fundamental to the expression of most eukaryotic genes, namely promotion, initiation, and polyadenylation of transcripts. The U_3 region contains the sequence TATAAA which is the "canonical" promoter for transcription of cellular and viral mRNAs. This sequence is 24-31 nucleotides away from the predicted initiation site (see Table 5).

In the U_3 region "enhancer" elements have been detected. The "enhancer" elements are short regions of DNA, found upstream of several transcriptional initiation sites, which have the capacity of increasing the activity of a wide variety of heterologous promoters in a nonpolar fashion and over a considerable distance (53).

The polyadenylation site is at the 3' end of U_3, but there is evidence that transcription of proviral DNA continue up to U_5. There are inconclusive data to explain how the viral messenger and genomic RNAs are polyadenylated. There are still many unresolved questions about the preference of cellular polymerase II in choosing the 5' LTR rather than the 3' LTR, a choice that may be dictated by the secondary structure of the RNA (53).

Synthesis of the *gag* gene products

The internal core proteins are synthesized as a single precursor polyprotein of molecular weight ranging from 70 to 80 kDa, depending on the particular virus strain. Once synthesized, the precursor polyproteins are associated to the plasmamembrane, most of the time phosphorylated, and then cleaved to generate the internal core proteins. In this context the organization and processing of avian C-type viral core proteins will be taken as example to present this processes (109) (Fig. 9).

The core proteins of avian retroviruses are synthesized from a precursor of 76 kDa ($Pr76^{gag}$) translated from a mRNA of genomic size (Fig. 9). Translation of the precursors initiates at an AUG codon 372 nucleotides away from the capped 5' end of genomic RNA. In murine retroviruses the *gag*-precursor is synthesized on membrane-bound ribosomes, whereas in avian retroviruses the $Pr76^{gag}$ *gag* precursor is synthesized on free cytoplasmic polyribosomes. The order of the core proteins within the precursor is NH_2-p19-p10-p27-p15-COOH (Fig. 9). Therefore, at the aminoterminal one finds first the coding sequences for lipid and RNA binding proteins, then those for the hydrophobic core-shell constituents followed by a nucleic acid-binding peptide. At the carboxyl terminus a protease (p15) specific for

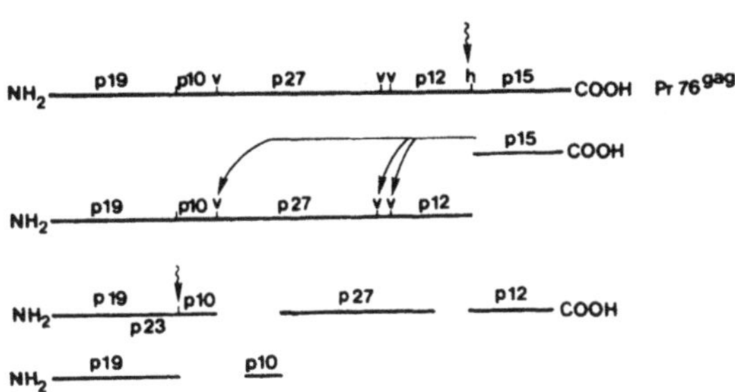

B

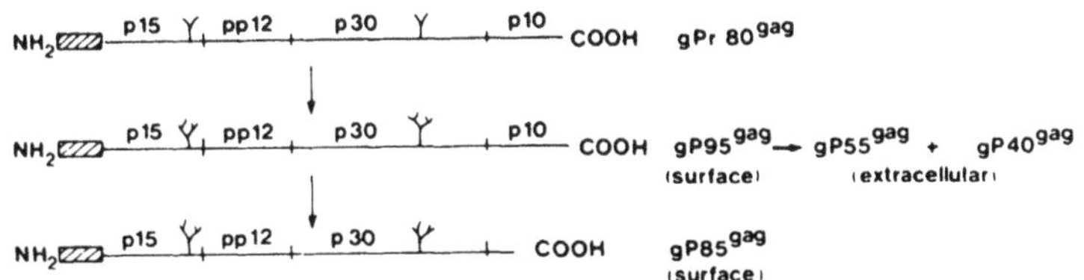

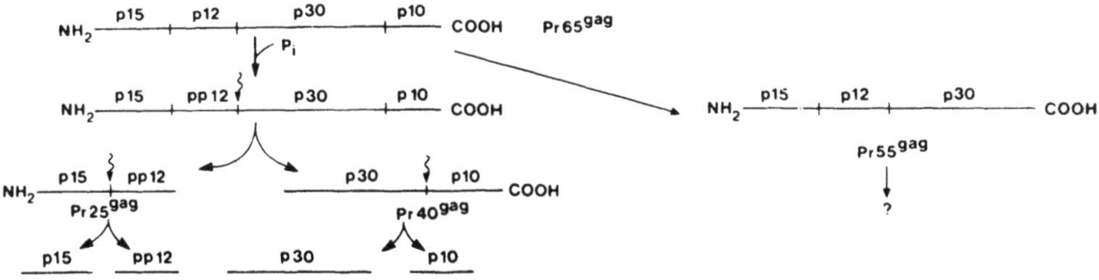

Figure 9. Processing of the gag and env precursor polyproteins of the avian replication competent c-type retroviruses.

A: The processing of the *gag* precursor $Pr76^{gag}$ begins with the removal of the p15 by a cellular protease which cleaves at the site marked h. p15 will then cleave the polyprotein at the sites marked V, giving rise to p12, p27, and p23. The latter is further processed to generate p19 and p10.

B: The maturation of *env* glycoproteins is described starting from the precursor $p63^{env}$. The cleavage site is indicated by an arrow. The glycosylation sites are indicate by branched lines.

(Reproduced with permission from Dickson, C., Eisenman, R., Fan, H., Hunter, E., & Teich, N: "Protein Biosynthesis and Assembly" In: *Molecular Biology of Tumor Viruses: RNA Tumor Virus*, eds. Weiss, R.A., Teich, N., Varmus, H.E., & Coffin, J.M., Cold Spring Harbor Laboratory, Cold Spring Harbor, NY (1982), 513-648, (18)).

the cleavage of the nearby aminoacids is present. A fairly similar arrangement of *gag* proteins exists within the MLV precursor $Pr65^{gag}$ (110). In fact, in MLV the lipid-binding (p15) and RNA-binding (pp12) proteins precede the core-shell protein (p30) at N-terminal end of $Pr65^{gag}$ and are followed by the nucleic-acid-binding protein (p10) (Fig. 10).

Pulse-chase experiments have shown that the avian $Pr76^{gag}$ has a half-life of 45 minutes. The precursor $Pr76^{gag}$ undergoes thereafter a series of cascade proteolytic cleavages that lead to the final core proteins. There is indirect evidence that a specific host cell protease is responsible for the first step, i.e. the cleavage and activation of the viral protease p15, which in turn will cleave the $Pr76^{gag}$ generating p12 and p27. p29 and p10 may thereafter be generated by the action of another protease (111).

The *pol* gene protein

Reverse transcriptase, the product of the *pol* gene was believed to be translated from a mRNA of *genomic* length that had been *spliced* to remove the terminator codon of the *gag* polyprotein. Recently, evidence has been provided that the precursor of the viral polymerase is translated from the same mRNA that codes for the *gag* precursor, but a frame shift mechanism forces the incorporation of a glycine in the protein product at the place of the terminator codon active in the *gag* mRNA. This control mechanism on protein synthesis was known in bacteria, but had never been shown to regulate protein synthesis in eukaryotic organisms.

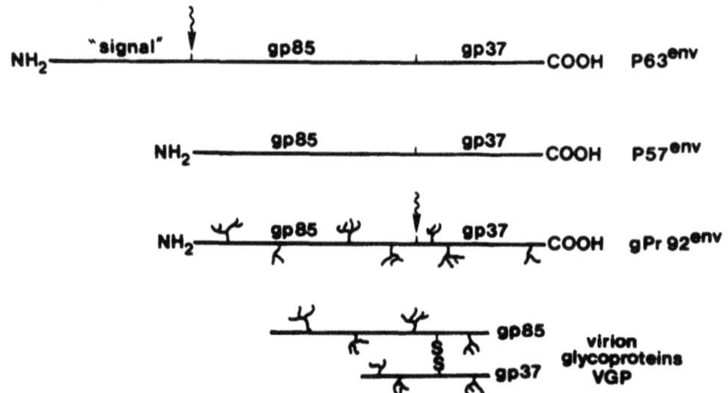

Figure 10. Synthesis and processing of Murine Leukemia Virus gag-polyprotein.
Initiation of translation from two sites in the *gag* mRNA may results in the synthesis of two primary translation products, $Pr65^{gag}$ and $gPr80^{gag}$. They have probably identical sequences, except for an extra peptides at, or near, the amino terminus in $gPr80^{gag}$ indicated by the hatched box. The treelike markes indicate the location of the two carbohydrate side chains. Phosphorylation of p12 and cleavage of the $Pr65^{gag}$ precursor at the sites indicated by arrow generate the mature internal structural proteins of the virion.

(Reproduced with permission from Dockson, C., Eisenman, R., Fan, H., Hunter, E., & Teich, N: "Protein Biosynthesis and Assembly" In: *Molecular Biology of Tumor Viruses: RNA Tumor Virus*, eds. Weiss, R.A., Teich, N., Varmus, H.E., & Coffin, J.M., Cold Spring Harbor Laboratory, Cold Spring Harbor, NY (1982), 513-648, (18))

The precursor of avian retroviral reverse transcriptase is a polyprotein ($Pr180^{gag\text{-}pol}$) containing both *gag* and *pol* determinants (Fig. 9), of about 180-200 kDa. The $Pr180^{gag\text{-}pol}$ is synthesized on cytoplasmatic polyribosomes. Even if the $Pr180^{gag\text{-}pol}$ contains the aminoacid sequences for the *gag* proteins, it is not their major progenitor. Its processing can be summarized as follows:

i) the precursor $Pr180^{gag\text{-}pol}$ is cleaved to give $Pr130^{gag\text{-}pol}$ which contains only p15 protease as *gag* determinant;
ii) the $Pr130^{gag\text{-}pol}$ is cleaved to remove the p15 protease. This process should occur during virus budding; and
iii) during (or shortly after) the budding the activated p15 brings about the final processing steps in maturation of reverse transcriptase giving rise to alpha, beta and p32 subunits. The alpha subunit must be phosphorylated at its carboxyl terminus before the cleavage which generates the p32 phosphoprotein (112).

The *env* gene proteins

The mRNA for the *env* polyprotein is generated by splicing out most of *gag* and *pol* sequences (see Fig. 8). The acceptor splice site and even the *env* coding sequences themselves are located upstream from the *pol* termination codon. In murine leukemia viruses the donor splice site for *env* mRNA is located upstream from *gag*, whereas in the Rous sarcoma virus is located downstream from the *gag* initiation codon (113).

The *env* gene products represent the glycosylated envelope proteins and are synthesized on subgenomic mRNA. These proteins are subjected to different maturation processes because of their structure and roles in virus morphogenesis. The *env* precursor is translated on membrane-bound polyribosomes. Recent experimental evidence suggests that *env* mRNA is translated to give the polypeptide $p63^{env}$. A first cleavage removes the hydrophobic signal peptide (64 N-terminal aminoacids) and the polypeptide is then glycosylated while inserted into the endoplasmic reticulum. These maturation steps generate the $gPr92^{env}$ precursor having both gp85 and gp37 in the order NH_2-gp85-gp37-COOH. The virion glycoproteins are thereafter generated by further cleavage, glycosylation and disulfide-bonds as shown in Fig 9.

v-onc

The *v-onc* genes are subjected to the same regulatory controls as the replication genes. The *v-onc* can be independently expressed from subgenomic mRNA, such as *v-src* and *v-myb* for instance. In the case of *v-myb* (that replaces part of the *env* gene), a transduced intron sequence is used as a splice acceptor site for generation of *v-myb* mRNA (114).

As for the *v-src* gene, its splice acceptor site derives from the cellular progenitor, and resides upstream the initiation codon.

Other *v-onc* such as *v-myc* or *v-erb B* are expressed from genomic length mRNA as *gag-onc* fusion protein, so that their expression is controlled in a manner analogous to that of *gag* (71).

C. Virion Assembly

Although much has been learned on the mechanisms regulating transcription, translation and processing of viral proteins, still much has to be learned on steps involved with virion morphogenesis.

In crude extracts the cleavage of the avian $Pr76^{gag}$ is inhibited by compounds which affect membrane protein association (115). For the assembly of viral proteins and virion budding it was suggested that *gag* and *gag-pol* precursors migrate under the plasma membrane where their amino termini interact with the transmembrane portion of envelope glycoproteins, while the carboxyl termini interact with the genomic RNA. There is experimental evidence that both avian and murine retroviral particles can be generated in the absence of envelope glycoproteins, a finding that underlines the importance of *gag* and *pol* products and their cleavage intermediates in the morphogenesis of virions. The interaction with the *env* glycoproteins apparently is not needed while the mere processing of *gag* and *gag-pol* polyproteins will activate the morphogenetic process (116).

Virus-specific genomic RNA is preferentially packaged into virions, even if also rRNA, tRNA, as well as cellular mRNAs and subgenomic viral mRNAs can be found within the viral particles. It was first proposed the *pol* product plays a key role in the selection of the viral genomic RNA over other available RNAs (117). Recent experimental evidences, however, points to pp19 and pp12 as the most important tools in the genome-selecting machinery. In fact, the viral RNA bound to be packaged into virions must be tightly associated with these phosphoproteins, an association that makes of it an easily selectable target which make it easily selectable among the pool of cytoplasmatic RNAs. The proposed model for virion assembly and genomic RNA selection have been formulated taking account of experimental data available, but there are still unresolved issues before making this working hypothesis the definitive scheme.

4. MOLECULAR BASIS OF PATHOLOGY

Retroviruses have been mainly associated with neoplasia development, but there are a variety of pathological syndromes with retroviral infections, which may not always result in overt disease.

Retroviruses can cause proliferative diseases, anemias, and slow degenerative syndromes sometimes resembling "autoimmune" diseases.

A. Non-Pathogenic Infections and Endogenous Retroviruses

The foamy viruses are the only retroviruses not yet associated with disease. These viruses are detectable in neural tissues, and when they infect cultured cells, vacuolated syncytia have been observed. On the other hand, almost all endogenous retroviruses are not pathogenic. The only pathogenic ones have been isolated from inbred strains of mice specially selected for high incidence of disease, like, for instance, the murine leukemia virus (MLV) genomes present in the AKR mice and the MMTV present in the GR mice.

The non-pathogenicity may have been generated under natural selection pressure which ensured a symbiotic relationship between virus and host. Xenotropism may be important in preventing disease, since pathogenic variants and recombinants are known to arise in mice in which replication of the ecotropic endogenous virus (AKR, C58) is allowed. On the other hand, infection by the AKR endogenous ecotropic virus does not induce disease in mouse strains susceptible to infection, and RAV-O seems to be completely nonpathogenic in chickens permissive for its replication. Therefore, xenotropism cannot entirely explain nonpathogenicity.

Endogenous viruses have also important roles in protecting the host from pathogenic sequelae of exogenous virus infection. For instance, chickens which express envelope-related gene(s) are protected from infection by avian leukosis viruses.

B. Acute Neoplasms

Some retroviruses induce tumors shortly after infection. For instance, chickens inoculated with RSV in the wing web develop tumors of 10 mm in diameter within 1 week; Friend erythroleukemia virus complex causes splenomegaly in mice 2 weeks after infection. The tumors are not clonal outgrowths but are formed by different clones of infected transformed cells.

Acute oncogenic retroviruses not only induce solid tumors and leukemias in animals but may transform cultured cells as well. All the acute oncogenic viruses which transform cells *in vitro*, bear genes (*v-onc*) responsible for their neoplastic activity in their genome. The pathogenesis of defective leukemia and sarcoma viruses carrying oncogenes is strictly related to the functions of the *v-onc* products in target cells (see above).

Retroviral oncogenes are derived from a limited set of cellular genes (118, 119). Usually, the transduction of these genes into the viral genome lead both to deletions in one or more viral structural genes and the appearance of point mutations and deletions in the captured genes, so that, with the exception of RSV, all the acutely transforming retroviruses are replication-defective and need a replication competent-virus (helper) to be able to replicate.

The conversion of a non-transforming proto-oncogene into the tumorigenic v-*onc* of a retroviruse is probably due to acquisition of an autonomous regulatory system provided by the viral promoter sequences present in the LTRs, that have also "enhancer-like" functions (12). However, as already mentioned, the proto-oncogenes can also be activated while still residing in the cellular genome by point mutation, translocation, or gene amplification.

Some acutely transforming retroviruses, such as the SFFV component of Friend virus complex, are quite different with regard to the oncogenic component of their genome; in fact, the transforming sequence is a rearranged viral *env* gene captured from an endogenous virus present in the host chromosomes. Unlike most *v-onc*'s, no sequences complementary to the *env* gene of SFFV have been found in eukaryotic genome (120).

The presence of *v-onc* genes seems crucial for the *in vivo* retroviral tumorigenesis; conversely, the role of the other viral sequences has not been clarified as yet. In fact, genetic constructs from RSV, that contain only the viral LTR sequences plus the *v-src* oncogene, are still able to induce sarcomas in chickens (121).

About 20 different *v-onc* genes have been discovered; they have been found in the genome of retroviruses isolated from mice, cats, chickens, rats and a few other animal species (see above); nevertheless, retroviruses carrying *v-onc* genes are isolated only occasionally in nature. They are not naturally transmitted from one host to another, either because they are defective (except RSV) or because they induce tumors so rapidly that there is a strong selection against vertical or horizontal transmission.

C. Chronic Neoplasms

Retroviruses, which do not contain transforming sequences, induce neoplasms only after a long period of incubation. For this reason they have been named *slow* transforming viruses. The slowness stays to indicate the long latency period between infection and manifestation of the disease, even if, once induced, the tumor grows very fast.

Tumors induced by *slow* retroviruses usually appear in viremic animals and the oncogenic growth is a multistage process. For instance, in newly hatched chicks one month after infection with ALV, a preneoplastic transformation can be observed: In fact, 10-100 follicles in the bursa, which are called transformed follicles, become filled with immature blast cells. Five months later, about 2% of the transformed follicles grow to form nodules. Most likely, the metastases which arise in the infected animals derive from these nodules, since the pattern of viral restriction fragments in their DNA is similar.

The ALV-induced leukosis represents the first and best understood example of viral oncogenesis caused by a non-defective retrovirus. In the tumor development the critical event is the site-specific integration of the ALV genome in the vicinity of the cellular proto-oncogene *c-myc*. Its subsequent activation apparently leads the target cells (B lymphocytes) to the neoplastic process. The activation is due to an enhanced transcription initiated by the ALV promoter of the U_3 region of LTR, located in the 3' end of the integrated provirus (122). Intriguingly, in a few cases (123) the proviral LTR is located either downstream of *c-myc*, or upstream but in the wrong orientation. To explain this inconsistence, the presence of an enhancer element that increases expression of the neighbouring gene in a non polar fashion has been hypothesized (123). In fact, such a type of element has been isolated in the RSV (124).

The activation of *c-myc* does not seem to be the only necessary event for the development of lymphomagenesis: In fact, additional genetic sequences able to induce transformation when transfected *in vitro* in NIH3T3 fibroblasts have been isolated from B cells transformed by ALV (125, 126). This new oncogene, named Blym-l, was subsequently cloned and sequenced, and the deduced aminoacid sequence predicts a peptide of 8 kDa with a partial homology with the aminoterminal domains of the secreted forms of the transferrin family of proteins (127). In contrast to most cellular proto-oncogenes, sequences homologous to Blym-l have not been found to date in the genome of any retrovirus.

The role played by Blym-l in lymphomagenesis is in keeping with the hypothesis that the neoplastic process can be a multistep phenomenon, although the exact relationship among the different events is still unknown. Hypothetically, different factors can act in neoplasia in a fixed order, so that one event is necessary for the occurrence of the subsequent one (subsequential model); alternatively, several factors can act without any order, neoplasia resulting from the cumulative effects of their properties (cumulative model). In fact, in most cases the conversion of a normal cell to a neoplastic one can be due to a process resembling a combination of the two above described models.

Very interesting results have been also obtained in the molecular studies of leukemogenesis induced by MLV. Studies carried out on an endogenous Mink Cell Focus forming (MCF) virus present in the spontaneous thymomas in AKR mice, and performed with recombinants containing only segments of viral genome, have shown that the U_3 region of the provirus seems to play the most important role in the development of leukemia (128).

Similar experiments aimed at analyzing target cell tropism of MLV have been performed with the construction of chimeric viral genomes from MLV with different target cell specificity (i.e. Friend-MLV, F-MLV, and Moloney-MLV, Mo-MLV). The results showed that the LTR are the principal determinants for the leukemogenic tropism; in fact, viruses with a chimeric genome containing the LTR of Mo-MLV and the structural genes of F-MLV are able to induce predominantly T-cell leukemias, as the prototype Mo-MLV. Reciprocally, recombinant viruses with the F-MLV LTR and the Mo-MLV structural genes induce almost exclusively the erythroleukemia typical of the prototype F-MLV virus (129).

At variance with ALV, MLV proviruses integrate in more than one single site; in fact, at least five preferred integration sites have been detected. Incidentally, the activation of *c-myc* proto-oncogene, typical of ALV infection, has been detected only in a low percentage of tumors. In these cases, as for exemple in the T cell lymphoma cell lines derived from mice inoculated with the Soule strain of MLV, rearrangements have been shown in *c-myc* proto-oncogene subsequently to the integration of MLV provirus (130).

In conclusion, in some cases the cellular transformation mediated by slow and acute transforming viruses may have a similar molecular basis: In fact, ALV can mediate cellular transformation by inserting its LTR near to the *c-myc* proto-oncogene, which then becomes activated and functions in a fashion analogous to that of an exogenous transforming *v-onc* .

The Human T lymphotropic virus family which, as ALV and MLV, induces neoplastic transformation without any oncogene sequences, will be dealt with in a separate section.

D. Tumors Induced by Mouse Mammary Tumor Viruses

The mouse mammary tumor viruses are replication-competent retroviruses, do not bear an oncogene, and induce tumors only after 4-9 months of latency. In strains of mice with a high mammary tumor incidence MMTV can be congenitally transmitted through the milk to the suckling offspring. In the mouse strains GR, DBAf and C3Hf MMTV is genetically transmitted (131).

MMTV has a life cycle similar to that of other well characterized retroviruses, but is unique in that its transcription is regulated by glucocorticoid hormones via interactions involving the LTR. The DNA sequence of the LTR of different endogenous and exogenous MMTV proviruses has been determined and they are similar but not identical (131). A region of 202 nucleotides before the LTR cap site is involved in the glucocorticoid response (132). Moreover, in the U_3 region there is an open reading frame leading to the expression of a protein with a molecular weight of 36 kDa. The role of this protein in the oncogenic process is not yet understood (131).

Molecular analysis of MMTV-induced tumors has shown new proviral sequences in their DNA. Moreover, by analyzing the virus-cell junction fragments with virus-specific probes, the clonal origin of the tumor cell population has been defined (133). Since ALV induces bursal lymphomas in chickens via its insertion next to the proto-oncogene *c-myc*, numerous experiments have been carried out in search of preferential integration sites which may be related to the neoplastic process induced by MMTV.

Although most MMTV-induced tumors contain multiple new proviral integration sites, it appears as if a specific integration site does exist in a region of about 30kb, defined as int-1 localized on chromosome 15 (134). In addition, 22/45 tumors detected in BR6 mice show a new proviral integration site in a different region about the same size and defined int-2 localized on chromosome 7; they are not related to any of the known oncogenes.

Analysis of the configuration of the proviruses next to the int-1 and int-2 loci suggests that the activation of the cellular genes is mediated by a mechanism other than that known as insertion-promotion. It has been proposed that MMTV integration provides via the LTRs an enhancer which activates preferentially the nearest cellular promoter (135).

5. HUMAN T LEUKEMIA VIRUSES

To date the human T-Lymphotropic virus (HTLV) family encompasses the only *bona fide* human retroviruses associated with disease.

The early difficulties in isolation of retrovirus from human tissues were in sharp contrast with the findings in some animal systems. The successful isolation of different strains of the HTLV family was the result of several developments in the fields of retroviral biochemistry and human T-cell biology during the last ten years, i.e. development of sensitive assay for detecting human retroviral reverse transcriptase (136) and, more recently, the discovery of T Cell Growth Factor also designated Interleukin-2 (IL-2) (137, 138).

Three major subgroups of HTLV viruses have been identified: The type I (HTLV-I) is associated with human adult T cell leukemia (ATL) (139-141); the type II (HTLV II) was originally isolated from a patient with a clinically benign hairy cell leukemia of the T-cell type (142). A third class of human retroviruses with similar properties was isolated from patients with immunodeficiency sindrome (AIDS) and was later named HIV (human immunodeficiency viruses) (143, 144, 145).

The viruses of HTLV family have the typical morphology of type-C retroviruses: The size (usually 100 nanometers) varies considerably from 90 to 140 nanometers in the different established cell cultures. All HTLV isolates share in common the following features: i) isolation from mature T cells; ii) selective tropism for $OKT4^+$ T cells; iii) an immunologically cross reactive internal core protein, and reverse transcriptase of similar size and biochemical properties (146-149).

A. HTLV Genome

The organization of the genome is similar to that of animal retroviruses (60): The provirus sequence is 9032 nucleotides long, is bounded by a terminal repeat of 754 nucleotides, and is flanked by LTR regions. LTRs of HTLV are unusually long (about 750 bp) as compared with other mammalian or avian type-C viruses (usually 320 to 580 bp). In this respect they closely resemble the bovine leukemia virus (BLV). The extra-length of the HTLV LTR regions is the result of unusually long R and U_5 regions, whereas U_3 regions are comparable in length to those of other type-C viruses. There are three structural genes in HTLV genome: *gag*, *pol* and *env* genes are present in the same order as in other replication-competent retroviruses. HTLV, however, are unique in that their genomic RNA presents a large open reading frame (X-LOR) between the *env* gene and the downstream LTR which, in different reading frame, codes for four proteins, designated pX proteins, the functions of which are still unknown. pX sequences are viral in origin as they are not found in uninfected cells. Indirect evidence implicates the X-LOR gene in a) the transforming activity of HTLV I and II, and b) the trans-acting transcriptional regulation (TAT) phenomenon: This is defined as a greatly increased rate of transcription directed by the viral LTR in infected as compared to uninfected cells. The high rate of transcription is induced by viral infection and acts in *trans*; the TAT phenomenon is not restricted to cell types that are *in vivo* targets of the transforming or cytopathic effects of these viruses.

Analysis of the genome of HIV reveals that, unlike HTLV-I and II, HTLV-III/ LAV isolates exhibit a high degree of polymorphism (150, 151): In fact, the widespread existence of variants of HIV is the main feature of this virus. It is of interest that the most divergent part of the genome lies within the major exterior glycoprotein portion of the *env* gene, as epitopes involved in virus neutralization generally reside in this region.

Comparison of prototypes HTLV III/LAV and HTLV I showed short stretches of significant homology in the *gag-pol* region and distant but still specific homology in the pX region (X-LOR).

The genomic organization of HIV viruses (Fig. 3) has some important differences from that of HTLV I and II. In fact, the genome of HIV contains, besides the *gag*, *pol* , *env* and *lor* regions, three additional open reading frames. The first one overlaps with the 3' end of the *pol* gene (short open reading frame or SOR) and is probably the vestige of on envelope gene. The second one, named 3' ORF-LTR, extends from the end of the *lor* region into LTR; the third is a gene, called tat-III, wich lies between the *sor* and *env* genes and is able to mediate activation, in a *trans* configuration, of the genes linked to HIV long terminal repeat (LTR) sequences (152-155).

B. Products of HTLV Genome

The *gag*, *pol* and *env* genes of HTLV code, as for all other retroviruses, for the core structural proteins, for the reverse transcriptase and for the envelope proteins, respectively. The *gag* gene product of HTLV I and II and, with minor differences, of HIV is a precursor polypeptide which is then processed into the protein p19 (at the NH_2 terminus), p24 and p15 (at the COOH terminus).

All HTLV viruses have a reverse transcriptase, biochemically different from that of most mammalian type-C retroviruses. In fact, the HTLV enzyme is larger (about 100 kDa) and prefers Mg^{++} as the divalent cation instead of Mn^{++} (156).

The product of *env* gene is a 61-68 kDa glycoprotein with two functional domains: An external glycoprotein part (gp46) and a transmembrane part (p21).

As already mentioned, one of the special features of HTLV I and II, as compared to other non-acute retroviruses that contain *gag*, *pol* and *env* genes, is that they have an additional sequence of approximately 1600 nucleotides located between the 3' end of the *env* gene and the 5' end of U3 region of the proviral LTR (157). The potential proteins (pX proteins) encoded by these long open reading frames (LOR) of HTLV I and II are approximately of the same length (357 aminoacids for HTLV I and 337 for HTLV II), and have the same aminoacid residues in 259 of the 337 positions. A protein of about 42 kDa that would correspond to the LOR protein has been observed (158).

As already described, besides the *gag*, *pol* and *env* genes, HIV viruses present three other open reading frames. The product of the SOR region is unknown. The length of the predicted 3' ORF products varies from about 13.5 kDa to about 35.5 kDa depending upon the HIV strain. The predicted protein products of the 3' ORF region differ from those predicted for other known retrovirus genomes. Finally, the tat-III gene encodes a 14 kDa protein which is likely responsible for the trans-acting transcriptional (TAT) phenomenon that greatly elevates the level of gene expression directed by the HIV LTR. The elements of this autostimulatory pathway include the above mentioned protein as effector and a responder element, the trans-activating responsive sequence (TAR) (155), located within the HIV LTR.

In cells that constitutively express the trans-activation protein, the level of synthesis of viral proteins and heterologous proteins under the control of the LTR of HIV increases dramatically. Increased levels of protein synthesis occur without a comparable increase in the levels of corresponding mRNAs. Probably the post-transcriptional event mediated by the HIV TAT protein accounts for positive regulation of HIV gene products in infected cells (159). Moreover, there is evidence that the product of tat-III gene is an absolute requirement for virus expression (160).

C. Cellular Pathology

T4 lymphocytes are the specific target of all HTLVs. The members of HTLV family can induce both tumors and immunosuppressive effects. Tumors induced by HTLV I occur after a long latent period and with a low incidence (161). Moreover, like the non-acute retroviruses, tumors induced by HTLV I are monoclonal with respect to integration of the provirus (162, 163), although the provirus can integrate at multiple sites. However, unlike the non-acute retroviruses, preferred sites for proviral integration have not yet been detected among different tumors. Furthermore, HTLV I and II can transform primary lymphocytes in culture, a property restricted to the acute transforming retroviruses containing oncogenes. HTLV I and II, as the HIV, can also induce immunosuppressive effects.

One of the distinctive properties of the HTLVs is their ability to reproduce *in vitro* some of their biological properties expressed *in vivo*. These include immortalization and transformation of normal T lymphocytes by HTLV I and II (164, 165), and cell killing by HIV.

When T cells are infected *in vitro* with HTLV I or II, some cells grow indefinitely, in contrast with the majority of T cells which enter a crisis period between days 30 and 40 of *in vitro* culture. A remarkable difference between the HTLV-I positive ATL cells and normal T4 cells is the constitutive expression of IL-2 receptors by ATL cells (166-168). Like the primary ATL cells, the T-cell lines immortalized by *in vitro* infection with HTLV-I or -II are predominantly OKT4$^+$ (only occasionally OKT8$^+$). They show partial or complete independence of exogenous IL-2, express high levels of receptors for IL-2 transferrin, and HLA-DR determinants, and often display lobulated nuclei and multinucleated giant cells. Although HTLV-I or -II infection of T cells is not able to transform *in vitro* the majority of cells, the immunological functions of the lymphocytes are dramatically impaired (169, 170); this explains both the lack of helper activity in the OKT4$^+$ ATL cells obtained from patients, and the increased susceptibility of HTLV-I-infected individuals to opportunistic infections (171). Furthermore, HTLV-I-infected T4 lymphocytes may cause polyclonal B-cell activation (169) and selective killing of certain cytotoxic T cell (170).

The T cells infected *in vitro* with HIV initially resemble the HTLV-I or -II-infected lymphocytes: In fact, numerous multinucleated giant cells appear but, unlike the indefinite growth of HTLV-I or -II-infected cells, a drastic loss of T lymphocyte viability is noted within 2 or 3 weeks; the onset of this premature cell death correlates with expression of virus. The specific target of HIV infection is the OKT4$^+$ T lymphocyte subset: The selective tropism of HIV for the T4 cells is correlated with the recent observation that T4 antigen is a component of the receptor for HIV. As a consequence of being selectively infected and killed, the OKT4$^+$ lymphocyte subset is depleted along with a decline and eventual cessation of virus production.

HTLV I, II or HIV can infect not only T lymphocytes but also other cell types: B cells can be infected by HTLV I, II or HIV, and in addition, fibroblasts (172) and vascular endothelial cells can be infected *in vitro* with HTLV I. Glial cells in the brain of AIDS patients have been found to be infected by HIV (173): Starting from these observations, experimental work is in progress to establish a direct correlation between some hitherto ununderstandable neurological pathology and HIV infection. Finally, also the macrophages have been shown to be sensitive to the infection by HIV.

6. CONCLUDING REMARKS

Studies on the replication cycle of retroviruses provided the first experimental evidence that RNA could be transcribed into DNA, thus opening new frontiers in the understanding of the molecular mechanisms involved in evolution and regulation of the genetic information. Moreover, nucleotide sequencing of retroviral proviruses and of a variety of mobile genetic elements (transposons) from many organisms has revealed that they are closely related on structural grounds as all of them contain structurally long terminal repeats or LTR.

These mobile genetic elements can be found in bacteria such as Tn 9 of *E. coli*, in yeasts (Ty 1) and in insects as well (copia of Drosophila). Like other kinds of transposable elements, they generate short duplications of host sequences at the site of insertion and carry short inverted terminal repeats. Furthemore, eukaryotic elements from Drosophila and yeasts often share with proviruses a sequence similar to tRNA adjacent to 5' long terminal repeat and a polypurine tract adjacent to the 3' LTR.

These structural homologies among transposable elements and proviruses do pose several interesting questions about their origin and functions. In fact, these transposable elements, once integrated in the host, can activate or switch off the flanking genes, playing, therefore, an important role in the regulation of gene expression. It is still open to debate whether transposable elements move around the chromosomes by mechanisms of non homologous recombination, or their RNA must be reverse transcribed prior to integration in the chromosome.

In Drosophila genes sandwiched between transposons could be reverse transcribed, and once inserted back into the chromosomes could be used as substrates to generate new functional genes.

Processed genes or pseudogenes may represent an example, since they are the intron-less versions of functional genes. Accordingly, the idea that they may have arisen by reverse transcription of processed mRNAs is receiving an ever stronger support. The human beta globin gene provides an excellent example.

The new ideas on the molecular mechanism of evolution of the eukaryotic genome appear to go hand in hand with our understanding of the biology of retroviruses, and their putative role as carriers of cellular genetic information: It is tempting to speculate as to whether transposons represent the degenerate decayed forms of pre-existing retroviruses or else they are their precursors.

Acknowledgements

We thank Dr. Raul Perez Bercoff for helpful discussions; we are indebted also to Mrss. Rosina Bellizzi and Catia Buschittari for the helpful assistance in typing the manuscript.

7. REFERENCES

1) ELLERMANN, V., & BANG, O: "Experimentelle leukamie bei Huhnern"Zentralbl. Bakteriol. (1908), **46**, 595-609.

2) ROUS, P: "A sarcoma of the fowl transmissible by an agent separable from the tumor cells" J. Exp. Med. (1911), **13**, 397-413.

3) BITTNER, J.J: "The milk-influence of breast tumors in mice" Science (1942), **95**, 462-463.

4) GROSS, L: "Spontaneous leukemia developing in C3H mice following inoculation in infancy, with AK-leukemic extracts, or AK-embryos" Proc. Soc. Exp. Biol. Med. (1951), **71**, 27-32.

5) FRIEND, C: "Cell-free transmission in adult Swiss mice of a disease having the character of a leukemia" J. Exp. Med. (1957), **105**, 307-318.

6) GARDNER, M.B: "Historical background" In: *Molecular Biology of RNA Tumor Viruses*, edited by J.R. Stephenson, pp. 2-46, Academic Press, New York (1980).

7) WONG-STAAL, F., & GALLO, R.C: "Human T-lymphotropic retroviruses" Nature (London) (1985), **317**, 395-403.

8) BOLOGNESI, D.P., MONTELARO, R.C., FRANK, H., & SCHAFER, W: "Assembly of c-oncornaviruses: A model" Science (1978), **199**, 183-186.

9) JAENISCH, R: "Endogenous retroviruses" Cell (1983), **32**, 5-6.

10) HARTLEY, J.W., ROWE, W.P., & HUEBNER, R.J: "Host-range restrictions of murine leukemia viruses in mouse embryo cell cultures" J. Virol. (1970), **5**, 221-225.

11) BISHOP, J.M: "Cancer genes comes of age" Cell (1983), **32**, 1018-1020.

12) WEISS, R., TEICH, N., VARMUS, H., & COFFINS, J: "Molecular Biology of Tumor Viruses" Cold Spring Harbor Laboratory (1984, 1985).

13) HAASE, A.T., BRAHIC, M., CARROLL, D., SCOTT, J., STOWING, L., TRAYNOR, B., & VENTURA, P: "Visna: An animal model for studies of virus persistence" In: *Persistent Viruses*, edited by J.G. Stevens et al., pp.643-654, Academic Press, New York (1978).

14) HOOKS, J.J., & GIBBS, C.J Jr: "The foamy viruses" Bacteriol. Rev.(1975), **39**, 169-185.

15) FINE, D., & SCHOCHETMAN, G: "Type D primate retroviruses: A review" Cancer Res. (1978), **38**, 3123-3139.

16) MASON, M.M., BODGEN, A.E., ILIEVSKI, V., ESBER, H.J., BAKER, J.R., & CHOPRA, H.C: "History of a rhesus monkey adenocarcinoma containing virus particles resembling oncogenic RNA viruses" J. Natl. Cancer Inst. (1972), **48**, 1323-1331.

17) AUGUST, J.T., BOLOGNESI, D.P., FLEISSNER, E., GILDEN, R.V., & NOWINSKI, R.C: "A proposed nomenclature for the virion proteins of oncogenic RNA viruses" Virology (1974), **60**, 595-601.

18) DICKSON, C., EISENMAN, R., FAN, H., HUNTER, E., & TEICH, N: "Protein Biosynthesis and Assembly" In: *Molecular Biology of Tumor Viruses: RNA Tumor Virus* edited by WEISS, R.A., TEICH, N., VARMUS, H.E., & COFFIN, J.M., Cold Spring Harbor Laboratory, Cold Spring Harbor, NY (1982), 513-648.

19) DICKSON, C., EISENMAN, R., & FAN, H: "Protein Biosynthesis and Assembly" In: *Molecular Biology of Tumor Viruses: RNA Tumor Virus Supplements and Appendixes* edited by WEISS, R.A., TEICH, N., VARMUS, H.E., & COFFIN, J.M., Cold Spring Harbor Laboratory, Cold Spring Harbor, NY (1985), 135-145.

20) BARBACID, M., & AARONSON, S.A: "Membrane properties of the gag gene-coded p15 protein of mouse type-C RNA tumor viruses" J. Biol. Chem. **253**, 1408-1414.

21) BARBACID, M., LONG, L.K., & AARONSON, S.A: "Major structural proteins of type B, type C, and type D oncoviruses share interspecies antigenic determinants" Proc. Natl. Acad. Sci. (1980), **77**, 72-76.

22) MERIC, C., DARLIX, J.L., & SPAHR, P.F: "It is Rous sarcoma virus protein p12 and not p19 that binds tightly to Rous sarcoma virus RNA" J. Mol. Biol. (1984), **173**, 531-538.

23) MASSEY, R.J., & SCHOCHETMAN, G: "Gene order of mouse mammary tumor virus precursor polyproteins and their interaction leading to the formation of a virus" Virology (1979), **99**, 358-371.

24) PEPINSKY, R.B., CAPPIELLO, D., WILKOWSKI, C., & VOGT, V.M: "Chemical cross-linking of proteins in avian sarcoma and leukemia viruses" Virology (1980), **102**, 205-210.

25) MOELLING, K., SCOTT, A., DITTMAR, K.E.J., & OWADA, M: "Effect of p15-associated protease from an avian RNA tumor virus on avian virus-specific polyprotein precursors" J. Virol. (1980) **33**, 680-688.

26) HUNTER, E., BENNETT, J.C., BHOWN, A., PEPINSKY, R.B., & VOGT, V.M: "Amino-terminal amino acid sequence of p10, the fifth major *gag* polypeptide of avian sarcoma and leukemia viruses" J. Virol. (1983), **45**, 885-888.

27) GAUTSCH, J.W., ELDER, J.H., SCHINDLER, J., JENSEN, F.C., & LERNER, R.A: "Structural markers on core protein p30 of murine leukemia virus: Functional correlation with Fv-l tropism" Proc. Natl. Acad. Sci. (1978), **75**, 4170-4174.

28) NUSSE, R., ASSELBERGS, F.A.M., SALDEN, M.H.L., MICHALIDES, R.J.A.M., & BLOEMENDAL, H: "Translation of mouse mammary tumor virus RNA: Precursor polypeptides are phosphorylated during processing" Virology (1978), **91**, 106-115.

29) BALTIMORE, D: "RNA-dependent DNA polymerase in virions of RNA tumour viruses" Nature (London) (1970), **226**, 1209-1211.

30) BEEMON, K.L., & KEITH, J.M: "Structure of Rous sarcoma virus RNA. 1. Localization of N^6-methyladenosine; 2. The sequence of 23 nucleotides following the 5' capped terminus m^7GpppG^mp. In *Animal virology* (ed. D. Baltimore et al.), pp. 97-105. Academic Press, New York.

31) PETERS, G., HARADA, F., DAHLBERG, J.E., PANET, A., HASELTINE, W.A., & BALTIMORE, D: Low-molecular-weight RNAs of Monoley murine leukemia virus: "Identification of the primer for RNA-directed DNA synthesis" J.Virol. (1977), **21**, 1031-1041.

32) GOLOMB, M., GRAND GENETT, P.D., & MASON, W: "Virus-coded DNA endonuclease from avian retrovirus" J. Virol. (1981), **38**, 548-555.

33) HAUSEN, P., & STEIN, H: "Ribonuclease H. An enzyme degrading the RNA moiety of DNA-RNA hybrids" Eur. J. Biochem. (1970), **14**, 278-283.

34) MOELLING, K., BOLOGNESI, D.P., BAUER, H., BUSEN, W., PLASSMANN, H.W., & HAUSEN, P: "Association of viral reverse transcriptase with an enzyme degrading the RNA moiety of RNA-DNA hybrids" Nature New Biol. (1971), **234**, 240-243.

35) GRANDGENETT, D.P., VORA, A.C., & SCHIFF, R.D: "A 32,000-dalton nucleic acid-binding protein from avian, retrovirus cores possesses DNA endonuclease activity" Virology (1978), **89**, 119-132.

36) GERARD, G.F: "Mechanism of action of Monoley murine leukemia virus RNA-directed DNA polymerase associated RNase H (RNase H I)" Biochemistry (1981), **20**, 256-264.

37) DION, A.S., WILLIAMS, C.J. & MOORE, D.H: "RNase H and RNA-directed DNA polymerase: Associated enzymatic activities of murine mammary tumor virus" J. Virol. (1977), **22**, 187-193.

38) PAWSON, T., MELLON, P., DUESBERG, P.H., & MARTIN, G.S: "*env* gene of Rous Sarcoma Virus: identication of the gene product by cell-free translation" J. Virol. (1980), **33**, 993-1003.

39) ANDERSON, K.B., & NEXO, B.A: "Entry of murine retrovirus into mouse fibroblasts" Virology (1983), **125**, 85-98.

40) REDMOND, S.M.S., & DICKSON, C: "Sequence and expression of the mouse mammary tumour virus *env* gene" EMBO J. (1983), **2**, 125-131.

41) COFFIN, J: "Structure of the Retroviral Genome" In: *Molecular Biology of Tumor Viruses: RNA Tumor Virus* edited by WEISS, R.A., TEICH, N., VARMUS, H.E., & COFFIN, J.M., Cold Spring Harbor Laboratory, Cold Spring Harbor, NY (1982), 261-368.

42) PETERS, G., & CLOVER, C: "tRNAs and priming of RNA-directed DNA synthesis in mouse mammary tumor virus" J. Virol. (1980), **35**, 31-40.

43) SCHWARTZ, D.E., TIZARD, R., & GILBERT, W: "Nucleotide sequence of Rous Sarcoma Virus" Cell (1983), **32**, 853-869.

44) FARAS, A.J., GARAPIN, A.C., LEVINSON, W.E., BISHOP, J.M., & GOODMAN, H.M: "Characterization of the low-molecular-weight RNAs associated with the 70S RNA of Rous sarcoma virus" J. Virol. (1973), **12**, 334-342.

45) FARAS, A.J., DAHLBERG, J.E., SAWYER, R.C., HARADA, F., TAYLOR, J.M., LEVINSON, W.E., BISHOP, J.M., & GOODMAN, H.M: "Transcription of DNA from the 70S RNA of Rous sarcoma virus. II. Structure of a 4S RNA primer" J. Virol. (1974), **13**, 1134-1142.

46) PETERS, G., & CLOVER, C: "Low-molecular-weight RNAs and initiation of RNA-directed DNA synthesis in avian reticuloendotheliosis virus" J. Virol. (1980), **33**, 708-716.

47) HUGHES, S.H: "Sequence of the long terminal repeat and adjacent segments of the endogenous avian virus Rous-associated virus O" J. Virol. (1982), **43**, 191-200.

48) DEVARE, S.G., REDDY, E.P., LAW, J.D., & AARONSON, S.A: "Nucleotide sequence analysis of the long terminal repeat of integrated simian sarcoma virus: Evolutionary relationship with other mammalian retroviral long terminal repeats" J. Virol. (1982), **42**, 1108-1113.

49) HASELTINE, W.A., & KLEID, D.G: "A method for classification of 5' termini of retroviruses" Nature (London) (1978), **273**, 358-364.

50) PALMITER, R.D., GAGNON, J., VOGT, V.M., RIPLEY, S., & EISENMAN, R.N: "The NH_2-terminal sequence of the avian oncovirus *gag* precursor polyprotein ($Pr76^{gag}$)" Virology (1978), **91**, 423-433.

51) PURCHIO, A.F., ERIKSON, E., & ERIKSON, R.L: "Translation of 35S and of subgenomic regions of avian sarcoma virus RNA" Proc. Natl. Acad. Sci. (1977), **74**, 4661-4665.

52) SWANSTROM, R., PARKER, R.C., VARMUS, H.E., & BISHOP, J.M: "Transduction of a cellular oncogene: The genesis of Rous sarcoma virus" Proc. Natl. Acad. Sci. (1983), **80**, 2519-2523.

53) VARMUS, H.E: "Form and function of retroviral proviruses" Science (1982), **216**, 812-820.

54) SRINIVASAN, A., REDDY, E.P., DUNN, C.Y., & AARONSON, S.A: "Molecular dissection of transcriptional control elements within the long terminal repeat of the retrovirus" Science (1984), **223**, 286-289.

55) DHAR, R., MCCLEMENTS, W.L., ENQUIST, L.W., & VANDE WOUDE, G.F: "Nucleotide sequences of integrated Moloney sarcoma provirus long terminal repeats and their host and viral junctions" Proc. Natl. Acad. Sci. (1980), **77**, 3937-3941.

56) WANG, L.H., SNYDER, P., HANAFUSA, T., MOSCOVICI, C., & HANAFUSA, H: "Comparative analysis of cellular and viral sequences related to sarcomagenic cell transformation" Cold Spring Harbor Symp. Quant. Biol. (1980), **44**, 755-764.

57) SAGATA, N., YASUNAGA, T., OGAWA, Y., TSUZUKU-KAWAMURA, J., & IKAWA, Y: "Bovine leukemia virus: Unique structural features of its long terminal repeats and its evolutionary relationship to human T-cell leukemia virus" Proc. Natl. Acad. Sci. (1984), **81**, 4741-4745.

58) PEREZ-BERCOFF, R., & BILLETER, M.A: "Characterization of the 3'-terminal region of the large molecular weight RNA subunits from normal and transformation-defective Rous sarcoma virus" Biochim. Biophys. Acta (1976), **454**, 383-388.

59) KOZAK, M: "Evaluation of the *scanning model* for initiation of protein synthesis in eukaryotes" Cell (1980), **22**, 7-8.

60) SEIKI, M., HATTORI, S., HIRAYAMA, Y., & YOSHIDA, M: "Human adult T-cell leukemia virus: complete nucleotide sequence of the provirus genome integrated in leukemia cell DNA" Proc. Natl. Acad. Sci. (1983), **80**, 3618-3622.

61) LOVINGER, G.G., & SCHOCHETMAN, G: "5' terminal nucleotide sequences of type C retroviruses: Features common to noncoding sequences of eukaryotic messenger RNAs" Cell (1980), **20**, 441-449.

62) ITIN, A., KESHET, E: "Nucleotide sequence analysis of the long terminal repeat of murine virus-like DNA (VL30) and its adjacent sequences: Resemblance to retrovirus proviruses" J. Virol. (1983), **47**, 656-659.

63) SHANK, P.R., & LINIAL, M: "Avian oncovirus mutant (SE21Q1b) deficient in genomic RNA: Characterization of a deletion in the provirus" J. Virol. (1980), **36**, 450-456.

64) DONEHOWER, L.A., HUAGER, A.L., & HAGER, G.L: "Regulatory and coding potential of the mouse mammary tumor virus long terminal redundancy" J. Virol. (1981), **37**, 226-238.

65) BESMER, P., MURPHY, J.E., GEORGE, P.C., QIU, F., BERGOLD, P.J., LEDERMAN, L., SNYDER, H.W., Jr., BRODEUR, D., ZUCKERMAN, E.E., & HARDY, W.D: "A new acute transforming feline retrovirus and relationship of its oncogene v-kit with the protein kinase gene family" Nature (London) (1986), **320**, 415-417.

66) DESGROSEILLERS, L., VILLEMUR, R., & JOLICOEUR, P: "The high leukemic potential of Gross passage A murine leukemia virus maps in the region of the genome corresponding to the long terminal repeat and to the 3' end of *env* " J. Virol. (1983), **47**, 24-32.

67) BISHOP, J.M: "Retroviruses and cancer genes" Adv. Cancer Res. (1982),37, 1-32.

68) HARVEY, J.J: "An unidentified virus which causes the rapid production of tumors in mice" Nature (London) (1964), **204**, 1104-1105.

69) MOLONEY, J.B: "Biological studies on a lymphoid-leukemia virus extracted from Sarcoma 37. I. Origin and introductory investigations" J. Natl. Cancer Inst. (1960), **24**, 933-951.

70) ABELSON, H.T., & RABSTEIN, L.S: "Influence of prednisolone on Moloney leukemogenic virus in Balb/mice" Cancer Research (1970), **30**, 2208-2212.

71) STEHELIN, D., GUNTAKA, R.V., VARMUS, H.E., & BISHOP, J.M: "Purification of DNA complementary to nucleotide sequences required for neoplastic transformation of fibroblasts by avian sarcoma viruses" J. Mol. Biol. (1976), **101**, 349-365.

72) STEHELIN, D., VARMUS, H.E., BISHOP, J.M., & VOGT, P.K: "DNA related to the transforming gene(s) of avian sarcoma viruses is present in normal avian DNA" Nature (London) (1976), **260**, 170-173.

73) LEVINSON, A.D., COURTNEIDGE, S.A., & BISHOP, J.M: "Structural and functional domains of the Rous sarcoma virus transforming protein ($pp60^{src}$)" Proc. Natl. Acad. Sci. (1981), **78**, 1624-1628.

74) COPELAND, N.G., ZELENETZ, A.D., & COOPER, G.M: "Transformation by subgenomic fragments of Rous sarcoma virus DNA" Cell (1980), **19**, 863-870.

75) HUEBNER, R.J., & TODARO, G.J: "Oncogenes of RNA tumor viruses as determinants of cancer" Proc. Natl. Acad. Sci. (1969), **64**, 1087-1094.

76) DE FEO-JONES, D., SCOLNICK, E., KOLLER, R., & DHAR, R: "*ras*-related gene sequences identified and isolated from *Saccharomyces cerevisiae*" Nature (London) (1983), **306**, 707-709.

77) VARMUS, H.E: "The molecular genetics of cellular oncogenes" Ann. Rev. Genet. (1984), **18**, 553-612.

78) WEINBERG, R.A: "The action of oncogenes in the cytoplasm and nucleus" Science (1985), **230**, 770-776.

79) LAND, H., PARADA, L.F., & WEINBERG, R.A: "Cellular oncogenes and multistep carcinogenesis" Science (1983), **222**, 771-778.

80) COOPER, G.M., & LANE, M.A: "Cellular transforming genes and oncogenesis" Biochimica and Biophisica Acta (1984), **738**, 9-20.

81) BERRIDGE, M.J: "Oncogenes, inositol lipids and cellular proliferation" Trends in Bioch. Sciences (1984), **5**, 542-546.

82) MARDON, G., & VARMUS, H.E: "Frameshift and intragenic suppressor mutations in a Rous sarcoma provirus suggest *src* encodes two proteins" Cell (1983), **32**, 871-879.

83) OPPERMANN, H., LEVINSON, A.D., & VARMUS, H.E: "The structure and protein kinase activity of proteins encoded by nonconditional mutants and back mutants in the *src* gene of avian sarcoma virus" Virology (1981),108, 47-70.

84) DOOLITTLE, R.F., HUNKAPILLER, M.W., HOOD, L.E., DEVARE, S.G., ROBBINS, K.C., AARONSON, S.A., & ANTONIADES, H.N: "Simian sarcoma virus *onc* gene, *v-sis*, is derived from the gene (or genes) encoding a platelet-derived growth factor" Science (1983), **221**, 275-277.

85) SHERR, C.J., RETTENMIER, C.W., SACCA, R., ROUSSEL, M.F., LOOK, A.T., & STANLEY, E.R: "The *c-fms* proto-oncogene product is related to the receptor for the mononuclear phagocyte growth factor, CSF-1" Cell (1985), **41**, 665-676.

86) DOWNWARD, J., YARDEN, Y., MAYES, E., SCRACE, G., TOTTY, N., STOCKWELL, P., ULLRICH, A., SCHLESSINGER, J., & WATERFIELD, M.D: "Close similarity of epidermal growth factor receptor and *v-erb-B* oncogene protein sequences" Nature (London) (1984), **307**, 521-527.

87) DUESBERG, P.M: "Retroviral transforming genes in normal cells?" Nature (London) (1983), **304**, 219-226.

88) HUNTER, T: "Oncogenes and growth control" Trends in Bioch. Sciences (1985), **10**, 275-280.

89) GROFFEN, J., STEPHENSON, J.R., HEISTERKAMP, N., de KLEIN, A., BARTRAM, C.R., & GROSVELD, G: "Philadelphia chromosomal breakpoints are clustered within a limited region,*bcr* on chromosome 22" Cell (1984),36, 93-99.

90) NOWELL, P., FINAN, J., DALLA FAVERA, R., GALLO, R.C., aR-RUSHDI, A., ROMANCZUK, H., SELDEN, J.R., EMANUEL, B.S., ROVERA, G., & CROCE, C.M: "Association of amplified oncogene *c-myc* with an abnormally banded chromosome 8 in a human leukaemia cell line" Nature (London) (1983), **306**, 494-497.

91) HUNG, M.C., & WEINBERG, R.A: "Multiple independent activations of the *neu* oncogene by a point mutation altering the transmembrane domain of p185" Cell (1986), **45**, 649-657.

92) PARK, M., DEAN, M., COOPER, C.S., SCHMIT, M., O'BRIEN, S.I., BLAIR, D.G., & VANDE WOUD, G.F: "Mechanism of *met* oncogene activation" Cell (1986), **45**, 895-A904.

93) MARTIN-ZANCA, D., HUGHES, S.H., & BARBACID, M: "A human oncogene formed by the fusion of truncated tropomyosin and protein tyrosine kinase sequences" Nature (London) (1986), **319**, 743-748.

94) EVA, A., & AARONSON, S.A: "Isolation of a new human oncogene from a diffuse B-cell lymphoma" Nature (London) (1985), **316**, 273-275.

95) SAKAMOTO, H., MORI, M., TAIRA, M., YOSHIDA, T., MATSUKAWA, S., SHIMIZU, K., SEKIGUCHI, M., TERADA, M., & SUGIMURA, T: "Transforming gene from human stomach cancers and a noncancerous portion of stomach mucosa" Proc. Natl. Acad. Sci. USA (1986), **83**, 3997-4001.

96) TAKAHASHI, M., & RITZ, J: "Activation of a novel human transforming gene, *ret*, by DNA rearrangement" Cell, (1985), **42**, 581-588.

97) KOHL, N.E., KANDA, N., SCHRECK, R.R., BURNS, G., LATT, S.A., GILBERT, F., & ALT, F.W: "Transposition and amplification of oncogene-related sequences in human neuroblastomas" Cell (1983), **35**, 359-367.

98) NAU, M.M., CARNEY, D.N., BATTEY, J., JOHNSON, B., LITTLE, C., GAZDAR, A., & MINNA, J.D: "Amplification, expression and rearrangement of *c-myc* and *N-myc* oncogenes in human lung cancer" Curr. Topics Microbiol. Immunol. (1984), **113**, 172-177.

99) YOUNG, D., WAITCHES, G., BIRCHMEIER, C., FASANO, O., & WIGLER, M: "Isolation and characterization of a new cellular oncogene encoding a protein with multiple potential transmembrane domains" Cell (1986),45, 711-719.

100) SPORN, M.B., & TODARO, G.J: "Autocrine secretion and malignant transformation of cells" N. Engl. J. Med. (1980), **303**, 878-880.

101) HANAFUSA, H., BALTIMORE, D., SMOLER, D., WATSON, K.F., YANIV A., & SPIEGELMAN, S: "Absence of polymerase protein in virions of alpha-type Rous sarcoma virus" Science (1972), **177**, 1188-1191.

102) PETERS, G.G., & HU, J: "Reverse transcriptase as the major determinant for selective packaging of tRNAs into avian sarcoma virus particles" J. Virol. (1980), **36**, 692-700.

103) CORDELL, B.R., SWANSTROM, R., GOODMAN, H.M., & BISHOP, J.M: "$tRNA^{Trp}$ as primer for RNA-directed DNA polymerase: Structural determinants of function" J. Biol. Chem. (1979), **254**, 1866-1874.

104) PANGANIBAN, A.T: "Retroviral DNA integration" Cell (1985), **42**, 5-6.

105) RESNICK, R., OMER, C.A., & FARAS, A.J: "Involvement of retrovirus reverse transcriptase-associated RNase H in the initiation of strong-stop (+) DNA synthesis and the generation of the long terminal repeat" J. Virol. (1984), **51**, 813-821.

106) TEMIN, H.M: "Structure, variation and synthesis of retrovirus long terminal repeat" Cell (1981), **27**, 1-3.

107) PANGANIBAN, A.T., & TEMIN, H.M: "Circles with two tandem LTRs are precursors to integrated retrovirus DNA" Cell (1984), **36**, 673-679.

108) BALTIMORE, D: "Retroviruses and retrotransposons: the role of reverse transcription in shaping the eukaryotic genome" Cell (1985), **40**, 481-482.

109) PALMITER, R.D., GAGNON, J., VOGT, V.M., RIPLEY, S., & EISENMAN, R.N: "The NH_2-terminal sequence of the avian oncovirus *gag* precursor polyprotein ($Pr76^{gag}$)" Virology (1978), **91**, 423-433.

110) BARBACID, M., STEPHENSON, J.R., & AARONSON, S.A: "*gag* gene of mammalian type-C RNA tumour viruses" Nature (London) (1976), **262**, 554-559.

111) VOGT, V.M., WIGHT, W., & EISENMAN, R: "*In vitro* cleavage of avian retrovirus *gag* proteins by viral protease p15" Virology (1979), **98**,154-167.

112) TYLER, J., & VARMUS, H.E: "Expression of the Rous Sarcoma Virus *pol* gene by Ribosomal Frameshifting" Science (1986), **230**, 1237-1242.

113) HUGHES, S.H: "Synthesis, integration, and trascription of the retroviral provirus" Curr. Top. Microbiol. Immunol. (1983), **103**, 23-49.

114) KLAMPNAVER, K.H., RAMSAY, G., BISHOP, J.M., MOSCOVICI, M.G., MOSCOVICI, C., Mc GRATH, J.P., & LAVINSON, A.O: "The product of the retroviral transforming gene *v-myb* is a truncated version of the protein encoded by the cellular oncogene *c-myb*" Cell (1983), **33**, 345-355.

115) VOGT, V.M., EISENMAN, R., & DIGGELMANN, H: "Generation of avian myeloblastosis virus structural proteins by proteolytic cleavage of a precursor polypeptide" J. Mol. Biol. (1975), **96**, 471-493.

116) HAYMAN, M.J., ROYER-POKORA, B., & GRAF, T: "Defectiveness of avian erythroblastosis virus: Synthesis of a 75k gag-related protein" Virology (1979), **92**, 31-45.

117) LEVIN, J.G., & SEIDMAN, J.G: "Selective packaging of host tRNA's by murine leukemia virus particles does not require genomic RNA" J. Virol. (1979), **29**, 328-335.

118) DUESBERG, P: "Activated proto-onc genes: sufficient or necessary for cancer?" Science (1985), **228**, 669-677.

119) BISHOP, J.M: "Viral oncogenes" Cell (1985), **42**, 23-38.

120) LINEMEYER, D.L., MENKE, J.G., RUSCETTI, S.K., EVANS, L.H., & SCOLNICK, E.M: "Envelope gene sequences which encode the gp52 protein of spleen focus-forming viurs are required for the induction of erythroid cell proliferation" J. Virol. (1982), **43**, 223-233.

121) FUNG, Y.T., CRITTENDEN, L.B., DADLY, A.M., & KUNG, H-J: "Tumor induction by direct injection of cloned v-*src* DNA into chickens" Proc. Natl. Acad. Sci. USA (1983), **80**, 353-357.

122) HAYWARD, W.S., BEEL, B.G., & ASTRIN, S.M: "Activation of a cellular *onc* gene by promoter insertion in ALV-induced lymphoid leukosis" Nature (London) (1981), **290**, 475-480.

123) PAYNE, G.S., BISHOP, J.M., & VARMUS, H.E: "Multiple arrangements of viral DNA and an activated host oncogene in bursal lymphomas" Nature (London) (1982), **295**, 209-214.

124) LUCIW, P.A., BISHOP, J.M., VARMUS, H.E., & CAPECCHI, M.R: "Location and function of retroviral and SV40 sequences that enhance biochemical transformation after microinjection of DNA" Cell (1983), **33**, 705-716.

125) COOPER, G.M., & NEIMAN, P.E: "Transforming genes of neoplasms induced by avian lymphoid leukosis viruses" Nature (London) (1980), **287**, 656-659.

126) COOPER, G.M., & NEIMAN, P.E: "Two distinct candidate transforming genes of neoplasms induced by avian lymphoid leukosis viruses" Nature (London) **292**, 857-858.

127) GOUBIN, G., GOLDMAN, D.S., LUCE, J., NEIMAN, P.E., & COOPER, G.M: "Molecular cloning and nucleotide sequence of a transforming gene detected by transfection of chicken B-cell lymphoma DNA" Nature (London) (1983), **302**, 114-119.

128) LENZ, J., CELANDER, D., CROWTHER, R.L., PATARCA, R., PERKINS, D.W., & HASELTINE, W.A: "Determination of the leukaemogenicity of a murine retrovirus by sequences within the long terminal repeat" Nature (London) (1984),308, 467-470.

129) CHATIS, P.A., HOLLAND, C.A., SILVER, J.E., FREDERICKSON, T.N., HOPKINS, N., & HARTLEY, J.W: "A 3' end fragment encompassing the transcriptional enhancers of nondefective Friend virus confers erythroleukemogenicity on Moloney leukemia virus" J. Virol. (1984), **52**, 248-254.

130) CORCORAN, L.M., ADAMS, J.M., DUNN, A.R., & CORY, S: "Murine T lymphomas in which the cellular *myc* oncogene has been activated by retroviral insertion" Cell (1984), **37**, 113-122.

131) NUSSE, R., & VARMUS, H.E: "Many tumors induced by the mouse mammary tumor virus contain a provirus integrated in the same region of the host genome" Cell (1982), **31**, 99-109.

132) PETERS, G., LEE, A.E., & DICKSON, C: "Activation of cellular gene by mouse mammary tumour virus may occur early in mammary tumour development" Nature (London) (1984), **309**, 273-275.

133) PETERS, G., BROOKES, S., SMITH, R., & DICKSON, C: "Tumorigenesis by mouse mammary tumor virus: Evidence for a common region for provirus integration in mammary tumors" Cell (1983), **33**, 369-377.

134) NUSSE, R., van OOYEN, A., COX, D., FUNG, Y.K.T., & VARMUS, H: "Mode of proviral activation of a putative mammary oncogene (*int-1*) on mouse chromosome 15" Nature (London) (1984), **307**, 131-136.

135) PETERS, G., KOZAK, C., & DICKSON, C: "Mouse mammary tumor virus integration regions *int-1* and *int-2* map on different mouse chromosomes" Mol. Cell. Biol. (1984), **4**, 375-378.

136) SARNGADHARAN, M.G., ROBERT-GUROFF, M., & GALLO, R.C: "DNA polymerases of normal and neoplastic mammalian cells" Biochim. Biophys. Acta (1978), **516**, 419-487.

137) MORGAN, D.A., RUSCETTI, F.W., & GALLO, R.C: "Selective *in vitro* growth of T-lymphocytes from normal human bone marrow" Science (1976), **193**, 1007-1008.

138) RUSCETTI, F.W., MORGAN, D.A., & GALLO, R.C: "Functional and morphological characterization of human T-cells continuously grown *in vitro*" J. Immunol. (1977), **119**, 131-138.

139) ESSEX, M., McLANE, M.F., LEE, T.H., FALK, L., HOWE, C.W.S., MULLINS, J.L., CABRADILLA, C., & FRANCIS, D.P: "Antibodies to cell membrane antigens associated with human T-cell leukemia virus in patients with AIDS" Science (1983), **220**, 859-862.

140) GALLO, R.C., KALYANARAMAN, V.S., SARNGADHARAN, M.G., SLISKI, A., VONDERHEID, E.C., MAEDA, M., NAKAO, Y., YAMADA, K., ITO, Y., GUTENSOHN, N., MURPHY, S., BUNN, P.A. Jr., CATOVSKY, K., GREAVES, M.F., BLAYNEY, D.W., BLATTNER, W., JARRETT, W.F., zur HAUSEN, H., SELIGMANN, M., BROUET, J.C., HAYNES, B.F., JEGASOTHY, B.V., JAFFE, E., COSSMAN, J., BRODER, S., FISHER, R.I., GOLDE, D.W., & ROBERT-GUROFF, M: "Association of the human type C retrovirus with a subset of adult T-cell cancers" Cancer Res. (1983), **43**, 3892-3899.

141) GALLO, R.C., SARIN, P.S., GELMANN, E.P., ROBERT-GUROFF, M., RICHARDSON, E., KALYANARAMAN, V.S., MANN, D., SIDHU, G.D., STAHL, R.E., ZOLLA-PAZNER, S., LEIBOWITCH, J. & POPOVIC, M: "Isolation of human T-cell leukemia virus in acquired immune deficiency syndrome (AIDS)" Science (1983), **220**, 865-867.

142) KALYANARAMAN, V.S., SARNGADHARAN, M.G., ROBERT-GUROFF, M., MIYOSHI, I., GOLDE, D., & GALLO, R.C: "A new subtype of human T-cell leukemia virus(HTLV-II) associated with a T-cell variant of hairy cell leukemia" Science (1982), **218**, 571-573.

143) SARNGADHARAN, M.G., POPOVIC, M., BRUCH, L., SCHPBACH, J., & GALLO, R.C: "Detection, isolation and continuous production of cytopathic retroviruses (HTLV III) from patients with AIDS and pre-AIDS" Science (1984), **224**, 497-508.

144) LEVY J.A., HOFFMAN, A.D., KRAMER, S.M., LANDIS, J.A., SHIMABUKURO, J.M., & OSHIRO, L.S: "Isolation of lymphocytopathic retroviruses from San Francisco patients with AIDS" Science (1984), **225**, 840-842.

145) BARR-SINOUSSI, F., MATHUR-WAGH, U., REY, F., BRUN VEZINET, F., YANCOVITZ, S.R., ROUZIOUX, C., MONTAGNIER, L., MILOVAN, D., & CHERMANN, J.C: "Isolation of lymphadenopathy-associated virus (LAV) and detection of LAV antibodies from US patients with AIDS" Jama (1985), **253**, 1737-1739.

146) MIYOSHI, I, KUBONISHI, I., YOSHIMOTO, S., AKAGI, T., OHTSUKI, Y., SHIRAISHI, Y., NAGATA, K., & HINUMA, Y: "Type C virus particles in a cord T-cell line derived by co-cultivating normal human cord leukocytes and human leukaemic T cells" Nature (London) (1981), **294**, 770-771.

147) POPOVIC, M., LANGE-WANTZIN, G., SARIN, P.S., MANN, D., & GALLO, R.C: "Transformation of human umbilical cord blood T cells by human T-cell leukemia/lymphoma virus" Proc. Natl. Acad. Sci. USA (1983), **80**, 5402-5406.

148) YAMAMOTO, N., OKADA, M., KOYANAGI, Y., KANNAGI, M., & HINUMA, Y: "Transformation of human leukocytes by cocultivation with an adult T cell leukemia virus producer cell line" Science (1982), **217**, 737-739.

149) CHEN, I.S., QUAN, S.G., & GOLDE, D.W: "Human T-cell leukemia virus type II transforms normal human lymphocytes" Proc. Natl. Acad. Sci. USA (1983), **80**, 7006-7009.

150) SHAW, G.M., HAHN, B.H., ARYA, S.K., POPOVIC, M., GROOPMAN, J.E., GALLO, R.C., & WONG-STAAL, F: "Molecular characterization of human T-cell leukemia (lymphotropic) virus type III in the acquired immunodeficiency syndrome" Science (1984), **225**, 1473-1476.

151) WONG-STAAL, F., HAHN, B., MANZARI, V., COLOMBINI, S., FRANCHINI, G., GELMANN, E.P., & GALLO, R.C: "A survey of human leukemias for sequences of a human retrovirus, HTLV" Nature (London) (1983), **302**, 626-628.

152) SODROSKI, J., ROSEN, C., WONG-STAAL, F., SALAHUDDIN, S.Z., POPOVIC, M., ARYA, S., GALLO, R.C., & HASELTINE, W.A: "Trans-acting transcriptional regulation of human T-cell leukemia virus type III long terminal repeat" Science (1985), **227**, 171-173.

153) ARYA, S.K., GUO, C., JOSEPHS, S.F., & WONG-STAAL, F: "Trans-activator gene of human T-lymphotropic virus type III (HTLV-III)" Science (1985), **229**, 69-73.

154) SODROSKI, J., PATARCA, R., WONG-STAAL, F., ROSEN, C., & HASELTINE, W.A: "Location of the trans-activating region on the genome of human T-cell lymphotropic virus type III" Science (1985), **229**, 74-77.

155) ROSEN, C.A., SODROSKI, J.G., & HASELTINE, W.A: "The location of cis-acting regulatory sequences in the human T cell lymphotropic virus type III (HTLV-III/LAV) long terminal repeat" Cell (1985), **41**, 813-823.

156) POIESZ, B.J., RUSCETTI, F.W., GAZDAR, A.F., BUNN, P.A., MINNA, J.D., & GALLO, R.C: "Detection and isolation of type-C retrovirus particles from fresh and cultured lymphocytes of a patient with cutaneous T-cell lymphoma" Proc. Natl. Acad. Sci. USA. (1980), **77**, 7415-7419.

157) SEIKI, M., HATTORI, S., HIRAYAMA, Y., & YOSHIDA, M: "Human adult T-cell leukemia virus: complete nucleotide sequence of the provirus genome integrated in leukemia cell DNA" Proc. Natl. Acad. Sci. USA (1983), **80**, 3618-3622.

158) LEE, T.H., COLIGAN, J.E., SODROSKI, J.A., HASELTINE, W.A., SALAHUDDIN, S.Z., WONG-STAAL, F., GALLO, R.C., & ESSEX, M: "Antigens encoded by the 3' terminal region of human T-cell leukemia virus: evidence for a functional gene" Science (1984), **226**, 57-61.

159) ROSEN, C.A., SODROSKI, J.G., GOH, W.C., DAYTON, A.I., LIPPKE, J., & HASELTINE, W.A: "Post-transcriptional regulation accounts for the trans-activation of the human T-lymphotropic virus type III" Nature (London) (1986), **319**, 555-559.

160) FISHER, A.G., FEINBERG, M.B., JOSEPHS, S.F., HARPER, M.E., MARSELLE, L.M., REYES, G., GONDA, M.A., ALDOVINI, A., DEBOUK, C., GALLO, R.C., & WONG-STAAL, F: "The **trans**-activator gene of HTLV-III is essential for virus replication" Nature (London) (1986), **320**, 367-370.

161) GALLO, R.C., & WONG-STALL, F: "Retroviruses as etiologic agents of some animal and human leukemias and lymphomas and as tools for elucidating the molecular mechanism of leukemogenesis" Blood (1982), **60**, 545-557.

162) KETTMANN, R., DESCHAMPS, J., CLEUTER, Y., COUEZ, D., BURNY, A., & MARBAIX, G: "Leukemogenesis by bovine leukemia virus: proviral DNA integration and lack of RNA expression of viral long terminal repeat and 3' proximate cellular sequences" Proc. Natl. Acad. Sci. USA (1982),79, 2465-2469.

163) SEIKI, M., EDDY, R., SHOWS, T.B., & YOSHIDA, M: "Nonspecific integration of the HTLV provirus genome into adult T-cell leukaemia cells" Nature (London) (1984), **309**, 640-642.

164) MIYOSHI, L., KUBONISHI, I., YOSHIMOTO, S., AKAGI, T., OHTSUKI, Y., SHIRAISHI, Y., NAGATA, K., & HINUMA, Y: "Type C virus particles in a cord T-cell line derived by co-cultivating normal human cord leukocytes and human leukaemic T cells" Nature (London) (1981), **294**, 770-771.

165) POPOVIC, M., LANGE-WANTZIN, G., SARIN, P.S., MANN, D., & GALLO, R.C: "Transformation of human umbilical cord blood T cells by human T-cell leukemia/lymphoma virus" Proc. Natl. Acad. Sci. USA (1983), **80**, 5402-5406.

166) GOOTENBERG, J.L., RUSCETTI, F.W., MIEI, J.W., GAZDAR, A.F., & GALLO, R.C: "Human cutaneous T cell lymphoma and leukemia cell lines produce and respond to T cell growth factor" J. Exp. Med. (1981), **154**, 1403-1418.

167) POIESZ, B.J., RUSCETTI, F.W., MIER, J.W., WOODS, A.M., & GALLO, R.C: "T-cell lines established from human T-lymphocytic neoplasias by direct response to T-cell growth factor" Proc. Natl. Acad. Sci. USA (1980), **77**, 6815-6819.

168) LEONARD, W.J., DEPPER, J.M., UCHIYAMA, T., SMITH, K.A., WALDMANN,T.A., & GREENE, W.C: "A monoclonal antibody that appears to recognize the receptor for human T-cell growth factor; partial characterization of the receptor" Nature (London) (1982), **300**, 267-269.

169) POPOVIC, M., FLOMENBERG, N., VOLKMAN, D.J., MANN, D., FAUCI, A.S., DUPONT, B., & GALLO, R.C: "Alteration of T-cell functions by infection with HTLV-I or HTLV-II" Science (1984), **226**, 459-462.

170) MITSUYA, H., GUO, H.G., COSSMAN, J., MEGSON, M., REITZ, M.S. Jr., & BRODER, S: "Functional properties of antigen-specific T cells infected by human T-cell leukemia-lymphoma virus (HTLV-I)" Science (1984), **225**, 1484-1486.

171) ESSEX, M.E., McLANE, M.F., TACHIBANA, N., FRANCIS, D.P., & LEE, T.H in "Human T-Cell Leukemia/Lymphoma Virus (eds. GALLO, R.C., ESSEX, M.E. & GROSS, L.) 355-362 (Cold Spring Harbor Laboratory, New York, 1984).

172) CLAPHAM, P., NAHY, K., CHEINGSONG-POPOV, R., EXLEY, M., & WEISS, R.A: "Productive infection and cell-free transmission of human T-cell leukemia virus in a nonlymphoid cell line" Science (1983), **222**, 1125-1127.

173) KLATZMANN, D., & GLUCKMAN, J.C: "HIV: facts and hypothesis" Immunology Today (1986), **7**,10, 291-295.

CHAPTER 17

THE MOLECULAR BIOLOGY OF HEPATITIS B VIRUS

DAVID A. SHAFRITZ

Departments of Medicine and Cell Biology and the Marion Bessin Liver Research Center, Albert Einstein College of Medicine, 1300 Morris Park Avenue, Bronx, NY 10461, U.S.A.

INTRODUCTION

Hepatitis B virus (originally termed serum hepatitis) has been known to exist for more than four decades and the virus particle was physically demonstrated by electron microscopy more than 15 years ago; yet until recently, progress has been very slow in understanding the nature and biology of this virus. The major limitation has been the lack of cell culture systems to effectively propagate hepatitis B virus (HBV) and study its replication cycle. However, with the advent of recombinant DNA methods in the 1970's, the cloning of the complete HBV genome, the information derived from sequencing this genome, and use of cloned HBV DNA as a molecular probe, significant advances have been made during the past six years. This review will deal with our current understanding of the HBV genome, the nature of its gene products and the virus replication mechanism, all which illustrate unique features. Finally, information will be presented concerning our present understanding of the possible role of HBV in hepatic oncogenesis for which a clear association has been shown by epidemiologic studies.

1. THE HBV CARRIER STATE

Hepatitis B virus infection represents a major health problem, as there are more than 750,000 carriers in the United States and more than 200,000,000 worldwide (1). Acute infection in adults is usually self-limited, but 10% become carriers. In long-term carriers, viral protein production and virus replication usually decrease over time, and about 10%/yr of adult carriers convert spontaneously to anti-HBs status (2). Infants who become infected have a much higher carrier rate (about 90%) and often remain as carriers for life (3). Clinically, most carriers show no evidence of liver disease, but as many as 10-25% develop mild chronic lobular or chronic persistent hepatitis and 2-10% chronic active hepatitis (2, 4); progression to cirrhosis is well documented (5).

Epidemiologic and other studies have established a relationship between the hepatitis B virus carrier state and development of primary liver cancer (1,6). The most convincing study is that of Beasley *et al.*. (6) in Taiwan in which 22,707 male government employees were divided in HBsAg positive or negative groups and were followed prospectively for development of cirrhosis and primary liver cancer. After a mean follow-up of 6.2 years, 113 $HBsAg^{+}$, but only 3 $HBsAg^{-}$, individuals developed liver cancer (6). The relative risk for liver cancer in carriers *vs.* non-carriers was 217/1. In Taiwanese males, primary liver cancer is the leading cause of death and in HBsAg positive individuals the lifetime attack rate has been estimated at ~40%. Similar conclusions have been reached from studies of railroad workers in Japan (7). What is most fascinating is that very high carrier rates for Hepatitis B virus (in excess of 10% of the total population) tend to occur in focal clusters and that the incidence of liver cancer is often very high in these areas. In China, there are more than 100,000 deaths per year from primary liver cancer and in Mozambique the incidence is 1 per 1000 total population per year. In nearly all ethnic populations around the world, an increased incidence of primary liver cancer has been associated with an increased HBsAg carrier rate. Exception include reports from Greenland and Egypt (8), but these findings have not been reconfirmed using recently developed more sensitive immunological or molecular methods to evaluate materials from these patients. Clearly, there is an association between hepatitis B virus carriage and development of primary liver cancer, but the precise basis for this association remains obscure.

2. THE VIRION PARTICLE

The *Hepatitis Associated DNA* (hepadna) virus family (9) is currently comprised of 4 members: (1) human (HBV), (2) woodchuck (WHV), (3) ground squirrel (GSHV), and (4) duck (DHBV). All have similar virion structure, genome organization, mechanism of virus replication, and predilection for persistent infection of the liver, although lymphoid and spleen cells have recently been recognized to harbor and replicate this virus (10-16). In ducks, pancreas and kidney also appear to support viral gene expression and replication (17). Man is the only known natural host for HBV, although chimpanzees can be infected experimentally. Attempts to propagate HBV in tissue culture have generally failed, but there is a recent report of HBV replication in a human hepatoblastoma cell line, HepG2, transfected with a recombinant cloned HBV dimer (18). This system may provide the first effective *in vitro* model of HBV replication.

HBV is found primarily in liver and serum. In serum, three different types of particles have been identified (Fig. 1). The presumed infectious virion is the Dane particle, which has a diameter of approximately 42 nm and consists of an outer envelope of protein and membrane lipids, surrounding an inner core structure of nucleocapsid (27 nm diameter) that contains the viral nucleic acid (schematically illustrated in figure 1A and demonstrated by electron microscopy of a serum concentrate in figure 1B). Found in far greater concentration are 16 to 25 nm spherical particles and tubular elements (22 nm in diameter and up to 1000 nm in length) that do not contain nucleic acid (Fig. 1B). Nucleocapsid particles can be identified in the nucleus of infected hepatocytes, but they are not found in serum.

For non-human hepadna viruses, virion sizes vary slightly from 40 nm in DHBV to 47 nm in GSHV. WHV and GSHV are associated with surface antigen-containing spheres and filaments similar to HBV, which partially cross-react with anti-HBs (27-29). WHsAg shares minor antigenic determinants with HBV (29), whereas one-third of the peptides produced by tryptic digestion of GSHsAg correspond to those produced by digestion of HBsAg. WHcAg is antigenically similar to

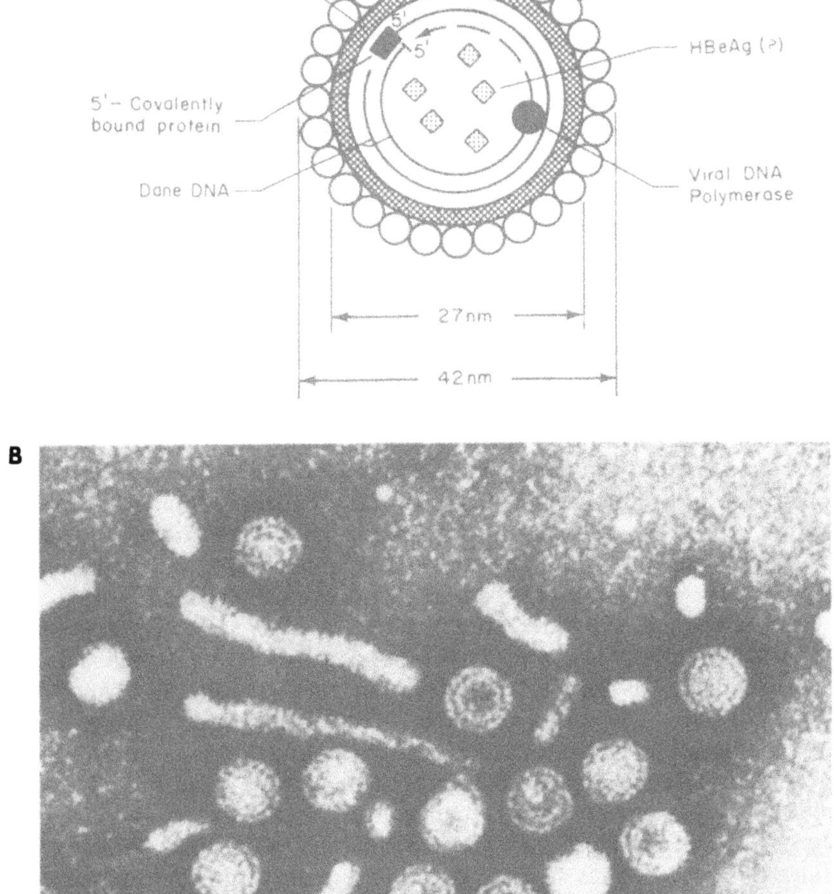

Figure 1. Hepatitis B virus particles as found in human serum. (A) Schematic diagram of the virion (Dane particle). (B) Electron micrograph of virion particles (42 nm) and both spherical and filamentous aggregates of HBsAg (22 nm diameter).

HBcAg when tested by immunodiffusion and shows 73% amino acid homology and 66% nucleic acid homology when tested by nucleic acid sequence analysis (30). In DHBV infections, the serum contains larger, 40 to 60 nm, spherical convoluted particles. These particles do contains larger, 40 to 60 nm, spherical convoluted particles. These particles do not cross-react with anti-HBs, but are presumed to be homologus to the other surface antigen-containing particles. WHV and GSHV core antigens cross-react with anti-HBc, but no similar cross-reactivity has been detected with DHBV.

Persistent infection has been documented in all three animal species. In woodchucks, persistent WHV infection is associated with acute and chronic inflammatory changes (that is, chronic hepatitis), but not with cirrhosis. A large proportion of persistently infected woodchucks also develop hepato cellular carcinoma (31, 32). Liver disease in DHBV-infected ducks has been reported from one area in China (33). Colonies of infected Pekin ducks from Japan and the United States generally do not show histologic evidence of liver disease. DHBV-infected animals have also been reported to develop hepato cellular carcinoma, but the relationship of this tumor to DHBV infection is not yet clearly established. Chronic liver disease does not generally develop in GSHV-infected ground squirrels, but hepato cellular carcinoma has been found in a substantial number of such animals (34).

A. Genome Organization

HBV has a small circular DNA genome of 3200 to 3300 base pairs (for recent review, see Tiollais et al. ref. 19). As isolated from Dane particles, it has a unique structure (see Figure 2). One strand, the long or L (−) strand, is complete except for a nick of one or a few nucleotides at a fixed point (1800 nucleotides from the Eco R1 site). The other strand, the S(+) or short strand, is incomplete, missing from 15 to 50% of its potential sequence complement. The 5'end of the S strand is located at nucleotide 1560 but the 3' end varies considerably. The circular configuration of the molecule is maintained by base pairing of the 5' extremities of both strands, which overlap by about 250 to 300 nucleotides, the so called "cohesive end" region (Fig. 2). An endogenous DNA polymerase activity associated with the nucleocapside can fill in the single-stranded region of the S strand. This polymerase is distinct from bacterial or other serum DNA polymerases in that HBV-DNA polymerase activity is retained or increased at high salt concentrations. *In vitro*, the DNA polymerase reaction does not join the 3' and 5'ends of the S strand, so that, although a complete circle is generated, it is not covalently closed. Nonetheless, particles containing covalently closed, circular double-stranded DNA molecules (so-called superhelical forms), have been identified in liver of chimpanzees persistently infected with HBV (20), as well as in ducks (21), woodchucks (22), and ground squirrels (23) infected with other hepadna viruses. Within the overlap or "cohesive end" region, there are also two 11 base pair direct repeats, called DR1 and DR2, which may have important functions in the virus life cycle or replication mechanism (for further details, see Fig. 4 and ref. 19).

Studies of cloned HBV DNA have shown significant heterogeneity in nucleotide sequence. Particles isolated from a single patient are homogeneous with respect to DNA sequence, as determined by restriction enzyme mapping (24), and within a single subtype only occasional differences are observed (25). However, different immunologic subtypes show considerable differences in restriction endonuclease maps. Nucleotide sequence analysis of cloned HBV DNA indicates up to 10% difference in nucleotide sequence between different clones, as well as small additions or deletions scattered throughout the genome (26).

B. Structural Genes

There are four translational open reading frames (ORF's) in HBV which code for polypeptides of substantial length (see Fig. 2). These are:

i) S, the surface antigen or *env* protein (map position 2848 to 833);
ii) C, the core antigen or nucleocapsid protein (map position 1836 to 2450);

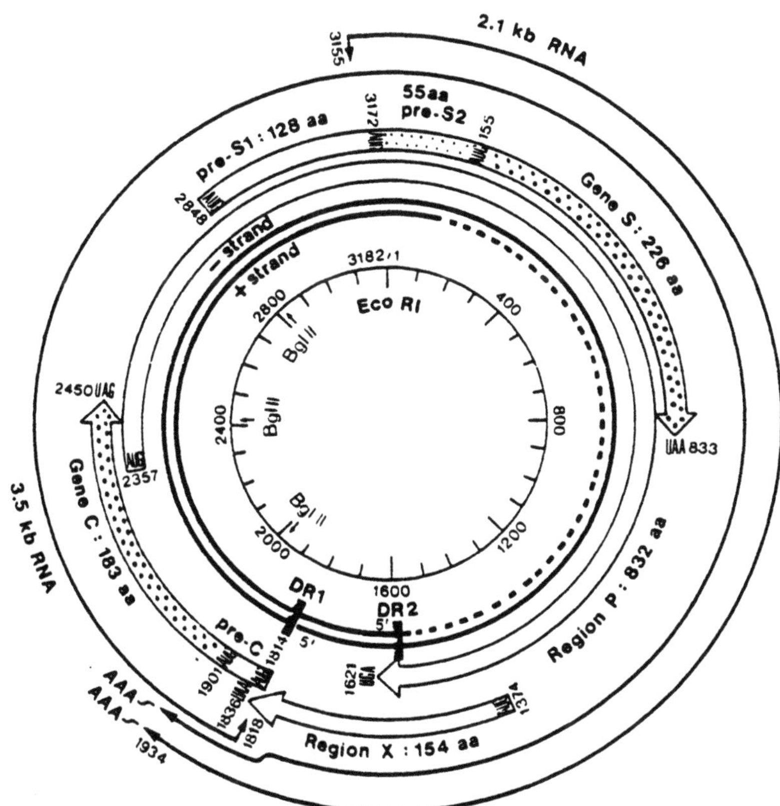

Figure 2. Structural and genetic organization of the Hepatitis B Virus genome. The four transcriptional open reading frames S, C, P and X and the position and orientation of the two major known HBV RNA transcripts are shown (Reproduced from Tiollais et al, ref. 19, with permission).

iii) P, the presumptive viral polymerase (map position 2357 to 1621); and
iv) X, a small protein of unknown function (map position 1374 to 1818) (19).

These ORF's are all located on the L(-) strand and are well conserved among the various members of the hepadna virus family. One exception is that the X ORF is not present in DHBV. There is also considerable nucleotide sequence homology between the human and animal viruses (~70,55 and 40% with WHV and GSHV (19).

The S(-) strand contains only very short ORF's, which could code for polypeptides of less than 10,000 daltons. The position of these ORF's on the S(-) strand are not well conserved in different HBV clones, suggesting that this strand is not involved in transcription of specific viral gene products.

C. HBV RNA Transcripts

Two major HBS-specific poly(A)$^+$ RNAs have been identified (35, 36). Both are L(-) strand transcripts and are referred to as the 2.1 kb and the 3.5 kb RNA (Fig. 2).

The 5' end of the 2.1 kb RNA has been mapped to ~20 bp up stream of the preS_2 region and the 3' end to the beginning of the C gene, where a specific polyadenylation signal is located. The 2.1 kb RNA covers the preS_2, S and X regions.

The 5' end of the 3.5 kb RNA has been mapped to the pre-C region and the 3' end to the identical position as that of the 2.1 kb RNA. It covers the complete genome plus an additional ~100 nucleotides. This implies that the 3.5 kb RNA is produced from covalently closed double-stranded DNA and that the polyadenilation signal is skipped over during the initial phase of transcription (i.e. during the first cycle around the genome).

Minor RNA species of larger and smaller sizes have also been detected. No splicing event has been identified in HBV transcription, at least for the two major RNAs. Cell transfection experiments with cloned HBV DNA fragments and oligomers of HBV DNA have shown clearly that the 2.1 kb RNA is the messenger for the major protein of the envelope(s) and the 3.5 kb RNA the messenger for the core protein (37, 38). The mRNAs encoding the other viral proteins remain unknown, but the 3.5 kb RNA probably encodes the DNA polymerase.

D. HBV Gene Products

As indicated above, there are four ORF's within the L(-) coding strand of the HBV genome which could code for polypeptides of substantial length S, C, P and X. However, at least 12 viral specific polypeptides have been identified and the S, C, and P ORF's have been shown to produce multiple gene products; six polypeptides from S, 2 from C, and 3 from P (for details, see ref 19).

The surface antigen(s) or envelop proteins are dimers of three size classes, S or major (226 amino acids), pre S_2 plus S or middle (281 amino acids) and pre S_1 plus pre S_2 plus S or large (389 or 400 amino acids, depending on viral subtype) (see

Figure 3). Each of these polypeptides can be glycosylated and the native proteins found in virions are dimers linked through disulfide bridges. The middle protein is found in two glycosylated forms ($GP33^s$ and $GP36^s$) depending on the degree of glycosylation, whereas the major protein is found in either unglycosylated ($P24^s$) or glycosylated ($GP27^s$) form, as is the large protein (unglycosylated, $P39^s$ and glycosylated, $GP42^s$). The 3' or S portion of these proteins is constant, whereas the 5' portion varies depending either on which S gene promoter-initiation site is used ($preS_1$ or $preS_2$, see below) or possibly on polypeptide processing events which have not yet been described.

HBsAg contains a major antigenic group specific determinant called "a" and subtype determinants d, y, w and r, producing four main serotypes adw, ayw, adr and ayr. Each of these virus subtypes has a separate worldwide distribution. An exposed hydrophilic segment of the major protein between residues 122 and 155 contains most of the information leading to this group and subtype antigenic specificity (39). The $preS_2$ region also appears to encode for a receptor for polymerized serum albumin (40-42). This receptor has been postulated to mediate attachment of virions to hepatocytes. The large $preS_1$ containing polypeptide may also be involved in attachment of HBV to hepatocytes.

The precise mechanism regulating the synthesis or level of these various surface antigen gene products has not been elucidated. However, development of specific antibodies to the $preS_1$, $preS_2$, and S regions, either using artificially synthesized oligopeptides or polypeptides generated from bacterial plasmid expression vectors containing specific $preS_1$, $preS_2$ or S sequences, has permitted additional analysis of the function of the various S gene polypeptide regions. From these studies, it is clear that 80-90% of the total S gene product on the surface of virions is the major or S polypeptide. Filamentous forms of HBsAg aggregates in serum have a similar composition, but the 22-nm spherical particles contain very little

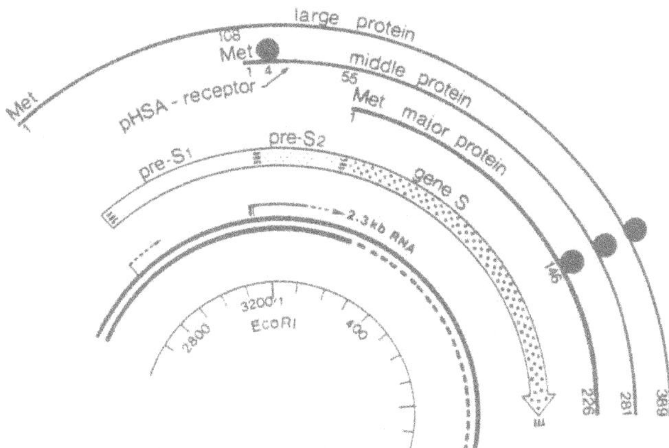

Figure 3. Detailed representation of the HBV surface antigen gene. The relationship of the large, middle and major polypeptides coded by this gene is shown in detail (Reproduced from Tiollais *et al.*, ref.19, with permission).

large protein. In the absence of viral replication, the spherical particles contain less than 1% middle protein and no large protein. Most recently, it has been recognized that a mRNA derived from the S region linked to the β-globin coding sequence directs synthesis of a secreted S-globin fusion polypeptide (43) and there are apparently 3 separate hydrophobic signal sequences present within the major S polypeptide (44). Presently unpublished studies also indicate that the S polypeptide is rapidly secreted from cells, whereas when $preS_1$ protein is also synthesized, significant quantities of the S and $preS_1$ protein are retained within intracellular membranous structures (45). Similar findings have been observed in reconstructed systems using frog oocytes or other eukaryotic cells into which plasmids containing expressable HBV sequences have been introduced (45), as well as in long-term hepatitis B virus carriers (S.J. Hadziyannis, D.A. Shafritz, and J. Sninsky, unpublished observations).

The core gene "C" codes for at least two major polypeptides, the core antigen of 22,000 daltons (nucleocapsid protein) and E antigen of 15,000 daltons (unknown function). The C-terminus of core antigen is rich in arginine, serine and proline residues and contains a tandem repeat of an octapeptide which is very similar to protein sequences of protamine and other DNA binding proteins (26). Core Antigen can also be phosphorylated by a protein kinase activity found within nucleocapsid particles (46, 47). The "E" antigen is a separate structural protein derived from the same reading frame of the core gene but uses a separate AUG initiation site ~80 nucleotides upstream to the core AUG. It also has a truncated 3'-end with absence of the last 33 core antigen amino acids (48). E is a crypt antigen, which is released from virion particles by detergents and is thought to serve in the mechanism of virus-membrane interaction and/or virus secretion by virtue of a hydrophobic signal sequence encoded within the extra 5' portion of the E antigen (49, 50).

The X gene (map position 1374 to 1836) is the ORF about which the least is known concerning its gene product(s). This region could encode for a polypeptide of 154 amino acids (~15,000 daltons), but no such protein has yet been identified. However, several types of evidence suggest that this region has specific gene function, including the presence of a transcriptional promoter at map position ~1300 (45), synthesis of a specific 16 kDa polypeptide when an expression vector containing this sequence is transfected into COS or HepG2 cells (45), and a putative X gene mRNA transcript of appropriate length (~750 nucleotides) and map location in rat 2 cells transfected with an HBV-DNA head-to-tail tetramer (38).

Evidence for X gene expression in natural HBV infections is limited to the finding of antibodies to synthetic X polypeptides in the serum of some patients (51, 52). In addition, antibodies directed against X synthetic peptides recognize a 28 kDa protein in Western blots of infected liver extracts (51). This suggests that X may be expressed as a fusion gene product, perhaps with a portion of the core or pol gene (51). Attempts to demonstrate either a nuclear or cytoplasmic subcellular location for the putative X gene product in either natural infections or viral DNA transfected eukaryotic cells have produced conflicting preliminary results (45).

Other virion products include a protein covalently bound to the 5' end of the L(-) DNA strand (which appears to serve as a primer for HBV DNA replication, see below) and a protein kinase activity which can phosphorylate HBcAg (47, 53). Presently, it is not known whether these two proteins are derived from viral or cellular sequences. Most recently, Will *et al.* (54) have identified at least three viral polypeptides containing sequence regions derived from both the core and pol ORF's. The major core-pol fusion polypeptides are 35 kDa, 38 kDa and 69 kDa and are found in HBV infected tumor or peritumor tissue.

E. Regulatory Sequences

Three major HBV gene promoters have been identified, $preS_1$ (map position 2786-2791, with a "TATA" sequence (55), $preS_2$ (map position 3122-3130, with an SV40 late promoter-like sequence) (35) and core (map position 1653-1660, and not containing a "TATA" sequence) (56,57). In addition, an enhancer element has been located at map position 1080-1234 (58). The HBV enhancer functions with both homologus (HBV) and heterologus (SV40) promoters. A glucocorticoid responsive element (GRE), which stimulates the HBV enhancer, also appears to be present in the HBV genome (59). This GRE, located at map position 355-366, contains all but the first nucleotide of the core consensus sequence ACAA--$TGT^{T}_{C}CT$ found in the 5'-regulatory region of other glucocorticoid responsive genes, e.g: the mouse mammary tumor virus LTR (60), human metallothionein IIA (61) and human growth hormone (62). The $TGT^{T}_{C}CT$ segment is also present in multiple tandem copies (2-3), as also observed for the mouse mammary tumor virus LTR (60).

Evidence has also been reported that the HBV enhancer may function in a tissue specific manner (58, 59), and recent studies suggest both viral and cellular transacting factors in HBV gene expression (63). Most recently, Shaul and coworkers have identified, by DNAse footprinting, regions within and adjacent to the HBV enhancer which serve as protein binding sites for factors which may be involved in tissue specific HBV gene expression (personal communication). Studies in other systems suggest that tissue specificity for eukaryotic gene transcription may also in part reside in sequences of the promoter (64, 65), and specific transcription factors are now being shown to bind to the CAAT or repeat sequence region of the promoter (66). In view of the obvious importance of these regulatory elements and factors, much current research on control of HBV gene expression is now focused on regulatory sequences in the HBV genome and how these sequences might function.

3. HEPADNAVIRUS REPLICATION

The hepadna virus replication mechanism, discovered originally by Summers and Mason for DHBV (67) and confirmed later in man (68), is strikingly different from that of other DNA viruses (Figure 4). The replication cycle involves a reverse transcription step, using a full length copy (3.5 kb) of virion RNA (+), transcribed from the L(-) strand of supercoiled DHBV (or HBV). This RNA, initiated at DR1 and referred to as "pregenome" RNA, is transported from the nucleus to the cytoplasm, where it binds with viral proteins to form core particles. Viral DNA

L(–) strand is synthesized within the cores using the presumed viral polymerase (perhaps a core pol fusion protein as recently identified by Will et al. (54)). The protein at the 5' end of the L(–) strand is thought to serve as a primer for L(–) strand synthesis. A ribonuclease H function of the polymerase degrades RNA sequences progressively as the DNA L(–) strand elongates. The S (+) strand is then transcribed from the L(–) strand beginning at DR2, using a 19 nucleotide capped oligoribonucleotide primer (69). The growing S(+) strand jumps the nick in the L(–) strand by an unknown mechanism and continues to elongate, but complete synthesis of the S (+) strand is not necessary for packaging and secretion of the virus.

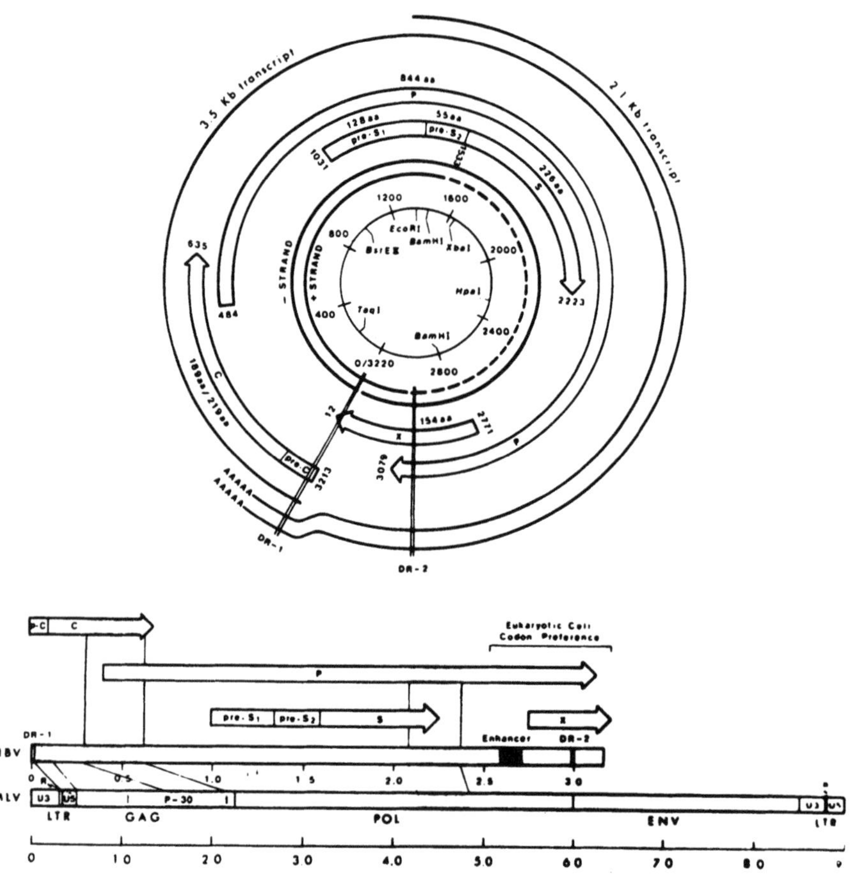

Figure 4. Linearized version of the HBV genome showing the orientation and overlapping nature of the known ORF's and the similarity between the genetic organization of HBV and that of retroviruses.

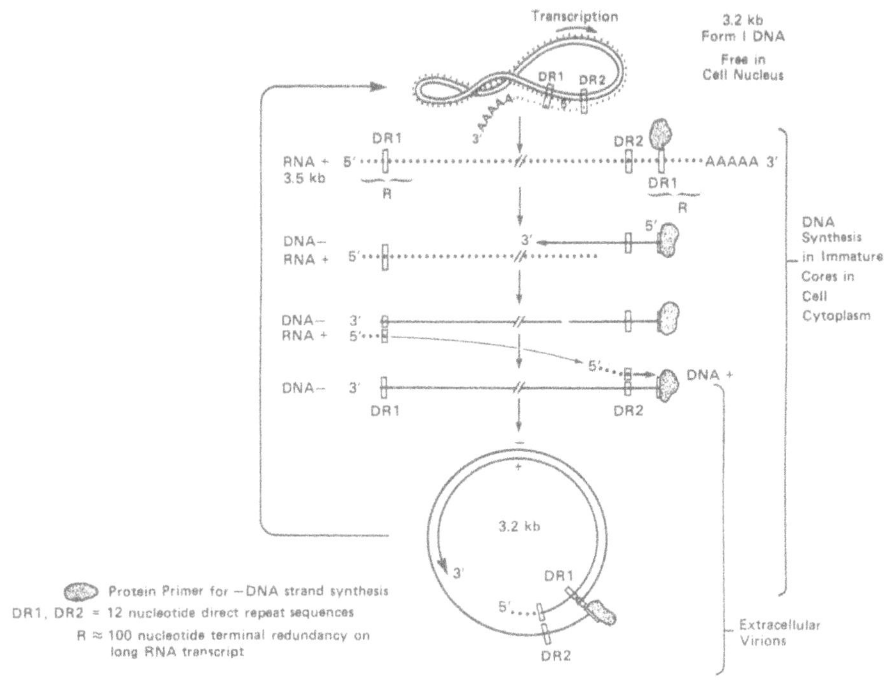

Figure 5. Replication scheme for various members of the Hepadna virus family.

A. Similarities between Hepadna Viruses and Retroviruses in Genome Organization and Replication Mechanisms

If the nicked, but repaired viral DNA were linearized and the overlapping ends filled, HBV would resemble a retrovirus proviral DNA with a large direct repeat at each end (Figure 5). The structural genes of HBV are also organized in a fashion similar to retroviruses in that they are all coded from the L(−) strand. However, in HBV, the pol gene ORF overlaps the other three coding segments. As indicated earlier, evidence for core-pol gene fusion products of 35, 38 and 69 kDa has also recently been found by Will *et al.* (54). Such polypeptides would have features consistent with the gag-pol fusion gene products of retroviruses. In this linearized form, the X gene is at the 3' end of the HBV genome and its location would suggest a possible function as a transactivating protein, similar to the Tat gene of retroviruses (70). Additional analogies have been made between hepadna viruses, cauliflower mosaic viruses and retroviruses, since they all replicate by reverse transcription (71). Miller and Robinson (72) have compared the sequence of HBV and retroviruses and conclude that the C-terminus of the core gene, the middle portion of pol and a large segment of X have considerable homology to retrovirus sequences.

Although the mechanism for hepadna virus replication has similarities to that observed for retroviruses, clear differences are evident. In contrast to hepadna viruses, retroviruses package a single stranded RNA genome, retroviruses use a tRNA as primer for DNA (–) strand synthesis, other steps in retrovirus replication are for more complex and, in contrast to retroviruses, integration of an intact hepadnavirus proviral DNA does not appear to be required for its replication (J. Summers, personal communication). Nonetheless, the members of the hepadna virus family are unique not only in terms of their unusual DNA structure, but also in the fact that these viruses share features with both DNA and RNA viruses.

B. Cis- and Trans-acting Factors in Viral Gene Expression

Evidence for both *cis* and *trans*-activation of viral and cellular gene expression is rapidly accumulating (70, 73-77) and it appears that viral gene products can influence cellular gene function (75-77). For some time, it has been known that the adenovirus early gene product E1A affects transcription of other adenovirus genes (74 and 75) and that Ad E1A also modifies expression of cellular genes (for a detailed discussion, see chapter 20 of this volume and ref.75-77). Similar regulatory phenomena have been observed with SV40, Epstein-Barr virus, and papilloma viruses. In the HTLV system, several mechanisms for viral transactivation have recently been proposed. In HTLV I and II, the viral pX (or Tat) gene produces a protein which transactivates the HTLV LTR enhancer-promoter to increase transcription (70). In contrast, the HTLV III Tat gene product does not increase transcriptional activity of HTLV III genes, but rather increases translation of HTLV III mRNAs, perhaps by increasing mRNA binding to the ribosome (Dr. W. Haseltine, personal communication).

By using a combination of plasmid expression vectors containing specific promoters, enhancers and/or other regulatory elements, it is now possible to explore the interaction of different viral and host gene products in regulating both virus and host cell functions. Such factors could play a major role in determining the liver tissue specificity for HBV infection and/or persistence by modulating expression of specific viral genes. Similar mechanisms could also produce interference in HBV gene expression, as observed during superinfection with other viruses (78, 79). Using competition experiments between co-transfected CAT vectors containing HBV vs. SV40 enhancers, Siddique *et al.*. have recently obtained preliminary evidence for a possible role of cellular factors ("transacting") in activating the HBV enhancer (63).

C. HBV DNA in Serum

Using molecular hybridization techniques, HBV DNA can be detected in the serum of many HBsAg positive carriers. Serum HBV DNA has not been well characterized, but is thought to represent virion DNA. Therefore, its presence is thought to correlate with infectivity and active virus replication (80). During acute viral hepatitis, when HBsAg and HBeAg are present, HBV DNA is always detectable, although in varying amounts (80-82). In immunosuppressed patients in whom high viremia would be expected, HBV DNA levels may be very high (82). HBeAg positive chronic carriers are also presumed to be infectious. Virtually all of these patients have circulating HBV DNA (83), as well as nuclear HBcAg in the liver, suggesting that they do indeed have active virus replication (83). Many such patients also have some form of chronic liver disease (chronic persistent or chronic active hepatitis). On the other hand, HBsAg carriers who are anti-HBe positive are though to have passed the stage of active viral replication and are presumably noninfectious, although exceptions have been described. In these patients, the finding of normal liver histology is generally associated with the absence of HBV DNA in serum (83). These patients, therefore, are probably noninfectious. However, if the liver biopsy in an HBsAg/anti-HBe positive patient shows chronic liver disease, HBV DNA is frequently found in the serum (83). This combination ($HBsAg^+$/anti-HBe^+, with active virus replication and active chronic liver disease) is apparently more common that previously suspected (83). Such patients are potentially infectious, despite being anti-HBe positive and show a very virulent form of chronic to active hepatitis, rapidly progressing to cirrhosis (84).

HBV DNA is frequently absent from the serum of patients with hepato cellular carcinoma (82). When present, it is usually found in a low concentration. Furthermore, HBV DNA can be detected in the serum of some patients who are anti-HBs positive and in patients in whom anti-HBc is the only marker of HBV infection. In addition, integrated HBV DNA has been found in hepato cellular carcinoma tissue from some patients who are serologically HBsAg negative but anti-HBs positive, as well as in patients with no serologic markers of present or previous HBV infection (85). Therefore, measurement of immunologic markers of HBV has certain obvious limitations, especially if viral antigens and their specific antibodies are present in serum at the same time. The formation of immune complexes may interfere with the competitive binding assays (radio-immunoassay or enzyme-linked immunosorbent assay) used for detection of viral antigens. In some instances, these immune complexes can be detected using monoclonal antibodies to HBsAg (85). However, with advances and simplification of molecular hybridization technology, it is likely that HBV DNA determinations will be used clinically as an adjunct to, or as a replacement for, some of these immunologic assays.

D. Replicating Versus Nonreplicating States of Persistent HBV Infection

From histologic, cellular, and molecular studies, it has been concluded that persistent HBV infection falls generally into two categories, permissive and nonpermissive (83, 86, 87). The features of replicating or permissive infection include: active inflammatory liver disease (chronic persistent or chronic active hepatitis), HBsAg and HBcAg in individual cells distributed randomly throughout the hepatic parenchyma, and free virions and lower molecular weight replicating forms of HBV DNA. The features of nonpermissive infection include: no active inflammatory liver disease (the liver may show no histologic evidence of HBV infection other than the presence of ground glass hepatocytes or inactive cirrhosis resulting from viral infection many years earlier), continued HBsAg, but not HBcAg, production, and integrated HBV DNA molecules.

At one extreme, cells continue to replicate virus and the patients show continued liver disease activity (permissive or actively replicating infection). In permissive infection, the bulk of HBV DNA in hepatocytes is in free virions or replicating forms, and integration into hosts genome is not observable by Southern blot analysis. However, as various studies have shown (81, 87-89), integration may be occurring randomly throughout the host genome during the infectious process. In nonpermissive infection, it is often possible to observe these random integrations as a smear of hybridization in the high molecular weight region on Southern blot analysis. In these cases, free virion DNA is not found in the liver and virus is not secreted into the bloodstream. In some nonpermissive infections, a single HBV DNA integration has been observed on Southern blots, suggesting that cell population from which the DNA was derived is of monoclonal origin (83, 86). In several of these cases, Hadziyannis *et al* have noted clusters of HBsAg-producing cells having the appearance of a focal clonal growth (83, 86). A mixed type of persistent infection may also occur in which features of replicating and nonreplicating infection are found in different cells or regions of the same liver. Raimondo *et al.* (90) have recently identified several tissue specimens (liver or tumor) in which substantial quantities of viral replicative intermediates and other extrachromosomal forms of HBV-DNA are present and yet these patients were not secreting virus into the serum. In these individuals, it appears that the virus replication, assembly or secretion pathway is defective or blocked. Under these circumstances, there is intracellular accumulation of HBV DNA replicative forms, which may augment the viral DNA integration process.

4. HBV-DNA INTEGRATION IN HEPATO CELLULAR CARCINOMA

Early studies of hepato cellular carcinoma DNA's using Southern blot analysis, showed that these tumors contained HBV-DNA integrated into unique sites, indicating that the tumors are clonally derived (81, 88, 89). Banding patterns of integrated HBV-DNA varied in different tumors and individual tumors often contained multiple viral DNA integrations (81, 88, 89). This suggested that individual hepatocytes can accumulate multiple integrations during chronic infection or that several integrations occurred simultaneously during transformation. It was also possible that integrations could duplicate themselves or translocate to new cellular locations subsequent to initial insertion (92-94). The fact that viral DNA integrations were also present in "normal" liver tissue in hepato cellular carcinoma patients suggested that these integrations may play a role in transformation (88, 89, 95, 96). This conclusion was based on the fact that most hepato cellular carcinomas in HBV carriers show clonally propagated integrations, whereas probably not more than 10-20% of hepatocytes in carrier liver tissue on the average contain an HBV integration, even after very long carriage of the virus. In addition, HBV-DNA integration into the liver cell genome was also found in patients with no evidence of hepato cellular carcinoma (81, 83, 87, 89).

HBV integrations from hepato cellular carcinomas often contain multiple rearrangements, including deletions, inversions and direct and inverted duplications (92-94, 97-100). However, some HBV integrations in hepato cellular carcinomas contain a greater than unit length HBV-DNA sequence with an apparently intact copy of the genome (97). Extrachromosomal high MW "novel" forms of viral DNA have also been found in woodchuck and ground squirrel carrier liver (101, 102). These molecules are comprised of oligomeric, greater than unit length viral

DNA's with deletions, duplications and rearrangements of viral sequences. These structural rearrangements are similar to those observed in WHV integrations in woodchuck hepato cellular carcinomas (103). It has been suggested that these "novel" forms may represent a by-product of defective genome replication during viral persistence (101) and could represent a precursor of integrations involved in hepatic oncogenesis. However, "novel" forms have not yet been identified in man.

The ability of HBV to integrate into cellular DNA during chronic infection is consistent with its potential function as a tumor initiator. For other viruses, this has been shown to occur either by direct transfer of oncogenes carried by the virus (104, 105) or by activation of cellular proto-oncogenes by insertion of viral promoters or enhancers into the cellular genome (106, 107). Hepadna viruses have not to date been shown to contain a viral oncogene, nor has an HBV-DNA integration been shown to occur immediately adjacent to any known cellular proto-oncogene (19, 92, 97, 108). Cellular sequences flanking HBV integrations have been used as hybridization probes to search for common viral integration sites on the host genome, but so far none have been identified.

Concerning the map location of the viral portion of the viral/cellular junctions of integrated HBV DNA molecules, wide variations have been noted. However, in ~30-40% of those viral/cell junctions which have been mapped and sequenced, integration at DR1 has been found at one junction (92, 94, 97, 99). This suggests that a specific sequence within DR1 might be involved in a recombinational event with an analogous sequence in the host cell genome, although the precise mechanism for this process will require further study.

Dejean *et al.* (109) have recently identified one HBV integration in an exon of a coding gene having partial sequence homology with the erb A oncogene and the glucocorticoid receptor which are structurally related (110). This integration is located on chromosome 3. Integrations that Rogler *et al.* have characterized are located on chromosomes 6, 9, and 11, with also a translocation between chromosome 17 and 18 (94, 100), and are associated with cellular gene deletions (100). Other HBV integrations in hepato cellular carcinomas or hepato cellular carcinoma cell lines have been identified on chromosome 2, 3, 7, 12, 15 and 18 (111). Therefore, many chromosomes are apparently involved in HBV integrations are located in random sites or are in specific regions associated with oncogenesis. In this light, it is interesting that one HBV integration located on chromosome 11 p 13-14 is in the same region that contains the Wilm's tumor locus (112, 113). It is also near the locus for Beckwith-Wiedmann Syndrome, which is associated with multiple system organomegaly, hepatoblastoma and hepato cellular carcinoma (114).

A. Possible Role of Chemical Carcinogens

Since many hepato cellular carcinomas, especially those observed in Western countries, occur in non-HBV infected individuals, (e.g., those associated with alcoholic cirrhosis, hemochromatosis, α_1-antitrypsin deficiency, mycotoxins, and chemical carcinogens), other factors are obviously involved in hepatic oncogenesis. Factors which stimulate chromosomal aberrations have been associated with increased risk of neoplasia (115), and agents, such as benzopyrene, acetaminofluorine, aflatoxin, etc. which damage DNA, clearly cause HCC (116). The ability of oncogenic viruses, probably through integration into the host genome, to damage DNA by random mutation of cellular genes and/or generation of chromosome aberrations is believed to be an important factor in viral mediated multi-stage carcinogenesis (117). Indeed, a similar process may be occurring with HBV, since HBV integrations are often associated with deletion of adjacent cellular sequences at the site of integration (93, 97, 99, 100).

B. Unifying Hypothesis Relating Viral and Nonviral Factors in Hepatic Oncogenesis

Evidence from molecular studies indicates that integration of HBV DNA occurs before or during the oncogenic process, but this does not prove that HBV itself is oncogenic. If neoplasia arises frequently in HBV-bearing livers, and if random integration of HBV DNA occurs frequently (i.e. in many cells in such livers), then tumors arising in those livers will often contain integrated HBV DNA, whether or not the integrated viral DNA itself is wholly or partly responsible for the oncogenic change. This does not exclude HBV from a role in oncogenicity, but merely implies that neoplastic change may occur by an epigenetic mechanism.

From information currently available, one can generate the following hypothesis (illustrated schematically in Figure 6): At some stage during the progression of the carrier state, or perhaps as early as the initial acute infection, HBV DNA integration occurs (Stage I). Hepatocytes expressing all, or at least some, specific viral proteins are efficiently removed by host defense mechanisms. However, integration of HBV DNA, which is probably a random event, may result in altered expression of viral antigens, so that the specific recognition factor or factors for host immune surveillance may not be expressed or may be expressed to a reduced degree in cells containing integrated versus free virion HBV DNA. Therefore, cells containing integrated HBV DNA might be preferentially retained within the liver parenchyma compared with cells actively replicating virus. With time, the normal process of cell division would therefore accelerate the accumulation of cells containing integrated HBV DNA. At this stage, HBV DNA would be integrated into multiple sites within the host genome and would appear as a diffuse smear on Souther blot hybridization analysis, reflecting its dispersed distribution. Cell division is essential to this stage, so that factors that stimulate cycles of hepatocyte necrosis and regeneration (such as ethanol, hepatotoxins, chemical carcinogens, infection with other viruses, etc.) may potentiate in these early selection events. Destabilization of the cellular genome during the mitotic process could also cause rearrangement, deletion or modification of integrated HBV-DNA and/or adjacent cellular sequences during these repeated cycles of liver regeneration.

Subsequently, a clone or several clones of hepatocytes containing integrated HBV DNA might become selected for preferential survival and/or multiplication by factors at present unknown, but which may include the state of the integrated HBV DNA. These cells are not yet neoplastic, although they are by definition transformed, since they contain integrated viral DNA (Stage II). On restriction analysis and Southern blotting, these cells would show discrete, integrated HBV DNA bands. Under the influence of still further factors (such as host immune surveillance, nutritional state, hormones, environmental toxins, or carcinogens), one or more of these clones may give rise to an eventual malignant clone (Stage III).

The above hypothesis implies that HBV does not act alone in the genesis of malignancy, but rather that hepatic oncogenesis is a multi-stage process involving both host and viral factors. Although there are still many uncertainties concerning the role of HBV in the pathogenesis of hepato cellular carcinoma, clearly the use of molecular biologic techniques will in great measure contribute to our eventual understanding of this process.

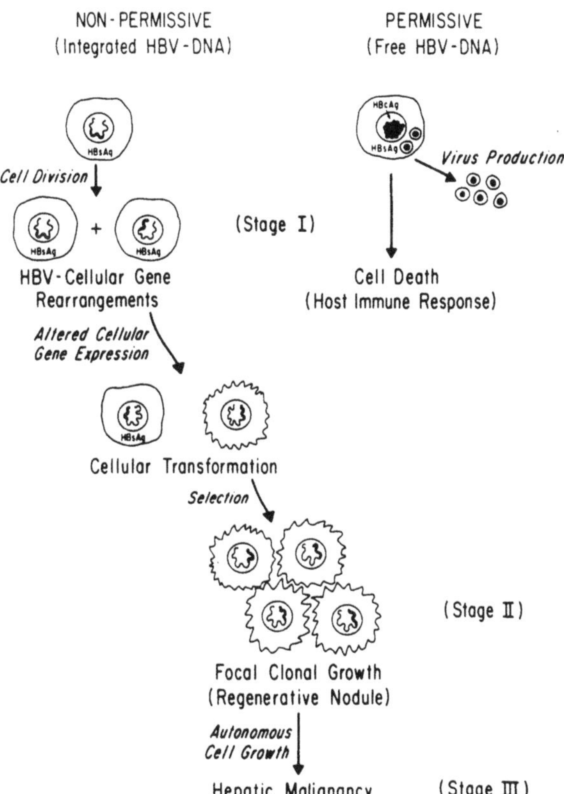

Figure 6. Schematic diagram illustrating various stages in persistent HBV infection and their potential relationship to the development of Hepatocellular Carcinoma. (Reprinted from ref. 86 with permission).

Acknowledgements

The author would like to thank his various colleagues, research associates, post doctoral fellows and predoctoral students for their contributions to studies reported from his laboratory, and to Ms. Anna Caponigro and Mr. Roy Forbes for typing the manuscript. Research supported in part by NIH Grants CA32605, P30-CA13330 and P50-AM17702, the Gail I. Wuckerman Foundation and the Marion Bessin Liver Center Fund.

5. REFERENCES

1) Smuzness W. 1978. Prog Med Virol **24**, 40-69.

2) Hoofnagle, JH, Alter HJ. In: *Viral Hepatitis and Liver Disease*, Vyas GN, Dienstag JL, Hoofnagle JH, eds., Grune et Stratton, 1984., pp. 97-113.

3) Stevens CE, Beasley RP, Tsui J and Lee WC. 1975. N Engl J Med **292**, 771-774.

4) Anderson, MG, Murray-Lyon IM. 1985. Gut **26**, 848-860.

5) Redeker Ag. 1975. Amer J Med Sci **270**, 9-16.

6) Basley RP and Hwang L-Y. 1984. In: *Epidemiology of Hepatocellular Carcinoma and Liver Cancer*, eds., Vyas, GN, Dienstag JL and Hoofnagle JH. New York: Grune and Stratton, pp. 209-224.

7) Obata K, Hatashi N, Motoike Y *et al.*. 1980. Int J Cancer **25**, 741-747.

8) Blumberg B, London WT. 1982. Current Probl Cancer **6**, 21-23.

9) Robinson WS. 1980. Ann NY Acad Sci **354**, 371-378.

10) Lie-Injo LE, Balasegaram M, Lopez CG, Herrera AR. 1983. DNA **2**, 301-308.

11) Elfassi E, Romet-Lemone J-L, Essex M, Francis-McLane M, Haseltine W. 1984. Proc Natl Acad Sci USA **81**, 3526-3568.

11) Chong-Jin O, Jeak-ling D. 1984. Lancet **ii**, 395-396.

12) Pontisso P, Poon MC, Tiollais P, Brechot C. 1984. Brit Med J **288**, 1563-1566.

13) Gu J-R, Chen Y-c, Jiang H-Q, Zhang Y-L, Wu S-M, Jiang W-L, Jian J. 1985. J Med Virol **17**, 73-81.

14) Korba BE, Wells F, Tennant BC, Yoakum GH, Purcell RH, Gerin JL .1986. J Virol **58**, 1-8.

15) Yoffe B, Noonan A, Melnick JL, Hollinger FB. 1986. J. Infec Dis **153**, 471-477.

16) Lieberman HM, Tung WW, Shafritz DA. 1987. J Med Virol **21**, in press.

17) Halpern MS, England JM, Deery DT, Petcu DJ, Mason WS, Molnar-Kimber KL. 1983. Proc Natl Acad Sci USA **80**, 4865-4869.

18) Sureau C, Romet-Lemonne J-L, Mullins JI, Essex M. 1986. Cell **46**, in press.

19) Tiollais P, Pourcel C, DeJean A. 1985. Nature (London) **317**, 489-495.

20) Ruiz-Opazo N, Chakraborty PR, Shafritz DA. 1982. Cell **29**, 129-138.

21) Mason WS, Aldrich C, Summers J, Taylor JM. 1982. Proc Natl Acad Sci USA **49**, 3997-4001.

22) Rogler CE, Summers J. 1982. J Virol **44**, 852-863.

23) Weiser W, Ganem D, Seeger C, Varmus HE. 1983. J Virol **48**, 1-9.

24) Sninsky JJ, Siddiqui A, Robinson WS, *et al.*. 1979. Nature (London) **279**, 236-348.

25) Siddiqui A, Sattler F, Robinson WS. 1979. Proc Natl Acad Sci USA **76**, 4664-4668.

26) Tiollais P, Charnay P, Vyas GN. 1981. Science **213**, 406-411.

27) Marion PL, Oshiro LS, Regnery DC, *et al.*. Proc Natl Acad Sci USA **77**, 2941-2945.

28) Gerlich WH, Feitelson MA, Marion PL, *et al.*. 1980. J. Virol **36**, 787-795.

29) Werner BG, Smolec JM, Snyder R, *et al.*. 1979. J Virol **32**, 314-322.

30) Galibert F, Chen TN, Mandart E. 1982. J Virol **41**, 51-65.

31) Snyder RL, Summers J. 1980. In: *Viruses in Naturally Occurring Cancers* Essex M, Todaro G, zur Hausen M (Eds): , **7**, Cold Spring Harbor Symposium on Cell Proliferation, pp. 447-457.

32) Summers J. 1981. Hepatology **1**, 91-98.

33) Omata M, Uchiumi K, Ito Y, et al. 1983. Gastroenterology **85**, 260-267.

34) Marion PL, Van Davelaar MJ, Knight SS, Salazar FH, Garcia G, Popper H, Robinson WS. 1986. Proc Natl Acad Sci USA **83**, 4543-4546.

35) Cattaneo R, Will H, Schaller H. 1983. Nature (London) **305**, 336-338.

36) Cattaneo R, Will H, Schaller H. 1984 EMBO J **3**, 2191-2196.

37) Pourcel C, *et al.*. 1982. J Virol **32**, 100-105.

38) Gough N. 1983. J Molec Biol **165**, 683-699.

39) Peterson DI, Nath N, Gavilanes F. 1982. J Biol Chem **257**, 10414-10420.

40) Neurath AR, Kent SBH, Strick N. 1984. Science **224**, 392-394.

41) Neurath AR, Kent SBH, Strick N, Taylor P, Stevens CE. 1985. Nature (London) **315**, 154-156.

42) Machida A, *et al.*. 1984. Gastroent **86**, 910-918.

43) Eble BE, Lingappa VR, Ganem D. 1986. Molec and Cell Biol **6**, 1454-1463.

44) Eble BE, Lingappa VR, Ganem D. 1986. Meeting on Molecular Biology of Hepatitis B Viruses. Cold Spring Harbor Laboratory, Abstract, p. 18.

45) Summers J, Tiollais P. (organizers) 1986. Meeting on Molecular Biology of Hepatitis B Viruses. Cold Spring Harbor Laboratory, Auga. 18-31, 1986.

46) Albin C, Robinson WS. 1980. J Virol **34**, 297-302.

47) Petit MA, Pillot J. 1985. J Virol **53**, 543-551.

48) Takahashi K, Machida A, Funatsu G, *et al.*. 1983. J. Immunol **130**, 2903-2907.

49) Ou J-H, Laub O, Rutter WJ. 1986. Proc Natl Acad Sci USA **83**, 1578-1582.

50) Bruss V, Gerlich W. 1986. Meeting on Molecular Biology of Hepatitis Viruses. Cold Spring Harbor Laboratory, Abstract, p. 14.

51) Moriarty AM, Alexander H, Lerner RA. 1985. Science **227**, 429-433.

52) Myers ML, Trepo LV, Nath N, Sninsky JJ. 1986. J Virol **37**, 103-109.

53) Gerlich W, Robinson WS. 1980. Cell **21**, 801-809.

54) Will H, Salfeld J, Pfaff E, Manso C, Thielman L, Schaller H. 1986. Science **231**, 594-596.

55) Pourcel C, Louise A, Gervais M, Chenciner N, Dubois MF, Tiollais P. 1982. J Virol **42**, 100-105.

56) Chakraborty PR, Ruiz-Opazo N, Shafritz DA. 1981. Virology **111**, 647-652.

57) Rall LB, Standring DN, Laub O, Rutter W. 1983. Molec Cell Biol **3**, 1766-1773.

58) Shaul Y, Rutter WJ, Laub O. 1985. EMBO J **4**, 427-430.

59) Tur-Kaspa R, Burk RD, Shaul Y, Shafritz DA. 1986. Proc Natl Acad Sci USA **83**, 1627-1631.

60) Hynes N, Van Oogen AJJ, Kennedy N, Hernlich P, Ponta H, Groner B. 1983. Proc Natl Acad Sci USA **80**, 3637-3641.

61) Krauter P, Westphal HM, Beato M. 1984. Nature (London) **308**, 513-519.

62) Moore DD, Marka AR, Buckley DI, Kaplan AG, Payvar F, Goodman HM. 1985. Proc Natl Acad Sci USA **82**, 699-702.

63) Jameel S, Siddique A. 1986. Molec Cell Biol **6**, 710-715.

64) Foster J, Stafford J, Queen C. 1985. Nature (London) **315**, 423-425.

65) Edlund T, Walker MD, Barr PJ, Rutter WJ. 1985. Science **230**, 912-914.

66) Carthen RW, Chodosh LA, Sharp PA. 1985. Cell **43**, 439-448.

67) Summers J, Mason WS. 1982. Cell **29**, 403-415.

68) Miller RH, Tran C-T, Robinson WS. 1984. Virology **139**, 53-63.

69) Lien J-M, Aldrich EC, Mason WS. 1986. J Virol **57**, 229-236.

70) Rosen CA, Sodroski JG, Haseltine WA. 1985. Proc Natl Acad Sci USA **82**, 6502-6507.

71) Mason WS, Taylor JM, Hull R. 1986. Advances in Virus Research, **32**, in press.

72) Miller RH, Robinson WS. 1986. Proc Natl Acad Sci USA **83**, 2531-2525, in press.

73) Berk AJ, Lee F, Harrison T, Williams J, Sharp PA. 1979. Cell **17**, 935-944.

74) Nevins JR. 1981. Cell **26**, 213-220.

75) Nevins JR. 1982. Cell **29**, 913-919.

76) Treisman R, Green R, Maniatis R. 1983. Proc Natl Acad Sci USA **80**, 7428-7432.

77) Tanaka K, Isselbacher KJ, Khoury G, Jay G. 1985. Science **228**, 26-30.

78) Smedile A, Dentico P, Zanetti A, Sagnelli E, Nordenfelt E, Actis GC, Rizzetto M. 1981. Gastroent **82**, 992-997.

79) Harrison TJ, Tsiquaye KN, Zuckerman AJ. 1983. J Virol Meth **6**, 295-302.

80) Bonino F, Hoyer B, Nelson J, et al. 1981. Hepatology **1**, 386-391.

81) Brechot C, Hadchouel M, Scotto J, *et al.*. 1981. Proc Natl Acad Sci USA **78**, 3906-3910.

82) Lieberman HM, LaBrecque DR, Kew ML, Hadziyannis SJ, Shafritz DA. 1983. Hepatology **3**, 285-291.

83) Hadziyannis SJ, Lieberman HM, Karvountzis GG, Shafritz DA. 1983. Hepatology **3**, 656-662.

84) Yokosuka O, Omata M, Imazechi F, Okuda K. 1985. Gastroent **89**, 610-616.

85) Shafritz DA, Lieberman HM, Isselbacher KJ, Wands JR. 1982. Proc Natl Acad Sci USA **79**, 5675-5679.

86) Shafritz DA, Hadziyannis SJ. 1984. In: *Advances in Hepatitis Research*, F.V. Chisari ed., Masson Press, New York, pp. 80-90.

87) Kam A, Rall LB, Smuckler EA, *et al.*. 1982. Proc Natl Acad Sci USA **79**, 7522-7526.

88) Shafritz DA, Shouval D, Sherman HI, *et al.*. 1981. N Engl J Med **305**, 1067-1073.

89) Koshy R, Maupas P, Muller R, Hofscneider PH. 1981. J Gen Virol **57**, 95-102.

90) Raimondo G, Burk RD, Lieberman HM, Muschel J, Kew MC, Shafritz DA. 1986. Meeting on Molecular Biology of Hepatitis B Viruses. Cold Spring Harbor Laboratory Abstract, p. 90.

91) Shafritz DA, Kew MC. 1981. Hepatology **1**, 1-8.

92) Koch S, von Loringhoven AF, Hofschneider PH, Koshy R. 1984. EMBO J **3**, 2185-2189.

93) Ziemer M, Garcia P, Shaul Y, Rutter. 1985. J Virol **53**, 885-892.

94) Hino O, Show TB, Rogler CE. 1986. Proc Natl Acad Sci USA, in press.

95) Louise A, Scotto J, Tiollais P. 1982. Hepatology **2**, 27s-35s.

96) Chen DS, Hayes BH, Nelson J, Purcell RH, Gerin JL. 1983. Hepatology **2**, 42s-45s.

97) Dejean A, Brechot C, Tiollais P, Wain-Hobson S. 1983. Proc Natl Acad Sci USA **80**, 2505-2509.

98) Koike K, Kobayashi M., Mizusawa H, Yoshida E, Yaginuma K, Taira M. 1983. Nucl Acid Res **11**, 5391-5402.

99) Koshy R, Koch S, von Loringhoven AF, Kahmann R, Murray K, Hofschneider PH. 1983. Cell **34**, 215-223.

100) Rogler CE, Sherman M, Su CY, Shafritz DA, Shows T, Henderson A, Kew MC. 1985. Science **230**, 329-332.

101) Rogler CE, Summers J. 1982. J Virol **44**, 852-863.

102) Marion PL, Robinson WS, Rogler CE, Topper D, Summers J. 1982. J Cell Biochem, Suppl 6, p 203.

103) Ogston CW, Jonak GJ, Rogler CE, Astrin SM, Summers J. 1982. Cell **29**, 385-394.

104) Land H, Parada LF, Weinberg RA. 1984. Science **222**, 771-778.

105) Bishop JM. 1985. Cell **42**, 23-38.

106) Hayward WS, Neel BG, Astrin SM. 1981. Nature (London) **290**, 475-480.

107) Payne GS, Bishop JM, Varmus HE. 1982. Nature (London) **295**, 209-212.

108) Mizusawa H, Mananori T, Yaginuma K, Kobayashi M, Yoshida E, Koike K. 1985. Proc Natl Acad Sci USA **82**, 208-212.

109) Dejean A, Bougueleret L, Grzeschik K-H, Tiollais P. 1986. Nature (London) **322**, 70-72.

110) Weinberger C, Hallenberg SM, Rosenfeld MG, Evans RM. 1985. Nature (London) **318**, 670-672.

111) Varmus H, Summers J. (organizers). 1985. Meeting on Hepatitis B viruses. Cold Spring Harbor Laboratory, abstracts. May 2-5, 1985.

112) Koufos A, Hansen MF, Lampkin BC, Workman ML, Copeland NG, Jenkins NA, Covenee WK. 1984. Nature (London) **309**, 170-172.

113) Orkin SH, Goldman DS, Sallan SE. 1984. Nature (London) **309**, 172-174.

114) Koufos A, Hansen MF, Copeland NG, Jenkins NA, Lampkin BC, Cavenee WK. 1985. Nature (London) **316**, 330-334.

115) Klein G. 1981. Nature (London) **294**, 313-318.

116) Pitot HC. 1977. Am J Pathol **89**, 401-412.

117) Geissler E, Theile M. 1983. Human Genetics **63**, 1-12.

CHAPTER 18

THE BIOLOGY OF THE PAPILLOMAVIRUSES

PETER M. HOWLEY
Laboratory of Tumor Virus Biology, Natl. Cancer Institute, Bethesda MD 20892, U.S.A.

INTRODUCTION

The papillomaviruses are grouped together with the polyomaviruses to form the papovavirus family (1). Each of these genera are physically distinguishable by capsid size (55 nanometers versus 40 nanometers) and by the size of their double-stranded DNA genomes (8000 bp versus 5000 bp). The biology of the polyomaviruses which includes SV40 and the human polyomaviruses BK and JC have been well-characterized because they are generally easily replicated in tissue culture. Because of the lack of a tissue culture system to propagate the papillomaviruses, they remained relatively refractory to study until the late 1970s when recombinant DNA technology permitted the cloning of the genomes of individual papillomaviruses which then permitted a systematic study of this group of viruses.

The papillomaviruses are widespread in nature. They have been detected or isolated from a variety of species of higher vertebrates, primarily mammalian, although the virus has been recognized in several avian species. A listing of the known papillomaviruses with the species in which the virus has been recognized is presented in Table 1.

Papillomaviruses induce benign squamous epithelial and fibroepithelial proliferations (warts and papillomas). The viral nature of warts was first suggested by Ciuffo by the cell-free filtrate transmission of warts in 1907 (2). The first papillomaviruses recognized was the cottontail rabbit papillomavirus (CRPV) which is the etiologic agent of rabbit papillomatosis (3). Subsequent studies by Peyton Rous established that benign rabbit papillomas could advance to malignant squamous cell carcinomas in the presence of certain non-viral cofactors such as coal tar or methylcholanthrene (4). This Shope papillomavirus system may still be the best system for the study of malignant progression of a benign virally induced papillomas into squamous cell carcinomas.

Table 1. - Papillomaviruses
Human papillomaviruses (types 1-46)
Bovine papillomaviruses (types 1-6)
Shope papillomavirus (CRPV)
European elk papillomavirus (EEPV)
Deer papillomavirus (DPV)
Ovine papillomavirus
Other species shown to host papilloma viruses: opossum, horse, dog, mast omys natalensis, chimpanzee, goat, elephant, giraffe, elk, parrot, and chaffinch.

1. BIOLOGICAL PROPERTIES

A. Virus Growth

The papillomaviruses are highly species-specific and induce squamous epithelial tumors and fibroepithelial tumors in their natural hosts. Histologically, the lesions induced by papillomaviruses share certain features. Generally, there is thickening of the epidermis with acanthosis and hyperkeratosis. Usually there is some degree of papillomatosis. Kerathohyalin granules are often prominent in the granular layer and basophilic intranuclear inclusions can sometimes be detected in the cells of the upper layer of the epidermis. These histologic features reflect the biologic properties of the papillomaviruses. The papillomaviruses have a specific tropism for squamous epithelial cells. Full viral replication, including vegetative viral DNA synthesis, production of viral capsid proteins, and the assembly of virions, occurs only in the more terminally differentiated squamous epithelial cells. It is generally believed by investigators (but not yet proven) that the papillomavirus DNA is present in a stable and latent form in the cells of the basal layer of the squamous epithelium. The acanthosis is thought to be due to cellular proliferation in the stratum spinosum and is believed to be secondary to the expression of some of the papillomavirus (early) genes which are expressed in the lower levels of the epithelium. These genes are expressed before the onset of vegetative viral DNA replication which by *in situ* hybridization experiments was shown to occur in the epithelial cells of the upper half of the stratum spinosum. Thus, vegetative viral DNA synthesis as a measure of a late virus function is squamous epithelial specific, and is linked to the state of differentiation of the squamous epithelial cell.

The wart specific (late) viral genes which encode the viral capsid proteins are expressed only in the terminally differentiated epithelial cells of the wart. In these same cells one can detect the major capsid protein (VP1) using antibodies to the genus specific antigen which recognizes epitopes of the major capsid viral protein (5).

This apparent absolute requirement for terminal differentiation of squamous epithelial cells for the expression of the late wart-virus specific genes for virus replication is probably the reason that virologists have not yet succeeded in propagating these viruses in tissue culture. Investigations continue in a number of laboratories to adapt epithelial cell culture systems using novel approaches with the hope of developing a system which will permit the *in vitro* replication of these viruses.

Until the mid-1970s, it was generally believed that there was a single human papillomavirus, and that the clinical and pathological characteristics of specific human wart virus associated lesions were a function of the nature of the squamous epithelium at the site of the lesion. The heterogeneity of HPV types became apparent in 1975 and 1976 through the use of restriction endonucleases to analyze the viral DNA from wart-virus associated lesions (6, 7). From these early studies it was clear that at least two distinct human papillomaviruses were associated with human plantar warts and that another unrelated papillomavirus was associated with common warts.

Although these early studies demonstrated that there were distinct human papillomaviruses associated with specific clinical lesions, nucleic acid hybridization experiments carried out under nonstringent conditions revealed that there were conserved sequences among the human papillomaviruses which were also shared with the animal papillomaviruses (8, 9).

The presence of these conserved sequences among all the genomes of the papillomaviruses made it possible to use a papillomavirus specific DNA probe to examine DNA form a clinical lesion for papillomavirus related sequences suspected of harboring papillomavirus genomes (10). This approach was successful used in demonstrating HPV DNA sequences of unknown types in DNA prepared from anogenital warts, juvenile laryngeal papillomas, and cervical carcinomas (11-14).

Using nonstringent hybridization screening techniques combined with molecular cloning, over 46 human papillomaviruses have now been identified. These types are referred to as genotypes rather than serotypes because they are distinguished by their DNA genomes rather than by antigenic determinants.

Type specific serologic agents have not yet been developed for the different papillomaviruses, thus a human papillomavirus isolate is considered a new human papillomavirus type if it shares less than 50% homology with other known HPV types when analyzed by standard hybridization conditions (15). The figure of 50% homology, however, is a relative figure. It does not represent the extent of actual homology at the DNA sequence level. Nucleic acid hybridization techniques tend to exaggerate differences since small amounts of base pair mismatch can destabilize DNA/DNA heteroduplexes. Table 2 lists the various HPVs that have been characterized to date. Each of these types tends to be preferentially, although not exclusively, associated with specific clinical lesions. Many of the types recognized have been found in patients with a rare disease called epidermodysplasia verruciformis (EV) which has been extensively studied by molecular biologists interested in HPVs.

B. Transforming Papillomaviruses

Despite the fact that to date there is no tissue culture system for the *in vitro* propagation of the papillomaviruses, cellular transformation of rodent cells by cer-

Table 2. – Human papillomaviruses

Virus Type	Clinical Association	Reference(s)
HPV-1	Plantar warts	(6,7)
HPV-2	Verruca vulgaris	(16)
HPV-3	Flat warts	(17)
HPV-4	Plantar warts	(9,18)
HPV-5	Macular lesions in EV[a]	(19)
HPV-6	Genital warts and laryngeal papillomas	(20)
HPV-7	Common warts in meat handlers	(21,22)
HPV-8	Macular lesions in EV	(19,23)
HPV-9	Macular lesions in EV	(24)
HPV-10	Flat warts	(25)
HPV-11	Laryngeal papillomas and genital warts	(26)
HPV-12	Macular lesions in EV	(25)
HPV-13	Oral focal epithelial hyperplasia	(27)
HPV-14	Macular lesions in EV	(28)
HPV-15	Macular lesions in EV	(29)
HPV-16	Cervical dysplasia, Bowenoid papulosis, and cervical carcinoma	(13)
HPV-17	Macular lesions in EV	(29)
HPV-18	Cervical dysplasia and carcinoma	(14)
HPV-19	Macular lesions in EV	(29,30)
HPV-20	Macular lesions in EV	(29,30)
HPV-21	Macular lesions in EV	(29)
HPV-22	Macular lesions in EV	(29)
HPV-23	Macular lesions in EV	(29)
HPV-24	Macular lesions in EV	(29)
HPV-25	Macular lesions in EV	(30)
HPV-26	Flat warts	(31)
HPV-27	Verruca vulgaris	K. Zachow et al, unpubl.
HPV-28	Flat warts	M. Favre et al, unpubl.
HPV-29	Verruca vulgaris	M. Favre et al, unpubl.
HPV-30	Genital warts and laryngeal carcinoma	(32)
HPV-31	Cervical dysplasia	(33)
HPV-33	Genital intraepithelial neoplasia and cervical carcinomas	(34)

Types 32, and types 34-46 have also been described at meetings but full descriptions have not appeared in press.

[a] EV is for epidermodysplasia verruciformis.

tain papillomaviruses has provided a system for researchers to study the viral functions involved in the induction of cellular proliferation, in cellular transformation, and in plasmid maintenance.

Although a few studies have now demonstrated transformation with certain of the HPV types, the analysis of papillomavirus transforming functions has largely been performed with a subgroup of papillomaviruses which are able to induce fibroblastic tumors in hamsters. These viruses include the bovine papillomavirus types 1 and 2, the deer papillomavirus, the European elk papillomavirus, and the ovine papillomavirus (35).

One characteristic of these viruses is that each is able to induce fibropapillomas in their natural host. These benign tumors contain a proliferative dermal fibroblastic component. Thus, the ability to induce fibroblastic tumors in heterologous hosts appears to correlate with the ability of a specific papillomavirus to induce fibroblastic proliferation in its natural host.

The most extensively studied of the transforming papillomaviruses has been the BPV-1. *In vitro* morphologic transformation was first described for BPV-1 in the early 1960s (36-38). A focus assay has been developed using mouse cells to study BPV-1 transformation using C127 cells and NIH 3T3 cells (39). In addition, a variety of other rodent cells including hamster and rat cells are susceptible to BPV-1 mediated transformation. Transformation by BPV-1 leads to an altered phenotype. The cells become anchorage independent, lose contact inhibition, and become tumorigenic in nude mice (39).

One characteristic of BPV-1 transformed mouse and rat cells is that *the integration of the viral DNA is not* required for either the initiation or the maintenance of the transformed state (40). The BPV-1 genomes exist as multiple copies (10 to 200 copies per diploid genome). In these initial studies, no evidence was found for integration of the viral genome in the host chromosomes. This property led to the development of BPV-1 as a mammalian cell plasmid cloning vector (41, 42). It seems likely that the viral functions involved in stable plasmid replication and faithful plasmid partitioning are manifestations of the normal biology of the virus and possibly representative of the non-productive, latent infection of cells by papillomaviruses. Thus, in addition to affording the opportunity to study the viral functions involved in the induction of cellular proliferation, the BPV-1 transformation system provides a model for the study of latent infection by papillomaviruses. Chromosomal integration of the viral DNA in these transformed cells, however, is not precluded. Several studies examining recombinant molecules containing BPV-1 have demonstrated the integration of BPV-1 sequences (43, 44). Allshire and Bostock (45) have reported evidence of integrated BPV-1 DNA molecules in mouse cells transformed with purified BPV-1 DNA and with recombinant molecules containing the BPV-1 genome.

The viral functions involved in BPV-1 transformation have been mapped and will be discussed later in this chapter. It has been shown, that the maintenance of the BPV-1 transformed state is dependent upon the continued presence and presumed expression of the BPV-1 genome (46). When BPV-1 transformed cells are treated during longterm passage with mouse L cell interferon, there is a decrease in the average number of copies of viral DNA genome per cell. Flat revertants can be observed in the treated cultures but not in the nontreated cultures after about 60 cell doublings. These flat revertants have lost the characteristic properties of transformed cells in that they grow to a decreased cell density similar to that of non-transformed C127 cells, and they have lost anchorage independence. The flat revertant cells had been "cured" of their BPV-1 DNA. Since the cells could be retransformed by BPV-1, they had not arisen from cellular mutation making them refractory to the BPV-1 transforming gene functions (46).

2. GENOMIC ORGANIZATION

The genomes of the papillomaviruses are double stranded circular DNA molecules containing about 8000 base pairs. To date, nine papillomavirus genomes have been completely sequenced (Table 3). The genomic organization of each of

Table 3. - Sequenced papillomavirus genomes

Virus	Host	Types of Lesions	Reference
HPV-1	Human	Deep plantar warts	(47)
HPV-6	Human	Condylomas	(48)
HPV-8	Human	Macular lesions in E.V.	(49)
HPV-11	Human	Condylomas and laryngeal papillomas	(50)
HPV-16	Human	Genital cancers	(51)
HPV-33	Human	Genital cancers	(52)
BPV-1	Cattle	Cutaneous fibropapillomas	(53)
CRPV	Rabbit	Papillomas	(54)
DPV	Deer	Fibropapillomas	(55)

the sequenced genomes is quite similar (Fig. 1). There are several large and relatively conserved open reading frames (ORFs) which are found in the same general vicinity and are *located on one strand* of the viral DNA. In each of the viruses studied to date, all of the RNAs found in either transformed cells, in productively infected cells, or in papillomavirus associated carcinomas have been transcribed from a single strand (56, 57, 58, 59). The other DNA strand for each of these viruses is presumed at this point to be non-coding. BPV-1 has served as the prototype for studying the molecular biology of the papillomaviruses. The genome for BPV-1 is depicted in Fig. 1.

The genome of BPV-1 can be functionally divided into two domains on the basis of its transformation capability. A fragment comprising 69% of the genome is capable of *in vitro* transformation of mouse fibroblasts and is indicated in Fig. 1 by the heavy line extending from the unique *Hind*III site to the unique *Bam*HI site (61). The corresponding region of the genome is termed the E region. This region contains all of the information required for the induction of cellular transformation as well as for stable plasmid replication in the transformed cells. This region contains eight ORFs, at least 400 bases in size, which could serve as potential coding exons. These are indicated in Fig. 1 as E1 through E8. It is possible that additional smaller ORFs are also contained in the genome.

The L region contains two additional ORFs, L1 and L2, which have been shown to code for *structural proteins* of the virus (Table 4). In general, most of these functions have been determined by studies of BPV-1. The assignment of L2 as a minor capsid proteins has been made by studies on HPV-1, and the assignment of E4 as a cytoplasmic protein has also been done for HPV-1 (Table 4).

Located 5' to the E ORFs is a region of the viral genome which has been referred to as the non-coding region (NCR) but more recently has been termed the long control region (LCR) because a small coding exon for a wart-specific late RNA has been mapped to this region (62). An analogous control region has been found in each of the other papillomavirus genomes that has been sequenced. This region ranges from approximately 500 bases in size (as is the case for HPV-8) to approximately 1000 bases in size. In general, this region is quite divergent at the level of nucleotide sequence from virus genome to another. This region contains the origin of DNA replication utilized by BPV-1 in rodent cells (63). It also contains transcriptional promoters and enhancers as described in the next section. Scattered through the LCRs of each of the papillomavirus sequences to date is the motif ACC(N)$_6$GGT. In the BPV-1 genome, this motif is repeated four times in a region which has been shown to contain the E2 conditional enhancer as discussed in the next section.

3. TRANSCRIPTION

BPV-1 transformed cells contain multiple transcripts varying in size from 1000 to 4000 bases (56). Because of the low abundance of these transcripts, precise mapping of the RNAs proved to be difficult.

Two recent studies involving the electron microscopic analysis of RNAs from transformed cells (64) and the direct sequencing of cloned cDNA copies of the viral RNAs from transformed cells (65) have revealed that there are multiple discrete species of RNAs generated by differential splicing. In addition, there are multiple viral transcriptional promoter which function in the transformed cells. At least five classes of RNAs can be distinguished by the locations of their 5' ends (62, 64). These are indicated in Fig. 2.

The 5' ends for the messages in transformed cells have been confirmed using a combination of exonuclease S1 analysis and primer extension analysis. The 5' ends of messages in transformed cells have been mapped in the vicinities of base 7185, base 7940, base 89, base 2440, and base 3090.

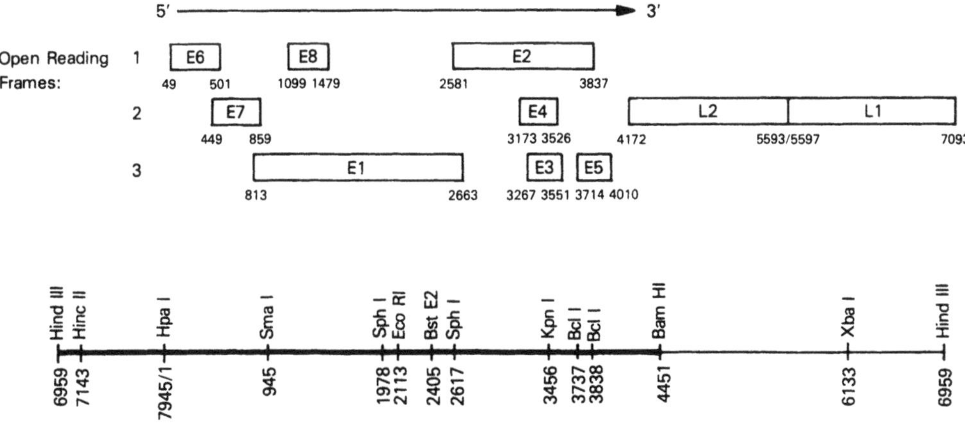

Figure 1. Genomic organization of BPV-1 DNA

The full-length molecule (7945 bp) of the BPV-1 genome opened at the unique *Hind*III site (base 6959) is marked off with restriction sites in bases noted at the bottom of the figure. The transforming segment from the *Hind*III site to the unique *Bam*HI is indicated by the heavy bar. The region transcribed in transformed cells and the direction of transcription are indicated by the arrow at the top of the figure (56). The open bars represent potential coding regions (ORFs) for the BPV-1 proteins (53). ORFs within the transforming region have been designated E1 to E8. Numbers beneath the ORFs represent the first and last bases of the ORF. Reprinted with permission from Sarver *et al.* (60).

All of the RNA species found in transformed cells are polyadenylated at the same site at base 4203 (65).

Direct sequence analysis of the viral cDNAs from transformed cells have indicated some of the precise splice donor and acceptor sites, and have provided information indicating combinations of how the various ORFs indicated in figures 1 and 2 can be spliced together at the level of RNA.

This information permits one to predict the amino acid composition of the viral proteins expressed in transformed cells. For instance, cDNAs were found in which the E2 ORF is intact and in which the E6 ORF is intact. Others were found in which a portion of the E6 ORF is spliced together with a portion of the E7 ORF and in which the same portion of the E6 ORF is spliced together with a segment of

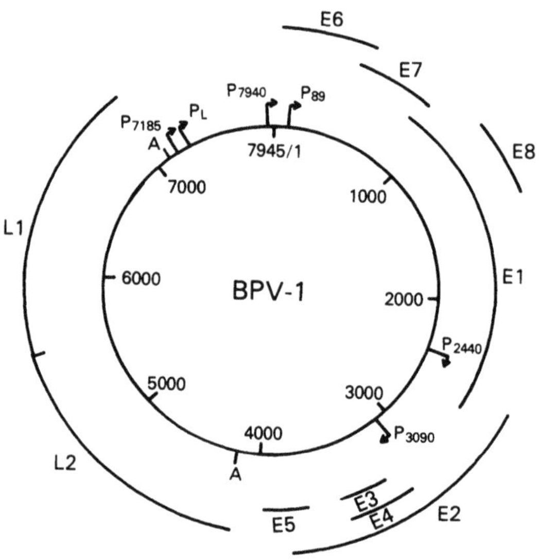

Figure 2. Transcriptional map of BPV-1 DNA

The entire 7945 bp circular genome has been sequenced (53) and the position of the base pairs in units of 1000 are noted on the inside of the genome. The 69% transforming region extends from the unique *Hind*III site at base 6959 to the unique *Bam*HI site at 4451 (61). All transcription is derived from the single strand and the ORFs on this strand are designated E1 through E8 for the transforming region. These are positioned outside the genomic map in the appropriate translation frames. The L1 and L2 ORFs are expressed only in productively infected bovine epithelial cells (57, 62). The small arrows and letters "p" correspond to the major 5' ends of the mRNA classes found in BPV-1 transformed cells (62, 64, 65, 80, 81). The five promoters found to be active thus far in transformed cells have 5' ends mapping to 7185, 7940, 89, 2440, and 3090. A wart-specific promoter which is only active in warts directs the synthesis of RNAs with heterogeneous 5' ends mapping around nucleotide 7250. The letters designate the sites of polyadenylation used in transformed cells (4203) and in warts for the L1 and L2 region in RNAs (7175).

Table 4. - Papillomavirus gene functions

Open Reading Frames	Function(s) Assigned	Virus	Reference
E1 (5′ portion)	Repressor, modulator	BPV-1	(66,67)
E1 (3′ portion)	Replication	BPV-1	(60,68)
E2	Transcriptional transactivation; probable indirect effect on replication and on trasformation	BPV-1 HPV-16	(69,70,71,72)
E3	—	—	
E4	Cytoplasmic protein in warts	HPV-1	(73)
E5	Transformation	BPV-1	(72,74,75)
E6	Transformation, plasmid copy control	BPV-1	(65,76,77)
E7	Plasmid copy control	BPV-1	(68,77)
E8	—	—	
L1	Major capsid protein	BPV-1	(57,78)
L2	Minor capsid protein	HPV-1	(79)

the E4 ORF (65). Thus, as has been shown for SV40 and polyoma virus early products, different proteins exist which share the same protein domains.

Studies have not yet been done to determine which viral genes are expressed from which promoters. It seems likely, however, that the E2 ORF is expressed from the promoter at base 2440. Spalholz *et al* have defined an E2 conditional enhancer located in the LCR (69) and have more finely mapped this enhancer between bases 7611 and 7805. Furthermore, this enhancer has been shown to increase transcription from the P_{7940} and P_{89} promoters (B.A. Spalholz, P.F. Lambert, and P.M. Howley, manuscript in preparation). It seems likely that the P_{7940} and P_{89} promoters may be involved in the expression of the E6, E7, and E1 ORFs. Interestingly, mutations in the E2 ORF dramatically decrease the transformation efficiency of BPV-1 and impair stable plasmid replication for BPV-1 (60, 70, 72). It seems likely that the effects of the E2 mutations are not direct but are rather indirect by eliminating the E2 mediated transcriptional activation of the promoters required for the E6, E7, and E1 genes necessary for transformation and for stable plasmid maintenance. Temporal studies to determine whether there is an ordered and regulated cascade of promoter activation for the papillomaviruses have not yet been done.

Late or wart-specific transcription has also best been studied for BPV-1. The L1 and L2 ORFs are only expressed in wart tissue and it has been presumed that they are only expressed in the more terminally differentiated of the epithelial cells (57, 82). Transcripts which could potentially encode the L1 and L2 ORFs are not found in transformed cells. Wart specific viral transcripts spanning the L1 and L2 ORFs are polyadenylated at the 3' end of the L1 ORF at base 7185 (57, 62). Analysis of cDNA clones from a productively infected bovine papilloma has revealed a cluster of 5' ends near base 7250 indicating a late promoter in the LCR in this vicinity. The DNA sequence in this region has a duplicated element which has some homology with a core element for the SV40 late promoter (62). The papillomaviruses contain transcriptional enhancer elements which may be involved in the control of viral gene expression in infected cells. Enhancers are *cis* regulatory elements which are able to increase the transcriptional activity of linked transcriptional promoters in an orientation and position independent manner. One such element maps 3' to the end of the transforming region near the *Bam*HI site (83, 85). The role of this enhancer element in the life cycle of BPV-1 is not clear at this point since it can be deleted without affecting the ability of the viral genome to transform or to remain as a stable plasmid in transformed cells (86). Additional enhancer elements map to the LCR. There is an enhancer element in the vicinity of the 7185 promoter, the activity of which is enhanced about 5-fold in the presence of the E2 gene product. In addition, there is the E2 conditional enhancer mapping between bases 7611 and 7805 which has been mentioned above (69; B.A. Spalholz, P.F. Lambert, and P.M. Howley, manuscript in preparation).

4. REPLICATION

Benign epithelial and fibroepithelial tumors induced by papillomaviruses contain the viral DNA as extrachromosomal plasmids. The DNA exists as a stable low copy number plasmid in the dermal fibroblasts of a bovine fibropapilloma (87). It is generally believed that the DNA exists as a low copy number plasmid in the basal cells of the epidermis in warts. Vegetative DNA replication only occurs in the cells of the upper portion of the epidermis (88). In the tumors induced in other species, such as hamsters and rabbits, the BPV-1 DNA remains extrachromosomal as a multicopy plasmid. The DNA is also present as a stable multicopy plasmid in transformed mouse cells and transformed rat cells (40, 89-91). As noted above, the integration of the viral genome is not required for the induction of BPV-1 transformation (40).

Plasmid replication of the BPV-1 genome in rodent cells provides excellent model to study the latent and persistent infection of cells by papillomaviruses. The DNA is maintained as a stable multicopy plasmid. Once established, each DNA molecule apparently replicates once per cell cycle (92). The BPV-1 plasmid copy number appears to be controlled within transformed cells indicating that there must be a mechanism whereby each plasmid is marked in a manner which insures its replication once and only once per cell cycle. Recently, evidence has been presented that a gene located in the 5' end of the E1 ORF has a modulator function involved in controlling plasmid replication (66, 67).

The mechanism by which BPV-1 plasmids become established in rodent cells is complex. BPV-1 transformation of mouse cells follows single hit kinetics (39) and yet the DNA replicates as a multicopy plasmid. Consequently, there must be an initial replication of the input BPV-1 DNA from a single copy to a level at which the DNA is maintained as a multicopy plasmid in the transformed cells. The pathway by which the viral DNA establishes and maintains copy number control has not yet been elucidated. Botchan has shown that BPV-1 plasmid replicates once per cell cycle as mentioned above, however, it is clear that immediately following infection or transfection the BPV-1 plasmid replication must not be tied to the cell cycle (67,

92). Both *cis* and *trans* acting functions must be involved in the stable replication of BPV-1 plasmids in mouse cells.

An electron microscopic analysis of BPV-1 replicative intermediates in transformed cells has mapped the center of replication bubbles to a discrete site within the BPV-1 LCR (63). This site maps closely, if not identically, with a plasmid maintenance sequence (PMS-1) mapped by Lusky and Botchan (93). A PMS is a minimal sequence required for the maintenance of BPV-1 plasmids and cells providing BPV-1 replication functions in *trans*. Lusky and Botchan mapped two regions referred to as PMS-1 and PMS-2, either of which is sufficient for the maintenance of plasmids in BPV-1 transformed cells when linked with an enhancer (93). The PMS-1 maps to the LCR in the vicinity of the origin of replication determined by Waldeck (63), and PMS-2 maps at a specific site within the E1 ORF. These PMS sequences, however, are by themselves not sufficient for maintaining plasmids in BPV-1 transformed cells in that each requires the presence of a functional enhancer element. The BPV-1 distal enhancer located adjacent to the *Bam*HI site is sufficient for providing this function as is the replication enhancer recently defined in the L1 ORF (94).

Several *trans* functions have been identified which are involved in BPV-1 plasmid replication and maintenance. Mutations in the 3' portion of E1 abolish plasmid replication leading to integration of the viral DNA (60, 68). In addition, mutations in E6 gene, as well as in the E6/E7 gene, result in DNAs which are able to be maintained as plasmids but only at low copy number. These viral plasmids cannot maintain a copy number of higher than five per cell (68, 77). Mutations in the 5' portion of the E1 ORF define an additional class of mutants which are defective in a modulator function, required for BPV-1 plasmids to become established (67). Mutations in this gene are lethal for the cell in that cotransfection of these mutants with a dominant selective marker such as the neomycin resistance gene leads to a decrease in the number of drug resistant colonies. In transient replication assays, there is no absolute requirement for this function for DNA replication; in stable replication assays, however, the mutant DNAs integrate (67).

In the context of the full viral genome, expression of the E2 gene product is also required for plasmid maintenance. Mutations in the E2 gene lead to integration of the viral DNA into the host chromosome (60, 71, 72, 70). Furthermore, Rabson *et al.* have shown the E2 function when provided in *trans*, allows plasmid replication (72). It seems likely that the role of E2 in plasmid replication (as well as in transformation) is, however, indirect through its effect on the activation of gene functions via the LCR E2 conditional enhancer (69, 74). Further evidence that it may have an indirect role comes from the observation that under certain experimental conditions some E2 mutants are able to be maintained extrachromosomally as stable plasmids (68).

5. TRANSFORMING FUNCTIONS

The genetics of BPV-1 transformation are complex. In addition to direct transforming viral oncogenes, there are functions encoded by the BPV-1 which have an indirect effect on transformation.

BPV-1 encodes two independent genes which are each capable of independently transforming rodent cells. It was originally shown that a specific 69% subgenomic fragment encompassed by the unique *Hind*III and unique *Bam*HI sites (see Fig. 2) was sufficient for inducing transformation of the C127 cells or NIH 3T3 cells (61). Mutagenesis studies of the full length viral genome and experiments utilizing subgenomic fragments expressed from surrogate promoters have mapped two segments of the viral genome involved directly in transformation.

One of these regions maps to the 3' ORFs which includes the E2, E3, E4 and E5 ORFs (60, 95, 96). The expression of these 3' ORFs, however, is not sufficient for the fully transformed phenotype.

Expression of the 3' ORFs from BPV-1 promoters located in the LCR or from a surrogate promoter lead to efficient focus formation, however, the resulting transformants have only a partially transformed phenotype. They are not anchorage independent nor are they highly tumorigenic in nude mice (60). Sarver *et al.* demonstrated that the integrity of the E6-E7 region of the BPV-1 genome was required for the full transformed phenotype (60). Further mutagenesis studies defined the E6 ORF as encoding the 5' transforming gene and the E5 ORF as encoding the 3' transforming gene. A cDNA containing the E6 ORF intact is able to transform C127 cells efficiently (74). In addition, the expression of the E6 ORF from a strong surrogate promoter, the Harvey sarcoma virus LTR, has been shown to be sufficient for morphologic transformation of C127 cells alone (76).

The BPV-1 E6 gene product has been identified in BPV-1 transformed cells (97). A fusion polypeptide containing the amino terminal end of the phage CII protein linked to the E6 protein was expressed in *E.coli* and used to generate antibodies. Antibodies to this fusion protein were capable of specifically immunoprecipitating a 15.5 kD BPV-1 E6 protein from mouse cells specifically transformed by the E6 gene. This gene product was identified by biochemical fractionation in both the nuclear as well as membrane fractions of transformed cells (97). The E6 ORF is conserved in the genomes of each of the papillomaviruses that have been sequenced thus far. The predicted amino acid composition of each of the E6 gene products contains the motifs CysXXCys, repeated four times; the spacing of these motifs is conserved. Based on the presence of this repeated motif and the high lysine and arginine content of the BPV-1 E6 protein, it has been suggested that the E6 protein may be a DNA binding protein (97). Direct evidence for such binding, however, has yet to be produced.

The role of the E6 protein in benign warts has not yet been defined. It should be pointed out, however, that this gene is generally expressed in papillomavirus associated carcinomas and carcinoma cell lines. The E6 gene, therefore, remains a potential candidate for being involved in malignant progression for those papillomaviruses that have been implicated in malignant transformation. The E6 gene has been selectively retained and is transcriptionally active in rabbit carcinomas induced by the CRPV (98, 99). In addition, in human cervical carcinomas, the E6 region of the associated HPV-16 and HPV-18 genomes are generally expressed (59, 100, 101).

The E5 gene is located in the extreme end of the 3' ORFs (Fig. 1) and is also capable of independently transforming mouse C127 cells and NIH 3T3 cells.

Mutagenesis studies have established that mutations downstream from a methionine codon at 3879 effectively eliminate transformation of mouse C127 cells or NIH 3T3 cells (71, 72, 75). Behind either the Harvey sarcoma virus LTR or the SV40 early region promoter, the E5 ORF downstream from the first methionine codon is sufficient for inducing efficient transformation (74, 96). The E5 transforming protein of BPV-1 has been identified (102). Antibodies generated against a synthetic peptide corresponding to the carboxy terminus of this ORF were used to identify this 7 kD protein in BPV-1 transformed cells. The amino acid composition of the 44 amino acid E5 protein has a strikingly hydrophobic structure; cell fractionation studies of these transforming cells have localized the polypeptide predominantly to cellular membranes (102).

The E5 ORF is strongly conserved among those papillomaviruses which induce fibropapillomas. These viruses include BPV-1, BPV-2, the deer papillomavirus, and the European elk papillomavirus. The putative E5 genes encoded by those viruses which produce purely epidermal cell proliferations (i.e., HPV-1 or CRPV) show little homology to the E5 gene of BPV-1, supporting the hypothesis that the normal role of the E5 protein is in the stimulation of the dermal fibroblasts in benign fibropapillomas.

It is not clear at this point how the papillomavirus transforming genes relate to other known viral or cellular oncogenes. They share no striking homology to other viral or cellular oncogenes. To date, attempts to use either of the two BPV-1 viral transforming genes to complement other known oncogenes in a transformation assay employing primary cells have not been reported.

6. VIRUSES AND CARCINOGENIC PROGRESSION

There is a subgroup of papillomaviruses which induces papillomas that can progress to squamous cell carcinomas. These viruses and their associated malignancies are listed in Table 5. The most extensively studied of these viruses has been the Shope or cottontail rabbit papillomavirus (CRPV). Studies with this virus date to the early 1930s when Shope identified the virus as the etiologic agent of rabbit cutaneous papillomatosis (3). The CRPV system has been the most extensively studied as a model for papillomavirus-induced carcinogenesis (4, 103-105).

One of the features of CRPV associated carcinogenic progression is the synergy between the virus and carcinogenic external factors. The synergism with external factors appears to be a general phenomenon involved in papillomavirus associated carcinogenic progression. In the case of CRPV, the carcinomas develop more rapidly and in a higher percentage of animals when the virally induced papillomas are painted with either methylcholanthrene or tar (103). These carcinomas contain the CRPV DNA which is often arranged in multiple head-to-tail copies (106, 107). In some tumors, integrated DNA can be detected. Transplantable carcinoma cell lines (VX 2 and VX 7) have been established and, the DNA is integrated in these cells (98, 108, 109).

Another animal papillomavirus system in which carcinogenic progression has been extensively studied is BPV-4. BPV-4 causes esophageal papillomatosis in cattle. Jarrett and his colleagues have found that the cattle from the Scottish highlands

Table 5. - Papillomaviruses and naturally occurring cancers

Papillomaviruses	Cancers	Other Factors
CRPV	Skin cancer	Methylcholanthrene, coal tar
BPV-4	Tongue, esophageal and foregut cancer	Bracken Fern
BPV (not typed)	Ocular cancers	Ultraviolet Light
Ovine papillomavirus	Skin cancer	Ultraviolet Light
HPV-5,8	Skin cancer in patients with EV	Ultraviolet Light
HPV-16,18,33	Anogenital cancers, some oral and laryngeal cancers	Smoking, ? herpes simplex, ? other factors

that are infected with BPV-4 and feed on Bracken fern, have a high incidence of squamous cell carcinomas of the foregut and of the esophagus (110). This Bracken fern is known to contain a radiomimetic substance.

In contrast to the CRPV associated carcinomas, in which the viral DNA is invariably present, BPV-4 DNA is usually not found in these alimentary tract carcinomas.

Extensive analysis of the squamous cell carcinomas of the upper alimentary tract in cattle infected with BPV-4 has revealed that these malignant tumors are generally negative for the viral DNA; in only one case out of 70 examined was BPV-4 DNA found in a carcinoma of the esophagus. In a second case, DNA was found in a carcinoma of the tongue (111). These data suggest that the continued presence of the papillomavirus genome may not be required for the maintenance of the carcinogenic state in these alimentary tract tumors. It is a formal possibility, however, that a different papillomavirus, genetically remote from BPV-4, may be associated with these carcinomas, and that the genomes of this alternate virus may not have been detected in the assays performed. It should be noted that BPV-4 has been reported to transform mouse fibroblasts in culture (112).

Additional animal papillomaviruses are also associated with carcinogenic progression. Ocular cancer in cattle have been associated with papillomaviruses (113). In these cases, ultraviolet light is thought to be the cofactor. In addition, skin cancer in sheep has been associated with an ovine papillomavirus and, ultraviolet light has been implicated in this carcinogenic progression (114).

In humans there are several entities which have been associated with papillomaviruses and carcinogenic progression. One is skin cancer in patients with epidermodysplasia verruciformis. This is a rare lifelong disease, usually beginning in infancy or early childhood. It is characterized by disseminated polymorphic skin lesions that resemble either flat warts or lesions that appear as reddish macules which are sometimes refered to as "pityriasis-like lesions" (115). About one-third of patients with epidermodysplasia verruciformis develop multiple skin carcinomas, usually during their third or fourth decade. In general, the carcinomas that develop in these patients arise in sun exposed areas and it has been proposed that UV irradiation plays a co-carcinogenic role with the specific HPVs in the etiology of these cancers. A variety of HPVs, as listed in Table 2, have been associated with this disease. A subset of these HPVs, particularly HPV-5 and HPV-8, have been associated with the carcinomas.

In one extensive study, HPV genomes were found in 27 or 28 tumor samples; 21 of these contained HPV-5 DNA and five contained HPV-8 DNA (116). Other investigators have found HPV-5 associated with primary carcinomas and metastatic carcinomas in patients with EV (117, 118).

One of the areas in which there has been significant advances over the last several years has been in the association of HPVs with anogenital carcinomas. Numerous epidemiologic studies have suggested that an infectious agent, possibly a virus, may be involved in the etiology of cervical carcinomas. Sexual promiscuity, an early age of onset of sexual activity, and poor sexual hygiene have been identified as risk factors. In the early 1970s the herpesviruses were considered as strong potential etiologic agents in cervical carcinoma but studies seeking molecular support for such a role failed to establish this association. Recent prospective epidemiologic studies by Vonka would tend to minimize the role of Herpes simplex virus 2 (HSV2) as an important etiologic factor in cervical carcinoma (119, 120).

Strong evidence linking an HPV with cervical carcinoma came from the recognition that the morphologic features previously interpreted on pap smears and biopsies as dysplasia were, in fact, manifestations of a papillomavirus infection of the cervix (121, 122, 123). This observation spurred the molecular analyses of genital wart lesions and cervical carcinomas for the presence of HPV DNAs. HPV-6 was cloned directly from a condyloma acuminata and using its DNA, along with that of the closely related HPV-11 genome cloned from a laryngeal papilloma, Gissmann and his colleagues were able to demonstrate a strong association of these HPV types with condyloma acuminata (124, 125). The majority of cervical carcinomas, however,were negative for HPV-6 and HPV-11 DNA.

Using non-stringent hybridization techniques described above, which are capable of detecting all HPV types, zur Hausen and his colleagues were able to clone HPV-16 and HPV-18 DNAs directly from human cervical carcinomas, and subsequently to detect these DNAs in approximately 70% of human cervical carcinomas examined (13, 14, 125). The cloning of HPV-16 and HPV-18 as well as the other HPVs now associated with genital tract lesions now provide the probes to permit an extensive analysis of specific HPV types with specific genital tract lesions.

Another area where HPVs have been associated with human carcinomas has been in the larynx and oral pharynx. In general, the same HPVs associated with genital tract papillomavirus lesions have been associated with some laryngeal and oral carcinomas. The percentage of the tumors in the larynx and oral cavity that are positive for HPVs, however, appears to be lower than that of the genital tract.

The genital tract papillomaviruses have been found in normal appearing epithelium of the genital tract as well as of the larynx (126, 127, 128). Thus, to interpret the associations with cervical and laryngeal carcinomas, a better understanding of the biology and of the natural history of papillomavirus infections in needed. The possible high prevalence of clinically inapparent HPV infections of the cervix in normal populations of women and in pregnant women (125) suggests that more extensive epidemiologic studies must be carried out in order to determine the prevalence of specific HPV infections and their consequent risk. In addition, functional studies establishing a mechanism for these viruses in carcinogenic progression must be done in order to determine whether or not the high degree of association of specific HPVs with specific human cancers is indicative of an etiologic role.

Acknowledgement

I am grateful to Nan Freas for her skilled assistance in preparing this chapter which was completed in October 1986.

7. REFERENCES

1) Melnick J.L. (1962) Science **135**, 1128-1130.
2) Ciuffo G. (1907) Giorn Ital Mal Venereol **48**, 12-17.
3) Shope R.E. (1933) J. Exp Med **58**, 607-624.
4) Rous P., Beard J.W. (1935) J. Exp Med **62**, 523-548.
5) Jenson A.B., Rosenthal J.D., Olson C., Pass F., Lancaster W.D., Shah K (1980) JNCI **64**, 495-500.
6) Favre M., Orth G., Croissant O., Yaniv M. (1975) Proc Natl Acad Sci USA **72**, 4810-4814.
7) Gissmann L., zur Hausen H. (1976) Proc Natl Acad Sci USA **73**, 1310-1313.
8) Law M.-F., Lancaster W.D., Howley P.M. (1979) J. Virol **32**, 199-207.
9) Heilman C.A., Law M.-F., Israel M.A., Howley P.M. (1980) J. Virol **35**, 395-407.
10) Howley P.M., Israel M.A., Law M.-F., Martin M.A. (1979) J. Biol Chem **254**, 4876-4883.
11) Kryzek R.A., Watts S.L., Anderson D.L., Faras A.J., Pass F. (1980) J. Virol **36**, 236-244.
12) Lancaster W.D., Jenson A.B. (1981) Intervirology **15**, 204-212.
13) Durst M., Gissmann L., Ikenberg K., zur Hausen H. (1983) Proc Natl Acad Sci USA **80**, 3812-3815.
14) Boshart M., Gissmann L., Ikenberg H., Zur Hausen H. (1984) EMBO J. **3**, 1151-1157.
15) Coggin J.R., zur Hausen H. (1979) Cancer Res **39**, 545-546.
16) Orth G., Favre M., Croissant O. (1977) J. Virol **24**, 108-120.
17) Orth G., Jablonska S., Favre M., Croissant O., Jarzabek-Chorzelska M., Rzesa G. (1978) Proc Natl Acad Sci USA **75**, 1537-1541.
18) Gissmann N.L., Pfister H., zur Hausen H. (1977) Virology **7**, 569-580.
19) Orth G., Favre M., Breitburd F., Croissant O., Jablonska S., Obalek S., Jarzabek-Chorzelska M., Rzesa G. (1980) Cold Spring Harbor Conference on Cell Proliferation **7**, 259-282.
20) Gissmann L., zur Hausen H. (1980) Int J. Cancer **25**, 605-609.
21) Ostrow R.S., Kryzyek R., Pass F., Faras A.J. (1981) Virology **108**, 21-27.
22) Orth G., Jablonska S., Favre M., Croissant O., Jarzabek-Chorzelska M., Jibard N. (1981) J. Invest Derm **76**, 97-102.
23) Pfister H., Nurnberg F., Gissmann L., zur Hausen H. (1981) Int J. Cancer **27**, 645-650.

24) Kremsdorf D., Jablonska S., Favre M., Orth G. (1982) J. Virol **43**, 436-447.

25) Kremsdorf D., Jablonska S., Favre M., Orth G. (1983) J. Virol **48**, 340-351.

26) Gissmann L., Diehl V., Schultz-Coulon H., zur Hausen H. (1982) J. Virol **44**, 393-400.

27) Pfister H., Hettich I., Runne U., Gissmann L., Chilf G.N. (1983) J. Virol **47**, 363-366.

28) Tsumori T., Yutsudo M., Nakano Y., Tanigaki T., Kitamura H., Hakura A. (1983) J. Gen Virol **64**, 967-969.

29) Kremsdorf D., Favre M., Jablonska S., Obalek S., Rueda L.A., Lutzner M., Blanchet-Bardon C., van Voorst Vader P.C., Orth G. (1984) J. Virol **52**, 1013-1018.

30) Gassenmaier A., Lammel M., Pfister H. (1984) J. Virol **52**, 1019-1023.

31) Ostrow Rs, Zachow K.R., Thompson O., Faras A.J. (1984) J. Invest Dermatol **82**, 362-366.

32) Kahn T., Schwarz E., zur Hausen H. (1986) Int J. Cancer (in press).

33) Lorincz A.T., Lancaster W.D., Temple G.F. (1986) J. Virol **58**, 225-229.

34) Beaudenon S., Kremsdorf D., Croissant O., Jablonska S., Wain-Hobson S., Orth G. (1986) Nature (London) **321**, 246-249.

35) Howley P.M,. & Schlegel R. (1987) In: *The Papillomaviruses* P.M. Howley, N. Salzman, eds., Plenum Press, New York (in press).

36) Black P.H., Hartley J.W., Rowe W.P., Huebner R.J. (1963) Nature (London) **199**, 1016-1018.

37) Thomas M., Boiron M., Tanzer J., Levy J.P., Bernard J. (1964) Nature (London) **202**, 709-710.

38) Boiron M., Levy J.P., Thomas M., Friedman J.C., Bernard J. (1964) Nature (London) **201**, 423-424.

39) Dvoretzky I., Shober R., Lowy D.R. (1980) Virology **103**, 369-375.

40) Law M.-F., Lowy D.R,. Dvoretzky I., Howley P.M. (1981) Proc Natl Acad Sci USA **78**, 2727-2731.

41) Sarver N., Gruss P., Law M.F., Khoury G., Howley P.M. (1981) Mol Cell Biol **1**, 486-496.

42) Sarver N., Byrne J.C., Howley P.M. (1982) Proc Natl Acad Sci USA **79**, 7147-7151.

43) DiMaio D., Metherall J., Neary K., Guralski D. (1985). In: *Papillomaviruses* P. Howley, T. Broker, eds., Alan R. Liss, New York, 437-456.

44) Sarver N., Muschel R., Byrne J.C., Khoury G., Howley P.M. (1985) Mol Cell Biol **5**, 3507-3516.

45) Allshire R.C., Bostock C.J. (1986) J. Mol Biol **188**, 1-13.

46) Turek L.P., Byrne J.C., Lowy D.R., Dvoretzky I., Friedman R.M., Howley P.M. (1982) Proc Natl Acad Sci USA **79**, 7914-7918.

47) Danos O., Katinka M., Yaniv M. (1982) EMBO J. **1**, 231-236.

48) Schwarz E., Durst M., Demankowski C., Lattermann O., Zech R., Wolfsperger L., Suhai S., zur Hausen H. (1983) EMBO J. **2**, 2341-2348.

49) Fuchs P.G., Iftner T., Weninger J., Pfister H. (1986) J. Virol **58**, 626-634.

50) Dartmann K., Schwarz E., Gissmann L., zur Hausen H. (1986) Virology **151**, 124-130.

51) Seedorf K., Krammer G., Durst M., Suhai S., Rowekamp W.G. (1985) Virology **145**, 181-185.

52) Cole S.T., Streeck R.E. (1986) J. Virol **58**, 991-995.

53) Chen E.Y., Howley P.M., Levenson A.D., Seeburg P.H. (1982) Nature (London) **299**, 529-534.

54) Giri I., Danos O., Yaniv M. (1985) Proc Natl Acad Sci USA **82**, 1580-1584.

55) Groff D.E., Lancaster W.D. (1985) J. Virol **56**, 85-91.

56) Heilman C.A., Engel L., Lowy D.R., Howley P.M. (1982) Virology **119**, 22-34.

57) Engel L.W., Heilman C.A., Howley P.M. (1983) J. Virol **47**, 516-528.

58) Georges E., Croissant O., Bonneaud N., Orth G. (1984) J. Virol **51**, 530-538.

59) Schneider-Gadicke A., Schwarz E. (1986) EMBO J. **5**, 2285-2292.

60) Sarver N., Rabson M.S., Yang Y.-C., Byrne J.C., Howley P.M. (1984) J. Virol **52**, 377-388.

61) Lowy D.R., Dvoretzky I., Shober R., Law M.-F., Engel L., Howley P.M. (1980). Nature (London) **287**, 72-74.

62) Baker C.C., Howley P.M. (1986) Submitted).

63) Waldeck W., Rosl F., Zentgraf H. (1984) EMBO J. **3**, 2173-2178.

64) Stenlund A., Zabielski J., Ahola H., Moreno-Lopez J., Pettersson U. (1985) J. Mol Cell **182**, 541-554.

65) Yang Y.-C., Okayama H., Howley P.M. (1985) Proc Natl Acad Sci USA **82**, 1030-1034.

66) Roberts J.M., Weintraub H. (1986) Cell **46**, 741-752.

67) Berg L., Lusky M., Stenlund A., Botchan M. (1986) Cell **46**, 753-762.

68) Lusky M., Botchan M.R. (1985) J. Virol **53**, 955-965.

69) Spalholz B.A., Yang Y.-C., Howley P.M. (1985) Cell **42**, 183-191.

70) DiMaio D. (1986) J. Virol **57**, 475-480.

71) Groff D.E., Lancaster W.D. (1986) Virology **150**, 221-230.

72) Rabson MS, Yee C., Yang Y.-C., Howley P.M. (1986) J. Virol **60**, 626-634.

73) Doorbar J., Campbell D., Grand R.J.A., Gallimore P.H. (1986) EMBO J. **5**, 355-362.

74) Yang Y.-C., Spalholz B.A., Rabson M.S., Howley P.M. (1985) Nature (London) **318**, 575-577.

75) DiMaio D., Guralski D., Schiller J.T. (1986) Proc Natl Acad Sci USA **83**, 1797-1801.

76) Schiller J.T., Vass W.C., Lowy D.R. (1984) Proc Natl Acad Sci USA **81**, 7880-7884.

77) Berg L.J., Singh K., Botchan M. (1986) Mol Cell Biol **6**, 859-869.

78) Pilacinski W.P., Glassman D.L., Kryzek R.A., Sadowski P.L., Robbins A.K. (1984) Biotechnology, **2**, 356-360.

79) Komly C.A., Breitburd F., Croissant O. and Streeck R.E. (1986) J. Virol (in press).

80) Ahola H., Stenlund A., Moreno-Lopez J., Pettersson U. (1983) Nucleic Acids Res **11**, 2639-2650.

81) Pettersson U., Ahola H., Stenlund A., Bergman P., Ustav M., Moreno-Lopez J. (1986). In: *Papillomaviruses* D. Evered, S. Clark, eds., John Wiley and Sons, Chichester, 23-34.

82) Amtmann E., Sauer G. (1982) J. Virol **43**, 59-66.

83) Lusky M., Berg L., Weiher H., Botchan M. (1983) Mol Cell Biol **3**, 1108-1122.

84) Campo M.S., Spandidos D.A., Lang J., Wilkie N.M. (1983) Nature (London) **303**, 77-80.

85) Weiher H., Botchan M.R. (1984) Nucleic Acids Res **12**, 2901-2916.

86) Howley P.M., Schenbron E.T., Lund E., Byrne J.C., Dahlberg J.E. (1985) Mol Cell Biol **5**, 3310-3315.

87) Lancaster W.D., Olson C. (1982) Microbiol. Reviews **46**, 191-207.

88) Howley P.M. (1985) Amer J. Path **113**, 414-421.

89) Lancaster W.D. (1981) Virology **108**, 251-255.

90) Binetruy B., Meneguzzi G., Breatnach R., Cuzin F. (1982) EMBO J. **1**, 621-628.

91) Babiss L.E., Fisher P.B. (1986) Virology **154**, 180-194.

92) Botchan M., Berg L., Reynolds J., Lusky M. (1986). In: *Papillomaviruses*, D. Evered, S. Clark, eds., John Wiley and Sons, Chichester, 53-63.

93) Lusky M., Botchan M.R. (1984) Cell **36**, 391-401.

94) Lusky M., Botchan M.R. (1986) Proc Natl Acad Sci USA **83**, 3609-3613.

95) Nakabayashi Y., Chattopadhyay S.K., Lowy D.R. (1983) Proc Natl Acad Sci USA **80**, 5832-5836.

96) Schiller J., Vousden K., Vass W.C., Lowy D.R. (1986) J. Virol **57**, 1-6.

97) Androphy E.J., Schiller J.T., Lowy D.R. (1985) Science **230**, 442-445.

98) Nasseri M., Wettstein F.O. (1984) J. Virol **51**, 706-712.

99) Danos O., Georges E., Orth G., Yaniv M. (1985) J. Virol **53**, 735-741.

100) Baker C.C., Phelps W.C., Lindgren V., Braun M., Gonda M., Howley P.M. (1986) Submitted.

101) Schwarz E., Freese U.K., Gissmann L., Mayer W., Roggenbuck B., Stremlau A., zur Hausen H. (1985) Nature (London) **314**, 111-114.

102) Schlegel R., Wade-Glass M., Rabson MS, Yang Y-C (1986) Science **233**, 464-466.

103) Rous P., Kidd J.G. (1936) Science **83**, 468-469.

104) Kidd J.G., Rous P. (1937) Proc Soc Exp Biol Med **37**, 518-520.

105) Rous P., Kidd J.G., Smith W.E. (1953) J. Exp Med **96**, 159-174.

106) Ito Y. (1960) Virology **12**, 596-601.

107) Stevens J.G., Wettstein F.O. (1979) J. Virol **30**, 891-898.

108) Sugawara K., Fujinaga K., Yamashita T., Ito Y. (1983) Virology **131**, 88-99.

109) Georges E., Croissant O., Bonneaud N., Orth G. (1984) J. Virol **51**, 530-538.

110) Jarrett W.F.H., McNeil P.E., Grimshaw W.I.R., Selman I.E., McIntyre W.I.M. (1978) Nature (London) **274**, 215-217.

111) Campo M.S., Moar M.H., Sartirana M.L., Kennedy I.M., Jarrett, W.F.H. (1985) EMBO J. **4**, 1819-1825.

112) Campo M.S., Spandidos D.A. (1983) J. Gen Virol **64**, 549-557.

113) Ford J.N., Jennings P.A., Spradbrow P.B., Francis J. (1982) Res Vet Sci **32**, 257-259.

114) Vanseow B.A., Spradbrow P.B., Jackson A.R.B. (1982) Vet Record **110**, 561-562.

115) Lutzner M. (1978) Bull. Cancer **65**, 169-182.

116) Orth G. (1986). In: *Papillomaviruses*, D. Evered and S. Clark, eds., John Wiley and Sons, Chichester, 157-174.

117) Ostrow R.S., Bender M., Niimura M., Seki T., Kawashima M., Pass F., Faras A.J. (1982) Proc Natl Acad Sci USA **79**, 1634-1638.

118) Pfister H., Gassenmaier A. and Nurnberger F. (1983) Cancer Res **43**, 1436-1441.

119) Vonka V., Kanda J., Hirsch I., Zavadeva H., Krcmar M., Suchankova A., Rezakova D., Broucek J., Press M., Domorazkova E., Svoboda B., Havrankova A., Jelinek J. (1984) Int J. Cancer **33**, 61-66.

120) Vonka V., Kanda J., Jelinek J., Subrt I., Sucharek A., Havrankova A., Vachal M., Hirsch I, Domorazkova E, Zavadova H, Richterova V., Naprstkova J., Dvorakova V., Svoboda B. (1984) Int J. Cancer **33**, 49-60.

121) Meisels A., Fortin R. (1976) Acta Cytol **20**, 505-509.

122) Purola E., Savia E. (1977) Acta Cytol **21**, 26-31.

123) Laverty C.R., Russell P., Hills E., Booth N. (1978) Acta Cytol **22**, 195-201.

124) Gissmann L., Wolnik L., Ikenberg H., Koldovsky U., Schnurch G. and zur Hausen H. (1983) Proc Natl Acad Sci USA **80**, 560-563.

125) Gissmann L., Schwarz E. (1986). In: *Papillomaviruses* D. Evered, S. Clark, eds., John Wiley and Sons, Chichester, 190-197.

126) Steinberg B.M., Topp W.C., Schneider P.S., Abramson A.L. (1983) N. Engl J. Med **308**, 1261-1264.

127) Ferenczy A., Mitao M., Nagai N., Silverstein S.J., Crum C.P. (1985) N. Engl J. Med **313**, 784-788.

128) Macnab J.C.M., Walkinshaw S.A., Cordiner J.W., Clements J.B. (1986) N. Engl J. Med **315**, 1052-1058.

CHAPTER 19

BIOLOGICAL AND MOLECULAR ASPECTS OF SIMIAN VIRUS 40 (SV-40) AND POLYOMAVIRUS REPLICATION

YOSEF ALONI

Department of Genetics, The Weizmann Institute of Science, Rehovot 76100, Israel

1. GENERAL OVERVIEW

A. Classification

SV40 and polyomavirus belong to the papova family. The term papova in an acronym of *Pa*pilloma (wart) viruses of humans and rabbits, *Po*lyomavirus of mice and *Va*cuolating agent (SV40) of rhesus monkeys (1). The main properties shared by the papova viruses are the double-strandness of their DNA genome surrounded by an icosahedral capsid, and the nucleus of the host cell as the site of multiplication.

Papillomaviruses are larger than the polyomaviruses. The genome of the papillomaviruses contain about 8000 bp and all the genetic formation is in one DNA strand. The genomes of the polyomaviruses contain about 5000 bp and the genetic information is separated almost evenly between the two DNA strands. For further comparison between Papillomavirus and Polyomaviruses see Table 1 and the preceding chapter.

Polyomaviruses are widely distributed in nature. The infection is specie-specific. Human, monkeys, cattle, rabbits, mice and hamsters can be infected by polyomaviruses (2). The best studied representatives of the polyomaviruses are SV40 and polyoma and this review article will focus on these two viruses only.

B. Isolation of Polyomaviruses

Polyomavirus was isolated and identified by Ludwick Gross in 1958 (3-5). The virus was a contaminant of extracts that contained the murine leukemia virus (MLV). Ludwick Gross was able to demonstrate that a virus different from MLV can cause salivary gland adenocarcinomas. Because the virus could cause a variety

Table 1. - Comparison of properties of papillomaviruses and polyomaviruses

	Papillomaviruses	Polyomaviruses
Virion diameter	55 nm	44 nm
MW of DNA	5.2×10^6	3.2×10^6
No. of base pairs in genome	8×10^3	5×10^3
Genetic information	On one strand	On both strands
Genome in transformed cells	Most often not integrated	Most often integrated
Target issue	Surface epithelia	Internal organs
Result of infection	Benign tumor	Inapparent, or organ pathol-
Viruses infecting humans	Human papilloma viruses, types 1 to 25	Bk virus, JC virus
Most significant human illness	Warts on skin, in oral tract, in oral cavity, and in respiratory tract	Progressive multifolcal leukoencephalopathy
Distribution in nature	Mammals, birds	Mammals, birds

(From ref. 2, with permisson)

of other tumors and transform different types of cell, Stewart and Eddy proposed in 1958 the name polyomavirus (6-8).

The discovery of SV40 was made during the screening for viruses of the monkey cells that were used to prepare the polio vaccine. Many simian viruses were isolated during this investigation and were given serial numbers. SV40 was discovered in 1960 by Sweet and Hilleman when they used African green monkeys of the genus *Cercopithecus* as their test cells (9). Immediately after this discovery it was found that SV40 can cause tumors in newborn hamsters and transform certain cells (10-12). There is no evidence that SV40 can cause tumors in its natural host. As a result of the immunization with the poliovirus vaccine millions of people between 1955 and 1961 were exposed to simian virus 40 and this raised considerable concern. There is no evidence however that the exposure to SV40 caused ill effects. Other known polyoma viruses that were isolated from tumor diseases and were named after the patients' names include: JC virus that was isolated from progressive multifocal leukoencephalopathy (PML) a fatal disorder of the central nervous system and BK virus that was isolated from the urine of a renal transplant (13-16). In nature polyoma is a virus of mice and SV40 is a virus of monkeys.

C. The Permissive and Nonpermissive Virus/Cell Interactions

In the laboratory, permissive cells are cells that can support virus multiplication and as a result the cell lyse and virus particles are released. In the nonpermissive interaction although the virus penetrates the cells it does not produce progeny particles but instead its DNA integrates into the cellular genome. As a result of the integration of the viral DNA, the cell can be transformed. (The mechanism of integration of the viral DNA into the cellular genome is discussed in a separate section of this review). Monkey kidney cells including the stable lines BSC-1, CV-1 and Vero are the permissive cells for SV40 infection; human cells are partially permissive. Mouse cells including stable lines such as 3T3 are the permissive cells for polyoma. It is believed that specific cellular factors are involved in rendering a cell permissive for virus multiplication and other cellular factors prevent multiplication (17).

D. SV40 and Polyoma Virions

The virions contains DNA molecules that are covalently closed, double-stranded circles in a superhelical form (see Fig. 1). The twisted superhelical DNA of SV40 and polyoma are called form-I DNA. Disruption of one phosphodiester bond in one DNA strand converts superhelical molecules into relaxed circles, called form-II DNA. Disruption of two phosphodiester bonds at the same site at both DNA strands converts superhelical molecules into linear molecules. The three forms of the viral DNA can be separated by migration through agarose gels and visualized by staining with ethidium bromide. Whereas forms I and II migrate through agarose gels at rates that are relatively constant with respect to each other regardless of the concentration of the gel, migration of form III is very sensitive to gel concentration (24,25).

Each virion contains one DNA molecule, which represents about 12% of the total mass and is associated with the host histones H2a, H2b, H3 and H4 but lacks H1 (21-23). The histones are assembled in 24-26 nucleosomes on the viral DNA giving rise to a structure like beads on a string known as the viral *minichromosome* (see Fig. 2).

Seventy two capsomers form the icosahedrical capsid that surrounds the minichromosome. Three proteins, VP1, VP2, and VP3 in decreasing order of size constitute the capsid. VP1, a 45 kDa phosphoprotein, is the major capsid component, while VP2 and VP3 are less prominent.

Mutations in VP1 appear to affect viral assembly and virion stability; mutations in VP2 and VP3 appear to affect the uncoating process of the virion following penetration into the cell (5, 26). Table 2 summarizes the structural features of polyoma and SV40.

E. The Viral Minichromosome

SV40 and polyoma have similar physical structures, both as naked DNA and as chromatin (27). The viral DNAs are found in the infected cells in the form of minichromosomes, composed of viral DNA and a maximum of 26 nucleosomes. Twenty-six is also the maximum number of superhelical turns observed in normal form-I DNA (31).

The nucleosomes are composed of the four cell histones H2a, H2b, H3 and H4 while H1 is found in the linker between nucleosomes (28-30). The superhelical turns of the viral DNA are constrained in the nucleosomes (32). Nevertheless, the molecule still contains an extra 1-2 superhelical turns that are not constrained in the nucleosomes and the molecule is therefore torsionally strained (33). It appears that both the bulk and the 1-2% of the actively transcribed minichromosomes contain the extra supercoils (34). The significance of the extra supercoils in the

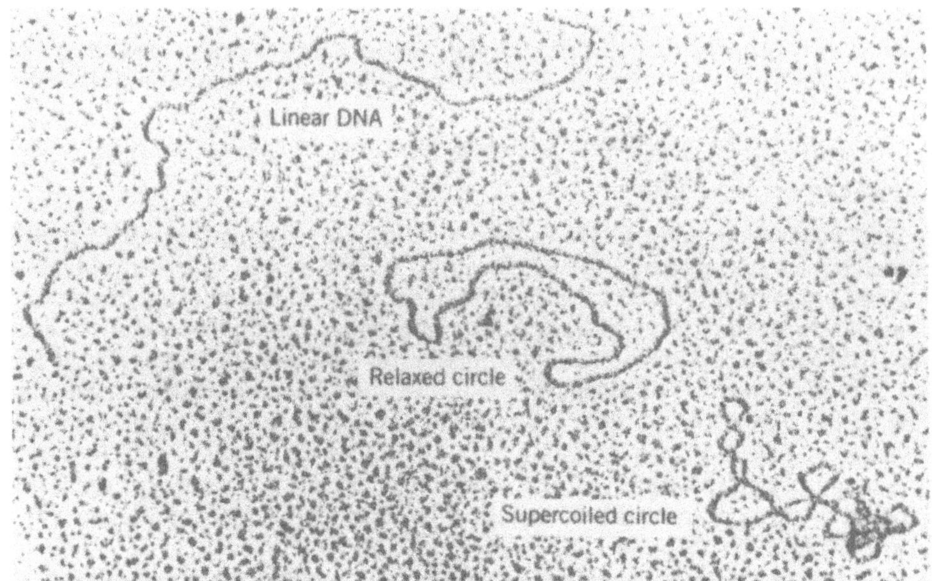

Figure 1. The three forms of SV40 DNA (from ref. 189 with permission).

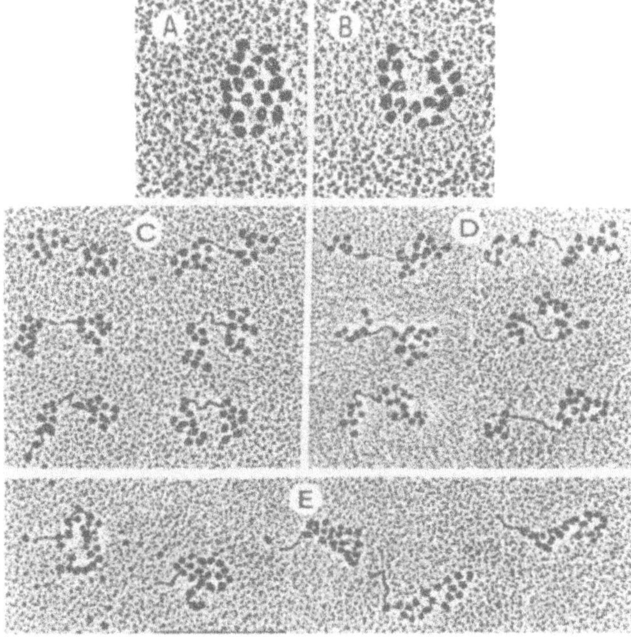

Figure 2. Electron microscope visualization of SV40 minichromosomes without (A) and with a gap (B), cleaved with *Bam* H-1 (C), *Eco* R1 (D) and *Bgl*-1 (E). The cleavage sites are: *Bam* H-1 15 map units (mu) *Eco* R1 0.0 mu and *Bgl*-1 67 mu. *Bgl*-1 cuts at the origin of replication. (With permission from ref. 39).

Table 2. - Structural features of polyoma and SV40

Feature	Polyomavirus	SV40
Size	45 nm	45 nm
Symmetry	icosahedral	icosahedral
Number of capsomers	72	72
DNA (m.w.)	3.4×10^6	3.4×10^6
DNA content (% w/w)	12.5	12.5
Virion (m.w.)	27×10^6	27×10^6
Protein composition		
major capsid protein (VP-1) (75% total protein)	47,00	45,00
minor capsid protein (VP-2)	35,000	42,000
(VP-3)	23,000	30,000
Histones (same composition in all viruses)	H3	15,300
	H2A	14,000
	H2B	13,800
	H4	11,300

(Adapted from ref. 5, with permission)

minichromosomes in regulating viral gene expression is currently under investigation in several laboratories.

A restricted region of about 400 bp around the origin of replication containing also the regulatory sequences for early and late transcription (see below), presents in 30% of the minichromosomes (like many cellular genes) an increased sensitivity to endogenous nucleases, to DNase I in 20% of the minichromosomes and to sequence specific endonucleases (35-38). In addition, electron microscopic analysis of isolated viral minichromosomes revealed in 10-25% of the molecules a distinctive non-nucleosomal region (gap) (39-41) (see Fig. 2). Further analyses have suggested that all transcribed SV40 minichromosomes possess a gap (42).

F. The Physical Maps of SV40 and Polyoma

The physical maps of SV40 and polyoma were established by cleavage with restriction enzymes (43-47). The cleavage site with *Eco* R1 was defined as zero coordinate and the viral DNAs were divided into 100 units. The origin of SV40 DNA replication has been located near 67.0 map units while that of polyoma virus at 70.8 map units (5).

Viral genes are expressed at different times during the infectious cycle. Those expressed *prior* to the onset of DNA replication are defined as *early* ones, and encode the viral T-antigens.

Early and late genes are located in symmetrical halves of the viral DNA.

SV40 has two T-antigens known as large T and small t. Polyoma has three T-antigens known as large T, middle T and small t (see below) (5, 26).

The late genes are those expressed after the onset of DNA replication and they encode the structural proteins VP1, VP2 and VP3 (5, 26). SV40 contain an additional late protein known as *agnoprotein* (48-53). The agnoprotein is a 61-amino acids basic protein. Several functions of the agnoprotein has been suggested (see below).

Maps of the early and late transcripts of SV40 and polyoma with locations of 5' ends, splice donors and acceptors sequences and the polyadenylation sites are presented in Fig. 3.

2. THE LYTIC CYCLE

Following the attachment of the viral particles to specific receptors on the cell membrane, the virions are internalized in the cell by pinocytosis (54-56). By 60 minutes postinfection they reach the nucleus of the cell, the viral DNA is uncoated and the half of the genome that contains the early region is transcribed (57-60).

Viral early products (T-antigens) are synthesized (see below). This is followed by the induction of cellular enzymes, and replication of the viral DNA (61).

Immediately afterwards, new mRNAs are transcribed from the other half and opposite strand of the viral DNA (Fig. 3). These transcripts known as late RNAs are translated to the three viral structural proteins VP1, VP2 and VP3 and to a small protein known as agnoprotein (26).

The early RNA is also synthesized following DNA replication but the amount of late RNA exceeds the amount of early RNA by about 20 fold (57).

A. The SV40 Early Promoter

The SV40 early promoter is one of the best studied examples of RNA polymerase II promoter and of regulation mediated by a known protein. Transcription initiated at the early promoter encodes information of the SV40 large and small T-antigens (26). Large T antigen regulates its own synthesis by decreasing the level of transcription from the early region (62-65). Dissection *in vivo* and *in vitro* of the SV40 early promoter region (see Fig. 4) has shown that it contains two overlapping promoters for RNA polymerase II, the early-early (EE) and late-early (LE) promoters (66-67).

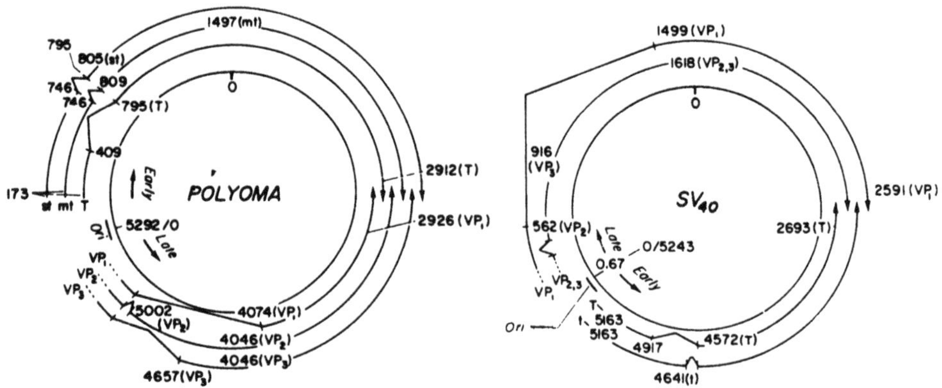

Figure 3. Physical, transcriptional, and translational maps of the polyoma and SV40 genomes. mt = middle-T antigen; st = small-T antigen; T = large-T antigen. (From ref. 26 with permission).

There are three major *cis*-elements that constitute the RNA polymerase II promoter.

The first one is the TATA box that appears to secure *in vitro* and *in vivo* the correct positioning at initiation of RNA synthesis. The TATA box is located at about 30 nucleotides from the EE transcription start site (66-69). Mutation within the TATA box results in transcription starting from heterogeneous sites. Deletions between the TATA box and the start site shift the initiation site downstream in such a way that the distance between the TATA box and the start site remains constant.

A second *cis*-element frequently called "upstream transcriptional regulatory sequence" influences the efficiency or RNA synthesis. In SV40 the upstream regulatory sequence is a GC-rich motif of the hexanucleotide GGGCGG that repeats twice in each of the three 21 bp tandem repeats (see Fig. 4); in total there are six repeats of this hexanucleotide (70-72).

The third *cis*-control element is the enhancer. In SV40 the enhancer is known as the 72 bp tandem repeat, located about 100 nucleotides upstream of the major EE start site (71, 73-75). Enhancer elements increases the efficiency of transcriptional initiation in a position- and orientation-independent fashion. It has been suggested that enhancers may serve as entry sites for RNA polymerase II or one of its associated subunits. Also the SV-40 enhancer functions in association with either

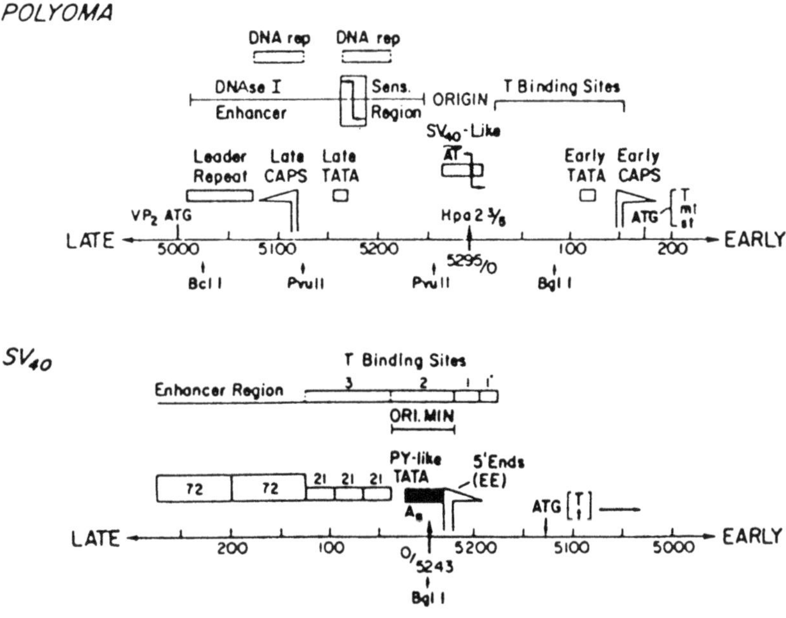

Figure 4. Replication origin and transcriptional control regions of polyoma and SV40 DNA. Symbols employed are: ⌐L, inverted repeat structure; ⊩>, locations of 5' ends of the relevant mRNAs; AT-rich region; [ ] indicates the approximate borders (hatched lines) of the relevant sequence(s). (From ref. 26 with permission).

early viral transcriptional unit or heterologous genes in most cells, a number of enhancers associated with viral or eukaryotic genes are highly tissue specific.

It is noteworthy that transcription from the late-early start sites (LES) also requires the presence of the 21 bp repeat region and of the enhancer but occurs in the apparent absence of a TATA box (70, 71, 77). *In vitro* transcription studies have shown that different *trans*-acting factors bind specifically to the three *cis*-acting regulatory sequences (77). The TATA box factor binds the TATA box to form a stable complex; this is a prerequisite for specific initiation of transcription (77-79). The transcription factor SP1 (80-82) binds specifically to the 21-bp repeat region and stimulate transcription from the EES and LES. Different *trans*-acting factor(s) bind(s) to the enhancer and stimulate(s) transcription from both the SV40 early and heterologous promoter elements (83, 84). Other factors that can bind to the enhancer are also able to down regulate transcription. In conclusion, at least three different proteins interacting with three different promoter elements, are required to allow the cellular RNA polymerase II to initiate transcription from the SV40 early promoter. Moreover, protein-protein interactions of the *trans*-acting elements should occur before transcription from SV40 early promoter can start. This interaction requires a stereospecific alignment of the *trans*-acting factors along the DNA template (77).

As noted above the large T-antigen regulates its own synthesis by down regulating transcription from the early promoter (62-65). T-antigen has three binding sites within the early promoter region. It is assumed that binding to the strongest binding site 1 results in the virtual suppression of transcription from the downstream early promoter but only in the partial inhibition of RNA polymerase II starts, which are under the control of the upstream promoter (LS) (26). Binding at both sites 1 and 2 inhibits transcription from both of the early promoters. The inhibition of transcription induced by T-antigen may follow the blockade of the elongation complex of RNA polymerase II, or else it can be the result of an alteration in the stereospecific alignments of the *trans*-acting factors (77).

Finally it is worth noting that all *cis*-controlling elements of the early promoter resides within the DNase-hypersensitive region that appears as a gap in the SV 40 minichromosomes isolated from the infected cells at late times after infection (35-42).

B. The Products of the Early Region

SV40

The products of the early region are the *large* T-antigen and *small* t-antigen. The large T is a 81-kDa phosphoprotein mainly confined to the nucleus but it is also found in the plasma membrane (136-139). The small-t antigen is a 20-kDa nonphosphorylated protein. It is found both in the nucleus and cytoplasm. Due to the partial overlapping of their coding regions large T and small t antigens share in common the first 82 amino acids (Figs. 3 and 4) (26, 140).

The large T antigen of SV40 has been studied extensively and detailed information is known. The *in vivo* and *in vitro* functions of SV40 large T are summarized in Table 3.

Table 3. - In vivo and in vitro functions of SV40 large T-antigen

DNA binding - origin of replication specific
DNA binding - to nonorigin DNA
Associated enzymatic activities
- ATPase
- Kinase (? Intrinsic)
- Self-adenylation (? Intrinsic)
p 53 Associatio
Activating SV40 late transcription
Inhibiting SV40 early transcription
Induction of host cell DNA synthesis in G-arrested cells
Functioning as part of an SV40 tumor-specific transplantation antigen
Initiating viral DNA replication
Establishing and maintaining SV40-induced transformation
Helping human adenoviruses to grow in monkey cells

(From ref. 26 with permission)

In the infected cells SV40 large T antigen is found as monomers, oligomers, or in a complex with the cellular phosphoprotein p53 (154-156). The binding of large T to p53 does not inhibit its binding activity to DNA. It appears that in transformed cells large T stabilizes p53 by protecting it from rapid degradation (154-159).

Much less is known about the biological significance of SV40 small T antigen: It appears not to be a DNA binding protein, and is not phosphorylated. Like large T the bulk of small-t antigen is also localized in the nucleus, but it has also be found in the cytoplasm (26). Although large T can function without small T in transformation both *in vivo* and *in vitro*, in certain resting established cell lines stable transformation requires both proteins. It is possible that in some cultures there are cellular proteins that can substitute small T in helping large T to transform the cell (26).

Polyoma

Polyoma has three early proteins known as large-T, middle-T and small-t. Due to the partial overlapping of their coding regions, the three proteins share in common the first 79 amino acids. Middle-T and small-t antigens share in common the first 191 amino acids (Fig. 3) (26). Polyoma large-T and small-t antigens have the same size and cellular localization as those of SV40 (26). Middle-T antigen is a phosphoprotein localized mainly in the plasma membrane but a small fraction can be found in the endoplasmic reticulum (26, 141).

Polyoma Middle-T

Polyoma Middle-T antigen is a 57-kDa membrane-associated phosphoprotein (160, 161). No analogous protein is found in SV40. Polyoma middle T has tyrosine kinase activity believed to be due to its association with the cellular pp-60 *c-src* the cellular progenitor of the transforming protein of Rous Sarcoma Virus (140, 161, 162). The C-terminus of middle T contains the transforming domain of the protein. This domain is characterized by 22 uncharged hydrophobic amino acids surrounded on both sides by clusters of basic residues. It has been suggested that this structure may play a role in the insertion of middle-T antigen in the plasma membrane, and may serve as substrate of pp60 *c-sarc* (163).

The Role of T Antigens

Large-T of SV40 and polyoma are essential for DNA replication (142) and indirectly enhance late transcription (91, 92, 143, 144). Small-t is not essential for virus growth in tissue culture (26). Host-range mutants of polyoma were found to be affected in the middle or small T (145).

Large T can indirectly promote the integration of SV40 or polyoma to the cellular genome, by an illegitimate mechanism of recombination and lead to a neoplastic transformation (see section 3 of this review). In polyoma, the middle- T and the amino terminal half of large T antigens are the transforming proteins (146-150). In cultured primary cells SV40 and polyoma large T immortalize the cells (151-153). The domain of SV40 large-T antigen responsible for the transforming ability is confined to the amino terminal half of the molecule (151, 154).

It is worth noting that neither SV-40 small-t antigen, nor polyoma middle- or small-t antigens are essential for virus growth in cell cultures; they may be required, however, for replication (or persistence) of the viruses in their natural hosts.

C. The SV40 Late Promoter

The *cis* and *trans*-acting elements that constitute the SV40 late promoter are less well characterized. The late region lacks the typical RNA polymerase II promoter elements such as the TATA box, the upstream regulating sequences and enhancer. Consequently, the late mRNAs have almost 20 heterogeneous 5' ends with the major initiation site located at residue 325 (Fig. 4) (85-88). At late time after infection more than 80% of transcription initiates from the major site. The current notion about the elements that constitute the late promoter is that sequences situated within a few base pairs of the various start sites (rather than upstream or downstream elements) play a primary role in determining the precise locations of these sites (89, 90).

Transcription from the late start sites are detectable also before the onset of DNA replication. However, before DNA replication the start sites closer to the origin of replication are preferred over the major start site at residue 325 (89-91). DNA replication is therefore required for the selection of the major start site at residue 325. Alternatively, transcriptional factors present at the late stages of the lytic cycle are needed for the optimal activity of the late start site.

Recent experiments have suggested that T-antigen induces the initiation of late transcription through an indirect mechanism which requires a cellular transcription factor (91, 92). The data available at the time of writing this review suggest that the topology, the torsional stress of the minichromosome and the "gap" in the minichromosome all may affect the selection of the start sites. In this regard it is interesting to note that E. coli RNA polymerase initiates transcription at the region of the late start sites and transcribe the late strand when viral minichromosome is the template for transcription. This is in contrast to transcription by *E.coli* polymerase of the naked viral DNA that initiates at map unit 15 and transcribes the early strand (93).

Attenuation

The late transcription unit is also negatively regulated. Experiments indicating that a mechanism resembling attenuation in prokaryotes regulates SV40 late gene expression have been described (94-102). The attenuation model suggests that agnoprotein (the 61- aminoacid protein encoded in the major leader of the late RNA) binds to the 5' end of the 16S mRNA and influences its conformation.

In its absence, transcription would proceed, and 16S mRNA would assume a conformation which favors production of agnoprotein.

In its presence, premature termination of transcription occurs and 16S mRNAs are translated to produced VP1. A similar mechanism has been proposed to regulate translation of 19S mRNAs to produce VP2 and VP3 (96, 101). Other experiments have suggested that agnoprotein may have a *trans*-active positive effect on late transcription (104) or that it is involved in the encapsidation process of the viral DNA (103).

D. Transcription Termination

For all the existing similarities between SV-40 and polyoma, termination of transcription of these two viruses may differ dramatically. Before discussing the molecular bases sustaining such differences, it would be proper briefly to review the mechanism of termination of transcription in eukaryotes.

Several groups have reported that RNA polymerase II transcribes several hundreds or even thousands of bases past the poly A site before chain elongation is arrested, the newly synthesized RNA molecules are released, and the enzyme quits the DNA template.

The termination sites of RNA polymerase II have not yet been defined at the nucleotide level (for review see 105). Nuclear run-on experiments indicate that in most genes studied so far termination by RNA polymerase II does not occur at one precise point but terminates heterogeneously within a defined region (105-109). It is conceivable, however, that there are several hot spots of transcription termination within the termination region. In SV40, adenovirus and the parvovirus minute virus of mice (MVM) it has been suggested that premature termination occurs within a U-stretch following a hairpin structure (94, 102-110).

In the iso-cytochrome C gene of yeasts (111), and in the ADE-8 gene of Drosophila (112) an 8 bp U-rich sequence, UUUUUAUA seems to be involved in termination of transcription. Similarly, it has been suggested that a U-tract is necessary (but not sufficient) for termination of the ovalbumin gene transcription (109).

The termination process of RNA polymerase III does not appear to involve auxiliary factors. The purified enzyme is capable of accurately and efficiently terminating transcription at the 3' end of the Xenopous 5S RNA genes *in vitro* (113). The information reguired for termination resides in the ability of RNA polymerase III to recognize a cluster of four or more Us in the RNA surrounded by GC-rich sequences at the end of the gene. The termination process of RNA polymerase I also occurs in front of a run of U-residues (114-118). It appears therefore that a U-stretch is an essential element in the termination process of the three classes of the eukaryotic RNA polymerases. It is interesting to note that the rho-independent transcription-termination process in prokaryotes also occurs in a stretch of Us (119). Martin and Tinoco (120) have suggested that termination within a stretch of Us is due to the marked instability of oligo (dA:rU) hybrids that facilitate the dissociation of the RNA from the DNA template.

It is tempting to speculate that once at the polyadenylation site the eukaryotic RNA polymerase II changes its allosteric conformation, or acquires (or loses) a certain factor and consequently can partially terminate transcription at any stretch of Us encountered during the elongation process. In some cases when termination should be avoided (e.g: during transcription of the adenovirus-2 at late time after infection) induced cellular or viral coded factors change the polymerase conformation in such a way that the enzyme would cross the Us stretch without terminating transcription. Similarly at the attenuation sites or at sites where termination should be regulated, viral or cellular factors as well as mutually exclusive RNA secondary structures are involved in the regulation of transcription-termination at these sites. Conceivably, the secondary structure of the RNA may slow down transcription, and facilitates termination at the U-stretch (101, 119, 120).

At late time after infection nuclear polyoma specific RNAs contain tandem repeats of the nucleotide sequences in an entire strand of the viral genome. Thus these giant transcripts are probably the result of multiple rounds of transcription of the circular polyoma virus DNA, implying that termination of transcription is inefficient on the L-strand (61, 121). Although some of SV40 late nuclear RNAs are also larger than the mRNAs, only a very small fraction is larger than genome size (121, 123).

Several explanations for this difference between the transcription pattern of polyoma and SV40 may be suggested:

i) processing of the polyoma primary transcripts is less efficient than those of SV40;
ii) transcription-termination factors are more abundant in polyoma infected mouse cells as compared to those present in SV40 infected monkey cells;
iii) anti-termination factors are less abundant in polyoma infected cells as compared to those present in SV40 infected monkey cells; and
iv) the polyoma primary transcripts contain may tracts of Us as compared to their presence in SV40 primary transcripts.

Table 4 shows the abundance of tracks of 5 or more Us in SV40 and polyoma L-strands. A dramatic difference is apparent. Whereas SV40 L-strand contains 24

Table 4. - Comparison of U-tracks in SV40 and polyoma late transcripts

A. SV40					
Residue No.	5 Us	Residue No.	6 Us	Residue No.	7 Us
415	CCUUUUUGC				
537	UCUUUUUGU	857	GAUUUUUUAG	1955	CAUUUUUUUGC
1026	ACUUUUUAA	2722	GGUUUUUUAA	3557	UCUUUUUUUAG
1048	GCUUUUUGG	2973	CAUUUUUUGA	5051	CAUUUUUUUAA
1142	GGUUUUUAG	2813	GGUUUUUUGG		
1654	GCUUUUUAA	3892	UCUUUUUUAA		
2413	CCUUUUUGU	4433	UCUUUUUUGG		
2878	CAUUUUUAU	5007	CAUUUUUUCU		
3773	GCUUUUUAG				
3822	GGUUUUUGC				
4174	AAUUUUUGA				
4282	UAUUUUUCC				
4315	GCUUUUUCC				
5173	GCUUUUUGC				
B. Polyoma					
Residue No.				Residue No.	
1962	GGUUUUUCC			2546	GCUUUUUUUCU
35	UCUUUUUCU				

such tracts, fourteen of 5 Us, seven of 6 Us and three of 7-Us, polyoma L-strand contains only three such tracks; two of 5 Us and one of seven Us.

It can be speculated that termination of transcription in each of the U-tracks is about 7% efficient while 93% of the RNA polymerase molecules readthrough a U-track. In such a situation, in the case of polyoma, following one round of transcription 20% of the RNA polymerase molecules will terminate transcription. Of the approximately 80% RNA polymerase molecules that will complete one cycle, 20% will terminate transcription in the second round of transcription. Termination of transcription will grind to a halt following 4-5 rounds of transcription. Such a mechanism will lead to the production of polyoma specific giant primary transcripts. In SV40 there are 16 U-tracks between the polyadenylation site and the transcription start site. If 7% of the RNA polymerase molecules will terminate transcription at each of these 16 sites complete termination of transcription will be achieved in one cycle. This will explain the near absence of SV40 specific giant primary transcripts.

E. Processing of the Late RNA and Generation of the mRNAs

SV40

As mentioned above transcription of the late RNA initiates at heterogeneous sites, the one at residue 325 is the major initiation site (85-90). The primary transcript (capped at the 5' end), contains a representation of sequences of nearly the entire L strand. This large RNA molecule is then processed around map unit 20 by two cleavages to give rise to a small RNA of 62-64 nucleotides known as SV40 associated small RNA (SAS-RNA) (130-132), and around map unit 17 to give rise to the 3' end of the two mRNAs: 16S and 19S. The 3' end is subsequently polyadenylated.

The signal for cleavage and polyadenylation is the hexanucleotide AAUAAA located 17 nucleotides upstream of the polyadenylation site (61). The polyadenylated RNA is spliced to generate the two mRNAs 19S and 16S and the two messages are transported to the cytoplasm (126, 127, 133).

The 19S codes for VP2 and VP3 while 16S codes for VP1 and agnoprotein (61). Figure 5 outlines the various events of the synthesis and processing of the 16S mRNA.

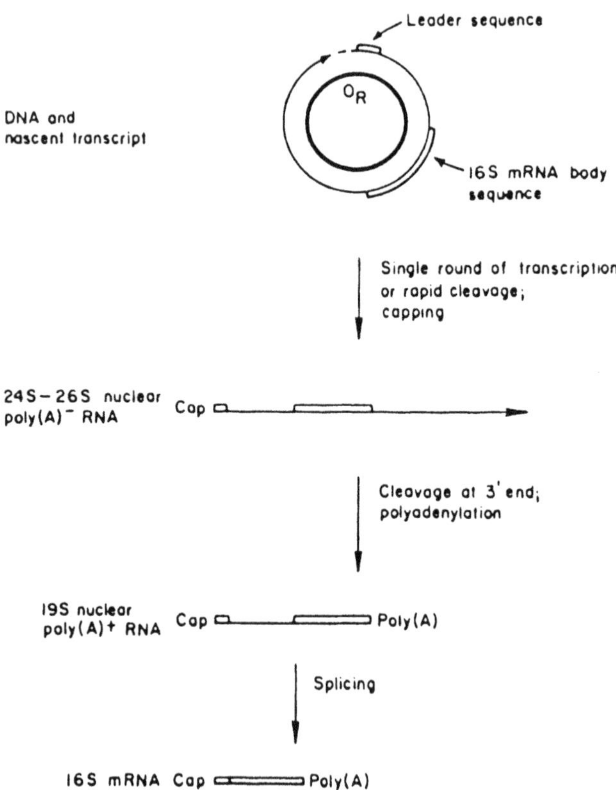

Figure 5. Model for synthesis and processing of SV40 16S L-strand mRNA O_R indicates origin of replication; (—) DNA; (—) RNA; (▭) mRNA body or leader sequences. See text for details. (From ref. 61 with permission).

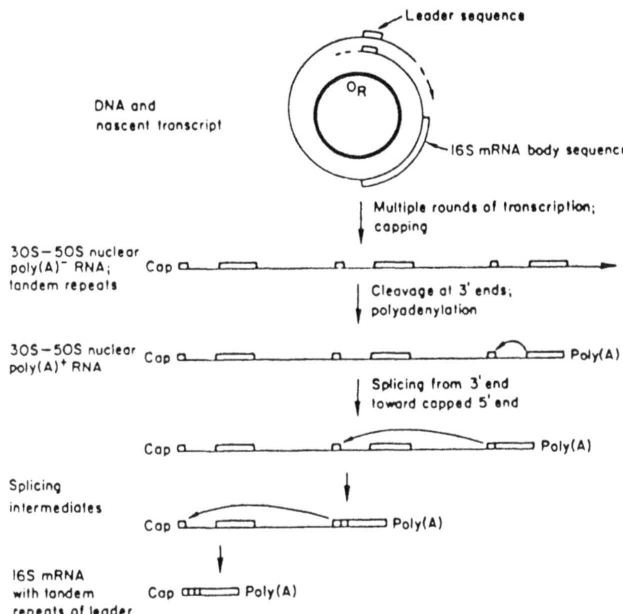

Figure 6. Model for synthesis and processing of polyoma virus 16S L-strand mRNA. See text for details. (From ref. 61 with permission).

Polyoma

Processing of polyoma primary transcript is complicated by the multiple rounds of transcription, which give rise to giant RNA containing tandem repeats of the sequences in the L-strand (61). Processing of polyoma primary transcript differ from that of SV40 in two main aspects:

i) there is no small RNA similar to SAS-RNA, and
ii) the primary transcript is processed to three mRNAs 16S, 18S and 19S that codes for VP1, VP3 and VP2, respectively (61, 134, 135).

Cleavage and polyadenylation occur at 26 map unit. Figure 6 outlines the various steps for the production of the 16S mRNA from the primary transcript (61).

Following polyadenylation the 16S body region is joined to the nearest upstream leader by splicing, giving rise to a splice intermediate still larger in size than the genomic sequences. Splicing continues upstream toward the 5' end of the RNA, joining the first leader to a second leader and then to a third leader until the capped 5' end is reached. This mechanism would explain the occurence of multiple leaders on polyoma virus L-strand RNAs although it appears only once in the genome (61, 135). It is interestingly to note that in comparison to SV40 the leader region of polyoma contains 57 nucleotides only and it does not code for any protein.

F. Replication of SV-40 DNA and the Minimum Replication Origin

Since most of the studies on the papova viral DNA replication have been carried out in the SV40 system, this section will deal with SV40 replication only. It is expected that knowledge on the mechanism of SV40 DNA replication will now advance at an accelerated pace. This is because *in vitro* SV40 DNA replication systems have recently been developed (164-166), a condition that should allow a detailed investigation of the enzymatic processes involved in the mechanism of DNA replication. It is assumed that DNA replication occurs in multiple separable ordered steps, each with its own enzymatic requirements (167). Below is a description of the overall mechanism of SV40 DNA replication.

The description of the *minimum replication origin* would be better followed on Figure 7. The minimum replication origin contains about 60 bp spanning from residue 5215 to 30. This region contains:

i) the start sites for early transcription (a);
ii) the 27-bp inverted repeat (b), and
iii) the 17 bp AT-rich region (c).

In addition there are four T-antigen binding sites 1,2,3 and 1', where 1' is a 15-bp extension of site 1. These sites are numbered according to their decreasing af-

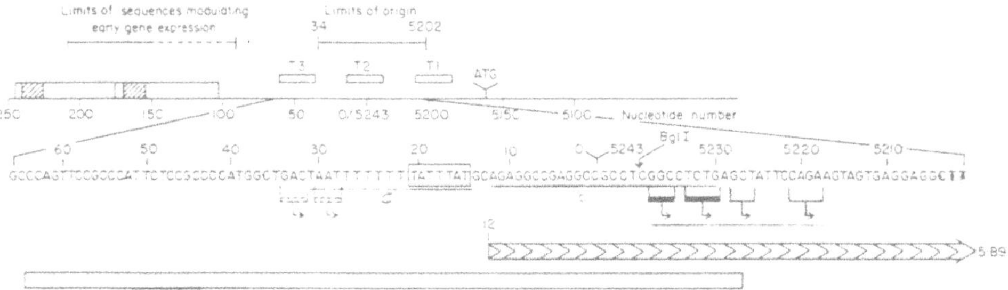

Figure 7. Structure and function of the region of SV40 DNA preceeding the early genes and including the origin of DNA replication. The T-antigen binding sites are indicates as T1, T2 and T3. The origin of DNA replication lies within the region spanned by nucleotides 34 and 5202.
The sequence between nucleotides 66 and 5202 is shown in detail. The AT-rich region related to the TATA box consensus sequence is indicated, as is the cleavage site of restriction enzyme *Bgl*I. Probable positions of the 5' ends of early mRNAs made *in vivo* and *in vitro* are shown by ↳ and ↳ (note that those detected *in vivo* only at late times are shown by the hatched marker). The 5' termini represent transcriptional start points.
The lower boxes indicate: (1) the extent of different deletion mutants that cause alternative 5' termini located at approxymately the same distances from the conserved AT-rich region ⪢, and (2) the extent of deletion mutants that remove the AT-rich region and cause heterogeneity of 5' termini *in vivo* and *in vitro* without significantly decreasing the overall efficiency of transcription▭.
The approximate extent of sequences that, when deleted, severely decrease the level of early gene expression *in vivo* but have no effect *in vitro* is indicated at the top left of the figure. This region is located in the vicinity of the 72-bp direct repeat indicated by the boxes. The shaded areas within the boxes indicate GC-rich tracts. (From ref. 190 with permission).

finity to T-antigen binding. Thus site 1 is the strongest binding site and site 3 is the weakest binding site. Site 1' approximates the affinity of site 3 (168-170). The limits of each site are indicated in Fig. 7.

Almost all DNA replication starts at site 2 (26). Following initiation of replication, two replicating forks are produced generating two branches of equal length (L1 and L2) representing the replicated portion of the molecule, and a supercoiled branch (L3) representing the unreplicated portion of the molecule (Fig. 8). The two replicating forks advance at about equal rates bidirectionally and terminate at about half way from the origin at around 17 map unit. No specific sequences are needed to terminate replication.

The strand growing in the 5' to 3' direction at each replication fork is synthesized in a continuous fashion; the strand growing in the 3'to 5' direction is synthesized as small pieces (Okazaki fragments) which are subsequently joined. Each Okazaki fragment is initiated with rAMP followed by 5-8 residues of ribonucleotide primer (141). This primer is then used to polymerize deoxyribonucleotides. A DNA primase is involved in the continuous and discontinuous DNA synthesis.

T-Antigen and DNA Replication

There is evidence that large-T antigen has to bind to site 2 and to the 27-bp inverted repeat structure as part of its role in initiating replication (103, 171). Similarly, some data suggest that large-T antigen might be involved also in post-initiation phases of DNA replication (26). The linkage between large T-antigen and the synthesis of the primer, however, is not clear. Some of the possible mechanisms are (a) T-antigen may direct the synthesis of a host encoded primase; (b) T-antigen might bind rAMP and serve it self as a primer, and (c) T-antigen can lead to a cleavage of small RNA whose synthesis starts upstream of the replication start site (26).

Termination of Replication

The mechanism by which replication terminates is poorly understood. There is evidence that the replicating forks pause after completing about 90% of the distance (172). Following the pause there is a gap-filling process and segregation of the daughter molecules. It is possible that segregation of the newly synthesized daughter molecules occurs through the formation of catenated dimers and the involvement of topoisomerase (141). Indeed catenated dimers have been identified during the lytic infection.

G. Virus Assembly

The SV40 minichromosomes are the central structure of the viral DNA that is used in replication, transcription and virion assembly. The arrangement of the nucleosomes (the viral minichromosomes) is changed during the virus life cycle (173). The "gap" found in the viral minichromosome purified from the infected cells late in the infection (35-42) is not found in the minichromosomes isolated from cells at early times after infection (173) or from disrupted virions (174, 175).

A model for the channeling of SV40 minichromosomes into the transcription, replication and encapsidation pathways has recently been suggested (176). According to this model the minichromosomes that function in replication and transcription contain the gap while those destined for the encapsidation process lose it. The VP1 directly or indirectly rearrange the SV40 minichromosome structure by sliding the nucleosomes towards the open region (176, 177).

Earlier reports concluded that pre-virions are formed first and the viral DNA is introduced into the empty shells (5, 17, 61). Recent studies illustrate a series of precursor-product relationships and define a clear pathway of viral DNA maturation beginning with the replicating form of the molecules and ending with fully packaged viral capsids. Thus, the viral minichromosomes (75S) are initially condensed into young virions, which mature into stable 250S virions (178-182).

The current idea is that the capsid proteins add to (and organize around) the 75S minichromosome. The prediction is that the assembly of previrions must be

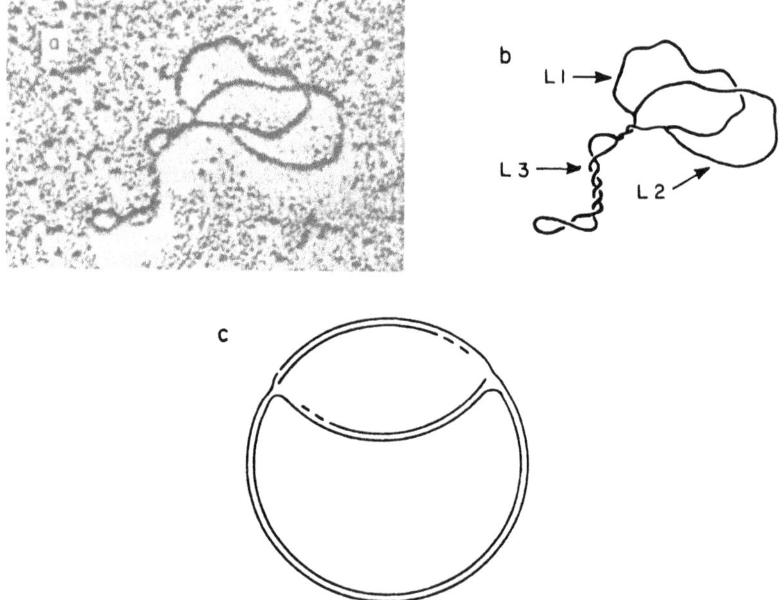

Figure 8. Replicative intermediate of viral DNA undergoing bidirectional replication. (a) Electron micrograph of SV40 replicative intermediate. (b) Interpretative drawing. Replicative intermediates contain two branch points (the replication forks), two branches of equal length (L1 and L2) representing the unreplicated portion of the molecule. (c) Summary of the basic features of the replicative intermediate. Both parental strands are covalently closed, and the daughter strands have free ends. The strand growing in the 5'→3' direction at each replication fork is synthesized in a continuous fashion; the strand growing in the 3'→5' direction is synthesized as small pieces, which are subsequently joined. This simplified diagram shows neither the double helical nature of the complementary strands nor the winding of the DNA around nucleosomes formed by cellular histones, which are present on replicating DNA. (From ref. 190 with permission).

cooperative, that is, the propagation step of shell assembly is much faster than the initiation event (176).

3. INTEGRATION OF VIRAL DNAs INTO THE CELLULAR GENOME

Integration of SV40 or polyoma DNAs into nonpermissive cells can render a small fraction of the cells stably transformed. In most cells the integration is not stable and the viral sequences are removed. The frequency of stable transformation is low (5). It appears that in those integration instances where the entire large T-antigen of the viruses are expressed the cells can lose the viral DNA presumably due to a process of *in situ* viral DNA replication initiating at the viral origin of replication. Such *in situ* replication leads to the release of form-I DNA containing viral and cellular sequences (186). By further intra and intermolecular recombination, intact free copies of viral DNA can be formed (187, 188). Such a mechanism can lead to the rescue of viruses from transformed cells by fusion with permissive cells which contain the factors needed for viral DNA replication. Stable integration can therefore be achieved if the origin of replication or the replicating activity of large T-antigen are inactivated (26).

The available information regarding the mechanism of integration of SV40 and polyoma DNAs into the cellular genomes suggest the following:

i) large-T antigen is needed for the integration of the viral DNAs into the cellular genomes. However, it does not seem that T-antigen has a direct function. It is more likely that T-antigen that stimulates cellular DNA synthesis in G1 arrested cells also triggers the induction of several enzymes, some of which are involved in DNA repair and recombination (183);

ii) there is no specific site of integration within the viral genome. However in order to obtain stable transformation the sequences coding for the large-T antigen of SV40 should remain intact. Similarly, in the case of polyoma the sequences encoding the whole middle-T antigen and 60% of the large-T are required. Therefore integration in stably transformed cells does not occur within these T-antigens coding regions (26, 61, 184);

iii) The cellular sequences where the viral DNA integrates are devoid of repeat, specialized sequences or sequences exhibiting homology with the viral DNA.

From these observations, it appears as if the mechanism of integration used by SV-40 or polyoma differs fundamentally from transposition or homologous recombination. It seems to be better defined as illegitimate, non-homologous, reciprocal recombination at random sites (26, 185).

Acknowledgements

The author's work described in this review was supported by USPHS Research grant CA 14995 and the Leo and Julia Forcheimer Center for Molecular Genetics.

4. REFERENCES

1) 1. Melnick, J. (1962) Science **135**, 1128-1130.

2) Shah, K.V. (1985) In: *Virology*, B.N. Fields ed. Raven Press, New York 371-391.

3) Gross, L. (1953) Proc. Soc. Exp. Biol. Med. **83**, 414.

4) Gross, L. (1953). Cancer **6**, 948.

5) Griffin, B.E. (1981) in: *DNA Tumor Viruses*, J. Tooze ed. Cold Spring Harbor Laboratory 61-123.

6) Stewart, S.E., Eddy, B.E. and Borgese (1958). J. Natl. Cancer. Inst. **20**, 1223.

7) Stewart, S.E. and Eddy, B.E. (1959) in: *Prospectives in Virology*, M. Pollard ed. Wiley, New York 245.

8) Stewart, S.E. (1960). Ad. Virus Res. **7**, 61.

9) Sweet, B.H. and Hilleman, M.R. (1960). Proc. Soc. Exp. Biol. Med. **105**, 420.

10) Eddy, B.E., Borman, G.S., Berkeley, W.H. and Young, R.D. (1961). Proc. Soc. Exp. Biol. Med. **107**, 191.

11) Eddy, B.E., Borman, G.S., Grubbs, G.E. and Young, R.D. (1962). Virology **17**, 65.

12) Shein, H.M. and Enders, J.F. (1962). Proc. Natl. Acad. Sci. USA **48**, 1164.

13) Gardner, S.D., Field, A.M., Coleman, D.V. and Hulme, B. (1971) Lancet i, 1253.

14) Padgett, B.L. and Walker, D.L. (1973). J. Infect. Dis. **127**, 467.

15) Jung, M., Krech, U., Price, P.C. and Pyndiah, M.N. (1975). Arch. Virol. **47**, 39.

16) Padgett, B. (1981). in: *DNA Tumor Viruses*, J. Tooze, ed. Cold Spring Harbor Laboratory 339-370.

17) Grodzicker, T. and Hopkins, N. (1981) in: *DNA Tumor Viruses, J. Tooze ed. Cold Spring Harbor Laboratory 1-60.*

18) Dulbecco, R. and Vogt, M. (1963). Proc. Natl. Acad. Sci. USA **50**, 236.

19) Weil, R. and Vinograd, J. (1963) Proc. Natl. Acad. Sci. USA **50**, 730.

20) Crawford, L.V. and Black, P.H. (1964) Virology **24**, 388.

21) Frearson, P.M. and Crawford, L.V. (1972). J. Gen. Virol. **14**, 141.

22) Lake, R.S., Barban, S. and Salzman, N.P. (1973) Biochem. Biophys. Res. Commun. **54**, 640.

23) Varshavsky, A.J., Bakayev, V.V., Chumackov, P.M. and Georgiev, G.P. (1976) Nucleic Acids Res. **3**, 2101.

24) Adler, S.P. and Nathans, D. (1973). Biochem. Biophys. Acta **299**, 177.

25) Griffin, B.E. and Fried, M. (1976). Methods Cancer Res. **12**, 49.

26) Livingston, D.M. and Bikel, I. (1985). Replication of Papovaviruses, B.N. Fields, ed. Raven Press, New York.

27) Griffin, J.D. (1975) Science 187, 1202-1203.

28) Felsenfeld, G. (1978) Nature (London) **271**, 115-122.

29) Korenberg, R.D. (1977). Ann. Rev. Biochem. **46**, 931-954.

30) Chambon, P. (1977) Cold Spring Harbor Symp. Quant. Biol. **42**, 1209-1234.

31) Keller, W. (1975). Proc. Natl. Acad. Sci. USA **72**, 4876-4880.

32) Muller, V., Zentgraf, H., Eicken, I. and Keller, W. (1978). Science **201**, 406-415.

33) Barsoum, J. and Berg, P. (1985). Mol. Cell. Biol. **5**, 3048-3057.

34) Choder, M. and Aloni, Y. (1986). unpublished results.

35) Varshavsky, A.J., Sundin, O.H. and Bohm, M.J. (1978) Nucleic Acids Res. **5**, 3469-3478.

36) Scott, W.A. and Wigmore, D.J. (1978). Cell **15**, 1511-1518.

37) Sundin, O. and Varshavsky, A.J. (1979). J. Mol. Biol. **132**, 535-546.

38) Waldeck, W., Fohring, B., Chowdhury, K., Gruss, P. and Sauer, G. (1978). Proc. Natl. Acad. Sci. USA **75**, 5964-5968.

39) Jakobovits, E., Bratosin, S. and Aloni, Y. (1980). Nature (London) **285**, 263-265

40) Jakobovits, E., Bratosin, S. and Aloni, Y. (1982). Virology **120**, 340-348.

41) Saragosti, S., Moyne, G. and Yaniv, M. (1980) Cell **20**, 65-73.

42) Choder, M., Bratosin, S. and Aloni, Y. (1984). EMBO J. **3**, 2929-2936.

43) Danna, K.J. and Nathans, D. (1971). Proc. Natl. Acad. Sci. USA 68, 2913.

44) Danna, K.J., Sack, G.H. and Nathans, D. (1973). J. Mol. Biol. **78**, 363.

45) Griffin, B.E., Fried, M. and Cowie, A. (1974). Proc. Natl. Acad. Sci. USA **71**, 2077.

46) Leibowitz, P., Kelly, T.J., Nathans, D., Lee, T.N.H. and Lewis, A.M. (1974). Proc. Natl. Acad. Sci. USA **71**, 441-445.

47) Griffin, B.E. (1977). J.Mol. Biol. 117, 447.

48) Jay, G., Nomura, S., Anderson, C.N. and Khoury, G. (1981). Nature, **291**, 346-349.

49) Jackson, V. and Chakley, R. (1981). Proc. Natl. Acad. Sci. USA **78**, 6081-6085.

50) Mertz, J.E., Murphy, A. and Barkan, A. (1983). J. Virol. **45**, 36-46.

51) Hay, N., Kessler, M. and Aloni, Y. (1984). Virology **137**, 160-170.

52) Nomura, S., Khoury, G. and Jay, G. (1983). J. Virol. **45**, 428-433.

53) Nomura, S., Jay, G. and Khoury, G. (1986). J. Virol. **58**, 165-172.

54) Mattern, C.F., Takemoto, K.K. and Daneil, W.A. (1966) Virology **30**, 242.

55) Hummeler, K., Tomassini, N. and Sokol, F. (1970) J. Virol. **6**, 87.

56) Oshiro, L.S., Rose, H.M., Margan, C. and Hsu, K.C. (1967) J. Virol. 1, 384.

57) Aloni, Y., Winocour, E. and Sachs, L. (1968). J. Mol. Biol. **31**, 415-429.

58) Weinberg, R.A., Ben-Ishai, Z. and Newbold, J.E. (1974). J. Virol. **13**, 1263.

59) Weil, R., Salomon, C., May, E. and May, P. (1975). Cold Spring Harbor Symp. Quant. Biol. **39**, 381.

60) Acheson, N.H. and Mieville, F. (1978). J. Virol. **28**, 885.

61) Acheson, N.H. (1981) in: *DNA Tumor Viruses*, J. Tooze ed. Cold Spring Harbor Laboratory, 125-204.

62) Cowan, K., Tegtmeyer, P. and Anthony, D.D. (1973) Proc. Natl. Acad. Sci. USA **70**, 1927-1930.

63) Tetmeyer, P., Schwartz, M., Collins, J.K. and Rundell, K. (1975). J. Virol. **16**, 168-178.

64) Reed, S.I., Stark, G.R. and Alwine, J.C. (1976) Proc. Natl. Acad. Sci. USA **73**, 3083-3088.

65) Parker, B.A. and Stark, G.R. (1979). J. Virol. **31**, 360-369.

66) Benoist, C. and Chambon, P. (1981) Nature (London) **290**, 304-309.

67) Ghosh, P.K. and Lebowitz, P. (1981) J. Virol. **40**, 224-240.

68) Ghosh, P., Lebowitz, P., Frisque, F. and Gluzman, Y. (1981). Proc. Natl. Acad. Sci. USA **78**, 100-104.

69) Rio, D. and Tjian, R. (1983) Cell **32**, 1227-1240.

70) Everett, R., Baty, D. and Chambon, P. (1983) Nucleic Acids Res. **11**, 2447-2464.

71) Fromm, M. and Berg, P. (1982). J. Mol. Appl. Genet. **1**, 457-481.

72) Hartzell, S., Yamaguchi, J. and Subramanian, K. (1983). Nucleic Acids. Res. **11**, 1601-1616.

73) Banerji, J., Rusconi, S. and Schaffner, W. (1981) Cell **27**, 299-308.

74) Gruss, P., Dhar, R. and Khoury, G. (1981) Proc. Natl. Acad. Sci. USA **78**, 943-947.

75) Moreau, P., Hen, R., Wasylyk, E., Everett, R., Gaub, M. and Chambon, P. (1981). Nucleic Acids Res. **9**, 6047-6068.

76) Khoury, G. and Gruss, P. (1983) Cell **33**, 313-314.

77) Takahashi, K., Vigneron, M., Matthes, H., Wildeman, A., Zenke, M. and Chambon, P. (1986) Nature (London) **319**, 121-126.

78) Davison, B.L., Egly, J.M., Mulvihill, E.R. and Chambon, P. (1983) Nature (London) 680-686.

79) Parker, C.S. and Topol, J. (1984) Cell **37**, 273-283.

80) Dynan, W.S. and Tjian, R. (1983) Cell **32**, 669-680.

81) Dynan, W.S. and Tjian, R. (1983) Cell **35**, 79-87.

82) Gidoni, D., Dynan, W.S. and Tjian, R. (1984) Nature (London) **312**, 409-413.

83) Sassone-Corsi, P., Doughetty, J.P., Wasylyk, B. and Chambon, P. (1984). Proc. Natl. Acad. Sci. USA **81**, 308-312.

84) Sassone-Corsi, P., Wildeman, A.G. and Chambon, P. (1985). Nature (London) **313**, 458-463.

85) Laub, O., Bratosin, S., Horowitz, M. and Aloni, Y. (1979). Virology **92**, 310-323.

86) Ghosh, P.K., Reddy, V.B., Swinscoe, J., Lebowitz, P. and Weissman, S.M. (1978). J. Mol. Biol. **126**, 813-846.

87) Haegeman, G., Van Henverswyin, H., Gheysen, D. and Fiers, W. (1979). J. Virol. **31**, 484-493.

88) Lai, C.J., Dhar, R. and Khoury, G. (1978) Cell **14**, 971-982.

89) Piatak, M., Ghosh, P.K., Norkin, L.C. and Weissman, S.M. (1983). J. Virol. **48**, 503-520.

90) Somasekhar, M.B. and Mertz, J.E. (1986). J. Virol. in press.

91) Brady, J. and Khoury, G. (1985). Mol. Cell. Biol. **5**, 1391-1399.

92) Keller, J.M. and Alwine, J.C. (1984) Cell **36**, 381-389.

93) Jakobovits, E.B., Saragosti, S., Yaniv, M. and Aloni, Y. (1980) Proc. Natl. Acad. Sci. USA **77**, 3297-3301.

94) Hay, N., Skolnik-David, H. and Aloni, Y. (1982) Cell **29**, 183-193.

95) Skolnik-David, H., Hay, N. and Aloni, Y. (1982) Proc. Natl. Acad. Sci. USA **79**, 2743-2747.

96) Aloni, Y. and Hay, N. (1983) Mol. Biol. Rep. **9**, 91-100.

97) Skolnik-David, H. and Aloni, Y. (1983) EMBO J. **2**, 179-184.

98) Pfeiffer, P., Hay, N., Pruzan, R., Jakobovits, E.B. and Aloni, Y. (1983) EMBO J. **2**, 185-191.

99) Abulafia, R., Ben-Ze'ev, A., Hay, N. and Aloni, Y. (1984). J. Mol. Biol. 172, 467-487.

100) Hay, N. and Aloni, Y. (1984) Nucleic Acids Res. 12, 1401-1414.

101) Aloni, Y. and Hay, N. (1985) Critical Reviews in Biochemistry, **18**, 327-383.

102) Hay, N. and Aloni, Y. (1985). Mol. Cell. Biol. **5**, 1327-1334.

103) Margolskee, R.F. and Nathans, D. (1983) J. Virol **48**, 405-409.

104) Alwine, J.C. (1982) J. Virol. **42**, 798-803.

105) Birnstiel, M.L., Busslinger, M. and Strub, K. (1985) Cell **41**, 349-359.

106) Hoefer, E., Hoefer-Warbinek, R. and Darnell, J.E. (1982) Cell **29**, 887-893.

107) Citron, B., Falk-Pedersen, E., Salditt-Georgieff and Darnell, J.E. (1984) Nucleic Acids Res. **12**, 8723-8731.

108) Hagenbuchle, O., Wellauer, P.K., Cribbs, D.L. and Schibler, U. (1984). Cell **38**, 737-744.

109) Le Meur, M.A., Galliot, B. and Gerlinger, P. (1984) EMBO J. **3**, 2779-2786.

110) Ben-Asher, E. and Aloni, Y. (1984). J. Virol. **52**, 266-276.

111) Zaret, K.S. and Sherman, F. (1982) Cell **28**, 563-573.

112) Henikoff, S., Kelly, J.E. and Cohen, E.H. (1983) Cell **33**, 607-614.

113) Cozzarelli, N.R., Gerrard, S.P., Schlissel, M., Brown, D.D. and Bogenhagen, D.F. (1983) Cell **34**, 829-835.

114) Grummt, I., Sorbaz, H., Ofmann, A. and Roth, E. (1985) Nucleic Acids Res. **13**, 2293-2304.

115) Niles, E.G., Cunningham, K. and Jain, R. (1981) J. Biol. Chem. **256**, 12856-12860.

116) Din, N., Engberg, J. and Gall, J.C. (1982) Nucleic Acids Res. **10**, 1503-1513.

117) Veldman, G.M., Klootwijk, J., de Jonge, P., Leer, R. and Planta, R.J. (1980) Nucleic Acids Res. **8**, 5179-5192.

118) Sollner-Webb, B. and Reeder, R.H. (1979). Cell **18**, 485-499.

119) Holmes, W.M., Platt, T. and Rosenberg, M. (1983). Cell **32**, 1029-1032.

120) Martin, F.H. and Tinoco, I. (1980) Nucleic Acids Res. **8**, 2295-2299.

121) Acheson, N.H. (1976) Cell **8**, 1-12.

122) Aloni, Y. (1974) Cold Spring Harb. Symp. Quant. Biol. **39**, 165-178.

123) Laub, O. and Aloni, Y. (1975). J. Virol. **16**, 1171-1183.

124) Aloni, Y. (1975) FEBS Lett. **54**, 363-367.

125) Groner, Y., Carmi, P. and Aloni, Y. (1977) Nucleic Acids Res. **4**, 3959-3968.

126) Aloni, Y., Dhar, R., Laub, O., Horowitz, M. and Khoury, G. (1978). Cold Spring Harbor Symp. Quant. Biol. **43**, 559-570.

127) Aloni, Y., Bratosin, S., Dhar, R., Laub, O., Horowitz, M. and Khoury, G. (1978). Cold Spring Harb. Symp. Quant. Biol. **43**, 559-570.

128) Aloni, Y. (1972) Proc. Natl. Acad. Sci. USA **69**, 2404-2410.

129) Aloni, Y. (1973) Nature, New Biology **243**, 2-6.

130) Alwine, J.C. (1982) J. Virol. **43**, 987-996.

131) Alwine, J.C. and Khoury, G. (1980). J. Virol. **36**, 701-708.

132) Hay, N., Amster-Choder, O. and Aloni, Y. (1986) J. Virol. **57**, 402-407.

133) Bratosin, S., Horowitz, M., Laub, O. and Aloni, Y. (1978) Cell **13**, 785-790.

134) Horowitz, M., Bratosin, S. and Aloni, Y. (1978) Nucleic Acids Res. **5**, 4663-4675.

135) Legon, S.A., Flavell, A.J., Cowie, A. and Kamen, R. (1979) Cell **16**, 373.

136) Pope, J.H. and Rowe, W.P. (1964). J. Exp. Med. **120**, 121-128.

137) Prives, C., Gilboa, E., Revel, M. and Winocour, E. (1977) Proc. Natl. Acad. Sci. USA **74**, 457-461.

138) Prives, C., Beck, Y., Gidoni, D., Oren, M. and Shure, E. (1981) Cold Spring Harb. Symp. Quant. Biol. **44**, 123-130.

139) Scheidtmann, K.H., Echle, B. and Walter, G. (1982). J. Virol. **44**, 116-133.

140) Ellman, M., Bikel, I., Figge, J., Roberts, T.M., Schlossman, R. and Livingston, D.M. (1984). J. Virol. **50**, 623-628.

141) DiPamphilis, M. and Wasserman, P. (1982) in: *Organization and Replication of Viral DNA*, A.S. Kaplan, ed. CRC Press, Boca Roton, Fla. 37-114.

142) Feunteun, J., Sompayrac, L., Fluck, M. and Benjamin, T.L. (1976) Proc. Natl. Acad. Sci. USA **73**, 4169-4173.

143) Brady, J. and Khoury, G. (1985). Mol. Cell. Biol. **5**, 1391-1399.

144) Aldwine, J.C. (1985) Mol. Cell. Biol. **5**, 1034-1042.

145) Staneloni, R., Fluck, M. and Benjamin, T.L. (1977). Virology **77**, 598-6089.

146) Brugge, J.S. and Butel, J.S. (1975) J. Virol. **15**, 619-635.

147) Fluck, M.M. and Benjamin, T.L. (1979) Virology **96**, 205-228.

148) Fried, M. (1965) Proc. Natl. Acad. Sci. USA **53**, 486-491.

149) Osborn, M. and Weber, K. (1975) J. Virol. **15**, 636-644.

150) Tetmeyer, P., Schwartz, M., Collins, J.K. and Rundnell, K. (1975). J. Virol. **15**, 613-618.

151) Colby, W.A., Shenk, T. (1982) Proc. Natl. Acad. Sci. USA **79**, 5189-5193.

152) Petit, C.A., Gardes, M. and Feunteun, J. (1983). Virology **127**, 74-82.

153) Rassoulzadegan, M., Cowie, A., Carr, A., Gliachenhaus, N., Kamen, R. and Cuzin, F. (1982) Nature (London) **300**, 713-718.

154) Lane, D.P. and Crawford, L.V. (1979) Nature (London) **278**, 261-263.

155) Linzer, D.I.H. and Levine, A.J. (1979) Cell **17**, 43-52.

156) McCormick, F. and Harlow, E. (1980) J. Virol. 34, 213-224.

157) Clark, R., Lane, D.P. and Tjian, R. (1981) J. Biol. Chem. **256**, 11854-11858.

158) Crawford, L.V., Pim, D.C., Gurnery, E.G., Goodfellow, P. and Taylor-Papadimitriou. (1981) Proc. Natl. Acad. Sci. USA **78**, 41-45.

159) Montenarh, M. and Henining, R. (1983) J. Virol. **45**, 531-538.

160) Ito, Y., Brocklehurst, J.R. and Dulbecco, R. (1977). Proc. Natl. Acad. Sci. USA **74**, 4666-4670.

161) Shaffhausen, B., Silver, J. and Benjamin, T.L. (1978) Proc. Natl Acad. Sci. USA **75**, 79-83.

162) Shaffhausen, B. and Benjamin, T.L. (1980). J. Virol. **40**, 184-196.

163) Carmichael, G., Dorsky, D., Oliver, D., Shaffhausen, B. and Benjamin, T.L. (1982) Proc. Natl. Acad. Sci. USA **79**, 3579-3583.

164) Li, J.J. and Kelly, T.J. (1984) Proc. Natl. Acad. Sci. USA 81, 6973-6977.

165) Stillman, B.W. and Gluzman, Y. (1985) Mol. Cell Biol. **5**, 2051-2060.

166) Wobbe, C.R., Dean, F., Weissbach, L. and Hurwitz, J. (1985). Proc. Natl. Acad. Sci. USA **82**, 5710-5714.

167) Wobbe, C.R., Dean, F.B., Murakami, Y., Weissbach, L. and Hurwitz, J. (1986) Proc. Natl. Acad. Sci. USA **83**, 4612-4616.

168) Delucia, A., Lewton, B., Tijan, R. and Tegtmeyer, P. (1983) J. Virol. **46**, 143-150.

169) Tenen, D., Haines, L. and Livingston, D.M. (1982) J. Mol. Biol. **157**, 473-492.

170) Tijan, R. (1978) Cell **13**, 165-179.

171) Shortle, D.R., Margolskee, R.F. and Nathans, D. (1979) Proc. Natl. Acad. Sci. USA **76**, 6128-6131.

172) Tapper, D. and DePamphilis, M. (1978) J. Mol. Biol. **120**, 401-412.

173) Cremisi, C. (1981) Nucleic Acids Res. **9**, 5949-5964.

174) Ambrose, C., Blasquez, J. and Bina, M. (1986) Proc. Natl. Acad. Sci. USA **83**, 3284-3291.

175) Hartman, J.P. and Scott, W.A. (1981) J. Virol. 37, 908-915.

176) Blasquez, V., Stein, A., Ambrose, C. and Bina, M. (1986) J. Mol. Biol. in press. 177. Coca-Prades, M., Yu, H. and Hsu, A.T. (1982) J. Virol. **44**, 603-609.

178) Garber, E., Seidman, M. and Levine, A.J. (1979) Virology **90**, 305-316.

179) Coca-Prades, N. and Hsu, M.T. (1979). J. Virol. **31**, 199-208.

180) Fanning, E. and Baumgartner, I. (1980) Virology **102**, 1-12.

181) Jakobovits, E.B. and Aloni, Y. (1980) Virology 107-118.

182) Jakobovits, E.B., Abulafia, R. and Aloni, Y. (1982) Virology **121**, 95-106.

183) Dulbecco, R., Hartwell, L. and Vogt, M. (1965) Proc. Natl. Acad. Sci. USA **53**, 403-408.

184) Gluzman, Y. and Ahrens, B. (1982) Virology **123**, 78-92.

185) Stringer, J.R. (1982) Nature (London) **296**, 363-366.

186) Aloni, Y., Winocour, E. and Sachs, L. (1969) J. Mol. Biol. **44**, 333-345.

187) Conrad, S.E., Lin, C.P. and Botchan, M. (1982) Science **218**, 1223-1225.

188) Sambrook, J., Greene, R., Stringer, J., Mitcheson, T., Hu, S.L. and Botchan, M. (1979) Cold Spring Harb. Symp. Quant. Biol. **44**, 568-583.

189) Lewin, B. (ed) (1985) *Genes*, John Wiley and Sons Inc. New York.

190) Tooze, J. (ed) (1981) *DNA Tumor Viruses, Molecular Biology of Tumor Viruses*, 2nd. Ed. Cold Spring Harbor Laboratory, Cold Spring Harbor, New York.

CHAPTER 20

THE MOLECULAR BIOLOGY OF ADENOVIRUSES

ARNOLD J. LEVINE

Department of Molecular Biology, Lewis Thomas Laboratory, Princeton University, Princeton NJ 08544, U.S.A.

INTRODUCTION

The adenoviruses were discovered in 1953-54 by two independent groups studying respiratory diseases (1, 2). W. Rowe and his collegues at the National Institute of Health in Bethesda, MD, USA, developed an experimental method for culturing adenoids that were removed from children. Cells derived from such cultured tissues subsequently liberated a viral agent which then destroyed these cells. The agent was called "the adenoid degenerating agent" or AD agent (1). At about the same time M. Hilleman and his research group at Merck Sharp and Dohme isolated a virus from the throat washings of military recruits (2). This virus replicated in and killed human HeLa cells in culture and was called "the respiratory illness agent" or RI agent (2). The RI agent and the AD agent were rapidly shown to be related to each other and the name adenoviruses was adopted to describe this group of viruses. Employing immunological reagents to classify viral antigens a large number of related adenoviruses, with a common group specific antigen, were detected in humans (3), monkies (4), dogs (5), mice (6), cattle (7), and birds (8). The adenovirus group now has over 90 members (9) with more than 40 distinct isolates of human origin (3). These viruses all share a common group specific antigen which is the exon virion subunit or virion surface capsid. Individual viruses are identified by a type specific antigen on the virion fiber protein which is responsible for absorption of viruses to cells.

A. Virion Structure

All of the adenoviruses are composed of protein (87%) and DNA (13%). The viruses are icosahedrial in shape (Fig. 1), with a diameter of 65-80 nm (10). The outer capsid is assembled form 252 subunits or capsomeres, 240 of which (the hexons) form the 20 triangular faces of the icosahedron. The 12 vertices are composed

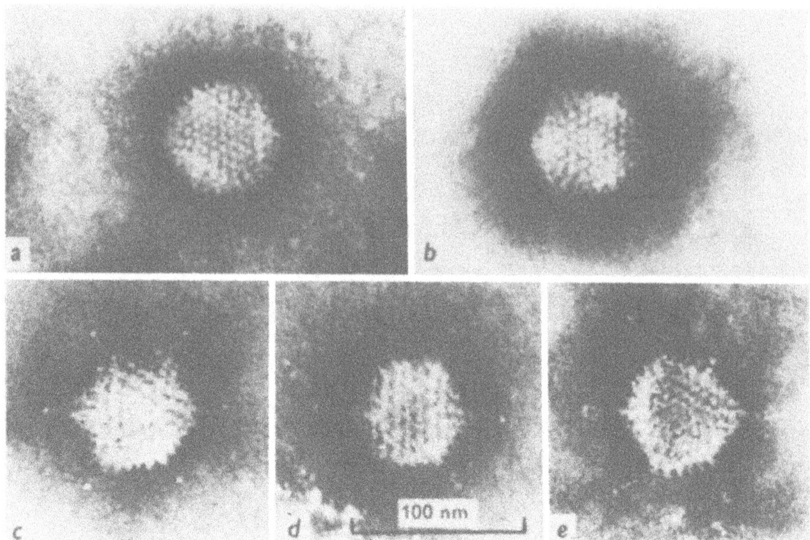

Figure 1. Electronmicrographs of Adenovirus (Reproduced from ref. 86)

Ultrastructure of virus particles of different serotype representatives of Rosen's subgroups I, type 3 (*a*); II, type 15 (*b*); and III, types 4 (*c*), 6 (*d*) and 2 (*e*). Negative contrasting with sodium tungstosilicate.

of penton bases (five nearest neighbors) to which is attached a fiber protein which is responsible for absorption of the virus to cells (11). The inner core of the virion is composed of two major proteins which are basic, histone-like peptides, associated with the condensed viral DNA (11, 12). A protein covalently liked to the 5' end of the two linear DNA strands (called pTP or terminal protein) is also present in the virion core (8, 11). About 11-14 distinct polypeptides are employed to form the virion particles (13). These proteins have been named (historically) by using roman numerals, where three polypeptides chains of protein II forms one hexon subunit or capsomer. Five or six peptides of protein III assemble to produce a penton base and three peptide IV molecules produce a fiber subunit (11, 13). Several additional polypeptides (VI, VII, IIIa, IX) are closely associated with the hexon subunits forming groups of hexons (nine at a time) during assembly of the virus particle (13). Peptides V and VII are the major constituents or polypeptides found in the core and are complexed with the viral DNA.

The adenovirus DNA has been isolated from virions as a double stranded linear molecule 20-23 $\times$ 10^6 daltons in size or about 33-37 kilobases long (14, 15). The DNA is bounded by terminally located, inverted repeated nucleotide sequences about 100-110 base pairs long (16, 17). Within this terminal repeat is a highly conserved nucleotide sequence, found in all adenovirus DNAs, located between nucleotides 9-22 (nucleotide 1 is at the end of the DNA molecule) (18, 19). These conserved sequences are the origin of viral DNA replication and are sites whose sequence is recognized by viral and/or cellular proteins involved in DNA replication

(18, 19). The 5' terminal nucleotide (number 1) of each DNA strand is covalently bound to a viral protein (8) which is involved in the initiation of newly synthesized viral DNA strands (see Chapter 3 of this volume and ref. 18,19).

B. Classification

The human adenoviruses have been divided into a number of subgroups based upon patterns of hemagglutination (20), oncogenic potential (21), DNA content and structure (15), the polypeptide composition of virion proteins (3), restriction site maps and polymorphism (3), DNA hybridization homologies (15) and more recently DNA sequence comparisons (17). At present, the 41 human adenoviruses are classified into seven subgroups (A-G) as summarized in Table 1. Adenoviruses 40 and 41 are recently isolated enteric adenoviruses and show a host range restricted ability to grow in only some cell cultures (3). Subgroups B, C and D are more commonly associated with epidemic conjunctivitis and respiratory disease with fever. Subgroup C adenoviruses are most commonly isolated from adenoids and tonsils and may be shed in feces over periods of years, representing an example of persistent infection in humans.

C. Questions under Study

The study of the adenovirus lytic or productive infectious cycle, transformation of cells in culture and tumorgenesis in animals has led to an understanding of a number of fundamental principles. The nucleotide signals regulating transcription in *cis* (enhancers) (22) and in *trans* (23, 24) (the E1A gene products) were among the first to be described using adenoviruses. The splicing of RNA transcripts into functional mRNAs was first elucidated in adenovirus infected cell systems (25, 26). Adenovirus gene functions have been shown to regulate transport of nuclear RNA

Table 1. - Properties of Human Adenovirus Subgroups

Subgroup	Serotypes	DNA Homology[a]	Oncogenicity[b]
A	12, 18, 31	48-69% with A 8-20% outside	100% tumors in 4 months
B	3, 7, 21, 34	89-94% with B 9-20% outside	10-50% tumors in 4-18 months
C	1, 2, 5, 6	99-100% with C 10-16% outside	None
D	8, 9, 24, 36	94-99% with D 4-17% outside	None
E	4	4-23 % outside	None
F	40	—	None
G	41	—	None

[a] DNA homology compared within a group (A vs A) or outside of a group (A vs B).

[b] Percent of animals (hamsters) with tumors after inoculation with purified viruses. The length of time it takes to detect a tumor.

into the cytoplasm (27), and affect the rate of translation of viral and cellular mRNAs by modulating the activity of a eukaryotic initiation factor (28, 29). The first *in vitro* DNA replication system initiating eukaryotic DNA synthesis (30) was described with adenoviruses. The modulation of gene functions regulating the immune response (class I major histocompatibility antigens) and its potential role in tumorgenesis was recently described with the adenoviruses (31, 32). Indeed, the early observation that only a subset of the closely related adenovirus serotypes (Table 1) produce tumors in animals (33), presented the challenge to investigate these molecular mechanisms of viral tumorgenesis. The past thirty years of research with these viruses have indeed uncovered a large body of knowledge central to life processes. As a model system for understanding mechanisms of gene expression and regulation, the adenoviruses have justified the intensive study and attention they have received. The goal of this review will be to detail the major concepts and ideas derived from the study of the molecular bases of adenovirus replication and transformation.

2. THE MOLECULAR BIOLOGY OF THE ADENOVIRUS REPLICATION CYCLE

A. The Early Events in Viral Gene Expression

After adsorption and penetration of the virus particle, the virion core becomes localized in the cell nucleus (11). The first viral gene transcribed by the cellular RNA polymerase II is the E1A gene (Table 2). There are at least two *cis*-acting nucleotide sites, located 5' to the E1A structural genes, that must be recognized by *cellular* gene products resulting in transcription of this viral gene (34). The E1A gene enhancer element has at least two functionally distinct domains (35), one, regulating E1A transcription, the other, early gene transcription. This signal has been localized at 200-350 bp from the left end of the molecule (Table 2). A E1A promotor whith a CAAT and TATA sequence motif located at 427 bp and 470 bp are also required to correctly begin the E1A transcript at 500 bp (17).

Splicing of the E1A primary transcript produces three mRNAs of 9S, 12S and 13S which share common sequences at their 5' and 3' ends (Fig. 2). The 12S message deletes sequences (46 amino acids) which are present in the middle of the 13S mRNA. While the primary translated products from these mRNAs are expected to be 13K, 26K and 32K proteins, in infected cells four to six polypeptides with molecular weights of 40-60K are observed (37). This is most likely the result of post-translational additions to these primary protein products. Phosphorylation of E1A proteins is known to occur (36, 37).

These E1A proteins, in association with cellular proteins, are responsible for enhancing the transcription of several viral early genes: E1B, E2A, E2B, E3 and E4 (see Table 2 and ref. 23, 24, 38). The E1A gene products can also positively regulate the transcription of its own gene (38, 39, 40) and the transcription of several cellular genes (41, 42). In addition to this enhancement of transcription, E1A gene products have been shown to be able to negatively regulate some viral promotors such as SV40 or polyoma virus (43, 44) and even be responsible for

Table 2. - Adenovirus Genes and Genetic Elements: The Early Gene Products

Gene or Element	Map Location[a] (base pairs)	Gene Product[b]	Function
1. Terminal repeat	1-103	cis-acting site	Origin of DNA replication Binds nuclear factor I (17-48 bp)
2. Packaging sequence	194-305 bp	cis-acting site	Required to package DNA into virion
3. E1A enhancer sequence	200-350 bp	cis-acting	Promotes transcription of E1A gene
CAAT and TATA promotor for E1A	427, 470 bp	cis-acting	Promotes transcription of E1A gene
4. E1A gene	500-1630 bp		
	12S mRNA	26K protein	Positive/negative
	13S mRNA (r-transcript)	32K protein	regulation of transcription in trans
5. E1B gene	1675-4070 bp		
	22S and 13S mRNA (r-transcript)	21K protein	DNA degradation, nuclear membrane location
	22S mRNA (r-transcript)	55K protein (complex with E4-34K protein or p 53 cellular protein)	Nuclear transport of m-RNAs
6. E2A	23.5-21.8 Kb (1-transcript)	72K	SS-DNA binding, DNA replication
E2B region			
1.	10.5-8.58 Kb (1-transcript)	87K	Terminal protein, (pTP), DNA replication
2.	8.4-5.2 Kb (1-transcript)	120K	DNA polymerase, DNA replication
7. E3	27.3-30.9 Kb (r.transcripts)	six proteins possible	Unknown function, deletion is viable
1.	28.6-29.1 Kb (r-transcript)	19K	Glycoprotein, alters glycosylation of cellular proteins
8. E4	35.7-33 Kb (1-transcripts)	seven possible proteins	
1.	34.8-34.4 Kb (1-transcript)	11K	Conserved, nuclear matrix protein
2.	34.1-33.2 Kb (1-transcript)	34K	Transport of mRNA from nucleus, bound to E1b-55K

[a] Map location is given in base pair (bp) or kilobases (Kb). Nucleotide 1 is at the left end of the molecule which is 36 Kb.

[b] Gene products are expressed by size times 1000 (K).

modulating down cellular gene activities (31, 45). The mechanisms mediating this E1A positive or negative regulation of transcription are unclear. The E1A proteins have short half lives, a nuclear location, and they exert a large impact upon viral and cellular gene expression in these infected cells.

The E1B genes, positively regulated by these E1A gene products, produce a 13S and a 22S-mRNA spliced from the primary transcript across this region (1675-4020 bp, Table 2). Two proteins are translated from these mRNAs, a 21K protein (from 13S and 22S) and a 55K protein (from 22S mRNA in a different reading frame) (37).

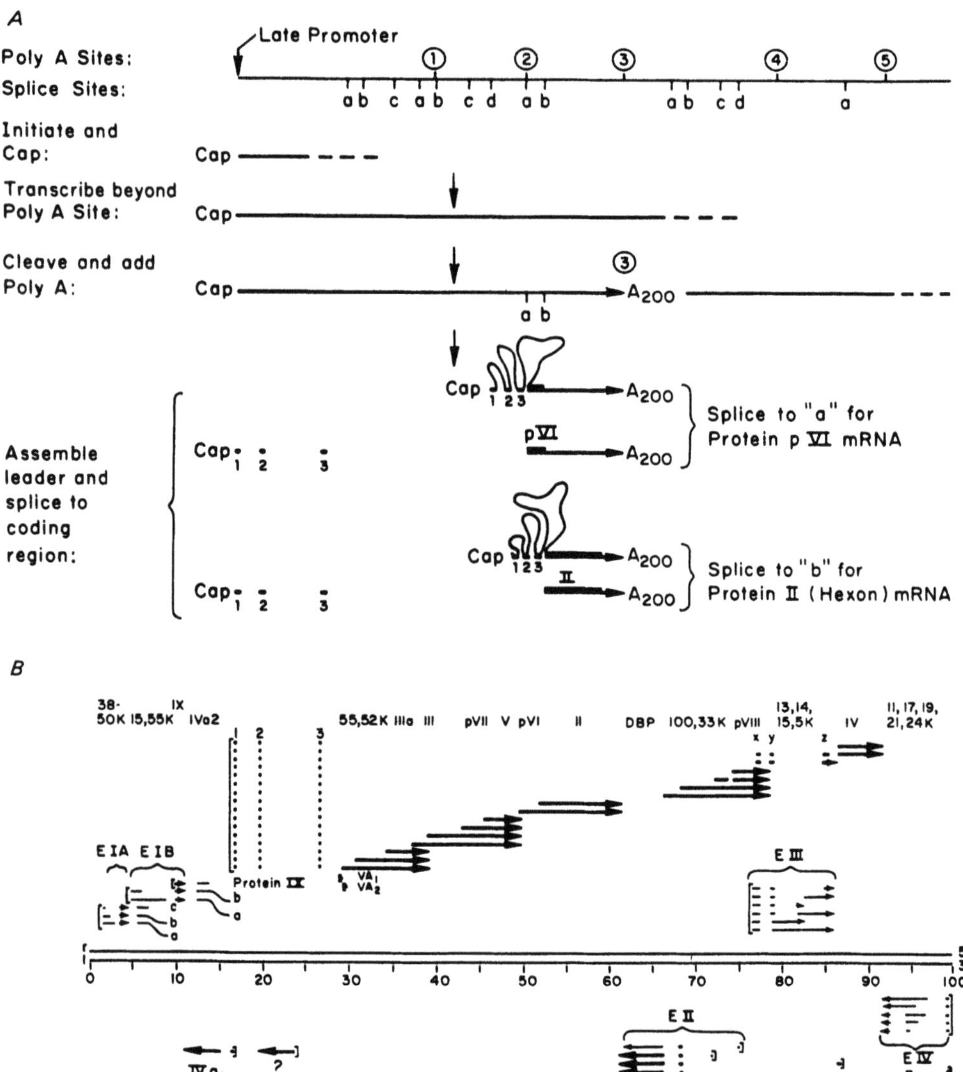

The 21K protein is localized in the nuclear membrane and mutants in this gene product often produce large plaques, have an enhanced cytolytic phenotype and degrade viral and cellular DNA in infected cells (46, 47, 48). These mutants replicate well in HeLa cells and are not defective for growth in cell culture.

The E1B 55K protein is found (in part) in virus infected cells to be associated in an oligomeric protein complex with the E4-34K viral protein (49). Mutants in either the E1B 55K protein or the E4-34K proteins (see Table 2) have similar phenotypes (37, 46). These mutants are defective for virus replication and fail to transport the late mRNAs out of the nucleus into che cytoplasm even through transcription and processing of late nuclear viral RNAs appears normal (27). Cellular transcripts, which are usually transported into the cytoplasm poorly at late times of infection (50), were not as severely inhibited and cellular protein synthesis continued in these E1B-55K mutant infected cells (27). The interpretation of these observations is that the E1B 55K-E4-34K protein complex efficiently transports out of the nucleus viral mRNAs with the late tripartite leader sequences (Table 3 and legend Fig. 2) but poorly transports cellular nuclear RNAs into the cytoplasm.

The E2A transcripts producing the 72K DNA binding protein (DBP) and the E2B mRNAs encoding the terminal protein (pTP) and DNA polymerase (Fig. 2 and Table 2) are all spliced from a single primary transcript reading leftward on the genome (from 23.5 Kb to 5.2 Kb, covering 18.3 Kb in length). The pTP is synthesized as a 87K precursor protein (Table 2) which reacts with dCTP to produce pTP-CMP, covalently linked via a phosphodiester bond to the OH group of serine on the protein and the 5' end of dCMP (19, 51). This leaves the 3'-OH group free to act as a primer for synthesis of new adenovirus DNA. The DNA polymerase associates with the pTP-CMP in a complex which recognizes the terminal repeat regions of the viral DNA (Table 2, origin of replication 1-103 bp, see also chapter 3 of this volume). A cellular protein termed nuclear factor I (19, 51) binds specifical-

Figure 2. Transcription and Processing of Adenovirus mRNAs.

Adenovirus DNA (~35 kb) has been conventionally divided into 100 map units 350 bp each. "Rightward" transcripts utilize the DNA r-strand (upper line in **A**), whereas "leftwards" transcripts are synthesized from right to left using the DNA l-strand as template.
A: Three separate transcriptional units on the r-strand (E1A, E1B, and EIII) and two on the l-strand (EII and EIV) encode the *early* adenovirus mRNAs. Different internal splices are used to originate several mRNA species from each transcriptional unit.
Late adenovirus mRNAs are transcribed from a major promotor site at map position 16.3 on the r-strand, and belong to five 3' co-terminal polyadenylation families. The untranslated tripartite leader share by late mRNAs is spliced from transcripts corresponding map positions 16.3, 17.6, and 26.6 (see below). *x, y,* and *z* indicate minor late promotors.
B: Schematic Representation of Transcription and Processing of Late Adenovirus mRNAs. Starting at the common late adenovirus promotor, the RNA transcript extends well beyond the polyadenylation site (circled 3 in this example), and reaches the 5' end of the genome. The nascent transcript is excised at the poly(A) site, and sequences from map positions 16.3, 17.6, and 26.6 are spliced into the tripartite leader which can be joined to the coding region. The diagram illustrates two alternative uses of the poly(A) site 3: in the first case the mRNA so produced would code for protein pVI, and for protein II (hexon) in the second one. Note that all late adenovirus mRNAs are 5' co-terminal (tripartite leader), and each family shares in common the poly(A) site. (Modified from E. Ziff, ref. 87).

Table 3. - Adenovirus Genes and Genetic Elements: The Late Gene Products from the Major Late Promoter

Gene or Element	Map Location[a] (map units)	Gene Product	Function
1. major late Promotor	16.3 mu	rightward start of transcript	
2. tripartite leader sequence	16.3, 19.6, 26.6 mu	200 bp RNA 5' end of all late m-RNAs	Translational efficiency
3. L1	30-38 mu	52K 55K IIIa	Expressed early, Function unknown Hexon associated
4. L2	38-50 mu	III pVII V	Penton base Core Core
5. L3	50-64 mu	pVI II 23K	Hexon associated Hexon Non-structural protease
6. L4	67-77 mu	100K 33K pVIII	Hexon assembly Non-structural Hexon associated
7. x, y, z leaders	79,86 mu	RNA-leader FOR L5	Unknow
8. L5	88-91 mu	IV	Fiber
Late Gene Products Not Under the Control of the Major Late Promotor			
Gene or Element	Map Location	Gene Product	Function
9. IX	10 mu (3.58 Kb-4.0 Kb) (r-transcript)	pIX	Virion protein, core and hexon associated
10. IVa2	5.7-4.1 Kb (1-transcript)	IVa2	Virion protein
Other Gene Products			
11. VA-I, II	30 mu (10.6 Kb)	160 bp RNA	Enhance translation Alter eIF-2 alpha Kinase activity

[a] Map units (mu) are given 1-100 as a percentage of genome length.

ly to nucleotide sequences (17-48 bp) in the terminal repeat regions and promotes the binding of pTP-CMP-DNA polymerase complex to the adenovirus DNA template (51). Polymerization of nucleotide triphosphates into adenovirus DNA proceeds displacing a single strand and copying one template strand (see Fig. 13, chapter 3). As this reaction proceeds (initiation of DNA replication), elongation of the DNA chain requires the adenovirus DBP which binds to single stranded DNA (displaced strand) and nuclear factor II which is a cellular topoisomerase (51). These proteins are required for the elongation step of DNA replication. After replication of one DNA strand is completed, the displaced single strand (with DBP) is employed as a template for double stranded synthesis. The 87K pTP is cleaved by a viral protease (52, 53) after DNA replication or during packaging of the DNA and assembly of virions, producing a terminal protein in the mature virion of 55K in size. Thus, only three viral proteins (pTP, polymerase, DBP) and two cellular proteins (nuclear factor I and II) are required to support adenovirus DNA replication *in vitro* (51).

The E3 transcripts are spliced into four major and four minor mRNAs (Fig. 2 and ref. 54). At least four E3 proteins have been detected and have been shown to be produced in infected cells (37). The most intensively studied of these proteins is the E3-19K glycoproteins (55, 56). This protein is found to be associated with a variety of membrane fractions in the infected cells. When in the endoplasmic reticulum and Golgi membranes, the E3-19K protein is found to be associated with the class I major histocompatibility antigens. As a result of this protein complex, these antigens fail to have proper terminal sugars added to the antigens (32). This then reduces the level of class I antigens on the infected cell surface and presumably decreases the efficiency of cytotoxic T-cell killer responses to infected cells (57). The E3-19K protein, like the other adenovirus E3 proteins, appears to be dispensable to virus replication in cell culture. Deletion mutants that eliminate the entire E3 region of the virus (46) replicate normally in vitro or in cell culture.

The E4 region of the genome is transcribed leftward (Fig. 2), and splicing produces some 9-14 transcripts (54). At least seven polypeptides can be produced from three open reading frames in the E4 region (17) and more proteins are possible. An E4-11K protein has been identified in infected cells (58) in the nuclear matrix. A frame shift mutation in this gene has no effect upon the viability of the virus in cell culture (58). An E4-34K protein has been isolated and shown to be in an oligomeric protein complex with the E1b-55K protein (49). Mutants in this gene, have similar phenotypes to E1b-55K mutants (27, 59) regulating the ability of late adenovirus mRNAs and cellular mRNAs to be transported from the nucleus to the cytoplasm. The other E4 gene produce have not been characterized in such detail (37, 46).

B. The Late Events

After the synthesis of the early viral proteins and the replication of viral DNA, viral gene expression shifts (Table 3), and the major late promotor becomes the most active transcription unit of the virus (60). Transcripts are initiated at 16.3 map units (mu) on the genome (Fig. 2), and continue across the chromosome in a rightward direction for 30Kb (61). Five overlapping families of mRNAs are processed from this primary transcript (25, 26, 54). Within each family a common 3'-terminal poly A addition site is shared. In all eighteen late mRNAs are generated and each bears the same tripartite leader sequence of 200 nucleotides derived from map coordinates 16.3, 17.6 and 26.6 mu (Table 3 and Fig. 2).

To date, eleven proteins have been shown to be synthesized in infected cells from these eighteen mRNAs (11,13). Three of these proteins, (Table 3) the L3-L23K, L4-100K and L4-33K are non-structural proteins and are not found in the mature virion. The L4-100K protein is involved in the assembly of hexon subunits or capsomers from protein II polypeptides (46). The 23K peptide is a protease that is required for protein processing of capsid polypeptides pVI, pVII and pVIII to mature VI, VII and VIII virion proteins. The same protease cleaves the 87K-pTP terminal protein to the 55K-pTP found in mature virions (53). Virus assembly requires the packaging of the viral DNA into virion particles. A *cis* acting recognition site on the DNA at nucleotides 194-305 (Table 2) is a packaging sequence required for the formation of mature virions (35, 62).

There are two other late virion structural proteins, IX and IVa_2 (Table 3) which are synthesized predominantly at late times during infection, but are not part of the major late transcript system (Fig. 2). It is not at present clear why these two structural proteins are treated differently, other than the different genomic positions of their genes (Table 3). Whether that is an important feature to the viral life cycle remains to be explored.

C. VA-RNA Translational Controls

A second gene family that is also regulated quite differently is the VA-I and VA-II genes of the virus (at 30 mu or 10.6Kb, Table 3). These genes are transcribed by the cellular RNA polymerase III enzyme and produce two closely related RNAs of 160 nucleotides in length. The VA genes contain both intragenic and extragenic (5') control regions for polymerase III transcription (63, 64). The VA-I RNA forms a ribonucleoprotein particle that prevents activation of an interferon induced eIF-2 alpha protein kinase (28, 29). The phosphorylation of the eukaryotic initiation factor of translation results in a decreased efficiency of translation of mRNA in infected cells (28, 29). This provides the adenoviruses with a mechanism to evade interferon action during persistant infection of animals (chapter 6 of this book).

3. THE MOLECULAR BIOLOGY OF ADENOVIRUS TRANSFORMATION AND TUMORIGENESIS

Three lines of evidence demonstrate that the adenovirus E1A and E1B genes and gene products are both necessary and sufficient for a transformation event to occur in at least a small percentage of cells.

First, the left end restriction fragment of the adenovirus genome, containing the E1A and E1B genes, produces transformed cell foci in culture (65). The E1A gene products alone, are sufficient to immortalize primary cells in culture (66) but these cells do not form foci, grow in agar or make tumors in animals. The addition of the E1B genes to these cells (E1A-positive cells) results in the phenotypes of transformation such as foci formation and enhanced tumorigenic potential. The overexpression of the E1A gene products in a cell can reduce the need for the E1B functions and result in a fully transformed cell line or formation of foci (67).

Second, mutations in the adenovirus E1A or E1B genes often result in the loss of the ability of these viruses to transform cells in culture (46).

Third, all adenovirus transformed cell lines retain the viral E1A and E1B genes (68) and express those genes in transformed cells (37, 68).

Taken together these observations strongly suggest that the viral E1A and E1B genes and gene products are required to initiate and probably maintain the transformed phenotype. Other viral gene products can enhance or alter the frequency of transformation in culture. Mutants in the E4-34K gene also enhance the frequency of transformation (69). This is probably due to the ability of the DBP to regulate the levels of E1A and/or E1B gene products (37). Mutations in the E4-34K gene also enhance the frequency of transformation (27, 59). In this case, the E4-34K protein is complexed with the E1B-55K protein and these gene products function together in some situations (49).

The mechanisms by which the E1A proteins and E1B proteins alter the properties of cells remains unclear. The E1A proteins have the ability to positively or negatively regulated viral and/or cellular gene promotors or enhancers (23, 24, 31, 38-45, 70). Among the cellular genes whose mRNA levels increase in response to E1A gene products are the family of heat shock genes (70). The E1A gene products thus increase the levels of two proteins, hsp 70 (a heat shock gene product) and the E1B-55K adenovirus protein, each of which has been shown to form an oligomeric protein complex with a third cellular protein, p53 (71, 72). The cellular p53 protein has been to shown to have oncogene-like activity in several assays:

i) p53 can function with the *ras* oncogene to produce foci with primary rat embryo fibroblasts (73, 74);
ii) p53 can immortalize murine cells in culture (75);
iii) p53 overproduced in already immortalized cell lines can increase their tumorigenic potential (76, 77, 78);
iv) p53 expression can result in the stimulation of cellular DNA synthesis in platelet-poor plasma (79) and antibodies directed against p53, microinjected into cells in culture, block the entry of cells into the S-phase (80);
v) p53 synthesis is regulated in the cell cycle (81, 82). Increased levels of p53 are commonly found in transformed cells and in SV40 (83, 84) and adenovirus transformed cells (72) the viral tumor antigens (SV40 T-antigen, E1B-55K) are complexed with this cellular p53 protein. This results in a longer half life for the p53 protein (83) and increased levels of this oncogene-like product.

Thus, one possible mechanism in which E1A and E1B-55K proteins may act to transform cell is through a p53 mediated alteration of cell regulation. The E1B-55K protein, complexed in infected cells with the E4-34K protein, appears to regulate the types of RNAs that can be transported from the nucleus into the cytoplasm (27). Thus, regulation at the transcriptional level via E1A and the nuclear transport level via E1B-55K could have a dramatic effect upon cellular gene expression. While the E1B-55K protein is complexed with the E4-34K protein during lytic infection (49), the E1B-55K protein is complexed with p53 in the transformed cell (72). This switch in one of the subunits of the E1B-55K complex, could well have functional consequences. It is likely that the E1B-55K-E4-34K complex selectively transports adenovirus mRNA into the cytoplasm (27) while the E1B-55K-p53 complex clearly permits cellular mRNAs to enter the cytoplasm. The implication of this reasoning is that p53 plays a role in differential transport of mRNAs into the cytoplasm. This could be done in a cell cycle dependent fashion (81, 82) and regulate entry of cells into the S-phase. That the hsp 70 gene family is also involved in regulating cell growth and the cell cycle has also been suggested before (71).

With the adenoviruses, the ability to form tumors in animals is not equivalent to the ability to transform cells in culture. The group C adenoviruses (Table 1) transform cells in culture just as efficiently as the group A adenoviruses, but the group C viruses fail to produce tumors in rats or hamsters while the group A viruses efficiently produce tumors in these animals (11, 15, 33). One reason that tumorigenesis in animals may differ from transformation *in vitro* is the role of the immune system *in vivo*. As noted, the E1A gene products can regulate a cellular or viral gene in a negative fashion turning off transcription (43, 44). The adenovirus type 12 E1A gene products (group A) modulate down the levels of class I major histocompatibility antigens in transformed cells while the adenovirus type 5 E1A gene products (group C) apparently do not exert this negative regulation (85). Adenovirus type 12 transformed cells in culture are usually more tumorigenic in syngeneric animals than adenovirus type 5 transformed cells (85). This is presumably due to the enhanced rejection of type 5 cells expressing higher levels of class I antigens on their cell surfaces. Adenovirus type 5 and 12 transformed cells produce tumors in nude mice (no rejection) with about equal efficiency (85). Presumably, adenovirus type 5 and 12 viruses can produce tumors in animals with roughly equal efficiences, but adenovirus type 5 tumors are usually rejected and the virus is then termed non-tumorigenic.

Two other gene functions of this virus play a role in virus replication *in vivo* and persistance. The E3-19K (Table 2) protein complexes with class I major histocompatibility antigens in the Golgi-endoplasmic reticulum membranes and reduces terminal glycosylation of these antigens (32). This also reduces the levels of class I antigens on the cell surface and presumably impairs the immune response of cytotoxic killer T-cells. The E3 genes are not found in most transformed cell lines and so, this mechanism may not play a large role in tumor formation. Rather, the E3-19K gene product probably aids the viral persistance and enhances virus titer in infected cells. Similarly, interferon, which can be elaborated in response to persistant virus infections, acts to both inhibit translation and raise the level of major histocompatibility antigens on the cell surface (chapter 6, and ref. 28). The eIF-2 alpha kinase is induced by interferon treatment and normally inhibits translation of viral mRNA. The VA RNAs reverse this and permit efficient translation of mRNA in virus infected cells and this enhances virus titer and persistance. These are good examples of how viruses have designed mechanisms to combat the host defenses of the immune system.

4. CONCLUSIONS

The adenovirus chromosome has an organization of genes and genetic elements that are programmed into transcription units that function at defined times in a productive infection cycle.

The only viral gene transcribed *solely* by cellular gene products, and therefore the first gene to be expressed, is the E1A gene. An enhancer-promotor element located 5' to the E1A gene is the *cis*-acting regulatory signal that must be recognized by the cellular RNA polymerase related activities, to permit an infectious cycle to begin.

The E1A gene products then activate a set of viral early genes, E1B, E2A and B, E3 and E4, at the transcriptional level. The E1B gene product selectively acts to transport viral nuclear RNA to the cytoplasm and the polymerase III VA gene product enhances translation of viral mRNAs. The E2 transcripts produce three viral proteins each required for viral DNA replication (along with two cellular gene products).

Following viral DNA replication, the transcript of the major late promotor is spliced into 5 families of late gene mRNAs comprising of eighteen different mRNAs and producing at least eleven proteins. Nine of these late functions, plus two others, are the virion structural proteins and assembly of the virus requires a specific recognition of the viral DNA by some of these proteins. The *cis*-acting viral packaging signal overlays and is contiguous with the E1A enhancer element. The first and last events in the cycle use identical signals. Viral functions that act positively or negatively upon both viral and cellular genes (E1A), or gene products (E1B-55K, VA-RNA, E3-19K) play a central role in virus replication. These same genes are involved in cellular transformation, tumorigenesis and viral persistance in the animal.

Two major features emerge from the data reviewed here: it is clear that a single transcriptional unit is involved in supplying functions for viral DNA replication (E2A and B) and a single transcript to produce the structural proteins of the virus. The *cis*-acting elements for the origin of DNA replication, packaging sequences and the E1A enhancer are all closely linked on the genome within the first 1% of the chromosome (350 base pairs). The adenoviruses thus represent a good example where the organization of genes and genetic elements reflect the temporal controls and functional linkage of transcription units.

The second conclusion to be drawn is the relationship between gene products that regulate viral functions and their related ability to regulate cellular gene functions resulting in transformation or tumorigenesis. This can be mediated by E1A and its associated proteins at the transcriptional level or E1B-55K switching its associated polypeptide from the E4-34K to the 53 protein. *In vivo*, in the whole animal, gene products like the E3-19K and VA-RNA play a role in persistance of the virus. As such, those gene products can affect viral mediated tumorigenesis or pathology.

Clearly, the study of the adenoviruses have served as a model system to understand how genes and their products are regulated. At the molecular, cellular and organismic level, the adenovirus group of agents continue to be a powerful probe to study fundamental questions about life processes.

5. REFERENCES

1) Rowe, W.P., R.J. Huebner, K.K. Gillmore, R.H. Parrott, and T.G. Ward, (1953). Proc. Soc. Exp. Biol. Med. **84**: 570.

2) Hillman, M.R. and J.H. Werner, (1954), Proc. Soc. Exp. Biol. Med. **85**. 183.

3) Wadell, G., (1984) in Curr. Top. Micro. Imm. **110**, Ed., W. Doerfler, The Mol. Biol. Adenoviruses 2, Springer-Verlag, New York, 191-220.

4) Hull, R.N., J.R. Minner and C.C. Mascoli, (1958), Am J. Hyg. **68**. 31.

5) Kapsenberg, J.G. (1959). Proc. Soc. Exp. Biol. Med. **101**: 611.

6) Hartley, J.W. and W.P. Rowe. (1960). Virol. **11**: 645.

7) Klein, M., E. Ealey and J. Zellat (1959). Proc. Soc. Exp. Biol. Med. **102**: 1.

8) Robinson, A.J., H.B. Younghusband and A.J.D. Bellett, (1973). Virol. **56**: 54.

9) Doerfler, W. (Ed.), Curr. Top. Micro. Imm., Vols: 109, 110 and 111, The Mol. Biol. Adenoviruses 1, 2, 3, Springer Verlag, New York.

10) Horne, R.W., S. Bonner, A.P. Waterson and P. Wildy, (1959). J. Mol. Biol., 1:84.

11) Tooze, J. (Ed.). Mol. Biol. Tumor Viruses, 2nd ed., revised. The DNA Tumor Viruses, Cold Spring Harbor Press, New York, Chapters 8, 9, 10, 11.

12) Brown, D.T., M. Westphal, B.T., Burlingham, U. Winterhoff and W. Doerfler, (1975). J. Virol. **16**: 366.

13) Philipson, L., (1983), Curr. Top. Micro. Imm. **109**: Ed., W. Doerfler, Springer-Verlag, New York., pp. 1.

14) Green, M., M. Pina, R. Kimes, P.C. Wensink, L.A. Machattie and C.A. Thomas, (1967). Proc. Natl. Acad. Sci. USA **57**: 1322.

15) Green, M. (1970). Ann. Rev. Biochem. **39**: 701.

16) Wolfson, J. and Dressler, D. (1972). Proc. Natl. Acad. Sci. **69**: 3054.

17) Van Ormondt, H. and F. Galibert, (1984). Curr. Top. Microbiol. Imm., Ed. W. Doerfler, Vol. **110**, Springer-Verlag, New York, 73-141.

18) Sussenbach, J.S. and P.C. vander Vleet, (1983), Curr. Top. Microbiol. Imm., **109**, Ed. W. Doerfler, Springer-Verlag, New York, 36.

19) Futterer J. and Winnocker, E.L. (1984). Curr. Top. Microbiol. Imm. **III**, Ed. W. Doerfler, Springer-Verlag, New York, 41-58.

20) Rosen, L. (1960). Am. J. Hyg. **71**: 120-128.

21) Huebner, R.J. (1967). M. Pollard, (Ed.), Perspect. in Virol., Vol. 5, Acad. Press, New York, pp. 147-166.

22) Hearing, P. and T. Shenk, (1983). Cell **33**: 695-703.

23) Berk, A.J., F. Lee, T. Harrison, J. Williams and P.A. Sharp. (1979) Cell **17**: 935-944.

24) Jones, N. and T. Shenk. (1979). Proc. Natl. Acad. Sci. USA **76**: 3665-3669.

25) Berget, S.M., C. Moore and P.A. Sharp. (1977). Proc. Natl. Acad. Sci. USA **74**: 3171.

26) Chow, L.T., R.E. Gelinas, T.R. Broker and R.J. Roberts. (1977). Cell **12**: 1.

27) Pilder, S., M. Moore, J. Logan and T. Shenk. (1986). Mol. Cell. Biol. **6**: 470-476.

28) Kitajewski, J., R.J. Schneider, B. Safer, S. Munemitsu, C. Samuels, B. Thimmappaya and T. Shenk. (1986). Cell **45**: 195-200.

29) Siekierka, J., T.M. Mariano, P.A. Reichel, and M.B. Mathews, (1985), Proc. Natl. Acad. Sci. USA **82**: 1959-1963.

30) Challberg, M.D. and T.J. Kelly. (1979). Proc. Natl. Acad. Sci. USA **76**: 655-659.

31) Schrier, P.T., R. Bernards, R.T. Vaessen, A., Houweling, A.J. vander Eb, (1983). Nature (London) **305**: 771-775.

32) Burgert, H.G. and S. Kvist, (1985). Cell **41**: 987-997.

33) Trentin, J.J., Y. Yabe, and G. Taylor (1962). Science **137**: 835-841.

34) Hearing, P. and T. Shenk, (1983). Cell **33**: 695-703.

35) Hearing, P. and T. Shenk. (1986). Cell **45**: 229-236.

36) Harter, N. and J. Lewis (1978). J. Virol. **26**: 736-744.

37) Levine, A.J., (1984). in Curr. Topic. Microbiol. Imm., Ed. W. Doerfler, Vol. **110**, Springer-Verlag, New York, pp 143-161.

38) Nevins, J., (1981). Cell **26**: 213-220.

39) Osborne, T.F., D.N. Avidson, E.S. Tyau, M. Dunsworth-Brown and A.J. Berk (1984). Mol. Cell. Biol. **4**: 1293-1305.

40) Hearing, P. and T. Shenk (1985). Mol. Cell. Biol. **5**: 3214-3221.

41) Stein, R. and E.B. Ziff. (1984). Mol. Cell. Biol. **4**: 2792-2801.

42. Treisman, R., M.R. Green and T. Maniatis (1983). Proc. Natl. Acad. Sci. USA **80**: 7428-7432.

43) Borrelli, E., R. Hen and P. Chambon (1984) Nature (London) **312**: 608-612.

44) Velcich, A. and E. Ziff (1985). Cell **40**: 705-716.

45) Bernards, R., A. Howveling, P.I., Schrier, J.L. Bos and A.J. vander Eb, (1982). Virology **120**: 422-432.

46) Shenk, T. and J. Williams (1984). Curr. Topic. Microbiol. Imm., Vol. **111**. Ed. W. Doerfler, Springer-Verlag, New York, pp. 1-30.

47) Chinnadurai, G., (1984) Cancer Cells, Vol. **2**, Eds. G. vande Wounde, A.J. Levine, W.C. Topp and J.D. Watson, Cold Spring Harbor Press, New York, pp. 511-517.

48) Logan, J. Pilder S. and T. Shenk. (1984). Cancer Cells, vol. **2**, Eds. G. vande Woude, A.J. Levine, W.C. Topp and J.D. Watson, Cold Spring Harbor Press, New York, pp. 527-532.

49) Sarnow, P., P. Hearing, C. Anderson, D. Halbert, T. Shenk and A.J. Levine, (1984). J. Virol. **49**: 672-700.

50) Beltz, J. and J. Flint. (1979). J. Mol. Biol. **131**: 353-373.

51) Friefeld, B.R., J.H. Lichey, J. Field, R.M. Gronostajski, R.A. Guggenheimer, M.D. Krevolin, K. Nagata, J. Hurwitz and M.S. Horwitz, (1984). Curr. Topic. Microbiol. Imm. vol. **110**, Ed. W. Doerfler, Springer-Verlag, New York, pp. 221-255.

52) Stillman, B.W. J.B. Lewis, L.T. Chow, M.B. Mathew, J.E. Smart (1981). Cell **23**: 497.

53) Stillman, B.W. (1981). J. Virol. **37**: 139.

54) Flint, J. (1982). Biochem. Biophys. Acta. **651**: 175-208.

55) Ross, S.R. and A.J. Levine (1979). Virology **99**: 427-430.

56) Persson, H., S. Kvist, L. Ostberg, P.A. Petersson and L. Philipson (1979). Cold Spring Harbor Symp. Quant. Biol. **44**: 509-517.

57) Zinkernagel, R.M. and P.C. Doherty. (1979). Adv. Immunol. **27**: 51-177.

58) Sarnow, P., P. Hearing, C.W. Anderson, N. Reich, A.J. Levine (1982). J. Mol. Biol. **162**: 565-583.

59) Halbert, D.N., J.R. Cutt and T. Shenk. (1985). J. Virol. **56**: 250-257.

60) Ziff, E. and R. Evans. (1978). Cell **15**: 1463-1475.

61) Fraser, N.W., J.R. Nevins, E. Ziff and J.R. Darnell (1979). J. Mol. Biol. **129**: 643-656.

62) Hammarskjold, M.L. and G. Wenberg. (1980). Cell **20**: 787-795.

63) Fowlkes, D. and T. Shenk. (1980). Cell **22**: 405-413.

64) Guilfoyle, R. and R. Weinmann (1981). Proc. Natl. Acad. Sci. **78**: 3378-3382.

65) Graham, F.L., P.J. Abraham, C. Mulder, H.L. Hajneker, S.O. Warner, F.A.J. de Vries, W. Fiers and A.J. vander Eb (1974). Cold Spring Harbor Symp. Quant. Biol. **39**: 637-650.

66) Houweling, A., P.J. vanden Eisen, A.J. vander Eb. (1980). Virology **105**: 537-550.

67) Senear, A.W. and J.B. Lewis. (1986). Mol. Cell. Biol. **6**: 1253-1260.

68) Sambrook, J., M. Botchchan, P. Gallimore, B. Ozanne, U. Peterson, J. Williams, P. Sharp (1974). Cold Spring Harbor Symp. Quant. Biol. **39**: 615-632.

69) Logan, J., J.C. Nicolas, W.C. Topp, M. Girard, T. Shenk, A.J. Levine (1981). Virology **115**: 419-422.

70) Nevins, J.R., M.J. Imperiale, H.T. Kao and L.T. Feldman (1984). Cancer Cells, vol. II, Eds. G. vande Woude, A.J. Levine, W.C. Topp and J.D. Watson, Cold Spring Harbor Press, New York pp. 533-537.

71) Pinhasi-Kimhi, O., D. Michalovity, A. Ben-Zeev and M. Oren. (1986). Nature (London) **320**: 182-185.

72) Sarnow, P., Y.S. Ho, J. Williams and A.J. Levine (1982). Cell **28**: 387-394.

73) Ehiyahu, D., A. Raz, P. Gruss, D. Givol, M. Oren. (1984). Nature (London) **312**: 646-649.

74) Parada, L.F., H. Land, R.A. Weinberg, D. Wolf and V. Rotter (1984). Nature (London) **312**: 649-651.

75) Jenkins, J.R., K. Rudge, and G.A. Currie. (1984). Nature (London) **312**: 651-654.

76) Eliyahu, D., D. Michalovitz and M. Oren. (1985). Nature (London) **316.**: 158-160.

77) Kelekar, A. and M.D. Cole. (1986). Mol. Cell. Biol. **6**: 7-14.

78) Wolf, D., Harris, N. and V. Rotter (1984). Cell **38**: 119-126.

79) Kaczmarek, L., M. Oren and R. Baserga (1986). Expt. Cell. Res. **162**: 268-272.

80) Mercer, W.E., C. Avignolo and R. Baserga (1984). Mol. Cell. Biol. **4**: 276-281.

81) Milner, J. (1984). Nature (London) **310**: 143-145.

82) Reich, N. and A.J. Levine. (1984). Nature (London) **308**: 199-201.

83) Oren, M., W. Maltzman and A.J. Levine. (1981). Mol. Cell. Biol. **1**: 101-110.

84) Linzer, D.J.H. and A.J. Levine. (1979). Cell **17**: 43-52.

85) Van der Eb, A.J., R. Bernards, P.I. Schrier, J.L. Bos, R.T.M.J. Vaesen, A.G. Jochemsen and C.J.M. Melief (1984). Cancer Cells, vol II, Eds. G. Vande Woude, A.J. Levine, W.C. Topp and J.D. Watson, Cold Spring Harbor Press, New York, pp. 501-510.

86) Norby, E. (1969) J. Gen. Virol, **5**, 221

87) Ziff, E. B. (1980), Nature (London), **287**, 491-499

CHAPTER 21

THE MOLECULAR BIOLOGY OF POXVIRUSES

BERNARD MOSS
Laboratory of Viral Diseases, Natl. Inst. Allergy and Infectious Diseases, N.I.H., Bethesda, MD, 20892, U.S.A.

INTRODUCTION

Distinctive lesions present on the face and torso of the Egyptian mummy Ramses V have led to speculation that he died of smallpox (1). If true, man has been on intimate terms with poxviruses for at least 3,000 years. The recognition, perhaps first in India or China, that recurrent infections with smallpox were rare, led to the wide-spread practice of variolation (1). This procedure consisted of inoculating a susceptible individual, by nasal or percutaneous routes, with vesicular material from a mild case of smallpox which provided immunity to severe natural infection. An equally effective, but considerably safer procedure, was developed in 1796 by Edward Jenner in England (1). He was aware of the belief held by country folk that the transmittal of cowpox to milkmaids protected them from smallpox. To test this hypothesis directly, Jenner inoculated several boys with material from cowpox (or vaccinia) vesicles and subsequently challenged them by variolation. The absence of a smallpox lesion convinced Jenner of the prophlylactic value of vaccination. Indeed, the global adoption of this procedure finally led to the eradication of smallpox 80 years after Jenner's demonstration. With the end of this disease and the nearly complete cessation of vaccination, one might think that poxviruses are now of only historical interest. On the contrary, current research on this family of viruses is more vigorous than ever before. The reasons for continued activity include: recent technological advances that permit, for the first time, the molecular dissection of the large poxvirus genome; perceived experimental advantages of studying poxvirus transcription and replication mechanisms; the demonstrated value of poxviruses as expression vectors and their potential as live recombinant vaccines for medical and veterinary purposes; and last, but not least, curiosity as to how this unique family of viruses replicate and express their genomes in the cytoplasm of cells whereas other DNA viruses carry out these vital processes in the nucleus. Some recent reviews on the biology of poxviruses are referenced (2-5). This chapter will summarize previous work and provide an update on the most active research areas.

1. CLASSIFICATION

Members of the Poxviridae family are large, complex, double-stranded DNA viruses that replicate in the cytoplasm of infected cells. There are 2 subfamilies:

(a) Chordopoxvirinae (poxviruses of vertebrates), and

(b) Entemopoxvirinae (poxviruses of insects).

The Chordopoxvirinae share a common NP antigen (6, 7) and can rescue other heat-inactivated members of this subfamily by a process called non-genetic reactivation (8). The Chordopoxvirinae are subdivided into the 5 genera listed in Table 1. Except for the Orthopoxviruses, members of each genus have a rather limited host range. Similarities in DNA restriction endonuclease cleavage maps and/or stringent DNA hybridization have been demonstrated for members within some genera but not between genera. The Entemopoxvirinae share the common features of Poxviridae e.g. large size, complex virion structure with associated enzymes and long double-stranded DNA genome but they are not well characterized in molecular terms.

Members of the Orthopoxvirus genus have been studied most intensively and the family prototype is clearly vaccinia virus. Since this volume is primarily concerned with the molecular basis of virus replication, results obtained with other poxviruses will be mentioned only occasionally and for comparative purposes.

2. BASIC VIRION STRUCTURE

Poxviruses are the largest of all animal viruses and exhibit a complex brick-shaped structure lacking obvious icosohedral symmetry (Fig. 1). The main structural features include a dumbell-shaped core, one or two more or less distinct lateral bodies, and a lipoprotein envelope (4). A second outer envelope is also found in particles that have been extruded from the cell.

Table 1. - Family poxviridae

Subfamily	Genus	Prototype	Host
Chordopoxvirinae	Orthopoxvirus	Vaccinia	Mammal
	Parapoxvirus	Orf	Ungulate
	Avipoxvirus	Fowlpox	Bird
	Capripoxvirus	Sheeppox	Ungulate
	Leporipoxvirus	Myxoma	Leporid
	Suipoxvirus	Swinepox	Swine
Entemopoxvirinae	A	M. melontha	Coleoptera
	B	A. moori	Lepidoptera
	C	C. luridus	Diptera

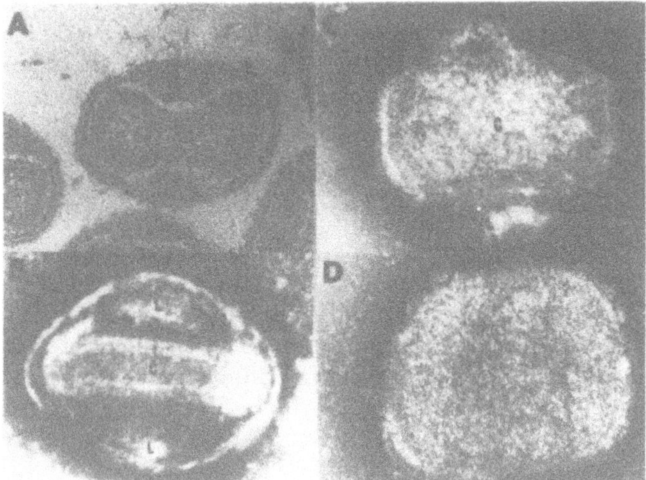

Figure 1. Electron microscopy of Vaccinia virus. Intact particle observed in thin section (A), or after negative stain (B). A virion deprived of its envelope (C), and the isolated core (D). L, lateral bodies; E, envelope, and C, the core. (Reproduced from ref. 183).

A. DNA Genome

Poxviruses contain a long double-stranded DNA genome that varies from 85 million daltons in Parapoxviruses to more than 200 million in Avipoxviruses (9). The genome of vaccinia virus is about 123 million daltons or 185,000 base pairs. Restriction endonuclease cleavage patterns of members of the Orthopoxvirus genus indicate a remarkable conservation of the central two-thirds of the genome (10, 11).

Genetic mapping indicates that the essential vaccinia virus genes including RNA polymerase subunits (12), DNA polymerase (13), and major structural proteins (14-17) as well as the sites of most temperature sensitive mutants (18, 19) are located in the central region. By contrast, the DNA near both ends of the genome is quite variable and spontaneous reiterations, transpositions and deletions readily arise under laboratory conditions (20-24).

For vaccinia virus and other poxviruses examined thus far, covalent links connect the two DNA strands (25). Nucleotide sequencing of the ends of vaccinia virus DNA demonstrated uninterrupted polydeoxynucleotide chains that take the form of incompletely base-paired hairpins (26). Interestingly, the hairpins exist in two isomeric forms that are inverted and complementary in sequence. This isomerism must arise during DNA replication and will be discussed later in this chapter. The hairpins of vaccinia (26) and Shope fibroma (27) viruses differ in sequence but they are both rich in adenylate and thymidylate residues and contain unpaired bases. There are similar sequences adjacent to the hairpins of vaccinia and Shope fibroma virus suggesting that this region may be important in replication or processing of the genome.

The presence of an inverted terminal repetition (28, 29) containing variable numbers of short direct tandem repeats (26, 30) as well as several early genes (31) is another feature of orthopoxvirus DNA. As might be expected because of their very close relationship, the repeated sequences of cowpox and vaccinia virus are very similar (32).

B. Virion Polypeptides

The complexity of the poxvirus particle is reflected in the large (greater than 100) number of polypeptides that can be resolved by polyacrylamide gel electrophoresis (33, 34). The majority of analyses have been carried out with the infectious intracellular form of vaccinia virus. Among the many proteins are an abundant phosphoprotein of 11,000 daltons (11 kDa) located in the core, a glycoprotein of 40 kDa apparently containing only N-acetylglucosamine and located near or at the surface of the particle, and 14 kDa and 58 kDa proteins on the surface which are capable of eliciting neutralizing antibody (35-39). At least 4 proteins are strongly bound to the genome and may help to maintain the DNA in a folded state (40).

The extracellular form of vaccinia virus contains an outer envelope with at least 8 additional polypeptides, seven of which are glycosylated (41-44). The genes for the hemagglutinin, which has a molecular weight of 85 kDa and both O- and N-linked oligosaccharide chains, and the abundant non-glycosylated 37 kDa protein, have been sequenced (45, 46).

Poxvirus particles contain a large number of enzymes many of which are involved in the synthesis and modification of mRNA (Table 2). The enzymes have been localized in the core as is needed for their interaction with the DNA genome. For the enzymes that have been purified, rough estimates of about 100 molecules per virus particle have been made. The properties and possible roles of some of the virion enzymes will be discussed in later sections.

3. THE INFECTIOUS CYCLE

A. Virus Entry into Cells

Both the intracellular form of vaccinia virus isolated following mechanical lysis of cells and the extracellular form harvested from the culture medium are infectious. Proteolytic treatment of vaccinia virus appears to enhance virus penetration into the cells (47). The outer coat of isolated intracellular virus has been observed to fuse with the plasma membrane at the surface of cells or within a vacuole formed by invagination of the membrane (48, 49). In either case, the virus core is released into the cytoplasmic matrix. More rapid and efficient adsorption of the extracellular form of vaccinia virus has been reported (50). The extracellular form of vaccinia virus may have an important role in the spread of poxvirus infections within animals (51, 52). A claim that the epidermal growth factor receptor is used by vaccinia virus to gain entry into the cell has been made (53) but this is clearly not obligatory since the virus can enter cells lacking the receptor (54). Moreover, a mutant virus that does not make vaccinia growth factor, which interacts with the

Table 2. - Family poxviridae

Enzyme	Subunit (kDa)	Ref.
RNA polymerase	17-140	78,79
Poly (A) polymerase	30,55	165
RNA guanylytransferase	95,31	166-168
RNA (guanine-7-) methyltransferase	95,31	166
RNA (nucleoside-2'-) methyltransferase	38	169,170
5'-phosphate polynucleotide kinase		171
DNA-dependent ATPase (NPH I)	61	172,173
DNA/RNA-dependent NTPase (NPH II)	68	172,173
Endoribonuclease		174
Deoxyribonuclease	50	175-178
DNA topoisomerase I	44	179
Protein kinase	62	180,181
Alkaline protease		182

receptor, is infectious (Chakrabarti, S. and Moss, B., unpublished). The vaccinia growth factor will be discussed further below.

B. Early Transcription

The strategy used by cytoplasmic DNA and RNA viruses to express their genes is so widely known now that it is difficult to appreciate the novelty of the finding, nearly 20 years ago, of a transcription system in vaccinia virus. At the time, virion proteins were thought to function as a protective casing for the genome and to facilitate virus adsorption to cells. The critical experiment demonstrated RNA synthesis when purified vaccinia virions were incubated with ribonucleoside triphosphates (55, 56). Evidence for transcriptase activities was subsequently obtained by similar methods with double-stranded and negative single-stranded RNA viruses as well as retroviruses (see other chapters in this volume).

The virus transcription systems provided a unique source of pure mRNA before general methods of isolating cellular mRNA had been developed. Indeed, poly(A) was first found at the 3' end of vaccinia virus mRNA (57) and this led to the development of oligo(dT) cellulose and poly(U) agarose chromatography methods for purifying mRNA. For similar technical reasons, the precise structure of the methylated cap at the 5' ends of mRNAs was initially determined for vaccinia virus (58) and reovirus (see chapter on the biology of reoviruses).

The presence of viral mRNA in the cytoplasm of infected cells within minutes after infection results from endogenous transcription within virus cores (Fig. 1). Addition of protein synthesis inhibitors, such as cycloheximide, prevents the uncoating of virus cores and actually enhances early RNA synthesis (59). Moreover, the polypeptides made by reticulocyte cell-free extracts programmed with vaccinia RNA made *in vitro* by permeabilized cores or in vivo in the presence of cycloheximide are virtually indistinguishable (60). Since vaccinia virus cores contain capping and methylating enzymes as well as poly(A) polymerase, viral mRNAs closely resemble their eukaryotic counterparts despite their cytoplasmic site of formation.

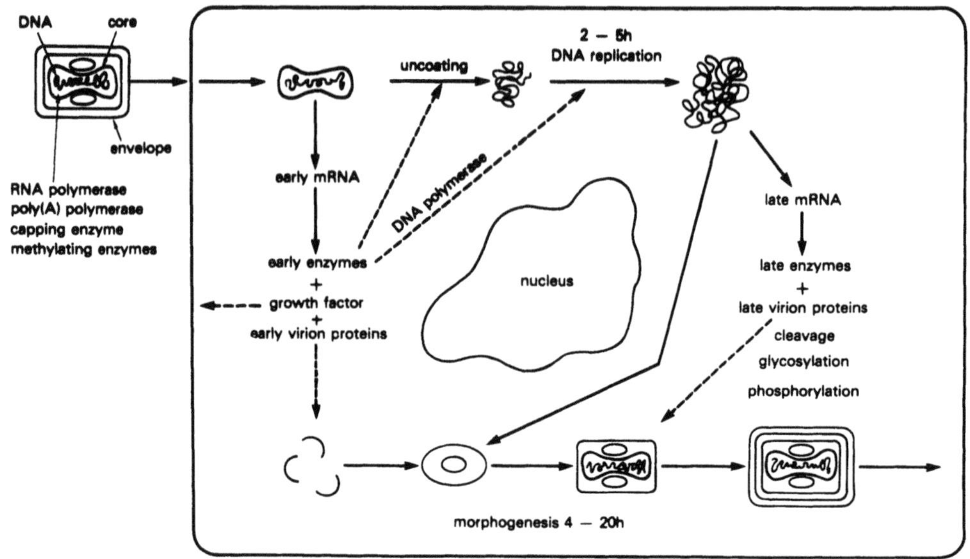

Figure 2. Vaccinia virus replication cycle.

DNA-RNA hybridization studies suggest that about half of the vaccinia genome is expressed prior to DNA replication (61-64). Further progress in analyzing vaccinia mRNAs followed the preparation of restriction endonuclease maps of the genome (65, 66). The physical location of early genes was determined by hybridization of early mRNAs to cloned restriction fragments (67). The specifically bound mRNAs were eluted and translated in reticulocyte extracts. These experiments established that early genes were distributed throughout the length of the genome and confirmed the large number of early mRNA species.

The next important steps in analysis of early transcription involved the mapping and sequencing of representative early genes and their flanking regions as well as the determination of the locations of the 5' and 3' ends of the corresponding transcripts (68-70). The following points were made:

i) none of the early genes examined had introns;
ii) the sequences just upstream of the coding regions were extremely rich in adenylate and thymidylate residues but did not closely resemble eukaryotic promoters; and
iii) the sequence AATAAA which forms part of the conserved eukaryotic RNA processing/polyadenylation signal was absent from the 3' terminal regions.

In order to identify the transcription signals, functional studies were needed. One approach involved the excision of putative promoter regions, e.g. the RNA start site and upstream sequences, and their ligation to test genes such as chloramphenicol acetyltransferase (CAT). These chimeric structures were inserted into the thymidine kinase (TK) locus of the vaccinia genome by homologous recombination and TK^- recombinant virus was isolated by plaque assay on TK^- cells in the presence of 5-bromodeoxyuridine. The latter drug is lethally incorporated into the DNA of wild-type TK^+ virus. When cells were infected with the recombinant virus, expression of CAT was regulated as a normal early vaccinia virus gene indicating that the promoter region was translocatable (71). By using shorter and shorter pieces of DNA, the signals were located within about 30 base pairs (bp) upstream of the RNA start sites (72). Mutagenesis of individual nucleotides is now required to precisely identify the signals.

Another recent approach involves the development of *in vitro* transcription systems. The permeabilized virus core was used to demonstrate that the 5' ends of early transcripts are unprocessed except for the addition of the cap. However, to study transcriptional regulatory signals, a soluble template dependent system was needed. Although methods of lysing the virus core with deoxycholate, removing endogenous DNA and isolation of active enzymes had been achieved many years ago, the cloning and identification of early genes were needed to prepare proper templates. Success in transcribing early genes was first achieved with extracts of infected cells (73, 74) and then with extracts prepared from deoxycholate-treated purified virus particles (75, 76). With the latter system, evidence was obtained that transcription was specific for vaccinia early genes and that the RNA start sites corresponded to those of in vivo mRNAs. Deletion mutagenesis also demonstrated that the same regulatory signals were required for *in vitro* and *in vivo* early transcription (Rohrmann, G. and Moss, B., unpublished).

The solubilized virion transcription system has been particularly useful for analyzing termination. When whole vaccinia virus early genes were used as templates, the major RNA species corresponded in size to mRNAs made *in vivo* and also were polyadenylated (77). Kinetic analysis suggested that the 3' ends were formed by termination rather than cleavage of run-off transcripts. The terminal addition of adenylate residues was non-specific and even unrelated RNAs added to the *in vitro* system were polyadenylated.

The signal for termination was located about 30 to 50 base pairs upstream of termination sites and the optimal sequence was shown by *in vitro* mutagenesis to be ATTTTTAT (Yuen, L. and Moss, B., unpublished). Deletion of a single T eliminated termination whereas substitution of other nucleotides for the A residues only lowered the efficiency. Remarkably, there appears to be no stringent sequence requirements at the actual site of termination.

The RNA polymerase has been isolated from vaccinia virus cores as a multisubunit protein of about 500 kDa (78, 79) and seven subunits have been shown to be virus encoded (Jones, E.V., Puckett, C. and Moss, B., unpublished). The large subunit of vaccinia RNA polymerase has extensive sequence homology with the large subunit of *E. coli* and yeast RNA polymerases (12). The homology, however, was greatest with the eukaryotic subunit. Further comparisons will be possible when the sequences of other eukaryotic RNA polymerase subunits are reported.

Despite the complexity of the vaccinia RNA polymerase, highly purified preparations can only transcribe single stranded DNA in a non-specific manner and require Mn^{++} instead of Mg^{++} (78, 79). Attempts to isolate other factors necessary for accurate transcription are in progress.

C. Late Transcription

If vaccinia virus infected cells are pulse-labeled with amino acids at intervals after infection, a changing pattern of polypeptides is revealed by autoradiographic analysis (80, 81). Within just a few hours, the appearance of new virus polypeptides and the disappearance of host cell polypeptides occurs. The shut off of host protein synthesis probably involves competition by excess viral mRNA (64, 82, 83) as well as specific mechanisms (84-86). Between 4 and 6 hours after infection, new labeled polypeptides appear and the majority, but not all early bands are diminished (80, 81). This early to late shift can be reversibly blocked by inhibitors of DNA replication.

There is evidence that many of the virion structural polypeptides are expressed late in infection (87). Translational mapping studies indicate that late genes are distributed throughout the genome but are concentrated in blocks within the central region (67).

A major question concerns whether early/late regulation occurs at transcriptional or translational levels. RNA-DNA hybridization studies, which one might have expected to readily resolve this problem, have been difficult to interpret. The presence of new RNA species at late times was indeed found but early RNA species continued to be expressed (61-64). The early sequences, however, were in RNAs that were considerably larger in size at late times. In addition, complementary RNA species were present such that duplex structures readily formed (88, 89).

In contrast to the confusing picture obtained by physical and hybridization analysis of RNA, translational studies were consistent with transcriptional control (82). When mRNAs isolated at early and late times after infection were translated in reticulocyte lysates, the autoradiographic patterns were similar to those obtained by analysis of in vivo pulse-labeled polypeptides. Differences between hybridization and *in vitro* translation data were partially reconciled by translation of size-fractionated RNAs (82, 90, 91). It was found that mRNAs coding for individual polypeptides varied in size over a range of several thousand nucleotides. The early to late switch could therefore result from changes in sites of transcription initiation and the anomalous findings could result from inefficient transcription termination at late times. In support of this interpretation, the method of nuclease S1 analysis revealed putative RNA start sites just upstream of coding segments (14, 15, 17, 92, 93). However, other explanations involving RNA processing need to be more thoroughly investigated.

Nucleotide sequencing has revealed some features of late genes that distinguish them from early genes. Most notable is the occurrence of the sequence TAAATG (or TAAAT followed within a few nucleotides by ATG) at the beginning of the open-reading-frames of late genes (14-17, 92, 93). This sequence was surprising since a purine is invariably present 3 nucleotides upstream of eukaryotic translation initiation codons. In addition, the putative RNA start sites, mapped by nuclease S1 analysis, are located just a few nucleotides upstream of the ATG.

Also, the early transcription termination sequence TTTTTNT is not present near the ends of late coding regions (Yuen, L. and Moss, B., unpublished).

Late transcriptional regulatory regions have been defined by deletion mutagenesis. Both a helper vaccinia virus-dependent transient expression assay, in which plasmids containing putative promoters linked to the CAT coding sequence are transfected into cells infected with vaccinia virus (72, 94), or recombinant virus expression (92, 95-98) have been used. The data indicate that the required transcriptional regulatory region is quite short and may be less than 18 base pairs. Nevertheless, it should be emphasized that the mechanism of regulation of late gene expression is still poorly understood and that some novel features may soon emerge.

D. Host Factors in Virus Expression

There have been several studies suggesting a possible role of the nucleus or the host RNA polymerase in vaccinia virus gene expression. For example, in cells that are enucleated (99, 100) or treated with alpha amanitin (101) an inhibitor of eukaryotic RNA polymerase II, infectious virus particles are not formed (although considerable viral RNA, DNA and protein synthesis occur). Paradoxically, it has been reported in one laboratory that wild-type vaccinia virus can propagate in alpha-amanitin-resistant cells in the presence of the drug (102) and in another that an alpha-amanitin-resistant vaccinia mutant can replicate in drug-sensitive cells in the presence of the drug (103).

Claims of poxvirus DNA and RNA in the cell nucleus have been made (104, 105) and refuted (106). The biological significance of a report that the large subunit of cellular RNA polymerase II is specifically transported out of the nucleus and that a large proteolytic cleavage product is associated with the viral RNA polymerase in virus particles (107) also needs to be determined.

E. DNA Replication

The cytoplasmic site of viral DNA replication is a distinctive feature of poxviruses. Autoradiographic studies revealed discrete cytoplasmic foci of DNA synthesis (108). A vaccinia virus-encoded DNA polymerase is synthesized as an early gene product. The enzyme has a molecular weight of about 110 kDa and has associated 3' exonuclease activity (109). The enzyme is sensitive to phosphonoacetate (PAA) and aphidicolin but drug-resistant virus mutants have been isolated (110, 111). Marker transfer of PAA-resistance was used to map the DNA polymerase gene (13). The nucleotide sequence of the gene confirmed the size of the enzyme and revealed extensive amino acid homology with DNA polymerases of members of the herpesvirus family (112). Interestingly, a short but highly conserved amino acid sequence is present in vaccinia virus, herpesvirus, and adenovirus DNA polymerases.

Although the vaccinia virus thymidine kinase (TK) may enhance DNA replication by increasing the thymidylate pool under certain conditions, the gene is not essential in tissue culture cells (113). Indeed, the non-essential nature of the TK provides a highly useful selection procedure for isolating mutants and recombinant viruses (71). The nucleotide sequence of the vaccinia TK revealed extensive homology with human and avian TK genes but little or none with the herpes simplex virus TK gene (114, 115).

Since replication of vaccinia virus DNA can occur in enucleated cells (99, 100), it seems likely that additional factors are virus encoded. In this regard, several DNA- temperature sensitive mutants in genes other than DNA polymerase have been isolated (18, 19). The functions of these genes, however, have not yet been defined.

DNA replication can be detected within 1.5 hours after infection, peaks at about 4 hours and then declines (116, 117). Preliminary evidence suggests that nicks occur in parental vaccinia DNA soon after infection (118), replication begins at each end of the genome (see: Figure 10, chapter 3 of this volume and ref. 119), strand displacement mechanisms are involved and small DNA fragments are intermediates (120, 121). The hairpin structure at the ends of poxvirus DNA suggests a self-priming model (26) similar to that proposed for parvovirus DNA replication. For a detailed discussion of this model, the reader is referred to section 4, chapter 3 of this volume. Concatemeric forms of rabbitpox (122) and vaccinia (123) DNA have been detected and may be replicative intermediates. The junction connecting unit genomes in concatemers has been cloned and sequenced (124). When circular plasmids containing the concatemeric junctions were transfected into vaccinia virus infected cells, they were resolved into linear molecules with vaccinia hairpins at the ends and vector DNA in the center (27, 124). Similar results were obtained with cloned synthetic concatemer junction fragments of Shope fibroma virus (27). Thus there appears to be a specific mechanism for resolving concatemers into linear unit genomes.

Apparently any plasmid, regardless of sequence, that is transfected into cells infected with vaccinia or Shope fibroma virus will be replicated although not resolved into linears (124, 125). Whether this experimental result signifies that any region on the vaccinia genome can serve as a replication origin remains to be determined.

F. Virus Assembly and Release

In view of the large number of polypeptides and the complex structure of poxviruses, assembly must be a very complex process. The first morphologically distinct structures visualized by electron microscopy appear in thin sections as crescents (4). These structures consist of a bilayer membrane with a brush-like border of spicules on the convex surface and associated granular material in the concavity. If the drug rifampicin is added to infected cells, then only irregular forms lacking the spicule layer accumulate (126, 127). Upon removal of the drug, the membranes are converted within minutes to spicule-coated crescent forms (128). Rifampicin-resistant vaccinia mutants have been isolated (129, 130) and marker transfer procedures were used to locate the gene conferring resistance (131-133). Whether this gene encodes the spicule protein has not been determined. Immature particles with varying degrees of internal complexity have been visualized either during normal infection (4) or after synchronization achieved by removal of rifampicin (130).

Images suggesting the entry of the nucleoprotein into the immature particle have been observed (134).

A double-layered membrane derived from golgi surrounds many of the immature particles (44, 135-138) and some appear to be drawn by actin-containing microfilaments into the tips of specialized microvilli from which they are released into the medium (136, 138). These extracellular forms of vaccinia virus retain an external envelope derived from golgi.

G. Cell Proliferation

Many poxviruses induce cellular proliferation. The effect is particularly dramatic in the case of rabbit fibroma virus (139), Yaba tumor virus (140, 141) and fowlpox virus (142) but can be seen with other poxviruses as well. An explanation for the cell proliferation caused by vaccinia virus has recently come in a serendipitous fashion. Computer searches of derived amino acid sequences (143-145) revealed homology between the active sites of epidermal growth factor (EGF) and transforming growth factor alpha and an early gene located in the inverted terminal repetition (69). Further work demonstrated that this vaccinia protein, now termed vaccinia growth factor or VGF, binds to EGF receptors on the surface of cells and stimulates tyrosine protein kinase activity which leads to cell proliferation (54, 146, 147). A mutant vaccinia virus with deletions inactivating both copies of VGF has been constructed (S. Chakrabarti, M. Buller, and B. Moss, unpublished). This mutant virus forms plaques and grows to normal titers in cell culture but has a markedly lower cell proliferative effect in rabbit skin and chick chorioallantoic membrane.

4. POXVIRUS EXPRESSION VECTORS

The development of vaccinia virus as an expression vector (148, 149) followed the application of marker transfer procedures for introducing DNA into the vaccinia virus genome by homologous recombination (150-152). Expression of foreign genes within the vaccinia virus genome may occur because of transcriptional regulatory elements at or near the site of insertion or by other means of genetic engineering. Plasmid vectors that facilitate insertion and expression of foreign genes have been constructed (71, 153, 154). These vectors contain an expression site, composed of a vaccinia virus promoter and one or more unique restriction endonuclease sites for insertion of the foreign coding sequence, flanked by DNA from a non-essential region of the vaccinia genome. The choice of promoter determines both the time and level of expression whereas the flanking DNA sequence determines site of homologous recombination. Although recombinant virus plaques can be identified by DNA hybridization, TK selection (71) as described in an earlier part of this chapter is more efficient. Plasmid vectors that contain the *E. coli* β-galactosidase gene, as well as an expression site for a second gene, permit an alternative or supplemental way of picking recombinant plaques (154). Plaques formed by such recombinants can be stained blue by addition of an appropriate indicator to the agar.

As a vector, vaccinia virus has a number of characteristics that may be used to advantage. These include a large capacity for added DNA (155), non-essential sites for insertion so that infectivity is retained (22, 23, 113, 156), and a wide host range. Experience expressing a variety of viral and mammalian proteins indicates that proper synthesis, folding, processing, glycosylation and transport occur (e.g. ref. 157-160). Vaccinia virus vectors have been exploited to greatest advantage for immunological studies involving the determination of targets of neutralizing antibody and cytotoxic T cells (161, 162, 163). Experimental animals vaccinated with recombinant vaccinia viruses expressing proteins from other agents have been protected against hepatitis B, influenza, herpes simplex, vesicular stomatitis, rabies, and respiratory syncytial viruses (reviewed in ref. 162). The success of these experiments make it logical to consider the use of vaccinia (or other poxviruses) as vaccines for veterinary and medical uses. Such vaccines would be economical to produce, *do not require refrigeration during transport*, and be simple to administer. Aside from establishing efficacy, the most important concern is with safety. Experience with vaccinia virus as a smallpox vaccine indicated that serious complications were uncommon. Nevertheless, present standards require that vaccinia virus be further attenuated. In this regard, evidence has already been obtained that disruption of the TK gene and other deletions lead to marked decreases in virulence for experimental animals (164). With further information regarding non-essential vaccinia virus genes, it should be possible to achieve a good compromise between retention of immunogenicity and safety.

5. CONCLUSIONS

The poxviruses are large DNA viruses that have the remarkable ability of replicating and expressing their genes in the cytoplasm of vertebrate and insect cells. Research on these viruses is intense and can be expected to reveal new information particularly with regard to mechanisms of transcription and replication. Because of its proven and potential use as expression vectors, poxviruses have become of interest to other virologists, molecular biologists, immunologists, and infectious disease specialists.

6. REFERENCES

1) Hopkins, D.R. (1983), *Princes & Peasants*, The University of Chicago Press, Chicago.

2) Moss, B. (1985), in: *Virology*, B.N. Fields, ed., Raven Press, N.Y., 685-703.

3) Fenner, F. (1985), in: *Virology*, B.N. Fields, ed., Raven Press, N.Y., 681-684.

4) Dales, S. & Pogo, B.G.T. (1981), *Biology of Poxviruses*, D.W. Kingsbury & H. Zur Hausen, ed., Springer-Verlag, N.Y.

5) Andrewes, C., Periera, H.G. & Wildy, P. (1978), *Viruses of Vertebrates*, Balliere Tindall, London, 356-389.

6) Takehashi, M., Kameyama, S., Kato, S. & Kamahora, J. (1959), Biken J **2**, 27-29.

7) Woodroofe, G.M. & Fenner, F. (1962), Virology **16**, 334-341.

8) Fenner, F. & Woodroofe, G.M. (1960), Virology **11**, 185-201.

9) Wittek, R. (1982), Experentia **38**, 285-410.

10) Mackett, M. & Archard, L.C. (1979), J. Gen. Virol. **45**, 683-701.

11) Esposito, J.J. & Knight, J.C. (1985), Virology **143**, 230-251.

12) Broyles, S. & Moss, B. (1986), Proc. Natl. Acad. Sci. USA, **83**, 3141-3145.

13) Jones, E.V. & Moss, B. (1983), J. Virol. **49**, 72-77.

14) Wittek, R. Hanggi, M., & Hiller, G. (1984), J. Virol. **49**, 371-378.

15) Wittek, R., Richner, B. & Hiller, G. (1984), Nucleic Acid Res. **12**, 4835-4848.

16) Weir, J.P. & Moss, B. (1985), J. Virol. **56**, 534-540.

17) Rosel, J. & Moss, B. (1985), J. Virol. **56**, 830-838.

18) Ensinger, M.J. & Rovinsky, M. (1983), J. Virol. **48**, 419-428.

19) Thompson, C.L. & Condit, R.C. (1986), Virology **150**, 10-20.

20) Moyer, R.W. & Rothe, C.T. (1980), Virology **102**, 119-132.

21) Moyer, R.W., Graves, R.L. & Rothe, C.T. (1980), Cell **22**, 545-553.

22) Panicali, D., Davis, S.W., Mercer, S.R. & Paoletti, E. (1981), J. Virol **37**, 1000-1010.

23) Moss, B., Winters, E. & Cooper, J.A. (1981), J. Virol. **40**, 387-395.

24) Moss, B., Winters, E. & Cooper, N. (1981), Proc. Natl. Acad. Sci. USA **78**, 1614-1618.

25) Geshelin, P. & Berns, K.I. (1974), J. Mol. Biol. **88**, 785-796.

26) Baroudy, B.M., Venkatesan, S. & Moss, B. (1982), Cell **28**, 315-324.

27) DeLange, A.M., Reddy, M., Scraba, D., Upton, C. & McFadden, G. (1986), **59**, 249-259.

28) Wittek, R., Menna, A., Muller, K., Schumperli, D., Bosley, P.G. & Wyler, R. (1978), J. Virol. **28**, 171-181.

29) Garon, C.F., Barbosa, E. & Moss, B. (1978), Proc. Natl. Acad. Sci. USA 75, 4863-4867.

30) Wittek, R. & Moss, B. (1980), Cell **21**, 277-284.

31) Cooper, J.A., Wittek, R. & Moss, B. (1981), J. Virol. **37**, 284-294.

32) Pickup, D.J., Bastia, D., Stone, H.O., & Joklik, W.K. (1982), Proc. Natl. Acad. Sci. USA **79**, 7112-7116.

33) Essani, K. & Dales, S. (1979), Virology **95**, 385-394.

34) Oie, M. & Ichihashi, Y. (1981), Virology **113**, 263-276.

35) Sarov, I. & Joklik, W.K. (1972), Virology **50**, 579-592.

36) Garon, C.F. & Moss, B. (1971), Virology **46**, 233-246.

37) Rosemond, H. & Moss, B. (1973), J. Virol. **11**, 961-970.

38) Stern, W. & Dales, S. (1976), Virology **75**, 232-241.

39) Rodriguez, J.F., Janeczko, R.A., & Esteban, M. (1985), J. Virol. **56**, 352-356.

40) Soloski, M.J. & Holowczak, J.A. (1981), J. Virol. **37**, 770-783.

41) Payne, L. (1978), J. Virol. **27**, 28-37.

42) Payne, L. (1979), J. Virol. **31**, 147-155.

43) Shida, H. & Dales, S. (1981), Virology **111**, 56-72.

44) Hiller, G. & Weber, K. (1985), J. Virol. **55**, 651-659.

45) Hirt, P., Hiller, G. & Wittek, R. (1986), J. Virol. **58**, 757-764.

46) Shida, H. (1986), Virology **150**, 451-462.

47) Ichihashi, Y. & Oie, M. (1982), Virology **116**, 297-305.

48) Chang, A. & Metz, D.H. (1976), J. Gen. Virol. **32**, 275-282.

49) Dales, S. & Kagioka, R. (1964), Virology **24**, 278-294.

50) Payne, L.G. & Norrby, E. (1978), J. Virol. **27**, 19-27.

51) Boulter, E.A. & Appleyard, G. (1973), Prob. Med. Virol. **16**, 86-108.

52) Payne, L.G. (1980), J. Gen Virol. **50**, 89-100.

53) Epstein, D., Marsh, Y.V., Schreiber, A.B., Newman, S.R., Todaro, G.J., & Nestor, J.J., Jr. (1985), Nature (London) **318**, 663-665.

54) Stroobant, P., Rice, A., Gullick, W.J., Cheng, D.J., Kerr, I.M., & Waterfield, M.D. (1985), Cell **42**, 383-393.

55) Kates, J.R. & McAuslan, B. (1967), Proc. Natl. Acad. Sci. USA **58**, 134-141.

56) Munyon, W.E., Paoletti, E. & Grace, J.T., Jr. (1967), Proc. Natl. Acad. Sci. USA **58**, 2280-2288.

57) Kates, J. & Beeson, J. (1970), J. Mol. Biol. **50**, 19-23.

58) Wei, C.M. & Moss, B. (1975), Proc. Natl. Acad. Sci. USA **72**, 318-322.

59) Woodson, B. (1967), Biochem. Biophys. Res, Commun. 27, 169-175.

60) Cooper, J.A. & Moss, B. (1978), Virology. **88**, 149-165.

61) Oda, K. & Joklik, W.K. (1967), J. Mol. Biol. **27**, 395-419.

62) Kaverin, N.V., Varich, N.L., Surgay, V.V. & Chernos, V.I. (1975), Virology **65**, 112-119.

63) Paoletti, E. & Grady, L.J. (1977), J. Virol. **23**, 608-615.

64) Boone, R.F. & Moss, B. (1978), J. Virol. **26**, 554-569.

65) Wittek, R., Menna, A., Schumperli, D., Stoffel, S., Muller, H.K. & Wyler, R. (1977), J. Virol. **23**, 669-678.

66) DeFilippes, F.M. (1982), J. Virol. **43**, 136-139.

67) Belle Isle, H., Venkatesan, S. & Moss, B. (1981), Virology **112**, 306-317.

68) Venkatesan, S., Baroudy, B.M. & Moss, B. (1981), Cell 805-813.

69) Venkatesan, S., Gershowitz, A. & Moss, B. (1982), J. Virol. **44**, 637-646.

70) Weir, J.P. & Moss, B. (1983), J. Virol. **46**, 530-537.

71) Mackett, M., Smith, G.L. & Moss, B. (1982), J. Virol. **49**, 857-864.

72) Cochran, M.A., Puckett, C. & Moss, B. (1985), J. Virol. **54**, 30-37.

73) Puckett, C. & Moss, B. (1983), Cell **35**, 441-448.

74) Fogelsong, P.D. (1985), J. Virol. **53**, 822-826.

75) Golini, F. & Kates, J.R. (1985), J. Virol. **52**, 205-213.

76) Rohrmann, G. & Moss, B. (1985), J. Virol. 56, 349-355.

77) Rohrmann, G., Yuen, L., & Moss, B. (1986), Cell in press.

78) Baroudy, B.M. & Moss, B. (1980), J. Biol. Chem. **225**, 4372-4380.

79) Spencer, E., Shuman, S. & Hurwitz, J. (1980), J. Biol. Chem. **225**, 5388-5395.

80) Moss, B. & Salzman, N.P. (1968), J. Virol. **2**, 1016-1027.

81) Pennington, T.H. (1974), J. Gen. Virol. **25**, 433-444.

82) Cooper, J.A. & Moss, B. (1970), Virology **96**, 368-380.

83) Rice, A.P. & Roberts, B.E. (1983), J. Virol. **14**, 704-708.

84) Moss, B. (1968), J. Virol. **2**, 1028-1037.

85) Bablanian, R., Coppola, G., Scribani, S. & Esteban, M. (1981), Virology **112**, 13-24.

86) Person, A., Ben-Hamida, F., & Beaud, G. (1980), Nature (London) **287**, 355-357.

87) Moss, B., Rosenblum, E.N. & Garon, C.F. (1973), Virology **55**, 143-156.

88) Colby, C., Jurale, C. & Kates, J.R. (1971), J. Virol. **7**, 71-76.

89) Boone, R., Parr, R.P. & Moss, B. (1981), J. Virol. **39**, 365-374.

90) Cooper, J.A., Wittek, R., & Moss, B. (1981), J. Virol. **39**, 733-745.

91) Mahr, A. & Roberts, B.E. (1984), J. Virol. 49, 510-520.

92) Weir, J.P. & Moss, B. (1984), J. Virol. **51**, 662-669.

93) Rosel, J. & Moss, B. (1986), J. Virol. in press.

94) Cochran, M.A., Mackett, M. & Moss, B. (1985), Proc. Natl. Acad. Sci. USA **82**, 19-23.

95) Bertholet, C., Drillien, R. & Wittek, R. (1985), Proc. Natl. Acad. Sci. USA **82**, 2096-2100.

96) Bertholet, C., Stocco, P., Van Meir, E. & Wittek, R. (1986), EMBO J. **5**, 1951-1958.

97) Hanggi, M., Bannwarth, W. & Stunnenberg, H.G. (1986), EMBO J. **5**, 1071-1076.

98) Weir, J.P. & Moss, B. J. Virol. in press.

99) Prescott, D.M., Kates, J. & Kirkpatrick, J.B. (1971), J. Mol. Biol. **59**, 505-508.

100) Pennington, T.H. & Follett, E.A. (1974), J. Virol. **13**, 488-493.

101) Hruby, D.E., Lynn, D.L. & Kates, J.R. (1979), Proc. Natl. Acad. Sci. USA **76**, 1887-1890.

102) ilver, M., McFadden, G., Wilton, S. & Dales, S. (1979), Proc. Natl. Acad. Sci. USA **76**, 4122-4125.

103) Villarreal, E.C. & Hruby, D.E. (1986), J. Virol. **57**, 65-70.

104) La Colla, B. & Weisbach, A. (1975), J. Virol. **15**, 305-315.

105) Gafford, L.G. & Randall, C.C. (1976), Virology **69**, 1-14.

106) Minnigan, H. & Moyer, R.W. (1985), J. Virol. **55**, 634-643.

107) Morrison, D.K. & Moyer, R.W. (1986), Cell **44**, 587-596.

108) Cairns, J. (1960), Virology **11**, 603-623.

109) Challberg, M.D. & England, P.T. (1979), J. Biol. Chem. **254**, 7812-7819.

110) Moss, B. & Cooper, N. (1982), J. Virol. **43**, 673-678.

111) De Filippes, F.M. (1984), J. Virol. **52**, 474-482.

112) Earl, P., Jones, E.V. & Moss, B. (1986), Proc. Natl. Acad. Sci. USA **83**, 3659-3664.

113) Dubbs, D.R. & Kit, S. (1964), Virology, **22**, 214-225.

114) Weir, J.P. & Moss, B. (1983), J. Virol. **46**, 530-537.

115) Hruby, D.E., Maki, R.A., Miller, D.B. & Ball, L.A. (1983), Proc. Natl. Acad. Sci. USA **80**, 3411-3415.

116) Salzman, N.P. (1960), Virology **10**, 150-152.

117) Joklik, W.K. & Becker, Y. (1964), J. Mol. Biol. 10, 452-474.

118) Pogo, B.G.T. (1980), Virology **101**, 520-524.

119) Pogo, B.G.T., O'Shea, M. & Freimuth, P. (1981), Virology **108**, 241-248.

120) Esteban, M. & Holowczak, J.A. (1977), Virology **78**, 57-75.

121) Holowczak, J.A. (1982), Curr. Topics Microbiol. Immunol. **97**, 27-79.

122) Moyer, R.W. & Graves, R.L. (1982), Cell **27**, 391-401.

123) Moss, B., Winters, E. & Jones, E.V. (1983), In: *Mechanisms of DNA Replication and Recombination*, N. Cozzarelli, ed., A. Liss, New York, 449-461.

124) Merchlinsky, M. & Moss, B. (1986), Cell **45**, 879-884.

125) DeLange, A.M. & McFadden, G. (1986), Proc. Natl. Acad. Sci. USA **83**, 614-618.

126) Moss, B., Rosenblum, E.N., Katz, E. & Grimley, P.M. (1969), Nature (London) **224**, 1280-1284.

127) Nagayama, A., Pogo, B.G.T. & Dales (1970), Virology **40**, 1039-1051.

128) Grimley, P.M., Rosenblum, E.N., Mims, S.J. & Moss, B. (1970), J. Virol. **6**, 519-533.

129) Subak-Sharpe, T.H., Timbury, M.C. & Williams, J.F. (1969), Nature (London) **222**, 341-345.

130) Moss, B., Rosenblum, E.N. & Grimley, P.M. (1971), Virology **45**, 135-148.

131) Tartaglia, J. & Paoletti, E. (1985), Virology **150**, 394-404.

132) Tartaglia, J., Piccini, A. & Paoletti, E. (1986), Virology **150**, 45-54.

133) Baldick, C.J. & Moss, B. (1987), Virology in press.

134) Morgan, C. (1976), Science **193**, 591-592.

135) Ichihashi, Y., Matsumoto, S. & Dales, S. (1971), Virology **46**, 507-532.

136) Stokes, G.V. (1976), J. Virol. **18**, 636-642.

137) Morgan, C. (1976), Virology 73, 43-58.

138) Hiller, G., Jungwirth, C. & Weber, K. (1981), Exp. Cell Research **132**, 81-87.

139) Shope, R.E. (1932), J. Exp. Med. **56**, 803-822.

140) Bearcroft, W.G.C. & Jamieson, M.F. (1958), Nature (London), **182**, 195-196.

141) Andrewes, C., Allison, A.C., Armstrong, J.A., Bearcroft, G., Niven, J.S.F. & Pereira, H.S. (1959), Acto Unio Int. Contra Cancrum **15**, 760-763.

142) Cheevers, W.P., O'Callaghan, D.J. & Randall, C.C. (1968), J. Virol. **2**, 421-429.

143) Blomquist, M.C., Hunt, L.T., & Barker, W.C. (1984), Proc. Natl. Acad. Sci. USA **81**, 7363-7367.

144) Brown, J.P., Twardzik, D.R., Marquardt, H. & Todaro, G.J. (1985), Nature (London) **313**, 491-492.

145) Reisner, A.H. (1985), Nature (London) **313**, 801-803.

146) Twardzik, D.R., Brown, J.P., Ranchalis, J.E., Todaro, G.J. & Moss, B. (1985), Proc. Natl. Acad. Sci. USA **82**, 5300-5304.

147) King, C.S., Cooper, J.A., Moss, B. & Twardzik, D.R. (1986), Mol. Cell. Biol. **6**, 332-336.

148) Panicali, D. & Paoletti, E. (1982), Proc. Natl. Acad. Sci. USA **79**, 4927-4931.

149) Mackett, M., Smith, G.L. & Moss, B. (1982), Proc. Natl. Acad. Sci. USA **79**, 7415-7419.

150) Sam, C.K. & Dumbell, K.R. (1981), Ann. Vir. **132E**, 135-150.

151) Nakano, E., Panicali, D., & Paoletti, E. (1982), Proc. Natl. Acad. Sci. USA 79, 1593-1596.

152) Weir, J.P., Bajszar, G. & Moss, B. (1982), J. Virol. **46**, 530-537.

153) Mackett, M., Smith, G.L. & Moss, B. (1985), in: *DNA Cloning, vol II*, D.M. Glover, ed., IRL Press, Oxford, 191-212.

154) Chakrabarti, S., Brechling, K. & Moss, B. (1985), Mol. Cell. Biol. **5**, 3403-3409.

155) Smith, G.L. & Moss, B. (1984), Gene **25**, 21-28.

156) Perkus, M.E., Panicali, D., Mercer, S. & Paoletti, E. (1986), Virology **152**, 285-297.

157) Smith, G.L., Mackett, M. & Moss, B. (1983), Nature (London) **302**, 490-495.

158) Stephens, E.B., Compans, R.W., Earl, P. & Moss, B. (1986), EMBO J. **5**, 237-245.

159) Ball, L.A., Young, K.K.Y., Anderson, K., Collins, P.L.,& Wertz, G.W. (1986), Proc. Natl. Acad. Sci. USA **83**, 246-250.

160) Elango, N., Prince, G.A., Murphy, B.R., Venkatesan, S., Chanock, R.M., & Moss, B. (1986), Proc. Natl. Acad. Sci. USA **83**, 1906-1910.

161) Moss, B. (1986), Immunol. Today **6**, 243-245.

162) Moss, B. & Flexner, C.W. (1987), Ann. Rev. Immunol., in press.

163) Yewdell, J.W., Bennink, J.R., Smith, G.L. & Moss, B. (1985), Proc. Natl. Acad. Sci. USA **82**, 1785-1789.

164) Buller, R.M.L., Smith, G.L., Cremer, K., Notkins, A.L. & Moss, B. (1985), Nature (London) **317**, 813-815.

165) Moss, B., Rosenblum, E.N., & Gershowitz, A. (1975), J. Biol. Chem. **250**, 4722-4729.

166) Martin, S.A., Paoletti, E. & Moss, B. (1975), J. Biol. Chem. **250**, 9322-9329.

167) Venkatesan, S., Gershowitz, A., & Moss, B. (1980), J. biol. Chem. **255**, 903-908.

168) Shuman, S. & Hurwitz, J. (1981), Proc. Natl. Acad. Sci. USA **78**, 187-191.

169) Barbosa, E. & Moss, B. (1978), J. Biol. Chem. **253**, 7692-7697.

170) Barbosa, E. & Moss, B. (1978), J. Biol. Chem. **253**, 7698-7702.

171) Spencer, E., Loring, D., Hurwitz, J., Monroy, G. (1978), Proc. Natl. Acad. Sci. USA, **75**, 4793-4797.

172) Paoletti, E., Rosemond-Hornbeak, H., & Moss, B. (1974), J. Biol. Chem. **249**, 3273-3280.

173) Paoletti, E. & Moss, B. (1974), J. Biol. Chem. **249**, 3281-3286.

174) Paoletti, E. & Lipinskas, B.R. (1978), J. Virol. **26**, 822-824.

175) Pogo, B.G.T. & O'Shea, M.T. (1977), Virology, **77**, 55-66.

176) Rosemond-Hornbeak, H., Paoletti, E. & Moss, B. (1974), J. Biol. Chem. **249**, 3287-3291.

177) Rosemond-Hornbeak, H. & Moss, B. (1974), J. Biol. Chem. **249**, 3292-3296.

178) Lakritz, N., Fogelsong, P.D., Reddy, M., Baum, S., Hurwitz, J., & Bauer, W.R. (1986), J. Virol. **53**, 953-943.

179) Bauer, W.R., Ressner, E.C., Kates, J. & Patzke, V.J. (1977), Proc. Natl. Acad. Sci. USA **74**, 1841-1845.

180) Kleiman, J.H. & Moss, B. (1975), J. Biol. Chem. **250**, 2420-2429.

181) Kleiman, J.H. & Moss, B. (1975), J. Biol. Chem. **250**, 2430-2437.

182) Arzoglou, P., Drillien, R. & Kirn, A. (1978), Virology **95**, 211-214.

183) Pogo, B.G.T., & Dales, S. (1969), Proc. Natl. Acad. Sci. USA, **63**, 820.

CHAPTER 22

HERPESVIRUSES: BIOLOGY, GENE REGULATION, LATENCY, AND GENETIC ENGINEERING

BERNARD ROIZMAN, FRANK J. KENKINS, AND THOMAS M. KRISTIE

The Marjorie B. Kovler Viral Oncology Laboratories, The University of Chicago, 910 East 58th Street, Chicago IL 60637, USA

1. THE FAMILY HERPESVIRIDAE

The family *Herpesviridae* includes a large number of structurally related DNA viruses classified on the basis of the architecture of the virion (1). Their hosts range from lower vertebrates to primates, and many species, including humans become infected with more than one Herpevirus. Members of this family have been known for over 60 years but only in the last decade have we begun to understand the full range of manifestations caused by these viruses as well as many fascinating aspects of their molecular biology. Although we know more about the human Herpesviruses (Herpes Simplex Viruses 1 and 2) (HSV-1 and HSV-2), Epstein-Barr Virus (EBV), Varicella-Zoster, and the Human Cytomegalovirus (HCMV) than about those infecting other species, the accrued knowledge is predictive of the properties of the family as a whole.

2. THE HERPESVIRION AND ITS COMPONENTS

A. Architecture, Composition, and Requirements for Infection

With few exceptions, the various members of the Herpesviridae cannot be differentiated with respect to the morphology. At the center of the virion is a torroidal structure consisting of DNA, polyamines, and proteins suspended by protein fibrils anchored in the underside of an icosadeltahedral protein capsid (2, 3).

The capsid is surrounded by an asymmetric structure designated as tegument and by a lipid envelope. The Herpesvirion has been reported to contain from 24 to as many as 34 species of proteins ranging from relatively few to thousand of copies of virion (4). Curiously, less than 10 of these proteins are contained in the capsid and at least 7 additional proteins, all glycosylated have been reported to protrude on the surface of the HSV virion (4, 5). The location and function of the remaining virion proteins is not known. The tegument varies in amount dependent on the host cell rather than the virus (3).

Infection occurs as a consequence of:

i) the attachment of surface glycoproteins to cell receptors,
ii) fusion of the viral envelope with the plasma membrane (6) which is mediated by a viral glycoprotein (5)
iii) transport of the capsids to the nuclear pores, and
iv) release of the DNA into the nucleus (7).

As discussed below, virion proteins play a significant role in initial gene expression. Nevertheless, viral DNA denuded of proteins is infectious (8-10).

B. The Polymorphism of Herpesvirus Genomes

In stark contrast to the uniformity of the virion is the polymorphism of the herpesvirus genomes. The herpesvirus genomes range from 80 to 150 million molecular weight in size (1). Their base composition ranges from 32 to 74 moles percent G + C. The most striking variability is in the structure of the different viral genomes. At least 6 different arrangements have been encountered and more are likely to be found (1, 11). The salient characteristic of the sequence arrangements is the presence of reiterations, both direct and inverted, in nearly all herpesvirus genomes examined to date. Probably the most striking example is the tandem reiterations of a relatively small sequence in herpesvirus saimiri DNA which accounts for nearly 30 percent of the length of the genome (12).

The most thoroughly studied reiterations are those of the terminal domains of the HSV genome which are present in an inverted orientation internally, splitting the genome into two components of unequal size (Figure 1). The reiterated domains of the *Long* (L) component designated as *ab* and *b'a'* contain several genes whereas those of the *Short* (S) designated as *c'a'* and *ca* contain a single gene (13-15). Together, the reiterated domains account for nearly 10% of the genome and the genes contained within these reiterations are diploid. One consequence of the internal reiteration is that the two components invert relative to each other giving rise to 4 isomeric arrangements of the HSV DNA (16).

The function of the inverted repeat sequences is not known. The diploidy, the presence of inverted repeats internally, and the inversion of the L and S components do not appear to be essential inasmuch as viable recombinants lacking all of these features have been constructed (Figure 1; 17-19).

The subset of the terminal sequence known as the *a* sequence is better understood. This sequence is 500 base pairs (bp) long in strains HSV-1(F), but somewhat smaller in other HSV-1 strains (20, 21). It is present in a single copy at the S component terminus and in one to several copies at the L component terminus and at the junction between the L and S components (22, 23). The *a* sequence appears to have at least 4 functions:

i) it mediates the inversions between the L and S components in the wild type genomes;
ii) it is the site of cleavage of unit length molecules from concatemers;
iii) it contains the recognition site for packaging of the genome into preformed capsids; and
iv) it contains the promoter-regulatory domain of a gene contained in the inverted repeats (20, 24-27).

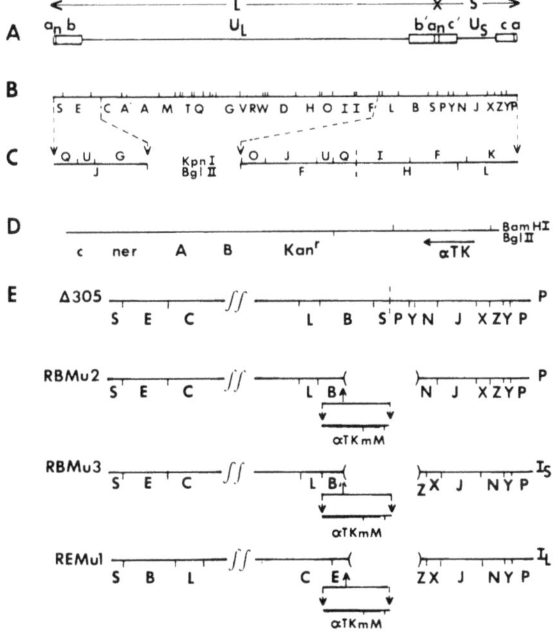

Figure 1. Restriction endonuclease maps and sequence arrangement of wild-type and recombinant HSV-1 viral genomes and of the alphaTK-mM genome.
(A) Sequence arrangement of HSV-1 DNA. The open boxes represent terminal sequences (*ab* and *ca*) of the L and S components, respectively, that are reiterated internally in an inverted orientation (*b'a'c'*).
(B) *Bam*HI restriction endonuclease map of wild-type HSV-1 DNA. The designations of the small fragments are not shown.
(C) Expanded portions of the termini of the L and S components of the HSV-1 DNA genome in prototype (P) arrangement showing the *Kpn*I and *Bgl*II cleavage sites. The dashed vertical line represents the junction of the L and S components.
(D) Sequence arrangement of the alphaTK-mM genome showing *Bam*HI and *Bgl*II cleavage sites. A, B, *ner* and *c* are mini-Mu genes; *Kan*^{r} refers to the gene for kanamycin resistance, and alphaTK refers to the TK structural gene fused to the regulatory and promoter domains of the alpha4 gene.
E) Sequence arrangement of the termini of the L and S components of HSV-1 305 DNA and of three recombinant HSV-1 viral genomes showing the *Bam*HI cleavage sites and the site of insertion of the alphaTK-mM. HSV-1 305 is shown in the P arrangement. The dashed vertical line represents the junction of the L and S components. The parentheses identify the domains of the internal inverted repeat sequences deleted in the recombinant viruses.

The *a* sequence in strain HSV -1(F) consists of a series of ordered structures described by the formula

$$DR1 - U_b - (DR2)_{23} - (DR4)_3 - U_c - DR1$$

where DR1 is a 20 bp sequence directly repeated at the ends, U_b and U_c are 65 and 59 bp unique sequences, DR2 is a 12 bp sequence directly repeated 19-23 times and DR4 is a 37bp sequence directly repeated 2-3 times (20).

Adjacent *a* sequences share the intervening DR1. The terminal DR1a are incomplete. The one at the terminus of S component contains one base pair and one 3' nucleotide extension whereas the L component terminus DR1 consists of 18 bp and a 3' nucleotide extension. Together, the two ends form a complete DR1 (28). It appears therefore that in the course of packaging the genomes into the virion, the head to tail junctions of HSV-1 DNA are cleaved asymmetrically within a DR1 shared by two *a* sequences, and the resulting portions of the DR1 are not further repaired (29). The role of the *a* sequence in inversion has been readily demonstrated by inserting it into the TK gene and demonstrating that segments of viral DNA bound by inverted copied of the *a* sequence invert (20, 29, 30). The inversion specific sequence appears to be primarily DR2 and DR4 (15).

While the focus of attention centers on relatively large reiterations, micro reiterations are a feature of the structural domains of many genes (e.g: the reiteration of 9 nucleotides 10 times in the HSV-1 $gamma_1$34.5 gene; ref. 26). The notion that they play an important role in the biology of the virus should not be dismissed out of hand.

C. Micropolymorphism of Herpesvirus Genomes: An Epidemiologic Tool

Early studies on fresh isolates revealed a micro heterogeneity in the sequence of HSV-1 and HSV-2 genomes (31). While the genomes are stable in both cell culture and in infected individuals and isolates from epidemiologically related individuals are identical, those from epidemiologically unrelated individuals were not (32-34). The significant variable is the presence or absence of restriction endonuclease cleavage sites although variability in the size of specific restriction endonuclease fragments has also been noted. Fingerprints of restriction endonuclease patterns have become an important tool in "molecular epidemiology" since it permits tracing the spread of infection from one individual to another. The significant conclusion from numerous studies is that HSV-1 and HSV-2, and very likely HCMV, are transmitted within epidemiologic "clusters", probably from reservoirs of latent virus infections in which the identity of the viruses is maintained (35, 36). The increased incidence of HSV-2 genital infections reflects a "frenzy" of transmission rather than an epidemic of HSV-2 sweeping the population.

3. REGULATION OF GENE EXPRESSION

A. The General Pattern of Gene Expression

HSV genes form at least 3 major groups, alpha, beta, and gamma, whose expression is coordinately regulated and sequentially ordered in a cascade fashion (37, 38).

There are 5 alpha genes designated alpha0, 4, 22, 27 and 47. The alpha4 gene product is required for the transition from alpha to beta and gamma gene transcription (39-41). The alpha4 and alpha0 gene products are involved in the overall expression of both β and gamma genes (42-46). The alpha27 gene product is required for the expression of some late genes (47) and the alpha 22 gene product is required for high level expression of gamma genes in some cells but not in others (48, 49). The alpha47 gene has been shown to be dispensible in cell culture (50). While the function of the alpha47 gene is not known, in transient expression systems it appears to stimulate indiscriminately the expression of the viral gene cotransfected with it.

The beta genes comprise a much larger and more heterogeneous group. The $beta_1$ genes are expressed earlier than the $beta_2$. Although both sets require the alpha4 protein for their transcription, the $beta_2$ genes may have additional requirements. The known functions of the beta genes concern the synthesis of viral DNA and of dXTP precursors. The rationale for the distribution of genes among the 2 groups is not clear. $Beta_1$ includes the major DNA binding protein ($beta_18$) and the ribonucleotide reductase ($beta_16$), whereas the $beta_2$ group includes the DNA polymerase, the thymidine kinase and most of the beta genes.

Like the beta genes, the gamma genes are heterogeneous with respect to the requirements for their expression. Whereas the expression of $gamma_1$ genes is reduced but not abolished by inhibitors of viral DNA synthesis, the expression of $gamma_2$ genes stringently requires viral DNA synthesis. The known gamma genes specify the structural proteins of the virus. The $gamma_1$ group includes the genes specifying the major capsid protein ($gamma_15$) and the glycoproteins gB and gD. The gene specifying the glycoprotein gC is an example of a $gamma_2$ gene.

In permissive cells, the herpesvirus DNA entering the nucleus is transcribed at all stages of infection by the host RNA polymerase II (51). It has become increasingly evident, however, that viral factors play a significant role in the transcription of the viral genes. A most significant finding, discussed in detail below, that signals an important difference between the expression of the herpesvirus genome and those of papovaviruses and adenoviruses is that a structural component of the virus, a $gamma_1$ protein, induces the expression of alpha genes (48, 52-57). This finding has significant implications with respect to the overall biology of these viruses.

B. Induction of Alpha Gene Expression by a Virion Structural Component

Early studies, designed to identify the *cis*-acting elements involved in alpha gene expression and regulation, utilized chimeric gene constructs consisting of the 5' nontranscribed and leader sequences of alpha genes fused to the leader and structural sequences of the thymidine kinase (TK) gene, normally a beta gene product (48, 52-54).

These constructs, designated as alpha-TK chimeras, were recombined into the viral genome and were also used to convert TK^- cells to the TK^+ phenotype. In the environment of the viral genome, the alpha-TK chimeric genes were expressed as alpha genes inasmuch as they were transcribed in the absence of *de novo* protein synthesis. Significant findings have emerged from analyses of alpha-TK gene expression in cells converted to the TK^+ phenotype. Specifically,

i) the alpha-TK genes were inducible by superinfection with HSV TK^- virus,
ii) *ts* mutants in the alpha4 protein efficiently induced the chimeric gene at the nonpermissive temperature, and
iii) the transcription of the alpha-TK gene was induced by infection of cells in the presence of an inhibitory concentration of cycloheximide.

The hypothesis that the alpha-TK gene was induced by a structural protein of the virus was reinforced by the observation that the expression of the alpha-TK gene resident in transformed cells was induced with high efficiency by HSV-1*ts*B7 at the non-permissive temperature (55). Previous studies have shown that this mutant is able to infect cells and the capsid is translocated normally to the nuclear pore. However, at the nonpermissive temperature, the capsid retains its integrity and does not release the viral DNA (7). Since viral genes are not transcribed at the nonpermissive temperature, the alpha gene *trans*-activating factor (alpha-TIF) had to be introduced into the cell during infection as either a component of the tegument structure surrounding the capsid or in the viral lipid envelope. Early evidence that alpha-TIF was a structural component of the virion, and not a product of the host genome induced by infection, was based on the observations that (a) induction of alpha-TK chimeras correlated with infectivity, and (b) infection with other herpesviruses (e.g: cytomegalovirus, bovine mammilitis virus, pseudorabies virus) or with adenovirus did not induce the expression the alpha-TK gene resident in the transformed cell lines (55).

The genes specifying alpha-TIF was mapped by cotransfection of cloned HSV-1 DNA fragments with an alpha-TK chimeric gene (56). The map location of the gene corresponds to that of the tegument protein designated as infected cell protein No. 25 (ICP25); a protein that is present in 500 to 1000 copies per virion (58). The nucleotide sequence of the gene predicts a protein 59,000 in translated molecular weight, but the predicted amino acid sequence does not have features that might suggest how alpha gene induction is effected (57).

C. A Specific Sequence Confers Inducibility upon Alpha Genes by Alpha-TIF

Studies designed to identify the *cis*-acting elements that determine alpha gene regulation have centered on the use of the alpha-TK chimeric genes as the indicator of alpha gene expression. Justification for the continued use of the alpha-TK chimeric gene for these studies rested on the observation that the ability of the alpha gene 5' nontranscribed and leader sequences to confer alpha regulation was not affected by the nature of the gene to which they were juxtaposed. The alpha promoter-regulatory domains could confer alpha regulation upon the chick oviduct ovalbumin gene, the hepatitis B S gene, and a variety of viral genes (59, 60). This key observation then led to systematic analyses of the alpha gene promoter-regulatory domains.

The initial studies on the sequences 5' to the alpha gene differentiated between the promoter domain (- 110 to + 33), that enabled efficient expression of the recipient gene, and the regulatory domain (- 110 to - 4500), that conferred the ability to be induced as an alpha gene by the infecting virus when fused to a promoter-indicator gene (54). Progressive deletions from the 5' end of the alpha gene sequences indicated that the elements required for response to alpha-TIF were reiterated within the alpha gene regulatory domain. Sequence analysis of the promoter-regulatory domains of alpha genes 4, 0 and 27 revealed that these genes contained numerous G + C rich repeats and from one (alpha27) to several (alpha0) homologs of an A + T rich element (52). Confirmation of the hypothesis that these elements play a role in the expression and regulation of alpha genes emerged from studies utilizing small cloned DNA fragments representative of these reiterated elements (61). Specifically, sequences derived from alpha genes 4, 0 and 27 (Fig. 2), were fused to a chimeric gene consisting of the alpha4 gene promoter (- 110 to + 33) linked to the leader and coding sequences of the TK gene. To control for possible complementing elements contained within the alpha4 promoter domain, these sequences were also fused to the beta promoter-TK gene which consisted of the natural beta TK sequences downstream from nucleotide - 80, relative to the transcription initiation site of the TK gene. The chimeric TK genes were tested for basal level expression and viral-directed regulation by infection of cells (i) transformed to the TK^+ phenotype; (ii) cells co-transfected with pSV_2Neo selected for resistance to G418 antibiotic; and (iii) cells transiently expressing the TK chimera.

These experiments demonstrated that the alpha4 regulatory domain (- 330 to - 110) contained all of the elements necessary to confer high basal level expression and induction by alpha-TIF upon both the alpha and beta promoter-TK genes. Sequence GC (Fig. 2), a 59bp element derived from the alpha4 regulatory domain conferred high basal level expression in an orientation semi-dependent manner. In

DESIGNATION		NUCLEOTIDE SEQUENCE
α Gene Donor Fragments		
		AT Cons.
AT_0	5'	CCGTGC ATGCTAATGATATTCTT TGGGGG 3' GGCACG TACGATTACTATAAGAA ACCCCC
		AT Cons.
AT_{27}	5'	CGGAAGCGGAACGGTGTATGTGAT ATGCTAATTAAATACAT GCCACGT 3' GCCTTCGCCTTGCCACATACACTA TACGATTAATTTATGTA CGGTGCA
		IR1 IR2 IR2 IR1
GC	5'	CGGATGGCGGG GCCGGGGG TTCGACCAACG GGCCG CGGCC ACGGG CCCCCGGC GTGCCG 3' GCCTACCCGCCC CGGCCCCC AAGCTGGTTGC CCGGC GCCGG TGCCC GGGGGCCG CACGGC

Figure 2. Nucleotide sequences of cloned alpha donor fragments.

AT_0 and AT_{27}, derived from the regulatory domains of the alpha0 and alpha27 gene, respectively, each contain a homologue of the 17-nucleotide A + T sequence reiterated within all the alpha gene regulatory domains. GC, derived from the regulatory domain of the alpha4 gene, contains numerosus G + C-rich inverted repeat sequences (IR) which are characteristic of the alpha gene regulatory-promoter domains.

addition, this G+C rich element complemented the beta class regulation of the truncated beta-TK promoter, but did not confer response to the alpha-TIF.

As will be discussed later, sequence GC contains an SP1 factor binding site (62) and a binding site for the major HSV-1 regulatory protein, alpha4. Sequence AT_{27}, a 47bp element from alpha27 gene and Sequence AT_0, a 29bp element from alpha0 gene each contain a homolog of the A+T element. Both sequences conferred alpha gene regulation upon the alpha4-TK chimeric gene in an orientation independent manner. However, only sequence AT_{27} contained all the components necessary to convert the beta-TK gene to an alpha-regulated gene. Consistent with these findings, transient expression systems also indicated a requirement for the presence of the A+T rich homolog for induction of alpha-TK gene expression by cotransfected DNA fragments encoding the alpha-TIF gene (63).

D. Protein Binding Sites in Alpha Promoter-regulatory Domains

To determine whether alpha-TIF binds to promoter-regulatory domains of alpha genes DNA-protein retardation assays were used to map and identify the viral proteins specifically binding to these DNA domains (64, 65). These studies detected the binding of numerous host cell proteins but failed to produce evidence that viral gene products bind to DNA fragments containing the alpha gene A+T homolog which is required for induction of chimeric genes by alpha-TIF. Of particular interest, however, was the observation that promoter-regulatory domains of HSV-1 and $gamma_2$ genes contain binding sites for the HSV-1 alpha4 protein.

In denaturing gels, the alpha4 protein forms 3 bands ranging from 160,000 to 172,000 in apparent molecular weight (14). In its native form, the protein is phosphorylated, exists as a homodimer, (66, 67) and has been reported to bind DNA (66, 68). The evidence that this protein binds to specific viral DNA sequences may be summarized as follows: Fig. 3 illustrates the use of the gel retardation assay (69, 70) to detect infected cell proteins bound to the alpha0-promoter (O-P) DNA (-110 to +72) in the presence of excess competitor nucleic acids. The identification of the HSV-1 specific protein in the major infected cell protein alphaO-P DNA complex was approached using protein extracts of cells infected at the nonpermissive temperature with viral *ts* mutants defective in the transition from alpha to beta or beta to $gamma_2$ protein synthesis. Since the complex was detected in all infected cell extracts, but was not found in extracts of cells infected and maintained in the presence of inhibitory concentrations of actinomycin D, the proteins in the complex were likely to belong to the alpha regulatory group.

The identification of the alpha4 protein as a component of the infected cell protein-alphaO-P DNA complex was determined by the use of monoclonal antibodies against several of the alpha gene products. Fig. 4 illustrates that only the alpha4 monoclonal antibody retarded the electrophoretic mobility of the alpha-O-P DNA-protein complex when added either before or after the binding reaction. Similar studies demonstrated that the alpha4 protein was present in complexes containing its own promoter and regulatory domains as well as in those formed by the $gamma_2$ promoter-regulatory region and the promoter-regulatory domains of the alpha27 gene. Although the alpha0 and 27 proteins have been shown to play a role in the regulation of beta and gamma gene expression, these proteins were not detected in any of the complexes with the monoclonal antibodies tested.

As the focus of these studies centered on the factors affecting the expression of alpha genes, the binding site of the alpha4 protein complex was mapped within the alpha0 promoter domain DNA and the alpha4 regulatory domain DNA. As shown in Fig. 5, exonuclease III was used to delineate the boundaries of the binding site of the alpha4 protein complex within the alpha0 promoter DNA. To control for the specific interaction of the alpha4 protein in the crude nuclear extracts used, the reactions were done in the presence or absence of monoclonal antibody H950. This antibody, directed against the alpha4 protein, significantly inhibited the formation of the alpha4 protein-alpha0 DNA complex. The DNA fragments specifically pro-

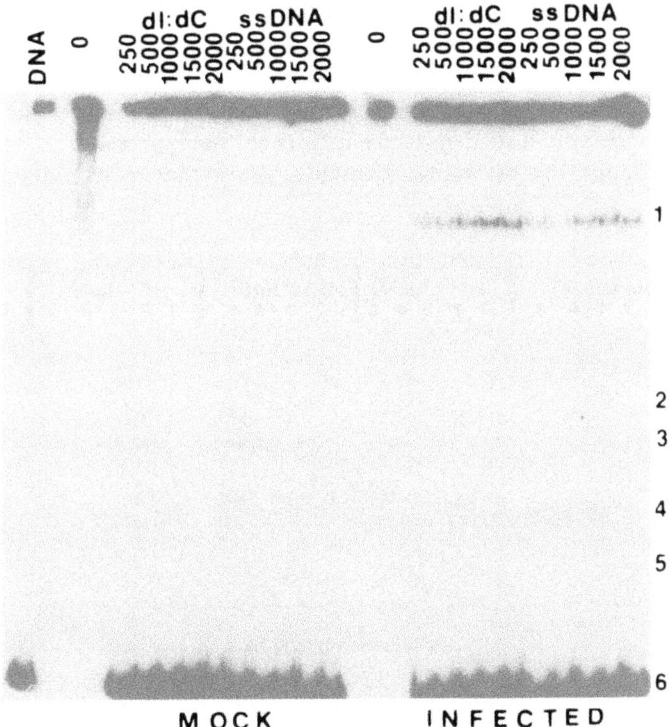

Figure 3. Autoradiographic images of labeled alpha0 promoter DNA complexed with protein in the presence of increasing amounts of competitor nucleic acids. The competitors, poly(dI) poly(dC), dI:dC, or salmon sperm DNA, ss DNA, were added prior to the addition of extracts of cells harvested 12 hours after mock-infection (MOCK) or HSV-1(F) infection (INFECTED). The quantity of competitor, in ng, is shown at the top of each lane. The lane marked DNA contained the alpha0 promoter probe DNA only. Lanes marked 0 contain the probe DNA and cell lysate in the absence of competitor nucleic acids. The DNA-protein complexes are numbered 1-5 while the unbound DNA migrates at position 6.

tected by the alpha4 protein from digestion with exonuclease III located the alpha4 protein binding site to - 46 to - 71, relative to the alpha0 transcription initiation site. The protected sequence element contains a portion of the dyad constituting the alpha0 promoter CCAAT box.

The binding site for the alpha4 protein within the alpha4 regulatory domain was localized using small cloned subfragments of the alpha4 regulatory domain. These studies mapped the binding site within the 59bp fragment (- 135 to - 194) designated Sequence GC in Fig. 2. Small deletions made in this 59bp fragment indicated that critical elements required for the alpha4 protein binding were contained within the promoter proximal region of Sequence GC. The SP1 binding site also containe in Sequence GC was not required for the binding of the alpha4 protein.

E. The Regulation of Alpha Genes: A Microcosm of the General Pattern of Viral Gene Regulation

Figure 6 illustrates the locations of the *cis*-acting elements mapped *in vitro* using the alpha4 gene as a model of the regulation of HSV-1 alpha genes. The promoter-regulatory domains of these genes consist of a complex mixture of elements which respond to host transcription components and viral *trans*-acting factors which positively and negatively influence the expression of these genes in the infected cell. Of the *cis*-acting elements, the evidence strongly supports the

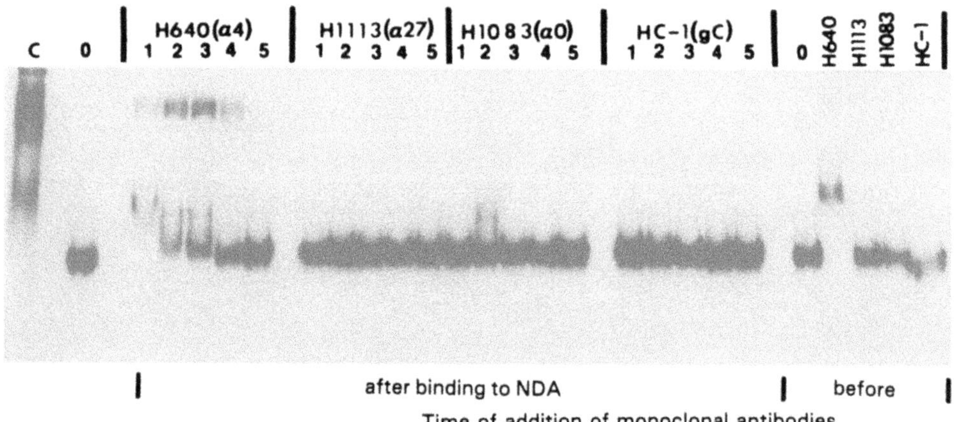

Figure 4. Autoradiographic images ot alpha0 promoter DNA-infected cell protein complex in the presence of murine monoclonal antibodies against alpha4 (H640), alpha27 (H1113), alpha0 (H1083) and Y_2gC (HC-1). MOCK, mock infected cell extract; INFECTED, HSV-1 (F) infected cell extract. The monoclonal antibodies were either added after protein-DNA binding or were preincubated with the protein extract for 0.5 hour prior to the incubation of the extract with labeled DNA. Lane C, DNA-protein complexes formed in the absence of poly(dI) poly (dC); lane 0, monoclonal antibody added. Lanes 1-5 differ with respect to the amount of monoclonal antibody added to the reaction as follows: 1, 500 ng; 2, 250 ng; 3, 100 ng; 4, 10 ng; 5, 1 ng. Preincubations were done with 500 ng antibody/1000 ng protein extract.

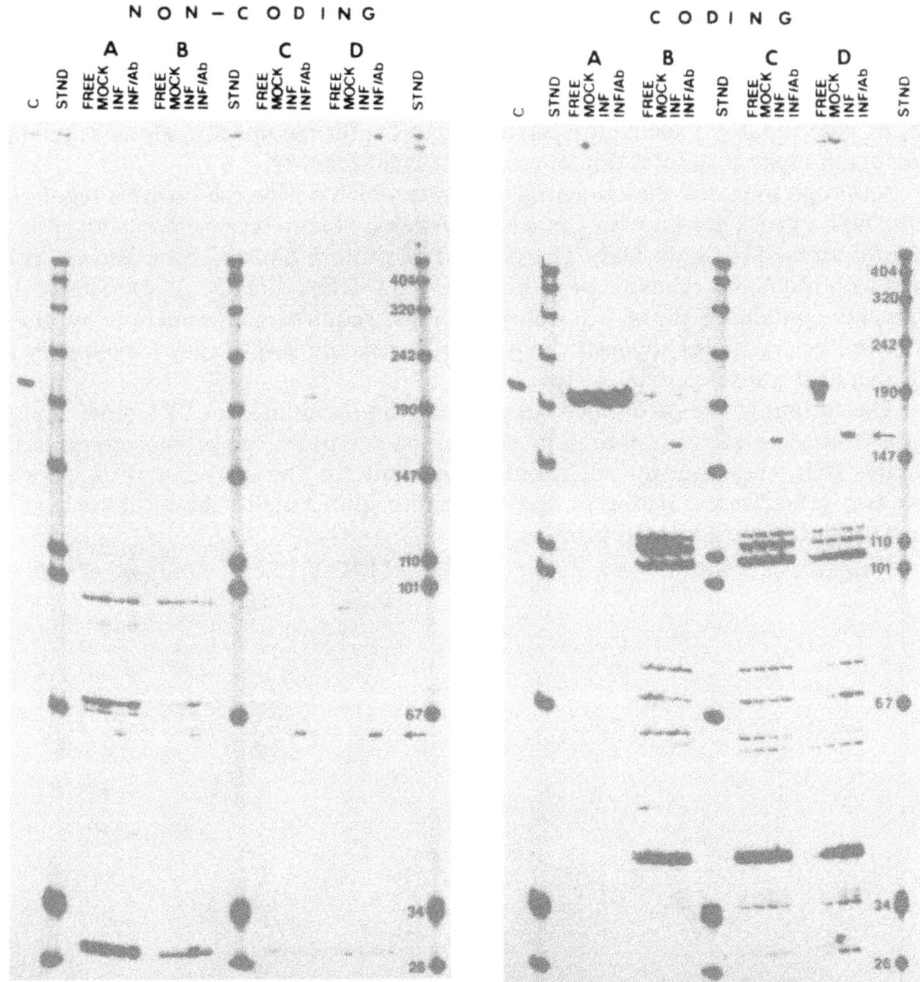

Figure 5. Autoradiogram of electrophoretically separated exonuclease III digests of alpha0 promoter DNA in the absence or presence of available alpha4 protein complex. C, non-digested alpha0 promoter DNA; FREE, exonuclease III digest after incubation without protein extract; MOCK, exonuclease III digest after incubation with mock-infected cell extract; INF, exonuclease III digest after incubation with infected cell extract; IFN/Ab, exonuclease III digest after incubation with infected cell extract that had been preincubated for 0.5 hour with monoclonal antibody H950. The left panel represents digests of alpha0 promoter DNA labeled at the 5' end of the noncoding strand. The exonuclease III digestions were: lane A, 100 units of enzyme for 15 minutes; lane B, 100 units of enzyme for 30 minutes; lane C, 100 units of enzyme for 45 minutes; lane D, 200 units of enzyme for 30 minutes. The right panel shows digests of alpha0 promoter DNA labeled at the 5' end of the coding strand. The exonuclease III digestions were for 15 minutes at the following enzyme concentrations: lane A, 1 unit; lane B, 25 units; lane C, 50 units; lane D, 100 units. pUC9 digested with *Hpa*II restriction enzyme and end-alabeled with [γ-^{32}P]-ATP (STND) served as size markers. The arrows indicate the position of bands representing DNA protected by the alpha 4 protein complex.

hypothesis that the reiterated A+T rich homolog present in the alpha gene domains is a component necessary for the positive response to the viral alpha-TIF. This element shares homology with a portion of the SV40 72bp enhancer as well as with other mammalian cell regulatory elements (71). Although necessary, it is unclear that the A+T element is sufficient to confer the specific alpha regulation *upon* a non-alpha promoter domain in an efficient manner.

Although many of the *cis*-acting elements which define the positive regulation of the alpha genes are known, the mechanism by which these elements respond to the viral alpha-TIF is unclear. The alpha-TIF protein has not been shown to interact with the A+T elements or to bind any viral DNA directly. The evidence that fragments containing the A+T rich homologs, required for induction by alpha-TIF, bind host proteins suggests the possibility that alpha-TIF may function by acting on a host transcriptional factor.

The second group of defined *cis* elements are the numerous SP1 binding sites. As these sites are a feature shared by both alpha and beta gene promoter-regulatory domains (62), they are not sufficient to account for the discrimination between these two gene classes. However, they probably contribute to the ultimate level of transcription of the activated promoters.

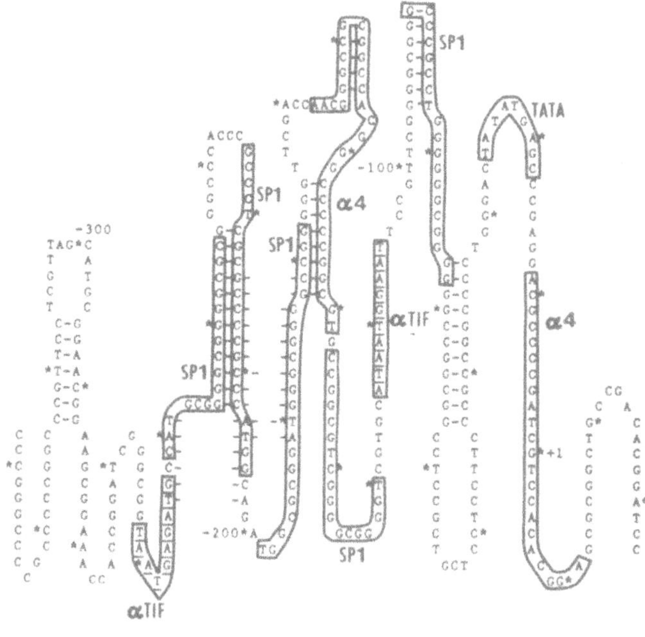

Figure 6. Summary of the *cis*-acting elements and *trans*-acting factors involved in alpha gene expression and regulation. The nucleotide sequence of the alpha4 promoter-regulatory domains is drawn to represent the numerous inverted repeat sequences. Alpha-TIF, sequences which enable response to the alpha-specific virion regulatory factor; SP1, location of the *in vitro* binding sites of the SP1 factor; ICP4, location of the mapped alpha4 protein binding sites. All number refer to the site of transcription initiation.

The alpha4 binding site constitutes the third group of *cis*-acting elements. In the case of the 59bp sequence from the regulatory domain of the alpha genes (Sequence GC in Fig. 2), the alpha4 protein binding site probably accounts for the strong beta regulation conferred by this element upon the - 80 beta promoter-TK gene. The function of the alpha4 binding site in its native location in the alpha4 regulatory domain is less certain. Although alpha gene expression appears to be reduced by DNA fragments containing the alpha4 gene in transient expression systems, the function of the alpha4 binding site in this particular location is not clear.

The location of the alpha4 protein binding sites and the evidence that the alpha4 protein regulates the expression of alpha genes offers a better insight into possible mechanism by which the alpha4 protein may regulate gene expression. The protein interacts with the promoter and regulatory domains of alpha, beta, and gamma genes. The alpha4 protein binding site in the alpha0 promoter domain (- 46 to - 71) and in the alpha4 regulatory domain is in close proximity to binding sites of host transcriptional components such as the SP1 and the CCAAT factors. Since the alpha4 protein may have both positive (beta and gamma genes) and negative (alphagenes) regulatory functions, the context of the binding site may play the most critical role in determining the net effect of the alpha4 protein upon the expression of that gene.

F. Herpesvirus Genome Arrangement and Gene Structure

The general impression of the HSV genome is that it consists of clusters of genes related with respect to function or regulation, but that strategy of the arrangement may reflect the evolutionary origin of the herpesvirus genome and not the strategy of its gene regulation.

For example, all alpha genes are in the reiterated sequences or in sequences immediately adjacent to the reiterated sequences flanking the L and S components (72-74). Unlike the alpha genes, the location of the beta and gamma genes are not predictable; they are located in both L and S components, singly and in clusters.

The HSV genomes contain at least 3 origins of replication, one in the middle of the L component and two in the reiterated sequences of the S component (24, 29, 75, 76). Flanking the S component origins are the alpha genes alpha4-alpha22 in one case, and alpha4-alpha47 in another. Flanking the L component origin are two beta genes specifying the major DNA binding protein and DNA polymerase (77, 78). Of the 6 major glycoproteins, 3 (gD, gE, gG) map in close proximity in the S component; the remaining 3 (gB, gC, gH) are dispersed in the L component (4,79,80). The segregation of *cis*-acting sites and of genes with related function in the two components underscores the hypothesis that the herpesviruses originated by the fusion of two genomes corresponding to the L and S components, respectively (81).

The herpesvirus genes vary in structure. Of the alpha genes, 3 are spliced (alpha0, alpha22 and alpha47). The introns of alpha 22 and alpha47 are in the 5' transcribed non-coding rather than coding domain (82-85), and the impression is that splicing is neither uniform nor predictable. Gene overlap is common, but not the general pattern of gene organization. The transcriptional signals generally conform to those found in eukaryotic genes, but variation from the generally derived consensus of the TATTAA sequence or the polyadenylation sequence YGTGTTYY (85) are not uncommon.

The promoter-regulatory domains of some genes have been studied in considerable detail, and the elegant studies on the sequence requirements for the expression of the thymidine kinase gene serve as models of the general structure of promoters functioning in eukaryotic environments. In the case of the thymidine kinase gene, the sequences which enable gene expression (promoter) and the regulatory sequences which enable the gene to respond to *trans*-acting regulatory signals appear to overlap (86). In the case of alpha genes, the overlap is minimal and each domain is independently movable to indicator genes to confer either expression (promoter) or regulation as an alpha gene (53).

As discussed earlier, regulatory alpha sequences require a promoter in order to confer alpha regulation upon the recipient gene. Of specifical interest is the observation that the regulatory domains of alpha genes contain at least two kinds of *cis*-acting signals. One, characterized by a high G + C content confers upon competent promoters a high level of gene expression and resembles enhancer sequences with respect to its mode of action. The other, A + T rich, confers upon the gene ability to be induced by the *trans*-acting structural protein but does not affect the basal level of expression in the absence of the inducer (61).

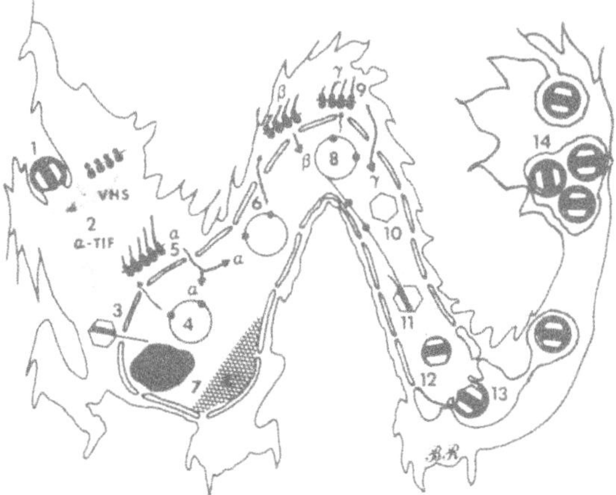

Figure 7. Schematic diagram of a productive HSV infection. (1) Fusion of the HSV virion envelope to the cell membrane. (2) Entry of nucleocapsid and tegument proteins into cytoplasm; VHS, virion associated host shutoff factor; alpha-TIF, alpha *trans*-inducing factor. (3) Transport of nucleocapsid to nuclear pores and release of viral DNA into nucleus. (4) Circularization of viral DNA by exonuclease activity on terminal *a* sequences. (5) Production of alpha mRNA and proteins. (6) Production of β mRNA and proteins. (7) Margination of cell chromatin along nuclear membrane. (8) Initiation of viral DNA synthesis by rolling circle mechanism. (9) Production of gamma mRNA and proteins. (10) Formation of empty capsids. (11) Packaging of viral DNA into nucleocapsids producing full capsids (12). (13) Envelopment of full capsids at the nuclear membrane. (14) Eggress of enveloped capsids through the cysternae of the endoplasmic reticulum.

G. *The Synthesis of Herpesvirus DNAs*

Unlike papova and adenoviruses, herpesviruses specify many of the gene products required for efficient production of dXTP pools and for DNA synthesis. Among these products are DNA polymerase, DNase (87), dUTPase (88, 89), ribonucleotide reductase (90, 91), and a nucleoside kinase (thymidine kinase; 92). Viral DNA synthesis occurs in the nucleus. At some point in the reproductive cycle the DNA synthesis satisfies some criteria for the rolling model of DNA replication. The replicated DNA is packaged from concatemers into preformed capsids (93). The capsids containing DNA differ from the preformed, empty capsids by the addition of a new protein (94, 95) and only capsids containing DNA and the additional protein are enveloped at the nuclear membranes by patches containing only viral membrane proteins (96, 97). The enveloped capsids accumulate in the space between the inner and outer lamellae of the nuclear membrane and in the cysternae of the endoplasmic reticulum through which the virus eggresses from the infected cell (97).

H. *Viral Replication and Cell Death*

Within the context of the systems studied to date, replication of viruses in the infected cells invariably results in cell death (98). Among the many manifestations of the effects of infection with HSV are the following:

- i) Shortly after infection, there occurs a disaggregation of host polyribosomes, decrease in host protein synthesis, degradation of host mRNA, and cessation of host DNA synthesis (99-107). A variety of studies involving inhibitors of transcription and viral mutants indicate that the early cessation in protein and DNA synthesis is caused by a structural component of the virus (103) but is not essential for virus multiplication in cell culture (108).
- ii) A more profound cessation of host macromolecular metabolism occurs with the initiation of the expression of beta and gamma genes (37). These changes include complete shut off of host ribosomal RNA synthesis, and the margination, breakage, and fragmentation of chromosomes and shut off of host DNA synthesis. The nucleus becomes distorted and may fragment (3).
- iii) Insertion of viral glycoproteins into host membranes coincides with and may cause changes in the shape of the infected cell and in the interaction of cells among themselves - the social behavior of cells (3).

4. LATENCY: AN INTERACTION OF HERPESVIRUSES WITH THEIR HOSTS

The capacity to remain latent, i.e. to rest in a specific tissue without apparent manifestations of multiplication and cell destruction, appears to be a general property of all herpesviruses examined to date. Relevant to an understanding of these latent infections are the following:

i) As a general rule, transmission of herpesviruses in the susceptible host population is characteristic for each virus. In the case of HSV, transmission is most often effected by physical contact of infected tissue with "wet" (i.e. mucous membranes, etc) tissue of susceptible individuals. The virus establishes latent infections by ascending through nerve endings at the portal of entry into the body to dorsal root ganglia and remains latent in neurons (109). Induction of virus multiplication causes progeny virus to descend by way of the axon to the portal of entry at or near the site of initial infection, multiply and be available for transmission. Thus, the latent virus is harbored at a site shielded from the immune system but accessible to the organ or tissue which serves as the donor for transmission to susceptibles. The pattern of transmission indicates that the latent virus is an important reservoir for the maintenance and dissemination of herpesviruses in their host populations.

ii) Each herpesvirus appears to be maintained in a latent state in a specific host cell. HSV is harbored in neurons of sensory and autonomic ganglia, but claims of involvement of other types of cells do occasionally surface (110). Other herpesviruses remain latent in salivary gland cells, in B lymphocytes, etc. These observations suggest that establishment of latency requires functions expressed by a specific cell. Infection of other cells (e.g. infection of CNS cells with HSV) while it may kill the host, does not contribute to the reservoir of latent or infectious virus of epidemiologic significance, since it is not transmitted.

iii) Latent herpesviruses are maintained in a "static" state (111), i.e. they do not multiply until multiplication is induced. The cell harboring the virus therefore restricts the extent of virus gene expression but is capable of supporting virus multiplication when induced. The factors terminating the restrictive state are not known. In the case of HSV, a plethora of factors related to physical and emotional trauma and changes in hormone balance are known to induce the termination of the latent state and initiate viral multiplication (109).

iv) The extent of viral gene expression required for establishment of the latent state is uncertain. As a general rule, initiation of viral DNA synthesis by viral enzymes signals an irreversible course leading to the destruction of the infected cell (98, 112). Moreover, it seems unlikely that functions required for the maintenance of the virus in the host population are totally those of the host and do not involve the products of the viral genome. It seems obvious that access to and infection of the cell harboring the latent virus involves viral functions, but nothing is known of the viral genes that might be required for establishment of HSV latency once the virus penetrates that cell. If viral genes do participate in this process, they are not likely to belong to beta or gamma regulatory groups inasmuch as the genes comprising these groups specify functions associated with DNA synthesis and virus assembly. The product of one alpha gene, alpha-4 has been reported in rabbit ganglia harboring latent HSV (113).

The human beta lymphocytes harboring EBV have been reported to express several genes, including EBNA1, EBNA2, and LYTMA, and it has been postulated that these genes are involved in the establishement and or maintenance of the latent state (114, 115). In contrast to neurons harboring HSV, the beta lymphocytes harboring EBV are capable of multiplying. Although EBV genomes integrated into host chromosomes have been reported (114), the bulk of the latent genomes is in the form of episomes (116, 117). It seems obvious that unless the episomal EBV DNA is replicated in tandem whith the multiplication of the beta lymphocyte, it would soon become lost. Therefore it appears that in the beta lymphocyte the functions required for the maintenance of EBV latency are the restriction of EBV gene expression and the maintenance of EBV episomes. Current studies suggest that at least EBNA1 is required for the maintenance of the episomal state (118). Whether these genes are involved in the restriction of EBV gene expression in the latently infected beta lymphocytes is still conjectural.

5. GENETIC ENGINEERING OF HSV GENOMES

HSV genomes encode a vast variety of information. In addition to coding sequences for proteins, these genomes encode specific signals for initiation and termination of transcription, for regulation of gene product abundance, and for modification of genome structure necessary for gene expression and genome replication. A central objective of the studies on the molecular biology of herpesviruses is the elucidation of the functions encoded in their genomes. A second objective, no less important, is the specific modification of the viral genomes for use as vaccines or as vectors of genes whose products protect against infectious disease.

Early studies designed to elucidate the function(s) of virus specific proteins and DNA sequences relied on the generation and characterization of temperature-sensitive (*ts*) mutants. Since the isolation of *ts* mutants is dependent on phenotypical changes or inhibition of viral replication at elevated temperatures, the majority of viral genes which can be studied by the aid of *ts* mutants are those that are essential for a productive virus infection in cell culture. Many herpesviruses are capable of a productive infection in many different cellular environments. For these viruses, it is conceivable that some of the viral genes are required for productive infection in the different cellular environments. These genes would be responsible for complementing a host specified protein that is diminished or absent in the cellular environment and therefore the detection of viruses containing *ts* mutations in these genes would depend on the growth properties of the viral mutants in the proper cellular environment.

A new approach to the elucidation of virus-specific functions involves site-specific mutagenesis of the HSV genome by the insertion or deletion of DNA sequences. Site-specific mutagenesis offers several advantages:

i) the procedures allow for either site-specific or target specific mutagenesis of the viral DNA genome, thereby interrupting a specific viral gene(s) or DNA sequence;

ii) mutants produced by the insertion or deletion of DNA sequences can be used for analysis of genes in which *ts* mutations would be undetected; and
iii) site-specific mutagenesis can be used in the modification of viral genomes for use as vaccines or as vectors of foreign genes.

A. Basic Strategy

As mentioned earlier, deproteinized HSV DNA is infectious. Co-transfection of intact DNA with a molar excess of a mutagenized DNA fragment recombines and replaces homologous DNA sequences (119). Viral recombinants can be isolated by screening for either the recombinant DNA sequences or expression of a gene contained in the insert (e.g. beta-galactosidase) but the work is tedious and time consuming. Another problem is that viruses modified by insertional mutagenesis may grow more slowly and yield less progeny than the wild type parent. In consequence, the slower growing recombinants may be overgrown by the parent, wild type virus. The isolation of desirable recombinants is greatly facilitated by the use of a selectable marker. Under selective pressure only the recombinants carrying the selectable marker are able to grow and are readily isolated by plaque purification.

One of the most widely used selectable markers is the HSV thymidine kinase (TK) gene because procedures are available to select both for and against the expression of the TK gene permiting the selection of TK^- and TK^+ recombinant progeny (29, 48).

B. Properties of the HSV TK

Cells in culture have two pathways for the synthesis of thymidine monophosphate (TMP). The major or *de novo* pathway involves the conversion of dUMP to TMP by thymidylate synthetase. The second, or scavenger pathway, involves phosphorylation of thymidine by TK. Several, but not all herpesviruses encode a TK, which has been claimed to be important for normal virus growth in experimental animals (120) but is not essential for growth in cell cultures since mutants lacking a functional TK gene have been known for many years (121). The host TK enzyme is also not essential for cell growth provided the primary pathway for TMP synthesis is not obstructed by metabolic inhibitors. The HSV TK differs in two important respects from its cellular counterpart. First, the viral enzyme phosphorylates deoxypurines and pyrimidines whereas the host TK does not phosphorylate deoxycytidine or deoxypurines (92). The second and more important characteristic is that the substrate range of the viral TK for deoxynucleotide analogues is rather wide and includes analogues that bear little resemblance to deoxynucleotides whereas the substrate range of the host TK is restricted to a small number of analogues (e.g: BudR) that closely resemble thymidine. In consequence, analogues that are phosphorylated by the viral TK (e.g: AraT, acyclovir, etc), but not by the host TK, will not affect uninfected TK^+ cells and will permit the multiplication of TK^- viruses but not that of TK^+ viruses in these cells. AraT and acyclovir can be used, therefore, to select TK^- viruses in both TK^+ and TK^- cell lines whereas BUdR can be used for this purpose only in TK^- cells inasmuch as this analog is phosphorylated by the host TK.

Obstruction of the thymidine synthetase pathway for TMP biosynthesis by such inhibitors as methotrexate or aminopterin is not deleterious in TK^- cells but will result in the depletion of the TTP pool and destruction of TK^- cells. Because of the depletion in the TTP pool, TK^- cells overlaid with medium containing methotrexate or aminopterin will permit the growth and plaque formation of TK^+ viruses but not of TK^- viruses. TK^+ viruses are therefore readily selected in TK^- cells overlaid in HAT medium containing methotrexate or aminopterin, thymidine and hypoxanthine (the latter two compensate for other metabolic products whose synthesis is blocked by these drugs).

It should be pointed out that the selection of TK^- recombinants is generally more efficient than that of TK^+ recombinants. Although spontaneous TK^- mutants are readily obtainable, the frequency of such mutations in a cloned viral population is 10^{-4} to 10^{-5}, i.e. lower than the recombination rate between viral DNA and a DNA fragment containing an insertion or deletion flanked by at least 2 Kbp of DNA homologous to contiguous DNA sequences in the viral genome. Furthermore, in cells doubly infected with a TK^+ parent and a TK^- recombinant, both viruses are destroyed by the analogue phosphorylated by the viral TK. The TK^+ selection suffers from two problems. First, the uninfected TK^- cells survive only a few days in the presence of methotrexate or aminopterin, and debilitated TK^+ recombinants may not have sufficient time to spread and form plaques. Furthermore in cells doubly infected with the TK^- parent and TK^+ recombinant, the TK^+ parent acts as a helper and both viruses grow.

C. Construction of Novel HSV Genomes

Figure 8 illustrates the strategy for construction of recombinant genomes based on the use of TK as a selectable marker. The first strategy (Figure 8-1) involves the insertion of a DNA sequence into the transcribed domain of the TK gene. The DNA fragment carrying the TK gene with the insert is amplified by cloning in *E. coli* and then co-transfected with intact HSV wild type DNA. TK^- recombinants carrying the insert are selected from among the progeny of transfection as described above. This technique allows rapid selection of recombinant progeny, but it restricts the location of the insertions to the domain of the TK gene.

The second strategy (Figure 8-2) permits the construction of recombinant viruses carrying insertions or deletions at specifically selected sites.

In the first step of this procedure, sequences are deleted near one end of the TK gene. The DNA fragment containing the deleted TK gene is co-transfected with intact wild type DNA and TK^- progeny carrying the deletion are then selected among the progeny of transfection. The TK^- recombinant constructed for these studies (HSV-1(F) 305) contains a 700 bp deletion extending downstram from the site of transcription initiation ($^+1$) beyond the *Sac*I site ($^-540$). Although this recombinant makes a truncated 19,000 molecular weight protein that reacts with antibody to TK (M. Ackermann, B. Roizman; unpublished observations), it is unable to convert thymidine to TMP.

In the second step a gene encoding a functional TK protein is inserted into a HSV DNA fragment at a specific site. Initial studies utilized the wild type TK gene (48). In subsequent studies, to reduce the probability that the TK gene will recom-

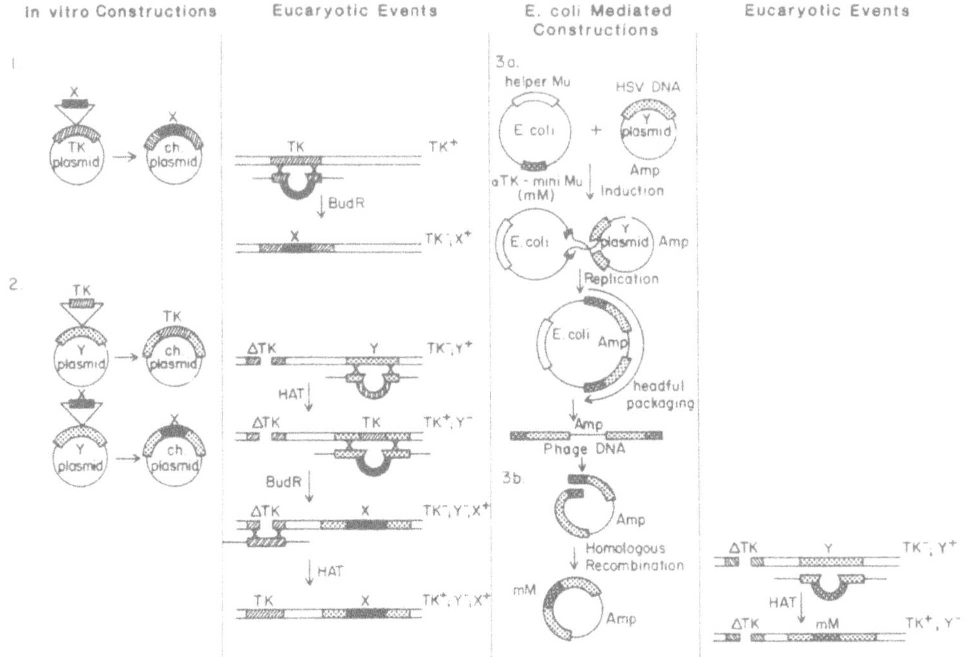

Figure 8. Schematic outline of procedures used to construct novel HSV genomes by site-specific and target-specific insertional mutagenesis as explained in the text.

bine at its natural location and restore the integrity of the wild type gene, a chimeric gene consisting of the coding sequences of the TK gene fused to the promoter-regulatory domain of alpha genes was used for this purpose (121, 123). Because the deletion in the natural TK gene extends both upstream and downstream of the *Bgl*II site whereas the new promoter regulatory domain extends upstream from the *Bgl*II site the probability that the chimeric gene will restore TK activity at the natural site is greatly reduced and in fact such recombinants have not been observed. Viral TK^+ recombinants carrying the TK insert at the desired site are selected by plating the progeny of transfection on TK^- cells maintained in HAT medium.

To effect a deletion or insert a specific HSV or foreign gene in the domain of the target sequence, an appropriate deletion or insertion is made in the target sequence at the site of insertion of the TK gene. This DNA fragment is then amplified and co-transfected with intact DNA of the recombinant carrying the TK gene inserted into the target sequence. Viral TK^- recombinants carrying the deletion or insertion in the target gene are then selected among the progeny of transfection as described above. Where it is desirable to assess the effect of the deletion in the target sequence only, the native TK gene can be restored by co-transfecting the DNA of this recombinant with DNA fragments carrying the entire domain of the TK gene. The TK^+ progeny of this transfection should differ from the wild type virus only with respect to the deletion or insertion in the target sequence.

This strategy permits the selection of viable recombinants carrying insertion or deletions at specific sites. The major disadvantages are that the location of the target sequence within the cloned DNA fragment must be known and furthermore, this sequence must contain a suitable restriction endonuclease cleavage site for the insertion of the TK gene. For example, if the objective is to inactivate by insertional mutagenesis a gene that overlaps in part with the domain of another gene either at its 3' or 5' termini, a suitable site would be upstream or downstream from the overlapping regions. In some instances no suitable sites are found or the precise domain of the target gene is not known.

The third strategy (Figure 8-3), a modification of the second, involves the use of the mini Mu derivative of the transducing bacteriophage Mu (123, 124). The mini-Mu prophage contains the left and right ends of the bacteriophage Mu plus the ner, A and B genes, a temperature-sensitive *c* repressor gene of Mu, and a selectable marker. The mini-Mu prophage is defective and requires a helper Mu for replication but the presence of the left and right ends plus the A and B genes of Mu allow for transposition within an *E. coli* cell. For use in insertional mutagenesis of HSV genomes, a functional TK gene under the control of the HSV alpha4 gene promoter and regulatory region was inserted into the mini-Mu to yield an alpha-TK mini-Mu (alphaTK-mM; 123). The alphaTK-mM was then used to lysogenize an *E. coli* strain containing a helper Mu prophage with a temperature-sensitive repressor *c* gene. Growth of the double lysogen *E. coli* strain at 30° C prevents replication and transposition of both the helper Mu and alphaTK-mM prophages.

The AlphaTK/miniMu System

The use of the alphaTK/mini-Mu system for insertional mutagenesis involves three steps.

In the first one (Figure 8-3a), the double lysogen *E. coli* strain is transfected with plasmid DNA containing the target HSV-1 DNA fragment. Induction by shift up to 42° C of the bacterial cells results in replication and transposition of both the helper Mu and alphaTK-mM. As a result of transposition, the alphaTK-mM will integrate into many different sites within the plasmid DNA producing a cointegrate structure consisting of the plasmid DNA flanked by single copies of the alphaTK-mM integrated into the *E.coli* DNA. Packaging of the cointegrate structures begins with the leftmost copy of the alphaTK-mM and proceeds through the plasmid DNA and into the second copy of the alphaTK-mM. If the cointegrate structure is less than 38 Kbp, then flanking *E. coli* DNA sequences will be packaged. Cointegrate structures greater than 38 Kbp results in deletions of the second alphaTK-mM copy. The alphaTK-mM is 9.7 Kbp in size. Therefore recombinant plasmids 18 Kbp or greater in length will cause deletion of part or all of the second copy of the alphaTK-mM. As will be evident below, at least a portion of the second copy of the alphaTK-mM is essential and therefore the recombinant plasmid must be less than 28 Kbp in size.

The lysates of the double lysogen transfected with plasmids containing HSV DNA and induced by shift up to 42° C should contain phages with defective genomes consisting of the recombinant plasmid DNAs linearized at random sites and flanked by direct copies of the alphaTK-mM. The second step (Figure 8-3b), involves infection of an *E. coli* Rec(A^+) strain lysogenized by a helper Mu prophage. The presence of the helper Mu prevents the cointegrate structure from transposing to the *E. coli* DNA. Homologous recombination between the flanking copies of alphaTK-mM results in the production of plasmid molecules in which the alphaTK-mM is randomly inserted into the recombinant plasmid. *E. coli* carrying the plasmid molecules can then be readily grown by selection for the antibiotic resistance marker contained in the alphaTK-mM. The isolated plasmid DNA consists of a pool of plasmid molecules containing an alphaTK-mM inserted at many different sites within the HSV DNA insert and non-essential vector sequences.

In the third step, the alphaTK-mM plasmids are co-transfected with intact TK^- HSV DNA and the progeny of the transfection is then plated on TK^- cells under HAT medium. The TK^+ progeny carrying the alphaTK-mM insert may be heterogeneous inasmuch as the alphaTK-mM may be expected to insert at any site within the target sequences of HSV DNA that is not essential for growth in the cells in which the selection is done.

The alphaTK-mM system offers several advantages. Foremost, it obviates the need for inserting the selectable marker at a specific and often inaccessible site. The production of recombinant plasmids containing randomly inserted alphaTK-mM is rapid —less than 48 hrs. The alphaTK-mM is an excellent probe for sequences and genes non-essential for growth in the cells used for the selection of recombinant viruses. Furthermore, as in the case of the inserted TK gene, the alphaTK-mM can be replicated by recombination with an insertion or deletion in the target sequence.

The techniques described above have been used to insert HSV and foreign sequences into HSV genomes, and to probe HSV-1 DNA for domains not essential for growth in cell culture. Among the central issues to be considered are the capacity of the genome as a vector for HSV or foreign genes, stability of the recombinants, and priviledged sites characterized by spontaneous rearrangements.

D. The Minimum and Maximum Sizes of HSV Genomes that Can Be Packaged

Current studies by N. Frenkel and associates (24, 25) and others (125) indicate that packaging of the genome requires the presence of terminal *a* sequences and is not accomplished by a simple headful mechanism. The capsids are capable of packaging small head to tail concatemers of defective genomes (25), but capsids containing concatemers which are significantly smaller than the wild type genome are not enveloped.

One hypothesis that could explain this observation is that capsids containing standard length DNA become modified and bind an additional protein (95) and that capsids packaging less than a minimum size DNA do not become modified and fail to bind this protein. The minimal size of the packaged genome is not known but the smallest non-defective genome packaged efficiently is that of the 1358 recombinant in which a 2 Kbp sequence containing the TK gene replaced 15 Kbp of DNA that included a portion of the unique sequences and nearly the entire internal inverted repeat sequence (17). The largest genomes packaged to date include an alphaTK-mM 9.7 Kbp in size (123). Taken together, these results indicate that the 2

Kbp sequence containing the TK gene in 1358 recombinant could be replaced by an insert at least 24 Kbp in size.

E. Sites and Sequences Not Essential for Growth in Cell Culture

The sites and sequences shown to date to be non-essential for viral replication include:

i) various portions of the domain of the TK gene (48),
ii) the sequences located between the 3' terminus of the alpha27 and alpha0 genes (17, 123),
iii) the sequences (approximately 15 Kbp) located between the 3' terminus of the alpha27 gene and the promoter-regulatory domain of the alpha4 gene, comprising unique sequences of the L component and nearly the entire internal inverted repeat sequence *b'a*$_n$'c' (17, 123),
iv) the sequences located between the terminal *a* sequence and the 3' terminus of the alpha4 gene at both ends of the S component (122),
v) the sequences flanking the origins of DNA synthesis of the S component (50),
vi) portions of the domains of the alpha22 gene (48), and
vii) sequences from (and including the) glycoprotein E gene contiguous to the terminus of the S component which includes the alpha47 gene and three additional HSV genes (50).

The recombinants carrying inserts or deletions in these regions of HSV DNA vary in their capacity to grow. The recombinants R325 and R328 carrying 500 and 100 bp deletions in the coding sequences of the alpha22 gene grow as well as the wild type parent in Vero and HEp-2 cells but poorly in human embryonic lung cells and various rodent cell lines. Studies on these recombinants suggest that a host factor is able to complement the deleted gene in Vero and HEp-2 cell lines but not in the restrictive cells (49). The 1358 recombinant described above yields approximately 10 fold less virus than the wild type in Vero and HEp-2 cells and considerably less virus in rodent cell lines or human embryonic cell lines. A characteristic of this virus is that its DNA is frozen in one arrangement, i.e. the L and S components do not invert relative to each other.

F. Insertion of HSV and Foreign Sequences into HSV Genome

The sequences and genes inserted into HSV-1 DNA to-date include: (a) HSV DNA sequences inserted for the purpose of identifying putative *cis*-acting sites, (b) HSV genes inserted for the purpose of identifying their function by complementing the authentic, mutagenized copy, and (c) foreign genes to determine whether the HSV genome can serve as a vector for expression of foreign genes.

Specifically, to identify the promoter regulatory domains of the alpha genes, a fragment containing the transcription initiation site and upstream sequences of the alpha4 gene was inserted into the *Bgl*II site (nucleotide +50) of the TK gene in the proper transcriptional orientation. In the resulting recombinants, the chimeric TK genes were regulated as an alpha gene (48). Similar insertion of a fragment carrying a putative $gamma_2$ promoter-regulator sequence converted the natural beta TK into a $gamma_2$ gene (126). Insertion of DNA fragments containing the *a* sequence into the *Bgl*II site of the TK gene led to the identification of this sequence as the site specific recombination site for the inversion of the L and S components (29). The HSV genes specifically inserted into the genome were designed to complement mutations in the native gene (127) or to express a mutated gene (128). The foreign genes inserted into the genome were the *S* gene of hepatitis B virus (60), the chick ovalbumen gene (M. Arsenakis, B. Roizman; manuscript in preparation), and the EBNA1 and EBNA2 genes of Epstein-Barr virus (129). The processing of products of foreign genes inserted into HSV-1 vectors has not been studied extensively, but normally secreted proteins are secreted from cells infected with HSV-1 vectors carrying their genes.

G. Requirements for the Expression of Foreign Genes

Host genes and genes of other viruses linked to their natural promoters are expressed poorly if at all in HSV vectors (130). When linked to promoter-regulatory domains of HSV genes, the foreign genes are expressed and regulated as the authentic HSV genes whose promoter was "borrowed" (60). HSV-2 genes are expressed in HSV-1 vectors under their own promoters, and it is conceivable that genes of other herpesviruses may also be expressed in HSV-1 vectors under their own promoters.

H. Stability of Genetically Engineered HSV-1 Genomes

Several major types of rearrangements in genetically engineered strains have been noted. These include the following:

Cell-specific restrictions

An example of this phenomenon is recombinant R316, which was constructed by the insertion of *Bam*HI N fragment of HSV-1 into the *Bgl*II site of the TK gene thus converting the natural (beta) TK into an alpha regulated gene (48). This recombinant is stable in Vero cells; in a variety of other cell lines and in experimental animals mutant carrying various size deletions in the inserted *Bam*HI N fragment emerge and rapidly overgrow the R316 recombinant (49). No explanation for this bizzare, cell-specific restriction has emerged so far.

Equalization of genetically related duplicated genes

Insertion of a gene closely related, but non-identical, to that resident in the HSV genome may result in the accumulation of identical recombinant sequences as a consequence of genetic recombination and segregation (131).

Insert-dependent defective genomes

Insertion of a second copy of a sequence in the same orientation relative to the first copy may result in the generation of head-to-tail reiterations of novel subsets of defective HSV genomes in which all of the sequences between the two copies of the reiterated sequence had been deleted. Defective genomes require an origin of DNA synthesis for amplification and a terminal *a* sequence for packaging by factors supplied by the helper virus. In addition, the length of the subset must be such that an integral number of tandem reiterations fall within the minimum and maximum size DNA that is both packaged and enveloped (18). The insertion of a second copy of sequences in the inverted orientation relative to the native copy has, in some instances, resulted in the inversion of the sequences flanked by the inverted repeats (29, 131). Whether a particular set of inverted repeats will cause inversions is totally unpredictable and may reflect recombinational "hot spots" within these sequences.

Privileged sites

These sites are defined as most likely to result in major rearrangements of the HSV genome. Thus insertion of sequences at or near the termini of the inverted repeats may result in the spontaneous deletion of most of the inverted repeat sequences $b'a_n'c'$ (Figure 1) (17, 123). Insertions within one of the inverted repeat sequences of the S component (*c'a'* or *ca*; Figure 1) results in the duplication of the insertion in the other repeat (122). The duplication is almost "obligatory" if the insert is close to the *a* sequence but diminishes in frequency with distance.

Site-specific or target-specific mutagenesis of viral DNA genomes using a selectable marker system is a powerful tool for the analysis of the function of specific regions of herpesviruses DNA genomes. Through these techniques both the determination of viral DNA sequence and gene function and the construction of vectors capable of delivering vaccines for the prevention of infectious disease in humans and animals is possible.

Acknowledgements

Our studies cited in this chapter were aided by United States Public Health Service grants CA-08494 and CA-19264 from the National Cancer Institute and grant MV-2T from the American Cancer Society. F.J.J. is a U.S. Public Health Service postdoctoral trainee (CA-09241) of the National Cancer Institute and T.M.K. was a National Cancer Institute predoctoral trainee (CA-192642).

6. REFERENCES

1) Roizman, B., Carmichael, L.E., Deinhardt, F., de The, G., Nahmias, A.J., Plowright, W., Rapp, F., Sheldrick, P., Takahashi, M., and Wolf, K. (1981), Intervirol. **16**, 201-217.

2) Furlong, D., Swift, H. and Roizman, B. (1972), J. Virol. **10**, 1071-1074.

3) Roizman, B. and Furlong, D. (1974) in: *Comprehensive Virology*, vol. **3**, H. Fraenkel-Conrat and R.R. Wagner eds., Plenum Press, New York, 229-403.

4) Spear, P.G. and Roizman, B. (1980) in: *DNA Tumor Viruses*, J. Tooze ed., Cold Spring Harbor Laboratory Press, Cold Spring Harbor, New York, 615-745.

5) Spear, P.G., (1985) in: *The Herpesviruses*, B. Roizman, ed., Plenum Press, New York, London, 315-356.

6) Morgan, C., Rose, H.M. and Mednis, B. (1968), J. Virol. **2**, 507-516.

7) Batterson, W., Furlong, D. and Roizman, B. (1983), J. Virol. **45**, 397-407.

8) Lando, D. and Ryhiner, M.L. (1969), C.R. Acad. Sci. **269**, 527.

9) Graham, F.L., Velihaisen, G., and Wilkie, N.M. (1973), Nature New Biology **245**, 265-266.

10) Sheldrik, P., Laithier, M., Larria, D. and Ryhinder, M.L. (1973), Proc. Nat. Acad. Sci. USA **70**, 3621-3625.

11) Koch, H.G., Delius, H., Matz, B., Flugel, R.M., Clarke, J., and Darai, G. (1985), J. Virol. **55**, 86-95.

12) Bornkamm, G.W., Delius, H., Fleckenstein, B., Werner, F.J. and Mulder, C. (1976), J. Virol. **19**, 154-161.

13) Wadsworth, S., Jacob, R.J. and Roizman, B. (1975), J. Virol. **15**, 1487-1497.

14) Morse, L.S., Pereira, L., Roizman, B. and Schaffer, P.A. (1978), J. Virol. **26**, 389-410.

15) Chou, J., and Roizman, B. (1985), Cell **41**, 803-811.

16) Hayward, G.S., Jacob, R.J., Wadsworth, S.C., and Roizman, B. (1975), Proc. Nat. Acad. Sci. USA **72**, 4243-4247.

17) Poffenberger, K.L., Tabares, E., and Roizman, B. (1983), Proc. Natl. Acad. Sci. USA **80**, 2690-2694.

18) Poffenberger, K.L. and Roizman, B. (1985), J. Virol. **53**, 587-595.

19) Jenkins, F.J., and Roizman, B. (1986), J. Virol. **59**, 494-499.

20) Mocarski, E.S. and Roizman, B. (1981), Proc. Nat. Acad. Sci. USA **78**, 7047-7051.

21) Davison, A.J. and Wilkie, N.M. (1981), J. Gen. Virol. **55**, 315-331.

22) Wagner, M.M. and Summers, W.C. (1978), J. Virol. **27**, 374-387.

23) Locker, H. and Frenkel, N. (1979), J. Virol. **32**, 424-441.

24) Vlazny, D.A. and Frenkel, N. (1981), Proc. Nat. Acad. Sci. USA **78**, 742-746.

25) Vlazny, D.A., Kwong, A., and Frenkel, N. (1982), Proc. Nat. Acad. Sci. USA **79**, 1423-1427.

26) Chou, J., and Roizman, B. (1986), J. Virol. **57**, 629-637.

27) Ackermann, M., Chou, J., Sarmiento, M., Lerner, R.A., and Roizman, B. (1986), J. Virol. **58**, 843-850.

28) Mocarski, E.S. and Roizman, B. (1982), Cell **31**, 89-97.

29) Mocarski, E.S. and Roizman, B. (1982), Proc. Nat. Acad. Sci. USA **79**, 5626-5630.

30) Mocarski, E.S., Post, L.E. and Roizman, B. (1980), Cell **22**, 243-255.

31) Hayward, G.S., Frenkel, N. and Roizman, B. (1975), Proc. Natl. Acad. Sci. USA **72**, 1768-1772.

32) Linnemann, C.C., Jr., Buchman, T.G., Light, I.J., Ballard, J.L., and Roizman, B. (1978), Lancet **i**, 964-966.

33) Buchmann, T.G., Roizman, B., Adams, G. and Stover, H. (1978), J. Infect. Dis. **138**, 488-498.

34) Buchman, T.G., Roizman, B. and Nahmias, A.J. (1979), J. Infect. Dis. **140**, 295-304.

35) Buchman, T.G., Simpson, T., Nosal, C., Roizman, B. and Nahmias, A.J. (1980), Ann. New York Acad. Sci. **354**, 279-290.

36) Huang, E.S. and Huong, S.M. (1980), Ann. New York Acad. Sci. **354**, 332-346.

37) Honess, R.W. and Roizman, B. (1974). J. Virol. **14**, 8-19.

38) Honess, R.W. and Roizman, B. (1975). Proc. Nat. Acad. Sci. USA **72**, 1276-1280.

39) Knipe, D.M., Ruyechan, W.T., Roizman, B. and Halliburton, I.W. (1978), Proc. Natl. Acad. Sci. USA **75**, 3896-3900.

40) Knipe, D.M., Ruyechan, W.T., Honess, R.W., and Roizman, B., (1979), Proc. Natl. Acad. Sci. USA **76**, 4534-4538.

41) Dixon, R.A.F. and Schaffer, P.A. (1980), J. Virol, **36**, 189-203.

42) O'Hare, P., and Hayward, G.S. (1985), J. Virol. **56**, 723-733.

43) O'Hare, P. and Hayward, G.S. (1985), J. Virol. **53**, 751-760.

44) Everett, R.D. (1984), EMBO J. **3**, 3135-3141.

45) Quinlan, M.P. and Knipe, D.M. (1985), Mol. Cell. Biol. **5**, 957-963.

46) Mavromara-Nazos, P., Silver, S., Hubenthal-Voss, J., McKnight, J.L.C. and Roizman, B. (1986), Virology **149**, 152-164.

47) Sacks, W.R., Greene, C.C., Aschman, D.P. and Schaffer, P.A. (1985), J. Virol. **55**, 796-805.

48) Post, L.E. and Roizman, B. (1981), Cell **25**, 227-232.

49) Sears, A.E., Halliburton, I.W., Meignier, B., Silver, S., and Roizman, B. (1985), J. Virol. **55**, 338-346.

50) Longnecker, R. and Roizman, B. (1986), J. Virol. **58**, 583-591.

51) Constanzo, F., Campadelli-Fiume, G., Foa-Tomas, L. and Cassai, E. (1977), J. Virol. **21**, 996-1001.

52) Mackem, S. and Roizman, B. (1982), J. Virol. **43**, 1015-1023.

53) Mackem, S. and Roizman, B. (1982), Proc. Nat. Acad. Sci. USA **79**, 4917-4921.

54) Mackem, S. and Roizman, B. (1982), J. Virol. **44**, 939-949.

55) Batterson, W. and Roizman, B. (1983), J. Virol. **46**, 371-377.

56) Campbell, M.E.M., Palfreyman, J.W. and Preston, C.M. (1984), J. Mol. Biol. **180**, 1-19.

57) Pellett, P., McKnight, J.L.C., Jenkins, F.J. and Roizman, B. (1985), Proc. Natl. Acad. Sci. USA **82**, 5870-5874.

58) Heine, J.W., Honess, R.W., Cassai, E. and Roizman, B. (1974), J. Virol. **14**, 640-651.

59) Herz, C. and Roizman, B. (1983), Cell **33**, 145-151.

60) Shih, M.F., Arsenakis, M., Tiollais, P. and Roizman, B. (1984), Proc. Natl. Acad. Sci. USA **81**, 5867-5870.

61) Kristie, T.M. and Roizman, B. (1984), Proc. Natl. Acad. Sci. USA **81**, 4065-4069.

62) Jones, K.A. and Tjian, R. (1985), Nature (London) **317**, 179-185.

63) McKnight, J.L.C., Kristie, T.M., Silver, S., Pellett, P.E., Mavromara-Nazos, P., Campadelli-Fiume, G., Arsenakis, M. and Roizman, B. (1986) in: *Cancer Cells 4, Control of gene expression and replication*, Botchan, M., Grodzicker, T., and Sharp, P. eds, Cold Spring Harbor Laboratories, Cold Spring Harbor.

64) Kristie, T.M. and Roizman, B. (1986), Proc. Natl. Acad. Sci. USA **83**, 3218-3222.

65) Kristie, T.M. and Roizman, B. (1986), Proc. Natl. Acad. Sci. USA **83**, 4700-4704.

66) Wilcox, K.W., Kohn, A., Sklyanskaya, E., and Roizman, B. (1980). J. Virol. **33**, 167-182.

67) Metzler, D.W. and Wilcox, K.W. (1985). J. Virol. **55**, 329-337.

68) Freeman, M.J. and Powell, K.L. (1982), J. Virol. **44**, 1084-1087.

69) Fried, M. and Crothers, D. (1981), Nucleic Acids Res. **9**, 6505-6525.

70) Garner, M.M. and Revzin, A. (1981), Nucleic Acids Res. **9**, 3047-3060.

71) Roizman, B., Kristle, T.M., Batterson, W. and Mackem, S. (1984) in: *Proceedings of the P and S Biomedical Sciences Symposium on Transfer and Expression of Eukaryotic Genes*, Vogel, H., and Ginsberg, H.S., eds., Academic Press, New York, 227-238.

72) Jones, P.C., Hayward, G.S. and Roizman, B. (1977), J. Virol. **21**, 268-276.

73) Watson, R.J., Preston, C.M. and Clements, B. (1979), J. Virol. **31**, 42-52.

74) Mackem, S. and Roizman, B. (1980). Proc. Natl. Acad. Sci. USA **77**, 7122-7126.

75) Frenkel, N., Locker, H., Batterson, W., Hayward, G. and Roizman, B. (1976), J. Virol. **20**, 527-531.

76) Stow, N.D. (1982), EMBO J. **1**, 863-867.

77) Chartrand, P., Crumpacker, C.S., Schaffer, P.A. and Wilkie, N.M. (1980), Virology **103**, 311-325.

78) Conley, A.F., Knipe, D.M., Jones, P.C. and Roizman, B. (1981), J. Virol. **37**, 191-206.

79) Roizman, B., Norrild, B., Chan, C. and Pereira, L. (1984), Virology **133**, 242-247.

80) Buckmaster, E.A., Gompels, U. and Minson, A. (1984), Virology **139**, 408-413.

81) Roizman, B. (1979), Cell **16**, 481-494.

82) Watson, R.J., Sullivan, M. and Vande Woude, G.F. (1981), J. Virol. **37**, 431-444.

83) Murchie, M.J. and McGeoch, D.J. (1982), J. Gen. Virol. **62**, 1-15.

84) Fontichiaro, K.L.E., Beck, T.W. and Millette, R.L. (1985), J. Virol. **53**, 235-242.

85) McLaughlan, J., Gaffney, D., Whitton, J.L. and Clements, J.B. (1985), Nucleic Acids Res. **13**, 1347-1368.

86) McKnight, S.L., Gavis, E.R., Kingsbury, R. and Axel, R. (1981), Cell **25**, 385-398.

87) Keir, H.M. and Gold, E. (1963), Biochimia et Biophysica Acta **72**, 263-276.

88) Wohlrab, F. and Francke, B. (1980). Proc. Nat. Acad. Sci. USA **80**, 100-104.

89) Preston, V.G. and Fisher, F.B. (1984), Virology **138**, 58-68.

90) Cohen, G.H. (1972), J. Virol. **9**, 408-418.

91) Huszar, D. and Bacchetti, S. (1981), J. Virol. **37**, 580-588.

92) Kit, S. and Dubbs, D.R. (1965), Virology **26**, 16-27.

93) Ben-Porat, T. and Tokazewski, S. (1977), Virology **79**, 292-301.

94) Gibson, W. and Roizman, B. (1972), J. Virol. **10**, 1044-1052.

95) Braun, D.K., Roizman, B. and Pereira, L. (1984), J. Virol. **49**, 142-153.

96) Darlington, R.W. and Moss, L.H., III (1969), Prog. Med. Virol. **11**, 16-45.

97) Schwartz, J. and Roizman, B. (1969), J. Virol. **4**, 879-889.

98) Roizman, B. (1972), in: *Oncogenesis and Herpesviruses*, Biggs, P.M., de The, G., and Payne, L.N., eds., I.A.R.C., Lyon, 1-21.

99) Roizman, B., Borman, G.S. and Kamali-Rousta, M. (1965), Nature (London) **206**, 1374-1375.

100) Sydiskis, R.J. and Roizman, B. (1968), Virology **32**, 678-686.

101) Wagner, E.K. and Roizman, B. (1969), J. Virol. **4**, 36-46.

102) Fenwick, M., Morse, L.S. and Roizman, B. (1979), J. Virol. **29**, 825-827.

103) Fenwick, M.L. and Walker, M.J. (1978), J. Gen. Virol. **41**, 37-51.

104) Nishioka, Y. and Silverstein, S. (1977), Proc. Natl. Acad. Sci. USA **74**, 2370-2374.

105) Nishioka, Y. and Silverstein, S. (1978), J. Virol. **25**, 422-426.

106 Nishioka, Y. and Silverstein, S. (1978), J. Virol. **27**, 619-627.

107) Silverstein, S. and Engelhardt, E.L. (1979). Virology **95**, 324-342.

108) Read, G.S. and Frenkel, N. (1983), J. Virol. **46**, 498-512.

109) Hill, T.J. (1985) in: *The Herpesviruses, Vol. 3*, B. Roizman ed., Plenum Press, New York/London, 175-240.

110) Scriba, M. (1975), Infect. Immun. **12**, 162-165.

111) Roizman, B. (1966) in: *Perspectives in Virology IV*, M. Pollard ed., Harper-Row, New York, 283-304.

112) Roizman, B. and Morse, L.S. (1978), in: *Herpesviruses and Oncogenisis*, Henle, W., Rapp, F., and de The, G., eds., I.A.R.C., Lyon, 269-297.

113) Green, M.T., Courtney, R.J. and Dunkel, E.G. (1981), Infect. Immun. **34**, 987-992.

114) Matsuo,T., Heller, M., Petti, L., O'Shiro, E. and Kieff, E. (1984), Science **226**, 1322-1325.

115) Reedman, B.M. and Klein, G. (1973), Int. J. Cancer **11**, 499-520.

116) Nonoyama, M. and Pagano, J. (1972), Nature (London) New Biol. **238**, 169-171.

117) Tanaka, A. and Nonoyama, M. (1974), Proc. Natl. Acad. Sci. USA **71**, 4658-4661.

118) Sugden, B., Marsh, K., and Yates, J. (1985), Mol. Cell. Biol. **5**, 410-413.

119) Ruyechan, W.T., Morse, L.S., Knipe, D.M. and Roizman, B. (1979), J. Virol. **29**, 677-697.

120) Tenser, R.B., and Dunstan, M.E. (1979), Virology **99**, 417-422.

121) Dubbs, D.R. and Kit, S. (1964), Virology **22**, 493-502.

122) Hubenthal-Voss, J., and Roizman, B. (1985), J. Virol. **54**, 509-514.

123) Jenkins, F.J., Casadaban, M.J. and Roizman, B. (1985), Proc. Natl. Acad. Sci. USA **82**, 4773-4777.

124) Castilho, B.A., Olfson, P. and Casadaban, M.J. (1984), J. Bacteriol. **158**, 488-495.

125) Stow, N.D., McMonagle, E.C. and Davison, A.J. (1983), Nucleic Acids Res. **11**, 8205-8220.

126) Silver, S. and Roizman, B. (1985), Mol. Cell. Biol. **5**, 518-528.

127) Lee, G.T.Y., Pogue-Geile, K.L., Pereira, L. and Spear, P.G. (1982), Proc. Natl. Acad. Sci. USA **79**, 6612-6616.

128) Gibson, M.G. and Spear, P.G. (1983). J. Virol. **48**, 396-404.

129) Hummel, M., Arsenakis, M., Marchini, A., Lee, L., Roizman, B. and Kieff, E. (1986), Virology **148**, 337-348.

130) Tackeny, C., Cachianes, G., and Silverstein, S. (1984), J. Virol. **52**, 606-614.

131) Pogue-Geile, K.L., Lee, G.T.Y. and Spear, P. (1985), J. Virol. **53**, 456-461.

SECTION V

The Martea Conference

THE MARATEA CONFERENCE

List of Participants

Guenter ADAM	Institut fuer Biologie der Universitaet Stuttgart D-7000 STUTTGART 60 (Germany)
Annika ALLARD	Dept. Virology University of Umea S-901 85 UMEA (Sweden)
Yosef ALONI	Dept. Genetics.- The Weizmann Institute of Sciences REHOVOT 76100 (Israel)
Simon BARTELING	Dept. Virology.- Central Veterinary Institute LELYSTADT 8221 RA (The Netherlands)
Massimo BATTAGLIA	Inst. Expl. Med. CNR Viale Regina Elena 324.- 00161 ROMA (Italy)
Ingrid BAUSE	Institut fuer Virologie der Freie Universitaet Berlin Norufer 20.- D-1000 BERLIN 65 (Germany)
Ali Osman BELUZ	Karadeniz University.- Fen-Ed. Fak. Biyoloji Bol. 61080 TRABZON (Turkey)
David H.L.	BISHOP NERC Institute of Virology OXFORD OX1 3SR (United Kingdom)
Dieter BLAAS	Institut fuer Biochemie der Universitaet Wien 1090-WIEN (Austria)
Jean M. BONNEVILLE	Friedrich-Miescher Institut CH-4002 BASEL (Switzerland)
Thomas BOEREN	Dept. Med. Biochemistry Univ. Goteborg S-400 33 GOTEBORG (Sweden)
Jonathan BRADLEY	Unite Virologie Moleculaire Institut Pasteur 75724 PARIS (France)
Hans-Joerg BUHK	Robert-Koch Institut des BGA Nordufer 20.- D-1000 BERLIN 65 (Germany)
Jozef BUJARSKI	Biophysics Laboratory University Wisconsin MADISON WI 53706 (U.S.A.)
Alfred T.H.BURNESS	Memorial University Newfoundland Faculty Medicine ST-JOHN'S, Newfoundland, (Canada)
Stefano BUTTO'	Dept. Virology Istituto Superiore Sanità 00161-ROMA (Italy)
G.CAMPADELLI-FIUME	Instituto Microbiologia e Virologia Università Bologna 40126-BOLOGNA (Italy)
W. CANAS-FERREIRA	Inst. Higiene & Med. Tropical.- Rua Juncal 96 1300-LISBOA (Portugal)
Zilda G. CARVALHO	Laboratorio Virologia.- Gulbekian Inst. Sciences 2781-OEIRAS Codex (Portugal)

Mehmet CEYHAN	Askeri Astane - Cocuk Sagligi ve Hastaliklari Uzmani ADANA (Turkey)
Phillips CHAMBERS	Dept. Biochemistry Univ. Newcastle upon Tyne NEWCASTLE UPON TYNE NE1 (United Kingdom)
Jean CONTENT	Institut Pasteur du Brabant B-1180 BRUSSELS (Belgium)
Julian COOPER	Microbiology Dept. Reading University READING RG1 5AQ (United Kingdom)
Roelf DATEMA	Dept. Antiviral Chemotherapy - Astra Lakemedel AB S-151 85 SODERTALJE (Sweden)
Anna M. DEGENER	Institute for Virology University of Rome 00185-ROME (Italy)
Suzanne U. EMERSON	Dept. Microbiology University of Virginia CHARLOTTESVILLE VA 22901 (U.S.A.)
Vincent C. EMERY	NERC Institute of Virology OXFORD OX1 3SR (United Kingdom)
Michele EQUESTRE	Institute for Virology University of Rome 00185-ROME (Italy)
Ludwig EVERAERT	Dept. Microbiol. & Hygiene Vrije Univ. Brussel B-1090 BRUSSELS (Belgium)
Pierre FERRER	Ctr. Biochimie & Genetique CNRS 31062-TOULOUSE Cedex (France)
Lucia FIORE	Laboratory Enteroviruses Istituto Superiore Sanità 00161-ROME (Italy)
Felix FILATOW	Iwanonowski Inst. Virology Acad. Medical Sciences D-98 MOSCOW (U.S.S.R.)
Kerstin GAEDICK	Institut fuer Biologie der Universitaet Stuttgart D-7000 STUTTGART 60 (Germany)
Jean Pierre GAGNER	Dept. Microbiol. & Mol. Genetics Harvard Med. Sch. BOSTON MA 02115 (U.S.A.)
Paola GALLINARI	Dept. Virology Istituto Superiore Sanità 00161-ROME (Italy)
Colomba GIORGI	Laboratory of Virology Instituto Superiore Sanità 00161-ROME (Italy)
Sergio GIUNTA	Laboratory Virology & Genetics I.N.R.C.A. 60100-ANCONA (Italy)
Karl GORDON	Inst. Biol. Mol. & Cellulaire CNRS 67084-STRASBOURG (France)
Giuseppina GRASSO	Institute for Virology University of Rome 00185-ROME (Italy)

Ludwig HAAS	Institute Virology Veterinary Sch. Hannover D-3000 HANNOVER 17 (Germany)
Anne-Lise HAENNI	Inst. Jacques Monod Université Paris VII 75251-PARIS Cedex 15 (France)
David L. HACKER	Dept. Microbiol. & Public Health Michigan State Univ. EAST LANSING MI 48824 (U.S.A.)
Howard V. HERSHEY	Department of Biology Indiana University BLOOMINGTON IN 47405 (U.S.A.)
M.J. HEWLETT	Departement Microbiologie Université Geneve 1211-GENEVE (Switzerland)
Elizabeth M. HOEY	Dept. Biochemistry Queen's University Belfast BELFAST BT9 7BL (United Kingdom)
Peter M. HOWLEY	Lab. Tumor Virus Natl. Cancer Inst.(Blg 41 Rm.D 201) BETHESDA MD 20892 (U.S.A.)
Alice S. HUANG	Div. Infectious Disease Children's Hospital BOSTON MA 02115 (U.S.A.)
Danny HUYLEBROECK	Eur. Mol. Biol. Org. Laboratory D-6900 HEIDELBERG (Germany)
Just JUSTESSEN	Institute Molecular Biology DK-8000 AARHUS (Denmark)
Walter KELLER	Div. Mol. Biol. German Cancer Res. Ctr. D-6900 HEIDELBERG (Germany)
Gila KOHEN	Laboratoire IFFA Institut Merieux 69342-LYON Cedex 07 (France)
Dan KOLAKOFSKY	Departement Microbiologie Université Geneve 1211-GENEVE 4 (Switzerland)
A. KOS	Primate Center TNO 2280-HV RIJSWIJK (The Netherlands)
Ernst KEUCHLER	Institut fuer Biochemie der Universitaet Wien A-1090 WIEN (Austria)
Jovi LARSON	Rabies Unit Wistar Institute PHILADELPHIA PA 19143 (U.S.A.)
Arnold J. LEVINE	Dept. Molecular Biology Princeton University PRINCETON NJ 08544 (U.S.A.)
LI Quan-ge	Dept. Virology Universty Umea S-901 85 UMEA (Sweden)
Florentyna LUSTIG	Dept. Med. Biochemistry University Goteborg S-400 33 GOTEBORG (Sweden)
Denis MACEJAK	Dept. Microbiol. & Immunol. Louisiana State Univ. NEW ORLEANS LA 70112-1393 (U.S.A.)
Malcolm McCRAE	Department Biological Sciences Univ. Warwick COVENTRY CV4 7AL (England)

Brian W.I. MAHY	Animal Virus Research Institute PIRBRIGHT GU24 ONF (England)
Maria L. MARCANTE	Laboratory Virology Istituto Tumori Regina Elena 00161-ROME (Italy)
Mario MIDULLA	Dept. Pediatrics Policlinico Umberto I Univ. Rome 00161-ROME (Italy)
Graziella MORACE	Laboratory Virology Istituto Superiore Sanità 00161-ROME (Italy)
Bernard MOSS	N.I.H. Blg. 5.- Rm.432 BETHESDA MD 20892 (U.S.A.)
B.N. NERSESSIAN	Poultry Dis. Research Ctr.- 953 College Rd. ATHENS GA 30605 (U.S.A.)
Loredana NICOLETTI	Dept. Virology Istituto Superiore Sanità 00161-ROME (Italy)
Yukihiro NISHIYAMA	Laboratory Virology Nagoya Univ. Sch. Med. Showa-Ku NAGOYA (Japan)
Tilman OLTERSDORF	Institute Virology German Cancer Res. Ctr. D-9000 HEIDELBERG (Germany)
David C. ORR	Dept. Microbial Biochemistry Glaxo Group Res. Ltd. GREENFORD Middx. (United Kingdom)
Hilary A. OVERTON	NERC Institute of Virology OXFORD OX1 3SR (United Kingdom)
Yoshiro OZAWA	Animal Health Service F.A.O. 00100-ROME (Italy)
Peter PALESE	Dept. Microbiol Mount Sinai Med. School NEW YORK NY 10029 (U.S.A.)
Alessandra PANI	Istituto Microbiol. Univ. Cagliari 09100-CAGLIARI (Italy)
Nathalie PARDIGON	Unite' Virologie Moleculaire Inst. Pateur 75724-PARIS Cedex 15 (France)
Jeffrey D. PARVIN	Dept. Microbiol. Mount Sinai Med. Sch. NEW YORK NY 10029 (U.S.A.)
Jean L. PATTERSON	Div. Infectious Dis. Children's Hospital BOSTON MA 02115 (U.S.A.)
Raul PEREZ BERCOFF	Institute for Virology University of Rome 00185-ROME (Italy)
Gerry PRODY	Department Chemistry Western Washington University BELINGHAM WA 98225 (U.S.A.)
Gianfranco RISULEO	Dip. Genetica & Biol. Mol. Univ. Rome 00161-ROME (Italy)
Bernard ROIZMAN	Marjorie Kovler Viral Oncol. Lab. Univ. Chicago CHICAGO IL 60637 (U.S.A.)

Maria G. ROMANELLI	Ist. Scienze Biologiche 37134-VERONA (Italy)
B. ROSENWIRTH	Sandoz Forschungsinstitut.- Brunnerstr. 58 A-1235-WIEN (Austria)
Giovanni B. ROSSI	Laboratorio Virologia Istituto Superiore Sanità 00161-ROME (Italy)
Michael ROSSMANN	Dept. Biological Sciences Purdue University WEST LAFAYETTE IN 47907 (U.S.A.)
Jeff ROTHENBERG	Dept. Virology Weizmann Institute for Sciences REHOVOT 76100 (Israel)
Erling W. RUD	Dept. Microbiol. & Immunol. Fac. Health Sci. OTTAWA KIH 8M5 (Canada)
C. SADZOT-DELVAUX	Lab. Microbiologie.- Tour Pathologie Univ. Liege B-4000 SART TILMAN (Belgium)
David A. SHAFRITZ	Albert Einstein College Med.- 1300 Morris Ave. BRONX NY 10461 (U.S.A.)
Milton SCHLESINGER	Dept. Microbiol. & Immunol. Washington University ST-LOUIS MO 63110 (U.S.A.)
Sondra SCHLESINGER	Dept. Microbiol. & Immunol. Washington University ST-LOUIS MO 63110 (U.S.A.)
D.A. SMALL	NERC Institute of Virology OXFORD OX1 3SR (United Kingdom)
Yuri SMIRNOW	Iwanowski Institute of Virology Acad. Med. Sci. D-98-MOSCOW (U.S.S.R.)
W.J.M. SPAAN	Institute Virology State Univ. Utrecht 3508 TD UTRECHT (The Netherlands)
Antonio TANGORRA	Institute for Virology University of Rome 00185-ROME (Italy)
Milton W. TAYLOR	Dept. Biology Indiana University BLOOMINGTON IN 47405 (U.S.A.)
Rainer UHSE	Inst. Wasser-Boden-Luft Hygiene BGA D-1000 BERLIN 33 (Germany)
John VAN EMELO	Plant Genetics Systems.- Jozef Plateaustr. 22 B-9000 GENT (Belgium)
Lucia VAN HOOF	Laboratory Microbiology Fac. Med. UIA B-2610 ANTWERP (Belgium)
D.A. VANDEN BERGHE	Laboratory Microbiology Fac. Med. UIA B-2620 ANTWERP (Belgium)
Aldo VENUTI	Institute for Virology University of Rome 00185-ROME (Italy)
B.J.M. VERDUIN	Dept. Virology Agricultural University NL-6709 PD WAGENINGEN (The Netherlands)
Heinz ZEICHARDT	Inst. Clin. & Exptl. Virology Freie Univ. Berlin D-1000 BERLIN 45 (Germany)

It's time for early breakfast for M. ROSSMANN...

B. NERSSESSIAN, and...

F. FILATOW & Y. SMIRNOW

D. KOLAKOFSKY, A. LEVINE, & M. HEWLETT

A-L. HAENNI (back), B. MOSS & H. HERSHEY

M. SCHLESINGER & P. PALESE

S.J. BARTELING & B. ROIZMAN

B. MAHY, D. KOLAKOFSKY, M. HEWLETT (back) & R. PEREZ BERCOFF

W. SPAAN, M. ROSSMANN (back) & D. HUYLEBROECK

Y. ALONI

F. RUGGERI & P. HOWLEY

J. PARVIN & W. SPAAN

Front row: P. ROY & D. H.L. BISHOP; sitting in the back row: J. CONTENT, W. SPAAN, & D. HUYLEBROECK

A-L. HAENNI, R. PEREZ BERCOFF & A. HUANG

B. ROIZMAN

S. SCHLESINGER, A. HUANG, D. KOLAKOFSKY (back) & J.M. BONNEVILLE

Front group: W. SPAAN, M. ROSSMANN (back), & D. HUYLEBROECK
In the rear: M. SCHLESINGER & P. PALESE

Y. NISHIYAMA & A. ALLARD

P. PALESE, S.J. BARTELING & B.ROIZMAN

D. HUYLEBROECK, D.H.L. BISHOP & P. ROY

S.J. BARTELING & P. PALESE

B. MOSS, A-L. HAENNI (back), & H. HERSHEY

J.P. GAGNER & E.W. RUD

A-L. HAENNI & B. MOSS

INDEX

www.ingramcontent.com/pod-product-compliance
Ingram Content Group UK Ltd.
Pitfield, Milton Keynes, MK11 3LW, UK
UKHW051325070726
13610UKWH00014B/87

9781468453515